WORKS ISSUED BY
THE HAKLUYT SOCIETY

Series Editors
Gloria Clifton
Joyce Lorimer

SIR JOSEPH BANKS, ICELAND AND THE NORTH ATLANTIC
1772–1820
JOURNALS, LETTERS AND DOCUMENTS

THIRD SERIES
NO. 30

THE HAKLUYT SOCIETY
Trustees of the Hakluyt Society 2015–2016

PRESIDENT
Captain Michael K. Barritt RN

VICE-PRESIDENTS

Peter Barber OBE
Professor Roy Bridges
Anthony Payne
Dr Nigel Rigby
Professor Will Ryan FBA
Dr Sarah Tyacke CB
Professor Glyndwr Williams

COUNCIL
(with date of election)

Professor Jim Bennett (2011)
Dr Jack Benson (co-opted)
Professor William Butler (2014)
Professor Daniel Carey (2014)
Professor Nandini Das (2014)
Dr John H. Hemming
  (Royal Geographical Society Representative)
Professor Claire Jowitt (2013)
Lionel Knight MBE (2012)
Dr John McAleer (2015)
Dr Guido van Meersbergen (2013)
Dr Roger Morriss (2015)
Roger Perry (2015)
Dr Maurice Raraty (2012)
Professor Suzanne Schwarz (co-opted)
Dr John Smedley (co-opted)
Professor Sebastian Sobecki (2015)
Dr Silke Strickrodt (2012)
Professor Charles Withers (2010)

HONORARY TREASURER
David Darbyshire FCA

HONORARY JOINT SERIES EDITORS
Dr Gloria Clifton   Professor Joyce Lorimer

HONORARY EDITOR (ONLINE PUBLICATIONS)
Raymond Howgego

HONORARY ARCHIVIST
Dr Margaret Makepeace

HONORARY ADVISOR (CONTRACTS)
Bruce Hunter

ADMINISTRATION
(to which enquiries and application for membership may be made)

Telephone: 0044 (0)1428 641850   Email: office@hakluyt.com   Fax: 0044 (0)1428 641933

Postal Address only
The Hakluyt Society, c/o Map Library, The British Library,
96 Euston Road, London NW1 2DB, UK

Website: http://www.hakluyt.com

Registered Charity No. 313168   VAT No. GB 233 4481 77

INTERNATIONAL REPRESENTATIVES OF THE HAKLUYT SOCIETY

Australia   Dr Martin Woods, Curator of Maps, National Library of Australia, Canberra, ACT 2601

Canada   Dr William Glover, 163 Churchill Crescent, Kingston, Ontario K7L 4N3

Central America   Dr Stewart D. Redwood, P.O. Box 0832-1784, World Trade Center, Panama City, Republic of Panama

France   Contre-amiral François Bellec, 1 place Henri Barbusse, F92300 Levallois

Germany   Monika Knaden, Lichtenradenstrasse 40, 12049 Berlin

Iceland   Professor Anna Agnarsdóttir, Department of History and Philosophy, University of Iceland, Reykjavík 101

Japan   Dr Derek Massarella, Faculty of Economics, Chuo University, Higashinakano 742–1, Hachioji-shi, Tokyo 192–03

Netherlands   Dr Anita van Dissel, Room number 2.66a, Johan Huizingagebouw, Doezensteeg 16, 2311 VL Leiden

New Zealand   John C. Robson, Map Librarian, University of Waikato Library, Private Bag 3105, Hamilton

Portugal   Dr Manuel Ramos, Av. Elias Garcia 187, 3Dt, 1050 Lisbon

Russia   Professor Alexei V. Postnikov, Institute of the History of Science and Technology, Russian Academy of Sciences, 1/5 Staropanskii per., Moscow 103012

Spain   Ambassador Dámaso de Lario, Glorieta López de Hoyos, 4, 28002 Madrid

Switzerland   Dr Tanja Bührer, Universität Bern, Historisches Institut, Unitobler, Länggasstrasse 49, 3000 Bern 9

USA   Professor Mary C. Fuller, Literature Section, 14N-405, Massachusetts Institute of Technology, 77 Massachusetts Avenue, Cambridge, MA, 02139-4207

Sir Joseph Banks in 1773, shortly after his return from Iceland, by Sir Joshua Reynolds. The globe refers to his travels. The inscription on the letter reads CRAS INGENS ITERABIMUS AEQUOR, which may be translated as 'Tomorrow we set out once more upon the boundless main'. NPG 5868. Courtesy of the Trustees of the National Portrait Gallery.

SIR JOSEPH BANKS, ICELAND AND THE NORTH ATLANTIC 1772–1820
JOURNALS, LETTERS AND DOCUMENTS

Edited and annotated with an introduction by
ANNA AGNARSDÓTTIR
University of Iceland

Published by
Routledge
for
THE HAKLUYT SOCIETY
LONDON
2016

Published 2016 for the Hakluyt Society by

Routledge
2 Park Square, Milton Park, Abingdon, Oxon OX14 4RN

and by Routledge
711 Third Avenue, New York, NY 10017

Routledge is an imprint of the Taylor & Francis Group, an informa business

© 2016 The Hakluyt Society

The right of Anna Agnarsdóttir to be identified as author of this work has been asserted by her in accordance with sections 77 and 78 of the Copyright, Designs and Patents Act 1988.

All rights reserved. No part of this book may be reprinted or reproduced or utilised in any form or by any electronic, mechanical, or other means, now known or hereafter invented, including photocopying and recording, or in any information storage or retrieval system, without permission in writing from the publishers.

Trademark notice: Product or corporate names may be trademarks or registered trademarks, and are used only for identification and explanation without intent to infringe.

British Library Cataloguing in Publication Data
A catalogue record for this book is available from the British Library

Library of Congress Cataloging in Publication Data
Sir Joseph Banks, Iceland, and the North Atlantic 1772–1820 : journals, letters, and documents / edited by Anna Agnarsdottir, Hakluyt Society.
  pages cm. – (Hakluyt Society, third series ; 30)
Includes index.
ISBN 978-1-908145-14-7 (hardcover) – ISBN 978-1-4724-7782-8 (ebook) 1. Banks, Joseph, 1743–1820. 2. Banks, Joseph, 1743–1820 – Correspondence. 3. Naturalists – Great Britain – Biography. 4. Natural history – Great Britain – History. 5. Natural history – North Atlantic Region – History. 6. Natural history – Iceland – History. I. Anna Agnarsdottir, editor.
 QH31.B19S57 2016
 508.092—dc23
  [B]
 2015029924

ISBN: 978-1-908145-14-7 (hbk)
ISBN: 978-1-4724-7782-8 (ebk)

Typeset in Garamond Premier Pro
by Waveney Typesetters, Wymondham, Norfolk

Routledge website: http://www.routledge.com
Hakuyt Society website: http://www.hakluyt.com

Printed in the United Kingdom by Henry Ling Limited,
at the Dorset Press, Dorchester, DT1 1HD

In memory of Harold B. Carter
and of my parents, Agnar Kl. Jónsson and Ólöf Bjarnadóttir

CONTENTS

List of Maps and Illustrations	xi
Preface	xv
Acknowledgements	xvii
List of Abbreviations	xix
A Note on Currency, Weights and Measures	xxi
Glossary of Icelandic and Danish terms	xxiii

INTRODUCTION: BANKS AND ICELAND — 1
1. Joseph Banks: Explorer and Naturalist prior to the Iceland Expedition, 1743–72 — 1
2. The Origins of the Iceland Expedition — 7
3. Previous Anglo-Icelandic Relations — 11
4. Iceland 1772–1815 — 13
5. The Scientific Exploration of Iceland before Banks's Visit — 15
6. The Banks Expedition in Iceland — 17
7. The Iceland Expedition: An Assessment — 20
8. Sir Joseph Banks, 1772–1820 — 25
9. Sir Joseph Banks as Protector of Iceland during the Napoleonic Wars — 29

TEXTUAL INTRODUCTION — 35
1. Banks as a Writer — 36
2. The Text — 37

INTRODUCTION TO THE JOURNALS — 39
The Scientific Papers of the Iceland Expedition — 42

THE ICELAND JOURNAL OF SIR JOSEPH BANKS PART I:
12 JULY – 6 SEPTEMBER 1772 — 45
Banks's Notes from 29 August–15 September 1772 — 90

THE ICELAND JOURNAL OF SIR JOSEPH BANKS PART II:
17 SEPTEMBER – 22 OCTOBER 1772 — 93
Banks's Notes from 28–31 October and 18–21 November 1772 — 111
Notes by Banks Regarding His Expedition to Iceland in 1772 — 113

THE ICELAND JOURNAL OF JAMES ROBERTS — 115

CALENDAR OF LETTERS AND DOCUMENTS — 141

THE ICELAND CORRESPONDENCE AND DOCUMENTS OF
SIR JOSEPH BANKS, 1772–1820   152

APPENDICES

The Iceland Expedition 1772
1. Biographical Details of the Members of Banks's Expedition to Iceland in 1772   593
2. Thomas Pennant and Banks   598
3. Sir Joseph Banks's Itinerary to Hekla, 18–29 September 1772   600
4. Checklist of the Actual Titles in the Banks Library Taken Ashore in Iceland from the Sir Lawrence Brig, 1 September 1772   602
5. Icelandic Contemporary Sources on the Banks Expedition to Iceland in 1772   604
6. Costs for Hiring the *Sir Lawrence*   608

Greenland
7. The Latitude and Longitude of the Danish Settlements in Davis's Streights   610

Trade
8. Abstract of the Trade Ordinance of 13 June 1787   612
9. The Trading Agreements of 16 June and 22 August 1809 Made in Iceland by Captains of the Royal Navy   615
10. The Order in Council of 7 February 1810   618
11. Admiralty Prize Court Bills against *Den Nye Prove*   620
12. Tonnage Regulations Affecting Trade between Iceland and the United Kingdom   624
13. Biographical Details of Banks's Correspondents   625

BIBLIOGRAPHY   643

INDEX   663

LIST OF MAPS AND ILLUSTRATIONS

Maps

Map 1. Map showing the places mentioned by Sir Joseph Banks on his journey from Gravesend up the western coast of the British Isles towards Iceland and on the return journey via the Orkney Islands. 49
Map 2. Detail of the Western Isles of Scotland, showing the places mentioned by Sir Joseph Banks. 56
Map 3 Map showing the route taken by Sir Joseph Banks on his excursion to Hekla, 18–28 September 1772. 105

Colour Plates

Frontispiece: Sir Joseph Banks in 1773, shortly after his return from Iceland, by Sir Joshua Reynolds. The globe refers to his travels. The inscription on the letter reads CRAS INGENS ITERABIMUS AEQUOR, which may be translated as 'Tomorrow we set out once more upon the boundless main'. NPG 5868. Courtesy of the Trustees of the National Portrait Gallery.

Between pages 000 and 000

Plate 1. Sir Joseph Banks, by Thomas Phillips, 1809. This portrait shows Banks as President of the Royal Society in 1809, the year of the Icelandic Revolution. By Courtesy of the Trustees of the Royal Society.
Plate 2. Iceland: View of the Danish merchant houses in Hafnarfjörður, by John Cleveley Jr. This was where Banks and his party stayed during their visit. The boat on the right is probably the one used in the chase to find a pilot, described by Banks in his journal. BL, Add MS 15511, f. 13. Courtesy of the Trustees of the British Library.
Plate 3. Iceland: Inside the merchants' house, unsigned, but probably by John Cleveley Jr. The illustration shows women and children in Icelandic dress, specimens on the table, a flute-player and Banks being handed a glass of wine. By permission of the Linnean Society, London.
Plate 4. Iceland: Banks and his travelling caravan, by John Cleveley Jr. The snow-covered Mount Hekla is in the background, with lava and a volcano. BL, Add MS 15511, f. 48. Courtesy of the Trustees of the British Library.
Plate 5. Iceland: The traditional farm of Þorsteinn Jónsson, Banks's guide, at Hvaleyri, by John Cleveley Jr. The image has been cropped to show the details in the

foreground, including an Icelandic pony with a woman's saddle and a woman in full riding habit to the left and to the right the artist's impresssion of people processing fish. BL, Add MS 15511, f. 19. Courtesy of the Trustees of the British Library.

Plate 6. Iceland: Woman in her bridal gown, with decorations in silver, by John Cleveley Jr. This is believed to be Sigríður Magnúsdóttir, the wife of Ólafur Stephensen, district governor in 1772. BL, Add MS 15511, f. 17. Courtesy of the Trustees of the British Library.

Plate 7. Iceland: Eruption of the Great Geysir, by John Cleveley Jr. The expedition made exact measurements of this 'volcano of water'. BL, Add MS 15511, f. 37. Courtesy of the Trustees of the British Library.

Plate 8. The frontispiece of 'A Journal of a Voyage to the Hebrides Iceland and the Orkneys by James Roberts. 1772'. A rather fanciful image with Hekla apparently looming in the background. By W. Brand, Boston, Lincolnshire, October 1796. A 1594. Courtesy of the State Library of New South Wales.

Plate 9. Jörgen Jörgensen (1780–1841), the 'Revolutionary Chief' or 'Protector of Iceland' in 1808, by Christoffer Wilhelm Eckersberg. Courtesy of the Museum of National History, Frederiksborg Castle.

Plate 10. Magnús Stephensen (1762–1833), Chief Justice of Iceland, by Andreas Flint. It was he who wrote to Banks in October 1807 asking him to come to the aid of the Icelanders, which led to Banks taking on the role of protecting Iceland. This is a copper engraving from c. 1800. Courtesy of the National Museum of Iceland, number 4978.

Plate 11. Bjarni Sívertsen (1762–1833), one of the leading Icelandic merchants, ascribed to Rafn Þorgrímsson Svarfdalín. He was among the first of the stranded Icelanders to visit Banks in London. Courtesy of the National Museum of Iceland, number 886.

Plate 12. Banks's visiting card. Banks had this visiting card made on his return from Iceland. Note the pride of place given to Mount Hekla. British Museum, C.1–740, Sarah Banks's collection of visiting cards. © The Trustees of the British Museum.

Plate 13. Contemporary Map of Iceland. This map was printed in Horrebow's *The Natural History of Iceland* in 1758, close to Banks's visit to Iceland. *District* here means *sýsla* and *quarter* is used for *amt*. Hekla is depicted. Courtesy of the National Library of Iceland.

Plate 14. A view of the town and harbour of Reykjavík c. 1820, by the Danish governor of Iceland, Count Moltke. The cathedral is on the left. The farmers bringing their produce into town, camping in tents, can be seen in the centre. Viðey is the island to the right and the Danish flags are a reminder that at that time Iceland was a dependency of Denmark. Courtesy of the National Library of Iceland.

Plate 15. 'A View of the prize ship Orion and the Margarite and Ann just at the moment her foremast caught fire Sixteen leagues southward of the island of Iceland', by Jörgen Jörgensen. Drawing with grey wash. © National Maritime Museum, Greenwich, London, object PAH8461, image B7482.

Plate 16. 'The Fly Catching Macaroni', satirical print from 1772 of Banks balancing on two globes, one lettered 'Antartick Circle' and the other 'Artick Circle', denoting his travels on the *Endeavour* voyage with Cook and his expedition to Iceland, respectively. 1868.0808.4476. © The Trustees of the British Museum.

Figures

Figure 1.	Page from Banks's Journal part I, for 28–29 August 1772. Courtesy of McGill University, Blacker-Wood Library, Montreal.	81
Figure 2.	Poor Man of Iceland. An unsigned pen and ink drawing, dated 4 September 1772. There is a description saying he is wearing a blue waistcoat and a light coat with.metal buttons. The cap he is holding is blue. Add MS 15512, f. 23. Courtesy of the Trustees of the British Library.	87
Figure 3.	Icelandic woman with two children. An unsigned pen and ink drawing of an Icelandic woman with a boy and girl. Banks described the Icelandic dress of the women in detail in his journal for 6 September. He found it 'not very pleasing to an European eye' and considered the headdress 'very unbecoming'. Add MS 15512, f. 3. Courtesy of the Trustees of the British Library.	89
Figure 4.	Page from Banks's Journal part II, for 23 September 1772. Courtesy of the Kent History and Library Centre, Maidstone.	103
Figure 5.	Page from James Roberts's Journal for 8–10 September 1772. Courtesy of State Library of New South Wales, A 1594.	131
Figure 6.	Count Trampe (1779–1832). Count Trampe was the governor of Iceland during the Icelandic Revolution of 1809. Banks met him when he came to London in the autumn and helped him gain redress for the Icelanders. Image PO–010379. Courtesy of Norges Teknisk-Naturvitenskapelige Universitet, Trondheim.	336
Figure 7.	Last Page of the Draft of the 1810 Order in Council (Document 144). Banks was instrumental in drafting the Order in Council of 7 February 1810. In the left margin in his well-known hand Banks has noted: 'as a voluntary acknowledgment on the Part of England that the Sovereignty of Iceland is still vested in the Crown of Denmark' and secondly 'as a Confession on the Part of these Countries /*that*/ they are under the Protection of England'. Courtesy of the University of Wisconsin Digital Collections.	385
Figure 8.	First page of 'Some Notes relative to the ancient State of Iceland', 1813. These notes were 'drawn up with a view to explain its importance as a Fishing Station at the present time, with comparative Statements relative to Newfoundland'. DTC 17, 140. Courtesy of the Natural History Museum.	527

PREFACE

The involvement of Sir Joseph Banks (1743–1820) with Iceland began in 1772 in a completely fortuitous manner. He was about to set out on the *Resolution* with Captain Cook, with whom he had sailed on the successful *Endeavour* voyage. Not finding the accommodation to his liking, he withdrew from the voyage. As he had assembled an impressive scientific party at his own expense, it was of prime necessity to engage them in a new project. The season was advancing and to Banks Iceland, a dependency of Denmark, relatively unknown and reasonably near, was an ideal objective. Thus Banks mounted the first scientific British expedition to Iceland, by way of the Western Isles. Subsequently, and especially during the Napoleonic Wars, he became Britain's leading expert on Iceland and protector of the Icelanders. As he wrote in one of the letters published here: 'The hospitable Reception I met with in Iceland made too much impression on me to allow me to be indifferent about any thing in which Icelanders are concernd.'[1]

This volume contains almost all extant documents regarding Banks and his relations with Iceland and the North Atlantic from the time of preparation for his voyage to Iceland in 1772 until his death in 1820. They comprise his journals and that of his servant James Roberts and letters from Icelanders he met during his visit. There are relatively few letters until Banks was called into action in 1807 by the Chief Justice of Iceland to come to the aid of Icelanders captured by the Royal Navy after the outbreak of hostilities with Denmark and brought to British ports. The bulk of the documents date from 1807 to 1815, covering the Napoleonic Wars.

Letters in this volume reveal that, during periods of hostilities between Denmark and Britain, in 1801 and 1807, the British government considered annexing Iceland as a retaliatory measure. And Banks was consulted. He was all for it but the politicians soon realized they could treat Iceland as they pleased without incurring the cost of adding the island to the British Empire. It was thanks to Banks, however, that Iceland and the other Danish North Atlantic Islands were taken under the protection of Britain in 1810. The fate of captured enemy ships, letters from captives and stranded Icelanders in Britain, Banks's horror at the Icelandic Revolution of 1809 and his efforts to have the revolutionaries punished, are just a few of the fascinating historical topics that come alive in the letters. Trade during wartime was fraught with difficulties as many letters here make amply clear. These documents are not only valuable as sources for Icelandic history but also for the history of Georgian Britain during the Napoleonic Wars. Last but not least, they illuminate how Banks acted as a powerful protector and benefactor. Without his unstinting and selfless help Iceland would have suffered greatly during the war.

This volume began as part of the Joseph Banks Archive Project, founded in 1989 by Harold B. Carter, an Australian research scientist, and the physicist Dr Desmond

---

[1] Document 41 in this volume.

King-Hele, FRS. It became a joint project under the auspices of the Royal Society and the Natural History Museum, London, with Carter as the first Executive Director and King-Hele as the first chairman of the Executive Committee. While in Australia Carter had become interested in Banks – Banks being of course very well-known there, he adorned the $5 bill at the time – and he went on to publish two volumes on Banks: *His Majesty's Spanish Flock: Sir Joseph Banks and the Merinos of George III of England* in 1964 and *The Sheep and Wool Correspondence of Sir Joseph Banks 1781–1820* in 1979. He had by this time moved to England where he began mapping out the extent of Banks's role in national life. Sir Joseph Banks had no legitimate heirs and during the nineteenth century his correspondence was sold at auction. Of this, 20,000 letters have survived and are to be found in repositories all over the world. Carter identified various themes in Banks's correspondence, one of them being the Iceland journals and correspondence, the extent of which justified a separate volume in his opinion. Harold Carter invited me to edit the Iceland correspondence, originally as a joint editor. However, more or less at the same time I was appointed to a tenured position in history at the University of Iceland and thus entered academic life full-time, with all that entails. Harold had already listed those letters he believed should be in the Iceland volume, thus laying the groundwork, and had begun transcribing them. The decision was taken at a later date to include the few letters dealing with Greenland and the Faroe Islands in the Iceland volume as the most fitting place for them in the Banks correspondence publications as a whole. As soon as the men trading with these islands heard about Banks coming to the aid of the Icelanders they were quick to approach him as well. Banks did not disappoint them. He was really interested in the North, as his sponsorship of many northern voyages indicated, and he soon added Greenland as one of the islands that might become a useful part of the British Empire. Carter then left the Iceland letters to me, turning his interest to other themes while directing the Banks Archive Project.

The aim of the Sir Joseph Banks Archive Project is to gather together copies of all the surviving documents, catalogue them in a database, transcribe, edit and finally publish them in similar scholarly editions. The Banks Archive Project is an independent research project with Dr Neil Chambers as the Executive Director. To date he has edited fifteen volumes of Banks correspondence, with a series on Africa currently ongoing. The editor of this volume has worked in close collaboration with Dr Chambers.

ACKNOWLEDGEMENTS

This volume has taken an extremely long time to complete because of academic commitments. It is also in essence inter-disciplinary and thus I am indebted to specialists in many fields. All geological material has been supplied by Professor Leifur Símonarson of the University of Iceland, botany has been elucidated by Dr David Hibberd as well as by Hörður Kristinsson, and meteorology by Trausti Jónsson of the Icelandic Meteorological Office.

At the beginning of this project Peter Warren of the Royal Society, Rex Banks and the late John Thackeray of the Natural History Museum were extremely enthusiastic and encouraging.

I would like to thank Dr Richard Virr of McGill University Library, the Linnean Society, the National Portrait Gallery, Margrét Hallgrímsdóttir of the National Museum and Örn Hrafnkelsson of the National Library here in Iceland for their generosity regarding illustrations. For the funding of the remainder of the illustrations I would like to thank the Institute of Humanities of the University of Iceland and especially the Rector of the university, Professor Jón Atli Benediktsson.

When the decision to include Greenlandic material was taken I was lucky enough to be ably assisted by Dr Thorkild Kjærgaard of the University of Greenland and Niels Frandsen. As to Faroese queries, Elin Súsanna Jacobsen was always quick to respond.

Not all the letters were in English. Special thanks go to Charles Burnett for translating the Latin letters, Professor Harald Gustafsson for help with Swedish, Professor Torfi Tulinius with French and Sigurður Pétursson and Svavar Hrafn Svavarsson for specific Latin inquiries. In the National Archives of Iceland I must especially thank Kristjana Kristinsdóttir for checking my transcriptions of the eighteenth-century Icelandic and Danish documents as well as Jón Torfason and Björk Ingimundardóttir. I am also indebted to all the librarians and archivists in the repositories of Banksian material worldwide. Special mention, however, must be made of Martha Whittaker of the old Sutro Library and Haleh Motiey-Payandehoo of the new Sutro Branch of the California State Library at the University of San Francisco; of John Tedeschi, who was so welcoming when I came to the Rare Book Library of University of Wisconsin, as was his successor Robin Rider; and of Andrea Hart of the Natural History Museum. I owe very special thanks to my friend the historian Dr Hrefna Róbertsdóttir who has urged me on during these long years spent in the company of Sir Joseph Banks.

My colleagues in the Department of History, Guðmundur Hálfdanarson, Guðmundur Jónsson, Gísli Gunnarsson and Sveinbjörn Rafnsson, have always been ready with help. I also thank Professor emeritus Andrew Wawn of the University of Leeds for helping with literary matters. I am also grateful for the help I have received from Colin Penman, Patrick Kaye, Andrew Cook and Margrét Eggertsdóttir. This work could not have been completed without my assistants, in particular Valgerður Johnsen, Rúnar Már Þráinsson,

Dr Ólafur Rastrick and Óðinn Melsted, and last but not least my doctoral student Margrét Gunnarsdóttir for a thorough read-through with positive results. I would like to give special thanks to Neil Chambers of the Banks Archive Project, my friend and collaborator, for simply always being there and responding to all my queries regarding our mutual friend, Joseph Banks. My editors from the Hakluyt Society, Will Ryan and Gloria Clifton, have been very helpful, their expertise saving me from serious mistakes. Finally thanks to the staff of Ashgate Publishing, especially John Smedley, and their freelance collaborators Mary Murphy and Barrie Fairhead.

LIST OF ABBREVIATIONS

Add MS	Additional Manuscripts, Department of Manuscripts, British Library, London
ADM	Admiralty (TNA)
Bar.	Barometer
BC	The Natural History Museum, London. Banks Correspondence, Botany Library, Special Collections
Bf. Rvk.	Bæjarfógetinn í Reykjavík, file in the National Archives of Iceland; see Glossary
BL	British Library, London
BM	British Museum, London
BT	Board of Trade (TNA)
CE	Board of Customs and Excise (National Archives of Scotland)
DBL	*Dansk Biografisk Leksikon*
DfdUA	Departement for de Udenlandske Anliggender [Danish Foreign Office] Rigsarkivet [Danish National Archives], Copenhagen
DI	*Diplomatarium Islandicum*
DNB	*Dictionary of National Biography*
DTC	Dawson Turner Collection, Natural History Museum, London
E	Exchequer (National Archives of Scotland)
Eg. MS	Egerton Manuscripts, Department of Manuscripts, British Library, London
FO	Foreign Office (TNA)
FRS	Fellow of the Royal Society
Ft	Foot or feet
Ges. Ark.	Gesandts Arkiver (archives of the ambassador or minister)
HCA	High Court of Admiralty (TNA)
HO	Home Office (TNA)
Hooker Corr.	Correspondence of Sir William Jackson Hooker, Archives of the Royal Botanic Gardens, Kew
JBK	Sir Joseph Banks Papers, Archives of the Royal Botanic Gardens, Kew
Kaye Coll.	Private Collection of Banks Correspondence belonging to Dr Patrick Kaye, York.
KK	Kommercekollegiet, Rigsarkivet, Copenhagen
km	kilometre (= 0.62 miles)
kmcl.	Commercial tons (see 'A Note on Currency, Weights and Measures')
Lbs.	Landsbókasafn Íslands-Háskólabókasafn (National and University Library of Iceland, Reykjavík)

L.S.	*Locus sigilli* (Latin) position of seal on document
ML	Mitchell Library, Sydney, Australia
MS	Manuscript
NAI	National Archives of Iceland, Þjóðskjalasafn Íslands, Reykjavík, Iceland
NAS	National Archives of Scotland
NHM	Natural History Museum, London
ODNB	*Oxford Dictionary of National Biography*
PC	Privy Council (TNA)
Phil. Trans.	*Philosophical Transactions of the Royal Society*
PRO	Public Record office (TNA)
RA	Danish National Archives, Danmarks Rigs Arkiv or Rigsarkivet, Copenhagen
Rd.	Rixdollar (see 'A Note on Currency, Weights and Measures')
RGO	Royal Greenwich Observatory (archives)
RS	The Royal Society, London
Rtk.	Rentekammer (Chamber of Rents, Rigsarkivet, Copenhagen)
SJB	Sir Joseph Banks
SL	The Sutro Library, University of San Francisco (The Banks Collection)
T	Treasury (TNA)
Th, Ther or Therm	Thermometer
TNA	The National Archives, Kew, London
TS	Treasury Solicitor and HM Procurator General (TNA)
Wisconsin MS	'Banks papers: Iceland, the Danish colonies & the Polar regions, 1772–1818.' Correspondence concerning Iceland: Written to Sir Joseph Banks. MS 3. Department of Special Collections, Memorial Library, University of Wisconsin-Madison, USA.
WO	War Office (TNA)

A NOTE ON CURRENCY, WEIGHTS AND MEASURES

These are highly complicated subjects with fluctuations and regional differences. Readers are further referred to Astrid Friis and Kristof Glamann, *A History of Prices and Wages in Denmark 1660–1800*, Copenhagen, 1958, especially I, pp. 119–40, and volume II by Dan H. Andersen and Erik Helmer Pedersen, Copenhagen, 2004. Also very useful is Guðmundur Jónsson and Magnús S. Magnússon, eds, *Hagskinna: Icelandic Historical Statistics*, Reykjavík, 1997, pp. 921–5.

Currency

Money was little used in Iceland during this period. The trade was one of barter. A standardized system of measurements had been introduced into the Danish realm at the end of the seventeenth century. In trade the weight of goods was measured in many different ways, the most common perhaps being *pund*, *lispund* and *skippund*.[1] It was a time of war and instability with high inflation and during the period 1772–1820 there were significant fluctuations, culminating in the actual bankruptcy of the Danish state in 1813.

Rigsdaler (in English always *rixdollar*) = 6 marks = 96 skillings. In 1772 the rixdollar equalled 4 English shillings and 6 pence.[2] According to Hooker, before the war (1807) the rixdollar was the equivalent of 4 shillings, thus a pound (20 shillings) then would have been equivalent to about 5 rixdollars. Inflation increased during the war years.

Distances

Danish mile = 7.5 kilometres (from 1683).
Fathom (Icelandic: *faðmur*): 4,000 fathoms = 1 Danish mile.

Weights and Measures

The basic weights were:
kommercelæst (commercial last) Danish weight (mostly used to measure capacity of ships) = 2.3 English imperial tons.

---

[1] Gunnarsson, *Fiskurinn sem munkunum þótti bestur*, pp. 99–100.
[2] See Banks's journal for 1 September 1772 (p. 85 below).

skippund (skp.) = 20 lispund = 320 pounds = 160 kilograms.
lispund (lp.) = 16 pounds (lb) = 8 kilograms.
pund = about 496 grams (in 1839 fixed at 500 grams as the Copenhagen pund).

These weights vary considerably according to what is being weighed. For example:
Klipfish	one last = 18 skippund
Stockfish	one last = 15 skippund
Sugar	one last = 10 skippund
A barrel of rye	10 lispund = 84 kilograms (1776)
A barrel of unground rye	96 kilograms
Butter, tallow, soap, liver oil, sulphur	14 lispund = 112 kilograms (1751).

Pot (quart) = 0.966 litres or $\frac{1}{32}$ cubic feet.

The *tønde* (barrel) was both a unit of capacity and a unit of weight. The *korn* (grain) barrel contained 144 *potter* (=139.121 litres), the salt barrel 176 potter, the *smör* (butter) barrel 224 pund, the *tjære* (tar) barrel 120 potter and the *öl* (beer) barrel 136 potter.

There are also references in the documents to the English hundredweight (cwt). This was equivalent to 112 pounds or 50.8 kilograms.

GLOSSARY OF ICELANDIC AND DANISH TERMS[1]

The Icelandic term and Danish equivalent, if applicable, are in italics. English visitors commonly adopted the Icelandic/Danish terms and anglicized them. The variations which appear in the text are noted here. The formal introduction of absolutism in Iceland in 1662 was followed with a great many changes in the administration.

Alþingi: the Althing, the Icelandic assembly with a legislative and judicial role, founded in 930 and convened every summer for two weeks at Þingvellir until 1798. It was abolished in 1800. By then it had lost all its legislative power to the absolute Danish king and its judicial role was taken over by a new High Court of Justice, named *Landsyfirréttur*, with a judge, two assessors and a secretary, situated in Reykjavík.

amt (Danish: *amt*): an administrative district. Originally four districts in 1662, from 1770 Iceland was divided into two *amts* – one for the south-west (to 1787) under the jurisdiction of the *stiftamtmaður* (q.v.), the other for the north-east – which was the situation when Banks came to Iceland in 1772. In 1787 the south and west *amts* were separated: the southern district continued under the jurisdiction of the *stiftamtmaður* while a new *amtmaður* (q.v.) was appointed to the western district. Thus there were three amts in Iceland during the period 1787–1820.

amtmaður (Danish: *amtmand*): district governor (in English variously written as ampman, amptman) and contemporary English translations include deputy-governor, bailiff, vice-governor, sub-governor, lieutenant-governor. The *amtmaður* had wide-ranging supervisory responsibilities regarding the general welfare of the country. In 1772 there was only one *amtmaður*, who governed the northern and eastern amts. During the period 1787–1820 there were strictly speaking only two *amtmenn* (one for the north and east and one for the west) as the *stiftamtmaður* (q.v.) also governed the southern amt. Until 1770, when it became obligatory for the *stiftamtmaður* to reside in the country, the *amtmaður* in the south was the de facto governor of the country. They served as a link between the governor and the *sýslumenn* (q.v.), ensuring that laws and proclamations reached the public and supervising the lesser royal officials.

bæjarfógeti (Danish: *byfoged*; English: *townfoged*): office established in Reykjavík in 1803. It combined the roles of recorder or judge and chief of police, responsible for keeping law and order in Reykjavík. In 1806 this office was merged with that of the *landfógeti* (q.v.), until 1874.

conferensråd: a high-ranking title in the Danish system of ranks, conferred by the king, now obsolete.

---

[1] I am indebted to my colleagues Professor Guðmundur Hálfdanarson and Dr Hrefna Róbertsdóttir for assistance with this glossary.

etatsråd: a title in the Danish system of ranks, conferred by the king, now obsolete. An *etatsråd* could eventually hope to be promoted to a *conferensråd*.

gjá: fissure.

gljúfur: ravine.

Handel: the Danish word for trade, a word used by the British visitors for the trading period during the summer when the farmers would bring their produce to one of the 24 designated trading ports. Called *kauptíð* in Icelandic.

hraun: lava or lava-field.

hver: hot spring.

justitsråd: the lowest title in the Danish system of ranks, now obsolete. A *justitsråd* could aspire to becoming an *etatsråd*.

jökull: glacier.

[*Dansk*] *Kancelli* (Danish): The Danish Chancery. It was organized as one of the main two ministries (*kollegium*) in the Danish administration (the other being the *Rentekammer*, see below), dealing with affairs of the interior and justice in Denmark, Norway and the Atlantic islands.

klipfish: dried and salted fish, most commonly cod.

Kommercekollegium: The Danish Board of Trade.

landfógeti (Danish: *landfoged*) (in English variously written as landfogt, landfogd or landvogt, and translated as King's Receiver or cashier, as well as treasurer): the treasurer of the island, who oversaw the country's finances, trade and the royal demesne. The office was founded in 1683.

landlæknir or *landfysicus*: the state physician. This office was founded in 1760, two more medical doctors being appointed in 1766. The state physician was also the apothecary until 1772.

Landsyfirréttur: The High Court of Justice in Reykjavík, established 1800 after the abolition of the *Alþingi* (q.v.) by amalgamating the two courts at Þingvellir, the *Yfirdómur* and *Lögrétta*. During the period 1800–1820 there was a Chief Justice or Justitarius, two assessors and one secretary. Cases could be appealed to the Supreme Court in Copenhagen. It continued until 1919.

lögmaður: (English: 'lagman'; Danish: *lovmand*, plural: *lögmenn*) lawman. There were two lawmen in Iceland from 1271, one for the North and West, the other for the South and East. They attended the *Alþingi* (q.v.) being the chief judges and thereby the leaders of the parliament, the de facto native leaders of the country (along with the bishops). Most were Icelanders. The office was abolished in 1800 along with the *Alþingi* itself.

lögréttumaður: a farmer appointed to accompany the *sýslumaður* (q.v.) to *Alþingi* (q.v.), their main function being to attend and participate in the *Lögrétta* (Law Council) or court of justice. They had tax privileges and were paid for their attendance. Originally 140 in number during medieval times, after 1764 this was reduced to twenty, ten of whom were obliged to attend *Alþingi* each year. This was a position of honour and respect.

Rentekammer: the Danish Exchequer, the ministry dealing with state finances.

Ridder af Dannebrog: Knight of the Dannebrog. Lowest class of the Order, instituted by Christian V. Until 1808 membership was limited to fifty members of noble or royal rank. The first Icelander to receive this honour was Bjarni Sívertsen (see Appendix 13).

söpas: Danish sea-licence.

stiftamtmaður (Danish: *stiftamtmand*), the governor (called variously stiftsamtman, stiftsampman, stiftsamptman, stiftsbefalingsmand), usually translated as governor but also grand bailiff. The position was founded in 1683. It was then largely a ceremonial post (the first holder of the title was the five-year-old illegitimate son of the Danish king, his successors being almost always Danish aristocrats). After 1770 the governor was obliged to reside in Iceland, being the highest royal official of the King of Denmark in Iceland and exercising his power as such.

stiftamtsskriver: secretary to the governor.

stiftsprófastur: the clergyman substituting for the bishop, one in each bishopric.

stockfish: dried fish, commonly cod.

sýsla (Danish: *syssel*) county. The traditional administrative and juridical district in Iceland. Iceland was divided into around twenty-three counties during the period of Banks's involvement with Iceland (1772–1820).

sýslumaður (plural: *sýslumenn*, Danish: *sysselmand-mænd*): county magistrate (variously translated as sheriff or bailiff or Anglicized as sysselman, sisselman, systelman or sytelman). His duties were collecting taxes, judging cases and meting out punishment to criminals. He also supervised trade, maintained law and order, and was obliged to attend the *Alþingi*. By the latter half of the eighteenth century these officials were required to have a law degree from the University of Copenhagen. The *sýslumaður* was the king's representative in the county. Other duties included being a notary and auctioneer.

Tatsroed: Derived from *etatsraad*, it was used by the British visitors when referring to Magnús Stephensen, the Chief Justice of Iceland (see Appendix 13), who indeed had been honoured with this title.

tómthúsmaður: a man with little or no land who made his living from the sea.

vaðmál: the traditional Icelandic woollen homespun cloth, resembling twill. In medieval times it was used as currency.

værtime (Danish), *vertíð* (Icelandic): fishing season.

INTRODUCTION: BANKS AND ICELAND

Sir Joseph Banks's relations with Iceland began in 1772 when he set off to explore the island, leading the first scientific expedition undertaken there by British naturalists. As a consequence of that visit, Banks became the acknowledged British expert on Iceland and a faithful friend of the Icelanders. Three decades later during the Napoleonic Wars, Banks assumed a crucial political role as self-appointed protector of Iceland, smoothing the way for their trade during the conflict and repeatedly urging the British government to annex the island for the benefit of the inhabitants. He became the architect of Britain's political and commercial policy towards the Atlantic dependencies of the Danish realm. This volume contains the journals of the Iceland expedition and almost all letters and papers worldwide in the Banks archives pertaining to Iceland, Greenland and the Faroe Islands.

1. Joseph Banks: Explorer and Naturalist prior to the Iceland Expedition, 1743–72

Joseph Banks[1] was born on 13 February 1743 to great wealth and a privileged position in society. His great-grandfather had bought the extensive estates of Revesby Abbey in Lincolnshire in 1714. His father, William Banks Hodgkinson (1719–61), had been a member of parliament, as had his father and grand-father before him, and Deputy-Lieutenant of Lincolnshire. Much land had been successively added and by Banks's time the family owned considerable estates, mostly in Lincolnshire, where he had 268 tenanted farms, Staffordshire and Derbyshire. He thus belonged to the landed gentry. Banks was educated first at Harrow and then Eton. In 1760 he entered Christ Church, Oxford, as a gentleman commoner, the first member of his family to attend university, where his main studies were in botany.[2] Four years later he left without a degree which in itself was quite common at the time for members of his class,[3] though there is little evidence that he was much of a scholar. His father died in 1761, when Banks was 18, and when he came of age three years later he was a very wealthy young man indeed, with an income of about £6,000 a year,[4] and thus able to pursue whatever kind of life he chose. He decided to devote his

---

[1] There is a vast bibliography on Banks, see especially: Carter, *Sir Joseph Banks (1743–1820). A Guide to Biographical and Bibliographical Sources*; Gascoigne, 'Banks, Sir Joseph, baronet (1743–1820)'; *ODNB*, III, pp. 691–6. He has been the subject of several biographies, the definitive one being Harold B. Carter's, *Sir Joseph Banks 1743–1820* from 1988. Gascoigne has also written a historiographical essay on Banks: 'The Scientist as Patron and Patriotic Symbol: the Changing Reputation of Sir Joseph Banks', in Shortland and Yeo, eds, *Telling Lives in Science*.

[2] Gentlemen commoners paid higher fees than other commoners and were permitted certain privileges such as wearing a velvet cap and silken gown.

[3] See Gascoigne, 'Banks, Sir Joseph', p. 691.

[4] Now about £708,000 (according to http://www.measuringworth.com for 2012).

life to the advancement of natural history with emphasis on botany: 'Botany has been my favourite Science since my childhood' he wrote in 1782.[1]

His passion for botany was famously kindled by the discovery of a copy of Gerard's *Herbal*,[2] a first edition, in his mother's dressing room at Revesby Abbey. At Oxford at that time botany was not rated highly and Humphrey Sibthorp (1713–97), the professor in that field, reputedly gave only one lecture. Banks resorted to his own measures, going to Cambridge where at his own expense he engaged the botanist Israel Lyons (1739–75).[3] Lyons gave Banks and his friends a series of lectures on botany in Oxford in 1764. Later Philip Miller (1691–1771), a near neighbour of Banks's mother and chief gardener of the Chelsea Physic Garden from 1722 to 1770, friend of Linnaeus and author of *The Gardener's and Florist's Dictionary*, played a significant part in Banks's botanical education.[4] Apart from his botanical studies, as Harold B. Carter, Banks's biographer, concluded, Banks was more or less 'self-educated in the natural sciences',[5] while Averil Lysaght, the editor of Banks's Newfoundland-Labrador journal, came to the conclusion that Banks was 'well-trained, fundamentally serious and extremely industrious in all his scientific undertakings'.[6]

In 1767 Banks bought a house in New Burlington Street in Mayfair. He became a well-known figure on the London social scene, attending scientific meetings and frequenting the British Museum. He made plenty of friends among the scientists of the age such as Thomas Pennant (1726–98) the zoologist and traveller and John Lightfoot (1735–88) the botanist. Neil Chambers has described his life at this point: 'Independent and affluent, gregarious and energetic, Banks had opportunities in London to indulge his taste for learning and for company, both of which he enjoyed to the full.'[7] Early in February 1766 Banks was nominated as a fellow to the Royal Society, barely 23 years of age, his qualifications being that he was 'versed in Natural History especially Botany and Other branches of Literature, and likely (if Chosen) to prove a Valuable Member',[8] and that same month he was also elected to the Society of Antiquaries.[9]

At the beginning of April 1766, Banks went off on his first scientific venture.[10] He joined his Etonian friend Lieutenant Constantine Phipps (later Lord Mulgrave)[11] as a

---

[1] London, Natural History Museum [hereafter NHM], Dawson Turner Collection [hereafter DTC], 2, Banks to Edward Hasted, [?February 1782], f. 97, quoted by Carter, *Sir Joseph Banks*, p. 177.

[2] Gerard's *Herbal* was first published in 1597 by John Gerard (1545?–1612) and has often been republished. It is well illustrated with more than 1,800 woodcuts (his illustration of the potato was the first to appear in any herbal) and it is easy to see why it would appeal to a boy like Banks.

[3] Israel Lyons was also a mathematician and astronomer and through Banks's help he was hired as the astronomer on the Arctic voyage of Constantine Phipps in 1773 (see below, p. 26, n. 1).

[4] Miller's book, first published in 1731, went through 24 editions in his lifetime.

[5] Carter, *Sir Joseph Banks*, p. 25.

[6] Lysaght, *Banks in Newfoundland*, p. 10.

[7] Chambers, 'General Introduction', *Scientific Correspondence of Sir Joseph Banks*, I, pp. ix–x.

[8] An illustration of Banks's certificate of election to the Royal Society is to be found in Carter, *Sir Joseph Banks*, p. 31. Chambers, *Indian and Pacific Correspondence of Sir Joseph Banks*, I, pp. 313–15.

[9] Appendix XXVII in Carter, *Sir Joseph Banks*, lists all his academic and civil honours and memberships of societies (pp. 585–7).

[10] See Lysaght, *Banks in Newfoundland*, for everything pertaining to this voyage.

[11] Constantine Phipps, 2nd Baron Mulgrave (1744–92), naval officer, explorer, future MP and Lord of the Admiralty.

supernumerary on an expedition to Labrador and Newfoundland on the *Niger*, a fisheries protection vessel. Banks set off well equipped for the collection of botanical and zoological specimens. With his assistant Peter Briscoe[1] from the Revesby estate he collected plants, shot birds, fished, and collected scientific specimens, as he would do later in the South Seas and in Iceland. Thus Banks became one of the first naturalist explorers; the nine-month Newfoundland experience would prove invaluable for his subsequent expeditions and he returned to England as a 'scientifically trained Linnaean naturalist'.[2] He was beginning to gain a reputation as 'a most expert naturalist, & a very Sensible worthy man'.[3]

Banks kept a diary on the *Niger* expedition, describing not only the flora, fauna and minerals but also the indigenous people he met. This would be a recurring theme of all his journals. From the late summer of 1767 until early 1768 Banks decided to explore Wales and the West Country from Cheshire south to Berkshire, collecting plants and investigating ancient monuments and antiquities.[4]

Banks had become acquainted with a Swedish botanist named Daniel Solander, a favourite student of the great Linnaeus at Uppsala, who had sent the young man to England in 1760 as one of his 'apostles'.[5] Solander is credited with consolidating the Linnaean system in Britain.[6] In 1763 he had been employed to catalogue the natural history collections of the British Museum and the following year he was elected a fellow of the Royal Society. Banks and Solander then began 'forging their working partnership and lasting friendship from which so many advances in natural history were to evolve'.[7] Solander was eventually engaged by Banks to put his Labrador specimens in order.[8] At this point in his life Banks was planning to go to Sweden and Lapland, accompanied by Solander who had already twice travelled to Lapland (in 1753 and 1755), with the intention of visiting 'our Master Linnæus' and 'Profiting by his Lecture before he dies'.[9] But before Banks could put the Swedish plan into action he heard about a proposed voyage around the world.[10]

In 1769 the transit of Venus across the Sun was due (and then not again until 1874), and the Royal Society wished to obtain good observations of the phenomenon which was expected to solve the problem of calculating the distance of the Sun from the Earth.

---

[1] Peter Briscoe: see Appendix 1.

[2] Gascoigne, 'Banks, Sir Joseph', p. 691. By chance, during that same voyage Banks actually met in St John's, Newfoundland, James Cook, who was surveying the coasts of Labrador and Newfoundland (Lysaght, *Banks in Newfoundland*, p. 41).

[3] Letter of 16 May 1767, Dr William Watson to Dr Richard Pulteney, quoted by Carter in 'The Royal Society and the Voyage of HMS *Endeavour* 1768–71', p. 247.

[4] Beaglehole, *Endeavour Journal of Joseph Banks*, I, pp. 16–17. Banks kept a journal from 13 August 1767 to 29 January 1768. See Carter, *Banks. The Sources*, pp. 53–4.

[5] Daniel Solander: see Appendix 1.

[6] Duyker, *Nature's Argonaut*, p. 258. The Linnean System is a taxonomic system, first set out by Linnaeus in his *Systema Naturæ* (1735). It consists of three kingdoms, divided into *classes*, and they, in turn, into *orders*, *families*, *genera* and *species*. This system still enjoys near universal acceptance.

[7] Carter, *Sir Joseph Banks*, p. 28.

[8] Now in the Natural History Museum, London.

[9] Banks to Pennant, late June 1767, Chambers, *Scientific Correspondence of Sir Joseph Banks*, I, pp. 17–18.

[10] On this occasion Banks wrote: 'Every blockhead does that [goes on a Grand Tour]; my Grand Tour shall be one round the whole globe' (Smith, *Life of Sir Joseph Banks*, pp. 15–16).

The government had been petitioned by the Royal Society to send observers to several parts of the world, one of the three locations being Tahiti,[1] and George III had been asked to sponsor it. In February 1768 the Royal Society's memorial was approved by the King who was prepared partly to finance a voyage to the South Seas. The Royal Navy would provide the ship *Endeavour* and a crew under the leadership of Lieutenant James Cook. Banks was determined to participate much in the same way as he had in the Labrador-Newfoundland voyage and on his own initiative, with the help of his old friend and fishing companion Lord Sandwich (1718–92), then Postmaster-General,[2] he succeeded in being accepted for the voyage as a supernumerary in natural history on the *Endeavour*, not only for himself but for a 'Suite' of eight.[3] In the minutes of the Royal Society of 9 June 1768 Banks was described as 'a Gentleman of large fortune, who is well versed in natural history' and it was 'very earnestly' requested that 'in regard to Mr. Banks's great personal merit, and for the Advancement of useful Knowledge', he be permitted to sail with Cook.[4] In fact his participation was to prove a godsend as it was he who recovered the stolen quadrant on which the observation of the transit of Venus was dependent, during a hazardous pursuit of some of the people of Otaheite (Tahiti).

The prime objectives of the *Endeavour* voyage were in the fields of astronomy and geographical discovery,[5] and originally there were no plans regarding the field of natural history. But Banks's participation in that capacity was welcome, especially when he proved more than willing to pay the costs of his expedition and that of his 'suite'.[6] Not only did he pay Solander's salary but those of his artists Sydney Parkinson[7] and Alexander Buchan;[8] his secretary Herman Spöring;[9] two field assistants from the Revesby estate: Peter Briscoe who had accompanied him to Labrador and Newfoundland and the sixteen-year-old

---

[1] Captain Samuel Wallis (1728–95), who circumnavigated the world in HM frigate *Dolphin*, had just arrived back and reported on this suitable island. Wallis named Tahiti 'King George the Third's Island'.

[2] John Montagu, 4th Earl of Sandwich (1718/19–95), FRS, statesman. He served in many capacities during his distinguished career, becoming First Lord of the Admiralty in 1748–51, again in 1763 and for the third time in 1771–82. There is a well-known portrait of Banks, Solander, Cook, Sandwich and Dr John Hawkesworth, the editor of Cook's papers, painted by John Hamilton Mortimer in 1771, now in the National Library of Australia.

[3] Admiralty to Cook, 22 July 1768, quoted by Carter, *Sir Joseph Banks*, p. 65.

[4] Quoted by Beaglehole, *Endeavour Journal of Joseph Banks*, I, p. 22.

[5] Carter, *Sir Joseph Banks*, p. 60.

[6] As Beaglehole has put it: 'What he [Banks] was proposing to do was to plant himself, a train of dependants and a mass of impedimenta on a small and already overcrowded vessel, commanded by a man he did not know ... and he was proposing it in the sure, certain, and unhesitating conviction that he had a right to be obliged, and would be made welcome. This it was that was so highly characteristic of the English gentleman of fortune of that age, so effortlessly superior, so candidly appropriative of privilege, upon his Grand Tour; this it was that was so completely the Banksian attitude to life' (*Endeavour Journal of Joseph Banks*, I, p. 23).

[7] Sydney Parkinson (1745?–71), draughtsman of natural history. In 1767 Banks engaged him to to copy a collection of drawings brought back from Ceylon by Governor Loten (see Document 13).

[8] Alexander Buchan, Scottish landscape artist, who died during the voyage in 1769.

[9] Herman Dietrich Spöring Jr (1733–71), naturalist, artist, surgeon and secretary to Banks. His father of the same name was a professor of medicine at Turku Academy. Spöring was a Swede from Finnish-speaking Turku. He studied medicine at Turku and surgery in Stockholm. He eventually came to London, working as a watchmaker or instrument maker. It was there that he met Solander and was employed by Banks as an 'Assistant Naturalist', as well as an artist and 'writer' on the *Endeavour* voyage. He was one of the casualties of the voyage. Beaglehole, *Endeavour Journal of Joseph Banks*, I, p. 27.

INTRODUCTION

James Roberts;[1] along with two servants, Thomas Richmond and George Dorlton. A reference library was assembled 'relating to the natural history of the Indies',[2] which would be diligently read during the voyage, and a vast quantity of equipment for collecting and storing specimens was sent on board ship.

The *Endeavour* expedition of 1768–71 is considered one of the most important voyages of discovery ever made. The ship sailed first to Madeira, across the Atlantic to Rio and south past Tierra del Fuego. Three months were spent in Tahiti, preparing for the transit of Venus. Subsequently the ship sailed on to New Zealand, previously visited by Europeans, but thought to be part of a large southern land mass, and found it to be two islands. These they claimed for Britain as New South Wales. Thus the belief that New Zealand was part of a large southern land mass was put to rest. They landed in Australia in April 1770, Cook naming the place Botany Bay. They subsequently sailed north along the coast of what Cook called New South Wales to Batavia (now Jakarta). He crossed the Indian Ocean to the Cape of Good Hope, stopping at Cape Town, then sailed up to St Helena, finally arriving back in England at Deal on the Downs on 12 July 1771. Though the voyage took a grim toll of Banks's party (only he himself, Solander, Briscoe and Roberts survived),[3] the *Endeavour* voyage was a triumph in the geographical and natural scientific sense, despite the fact that the object of the voyage was not realized. Though no one was aware of it at the time, it was impossible to make accurate observations of Venus in the way intended.

Banks and his assistants had collected some 30,000 botanical specimens, including 110 new genera and 1,400 new species. Among the zoological collections were over 1,000 animal specimens, famously including the kangaroo. Besides which Banks had shown great interest in ethnology,[4] actually learning Polynesian, making word-lists, and recording native customs in his journal.[5] Banks's journal of 200,000 words provided 'useful accounts ... and in a greater measure than Cook had either time or inclination to report'.[6] It proved invaluable to Cook when he had to send in his official report to the Admiralty.

These were the first collections from the South Pacific to be seen in Britain. On their return it was the gentlemen Banks and Dr Solander[7] who were fêted, Cook being somewhat in their shadow at that time. Lady Mary Coke, daughter of the Duke of Argyll, wrote on 9 August 1771:

---

[1] James Roberts (Mitchell Library) and Briscoe (Dixson Library) both wrote journals which Beaglehole dismissed as having 'no particular value as journals' (*Endeavour Journal of Joseph Banks*, I, p. 28n).

[2] Beaglehole, *Endeavour Journal of Joseph Banks*, I, p. 33.

[3] All four survivors were to go together to Iceland.

[4] Beaglehole calls him 'the founder of Pacific ethnology': *Endeavour Journal of Joseph Banks*, I, p. 40. See Lysaght on this aspect of Banks and his Lockeian emphasis on observation, 'Banksian Reflections', *The Journal of Joseph Banks in the Endeavour*, I, pp. 13–28.

[5] For further on the natural history results of the voyage, see Chambers, 'General Introduction', *Indian and Pacific Correspondence of Sir Joseph Banks,* I, p. xv.

[6] Carter, *Sir Joseph Banks,* p. 93. Banks's *Endeavour* journal was not published until 1896 in London by Sir Joseph D. Hooker. It was a heavily edited version. A scholarly edition was published by J. C. Beaglehole in 1962, with an extremely useful introduction.

[7] These devoted Linnaeans, however, did not send any plants to Linnaeus, who was very hurt, especially by the behaviour of Solander, his 'apostle' (Duyker, *Nature's Argonaut*, pp. 223–4).

the people who are most talk'd of at present are Mr Banks & Doctor Solander: I saw them at Court & afterwards at *Lady* Hertford's, but did not hear them give any account of their Voyage round the world, which I am told is very amusing.[1]

In early August 1771 George III and Queen Charlotte[2] formally received Banks and Solander at Kew, where Banks gave the king 'a coronet of gold, set around with feathers', originally a gift to Banks from a Chilean chief.[3] Banks's natural history collections were inspected and subsequently he became a close friend of George III. Banks and Solander dined with Johnson[4] and Boswell,[5] and both received honorary doctorates from Oxford in November 1771, the only academic degree Banks ever attained.[6] Needless to say, the *Endeavour* voyage was of great importance to Banks's career. In Professor John Gascoigne's estimation it elevated him to 'a figure of international scientific significance'.[7] As befitted his position he had two portraits painted at that time.[8]

As a result of the success of the *Endeavour* voyage another expedition to the South Pacific was planned for 1772. There was intense rivalry between the French and British regarding exploration and the acquisition of colonies. The French had preceded the latter in the Pacific to Tahiti, though it later came to light that Samuel Wallis[9] in HMS *Dolphin* had been the first European to claim the island, in June 1767. However, Bougainville,[10] the first Frenchman to circumnavigate the globe in 1766–9 had then, in 1771, published his sensational book *Le voyage autour du monde, par la frégate La Boudeuse, et la flûte L'Étoile*, his description of Tahiti attracting particular attention.

The prime aim of the second Cook voyage on the *Resolution* was to search for the existence of an Antarctic continent, the mythical *Terra Australis*. Banks, convinced that a 'Southern' continent existed, was overjoyed when Lord Sandwich, by now First Lord of the Admiralty, invited him to be the scientific leader of the expedition. He was ecstatic: 'O how Glorious would it be to set my heel upon ye Pole!', he wrote.[11]

Throughout the winter of 1771–2 Banks and Solander were sorting the *Endeavour* collections to make them 'usefull to the world even in Case we should perish in this', the

---

[1] *The Letters and Journals of Lady Mary Coke*, III, p. 435. Lady Mary (1726–1811) was married to Edward, Viscount Coke.

[2] The queen was interested in botany and with her two daughters, the Princesses Augusta and Elizabeth, was privately tutored in the subject by the president of the Linnean Society, James Edward Smith. Banks introduced a plant found in the Cape of Good Hope, and it was named after her *Strelitzia reginae,* the Queen being a member of the family Mecklenberg-Strelitz.

[3] Quoted by Beaglehole, *Endeavour Journal of Joseph Banks*, I, p. 51.

[4] Samuel Johnson (1709–84), author and lexicographer. Banks was invited to join the Literary Club or the Society of Dilettanti, a dining club and one of the most influential societies of the British Enlightenment. Johnson was a leading member and they became friends, Banks being one of Johnson's pallbearers in 1784.

[5] James Boswell (1740–95) was a Scotsman and great friend of Johnson. They visited Scotland together in 1773 and Boswell penned the famous *Life of Johnson*, first published in 1791.

[6] Carter, *Sir Joseph Banks*, pp. 96–7.

[7] Gascoigne, 'Banks, Sir Joseph', p. 692.

[8] One was by Sir Joshua Reynolds (1772–3) and the other by Benjamin West (1771–2). On the Reynolds portrait (see frontispiece) there is the saying of Horace *Cras ingens iterabimus aequor*: 'Tomorrow we set out once more upon the boundless main.'

[9] See above, p. 4, n. 1.

[10] Louis-Antoine, Comte de Bougainville (1729–1811), French admiral and explorer.

[11] 6 December 1771, Banks to the Comte de Lauraguais, published in Chambers, *Indian and Pacific Correspondence of Sir Joseph Banks*, I, pp. 52–8.

coming voyage of the *Resolution*.[1] As before, Banks was assembling a party of scientists, draughtsmen and assistants at his own expense. Because of the high rate of mortality among Banks's party on the *Endeavour* he was taking no chances and organized a larger party than before. All in all they were eighteen. Of course Solander would accompany him as well as his assistants Briscoe and Roberts, veterans of the *Endeavour* voyage. Four artists or draughtsmen were now hired including the celebrated Johann Zoffany (1733–1810),[2] the others being John Cleveley Jr, James Miller and his brother John Frederick Miller. The two secretaries were Sigismund Bacstrom and Frederick Herman Walden. Other members of the party were Dr James Lind, astronomer and physician; Lieutenant John Gore who had circumnavigated the globe three times; John Riddell, a young seaman and traveller; and a further six servants and assistants.[3] All was progressing well until Banks saw the shipboard facilities for himself and his party. He became famously displeased and threatened to abandon the voyage. As Banks wrote to Lord Sandwich on 30 May:[4]

> tho my services are upon this occasion refusd, I shall always hold myself ready to go upon this, or any other undertaking of the same nature, whenever I shall be furnished with proper accomodations for myself & my people to Exert their full abilities to Explore ...

The Navy Board was not impressed. In a memorandum drafted for Lord Sandwich, it stated:

> M^r. Banks seems throughout to consider the Ships as fitted out wholly for his use; the whole undertaking to depend on him and his People; and himself as the Director and Conductor of the whole; for which he is not qualified ...[5]

Banks withdrew 'in high dudgeon' from the expedition.[6] Despite the fact that he did not sail on the *Resolution* he made sure that a botanist did. This was Francis Masson (1741–1805), the first plant collector to be sent from what is now the Royal Botanic Gardens at Kew, who sailed to the Cape of Good Hope in 1772 sending back more than 500 plant specimens.

2. The Origins of the Iceland Expedition

But Banks was not one to give up. Though 'disagreeably disappointed', he had already assembled an impressive scientific party at his own expense and it was of prime necessity to engage them in a new project. He thus quickly 'resolved upon another excursion'.[7] By early June Banks had settled on his new destination. Instead of searching for a massive

---

[1] Carter, *Sir Joseph Banks*, p. 99.

[2] Johann Zoffany (1733–1810), German painter. He was born in Ratisbon, studied in Rome and emigrated to England in 1758. Among his subjects were King George III and Queen Charlotte.

[3] For biographical notes of those who actually sailed on the *Sir Lawrence* to Iceland, see Appendix 1.

[4] Chambers, *Indian and Pacific Correspondence of Sir Joseph Banks*, I, pp. 116–21.

[5] Ibid., pp. 125–6, 3 June 1772.

[6] Gascoigne, 'Banks, Sir Joseph', p. 692. In Banks's *Endeavour Journal* Beaglehole has an appendix with all the relevant correspondence dealing with Banks's withdrawal from the voyage of the *Resolution* (II, Appendix V, pp. 335–55).

[7] Von Troil, *Letters on Iceland*, pp. 1–2.

continent south of Australia, he decided to head north, his choice falling on Iceland.[1] The question that demands to be asked is – why did he choose this destination?

Lysaght suggested that this was a follow-up voyage to Phipps's expedition of 1766,[2] while Beaglehole, editor of the *Endeavour* journal, believed Banks had already given some thought to a northern voyage which now became a real option.[3] Halldór Hermannsson, the distinguished Icelandic scholar who was the first to write on the subject of Banks and Iceland, mentioned the English interest in Iceland's medieval literature emerging in the second half of the eighteenth century.[4] Samuel Johnson had apparently been on the brink of visiting Iceland for this reason.[5] It must be said that it is rather difficult to reconcile Banks with an interest in the Sagas.[6] Two Icelandic scholars have firmly suggested that Solander influenced his decision, because he was almost certainly acquainted with the botanist Johan Gerhard König,[7] both being pupils of Linnaeus, and Solander would have known that König had been sent to Iceland to collect plants in 1764–5 for the *Flora Danica*.[8] Furthermore, a couple of months earlier, in April 1772, Banks had been given some specimens of Icelandic lignite (*surtarbrandur*).[9] This would have strengthened his belief that Iceland had much to interest a naturalist.[10]

At the time there was a growing interest in vulcanology. When scientific travel first took off around the mid-eighteenth century the main interest was in botany and zoology, with geology (called mineralogy at the time) coming a poor third. At first travel was regional, the first instructions for travellers regarding geological observations appearing in Italy in 1751. In chapters 8 'Physica' and 9 'Lithologica' of Linnaeus's *Instructio peregrinatoris* (1759), measuring the height of mountains and collecting minerals and fossils were prescribed – and Banks and Solander were nothing if not devoted Linnaeans.[11] Geology was gaining ground and, as Edward Duyker, Solander's biographer, has remarked, Banks seems to have been more attracted to Iceland's geology than to its flora and fauna. Solander himself was also definitely interested in mineralogy. Before coming to London he had 'amassed an impressive collection of 2000 Swedish and foreign mineral specimens'.[12]

---

[1] Carter, *Sir Joseph Banks*, pp. 101–2.

[2] Lysaght, 'Joseph Banks at Skara Brae and Stennis', p. 221.

[3] Beaglehole, *Endeavour Journal of Joseph Banks*, I, p. 82. He quotes a letter from Banks to Thomas Falconer. The letter is undated but believed to be from late January 1771.

[4] Hermannsson, 'Banks and Iceland', pp. 1, 4.

[5] The year was 1752 and Boswell states: 'There was a talk of his going to Iceland with him, which would probably have happened had he lived' (*Life of Samuel Johnson*, I, p. 281). See also Beaglehole, *Endeavour Journal of Joseph Banks*, pp. 83–4.

[6] See Sigurðsson, 'Inngangur', *Bréf frá Íslandi*, p. 21.

[7] Johan Gerhard König (1728–85), a German pupil of Linnaeus. He lived in Denmark 1759–67 and spent a year in Iceland in 1764–5 making a valuable collection of plants for the *Flora Danica*. He later took a medical degree, went as a doctor to Tranquebar, the Danish colony in India, where he stayed 1773–85. From there he travelled widely to Madras, Ceylon, Siam and the Malacca Straits, collecting plants, corresponding with Banks and eventually bequeathing his collections to him. See also Dawson, *The Banks Letters*, p. 509; and Thoroddsen, *Landfræðissaga Íslands*, III, pp. 46–7. The *Koenigia islandica*, Iceland purslane (*naflagras*), is named after him.

[8] The *Flora Danica* is the extraordinary Danish botanical work, begun in 1761 and only completed in 1883, which consists of 3,240 engravings in folio of all wild plants that grew in the kingdom of Denmark.

[9] See Documents 1 and 2.

[10] Hermannsson, 'Banks and Iceland', pp. 4–5; Sigurðsson, 'Inngangur', *Bréf frá Íslandi*, pp. 22–3.

[11] Vaccari, 'The Organized Traveller', pp. 7–10.

[12] Duyker, *Nature's Argonaut*, p. 231.

The documentary evidence certainly points to the fact that seeing 'burning mountains'[1] was the major aim of the voyage. In his passport, quickly issued on 2 July by Count von Diede the Danish envoy in London, the main purpose of Banks's visit was recorded as 'observing Mount Hekla',[2] the most famous of the Icelandic volcanoes. And Claus Heide, a Dane resident in London, promised to write to Denmark to find the best advice on 'the most Proper places for you [Banks] to go to Vieuw [*sic*] Mount Heckla, or what place is burning at present'.[3] In the only contemporary printed account of the expedition, *Letters on Iceland*, volcanoes figure prominently in the introduction and the first section is entitled 'On the Effects of Fire in Iceland'. Von Troil wrote: 'The subject of volcanos, and of the origin and certain kinds of stones and fossils, have of late attracted the attention of philosophers', remarking the 'whole island of Iceland is a chain of volcanos'.[4] The ascent of Hekla is the highlight of the Icelandic part of Banks's 1772 journal, the measurements of the spouting hot springs coming a close second. On their return *The Scots Magazine* reported that they had 'applied themselves in a particular manner to the study of volcanoes'.[5] Despite Banks's undoubted passion for botany he would have been well aware that the flora of Iceland would have had comparatively little to offer in late August, as his leisurely travels up the coast of Scotland to Iceland suggest. He was in no hurry to catch the flowers blooming.

In his journal Banks explained the reasons for his decision: As the sailing season was much advanced he:

> saw no place at all within the Compass of my time so likely to furnish me with an opportunity as Iceland, a countrey which ... has been visited but seldom & never at all by any good naturalist to my Knowledge. The whole face of the countrey new to the Botanist & Zoologist as well as the many Volcanoes with which it is said to abound made it very desirable to Explore ...[6]

Accordingly, Iceland had the advantages of being relatively near and unexplored, its chief attraction being that it was full of volcanoes.

Banks prepared his voyage as best he could within the limited period of time he had. Understandably he found no-one in London who had been to Iceland but Claus Heide had also promised Banks information 'Chiefly out of books' and tried to gain as much information as he could from his friend Andreas Holt, regarding 'the names of People of the greatest note, likewise Letters of recommendation'.[7] Holt had first-hand knowledge of the island, having actually travelled around Iceland in 1770–71 as the leading member of the Royal Commission. The King of Denmark was notified of their wish to visit Iceland and was only too happy to sanction the journey to Iceland and the Faroes of the 'celebrated English Lords' (*berömte Engelske*

---

[1] Von Troil, *Letters on Iceland*, p. 220. Edward Smith, a biographer of Banks, attributed the decision to visit Iceland to the influence of von Troil, (*Life of Sir Joseph Banks*, p. 32), but Hermannson (rightly) considered this improbable. Von Troil was a latecomer to the expedition (see below p. 10).

[2] See Document 3.

[3] See Document 2.

[4] Von Troil, *Letters on Iceland*, p. xii.

[5] *The Scots Magazine*, 34, November 1772, p. 638

[6] See the Banks journal, p. 47 below.

[7] See Document 2.

Lords).[1] The Governor of Iceland was enjoined to show them politeness and do everything in his power to assist them. As all merchant ships had already left for Iceland, the letter would be sent to the governor himself to take with him.[2] On their departure *The Annual Register* noted that they 'carried every thing that can give them assistance in examining the natural history' of the places they proposed to visit[3] and off they set for Iceland 'at their own private expence', 'to prosecute new discoveries in the science of botany' as *The London Magazine* worded it.[4]

The Banks Expedition was about twenty strong. Only Zoffany was not among the original group Banks had engaged to sail with him on the *Resolution*. In Harold Carter's opinion Zoffany had 'perhaps been the chief sufferer' when Banks withdrew from the *Resolution* expedition, but Banks paid him one year's salary, a sum of £300,[5] and he went off to Florence to copy pictures for the King. But another Swede had joined Banks's party. This was the 26-year-old Uno von Troil, a friend of Solander's, described as 'a gentleman well acquainted with the northern languages and antiquities'.[6] He was invited by Banks to join the expedition at the last moment. He had come to England directly from Paris, where he had met such luminaries of the Enlightenment as Rousseau, d'Alembert and Diderot, and attended King Louis XVI's levée at Versailles.[7]

Banks chartered a ship, the *Sir Lawrence*, a brig of 190 tons, captained by James Hunter with a crew of twelve, at a cost of £100 a month.[8] The *Sir Lawrence* eventually left Gravesend on 12 July, ironically the same day as Cook started on his second voyage of exploration with his two ships *Resolution* and *Adventure*.

Thus Banks set off to explore Iceland, the coast of which had been frequented by British seamen between the fifteenth and seventeenth centuries, after which British contact had all but ceased until his visit in 1772. Before progressing further a brief account of previous Anglo-Icelandic relations, so obviously of interest to Banks,[9] will be given.

---

[1] MS Copenhagen, Danish National Archives, Rigsarkivet [hereafter RA], Departement for de Udenlandske Anliggender [hereafter DfdUA] 892, Count von der Osten [see below] to Baron de Diede, 11 July 1772. Banks would have been on his way by the time this letter reached London. It appears that a visit to the Faroe Islands was part of the original plan but time did not permit it.

[2] RA, DfdUA, 892, The Danish Chancellery (*Kanselli*) to the Foreign Minister Count Adolph Sigfried von der Osten, 10 July 1772. The governor always sailed on a special ship from the Danish Admiralty.

[3] *The Annual Register for the Year 1772*, London, 1773, p. 116. On pp. 139–40 there is a news item regarding Banks's visit to Staffa.

[4] *The London Magazine*, 1772, p. 342.

[5] Carter, *Sir Joseph Banks*, p. 123.

[6] *The London Magazine*, 1772, p. 342.

[7] Von Troil, 'Själfbiografi', pp. 166–7, 209, 212, 215, 217, 224. Von Troil had arrived in London on 21 April 1772. He was very pleased to make Banks's acquaintance, joining his circle where he met, among others, Benjamin Franklin. There has been some debate about how he came to join Banks's Iceland party and as he recounts himself in his autobiography he was already on board ship ready to sail to Gothenburg when the invitation from Banks arrived. On their return Banks invited von Troil to stay on in London until the spring at his expense, but the Swede felt he could not accept this generous offer, leaving in early January 1773 for Sweden ('Själfbiografi', pp. 167–8). On his stay in England see also Bergström, 'Indledning' (introduction to the Swedish edition of von Troil's letters), *Brev om Island*, pp. 10–12.

[8] Carter, *Sir Joseph Banks*, p. 104.

[9] See his memoranda of 1801 and especially 1813 (Documents 38 and 258).

3. Previous Anglo-Icelandic Relations

Anglo-Icelandic relations date back to the period of colonization in Iceland in the ninth century.[1] Most of the Scandinavian settlers came from Norway, but some had lived for a long time in the British Isles, bringing their Celtic slaves with them. A commonwealth until 1262, Iceland first came under the rule of the King of Norway, and then of Norway-Denmark, when the two monarchies were united in 1380. The superb fishing banks off Iceland were attracting English seamen by the beginning of the fifteenth century, if not earlier.[2] In about 1436–8 *The Libelle of Englyshe Polycye*, one of the earliest English political poems, had this to say of Iceland:

> Of Yseland to wryte is lytill need
> Save of stokfische[3] ...

and mentioned that men from Bristol and Scarborough had sailed 'unto the costes colde'.[4]

The Iceland cod fishery remained one of England's major fisheries until the 1650s. One successful voyage to Iceland could almost cover the cost of the ship and the expenses for its outfitting. It is estimated that on average about 100 English fishing vessels sailed annually to Iceland.[5] The Iceland fleet came mainly from East Anglian ports such as King's Lynn and Yarmouth,[6] but also from Bristol on the west coast. The English also engaged in trade, which was welcomed by the Icelanders as they offered a much better price for the fish than the Norwegian merchants (who had a monopoly) were willing to pay. So significant was the English presence during the fifteenth and early sixteenth centuries, that in Icelandic history the period is known as 'the English Century' (*Enska öldin*), and Edward IV (1442–83) even went so far as to speak in 1461 of 'terra nostra Island'.[7] Forts were erected by the English in the Westmann Islands and Grindavík on the south-west coast of Iceland.

The Danes objected strongly to the English seamen fishing off Iceland and trading with the inhabitants, but were powerless to curb the practice. The English were required to buy licences but usually dispensed with that. Disputes between the two countries were frequent as were clashes between merchants and royal officials on the island. This often led to violence, perhaps the most serious incident being the killing of one of the royal governors Björn *ríki* (the rich) Þorleifsson in 1467 by English merchants from Lynn. In revenge King Christian I had the Sound, which separates the Danish island of Zealand from Scania, closed to English ships in 1468 and in 1484 Richard III ordered naval protection for those sailing from Norfolk and Suffolk to Iceland, the first extant example of convoying.[8] By 1480 about 50 Icelanders were employed in Bristol and many in other

---

[1] The date of consensus for the settlement of Iceland is now 874 but Banks chose 860 (see below p. 526).
[2] Professor Björn Þorsteinsson of the University of Iceland argued that Icelandic waters became the 'practice ground' for the seamen who would in time create the British Empire (*Tíu þorskastríð*, p. 7). See also Jones, 'England's Iceland Fishery in the Early Modern Period', pp. 105–10.
[3] Stockfish was dried cod and ling.
[4] Edited by George Warner, 1926, p. 41.
[5] Þorsteinsson, *Tíu þorskastríð*, pp. 11, 16.
[6] See further Agnarsdóttir, 'Iceland's "English Century"'.
[7] Þorsteinsson, *Enska öldin í sögu Íslendinga*, p. 178n.
[8] Ibid., pp. 204–22, 243–4.

towns.¹ In 1490 a treaty was finally signed allowing the English to sail to Iceland 'in all perpetuity' if they bought licences and paid customs dues and taxes.²

In times of financial straits the Danish king attempted to sell or pawn Iceland to the English Crown; thus Henry VIII was three times offered Iceland in this manner, but refused to be tempted.³ In the late fifteenth century Hanseatic merchants began competing with the English for the Icelandic market. They were encouraged to do so by the Danish king in a bid to oust the English. Despite the King of Denmark, buoyed up by a more powerful navy, establishing a trade monopoly in Iceland in 1602, English, German, Dutch, Scottish, Spanish and French seamen all continued to fish in Icelandic waters, illicitly trading with the inhabitants.

After the Reformation in Iceland in 1550 royal power increased greatly at the expense of the church, parliament (*Alþingi*) and the Icelandic landowning class. In 1662, following the European trend, absolutism was introduced and centralized administration strengthened.⁴

Little reliable information on the island in the north was available in Europe in published form. There were, however, numerous accounts with amazing and fanciful stories to be found: for instance, in Sebastian Münster's *Cosmographia*, first published in 1544,⁵ and in *Islandia* by the infamous Dutchman Dithmar Blefken in 1607. Bishop Guðbrandur Þorláksson (1541–1627) of Hólar in northern Iceland, famous for translating the Bible into Icelandic in 1584, was incensed by these accounts and commissioned a young clergyman, Arngrímur Jónsson (1568–1648), to write a treatise refuting the calumnies against Iceland. Jónsson attempted to dispel the travel lies in various works. His *Brevis Commentarius de Islandia*, published in Copenhagen in 1593, received the widest circulation. It was the first Icelandic work to be written for foreigners, in Latin of course, and was subsequently published five years later in 1598 in an English translation in the first volume of the second edition of Richard Hakluyt's *The Principal Navigations, Voyages, Traffiqves, and Discoveries of the English Nation*. The fact that it was included can only confirm the importance of the Iceland fisheries for England. However, ignorance at the highest levels was still to be found, and in 1632 Mr Secretary Coke, one of the two secretaries of state of Charles I, remarked in a memorandum on the fisheries off Iceland: 'Iceland itself is a great territory, and unknown whether it be a main continent with Newfoundland or no'.⁶

Cultural contacts, especially among antiquarians, were frequent and it must be mentioned, especially with Banks's future lengthy presidency in mind, that the Royal Society (founded in 1662) was extremely active in collecting news of Iceland and its phenomena right from its year of foundation asking that a 'set of queries' be prepared for Iceland.⁷ There was in fact quite a lot of correspondence between the Society and learned

---

¹ Þorláksson, 'Útflutningur íslenskra barna', p. 47.
² Þorsteinsson and Jónsson, *Íslandssaga til okkar daga*, p. 167.
³ Þorsteinsson, 'Henry VIII and Iceland', pp. 67–101.
⁴ The only available general history of Iceland in English is Karlsson, *Iceland's 1100 Years*.
⁵ Around 40 editions of the *Cosmographia* were published between the years 1544 and 1628. For Blefken, see Johann Anderson, *Frásagnir af Íslandi*.
⁶ Quoted by Seaton, *Literary Relations of England and Scandinavia in the Seventeenth Century*, p. 221, from State Papers, Domestic.
⁷ In 1662/3 Robert Hooke (1635–1703), one of the luminaries of the Scientific Revolution, prepared his 'Enquiries for Iceland' for the Royal Society, which appeared in print at various points in the 17th and 18th centuries (see Hooke, *Philosophical Experiments and Observations*, pp. 19–21).

Icelanders: for example, the Reverend Páll Björnsson in Selárdalur, a famous witch-hunter who answered some of the queries in 1671, had a paper published in *The Philosophical Transactions of the Royal Society* in 1674.[1]

For the year 1702 the following item is to be found in one of the Icelandic annals: 'Danish merchant ships visited all the harbours in the country, also Dutch fishermen and French and Spanish whalers, but no English doggers.'[2] Indeed, none came that year, but they were certainly expected and their non-appearance was a sign of the times. By the eighteenth century, the illicit activities of foreigners were down to a trickle, as indeed mentioned by Banks in one of his memoranda.[3]

The Danish Royal Society sent Niels Horrebow to Iceland in 1749–51 to make astronomical and meteorological observations. His book *Tilforladelige Efterrretninger om Island*, published in 1752, was the first general account of Iceland that could be considered as reliable and would be translated into the major European languages (see Plate 13).[4] It was a first-hand account, used as the main source for the article on *Islande* in Diderot and d'Alembert's *Encyclopédie*, published in 1751–72, where it says that most accounts of Iceland until now 'ont donné des notions très peu exactes'.[5] Thus Iceland remained to all intents and purposes a *terra incognita* to most Europeans, a country *très peu connue*,[6] when Banks set off to explore it in 1772.

4. Iceland 1772–1815

The *Sir Lawrence* sailed to Iceland by way of the Western Isles, making many leisurely stops on the way, and, after suffering extreme bouts of seasickness, the Banks expedition finally arrived in Hafnarfjörður in south-west Iceland on 28 August.[7]

What was Iceland like in 1772 when Banks arrived? In the late eighteenth century Iceland was ruled by Denmark. The King's chief representative, the *stiftamtmaður* or governor, had only been resident in the island since 1770.[8] The country was divided into

---

[1] Björnsson, Páll, (Biornonius, D. Paulus), 'An Accompt of D. Paulus Biornonius, Residing in Iceland, Given to Some Philosophical Inquiries Concerning That Country, Formerly Recommended to Him from Hence: The Narrative being in Latine, 'tis Thus English'd by the Publisher', *Phil. Trans.*, 1 January 1674, IX, pp. 101–11, 238–40. He also answered some of the queries on Iceland. The Royal Society was, as Ethel Seaton wrote (p. 184), 'extraordinarily active' in gathering information on Iceland, which she describes in *Literary Relations of England and Scandinavia in the Seventeenth Century*, pp. 184–8.

[2] 'Eyrarannáll', *Annálar 1400–1800*, III, p. 413.

[3] See Document 258.

[4] Horrebow's book was published in Danish in 1752 and translated into German (1753), Dutch (1754), English (*The Natural History of Iceland* in 1758) and French (1764). Banks's copy of the book is in the British Library. It is a very handsome volume 'and illustrated with a New General Map of the Island', an excellent source for Banks's expedition. Not until 1966 was an Icelandic translation published.

[5] Translation: 'have not given very exact information'. *L'Encyclopédie*, VIII, p. 915.

[6] Translation: 'very little known'. Kerguelen-Trémarec, *Relation d'un voyage ... aux côtes d'Islande*, p. 34.

[7] The most important primary sources for the Iceland expedition are the journals of Joseph Banks and of James Roberts (both published here), and thirdly the above-mentioned von Troil, *Letters on Iceland*. For more detailed accounts of the Iceland voyage see: Carter, *Sir Joseph Banks*, pp. 101–12; and two useful modern introductions to von Troil's book by Sigurðsson, the editor of the Icelandic edition *Bréf frá Íslandi*, and Bergström, the editor of the 1933 Swedish edition.

[8] Before that the royal governor would visit Iceland during the summer, though far from every year.

four districts governed by district governors and sub-divided into just over 20 counties, administered by county magistrates. The treasurer of the island, the *landfógeti*, oversaw the royal finances. The ancient Icelandic general assembly, the *Alþingi* (Althing) founded in 930, was still in existence when Banks visited the island, but its law-making role was long over and by then it was little more than a judicial court. It would be abolished in 1800, a High Court of Justice (*Landsyfirréttur*) taking its place.[1]

During the period under review Iceland had a population of just under 50,000, though there were violent fluctuations due to natural catastrophes. Society was made up of a small landowning class and a large tenant peasantry (about 95 per cent). The landowning elite also supplied the officials appointed by the King. These royal officials were, however, also farmers, as were the clergy. As in western Europe, there were huge discrepancies between the rich farmer who owned a great deal of land and landless labourers. Those who did not own or rent a farm were in general servants (*vinnuhjú*).[2] From 1746 every Icelander had to be registered annually at a farm. This so-called 'labor bondage' (*vistarband*) was strictly regulated in pre-industrial Iceland in order to control the number of people settling by the coast, both to prevent the drain from agriculture to fisheries and to combat pauperism.[3] The precondition for marriage was a household, i.e. a farm of some sort where the farmer, his wife and children lived, together with the bonded servants and perhaps a pauper or two. This meant that during the eighteenth century about 50 per cent of Icelandic women never married.

The farms were isolated and the Icelanders were primarily engaged in animal husbandry: sheep farming, with some cattle and horses. Fishing, a subsidiary occupation, was carried on in a very primitive manner. There were hardly any fishermen as such and the farmers sent their servants to sea in open rowing-boats. In spite of this fish products were the major exports. Trading stations were dotted around the coast. It was not until the turn of the century that Reykjavík began to emerge as the nascent capital, boasting a population of 307 according to the census of 1801, including two policemen (see Plate 14).[4] When Banks visited the island, there were no villages or towns and thus no villagers or 'middling class'.

In 1772 the Iceland trade was still conducted as a monopoly with Danish merchants sailing to Iceland in the spring (March to June) bringing necessities to an island with limited resources. During the summer the ships would be used for fishing, the merchants subsequently returning to Denmark in August or September. Most merchants resided in Denmark for the best part of the year, employing factors, who were often native Icelanders, to supervise their trade. The Iceland trade was mainly a barter trade, as little money was in circulation. The principal exports were: klipfish (dried salted cod), stockfish (dried cod), fish liver oil (train-oil from cod, catfish and shark),[5] tallow, salted mutton,

---

[1] See Glossary.

[2] Servants did not regain full legal status until 1863.

[3] See Hálfdanarson, *The A to Z of Iceland*, pp. 125–6.

[4] Workshops for spinning and weaving wool had been established in Reykjavík in the 1750s. By 1759 there were 16 houses forming a street there. In 1764 some of them burned down and by 1768 the woollen production had decreased by half. In 1772 about 40 people were still employed who lived with their families on small farms in the surroundings. See further Róbertsdóttir, *Wool and Society*.

[5] Fish-liver oil was principally used for street lighting and lamps, and sometimes by curriers and soap manufacturers. For contemporary information on fish, fisheries and fish-liver oil, see Documents 178–9, and 196.

eiderdown, sulphur, salted salmon, wool and woollen products of various kinds (e.g. stockings, mittens), swansdown and feathers, sheep-, fox- and swanskins. The main imports were: grain (chiefly rye), timber, iron, fishing tackle, salt, linen and cotton goods, alcohol and tobacco. From about 1770 coffee, tea and sugar became valued imports.[1]

The Iceland trade was of considerable importance to the Danish economy, the fees for it constituting the King's major source of revenue from the island, and was highly prized by the Copenhagen merchants. Commercial profits were shipped to the Danish capital, with little capital investment in Iceland. However, by the latter half of the eighteenth century the Danish government, imbued with the spirit of enlightenment, had a clear policy of improving conditions in Iceland, especially in the economic sphere. Official scientific expeditions were sent to gather information on the situation, such as the aforementioned König botanical expedition and Horrebow's visit, and eventually, from the 1750s onwards, the Danes invested considerable sums in their dependency, introducing reforms in farming, the woollen industry and the fisheries.[2]

In 1784, with a change of government in Denmark, matters improved still further. The Bernstorff administration was enlightened, favouring liberal policies in the spirit of Adam Smith and the physiocrats. In 1785 a second Royal Commission was appointed to examine the commercial state of Iceland. It concluded that the trade monopoly was damaging to the Icelanders and successfully proposed the trade be thrown open to most Danish subjects. The subsequent 'free trade' (*fríverslun*) took effect from the beginning of 1788, the major change being that individual independent Danish merchants throughout the kingdom, with the exception of the inhabitants of the Faroes and Greenland, but including those of Iceland, could participate in the Iceland trade. All non-native merchants had to become 'burghers' (*borgarar*) of their trading districts and were obliged to own property there worth at least 3,000 rixdollars. There was, however, a strict ban in force against commercial dealings with 'foreigners' (i.e. those not subjects of the King of Denmark). After its introduction the Free Trade Charter was in full force throughout the rest of the period in question.

The level of education, as Banks was to note, was higher in Iceland than in most other European countries at the time,[3] literacy being widespread (confirmation in the Lutheran church insisted on this), and the sons of the elite were educated at the University of Copenhagen.[4]

5. The Scientific Exploration of Iceland before Banks's Visit

Banks was *not* the first 'good naturalist' to visit Iceland, as he believed himself to be. His expedition to Iceland must be put in context. The Danish authorities had not only been

---

[1] There is an excellent list of necessaries in Document 102.

[2] See Róbertsdóttir, *Wool and Society*, *passim*.

[3] See Document 63.

[4] *The London Magazine* reported: 'It is remarkable that Iceland was one of the earliest seats of learning in Europe. They have long had printing among them and their *Gymnasium* or college still flourishes. Much of the ancient history of the northern parts of Europe, in particular of Sweden and Denmark, is preserved by them, and they have several hundred books in their language, and some translations, *The Whole duty of Man* is translated into Icelandick.' 'For the London Magazine. Some Anecdotes of the late Voyage of Mr. Banks and Dr. Solander in the Northern Seas', 1772, pp. 508–9, at p. 509.

sending scientists to Iceland since the middle of the eighteenth century but had also financed exploration to exotic places outside the Danish realm. For example, Frederik V of Denmark sent a Danish expedition to Egypt, Syria and Arabia in 1761–7, the members of which were the first Europeans to map the Red Sea.[1]

In the late 1760s the French government was the first foreign nation to show political interest in Iceland. At the end of the Seven Years War, the loss of Canada to Britain was a great disappointment. France was intent on regaining her former colony. One plan suggested exchanging Iceland for the territorially large but unproductive French colony Louisiana, the idea being to establish a French naval station in Iceland from where ships could be sent to regain Canada. Étienne François de Choiseul, chief minister of Louis XV, received detailed plans to that effect.[2] At the beginning of 1767 Yves Joseph de Kerguelen-Trémarec (1734–97) was summoned to Versailles and commanded to set sail to explore the islands of the North Atlantic. He published an account of his voyage in 1771.[3] Though his book is in many ways valuable, he claimed to have found marble, crystal and mines of copper and iron in Iceland – all of which are yet to be discovered. Kerguelen-Trémarec was on the whole fairly impressed by the Icelanders: they were keen chess players, well-proportioned with superb teeth, but lazy and prone to drink.

A few years later, in 1771, a more ambitious expedition under the auspices of the King of France set off to explore the wider Atlantic region. It was led by Jean-René Antoine, Marquis de Verdun de la Crenne (1741–1805), and included a party of scientists. They were to correct maps and use new instruments and methods for navigation. In June 1772 they arrived at Patreksfjörður in western Iceland where they erected an observatory, planning to take various astronomical measurements. Unfortunately, they had no idea that during an Icelandic summer there is little difference between night and day, besides which torrents of rain and dense fog understandably hampered their work. They kept themselves to the Western Fjords and left on 20 July, eventually publishing an account of the voyage in 1778.[4]

Also in 1772, on 28 August, the first British expedition arrived in Hafnarfjörður in south-west Iceland, finding that the prospect before them 'though not pleasing, was uncommon and surprizing'.[5] Banks was in total ignorance of the French presence, which was understandable as they kept to different parts of a country with very primitive communications.

---

[1] See Hansen, *Arabia Felix: the Danish Expedition of 1761–1767*; and Rasmussen, *Den Arabiske Rejse*. There was only one survivor.

[2] The documents are preserved in the archives of the Ministère de la Défence at the Château de Vincennes and the research is ongoing.

[3] *Relation d'un voyage dans la mer du nord, aux côte d'Islande, du Groenland, de Ferro, de Schettland, des Orcades et de Norwége*, which was subsequently published in Amsterdam, Leipzig and Berlin. The book was translated into English and published in 1808, in the first volume of Pinkerton's *A General Collection of the Best and Most Interesting Voyages*.

[4] Admiral Verdun de la Crenne (1741–1805) in 1771–2, who published two volumes in Paris in 1778, *Voyage fait par ordre du roi en 1771–1772, en diverses parties de l'Europe, de l'Afrique et de l'Amérique...* The aim, as the title goes on to explain, was to verify the utility of several methods and instruments to determine latitude and longitude and to explore with a view to rectifying hydrographic maps. His co-authors were the mathematician Jean-Charles Chevalier de Borda (1733–99) and the astronomer Alexandre Guy Pingré (1711–96).

[5] Von Troil, *Letters on Iceland*, p. 3. For a fuller account, see Agnarsdóttir, 'Sir Joseph Banks and the Exploration of Iceland'.

6. The Banks Expedition in Iceland

The arrival of the *Sir Lawrence* initially caused great alarm. Many local boats were out fishing, but when the visitors lowered a boat, a chase began, with the Icelanders rowing furiously away. However, the British managed to overtake them and three Icelanders, coaxed by Solander, were induced to come aboard the *Sir Lawrence*, visibly trembling even after a large glass of brandy each. Having eaten and drunk 'plentifully' and having received Solander's guarantee that they were indeed Christians they recovered – one of them agreeing to pilot the ship into the port of Hafnarfjörður,[1] where Banks found to his 'great Joy … that the sides of the Harbour were constituted of the Surface of an ancient flow of Lava'.[2]

The governor of Iceland, Lauritz Andreas Thodal (1718–1808), received them with great courtesy and friendliness, the Icelanders being very happy to find they were 'Peaceable people'. On hearing the visitors were British, the Icelanders had first been 'much alarmd'. Banks noted in his journal: 'At first they thought that we were come with a hostile intention … They thought that we were the Prelude of an English fleet sent to take possession of the Island …'[3] Diplomatic relations between the Danish and British courts were strained at the time because of the marital difficulties of Christian VII and his wife Caroline Matilda, sister of George III. Nevertheless, empty warehouses, belonging to departed Danish merchants, were opened for the benefit of the English visitors, where they settled relatively comfortably for the duration of their visit. There is not much to be added to the journals of Banks and Roberts in this volume about their stay in Iceland. However, there are some contemporary Icelandic sources.[4] The only autobiographical account is that of a famous Icelandic clergyman, the Reverend Jón Steingrímsson, subsequently dubbed 'the fire-priest' (*eldklerkurinn*), for supposedly stopping the lava flow of the terrible Laki eruption in 1783, the greatest eruption in the history of Iceland.[5] The account is as follows:

> In 1777 [*sic*] a fine English ship arrived in Iceland. On board were some very learned naturalists, come to investigate Hekla and other rarities of nature. I was then on a journey, and met one of them in Hafnarfjörður; the other two were not there at the time. I was invited into their reception room, on one side of which there was a long table where I could partake of wine or whatever else I wished; for they were gracious hosts. On the table lay an open book;

---

[1] See the Banks journal, p. 82 below.

[2] MS London, British Library [hereafter BL], Additional [hereafter Add] MS 56301 (7), Banks to Thomas Falconer, 2 April 1773, ff. 20–24.

[3] See the Banks journal, p. 84 below.

[4] Other contemporary accounts from the annals are be found with an English translation in Appendix 5.

[5] This autobiography has recently been translated into English: *A Very Present Help in Trouble: the Autobiography of the Fire-priest*, pp. 289–90. The Laki eruption of June 1783 was the most destructive in Iceland in historical times, with the greatest lava flow recorded in the world, 223.9 sq. miles (580 km^2). About 70 per cent of the livestock perished. Ash and poisonous gases spread across western Europe resulting in thousands of deaths in 1783–4. Effects were also felt in North America where the winter of 1783–4 was the coldest on record on the east coast of the United States. The eruption ended in February 1784 but the Icelandic economy was devastated. It was followed by what has been termed *Móðuharðindi* (Famine of the Mist), resulting in the most serious population decline recorded in Icelandic demographic history, the population dropping by about a fifth from 49,000 down to 39,000 in 1786.

and what should there be on its pages but musical notation to sing from? I looked it over; and the interpreter, an Icelander who spoke Danish,[1] asked me whether I would like to hear how some of the tunes sounded, to which I replied that I would indeed like to hear them. He turned the pages of the book, and pointed out one dance melody. The maestro, whose name was Müller,[2] took his seat at the upper end of the table and played the melody on his musical instrument, while eight Englishmen sat on the bench opposite me and sang along with it, swaying back and forth, and accompanying the music with their hands and feet, stamping on the floor roughly or gently according as the music was loud or soft, so that the whole house seemed to be playing along with them. When this had been going on for some time, I saw that they were looking at me and starting to laugh. I looked round to find the cause, and finally realized that the dance music had so stirred my blood and senses that I had begun, entirely unwittingly, to get excited and sway back and forth. At that I controlled myself and kept still, whereupon their merriment ceased. The interpreter told me that they had been showing off their skill and testing my reactions.

Unfortunately there was no volcanic activity taking place, but the governor assured them they could examine effects of former eruptions everywhere. Banks spent exactly six weeks in all in Iceland, most of the time making short excursions from his base at Hafnarfjörður which, situated in the middle of a lava field, boasted much to explore. That is where he picked up his famous ballast of lava, which can be found at Kew Gardens and in the Chelsea Physic Garden of the Society of Apothecaries.[3] One of the excursions was to see hot springs 'at a place called Reikavik'.[4] After exhausting the possibilities of their neighbourhood they eventually set off on their main excursion to Mount Hekla. Guides were hired and nineteen packhorses loaded with supplies for the twelve-day journey, which would cover 365 English miles.[5] Their route lay first to Þingvellir, the site of the old Icelandic general assembly – the *Alþingi* – then on to view the warm springs at Laugarvatn. From there they rode to the Great Geysir. Here Banks found his volcano in eruption, but it was a 'volcano of water' not of lava. A lot of time was spent measuring the frequency, height and temperature of Geysir.[6] The novelty of boiling meat and fish in the geysers encouraged them to do some cooking, each member of the party having a taste of a ptarmigan, shot by Banks, and almost 'boiled to pieces in six minutes' but tasting 'excellently' according to von Troil.[7]

Then on to Skálholt, the bishopric of southern Iceland, where they met the learned bishop Finnur Jónsson and Bjarni Jónsson, the rector of the cathedral school who

---

[1] Probably the county magistrate Guðmundur Runólfsson (1709?–80); though he never actually studied in Copenhagen he could certainly write Danish.

[2] One of the Miller brothers of Banks's entourage – John Frederick or James.

[3] A French mineralogist, B. Faujas de Saint Fond (1741–1819), wrote about the lava Banks had collected in Iceland, a plentiful supply of which he had brought from Mount Hekla, saying that solely this lava had provided the ship's ballast. See *A Journey through England and Scotland*, I, pp. 82–3. For the refutation of this myth, see Meynell and Pulvertaft, 'The Hekla Lava Myth', pp. 433–6. Faujas was, however, the first to recognize that the basaltic columns in Fingal's Cave on Staffa were volcanic. He conducted a lifelong correspondence with Banks (1779–1819).

[4] The Roberts journal, p. 132 below.

[5] Von Troil, *Letters on Iceland*, p. 5n.

[6] Described by Banks in his journal (pp. 98–101 below) and in his letter to Thomas Falconer of 2 April 1773 (Document 12). The Great Geysir has given its name to all other geysers.

[7] Von Troil, *Letters on Iceland*, p. 10; the Roberts journal has a detailed description of this operation (pp. 133–4 below).

composed odes in Latin in their honour.[1] On his departure Banks presented the bishop with a silver razor and the schoolmasters with silver watches.[2] And thus finally on to Hekla and a very cold and difficult ascent. Banks was elated, believing they were the first to reach the summit.[3] Having accomplished the main purpose of their visit, the 'Philosophical Adventurers'[4] returned to their base in Hafnarfjörður on 28 September, inspecting some more hot springs on the way.

During their stay in Iceland they also did some fishing and botanizing, and led an active social life, dining with the governor and other local notables. The district governor Ólafur Stephensen became a particular friend of Banks, as attested by the correspondence between them in this volume. Banks made a point of entertaining all the leading men of Iceland to exquisite meals 'greater than anyone had seen before in Iceland'[5] prepared by his French chef Antoine Douvez. The Icelandic guests were surprised at the variety of wines and were amazed by the music of the French horns which accompanied these dinners. Solander and von Troil must have acted as interpreters. Latin was perhaps sometimes used, but Banks, apart from his grasp of Tahitian, has not been considered much of a linguist.

Their hospitality was amply returned by the Icelanders. Especially memorable must have been a dinner given by Bjarni Pálsson, Iceland's first state physician, whom they had requested to entertain them 'after the Icelandic manner'.[6] They spent a whole day at his official home in Seltjarnarnes learning much from him. After starting off with a glass of Danish akvavit (corn-brandy) they had biscuit, cheese, sour butter and dried fish, followed by roast mutton, meat-broth and trout.[7] Pálsson's guests ate with an excellent appetite, but 'the sour butter and dried fish were not often applied to' until the last course was served. Uno von Troil described it thus:

> So elegant an entertainment could not be without a desert; and for this purpose some flesh of whale and shark ... was served. This ... looks very much like rusty bacon, and had so disagreeable a taste, that the small quantity we took of it, drove us from the table long before our intention.[8]

Pálsson gave them gifts of mineral specimens, manuscripts and antiquities and in return received a magnificent magnifying-glass.[9] Rector Bjarni Jónsson later sent them a poem about Hekla,[10] Icelandic books the *Fingrarím*[11] and *Festa Mobilia*,[12] a snuffbox, a spoon

---

[1] Printed in Hooker, *A Tour in Iceland*, II, pp. 279–92.

[2] Espólín, *Íslands Árbækur*, XI, p. 3.

[3] See the Banks journal, pp. 104–6 below. They all had a tendency to believe this: e.g. Uno von Troil wrote in his *Letters on Iceland* that they had 'at last the pleasure of being the first who ever reached the summit of this celebrated volcano' (p. 5). Of course nobody knows who was the first person to climb Hekla but the honour of the first recorded ascent goes to Eggert Ólafsson in 1750.

[4] So called by Roberts (p. 135 below).

[5] Espólín, *Íslands Árbækur*, XI, p. 3.

[6] Von Troil, *Letters on Iceland*, p. 111.

[7] Pálsson, *Æfisaga Bjarna Pálssonar*, p. 72. 'They lost their appetite', writes Pálsson's biographer diplomatically.

[8] Von Troil, *Letters on Iceland*, p. 112. Pálsson, *Æfisaga Bjarna Pálssonar*, p. 72. It might be mentioned that Banks and Solander ate stewed shark on the *Endeavour* voyage 'and very good meat he was': Beaglehole, *Endeavour Journal of Joseph Banks*, I, p. 168.

[9] Pálsson, *Æfisaga Bjarna Pálssonar*, p. 72.

[10] See Hooker, *A Tour in Iceland*, II, pp. 273–6.

[11] *Dactylismus ecclesiasticus: edur Fingra-Rijm* ... by Bishop Jón Árnason (1665–1743), first published in 1739.

[12] This title is not in the National Library of Iceland and experts have failed to identify it. The title means moveable feasts, feasts that do not take place on the same date every year, e.g. Easter and Whitsun.

carved from whale tooth and a couple of delicacies, i.e. lumpfish and shark aged for three years to be eaten 'by itself with bread'. His wife also sent a silver chain in the hope they could have it repaired and gilded in Copenhagen. Finally, he asked for a *globum astronicum* (celestial globe) in exchange for 'Icelandic stories or antiquities'.[1] To the Icelanders, Banks and Solander were 'Lords'[2] and they received excellent reviews in contemporary Icelandic sources.[3]

They left Iceland on 8 October.[4] On the way back to Scotland they made a stop in Stromness in the Orkneys on 15 October, where they remained for a week. Here they visited Skara Brae and Stennis to see the semicircle of stones and spent some time in opening up ancient tombs.[5] On 22 October they left for Leith and arrived in Edinburgh a week later.[6] *The Scots Magazine* published an item on the voyage, reporting that: 'The gentlemen were very hospitably received by the inhabitants of Iceland, who are strong, simple, honest, and industrious, in general excellent players at chess, and exceedingly fond of spirits and tobacco.'[7] After a three-week stay, Banks, Solander and Lind left Edinburgh for London on 19 November 1772.[8]

7. The Iceland Expedition: An Assessment

Was the Icelandic voyage a success or was it a scientific fiasco? In the *New Oxford History of England*, published in 1989, Banks is portrayed as an arrogant 'primadonna' having 'initially outshone Cook in fame' then being forced to go off on a 'less than successful Icelandic voyage'.[9]

There is of course no comparison to the *Endeavour* voyage. From Banks's point of view the expedition must in some ways have been disappointing. No volcanoes were in eruption – Hekla erupted on average only twice a century, the last time being in 1766 – a major eruption which lasted for two years leaving a field of lava of some 25.1 square miles (65km^2) – so Banks had unfortunately just missed one. These keen botanists found 'but few plants from the lateness of the season',[10] though they could hardly have expected much. As far as is known only two species were directly introduced by Banks from Iceland: the *Koenigia Islandica* (Iceland purslane) and *Salix myrtilloides* (bog willow). And Bjarni Pálsson – who served the shark and whale – may well have told them that he himself, along with Eggert Ólafsson, had been the first recorded persons to climb Hekla some twenty years earlier. Since then he had even been up a second time. In fact,

---

[1] 21 October 1722, Bjarni Jónsson to Solander, in Duyker and Tingbrand, *Solander*, pp. 302–5.
[2] Espólín, *Íslands Árbækur*, XI, p. 3.
[3] See Appendix 5.
[4] According to Roberts. Banks did not note the date.
[5] The excavations of Gordon Childe in the 20th century revealed the best-preserved Neolithic village in the western world. See a detailed account of Banks's visit in Lysaght, 'Joseph Banks at Skara Brae and Stennis', pp. 221–34.
[6] *The Scots Magazine*, 34, November 1772, pp. 637–8.
[7] Loc. cit.
[8] A news item on Staffa in *The Annual Register for the Year 1772*, pp. 139–40. *The Gentleman's Magazine*, November 1772, p. 540, also had an item about this expedition, the emphasis again being on Staffa.
[9] Langford, *A Polite and Commercial People*, p. 511.
[10] The Banks journal, p. 86 below.

along with another Icelander, the naturalist Eggert Ólafsson, he had been sent all over Iceland during the years 1752–7 on an official expedition by 'Order of His Danish Majesty', sponsored by *Det Kongelige Danske Videnskabernes Selskab*, the Danish equivalent of the Royal Society, to explore Iceland's nature in all its aspects (flora, fauna, geology, meteorological measurements, etc.).[1] They, too, were familiar with Linnaeus and Pálsson had published works on botany. Both had degrees from the University of Copenhagen. Their description of Iceland, *Reise igiennem Island*, over 1,100 pages long, was published in two massive volumes, the first of which appeared that same year of 1772 in Danish, to be translated into German in 1774 and into French 1802. An abridged version of their *Travels in Iceland* was finally published in English in 1805, where they are known as Olafsen and Povelsen. The two quarto volumes were described as 'a faithful and ample account of all that deserves the attention of the learned and curious', besides which they were adorned with many engravings. Even though they were deemed by some as 'long and dull' and 'clogged with repititions',[2] this was far superior to what Banks could ever have hoped to achieve after a relatively short six-week stay in only the south-west of Iceland.

Banks did not, however, come back empty-handed. He collected mineral specimens, farming implements, riding tackle and examples of Icelandic dress, both male and female, even though he found the Icelandic female costume 'certainly not very pleasing to an European eye'.[3] He even purchased two Icelandic dogs, appropriately named Hekla and Geysir.[4] The ballast of lava taken in Hafnarfjörður where they stayed went to form the moss garden at Kew and the rockeries of the Apothecaries' Garden at Chelsea.

Banks was a diligent collector of Icelandic manuscripts and books which are now in the British Library (121 books and 31 manuscripts), including copies of the first Icelandic version of the Bible noted above, translated by Bishop Guðbrandur Þorláksson in 1584, Snorri Sturluson's *Edda*,[5] and a couple of the Icelandic Sagas, including *Njal's Saga*.[6] Hermannsson is disparaging, the manuscripts in his opinion being 'not very important', most being fairly recent, from the late seventeenth and eighteenth centuries. Some had even been copied expressly for Banks.[7] But Banks should receive credit for having acquired these. He even had men sent to Hólar in northern Iceland, where the only printing press in the island was located, to buy copies of the books printed there.[8] The magnificent drawings and watercolours made by the Miller brothers and Cleveley are invaluable

---

[1] Thoroddsen, *Landfræðissaga Íslands*, III, pp. 18–46.
[2] Anonymous, 'Introduction', in von Troil, *Letters on Iceland*, p. viii.
[3] The Banks journal, p. 88 below. The costume, brought back by Banks, has completely disappeared.
[4] Espólín, *Íslands Árbækur*, XI, p. 3.
[5] The *Edda* in Banks's possession was *Edda Islandorum An. Chr. MCCXV. Islandice conscripta per Snorronem Sturlæ ... nunc primum Islandice, Danice, et Latine ex antiquissimis codicibus ... in lucem prodit opera ... P. J. Resenii*, Copenhagen, 1665.
[6] The *Njal's Saga* in Banks's possession was a manuscript (BL, Add MS 4867). See further Hermannsson, 'Banks and Iceland', pp. 15–17. There is a catalogue of the Icelandic books Banks presented, preserved in the King's Library (BL, Add MS 45712). Another identical manuscript copy is owned by the author of this introduction. See also BL, Add MS 4857–96, where the Iceland books and manuscripts collected by Banks 1772–81 are to be found.
[7] Hermannsson, 'Banks and Iceland', p. 17. On the manuscripts see Helgason, 'Íslenzk handrit í British Museum', pp. 109–12; and Thomas Lidderdale's catalogue of Icelandic books in the British Museum.
[8] Espólín, *Íslands Árbækur*, XI, p. 3.

sources.[1] These illustrations, over seventy of them, are now deposited in the British Museum and in steady use.[2]

Solander compiled lists of Iceland's flora (112 pages), and of the fauna and rocks and minerals (48 pages), now preserved in the Botany Library of the Natural History Museum, but he died in 1782 before any could be published.[3] These papers make it abundantly clear that volcanoes and geysers were not the only fascinating phenomena examined by the party.[4]

On the return from Iceland Solander was promoted to Keeper of the Natural History Department of the British Museum as well as continuing to catalogue Banks's collection. He also assisted in the supervision of artists and engravers to complete illustrations from the *Endeavour* voyage for their projected *Florilegium*.[5] Though Banks himself published nothing on his Iceland expedition, he was generous in lending his journal to others. His companion Uno von Troil, later archbishop of Uppsala, took on the task of publication. His book *Bref rörande en resa til Island MDCCLXXII* (*Letters on Iceland*) was published in Uppsala in 1777, dedicated to the King and Queen of Sweden.[6] It was based on the expedition but is disappointing in that it adds little information on their stay in Iceland. Von Troil had consulted books on Iceland, particularly the above-mentioned massive work by Eggert Ólafsson and Bjarni Pálsson and Horrebow's account, to which he refers. To von Troil's credit, however, he corresponded with three learned Icelanders, bishop Hannes Finnsson, the scholar the Reverend Gunnar Pálsson (1714–91) and Hálfdán Einarsson (1732–85), the rector of Hólar Cathedral School, to gain further information.[7] Von Troil's book was promptly translated into German (Leipzig, 1779; Nuremberg 1789) and no less than three editions were published in English in 1780 (two in London and one in Dublin; followed by a third London edition in 1783), with subsequent translations into French in 1781 and into Dutch (Leiden and Amsterdam) in 1784. Thus it received a wide circulation in Europe,[8] awakening 'the curiosity of science to that

---

[1] For example, without the drawings we would have no idea what the 1772 cathedral in Skalholt had looked like.

[2] BL, Add MS 15511–12. In Add MS 15509–10 are drawings of the first part of the voyage to the Western Isles. These are described as 'Drawings, partly coloured, illustrative of Sir Joseph Banks's voyage to the Hebrides, Orkneys and Iceland, in 1772, by John F. Miller, J. Cleveley jun. and James Miller. 4 vols. Large folio. Bequeathed by Sir Joseph Banks, Bart.' Many have been published in Frank Ponzi, *Eighteenth-century Iceland*, and most of them in black and white by Sigurðsson, *Bréf frá Íslandi*.

[3] On Solander's illness and death see Carter, *Sir Joseph Banks*, pp. 181–2.

[4] See 'Introduction to the Journals' on this (p. 22, n. 4 below) and Diment and Wheeler, 'Catalogue of the natural history manuscripts and letters by Daniel Solander (1733–1782)', pp. 468–9.

[5] The plan was to publish a 14-volume Florilegium of new plant species discovered on the *Endeavour* voyage. This project was not completed until the 1980s.

[6] The first accounts of the Iceland expedition were actually published in the Uppsala newspapers in 1773 (see von Troil, *Letters on Iceland*, p. 1n) but this was the first full-length edition. The copy of von Troil in the British Library is Banks's own, with a dedication from the author. Von Troil used drawings from Iceland and Staffa and a map drawn by Friderik Ekmansson. However, the map in the English (Dublin and London) edition is touted as 'An accurate and correct Map of Iceland compiled from Surveys and authentic Memoirs of Mssrs. Erichsen & Schoonning'. They are not identical but the one in the Swedish edition is better printed. See further Hermannsson, 'Banks and Iceland', p. 8n. Banks's notes on the 1780 English edition are preserved in the Sutro Library, San Francisco, the Banks Collection [hereafter SL, Banks Collection], I, 1:42.

[7] Thoroddsen, *Landfræðissaga Íslands*, III, p. 107.

[8] It was reprinted in London in 1808 in the first volume of Pinkerton's, *A General Collection of the Best and Most Interesting Voyages*. See further on all these editions in Sigurðsson, *Writings of Foreigners Relating to the Nature and People of Iceland*, pp. 147–8.

neglected, but remarkable country'.¹ In the opinion of one modern expert, von Troil's book is one of the best accounts written on Iceland² and, along with Banks's and Roberts's journals, is an important source for Icelandic history today. Banks also lent his journal to the botanist William Jackson Hooker, who made use of his description of the Hekla ascent. To his antiquarian friend Dawson Turner he generously wrote: 'Pray keep my Icelandic Drawings as Long as you Please & have Copies taken of all you Chuse.'³

As already mentioned, before reaching Iceland Banks and his party had made a leisurely exploration of the Western Isles. Banks's first description of Staffa won instant fame.⁴ *The Scots Magazine*, naturally enough, considered the island 'one of the greatest natural curiosities in the world', the wonders of Stonehenge paling into comparison as 'trifles when compared to this island'.⁵ And Thomas Pennant at once incorporated Banks's journal description of Staffa into his *Tour in Scotland and Voyage to the Hebrides* in 1774.⁶

On their return to the capital the voyage attracted immediate attention in London, an account appearing in the *Annual Register*⁷ and *The London Magazine* reporting the highlights of the journey. Banks and his retinue were 'the first human beings who have been upon the top of Mount Hecla in Iceland, that most extraordinary burning mountain, whose bowels are on fire while it is covered in snow' and as to the Icelanders themselves they were:⁸

> much depressed by the inclemency of their climate and other causes, and have no encouragement to industry. They are an honest, plain, pious race of men, unaccustomed to see strangers, and therefore not expert at entertaining them; but withal abundantly hospitable, and ready to do everything to oblige them, so soon as the wants and wishes of strangers are made known to them.

As with the *Endeavour* voyage, British polite society was fascinated. Mrs Delany reported on Staffa and that the expedition had 'met with a mountain called Hecla' and 'a fountain called Geyser' in which they had boiled a 'dead partridge' in seven minutes.⁹

Banks was well aware of the shortcomings of the expedition and wrote in his journal that, despite the late season, he had hoped 'that something might be done at least hints

---

¹ Mackenzie, *Travels in Iceland*, p. vii.

² Sigurðsson, 'Inngangur', *Bréf frá Íslandi*, p. 30.

³ Banks to Dawson Turner, 31 March 1811, printed in Chambers, *Scientific Correspondence of Sir Joseph Banks*, VI, p. 53.

⁴ Lind suggested Banks buy Staffa: 'the annual rent is £10, and it is supposed £200 will buy it'. Lind to Banks, 2 March 1775, quoted by Beaglehole, *Endeavour Journal of Joseph Banks*, p. 94n.

⁵ *The Scots Magazine*, 34, November 1772, p. 637.

⁶ Published in Chester, pp. 261–9. Thomas Pennant (1726–98), naturalist and antiquarian, stuck closely to the journal, see changes made in footnotes to the Banks journal. See further on Pennant in Appendix 2. The edition published in London in 1776 was dedicated to Banks. See also Johnson, *Journey to the Western Islands of Scotland*, p. 93.

⁷ See above, p. 10, n. 3 and p. 20, n. 8.

⁸ *The London Magazine*, 1772, p. 509.

⁹ Mrs Delany to Mrs Port of Ilam, 2 January 1773, *The Autobiography and Correspondence of Mary Granville, Mrs Delany*, I, p. 491. Mrs Delaney, née Granville (1700–88), was an artist and prolific letter-writer, with an interest in botany. She was introduced to Sir Joseph Banks by her friend Margaret Bentinck, Dowager Duchess of Portland.

might be gathered which might promote examination of it by some others'.[1] In this he was successful, as other distinguished Englishmen were to follow in his footsteps to Iceland: Sir John Thomas Stanley (1766–1850), son of one of the richest men in England and a student at the University of Edinburgh, sailed in 1789. Aged twenty-two at the time, he had already been on his Grand Tour during which he had ascended both Etna and Vesuvius.[2] William Jackson Hooker (1785–1865), later director of Kew Gardens, was sent by Banks as his protégé in 1809 to collect plants.[3] The following year it was the turn of Sir George Steuart Mackenzie (1780–1848), the Scottish mineralogist, accompanied by the doctors Henry Holland (1788–1873) and Richard Bright (1789–1858).[4] And in 1814–15 the Scottish clergyman Ebenezer Henderson (1784–1858) became the first Englishman to stay for a lengthy period of time and travel all around the island, distributing bibles from the British and Foreign Bible Society.[5] With the exception of Henderson, they sought Banks's advice and assistance, acknowledging their debt to his pioneering effort, and all wrote books about their adventures in Iceland. These British visitors kept in touch with each other regarding Icelandic affairs. In 1813 Hooker, for example, wrote of a dinner to which Stanley had invited him. There he met Banks and Bright and they had all been able to compare their experiences, especially the feasts they had respectively enjoyed in the home of the former governor of Iceland, Ólafur Stephensen.[6] As will be seen in the correspondence, Hooker, Mackenzie and Stanley were to take on an active role in Anglo-Icelandic relations during the Napoleonic Wars.

So what was the significance of Banks's expedition to Iceland? It was the first British scientific expedition to explore the island. For Banks personally the importance of this voyage to the north must have been considerable – as Harold Carter has pointed out this was 'his first and only expedition as indisputable leader'.[7] Banks became so fond of the island that he had a map of Iceland engraved on his visiting card, with Mount Hekla of course in pride of place (see Plate 12).[8]

From the Icelandic point of view the voyage was a great success.[9] In contemporary Icelandic sources Banks, who was considered a Lord and called Baron Banks, received

---

[1] The Banks journal, Introduction, p. 47 below. A good example is Joseph Black (1728–99), Professor of Medicine and Chemistry at the University of Edinburgh, who presented a paper entitled 'An Analysis of the Waters of some Hot Springs in Iceland' to the Royal Society of Edinburgh on 4 July 1791 based on samples brought back by Banks and Stanley. Further readings on hot springs took place on 7 November 1791 and on 30 April 1792, all published in the *Transactions of the Royal Society of Edinburgh*, vol. 3, 1794, pp. 95–153.

[2] Sir John Thomas Stanley (1766–1850), 7th Baronet of Alderly Hall, 1st Baron Stanley of Alderley, 1839, FRS. See Wawn, 'John Thomas Stanley and Iceland'; and West, *The Journals of the Stanley Expedition to the Faroe Islands and Iceland in 1789*, 3 vols.

[3] See Hooker, *A Tour in Iceland*.

[4] A review of Mackenzie's *Travels in the Island of Iceland during the Summer of 1810* appeared in *The Scots Magazine and Edinburgh Literary Miscellany*, 74, pp. 43–53. Many excerpts from Mackenzie's book were published in the *Annual Register* of 1811.

[5] Henderson, *Iceland; or the Journal of a Residence in that Island, during the Years 1814 and 1815*, first published in 1818.

[6] Agnarsdóttir, 'This Wonderful Volcano of Water', pp. 19–20.

[7] Carter, *Sir Joseph Banks*, p. 104.

[8] British Museum, London [hereafter BM], Department of Prints & Drawings, Sarah Banks's Collection of Visiting Cards: C.1–740.

[9] The heyday of writing annals (contemporary chronicles) was largely over, but Banks's expedition is mentioned in five of them. See Appendix 5.

nothing but praise for his scholarly pursuits, his affability and generosity. The following year the first periodical ever to be published in Iceland (albeit in Danish), the *Islandske Maaneds-Tidender*,[1] adorned its front page with an account of Banks and Solander's visit, commending their generosity and humanity and filling the following pages with the ode composed by Rector Bjarni Jónsson.

Finally, as a result of his visit to Iceland and because of his subsequent prominent position in society, Banks was recognized by the British government and individuals alike to be *the* expert on Icelandic affairs. And his correspondence amply attests to this fact.

8. Sir Joseph Banks, 1772–1820

As he looked back on his life Banks was quite sensible of the fact that his travels had 'made' his life. He wrote the following in 1813 to William Jackson Hooker, whom he was advising to undertake a journey to Java:[2]

> Let me hear from you how you feel inclind to prefer Care & indulgence to Hardship and activity. I was about 23 when I began my Peregrinations you are some what older but you may be observed that if I had Listend to a multitude of voices that were raisd up to dissuade me from my Enterprice I should have been now a Quiet Countrey Gentleman ignorant of a multitude of matters I am now acquainted with & probably never attain to no higher Rank in Life than that of county Justice of the Peace.

On his return from Iceland Banks lived 'in no particular station' until his election to the presidency of the Royal Society in 1778.[3] For a man in Banks's position, a landowner with Tory sympathies, a parliamentary seat would have been the natural course. All his immediate forebears had been MPs but his biographers stress that he was not interested in politics, refusing to stand for parliament.[4] Banks never sought public office though he certainly used his position to deal with politicians in matters he held dear, managing to remain above party politics. 'I have never entered the doors of the house of Commons' he wrote to Benjamin Franklin and so 'I have escaped a million of unpleasant hours & preserved no small proportion of Friends of both parties'.[5] Banks had many friends among the major politicians of his age, particularly Charles Jenkinson, Lord Hawkesbury (Earl of Liverpool from 1796), his son Robert Banks Jenkinson, Lord Hawkesbury, and Earl Bathurst.[6] The latter two would receive correspondence dealing with Iceland.

Still he found much with which to occupy himself. In February 1773 he went with his friend Charles Greville[7] to Holland. In the Hague he used the opportunity to meet with Dutch whaling captains familiar with the Greenland waters to gain information for his

---

[1] *Islandske Maaneds-Tidender*, Hrappsey, Iceland, 1774.
[2] Chambers, *Scientific Correspondence of Sir Joseph Banks*, II, p. 5.
[3] Gascoigne, *Science in the Service of Empire*, p. 48.
[4] Carter, *Sir Joseph Banks*, pp. 132, 147.
[5] Chambers, *Scientific Correspondence of Sir Joseph Banks*, II, p. 5.
[6] On this aspect of Banks see Gascoigne, *Science in the Service of Empire*, passim.
[7] Charles Francis Greville (1749–1809), FRS. An antiquarian and collector, especially interested in minerals and horticulture, he was an MP for Warwick 1773–90 and held various governmental posts.

friend Constantine Phipps's proposed voyage to the North Pole.[1] It seems that he was thinking of joining Phipps on this voyage, as he had to Labrador and Newfoundland, and there were rumours that he was even going on another Pacific voyage.[2] But the visit to Holland was the last journey abroad for this intrepid traveller.[3] Instead he turned to exploring his own country.

During the summer of 1773 Banks went to Wales accompanied by Solander, Dr John Lightfoot (1735–88) the botanist and Dr Charles Blagden (1748–1820),[4] a physician trained at the University of Edinburgh who would become one of Banks's closest friends. A veteran of climbing Hekla, Banks now climbed the highest peak in Wales, Snowdon.

At the end of July 1774 the *Resolution* returned. Banks probably greeted this with mixed feelings. He must have been disappointed at not participating in this voyage. Banks's biographers agree that he had 'behaved most foolishly', missing out on this remarkable voyage through his own arrogance and a fit of pique.[5] But Banks and Cook (the dispute after all had been with the Navy Board) had remained on friendly terms. Solander, who met Cook shortly after his return, informed Banks that Cook had 'expressed himself in the most friendly manner towards you, that could be; he said: nothing could have added to the satisfaction he has had, in making this tour but having had your company'.[6] Banks was subsequently involved in supervising the sorting of the natural history specimens of the *Resolution* voyage and the planning of Cook's last voyage, as well as overseeing the publication of the accounts of the two preceding voyages.[7] On board the *Resolution* was Omai[8] from Tahiti, the first visitor to England from the Pacific islands. He was promptly handed over to Banks, becoming 'the darling of social London'. It turned out that Banks and James Roberts had not lost their command of the Tahitian language and could communicate with him.[9]

Banks's reputation was continually growing. As already mentioned he had become a close friend of George III, who in 1773 appointed Banks as his special advisor or unofficial director to the Royal Botanic Gardens at Kew, which included supervising all matters regarding Kew, botany and agriculture.[10] Under his guidance Kew became the major

---

[1] This was sponsored by the Royal Society and the Admiralty. Phipps would sail with two ships, the *Racehorse* with a complement of 90 men and the *Carcass* with 80, not counting the officers. Two masters of Greenland ships were hired as pilots. See Weld, *A History of the Royal Society*, II, pp. 70–73. Fifteen-year old Horatio Nelson took part in the expedition. Phipps was the first European to describe a polar bear. He wrote an account of this expedition, *A Voyage towards the North Pole undertaken by His Majesty's Command, 1773*, first published in 1774. See van Strien, 'Banks, Holland Journal', p. 179.

[2] Barrow, *Sketches of the Royal Society*, p. xxix.

[3] Banks's journal of the Holland trip has been published by Kees van Strien.

[4] Sir Charles Blagden (1748–1820), British physician and scientist, FRS. He worked closely with Banks as the Secretary of the Royal Society (1784–97).

[5] Lysaght, *Banks in Newfoundland*, p. 48; Beaglehole, *Endeavour Journal of Joseph Banks*, I, pp. 64, 71–81.

[6] Quoted by Beaglehole, *Endeavour Journal of Joseph Banks*, I, p. 105.

[7] See further Cook, 'James Cook and the Royal Society', pp. 37–55.

[8] Omai was painted by Sir Joshua Reynolds, the most distinguished portrait artist of his day, and by William Parry (1743–91) with Banks and Solander. He was an interpreter on Cook's second and third voyages. After being lionized by English society, he returned to Tahiti, where he died 18 months later.

[9] Beaglehole, *Endeavour Journal of Joseph Banks*, I, pp. 102–3.

[10] As attested by Harold Carter's *The Sheep and Wool Correspondence of Sir Joseph Banks*, Banks was very interested in breeding Spanish merino sheep, both at his experimental farm at Spring Grove, Middlesex, and with the royal flock on the King's farm at Kew, the aim being to produce fine wool.

botanical research centre in the world, sending out collectors worldwide and facilitating botanical exchange between continents.

In 1777 Banks took up residence, with a staff of twenty servants, in a large house at 32 Soho Square which quickly became the London centre for natural science. This 'Academy of Natural History'[1] housed the 'finest collection of books, pamphlets, and journals on natural science ever collected by one man',[2] not to mention Banks's herbarium and other collections of natural history specimens. It was indeed a research institute welcoming scientists, scholars and students alike.[3] All the most famous English and foreign naturalists visited him there, among them the Icelandic antiquarian Grímur Thorkelín.[4] Banks had an able staff in Soho Square to sort, catalogue and manage his own collections and those made on the voyages and missions he sponsored later in life. Solander was his first invaluable collaborator. After his death Jonas Dryander[5] was his curator and secretary and Robert Brown (1773–1858) his librarian. Banks's library was catalogued 1796–1802.[6]

Banks finally married in 1779. His wife was Dorothea Hugessen, eldest daughter of a wealthy Kent landowner. They were childless, but Banks's sister Sarah Sophia (1744–1818) lived with them all her life. She has also earned an entry in the *Oxford Dictionary of National Biography* as a collector of antiquarian items such as coins, medals, visiting cards and playbills.[7]

In 1778 Banks was elected President of the Royal Society and from then on his position in society was assured, becoming the longest-serving president ever in the history of the society (remaining in office until his death) (see Plate 1). This was both a great honour, Isaac Newton being among the former presidents, and an influential post wielding both scientific and political power, giving Banks an unrivalled opportunity to participate in everything relating to natural science. His term of office, 1778–1820, was a period of far-reaching geographical exploration, war and revolution. His correspondence shows that he managed to maintain relations with foreign scientists during the wars.[8] On many occasions collections of foreign naturalists captured by British vessels were restored to their owners through the direct intervention of Banks with the Lords of the Admiralty and Treasury.[9] Even Napoleon expressed a wish to meet him.[10]

The Royal Society had sponsored all Cook's voyages and needless to say that tradition of exploration was continued under Banks. The veteran of the epic *Endeavour* circumnavigation had a hand in almost all the British voyages of discovery in the late

---

[1] Carter, 'Introduction', in Chambers, *Letters of Sir Joseph Banks*, p. xiv.

[2] Henrey, *British Botanical and Horticultural Literature before 1800*, II, p. 659.

[3] For a contemporary description of the welcome received by foreign visitors, see Faujas de Saint Fond, *A Journey through England and Scotland*, I, pp. 2–7, 46–50, 59–61.

[4] See Documents 35 and 36.

[5] Jonas Dryander: see Appendix 13.

[6] Part of which was published in 1798–1800: see Dryander, *Catalogus bibliothecae historico-naturalis Josephi Banks*. See also 'A Manuscript Inventory of the Library of Sir Joseph Banks as received by the British Museum' (shelf-mark 460.g.1).

[7] Gascoigne, 'Banks, Sarah Sophia', pp. 697–8. 'Her unusual status was acknowledged in a glowing obituary in the *Gentleman's Magazine*', 1818, part 2, p. 472. Her collections went to the British Museum.

[8] See e.g. Document 235.

[9] Hooker, Joseph, *Banks's Endeavour Journal*, p. xxxiii.

[10] Cameron, *Sir Joseph Banks*, p. 145.

eighteenth and early nineteenth centuries; here only a few will be mentioned. Banks sent collectors and observers on naval missions, 126 of them, all over the world,[1] often at his own expense. One of his projects was the transportation of breadfruit from Tahiti to the West Indies, including the voyage which ended with the mutiny on the *Bounty* in April 1789.[2] He was involved with Sir George Macartney's embassy to China (1792–4) and with George Vancouver's surveying voyage to the north-west coast of America (1791–4). Banks organized Mathew Flinders's coastal surveys of Australia in 1795 and his later voyage on HMS *Investigator* (1801–3), which circumnavigated Australia for the first time, mapping the coast. Mungo Park (1771–1806) was sent by Banks to West Africa and became the first European to see inland stretches of the Niger River in 1796. Banks was always interested in the north and the perennial search for the Northwest Passage, supporting the Arctic explorations of Constantine Phipps in 1773, William Scoresby Jr (1789–1857) during 1807–23, John Ross (1777–1856) in 1818, John Franklin's (1786–1847) first expedition of 1819 and William Edward Parry (1790–1855) in 1819–20 on HMS *Hecla*, which finally solved the question of the Northwest Passage passage: it did not then exist. Among Banks's correspondents were scientists and scholars from all the Nordic countries.

As president of the Royal Society, Banks worked closely with government offices, especially the Privy Council Committee of Trade and Plantations and was involved in almost all the governmental scientific bodies of the time such as the Board of Longitude, the Royal Observatory at Greenwich and the Board of Agriculture, as well as being a trustee of the British Museum.[3] As Gascoigne has pointed out, Banks applied science to imperial policy, he was an important adviser on imperial and scientific affairs, the promotion of science and empire often complementing each other.[4] He was invariably consulted and involved in governmental projects and colonial policy, especially regarding Australia. As early as 1779 he proposed a settlement on the east coast of Australia, the region he considered best adapted to European settlement, which resulted in the founding of the convict colony in 1788 at Sydney Cove, near Botany Bay. Thus he was the acknowledged expert not only on Iceland but on New South Wales. Although Linnaeus's suggestion of naming the new country 'Banksia' was not adopted,[5] Banks's name was bestowed upon a genus of Australian plants. Not for nothing did he gain the epithet of 'Father of Australia', as one of his biographers, Joseph Henry Maiden, called him.[6]

He was an active participant in the founding of scientific societies. In 1788 he became the first president of the African Association (which later merged with the Royal Geographical Society), the aim of which was to discover the interior of Africa. With Francis Masson, Kew's first plant collector, and the explorer Mungo Park he was a central figure in founding the Linnean Society that same year and the Horticultural Society in 1804. Banks also played his part in establishing botanical gardens in places such as

---

[1] Mackay, 'Agents of Empire', p. 39.

[2] Banks remained a staunch ally of Captain William Bligh of the *Bounty*.

[3] See e.g. Chambers, 'Joseph Banks, the British Museum and Collections', pp. 99–113.

[4] Gascoigne, 'Banks, Sir Joseph', p. 963; Gascoigne, 'Joseph Banks and the Expansion of Empire', p. 39.

[5] Duyker, *Nature's Argonaut*, p. 223.

[6] The title of the biography is *Sir Joseph Banks: 'The Father of Australia'*, published in 1909. Banks's portrait adorned the Australian $5 bill for many years.

Jamaica, St Vincent, Calcutta and Ceylon. In addition, he was concerned in smaller scientific projects, such as William Herschel's telescope, to name a famous instance. On the whole, as David Mackay has pointed out, Banks 'believed that science, and botany in particular, could be applied in ways which would materially benefit Britain, and in this sense he was following directly in the Linnaean tradition. His imperial vision was a profoundly mercantilist one ...'[1]

On 23 March 1781 Banks was created a baronet, the first and only baronet of Revesby Abbey, and in 1795 he became a Knight Commander of the Order of the Bath, these being the only honours accepted by him. In 1797 he was appointed to the Privy Council and served as a member of the committees on trade and coinage. It was especially through his place on the Privy Council Committee for Trade and Plantations that he managed to help the Icelanders throughout the war years.

Banks is quite often described as being a dilettante 'who published very little and left no indelible mark on any scientific discipline'.[2] However, Harold B. Carter managed to compile a list of sixty publications,[3] the most important being *A Short Account of the Cause of the Disease in Corn Called by Farmers the Blight, the Mildew and the Rust* in 1805 and a paper on the introduction of the potato published in the *Transactions of the Horticultural Society of London* in 1807. As David Philip Miller has concluded, Banks's social station, his fame following the *Endeavour* voyage as 'an intrepid voyager', his connections with George III and the leading politicians, and his 'ubiquity in the scientific and learned institutions' of London, made him a powerful figure.[4] He was certainly more of an administrator than a scientist, a protector and patron of science. The term 'statesman of science' perhaps best sums him up.[5]

Banks's fame and influence was worldwide and he was respected by the scientific world, as his election to at least seventy-two learned societies demonstrates. In his obituary in the *The New Monthly Magazine* he was compared favourably to Linnaeus: 'Not even excepting the great Swedish Naturalist [Linnaeus], it may with justice be asserted, that Sir Joseph Banks was the most active philosopher of modern times'.[6]

9. Sir Joseph Banks as Protector of Iceland during the Napoleonic Wars

In the years following the Iceland expedition Banks continued to take an interest in Icelandic affairs.[7] Sir John Thomas Stanley, for example, on his return from Iceland in 1789, gave Banks seeds and plants from Iceland, which Banks promised to give to the gardener at Kew. As Stanley wrote in January 1790: 'Sir Joseph Banks has been very civil with visits, messages, and invitations, and I dine with him next week.'[8] Banks initially kept up a correspondence with Ólafur Stephensen, who was to become the only native

---

[1] Mackay, 'Agents of Empire', p. 49.
[2] Gascoigne, 'The Scientist as Patron', p. 243.
[3] Carter, *Banks, the Sources*, pp. 168–74.
[4] Miller, 'Joseph Banks, Empire and "Centres of Calculation" in Late Hanoverian London', p. 22.
[5] Chambers, 'General introduction', *Scientific Correspondence of Sir Joseph Banks*, I, p. xxxv.
[6] 'Memoir of Sir Joseph Banks', *The New Monthly Magazine*, 14 (2), 1820, pp. 185–94.
[7] Hermannsson, 'Banks and Iceland', p. 22 *passim*.
[8] Adeane, *The Early Married Life of Maria Josepha, Lady Stanley*, p. 89.

governor of Iceland ever (1790–1806), and who continued to collect and send him manuscripts and scientific specimens, until their correspondence petered out.[1]

Banks's quasi-official position as the expert on Iceland began, according to the available documentary evidence, when the government first called upon him regarding Icelandic matters in 1801. The northern powers of Europe, including Denmark, had combined against England in the League of Armed Neutrality. To combat this very real threat to British naval supremacy, the British Cabinet prepared to break up the League. All Danish colonies in the West Indies and India were seized and, as a further retaliatory measure against Denmark, the Pitt administration sought Sir Joseph Banks's opinion on a proposal to annex Iceland. In a lengthy memorandum entitled *Remarks concerning Iceland* and dated 30 January 1801, Banks offered practical information based on his visit and professed himself in favour of 'the conquest of Iceland', which could easily be undertaken by a force of 500 soldiers.[2] During his visit in 1772, Banks had found that the unhappy Icelanders would be 'much rejoiced in a change of masters that promised them any portion of liberty'. Thus the proposed annexation would be a benevolent gesture while Britain's gain would be control of the potentially important cod fishery. The Danes would suffer political humiliation and the possession of Iceland would in the future benefit England's trade, revenues and nautical strength. However, there is no indication that the Pitt administration took any steps to act on Banks's advice and shortly afterwards the League of the Armed Neutrality was dissolved.

Six years later, in 1807, Banks began his active role as patron and benefactor of the Icelanders. Neutral Denmark, following the British bombardment of Copenhagen, had reluctantly entered into an alliance with Napoleon. Britain and Denmark were thus at war again and the Royal Navy captured many of the Icelandic merchant ships on their way to Denmark, bringing them to British ports. Among the passengers travelling on these vessels was Magnús Stephensen, Chief Justice of Iceland and the leading Icelander of his day (see Plate 10). He was the eldest son of Banks's old friend Ólafur Stephensen. The younger Stephensen now sought Banks's assistance. The situation facing the Icelanders was desperate: if the navigation between Iceland and Denmark became impossible the Icelanders would face certain starvation.[3]

Banks went immediately to the rescue of the Icelanders. As he himself wrote: 'The hospitable Reception I met with in Iceland made too much impression on me to allow me to be indifferent about any thing in which Icelanders are concernd.'[4] Banks was now a privy councillor and ideally placed to help the Icelanders. In his opinion a British annexation was the only solution to the wartime plight of the Icelanders. He was also convinced of its appropriateness. Banks the imperialist felt that Iceland quite simply 'ought to be a part of the British Empire'.[5]

As soon as Banks received Magnús Stephensen's letter of October 1807 he sent it to his friend Lord Hawkesbury, the Home Secretary. Hawkesbury, after discussing the matter with the other ministers, asked Banks to find a way that, in his words, 'Iceland could be

---

[1] See e.g. letters in BL, Add MS 8094.
[2] See Document 38.
[3] See Document 40.
[4] See Document 41.
[5] See Document 61.

secured to His Majesty, at least during the continuance of the present war.'[1] Thus encouraged, Banks set to work planning in minute detail the annexation of Iceland. By December 1807 his plans were ready. Banks was now unwilling to subject Iceland to what he called 'the horrors of conquest'. Instead a warship should be sent to Iceland with a negotiator explaining the benefits of British rule and offering the inhabitants the splendid option of voluntarily becoming British subjects. A simple transfer of governors would take place. He had no qualms about the attitude of the Icelanders: they would be overjoyed. Expecting perhaps, however, to meet with some hesitation on the part of the ministers, Banks pointed out that once the Icelandic sailors had been 'active and adventurous' to the point of discovering America before Columbus but under the Danes they had degenerated to 'their present torpid character'. Under British rule, however, they would become 'animated' and 'zealous', eventually providing plenty of hardy sailors for the Royal Navy.[2] Manning the navy was a perennial problem.

Though the British government ultimately decided against annexing Iceland, Banks managed to persuade the authorities to release the Iceland merchant ships, which were of course technically enemy vessels. They were, moreover, issued with British licences and permitted to carry on their traditional trade with Denmark in relative safety. The ships were released by the government ostensibly on humanitarian grounds, and this was certainly Banks's motive, although the politicians may have had more pragmatic aims. Napoleon's blockade attempted to close the Continent, one of England's major markets, to British trade. England was to find it most useful to issue licences to neutral ships (which was to be the acknowledged status of the Iceland ships) to evade the Continental System. As the government was encouraging the search for new markets for British manufactures to compensate for those now closed on mainland Europe, the possibility of an advantageous trade with Iceland may have been an important consideration.[3] As soon as news leaked out about Banks's generous help to the Icelanders, the native leaders of the Faroes and merchants trading with the Faroes and Greenland sought his help, successfully. As Banks's policy on trade comprised all the Danish dependencies in the North Atlantic their correspondence has been included in this volume.

As will be seen in the correspondence, Banks frequently communicated with His Majesty's ministers during the Napoleonic Wars, presenting detailed plans for the conquest of Iceland.[4] He repeatedly urged the government – unsuccessfully – to this course of action. Banks had to admit defeat even though he always believed that a British annexation of Iceland would have been a wise political act, both for Iceland and for Britain. Banks's invaluable help to the Icelanders was not only acknowledged by the Icelanders themselves but also by men like John Barrow (1764–1848), Secretary to the Admiralty, who was heavily involved in Lord Macartney's embassy and the Arctic voyages of discovery, all Banksian projects. He wrote that Banks's humanity[5]

> was of signal service to these poor creatures; for whom some years afterwards, they were in a state of famine, the benevolence and powerful interest of this kind-hearted man brought about

---

[1] See Document 46.
[2] See Document 63.
[3] See Agnarsdóttir, 'Great Britain and Iceland', pp. 54–6.
[4] See e.g. Document 258.
[5] Barrow, *Sketches of the Royal Society*, p. 29.

the adoption of measures which absolutely saved the inhabitants from starvation. We were at war with Denmark, and had captured the Danish ships, and no provisions could be received into Iceland. Clausen,[1] a merchant, was sent to England to implore the granting of licences for ships to enter the island, and through the active intervention of Sir Joseph, who, as a Privy Councillor, was an honorary member of the Board of Trade, the indulgence was granted.

In 1809 a British trading expedition, granted a licence to trade in Iceland by the Privy Council, and led by a London soap merchant named Samuel Phelps, seized power in Iceland.[2] The Danish governor was taken prisoner and Iceland was proclaimed an independent country, under the protection of Great Britain. Phelps's interpreter, a Danish adventurer named Jörgen Jörgensen (see Plate 9), took up the reins of government as 'protector' of Iceland, issuing radical proclamations in the French revolutionary spirit. It was stated that the 'poor and common' had the same rights as 'the rich and powerful'. An Icelandic flag was designed – three white codfish on a blue background – and elected representatives were to meet the following year to draw up a constitution. This colourful episode has been called 'the Icelandic Revolution' and Banks has been cast in the role of 'a conspirator and revolutionary'[3] and the brain behind the event. Until recently he has been portrayed as such in Icelandic history-books.[4] There is no truth in this. Banks was admittedly involved in encouraging this trading venture, but its aim from his point of view was to bring much-needed supplies to Iceland. In Banks's opinion the expansion of British commerce should be the guiding motive of British imperial policy.[5] Banks used the opportunity to send William Jackson Hooker along on the voyage to gather plants.[6] Banks did know the revolutionaries personally – Jörgensen had made his acquaintance in 1806, both having been on voyages of discovery in the Pacific,[7] and he had even, in April 1809, discussed the question of annexation with Phelps and Jörgensen.[8] Both are among his correspondents.

Banks, however, was certainly innocent of any part in planning this event. The Revolution was simply a spontaneous reaction to the Danish governor's refusal to permit the British soap merchant to trade with the Icelanders. At the time it was the general belief in Iceland that the Revolution had the backing of the British government, and certainly that was implied by the revolutionaries themselves with Jörgensen strutting about in the undress uniform of a post-captain. But there is no evidence to support this. In fact the Revolution was ended after only two months through the intervention of the Royal Navy.

When Banks discovered what had happened, he was deeply shocked, condemning this attempt to involve his innocent Icelanders in what he called the 'horrors of a revolution'. The Icelandic Revolution was a far cry from the sort of proper peaceful annexation, desired and supported by the natives, Banks had had in mind. Instead the revolutionaries

---

[1] Holger P. Clausen: see Appendix 13.
[2] See Agnarsdóttir, 'Great Britain and Iceland', chaps 5 and 6; McKay, 'Great Britain and Iceland in 1809', pp. 85–95.
[3] Briem, '"King" Jörgen Jörgensen', p. 124; Briem, *Sjálfstæði Íslands 1809*, chap. 48.
[4] Jóhannesson, *Saga Íslendinga 1770–1830*, pp. 304–26.
[5] Gascoigne, 'Joseph Banks and the Expansion of Empire', p. 42.
[6] William Jackson Hooker wrote an account, 'Detail of the Icelandic Revolution in 1809', in *A Tour in Iceland*, II, Appendix A.
[7] Document 39.
[8] Document 113.

had actually dared to inflict a Jacobin republican regime on the Icelanders and declared the island independent. As will be seen, Banks collected all the information he could on the revolution, to the point of obsession, sending accounts to the British government of 'the atrocities' committed in Iceland.[1] Perhaps Banks's reaction to the affair was so violent, because of a twinge of conscience. After all, he had been instrumental in sending these men to the island, albeit from humanitarian motives.

Banks was of the opinion that it was time Anglo-Icelandic relations were placed on a firm official footing. In this he was successful. On 7 February 1810 the Privy Council issued an Order in Council stating the British government's official policy towards Denmark's three Atlantic dependencies: Iceland, Greenland and the Faroe Islands.[2] Banks was the principal author. Drafts of it, in his unmistakable handwriting, are to be found in the Rare Book Library of the University of Wisconsin in Madison.[3] The Danish dependencies were placed in a state of neutrality and amity with England and free trade was established between Iceland and Great Britain. The Order acknowledged that the sovereignty of Iceland was still vested in the Crown of Denmark but that Iceland was under the protection of Britain. A British consul, John Parke, was subsequently sent to Iceland, being thoroughly briefed by Banks before his departure, as the correspondence shows.[4]

Throughout the war years Banks was always ready to approach the proper government offices on the Icelanders' behalf. He bombarded government ministers including Lord Hawkesbury at the Home Office, Lord Castlereagh at the War Office, Earl Bathurst at the Board of Trade and the Marquis Wellesley at the Foreign Office, with letters and visits on Icelandic affairs. In all this Banks seems genuinely to have been motivated by Icelandic interests. When, for example, British fishing interests opposed the licensed Iceland trade and the British government decided to restrict the major Icelandic fish exports, Banks – the Iceland expert – was called in to draft the regulations. This he endeavoured to do as near as possible to the interests of the Icelanders themselves, solicitously enquiring of them the total export figures and fixing the regulations at that level.[5] Banks was thus the architect of the commercial policy adopted by the British government towards Iceland, as well as the political one. Letters dealing with trade are a major feature of his Iceland correspondence.

In summary, during the Napoleonic Wars Banks acted as a powerful protector and benefactor of the Icelanders. Without his unstinting and selfless help Iceland would have suffered greatly during this period. Throughout the war years he continuously watched over Iceland's welfare. There is no evidence that Banks ever refused to help an Icelander in need, even to the point of lending them money from his own pocket.[6] Even the Anglophobe King of Denmark was forced to acknowledge Banks's role as protector of his North Atlantic dependencies. At the end of the war a grateful Frederik VI sent a

---

[1] See e.g. Document 134.

[2] The original is in the National Archives, Kew, London [hereafter TNA], Privy Council [hereafter PC] 1/3901; *The London Gazette*, 10 February 1810.

[3] See Document 144.

[4] See Documents 202–4, 208. Parke did his part in keeping Banks informed. Near identical letters were sent to Banks and the Foreign Office.

[5] See Document 178.

[6] See e.g. Document 54.

personal letter of thanks to Banks.[1] Banks refused the Order of the Dannebrog, Denmark's highest honour, as a matter of principle,[2] but later accepted three cases of books, including the *Flora Danica* with illustrations of the flowers he missed seeing in bloom when in Iceland in 1772.[3] Banks's fame as a protector was not forgotten in Iceland. At the beginning of the Second World War, Jónas Jónsson, a former government minister and leader of the Progressive Party, who was rightly worried about the state of commerce because of the war, wrote to the British Foreign Office:

> If this branch of commerce [the Iceland trade] is too small for the English, we in Iceland will drift away, as we did in the Napoleonic war. We are in many ways better off than then. We have now the potatoes, and many other things which were then unknown. You remember our hot springs, which are now our very good friends ... During the Napoleonic war there was in England a gentleman and a scientist with the name of Joseph Banks. He remembered Iceland, which might else have been quite forgotten in England. Would you read about Mr. Banks, and see if you could be his successor now![4]

---

[1] See Document 293.

[2] RA, DfdUA, 1990, Bourke to Rosenkrantz, 26 August 1814.

[3] Later, after Banks's death in 1820, as further volumes of the *Flora Danica* were published, copies were sent to the British Museum, the Danish authorities being aware that Banks had bequeathed his books to that institution. See RA, DfdUA, 817, Clausewitz to Département Royal des Affaires Etrangéres á Copenhague, 29 and 30 September 1820. This continued at least until 1833: 'Sir, I am directed by the Trustees of the British Museum to acknowledge the receipt of your letter of 27 Dec accompanying the obliging present made to the Library of this Institution in the name of His Majesty the King of Denmark of the 33rd, 34th and 35th Numbers of the "Flora Danica"' (RA, DfdUA, 817, J. Forshall to Chevalier de Bourke, 16 January 1833).

[4] TNA, FO 371/23639, Jónas Jónsson to Mr Gage, 3 October 1939, ff. 257–258. I am indebted to Professor Guðmundur Jónsson for this quotation.

TEXTUAL INTRODUCTION

Sir Joseph Banks died in 1820 without a legitimate heir. According to his will, his librarian, the botanist Robert Brown, was to be permitted the use of all Banks's resources, including his library and manuscripts, during his lifetime. Subsequently, the great Banks herbarium, which was later to form part of the Natural History Museum, and his natural history library of 22,000 titles, went to the British Museum. His papers were not really dealt with until they became the province of one of his heirs, Edward Knatchbull-Hugessen (1828–93), later the first Lord Brabourne, who decided to sell them. To cut a long story short, after the Colonial government of Australia had bought a large collection of documents relating to Australia in 1884, preserved in the Mitchell Library in Sydney, and an effort had been made, unsuccessfully, by Brabourne to sell most of the rest to the British Museum, he resorted to auctioning them off. The greatest part of the manuscripts that Banks had classified, bound and filed away were sold in lots by Sotheby, Wilkinson & Hodge in March and April 1886, for a derisory sum of less than £200, and 'flung away and dissipated all over the earth'.[1] Some indeed were bought by manuscript dealers who were only interested in the autographs, destroying the rest of the letter. A final auction was held at Sotheby's in 1929.

Luckily, before the great dispersal, the invaluable Dawson Turner copies had been transcribed. Dawson Turner was to write Banks's biography and endless boxes of documents were sent to his residence. Turner never wrote the book but his daughters Hannah and Mary and his clerks produced no less than twenty-three volumes of transcripts, now preserved in the Botany Library of the Natural History Museum. In many cases the originals have survived and comparison has shown that the copies are 'accurate and reliable'.[2] Many Iceland letters have been saved in this manner.

It is estimated that more than 20,000 letters of Banks's correspondence have survived, 14,000 of them to Banks, 6,000 from him, some autograph, others copies made by an amanuensis. The number of correspondents is more than 3,000. These documents are preserved in at least 150 repositories throughout the world, private and public. As a Banks specialist has remarked, 'Sir Joseph's correspondence was global in circulation and global in concern'.[3] Despite that, much of Banks's work was done 'in conversation'.

A rescue mission to reunite the Banks correspondence was mounted in 1989 by the Australian scientist Harold B. Carter. He founded *The Sir Joseph Banks Archive Project* under the auspices of the Royal Society and the Natural History Museum in London.

---

[1] For the fate of the Banks papers, see Carter, 'Introduction', *Banks. The Sources*; and Dawson, *The Banks Letters*, pp. xiii–xvii. This quotation is by Beaglehole in Carter, *Banks. The Sources*, p.18. See also below, Document 123 and associated notes.

[2] Dawson, *The Banks Letters*, p. xxix.

[3] On Banks as a letter-writer, and on his correspondence in general, see Chambers, 'Letters from the President'. This quotation is on p. 38.

Dr Desmond King-Hele was the first chairman of the Executive Committee. The aim has been to gather together copies of all the surviving documents, catalogue them in a database, transcribe, edit and finally publish them. The Archive was first housed in the Natural History Museum in London and was then transferred to Nottingham Trent University. The Banks Archive Project is now an independent research project with Dr Neil Chambers as the Executive Director and editor of the volumes of correspondence.

In England, the major repositories for Banks's papers are the British Library, where, for example, his seven volumes of Foreign Correspondence are now to be found in the Department of Manuscripts (Add MS 8094–8100); the Natural History Museum; and the library of the Royal Botanic Gardens at Kew – all yielding Iceland correspondence. In the wider world, the major collections of Banks's Iceland correspondence are to be found in the United States, particularly in the Sutro Library at the University of San Francisco and the Rare Book Library of the University of Wisconsin. Other major repositories of material on Iceland are at McGill University in Montreal, and in various archives and libraries in Australia. The editor of this volume has travelled to all repositories except for a couple in Australia, comparing the transcriptions with the originals. Ironically there is only a single document relating to Banks in Iceland itself (Document 297). Though many Icelanders were among his correspondents none of his letters to them appears to have survived.

Banks kept all letters however trivial, made notes, and wrote lengthy reports and memoranda, some in answer to government requests, others on his own initiative. Among the Banks papers are also documents relating to his fields of interest which were sent to him and he kept: for example, treaties and bills. Almost all documents that can be classified under the headings 'Iceland' or 'North Atlantic', dealing with Norway, Greenland and the Faroe Islands, are included in this volume. They are arranged chronologically and numbered accordingly.

1. Banks as a Writer[1]

Banks's handwriting was quite legible when he was young – for example in his Iceland journal, which he wrote at the age of 29 – but steadily deteriorated as his life progressed, due to gout and the natural processes of age. His letters thus become more difficult to decipher. Banks's spelling was idiosyncratic, but the words are invariably recognizable. He was sometimes consistent, such as almost always writing *ei* instead of *ie*, and dispensing with the *e* when using a past tense ending in *–ed*; but then this was common in the eighteenth century. And he often dispensed with double consonants, for example writing *Mul* instead of *Mull*.

But what makes his writing more bizarre is, firstly, his use of capital letters, either sprinkling them without any discernible pattern or using small letters at the beginning of sentences and paragraphs as well as quite often in proper names, such as denmark; and secondly, his lack or abuse of punctuation. It is said that he was fonder of commas than full stops but in his Iceland journal, for instance, there is little or no punctuation. As the

---

[1] See Chambers, 'Letters from the President', pp. 25–57.

editor of the *Endeavour* journals, J. C. Beaglehole, so aptly put it, Banks wrote with a 'punctuation-free pen'.[1]

2. The Text

In the first place the spelling and capitalization of Banks and his correspondents has been almost entirely preserved, though occasional missing letters have been added in italics and words in brackets. The transcriptions are as near to the original as possible. Serious misspellings are explained in footnotes. The conventions regarding deletions and insertions established by Harold B. Carter, and also used by Neil Chambers, have been employed in transcribing the manuscripts. Carter showed deletions in italics between obliques thus: /*Deleted text*/. He showed insertions in plain text in obliques thus: /Inserted text/. Carter also felt that it was important to denote whether a correspondent was an FRS to show Banks's wide network of scientists. This convention is adhered to here. Otherwise, normal Hakluyt Society style has been followed which, for example, does not aim for a facsimile appearance. All underlinings are those of the original documents. Complete words which have been added for the sake of clarity have been put in square brackets, while abbreviations have been expanded by the use of italics. Square brackets have also been used for editorial insertions. Banks generally endorsed upon letters the dates of receipt and when he replied to them. The date of receipt might be jotted down at the foot of the first page or on the backs of letters. Those have been added as footnotes.

For the convenience of the reader, it has sometimes been necessary to insert capital letters in Banks's documents and full stops to mark the end of sentences. This has been done in his journals and when done in individual documents it is generally noted in a footnote. Banks used commas and they will be found where he chose to put them – no attempt has been made to make any changes in his use of commas. Superscripts were common during Banks's day and are reproduced here.

Icelandic place-names and the proper names of Icelanders are usually garbled and have been put into modern Icelandic in footnotes as they occur. Banks's correspondents have been identified and short biographies are to be found in the list of correspondents in Appendix 13.

Marginal notes in Banks's documents are either included as insertions or in footnotes. Ships' names are in italics throughout. Some of the letters are written in languages other than English, namely Latin, French, Danish, Swedish and Icelandic. These have been translated, but the text in the original language is included at the end of each letter. A couple of Danish letters have contemporary translations attached. In that case, as they would have been the ones read by Banks, they have not been translated anew. The letters from Danes and Icelanders in English are usually extensively dotted with umlauts and accents. These have been omitted but are mentioned in footnotes.

Some letters are extant in more than one copy. Obviously, if autograph letters are available they have been transcribed. As his life progressed Banks employed amanuenses, especially if the letter was intended for a government office.[2] One, William Cartlich,

---

[1] Beaglehole, *Endeavour Journal of Joseph Banks*, I, p. 14.
[2] See Marshall, 'The Handwriting of Joseph Banks'.

served him faithfully for thirty-five years. Drafts have only been used if nothing else is available. Banks usually wrote his replies at once; these are commonly to be found on the backs, the margins or any other free space on the letters. When the original and the draft reply both exist they are found to be 'verbally identical', the main difference being that the abbreviations jotted down in the draft have been written out in full.[1] The fact that the letters were auctioned off in lots is obvious when inspecting the calendar of documents in this volume: they are concentrated in 'blocks'. Several letters are undated and attempts have been made to date them and place them in their chronological order.

Many of Banks's letters concerning Iceland and the North Atlantic are clearly lost. One must hope that the publication of this volume will bring more to light.

---

[1] Dawson, *The Banks Letters,* p. xix.

INTRODUCTION TO THE JOURNALS

Banks's habit was to write journals during his travels, but he left their publication to others. The first ones were not published until the latter half of the twentieth century and some of the eight remain unpublished.[1] Banks himself wrote that he had not been able to publish his *Endeavour* journal both because his time was being spent on preparing for his second Cook voyage in 1772 and because Dr John Hawkesworth had been engaged to do so. Perhaps he lacked self-confidence, as has been suggested, or as he confessed in 1807, so truthfully: 'I am scarce able to write my own Language with Correctness, & never presumd to attempt Elegant Composition, Either in Verse or in Prose in that or in any other tongue.'[2]

The latter half of the twentieth century has seen the publication of his major journals. In 1962 the New Zealand historian Professor John Cawte Beaglehole (1901–71) published a superb edition of the *Endeavour* journal in two volumes.[3] Banks's journal from his voyage to Labrador and Newfoundland was published by another New Zealand scholar, Averil Margaret Lysaght (1905–81), in 1971, along with manuscripts and collections.[4] In 1980 Lysaght published a magnificent two-volume facsimile edition of *The Journal of Joseph Banks in the Endeavour* with a commentary, in a limited edition of 500 numbered copies, with a preface signed by Prince Philip (autograph). She also compiled a catalogue of the drawings of Alexander Buchan, Herman Diedrich Spöring, Sydney Parkinson and others. In addition, she discovered an unknown transcript of Banks's holograph *Endeavour* journal, attributed to Princess Elizabeth, second daughter of George III, and corrected by Sarah Sophia Banks.[5] Most recently, in 2005, Banks's Holland journal of 1773 has been published in a scholarly edition by Kees van Strien.

In the 1770s it was rumoured that a publication on the Iceland voyage by Daniel Solander was in the pipeline, but this proved to be untrue and it was left to the American historian Roy Anthony Rauschenberg (1929–2011) to publish his edition of the Iceland journals in *Proceedings of the American Philosophical Society* in 1973.[6] A short extract, from 28 August until 7 September, was translated into Icelandic by Dr Jakob Benediktsson and published in the Icelandic journal *Skírnir* in 1950.[7]

---

[1] On the journals written by Banks see Carter, *Banks. The Sources*, pp. 51–7.

[2] Beaglehole, 'Introduction', *Endeavour Journal of Joseph Banks*, I, pp. 120–22.

[3] Beaglehole, *Endeavour Journal of Joseph Banks 1768–1771*, 2 vols, Sydney, 1962. It is a scholarly version, heavily annotated with detailed introductory chapters, for instance on the extant transcripts.

[4] Lysaght, *Joseph Banks in Newfoundland and Labrador, 1766*.

[5] Lysaght, 'Textual note', *The Journal of Joseph Banks in the Endeavour*, pp. 31–2.

[6] Rauschenberg, 'The Journals of Joseph Banks's Voyage up Great Britain's West Coast to Iceland and to the Orkney Isles July to October 1772', pp. 186–226.

[7] Benediktsson, 'Sir Joseph Banks: Dagbókarbrot úr Íslandsferð 1772', *Skírnir*, 124, 1950, pp. 210–21. It is lightly annotated.

The Iceland journal is not a single document. There are two main manuscripts in Banks's autograph.[1] The first is a fair copy (and thus easily decipherable), in a notebook, ninety-six pages in length. Banks begins with an introduction where he explains his reasons for going to Iceland. The bulk of it deals with his leisurely visit to the Western Isles, only the last fourteen pages being devoted to the arrival and first days in Iceland up to and including 6 September.[2] This manuscript is preserved in the Rare Book Room of Blacker-Wood Library, McGill University, Montreal.[3] This notebook also contains a copy of Banks's passport in French and Latin by a different hand.[4] In addition, there is a fair copy made by Banks's sister, Sarah Sophia, and preserved in the Kent History and Library Centre in Maidstone, Kent.[5] According to Guy Meynell, there is yet another copy in the Hawley Collection, an exact copy of Sarah Sophia's. However, the present whereabouts of the Hawley Collection are unknown.[6] It does seem that Banks was at some point contemplating publishing his Iceland journal, given the existence of the fair copy and the fact that in 1783 Banks refused to permit Thomas Pennant to publish some of the drawings of Iceland, writing: 'I trust that you will excuse me for refusing my consent ... as I must consider such a measure a material injury to the journal of my Icelandic voyage ... if leisure allows me hereafter to print it.'[7]

The second manuscript, in a notebook with Lord Brabourne's binding,[8] covers the period 17–30 September 1772, the highlight being the excursion to Hekla, followed by two blank half-pages, suggesting that Banks meant to fill them in at a later date. He continues with entries for 16–22 October, after his arrival in the Orkneys on his return to Britain. This manuscript is very different from the first, being in the form of dated but rough notes and is at times very difficult to decipher. It is nineteen pages long but there is no page numbering. It is preserved in the Kent History and Library Centre in Maidstone.[9]

Banks's Iceland journal has thus been considered incomplete, in the sense that a journal for the days 7–17 September is missing.[10] However, Harold B. Carter discovered some

---

[1] For further information on the Iceland journals, including transcripts, see Carter, *Banks. The Sources*, p. 54.

[2] For a very detailed account of the manuscripts, see Guy Meynell, 'Banks Papers in the Kent Archives Office, including notebooks by Joseph Banks and Francis Bauer'; on the Iceland manuscripts see pp. 78–9.

[3] Meynell suggested that this notebook may have been lot 21 of the auction of 14 April 1886, 'Notes and brief history of Iceland in the autograph of Sir Joseph Banks', and was purchased by McGill University from Messrs Wheldon & Wesley on 10 December 1924.

[4] See Document 3.

[5] Kent History and Library Centre, Banks MS, U1590 S1/2. It includes a copy of Banks's letter to Lord Sandwich.

[6] The Hawley Collection was a collection of Banks papers in the possession of Major Sir David Henry Hawley, 7th Bart., of Mareham-Le-Fen, Lincolnshire. Sir Henry Hawley was one of Banks's executors. See further, p. 174, n. 1.

[7] 4 May 1783, Chambers, *Scientific Correspondence of Sir Joseph Banks*, II, p. 81.

[8] Lord Brabourne or Edward Knatchbull-Hugessen was the son of Sir Edward Knatchbull, 9th Bart., who was one of Banks's executors. Lady Banks was a member of the Hugessen family.

[9] It is bound in a small notebook, JB holograph memoranda and notes to 22 October 1772, Kent History and Library Centre, Banks MS, U951 Z31.

[10] Meynell suggested that Banks took with him a series of pocket-sized notebooks and then lost one for the period 6–18 September. He added: 'Even the Kent notebook may have been temporarily mislaid in 1772–73 when he and his sister were preparing their fair copies, for it is otherwise hard to see why they chose to finish with the passport and the letter to Sandwich and to leave out one of the most exciting incidents of the entire voyage, the ascent of Hecla' (Meynell, 'Banks Papers in Kent', p. 79).

memoranda and notes on loose leaves regarding the Iceland expedition in Banks's autograph in the Natural History Museum, London.[1] They are dated and cover the periods 29 August–15 September, 28–31 October and 18–21 November 1772, and have never been published before. Thus, regarding the Iceland expedition, in fact only an entry in Banks's hand for 16 September 1772 is lacking. However, from the journal of his servant James Roberts we know what took place on that day – in fact not much to report.[2]

The James Roberts journal, preserved in the Mitchell Library, State Library of New South Wales, in Sydney, Australia, has never been published before in an annotated edition.[3] It covers the period 11 July–29 October 1772. Roberts's journal is a fair copy written in a very clear hand. Roberts was Banks's collector and servant and had accompanied him on the *Endeavour*.[4] No doubt he made notes during the voyage, though they have not survived. According to information given by the Library website it 'appears to be a copy prepared for publication'. The manuscript has a title-page and opposite it is an illustration in watercolour containing an inscription 'W. Brand invt. deln., Boston, October 1796', which is difficult to decipher but doubtless indicates the date of the fair copy. This watercolour is Plate 8. Averil Lysaght believed that Roberts had written his journal some years later, no doubt based on this date.[5] There is quite a lot of material here which is a welcome addition to the Banks journal. It is quite clear that Roberts was not consulting Banks's journal because of the differences in the spelling of proper and place names.

The two journals are very different. Banks is busy making scientific observations; for instance noting all the plants and animal life he finds, describing Staffa in detail and measuring Geysir with great exactitude. Roberts tells us more about the day-to-day happenings, such as the social life and the weather.

Rauschenberg wrote a useful introduction to his edition of the Iceland journal but it is not heavily annotated. He put in punctuation for the sake of clarity and corrected Banks's use of capitals to twentieth-century usage.[6] In this edition capital letters at the beginning of sentences and full stops have been added to facilitate the reading of the journal, and Banks's spelling has occasionally been amended with the addition of italicized letters. Roberts's punctuation is also idiosyncratic. He has, for instance, a strange habit of sometimes placing a full stop and then continuing without a capital letter. No changes have been made to his spelling, which in general conforms to modern English, with a few easily-understood exceptions. Roberts always repeats the last word of the previous page on the following page, as was common at the time. The second appearance of the same word has been omitted here. For the sake of clarity the dates are now in bold.

---

[1] Unfortunately the staff of the Botany Library of the Natural History Museum have not been able to locate these pages and thus Harold Carter's transcriptions, preserved by The Banks Archive Project, are used here.

[2] See the journal of James Roberts, p. 133 below.

[3] Mitchell Library, A 1594. There are digital images on-line at: <http://www.acmssearch.sl.nsw.gov.au/search/itemDetailPaged.cgi?itemID=456959>.

[4] See Appendix 1.

[5] Lysaght, 'Joseph Banks at Skara Brae and Stennis', p. 223.

[6] Rauschenberg, 'Iceland Journals', p. 195.

The Scientific Papers of the Iceland Expedition

All the scientific papers, compiled by Joseph Banks and Daniel Solander before and during their Iceland expedition, are preserved as Solander MSS in the Botany Library of the Natural History Museum.[1] These have never been published. Judith A. Diment and Alwyne Wheeler catalogued these papers in 1984. They listed them as follows:
The *Flora Islandica*,[2] 112 pages in all, a checklist of Icelandic plants, which Banks and Solander collected and/or observed. There is a list of thirteen places[3] where the plants were observed with their abbreviations and the plants they found in the *herbaria* of Bjarni Pálsson, the state physician of Iceland and Bjarni Jónsson, the rector of the Skálholt school, have also been noted.[4] Bacstrom wrote the scientific names and Solander the popular names as well as numbering them.[5]

The other scientific papers are as follows:
[25] *D. C. Solander's Plantae Islandicae et Notulae Itinerariae together with a catalogue of the contents of the numbered bundles as they were collected during his visit with Joseph Banks to that island in 1772*, 76 pages.
[25a] List of seeds collected, 1 page. Only four are listed: the *Koenigia islandica* and three types of *Gentiana*.[6]
[25b] Itinerary, 1 page.[7]
[25c] Plants collected in various localities [4 pages]. This is a list of plant names.
[25d] *Plantae*, 11 pages. These include detailed descriptions of 21 plants.
[25e] Authors treating on, or referring to Iceland[8]
[25f] Invitation card from Banks to Governor Lauritz Thodal[9]
[25g] List of plants, 1 p.
[25h] Memoranda for the visit to Iceland including notes on the plants, animals, rocks and minerals, 48 pages.

This is of specific interest as here is to be found information under such headings as *oeconomica animalis, vegetabilis and lapidae, zoological, botanica, physica, geographica, plantae, Pisces Islandica* and *Plantae Islandica*.

---

[1] Diment and Wheeler, 'Catalogue of the Natural History Manuscripts and Letters by Daniel Solander', pp. 468–9.

[2] Solander MSS, No. 24 in the catalogue; hereafter the catalogue numbers will be placed in brackets. Marshall says that this is '*not* a list of the plants collected' (p. 13). That list is [25i].

[3] These (the place-names are not geographically in the right order) are today: Hafnarfjörður, Laugarnes, Múli, Gráfell, a lava field Kárastaðahraun, Skálholt, Þjórsá, Vaðeyri, Þjórsárholt, Hekla, Geysir, Reykir, Skarð, Þingvellir.

[4] See further Marshall, 'The Handwriting of Joseph Banks', p. 13; and Babington, 'The Flora of Iceland', p. 285.

[5] Marshall, 'The Handwriting of Joseph Banks', p. 16

[6] *Gentiana* with red, white and small flowers. See also the *Specimina Plantarum Islandiae* where Banks and Solander noted the plants they collected and brought back to England and illustrations of some plants were made by John Frederick Miller, whose watercolour drawings are to be found in the Botany Library of the NHM. One is of the *Hieracium praemorsum,* which Britten (1907) has identified correctly as *Hieracium caesiumne*.

[7] Appendix 3.

[8] Appendix 4. A very faint note at the top of the first page reads 'Books taken on shore'.

[9] Document 4.

[25i] *Specimina Plantarum Islandica* [Catalogue of the contents of the numbered bundles of specimens.] 6 pages. According to Marshall this is probably a list of the plants collected in Iceland, 153 plants contained in four bundles.[1]

[25j] Catalogue of the contents of the numbered bundles of specimens, 4 pages.[2]

Of the above only 25 *b, e,* and *f* have been published in this volume, as footnoted. The lists of plants have been useful in identifying some of the flora. Thus the great majority of the scientific papers await publication, a project that will require the collaboration of scholars and scientists in several fields of study. Plans for such a project are in their very early stages among academic staff of the University of Iceland.

---

[1] Marshall, 'The Handwriting of Joseph Banks', p. 13.
[2] This is a shorter list than [25i] and Diment and Wheeler suggest this may be an earlier version of it (p. 469).

THE ICELAND JOURNAL OF SIR JOSEPH BANKS
PART I: 12 JULY–6 SEPTEMBER 1772

Introduction[1]

after my return from my Voyage round the world I was sollicited by Ld Sandwich[2] the first Lord of the admiralty to undertake another voyage of the same nature. His sollicitation was couchd in the following words viz (if you will go we will send other ships). So strong a sollicitation /*acceding*/ agreeing exactly with my own desires was not to be neglected. I accordingly answerd that I was ready & willing. The navy board was then orderd to provide two ships proper for the service. This they did & gave me notice when it was done. I immediately went on board the principal ship[3] & found her very improper for our purpose. Instead of having provided a ship in which an extrordinary number of people might be accommodated they had chose one with a low & small cabbin & remarkably low between decks. This I objected to & was answerd that it could not nor should it be remedied.

With this answer I went immediately to Ld Sandwich who having advisd with several people orderd the Cabbin to be raisd 3 inches **[2]** for our convenience & a Spar deck[4] to be laid the whole length of the Ship for the accommodation of the people. This order I suppposed vex'd[5] the navy board for from that time they never ceasd to pursue me with every Obstacle they could throw in my way & at last Overthrew my designs.

First to the proposd alterations. They added a round house for the Captain[6] to be built over all. This & all other alterations they made with timber so heavy & strong that the top of the round house was literaly thicker than the gundeck of the Ship. This tho I saw, I could not remedy. The Ship was made so Crank[7] by it that she could not go to Sea. Some of the oldest sea officers who I beleive were jealous that discovery should go out of their line procurd an order that the Ship might be reducd to her original state. In this situation then I was again offerd the alternative to go or let it alone with a great deal of Coolness

---

[1] Banks's Iceland journal begins with this introductory chapter explaining the reasons for the voyage and then continues with a new pagination, for the period 12 July 1772 to 6 September 1772. The dates and folio pages are here printed in bold for the benefit of the reader. In the manuscript each page has a heading denoting a place, which more or less corresponds to where they are at the time.

[2] This was the Earl of Sandwich's third term as First Lord of the Admiralty, this time from 1771–83. See p. 4, n. 2.

[3] The *Resolution*. The other ship was the *Adventure*.

[4] A spar deck is a light upper deck in a vessel.

[5] This is difficult to decipher, but this is Carter's reading (in a transcript preserved at the Banks Archive Project) and also this editor's, while Rauschenberg read this as *hurt* ('Iceland Journals', p. 190).

[6] Captain James Cook (1728–99), the celebrated explorer and navigator.

[7] A nautical term, used of a ship when she is built too deep or narrow, and is therefore liable to lean over or capsize.

however for I now had inadvertently opend to them Every [3] Idea of discovery which my last voyage had suggested to me & these they thought themselves able to follow without my assistance now they had once got possession of them.

As the alterations which they had made renderd it impossible for my people to be Lodgd or to do their respective dutys I resolvd to refuse to go & wrote a letter to Ld Sandwich a copy of which is inserted in the appendix stating my reasons.[1] I shall now give a list of the People who I had at my own expence Engagd as assistants in this undertaking

D^r. Solander[2]	now well Known in the learned world as my assistant in Nat*ural* Hist*ory*
M^r. Zoffanii[3]	Painter of Figures & Landscapes
M^r. Jno. Fred^k. Miller[4]	
M^r. Ja^s. Miller[5]	Draughtsmen for Nat*ural* Hist*ory*
M^r. Cleveley[6]	
M^r. Walden[7]	
M^r. Backstrom[8]	Secretaries

besides 9 Servants all practisd & taught by myself to Collect & preserve such Objects [4] of Natural History as might occur three of whoom [*sic*] had already been with me on my last voyage.[9] Besides this I had had influence Enough to to [*sic*] prevail upon the board of Longitude to Send with us Mes^s Baily[10] & Wales[11] as Astronomers & also with the House of Commons to give 4000 pound to Enable Dr Lind[12] of Gorgie[13] remarkable for his Knowledge in Nat*ural* Philosophy & mechanicks to accompany us.

These Gentlemen except only the Astronomers who did not at all belong to me were to a man so well convincd of the impossibility of our going out in the state the ship was now reducd to that they all to a man refusd with me & so well were they satisfied with my conduct that tho I beleive every one but Dr Solander were separately tamperd with to embark without me not one would at all listen to any proposals which could be offerd to them.

---

[1] He did not do so. However, the letter in question is published in Chambers, *Letters of Sir Joseph Banks*, pp. 25–9.

[2] Daniel Carl Solander (1733–82), Swedish botanist. He left much manuscript material on the Iceland voyage, now in the Natural History Museum, London. See further p. 22 above. See also Appendix 1 for details of the members of Banks's entourage on the expedition to Iceland.

[3] Johann Zoffany (1733–1810): see Introduction, pp. 7 and 10 above.

[4] John Frederick Miller (fl. 1772–85), draughtsman and engraver.

[5] James Miller (fl. 1772–82), draughtsman and engraver.

[6] John Cleveley (often spelt Clevely) Jr (1747–86), artist and draughtsman.

[7] Frederick Herman Walden, Swedish secretary to Banks.

[8] Sigismund Bacstrom (fl. 1770–99), ship's surgeon and naturalist, as well as secretary to Banks.

[9] That is to say James Roberts, Peter Briscoe and Nicholas Young.

[10] William Bayly (1737–1810), astronomer.

[11] William Wales (c. 1734–98), FRS, mathematician and astronomer.

[12] James Lind (1736–1812), Scottish physician and astronomer. Such was Lind's reputation that Parliament had agreed, on the suggestion of Banks, to award him a grant of £4,000 for sailing in the *Resolution* (Beaglehole, 'Introduction', *Endeavour Journal of Joseph Banks*, I, p. 72 and see further below in Document 15). Later Banks would have Lind in mind when planning the Macartney Embassy in 1792 and be 'mortified' to discover that he would only consider a fee of not less than £6,000. See further Carter, *Sir Joseph Banks*, p. 292.

[13] Gorgie is a district of Edinburgh.

Upon my refusal to go out the ships were orderd to proceed & in order to do [5] as much as possible even in the branch of Natural history M^r. Forster[1] a gentleman Known to the learned world by his translations of several Books was Engagd under the immediate protection of the King & soon after M^r [Hodges][2] a young man who had cheifly studied architecture was joind to him as Lan*d*scape & figure painter. This young man was so much in debt that he was obligd to leave town without acquainting a single soul of where he intended to go & no sooner was it Known that he was at Plimouth than Baylifs were sent down to apprehend him whoom he Escapd by keeping continualy on board the Ship. With these gentlemen on board the Ships *Resolution* & *adventure* saild from Plymouth on the 12th. of July 1772.

In the Mean time I had receivd several Overtures from the East india Company who [6] seemd inclind to send me on the same Kind of Voyage the next Spring as our Adventurers had now set out upon.[3]

My People all continued faithfull to me. Even Mr Zoffani tho he was, the moment I refusd to proceed, sent by the King to Copy some pictures for him in the Florentine Gallery engagd to leave that business & return to me at a fortnights warning.

The rest were all left upon my hands & as they were a considerable running expence I thought it prudent to employ them in some way or other to the advancement of Science. A voy*ag*e of some Kind or other I wishd to undertake & saw /no*ne*/ place at all within the Compass of my time so likely to furnish me with an opportunity as Iceland, a countrey which from its being in some measure [7] the property of a danish trading company[4] has been visited but seldom & never at all by any good naturalist to my Knowledge.[5] The whole face of the countrey new to the Botanist & Zoologist as well as the many Volcanoes with which it is said to abound made it very desirable to Explore it & tho the Season was far advancd yet something might be done at least hints might be gatherd which might promote the farther Examination of it by some others.[6]

---

[1] Both Johann Reinhold Forster (1729–98), FRS, German naturalist, and his son Johann Georg Adam (1754–94), FRS, German botanist, were engaged as naturalists on the *Resolution* voyage. They were 'disastrous' according to Beaglehole (*Endeavour Journal of Joseph Banks*, I, p. 80).

[2] This is blank but according to Rauschenberg the missing name is that of William Hodges (1744–97), a London-born artist. Through the influence of Lord Palmerston he was appointed the official artist on the *Resolution*. He was subsequently employed by the Admiralty and he exhibited at the Royal Academy 1776–94. He committed suicide after an unsuccessful attempt at banking in Dartmouth (see further Rauschenberg, 'Iceland Journals', p. 197; Beaglehole, *Journals of Captain Cook*, II, pp. 885–6.)

[3] Lord Sandwich said that he hoped Banks would consider alternative voyages 'for the advantage of the curious part of Mankind' and he also hoped that Banks's 'zeal for distant voyages will not yet cease'. Furthermore Sandwich wrote: 'I would advise you in order to insure that success to fit out a ship yourself; that and only that can give you the absolute command of the whole Expedition' (Sandwich to Banks, 22 June 1772, printed in Beaglehole, *Endeavour Journal of Joseph Banks*, II, p. 355). And it seems that Sandwich had considerable influence in the East India Company at this time (see Rodger, *The Insatiable Earl*, p. 92, and Sutherland, *The East India Company*, p. 124). The company was certainly the only organization outside the Admiralty to which Banks might plausibly have turned in his search for a vessel to command on a Pacific voyage. Perhaps Sandwich is here hinting at an East India Company – sponsored mission and this therefore was what Banks had in mind when writing this. (I am indebted to Neil Chambers for this information.)

[4] At this time the Iceland trade was conducted as a monopoly of the Danish Crown. See Introduction, pp. 12 and 14–15 above.

[5] In this Banks was wrong. Horrebow's *The Natural History of Iceland* had been translated into English in 1758, and was among the books taken to Iceland. See Appendix 4.

[6] In this Banks was successful. See Introduction, p. 24, n. 1 above.

Influenced by these reasons I Applied to Baron Diede[1] the Danish Envoy who readily granted me a pasport[2] & having added to my people a Gardiner whose care was to be the preservation of live plants in tubs & seeds I Engagd a Brig of 190 tons *The Sir Lawrence* Captain Hunter with twelve men to **[8]** proceed according to my Directions at the Rate of 100 pounds a month for four Months Certain.

List of the People who embarkd with me on board the *Sir Lawrence*

Capt*ain* Hunter[3]	Alex^r. Scot[7]
Dr Solander	Peter Briscoe[8]
D^r. Lind	Jam^s. Roberts[9]
M^r. Troil[4]	J^{no}. Asquith[10]
M^r. Gore[5]	Peter Sidserf[11]
M^r. Riddel[6]	Nick Young[12]
M^r. J:F Miller	J^{no}. Marchant[13]
M^r. J^s. Miller	Rob Holbrook[14]
M^r. Clevely	Anthony Douez Cook[15]
M^r. Walden	Mr Moreland a Gardiner
M^r. Backstrom	Alexander a Malay who came with me from Batavia.

*O*f these M^r. Troil M^r. Gore & M^r. Riddel are independent of my original plan. I shall say a word or two of them. M^r. Troil a young sweedish Gentleman who has made the Islandich Language his Study wishd to Embark with me to make observations upon it. M^r. Gore Lieu^t. in the navy who now has 3 times circumnavigated the Globe out of mere freindship chose to take the the trip. M^r. Riddel a young Gentleman intended for the Sea still intends to embark with me if I get a ship from the Company.

[1][16] July [17]72

from Gravesend[17]

12.

At 11 at night saild down the river from this place with a fair breeze of wind.

---

[1] Baron Vilhelm Christof Diede von Fürstenstein, the Danish envoy to London: see Appendix 13.
[2] The passport is Document 3.
[3] For biographical details of Banks's entourage, see Appendix 1.
[4] Uno von Troil (1746–1803): see Introduction, p. 10.
[5] John Gore (c. 1730–90), captain in the Royal Navy, soldier, seaman and explorer.
[6] John Riddel, a midshipman.
[7] Alexander Scot, servant.
[8] Peter Briscoe (1747–1810), Banks's servant who also sailed on the *Endeavour*.
[9] James Roberts (1752–1826), Banks's servant who also sailed on the *Endeavour*. He wrote an Iceland journal, see pp. 115–40 below.
[10] John Asquith, servant.
[11] Peter Sidserf, servant.
[12] Nicholas Young (b. 1757), servant. He had sailed on the *Endeavour*, subsequently entering Banks's service.
[13] Jonathan Marchant, servant.
[14] Rob Holbrooke, servant.
[15] Antoine(?) Douez, Banks's French cook.
[16] Here Banks starts numbering the pages of his manuscript from 1 again, denoted in bold and in brackets. The other numbers in bold, but not in brackets, are the days of the month.
[17] Gravesend was at the time a major town on the south bank of the River Thames, the common landing place for those coming to London and those leaving for France.

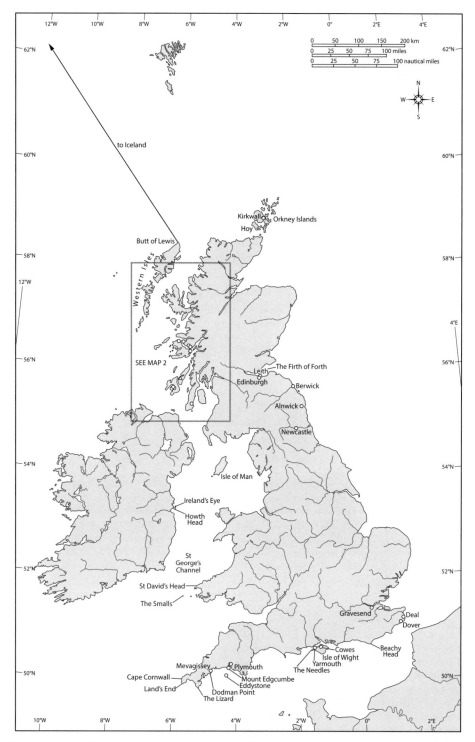

Map 1. Map showing the places mentioned by Sir Joseph Banks on his journey from Gravesend up the western coast of the British Isles towards Iceland and on the return journey via the Orkney Islands.

13.
This morn were at an anchor on the warp about 8 got under way & soon after passd the *Augusta* Yatch with Ld Sandwich on board who was just returning from his visitation of all the southern dock yards. About noon passd by deal[1] it being the day twelvemonth since I landed there from my voyage round the world. From hence we proceeded to Dover where I had promisd to set Count Lauragais[2] ashore. About 4 made a signal for a shore boat which immediately came off & carried us ashore. Here we were fortunate enough to meet an acquaintance Mr. Hatsell[3] who was going over to Calais & undertook at my desire to carry a bird I had with me, Columba coronata Linn.[4] to Calais from which place the Duc de Croÿ[5] had undertaken to forward it to M^r. De Buffon[6] for whoom it was intended. In the even*ing* walkd up to the Castle & observd the Great Brass Cannon which lies on the Cliff it is said to have been a present from the states to Queen Elisabeth.[7]

[2] Dover

& certainly is the hansomest & I beleive the Longest gun I have seen.

14.
The wind last night having blown too fresh for the Ship to anchor at Dover I had sent her back to Deal. So in the Morning we followd her but had the mortification to find that the wind was settled at west so judgd it more prudent to remain at anchor in the Downs than to attempt to proceed.

15.
This morn spent Botanizing about Sandown Castle[8] where we Observd nothing remarkable. Found however Salix arenaria?[9] & Silene conica[10] as they are said to grow in Rays Synopsis. After dinner a small breeze springing up at East we hurried on board & immediately Set Sail. Before however we could get the lenght [*sic*] of Beachy head[11] the wind came to west & blew so fresh as to make me very sick

---

[1] Deal is a seaport in Kent.

[2] Louis Léon Félicité, Comte de Lauraguais and Duc de Brancas (1733–1824). In 1771 he became a member of the Académie des Sciences. He was a confirmed Anglophile making frequent trips to England. There are three extant letters between the two. Banks wrote one dated 6 December 1771 (published both in Chambers, *Letters of Sir Joseph Banks*, pp. 17–24, and Beaglehole, *Endeavour Journal of Joseph Banks*, II, pp. 323–9) which was written as a private account of the *Endeavour* voyage. But evidently Lauraguais copied it for D'Alembert, had it printed and Banks stopped it in the press. A printed version in the Mitchell Library is therefore marked: 'Abstract of Endeavour's voyage written for Count Lauraguais who printed it. I seized the impression and burn'd it.' MS Sydney, State Library of New South Wales, Mitchell and Dixson libraries, Banks Paper Series 05.01.

[3] Possibly John Hassell, the author of *A Tour of the Isle of Wight*, 1790.

[4] *Columba coronata Linn*: the great crowned Indian pigeon (Turton, *A General System of Nature*, I, p. 468).

[5] Anne Emmanuel, the 8th Duc de Croÿ, Prince de Solre (1718–84).

[6] Georges-Louis Leclerc, Comte de Buffon (1707–88), French naturalist and mathematician. He was appointed director of the *Jardin du Roi* (now the *Jardin des Plantes*). He is most famous for his monumental *Histoire naturelle, générale et particulière* in 36 volumes (1749–88).

[7] The cannon, known as 'Queen Elizabeth's pocket pistol', was presented to Elizabeth I by the States of Holland in thanks for her assistance in their war of independence against Spain.

[8] Sandown Castle stood to the north-east of Deal. It was built by Henry VIII as a defence against the French.

[9] *Salix arenaria*: the sand willow, found on sea shores in the loose blowing sand (John Lightfoot, *Flora Scotica*, II, pp. 604–5). Banks himself inserted the question mark.

[10] *Silene conica*: the sand catchfly, presumably Ray's 'Lychnis sylvestris angustifolia caliculis turgidis striatis … A little to the North of *Sandown* Castle, plentifully; Mr. *J. Sherard* in Company with Mr. *Rand*' (*Synopsis stirpium*, 1724, p. 341). (Rauschenberg reads *Silene cuien* and had difficulty identifying it, 'Iceland Journals', p. 198).

[11] Beachy Head in Sussex is the highest headland on England's south coast, rising to 532 ft (162.2 m).

19.

The Wind has been in our teeth ever since the 15th. & myself too sick to write. Now for the first time the weather is rather more

[3] Isle of Wight[1]

moderate & we hope to anchor within the Isle of Wight tonight in order to stop tide.

20.

Late last night we anchord in Cowes[2] road & as we had now expended our small refreshments as Butter Eggs &c we resolvd to go ashore. In the morn accordingly at five we set out & landed in Cowes before the Shops were open as the flood was to run till twelve however we had time to walk about a good deal.

Cowes is a pleasant town situate on the North side of the Isle of Wight. Its road is open but as the distance between the Island & the Main is very short sufficiently well shelterd. Tho it has no trade yet many ships touch here as their last port & clear themselves out. Also such American ships as are obligd to take England in their way /home/ to dispose of their Cargoes in Holland or Elsewhere chuse this as a place likely to cause very little delay as they can sail from it with any wind.

Here is a small Fort[3] of no kind of

[4] Cowes

use I beleive except its name may fright small privateers who /would/ might insult the inhabitants in war time.

On the South side of the town is a pretty large Salt work[4] where salt is made from the Sea water which is pumpd up into flat pans made in the mud 4 or 5 inches deep where the sun evaporates a great part of the water leaving the brine strong Enough to be boild down without any addition of rock salt as is usual at the salt springs. The profits of the work seem however to be very small as the Greatest quantity of Salt that has ever been made there did not exceed 150 Tons in a year & in general was under 100.

In the course of our walks we observd Rubia Anglica[5] growing plentifully in the Hedges near the Road leading to Newport[6] Anthemis maritima[7] near the Salt works linum glabellum[8] Linum sylv. cœrul. &c. Raii synop. Edit 3 p. 362. N. 4[9] which upon examination we were of opinion to [5] be a very distinct species. On the Sea beach about 200 yards to the westward of the Fort a vein of very fine red ochre[10] appeard at the surface

---

[1] An island in the English Channel off Hampshire.

[2] Cowes, seaport and market town on the Isle of Wight was described in *A Tour of the Isle of Wight* by John Hassell in 1790. This description was reprinted in the second volume of Pinkerton's *A General Collection of the Best and Most Interesting Voyages*, in 1808.

[3] Cowes Castle is one of the circular forts built by Henry VIII in 1540.

[4] Salt-making was an ancient Hampshire industry. Hassell makes no mention of it in his description of Cowes in 1790 in *A Tour of the Isle of Wight*.

[5] *Rubia anglica*: the wild madder, bedstraw family (*Rubiaceae*).

[6] The isle's main town, situated inland, south of Cowes.

[7] *Anthemis maritime*: the sea chamomile.

[8] *Linum glabellum*: the perennial blue flax.

[9] Banks's own reference to the 1724 edition of Ray's *Synopsis stirpium*, p. 362, which listed: 'Linum sylvestre cœruleum perenne procumbens, flore & capitulo minore. Linum perenne minus cæruleum; capitulo minore ... Wild perennial blue Flax, the lesser.'

[10] Red clay. Ochre is clay coloured by iron oxides, used in ground form as a pigment.

of the Earth below high water mark. In all probability it runs under the hill which is common & might there be got to advantage.

At 3 O'Clock, the tide of Ebb making we got under way & proceeded but the wind being contrary & our people very much tird we agreed to go no farther than Yarmouth[1] at which place we arrivd about 6 & as my servants &c. &c. almost all Landmen seemd desirous of a Landing I resolvd to carry Every body ashore while the Crew of the Ship slept. Accordingly we landed with French Horns[2] to the no small surprise of the people who little expected to see such a motley crew issue from so small a vessel.

The town is small & ill built. The people seem much less humanisd than those of Cowes much less usd I suppose to see strangers. The children followd us about the Streets begging for halfpence.

[6] Yarmouth

The town is very small & ill built yet here is a little fort which from the Sea makes rather a formidable tho an old fashiond appearance.

Near the town is a small salt work near which we gatherd Frankenia laevis.[3] In the meadows was plenty of Linum glabellum & here & there a little of the Agrostis pallida[4] by the See shore. To the Eastward about a mile from the town were many flat plates of Stone a little impregnated with Iron in which were many fine Casts of Shells.

21.

At 3 O'Clock this morn we were calld up to see the needles[5] by which we passd & admird the small perpendicular rock resembling indeed a needle from which they probably had their name. That very stone would certainly be an excellent situation for Observations to be made to prove whether or not the theory of the Seas decrease is founded upon facts. At present our pilot told me that at low water there was not more than three feet [of] water between it & the larger rock which lays near

[7] Plymouth

it. If so & this paper should be read a thousand years hence they will probably be united if our present philosophers build upon good grounds.

23

After a variety of winds & a total dearth of adventures we this day at noon arrivd at Plymouth[6] where to my great disapointment I learnd that M[r]. Arnold[7] had carried my time Keeper to London with him whether to evade a trial or through thoughtlessness I cannot say however. To complain was needless so we spent our Evening at the Long

---

[1] Yarmouth (not to be confused with Great Yarmouth) is 12 miles (19.3 km) south-west of Cowes with yet another fort built by Henry VIII.

[2] During his stay in Newfoundland Banks spent his time 'Very Musically', as he wrote to his sister Sarah Sophia, rejecting the flute and playing the guitar 'with Great success'. Quoted by Beaglehole, *Endeavour Journal of Joseph Banks,* I, p. 12; Carter, *Sir Joseph Banks,* p. 35.

[3] *Frankenia laevis*: sea-heath.

[4] *Agrostis pallida*: panic millet grass.

[5] Three detached masses of Mucronata chalk (very hard chalk) off the westernmost point of the island, rising to c. 100 ft (30.5 m). The original 'Needle' was a pinnacle, 120 ft high (36.6 m), which collapsed in 1764. There is a drawing of them by J. F. Miller in BL, Add MS 15509, f. 1.

[6] Plymouth was a major naval port, the starting point of many expeditions, e.g. against the Armada and the attack on Cadiz.

[7] John Arnold (1735/6–99), a London watchmaker and one of the first makers of chronometers in England. He made chronometers for George III, the government and the East India Company.

rooms[1] which are neat & well situated for a beautifull prospect tho I beleive but ill frequented. Here is however every convenience to make Sea bathing convenient. Baths either hot or cold & a machine for fine weather in which you may bathe at any time of tide.

24

This morn we set out for mount Edgecumb[2] which is certainly a fine thing tho I cannnot help calling it even now a place of great capabilities was its noble owner[3]

[8] Mount Edgecumb

a man of refind taste for Laying out ground it certainly might be made a most Elegant place. At present nature is vast but absolutely naked was the view it commands of the immence ocean with all its inhabitants Plymouth sound the Legions of Vast ships lying there the dock the town & a fine countrey. Was this I say releivd by some internal beauties were there some vales where the mind might be releivd[4] by bounded prospects how fine would it be but at present nature magnificent as she certainly is fatigues the mind with that very magnificence & wheresoever she roves she can find no releif but is almost persecuted by repeated views of the sound of the dock of the town of the Ships all which she at first comprehended in one magnificent peice of Scenery. That famous & facetious voluptuary Juin[5] in speaking of fish divided the good ones [9] into two classes those which were good with sauce he calld <u>Fish of merit</u> those on the other hand which were good without sauce were fish of <u>Personal</u>[6] <u>merit</u> if the same distinction was to be applied to places Mount Edgecumb would certainly be a place of merit for it derives I may say all its merit from the sauce that is the accompanyments with which nature not art has furnishd it.

From mount Edgecumb we proceeded to the dock[7] which is truly magnificent certainly the first in England of course the first in Europe. Tho the intended improvements which were plannd by S^r. Thos. Slade[8] are not yet half executed they prosecute them however with great spirit. The rope walks which are now finishd are Six rows each twelve hundred feet in lenght [sic] the prettiest examples of Perspective that certainly can be seen. At Eleven at night we came on board & departed from this place with very little

[10] off the Eddystone[9]

wind but before morning got the wind at west.

---

[1] Presumably the Royal Clarence Baths, situated on a beach directly opposite Mount Edgcumbe.

[2] Seat of the Earls of Mount Edgcumbe, built c. 1550 by Sir Richard Edgcumbe. It is situated across Plymouth Sound on a peninsula which juts out into the west side of the harbour. Banks had explored the estate with Phipps while waiting for favourable winds for the *Niger* to sail in April 1766 and describes it in great detail. See Lysaght, *Banks in Newfoundland*, pp. 115–17.

[3] George Edgcumbe (1720–95) 3rd Baron Edgcumbe, later (1789) 1st Earl of Mount Edgcumbe, FRS. In 1772 he was Lord Lieutenant of Cornwall.

[4] Rauschenberg omits 'releivd by some internal beauties were there some vales where the mind might be' ('Iceland Journals', p. 200).

[5] Probably Nicolas Jouin (1684–1757), French satirical poet, author of anti-Jesuit pamphlets and Jansenist.

[6] Difficult to decipher.

[7] The royal dockyard was established by William of Orange in 1689. The town which grew up around it was known as Plymouth Dock or simply Dock, until it was renamed Devonport in 1824.

[8] Sir Thomas Slade (1703/4–71), shipbuilder.

[9] Lighthouse on the Eddystone Rocks 14 miles (22.5 km) south-west of Plymouth. This reef was the scene of numerous shipwrecks, and the lighthouse was the third on the site. Designed by John Smeaton, it lasted until 1882.

25

Which blew very fresh & not only prevented our proceeding but made me sicker than I have been since we saild.

26

This morn we were very little to windward of Plymouth sound quite calm. We fishd but caught nothing except 4 dogfish (Sqalus Acanth[1]) on whose fins were however a new Species of Oniscus.[2] The wind very soon breezd up as foul as Ever & we spent the day advancing very little.

27.

This morn we were about 2 leagues from the Deadmen abreast of a town calld Maragise[3] in the Channel draught; & seeing many fishing boats along shore we hoisted our colours on which a legion of small boats put off all however intended to buy of us any smugling commodity we might have. So our treaty ended not much to any of our satisfactions. At twelve got the wind fair & with a fresh breeze proceeded along Shore. Some Gannets[4] or Solan geese[5] were about the

[11] Lands End

ship probably bred on some rock in this neighbourhood as those whose nests are once disturbd (as is probably the case with most of those which breed on the coast of Scotland) are said not to fly till the month of September.

About noon we got round the lizzard[6] & met a large sea from the South westward which very soon incapacitated me from writing by making me more sick than I had been during my whole excursion.

28.

This morn saw the fleet of Observation ten Sail in all returning. They seemd to stand in to Plymouth but we were too far off to attempt speaking to them. At night we were near the Lands End[7] & in the night got round it. Soon after got the wind at south which determind us to Sail up the Irish channel.[8]

29.

Hazey weather & Strong breeze at S /SW/ we had steerd from Cape Cornwal[9] NNE for a short time & then /N/NE which by our draughts should have carried us 5[10] leagues without the Smalls rocks[11] lying off St Davids head.[12]

---

[1] *Squalus acanthias*: piked dogfish or *Picked Dog-fish* (Turton, *A General System of Nature*, I, p. 921).

[2] *Oniscus*: a genus of crustacean, Oniscidae family.

[3] Now Mevagissey near Dodman Point. An English league was 3 nautical miles.

[4] *Morus bassanus*: the northern gannet.

[5] The common gannet is the solan goose.

[6] Lizard Point, or simply The Lizard, a headland about 10 miles (16.1 km) south-east of Helston, the southernmost point of the British mainland. A reef running out from the Point makes passage round it extremely dangerous.

[7] Land's End is a promontory about 9 miles (14.5 km) south-west of Penzance, the westernmost point of England. Dangerous reefs lie off it.

[8] The Irish Sea, that part of the sea between northern England and the north of Ireland.

[9] Cape Cornwall is a headland 4 miles (6.4 km) north of Land's End.

[10] Rauschenberg read 4 ('Iceland Journals', p. 200).

[11] The Smalls are a cluster of rocks off the coast of Pembroke.

[12] St David's Head is a precipice about 100 ft (30.5 m) high, 3 miles (4.8 km) north-west of St David's, Pembrokeshire, the westernmost point of Wales.

[12] Irish Channel

At dinner time however we were surprisd by the sight of breakers ahead no more than two miles off on which we hauld our wind & stood W. by which course we soon cleard them whether this was causd by the false position of these rocks in our charts those of Capt[n]. G Collins[1] & Mons[r.] Dapres de Manivillette[2] or by the indraught of the Bristol channel[3] I cannot with certainty affirm but suppose rather the latter to have been the occasion of it.

In the Evening the weather cleard up & we steerd again N N E the wind still blowing strong on the Southern bord. In the morn we we [sic] were abreast of Dublin & saw plainly

30

the head of Hoath[4] & Irelands Eye. The weather fine & wind S W a fresh breeze

Many birds about the Ship young & old Solan Geese puffins[5] Guilemots & one large flock of Gulls. At noon the high Land of Dundrum[6] in sight. Soon after we had a sight of Man.[7]

[13] Mul of Cantire[8]

31.

This morn we were oft the mul of Cantire. The tide set so strong against us that instead of gaining we lost ground very considerably.

The Shore of Scotland which we were very near to gives as wild an idea as any that can easily be conceivd. Bare rock & heath constitute the greatest part. Yet here & there a miserable hut with a small enclosure seldom more than $\frac{1}{4}$ of an acre seems to tell you that the neighbour Countrey is not better /else/. A rational being would never fix upon so wretched a spot to continue his existence if a better could be found which he might get posession of.

---

[1] Greenvile Collins (fl. 1679–93), Captain in the Royal Navy and hydrographer. Between 1681 and 1688 he charted the whole of the coast of Britain. The charts were published collectively in 1693 under the title *Great Britain's Coasting Pilot*.

[2] Jean-Baptiste d'Après de Mannevillette (1707–80), French hydrographer. He was a captain on board the ships of the French Compagnie des Indes. He travelled the world and published a great deal including *Description et usage d'un nouvel instrument pour observer la longitude sur mer, appelé le quartier anglais* in Paris in 1739.

[3] The inlet separating south Wales from south-west England.

[4] Howth Head, promontory 560 ft (170.7 m) high on the north side of Dublin Bay. Ireland's Eye is a small island one mile (1.6 km) to the north. It was formerly called Inis-Ereann, 'Island of Erin'.

[5] *Fratercula arctica*: the Atlantic puffin.

[6] Small town on the coast of County Down, 7 miles (11.3 km) south-west of Downpatrick. On a hill close by the town are the remains of a medieval castle.

[7] The Isle of Man.

[8] The Mull of Cantyre (now Kintyre) is the nearest point of land in Britain to Ireland, Tor Point in Antrim, and the most southern part of Argyleshire. The Mull of Cantyre was 'noted for the violence of the adverse tides'. King Magnus Barelegs or Barefoot (*Magnús berfættur*) of Norway conquered it, and added it to the Hebrides (see Pennant, *Tour in Scotland and the Hebrides*, p. 220). Thomas Pennant was the first Englishman to publish a book on Scotland and the Hebrides. This was in 1769, so he preceded Banks, a point that Banks was very aware of (see Document 10). In later editions Pennant included Banks's account of his visit to Staffa. There are many editions of Pennant which differ in pagination; here the 1790 edition, published in London, is used throughout. See further on Banks and Pennant in Appendix 2.

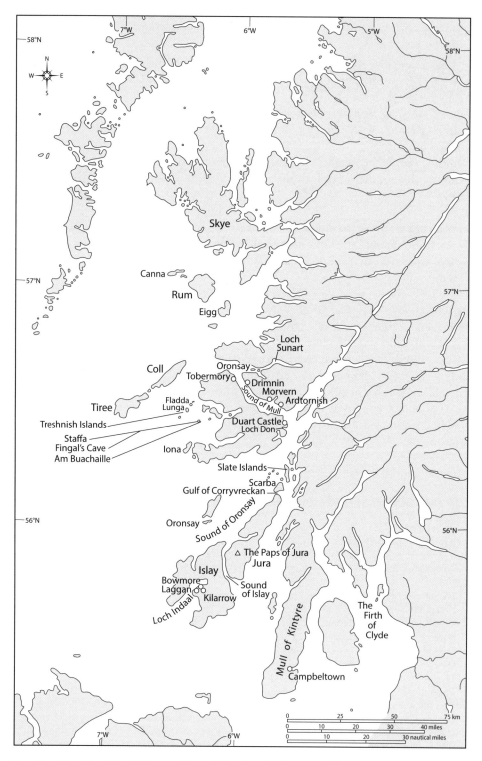

Map 2. Detail of the Western Isles of Scotland, showing the places mentioned by Sir Joseph Banks.

In the bay towards Camelstown¹ the countrey indeed wears a more agreable appearance. Pasture mixd with corn is to be seen on the face of some of the hills but we were at too great a distance to see any thing distinctly.

About noon the tide hove us so strong into the mouth of [the] Clyde that we were obligd to tack

[14] off Ilay²

Aug^st.³

& stand over to the Coast of Ireland which here cannot be above 8 Leagues from Scotland. Tho the wind was very slight yet we soon ran over & saw on that side a much better looking countrey than on the scotch. The Cultivations were larger & the houses surrounded by an uncommon quantity of outhouses so that each farm seemd a little village /at night/ in the Evening it fell calm & I sent out a party to Shoot who shot Larus canus the Common Gull & Alca arctica the Puffin. Islay was now seen plain but we had no hopes of arriving there at night on account of the want of Wind. At Sunset a Seal was seen swimming in the water.

1st. [August]

At day break this morn we anchord in Lochindale⁴ & went ashore immediately. The town of Bomore⁵ we found to consist of but few houses. Among them however were two publick ones. These could supply us with victuals but by no means with Lodging or even a room to Eat in so it became necessary

[15] Ilay

to pitch our tents which was finishd about 4 O'Clock in the midst of an immence /crowd/ of People who had been brought together on account of preparation for the Sacrament which is here administerd only once a year & seems to be receivd with much more respect & much more generaly than in England.

The Evening provd rainy so we were obligd to amuse ourselves with a plentifull Highland dinner composd of various legs of mutton & puddings which shewd the plentifullness of the countrey & that Luxury had yet made few advances in it. Some Gentlemen of the Countrey dind with us & after dinner introducd us to some of the Ladies who gave us tea & thus we at once commenced an acquaintance in the countrey.

---

¹ Presumably Campbeltown in Argyll, on the shore of Campbeltown Bay, by a loch of same name. Formerly it was the seat of the ancient Lords of the Isles, the rulers of the Inner Hebrides. It was founded in 1607 as part of the 'plantation' policy to bring Protestant settlers to the Hebrides, orchestrated by the 7th Earl of Argyll. Pennant found it 'a very considerable place'. In 1744 it had only two or three small vessels, but by 1772 there were 78, with a population of 7,000. According to Pennant, 'the oppressed natives of lowlands' had been encouraged to settle and 'are esteemed the most industrious people in the country' (Pennant, *Tour in Scotland and the Hebrides*, pp. 219–20).

² Islay (often spelled Ilay) is the most southerly of the Inner Hebrides and third in size. Its history dates back to Neolithic times and reached a peak when it was the seat of the Lords of the Isles, who were crowned here by the bishops of Argyll. In 1726 the island was bought by Daniel Campbell of Shawfield. There are numerous ruins of castles and forts in the island. It was then as now home to numerous species of birds and a wide variety of flora.

³ So placed in manuscript. The events of 1 August begin further down the page.

⁴ Loch Indaal.

⁵ Bowmore, a village and small seaport, lies near the head of Loch Indaal, and is now the geographical and administrative centre of Islay. When Banks was there the village was just coming into existence, the first houses being built in 1768.

2.

Being Sunday an immence concourse of people came to receive the Sacrament so many that tho three Clergymen Officiated by turns the Communicants passd & repassd till after Six O'Clock. This whole day it raind immoderately /from/ to which circumstance cheifly we owe

[16] Ila[1]

the preservation of our Characters for had we done any Kind of Work even walkd out Botanizing on a day held so sacred in this countrey the black seal[2] would have been irreversibly set upon us.

3ᵈ.

Tho it still raind our patience was so far wore out that we /began/ set out this morn & scarcely had we proceeded a mile towards Kilara[3] when we met an object that atracted our attention. A highland house so miserably constructed that it tempted us to have drawings made of every particular in it. Twas built of Stones so loosely laid together that wind & rain could scarcely be stopd in their Course by them. There were two doorways one of which servd at all times for a window for the house was furnishd with only one door or rather substitute for one a faggot of sticks not more closely tied up than faggots in general are which was occasionaly plac'd in one or the other doorway as the family found it most convenient.

In the middle of the house was the fire over which hung a pothook not in the chimney [17] but under that hole which was made in the roof as an expedient to let out a part of the smoak which it did but not till after the house was full so that none seemd to be lookd upon as superfluous but the mere overflowings. Round this upon miserable benches sat the family consisting of a Weaver his wife her mother a Stranger woman & Six children. These had two beds to accomodate them. The rest of the furniture consisted of a Loom & a lamp.[4]

Few as these conveniencies were to be allotted to the use of ten people yet they all appeard chearfull & content rather more so than common & the man in particular answerd all our questions with that becoming ease that total absence of Mauvais honte[5] that the whole scotch nation are blessd with in a degree so superior to the English to which cheifly I am much inclind to attribute the great success that their adventurers meet with in our capital.

[1] i.e. Islay.

[2] The black seal means that he would have been ostracized by the populace who frowned on any kind of non-religious activity on the Sabbath.

[3] Now Kilarrow. 'Kilarow, a village seated on *Loch-in-Daal*, a vast bay that penetrates very deeply into the island. Opposite *Bomore*, ships of three hundred tuns may ride with safety … Near *Kilarow* is the seat of the proprietor of the island.' There were tombs to be found there with 'most remarkable grave-stones' (Pennant, *Tour in Scotland and the Hebrides*, p. 251). James Miller drew a sketch entitled: 'View in the town of Killaru on the Island of Ila' (BL, Add MS 15509, f. 6). The parish church built there in 1767–9 by Daniel Campbell of Shawfield and Islay is the only circular church of this period in Scotland still in use. Legend has it that the church was built as a round so there would be no corner for the Devil to hide in.

[4] There are several illustrations from here, all in BL, Add MS 15509: 'Forefront of the Weaver's House near Bomore on the Island of Ila', by J. F. Miller (f. 24), with children playing ouside; 'The Backfront of the Weaver's House near Bomore on the Island of Ila', by James Miller (f. 26); 'The inside of the Weaver's House near Bomore on the Island of Ila', by J. F. Miller, with the mother breastfeeding with a further seven members of the family looking on, (f. 28). Lastly there are pencil sketches of the weaver and his daughter dated 4 August 1772 (f. 29).

[5] *mauvais honte*: self-consciousness, bashfulness.

From hence we proceeded to Killara[1] a small town at the head of the bay the residence of M[r]. Campbel of Shawfeild[2] the principal proprietor of [18] the Island. A very bad house is the best in the Island but as he very seldom resides here it is very sufficient.

In the Town of Killara is the remains of a religious foundation. In the neighbourhood are several tombstones said to have been brought from Lough Finlagan where M[c]Donald King of the Isles[3] had his residence. Of them I orderd my draughtsmen to Copy several & then proceeded to some lead mines about four miles in the Countrey. The stratum in which the lead lies Is found to be limestone exactly similar to that of the peak of Derbyshire & like it having shale for the stratum above it. These strata I was told lay in one lump in the middle of the Island & do not reach the sea in any direction. The mines at present bear a promising appearance. They have been originaly workd open cast by the Danes. As the people say this gives a great advantage to a speculative miner for the tops of the veins being laid bare it is easy to judge by their directions where they [19] will meet. At present the Lessee M[r]. Freebairn[4] is working on with spirit & will soon arrive at a place where five of them /*probably*/ run together in one point so that probably they may turn out very rich. At present his far feild carries 4 inches of ore.

4.

Weather again so rainy that it was impossible to see any thing with pleasure. We went however to see a Cave near Laggan of which we had heard a very pompous account but found it a dirty nasty hollow in a rock about 100 feet in depth from whence we receivd no satisfaction but from the following experiment.

Having when we came to the cave no fire we attempted to light one by firing gunpowder just within the mouth of the cave. When we had /*litghet*/ lighted our candles we were surprizd to find that the smoak had penetrated to the very bottom of the cave & was there so thick that we could scarce see our way.

5

The weather still continued rainy tho the Barometer stood at $29 \frac{28}{100}$ however we resolvd [20] to remove from a place where in reality we could see nothing at all. We resolvd to remove to the other side of the Island & no sooner had we signified that intention than the Gentlemen of the Island sent in above 20 horses to induce us to travel by Land. We accepted their invitation & orderd the Ship to come round with all Expedition to the Sound of Ila. Our journey was about 12 miles over a Countrey in which cultivation has yet made but a small progress. The Soil of the whole Island is a Kind of Turf under which lies Marl[5] of different Kinds which being got up & only thrown upon the surface of the

---

[1] Kilarrow in Islay.

[2] Daniel Campbell (d. 1777) of Shawfield, MP, was the proprietor of the island at that time.

[3] McDonald King of the Isles: probably the second Lord of the Isles of the MacDonald clan. The MacDonalds were instrumental in expelling the Norsemen from the Western Isles in the 12th century. Loch Finlaggan is towards the north of Islay.

[4] Charles Freebairn's letter to Banks, dated 8 July 1773 (Archives of the Royal Botanical Gardens, Kew, JBK 1/2/37), contains a description of mining techniques and ores. Pennant says he had worked the lead mines since 1763. He had smelted some £6,000 worth of ore in his furnace at Freeport in the nine previous years (Pennant, *Tour in Scotland and the Hebrides*, pp. 249–50). An illustration of his house by J. F. Miller is in BL, Add MS 15509, ff. 2–3. For lead mining on Islay, see Callender's pamphlet, *The Ancient Lead Mining Industry of Islay*.

[5] Marl is a kind of soil consisting principally of clay mixed with carbonate of lime, forming a loose unconsolidated mass, valuable as a fertilizer.

land under which it lies immediately destroys the ling[1] & brings the Land to a fine turf which will yeild very good crops of Corn as experience has shewn particularly in the farm of M[r]. Graham[2] who improves his farm with great spirit & as great success. The first principle of improvement is however in this Island totaly neglected I mean that of dividing the Lands. In [the] Whole Island is scarcely **[21]** a hedge or a wall. This at once prevents them from having any wood, winter corn, or hay in the Island. For in winter time the Black cattle of which they rear great numbers run all over the Island without the least check feeding upon whatever they can find & tho there are many valleys in the Island full of Brush wood Oak Ash birch &c they are never allowd to rise above man hight.

At [*blank*] in the Evening we arrivd at M[r]. Freebairns house on the sound of Ila where we were entertaind. The house is situated in a very romantick spot under an almost overhanging clift close by the sea side here not above an English mile in breadth. But on one side a small brook tumbles down from the hill in a pretty cascade. In front is the Island of Jurah[3] barren indeed but rising into two hills higher than any in that neighbourhood. Near his house is a very good air Furnace at which he smelts the lead producd by his mines.

6.

The weather in the morning being tolerably fine I resolvd not to lose that opportunity of going upon the paps[4] /of Jura/ high hils on the Island of that name

[22] Jura

lying opposite & about 8 miles from the place where we were. Accordingly we set out carrying a barometer with us in order to measure the hight of the hill.[5] About half way up the hill we met a covey of Ptarmigans[6] & I was fortunate enough to Shoot two. They are clearly the same bird as the white Partriges of the Northern countries but differ from them in that they never come down to the low countrey but Keep always upon the highest hills so at least the gentlemen who were with us universaly asserted. We found them in a place coverd with large stones so thick that not the least vegetable was to be seen. The countrey people went so far as to assert that they were found in no other places & that they verily beleivd the birds to feed upon nothing but stones as they had never found any other thing in their crops which is not difficult to account for as people cannot be supposd to arrive very early at the places they frequent & towards the middle of the day the food of the morning being digested leaves nothing in their maws[7] **[23]** but the stones of which these birds use a large quantity.

About 12 we arrivd at the top of the Southernmost of the hills & immediately set up our barometer & Observd every /hour/ quarter of an hour. The medium hight we found about 27 $^{inch}\cdot\frac{7}{10}$. The mist was now thick upon the hill & the thermometer about 56°. We observd that /if/ the Columns of mist which passd quick over us sometimes thicker & at

---

[1] 'Ling' refers to heather, which the locals used 'marl' to destroy, leaving grassland.

[2] Unidentified.

[3] Jura is an island in the Southern Inner Hebrides. Pennant wrote much about Jura calling it 'this mass of weather-beaten barrenness' (Pennant, *Tour in Scotland and the Hebrides*, pp. 243–8).

[4] The Paps (hills) of Jura are formed of quartzite. There are three of them rising to 2,575 feet (785 m). On the summit were several 'lofty cairns' as Pennant, who climbed the Paps, mentioned (*Tour in Scotland and the Hebrides*, pp. 247–8).

[5] Probably Beinn a' Chaolais, today calculated as being 2,408 ft (734 m).

[6] Pennant also met with: 'a brace of *Ptarmigans* [that] often favored us with their appearance' (*Tour in Scotland and the Hebrides*, p. 248).

[7] Rauschenberg says 'craws' ('Iceland Journals', p. 203).

others of a thinner substance alterd the hight of the mercury very fast the dense ones Lowering the mercury & the thinner ones on the contrary raising it. Indeed it seldom remaind many minutes of the same hight. These variations however were but small never arising to $\frac{1}{20}$ of an inch in all.

The hill itself does not raise high enough to produce Alpine plants. We found not one species of Saxifraga[1] nor indeed any one plant that inhabits the regions near perpetual snow many of which are found upon Snowden[2] Cader Idris[3] & even the Van[4] in Caermarthenshire. The sides near the top & for $\frac{1}{3}$ down are frequently Coverd /with/ for large tracts with /vast/ stones of [24] all Kinds of dimensions the sides & angles of which were perfectly sharp so that they must have been laid there since any great Power of water has acted so high. By what operation of nature I cannot guess possibly by fire but no remaining signs of that Element which generaly leaves conspicuous enough traces of its operations occurd to me. The fog indeed being thick upon the hill prevented our seeing 100 yards before us in any direction so we might omit to Observe things which may be very palpable to those who come after us.

Having finishd our Observations we dind & then erected a heap of Stones about 7 feet high on which we Erected an upright one as a token of the place we had chose. We meant then to have proceeded to the middlemost of the paps but having no one in Company who had been there & the mist being too thick to allow us at all to see the road we thought it more prudent to desist from the attempt. Accordingly we set out on our descent in the course of [25] which we again saw the Ptarmigans & when we got so near the foot of the hill as to be out of the mist a few Moor fowl. About 7 we arrivd at M^r. Frebairns when on comparing the Barometer which we had left with him with ours we found it had stood at about 30°: $\frac{2}{10}$ $\frac{34}{500}$ the difference between the two observations after the usual corrections gives about 2359 feet which I suppose to be near about the hight of the mountain.
7.
This morn the Ship came round to us. The weather being fine & Clear Dr Lind got ashore the Equatorial[5] in order to fix the Latitude of Freeport[6] the place we were at which he found to be /55°:52′:32″/. In the mean time I set the draughtsmen to work more to inure them to drawing than from any thing curious which I had to propose. We then dind with Maj^r Donald Campbell from whoom we had receivd many civilities & at night proceeded to the Ship in order to go to Sea in the Morn.
8.
The wind coming so criticaly foul that we could not break lose the ship I resolvd to take this

---

[1] *Saxifraga* refers to perennial herbs found in northern temperate, Arctic and Andean zones. According to Mabey, the purple saxifrage is chiefly to be found in Scotland (*Flora Britannica*, pp. 179–81). According to Lightfoot's *Flora Scotica*, there are various types of saxifraga but the Alpine Saxifrage can be found 'Upon the summits of the highland mountains, but not common' (I, pp. 220–25).

[2] Snowdon is the highest mountain in Wales, 3,560 ft (1,085 m) in height.

[3] The Cader Idris or Cadair Idris, meaning the chair of the Welsh giant Idris, is a mountain, 2,930 ft (893 m) high, in Merioneth, north-west Wales, at the southern end of the Snowdonia National Park. Banks was to see it in 1773 on his tour of Wales with John Lightfoot but did not climb it, though he climbed Snowdon.

[4] The Carmarthen Van or Fan is a peak in the Black Mountains, sometimes called the Carmarthenshire Beacons, and is 2,631 ft (802 m) high.

[5] There is a picture of this type of astronomical telescope in Carter, *Sir Joseph Banks,* p. 105.

[6] Freeport was Charles Freebairn's home.

[26] Oransay

opportunity of visiting Oransay.[1] An Island in the neighbourhood where I had been told were some considerable remains of an ancient Monastery.[2] Accordinly we set out in the boat & tho it blew very fresh & raind arrivd there in about 2 hours. Dripping wet as we were we immediately made towards the only good house in the Island. The Master M[r]. Macneal[3] was not at home but a relation of his Cap[tn]. Macdougal[4] receivd us immediately with all those marks of hospitality which tho not to be met with in England are yet so common in these unpolishd countries. Every Kind of refreshment was producd in a moment & every assistance to enable us with greater ease to examine & take drawings of such things as we might think worth our attention. We did not indeed find much. The church & buildings about it were all in a very ruinous state. We made draughts however of Every thing remarkable.[5]

In one of the Chapels was a singular [27] instance of superstition. It was the burying place of the Macduffies or M[c]Fees[6] as they are calld. In one corner of this was a wand about 8 feet long supported by a stone through which a hole had been made for that purpose. This we were told was the flagstaff of a M[c]fee who had been buried above 200 years. On this the people here beleivd the fate of the M[c]fee family to depend. They are to last sayd they as long as this staff but will be extinct as soon as it is taken away or destroyd. We behavd to it with the utmost respect tho we could not help fancying that the Macfees had renewd it several times since the death of their great Predecessor.

At night we returnd to the ship & the

9

next morning the wind being fair took leave of the Sounds of Ila famous for having been chose as an anchoring place by Thurot[7] when he visited scotland during the last war.

---

[1] Now Oronsay, an island of the Inner Hebrides, just over 2 sq. miles (5.2 km^2), immediately south of Colonsay and inhabited since prehistoric times. Ruins of the church and cloisters of a 14th-century Augustinian priory still survive. Founded under the patronage of the Lords of the Isles between 1330 and 1350, intricately carved graveslabs and stone crosses were produced there until 1500. John Cleveley made a sketch of the tomb in the McFee Chapel at Oransay (BL, Add MS 15509, f. 49). On the back of the sketch in Banks's autograph it says: 'We were told this M[c]Fee or McDufie was a factor or manager for Macdonald King of the Isles upon these Islands of Oransay & Colonsay & that for his mismanagement & Tyranny he was executed by order of that prince. This stone lies close to the South wall of M[c]Duffies (or as he is generally calld) M[c]Fees Chapel at Oransay the inscription may be: "Hic jacet Murchardus Macdufie de Collonsay anno. Do. 1539 mense mart. Ora med. ille ammen" [Here lies Murchard Macdufie of Colonsay [died] in the year of our Lord 1539 in the month of March. Pray for his soul Amen.].'

[2] Pennant wrote: 'the antient monastery, founded (as some say) by St. *Columba*, but with more probability by one of the *Lords of the Isles*, who fixed here a priory of *Canons regular of Augustine*, dependent on the abbey of *Holyrood* in *Edinburgh*'. A description follows (*Tour in Scotland and the Hebrides*, pp. 269-71).

[3] 'This island is rented by Mr. *Mac-Neile*, brother to the proprietor of both islands [Colonsay and Oransay]' wrote Pennant (ibid., p. 272).

[4] Alexander McNeil was married to Mary, the daughter of Alexander Macdougall of Macdougall.

[5] The illustrations from Oronsay are in BL, Add MS 15509, ff. 38-55.

[6] Pennant does not mention the McFees. The MacDuffies or MacFies of Colonsay were replaced as the ruling family in 1701 by the McNeils.

[7] François Thurot (1727-60), corsair. A famed French naval officer who scored several victories against the English until mortally wounded in action against three frigates in 1760. Born in Burgundy, he commanded a privateer taking many English prizes, eventually becoming a captain of a frigate in the French Royal Navy. When anchored off Islay, he carefully avoided plundering the islanders to supply his ship and crew, thus earning their affection. Thurot was one 'of the great Corsairs of France' (see Norman, *The Corsairs of France*, pp. 240-86.)

He anchord at the N W mouth of the Sound in a place where it was not usual for Ships to lie. The inhabitants went out to him & told him

[28] Ila

so but he in return told them that he knew its conveniencies far better than they did which in reality was true. He was in such a position that was he pursued by a superior force by taking the advantage of the tides which he was perfectly acquainted with he could run out either at one or the other passage & avoid his enemy.

Whether or not his orders were to enquire if any remains of the spirit of Rebellion was still to be found among these people he certainly took every means in his power to please them. They now speak of him with the utmost regret among the instances of his lenity I will relate some.

While he lay at anchor in the Sound there were several sail of West Indiamen in Lochindale[1] a few leagues only from them. Instead of making them a prize he sent round to them to advise them to go to sea for said he should any bad weather oblige me to come into Lochindale I shall be Obligd to take them all.

Some of the Soldiers that he had on board having been landed for refreshment began with their bayonets to dig up the potatoes thinking [29] them fair plunder in an enemies countrey. He seeing this from the Ship sent word ashore that if they did not desist even from that small mischeif he would instantly fire upon them from the ship.

He had while he lay here furnishd himself with all Kinds of refreshments which the people willingly supplied him with to prevent his people from plundering. Among other things he had met a vessel loaded with meal intended for one of our garrisons in the highlands. For all these things he was preparing to pay at the market price when his Officers remonstrated saying that it was wrong to pay an enemy for what was fair plunder. He persisted & calld a council of war in which his opinion was overruld. On this say the Islanders he producd an order signd by the King of Frances sign manual that no Kind of Damage should be done to the Scotch in any shape or Kind. This done he paid for every thing at the prices set upon them & to the Meal vessel he gave as well as the price of this meal the freight & profit which he would have made had he continued his voyage.

[30] Scarba[2]

Having heard much from all Kinds of people of a whirlpool between the Islands of Jura & Scarba which they represented almost if not quite as remarkable as the famous Maelstrom[3] of Norway I thought it incumbent upon me to see it tho the tides being at present very low gave me little hopes of seeing any thing very extraordinary. Accordingly I orderd the ship round to Scarba where we were told it might be seen with the greatest conveniencey & about 12 landed with provisions & a little Tent. This we pitched but having waited with impatience the whole tide saw nothing at all remarkable. There was it is true a strong current & a few whirlpools made I beleive by the meeting of two tides but not enough to have endangerd the smallest wherry that ever swam.

M\. MacNeal the principal gentleman in the Island was polite enough to walk over a very rough road to the uncultivated place which (it being Sunday) we had purposely chose

---

[1] Lochindale in Islay, now Loch Indaal, had a busy spacious harbour.
[2] Scarba is an island by the Gulf of Corryvreckan, 8 sq. miles (21 km^2) in size. Scarba has the ruins of a chapel but is now uninhabited.
[3] The Maelström is a whirlpool off the north-west coast of Norway. The whirlpool Banks sought was called Corryvreckan (the cauldron of Breckan), and is in the Gulf of Corryvreckan between Jura and Scarba.

to avoid scandal. He askd us home to his house which we readily accepted as it gave us an opportunity of asking questions from a man who living upon the spot must have a perfect Knowledge of the Whirlpool. [31] It is calld in the Galick[1] Cory Vrehan[2] & much feard by navigators in general who tell wonderfull stories of force & violence & of the Ships that have at different times been destroyd suckd in by its violence only in Passing by the mouth of it.

Mr. MacNeal told us that indeed at Spring tides & especialy with a N W wind there was a very great rippling & dashing together of the waves yet he who had livd there many years never Knew of more than one boat Lost in it & that a small one carrying only two men. He had Known however of several that had been in it during the time of its raging, which tho supposd to have been in great danger had Escapd.

At night we went on board the wonders of Cory vrehan being much sunk in the opinions of Every one of us. I can say no more of it however than just advise any traveler who may come after us to chuse a spring tide & NW wind when he shall go to See it or expect very little amusement.

[32] Lough Don

10.
From hence I was desirous of proceeding to Y Columb Kill[3] the only thing in this part of the highlands that I Knew of & wishd to see. I therefore attempted to persuade the pilot to go strait to it. He refusd & insisted on going through the sounds of Mull.[4] This tho tedious I was oblig'd to submit to & accordingly weighd with the first of the flood & proceeded towards them. Before noon we passd the Slate Isles[5] two small rocks neither of them more than a mile in circumference. Many Ships are however every year loaded from them. A little after noon we arrivd at Lough Don[6] where we were to Stay till morn for the tide.

As I had no other occupation to atract my attention & the Shore was almost without either houses or cultivation I employd the even*ing* in a manner of fishing quite new to me. We had rods about ten feet long to which were fas*t*ened hair lines a little longer than the rods. The hooks which were of the size Commonly usd for trout were baited with a small white feather. When we fishd we rowd our boat very gently through the water &

---

[1] Gaelic.

[2] Though Banks was not impressed, the Corryvreckan whirlpool is the third largest in the world. In the summer of 1947 George Orwell nearly drowned in it. He was living on Jura at the time and writing *Nineteen Eighty-Four*. There is an illustration of 'Whirlpool of Corivrachan near the Island of Jura' by James Miller in BL, Add MS 15509, f. 35.

[3] Pennant calls it Jona and not surprisingly offers a detailed account (*Tour in Scotland and the Hebrides*, p. 276).

[4] The Isle of Mull is the second largest island in the Inner Hebrides, about 337.8 sq. miles (875 km²) in size, and is the fourth largest island off the British mainland. In the 14th century Mull came under the rule of the Lords of the Isles who, holding their territories under the kings of Norway, exercised a kind of sovereignty independent of the Scottish monarchs. After the collapse of the Lordship in 1493 the island was taken over by the Clan MacLean, and in 1681 by the Clan Campbell.

[5] The Slate Islands are a group of seven main islands, few inhabited in present times. Dalradian slate was quarried from 1630 until the 20th century. This was the centre of the Scottish slate industry, the slate being used for roofing.

[6] Loch Don, a tidal sea loch to the south of Craignure, close to Mull's easternmost point, the main ferry port on Mull.

[33] Sound of Mull

immersd the points of our rods about 3 feet under the water behind her. By this uncommon method of fishing we caugh*t* a tolerable plenty of Fish cald here Grey fish (Gadus carbonarius Lin)[1] Calld on the yorkshire coast Coal fish. These were a little larger than a herring & I found took our feather bait supposing it to be young herrings of which we saw innumerable shoals about 2 inches long.

11.
This morn while the ship waited for the tide I went out with my gun & among the numbers of Guls which I killd as all our gentlemen think them excellent meat was an Arctick Gull Larus Parasiticus Linn. the first I ever saw. A bird I beleive scarce in this countrey as some of the countrey gentlemen who were on board did not know it. With the tide of Flood we proceeded & soon came in sight of Castle Duart[2] upon Mull the last of the line of forts intended in case of rebellion to cut off the highlands from the Lowlands. A more miserable remains of an ancient fort I never saw. It appeard scarce wind tight & water

[34] Morven

tight. It lookd picturesque however & had it drawn.[3] The Garrison I was told consisted of sixteen private*s* & an Ensign the greater part of whoom I saw lying upon a hillock just by their door sunning themselves.

Mull was now on our left hand & Morven[4] on our right. The former shewd but a barren appearance the latter lookd much more fertile. Here & there were pretty banks of Wood particularly in the neighbourhood of a small ruinous Castle Calld Artaurinish[5] a most Elegant one through which two pretty considerable brooks Came foaming down to the Sea.

Morven the Land of Heroes /once/ the seat of the Exploits of Fingal /&/ the mother of the romantick scenery of Ossian.[6] I could not even sail past it without a touch of

---

[1] *Gadus carbonarius Linn.*: now better known as pollock or coley.

[2] 'once the seat of the *Macleanes*, lords of the island; but now garrisoned by a lieutenant and a detachment from *Fort-William*' (Pennant, *Tour in Scotland and the Hebrides*, p. 408). The castle dates back to the 13th century and was the seat of Clan MacLean and the base of their sea-borne power for 400 years. The Castle Duart is built upon a cliff. In 1691 the MacLeans surrendered Duart and all their lands on Mull to the Duke of Argyll. In 1751 the castle was abandoned but has been restored. See *An Historical and Genealogical Account of the Clan Maclean* written by 'a Seneachie'. See also Martin, *A Description of the Western Isles*, p. 285.

[3] There is an illustration of 'Castle Douart' on the Isle of Mull by J. F. Miller, BL, Add MS 15510, f. 14.

[4] Morvern on the Sound of Mull had been yet another stronghold of the Lords of the Isles, belonging to the MacLeans since the 17th century.

[5] There are two illustrations of this castle by Cleveley in BL, Add MS 15510, ff. 17 and 18. It is on the north side of the Sound of Mull in Argyleshire (note by Banks): 'View of the Castle of Artaurinish at Ardtornish Point in Argyllshire, Scotland, overlooking the Isle of Mull', drawn by J. F. Miller in 1775 after sketches made in 1772.

[6] Ossian was a poet, supposedly discovered by the Scottish poet James MacPherson (1736–96) who claimed he had translated his epic poetry describing the heroic age of the Highlands from ancient sources in Scottish Gaelic. His most famous poem was 'Fingal' published in 1762. The publication was an immediate success and a subject of controversy. Ossian was a figment of MacPherson's imagination. MacPherson never denied allegations of forgery directed against him; however, his poems seem to have some basis in fact. He subsequently embarked on a political career serving as secretary to the governor of West Florida, as a publicist for Lord North's administration's North American policy and as an MP for Camelford 1780–96. See Banks's remarks on Ossian in this journal, pp. 66 and 69.

Enthusiasm, sweet affection of the mind which can gather pleasures from the Empty Elements & realise substantial pleasure which three fourths of mankind are ignorant of. I lamented the busy bustle of the Ship & had I dard to venture the Censure of my Companions would certainly [35] have brought her to an anchor. To have read ten pages of Ossian under the shades of these woods would have been Luxury above the reach of Kings.

We soon after passd by the mouth of a beautifull little inlet. Tho the tide was not half spent I venturd to propose a wish to go in there but the Cruel pilot declard that it was a bar harbour into which we could not go but at high water.

Evening came on & the tide became unfavourable. We anchord as fate directed in as ugly a spot as we could have chose along the whole coast sufficiently so I think to have destroyd the Enthusiasm of even an Ossian. The Master of a pretty little house however came off to us & as we wishd to see the burning of Kelp[1] an operation which was then going on along the Coast offerd to accompany us.

The Kelpers were at work by the sea side. They had got together many little heaps of sea weed pild up like small haycocks witherd pretty much by the sun yet by no means thoroughly dry. This material they continualy heapd [36] upon a fire made in a frame of Stone about 20 feet long 4 feet broad & three deep. Mr Morison our host informd us that this crop of Sea weed was regularly cut from the rocks on which it grows above low water mark once in three years & that it does not grow to perfection in a less time. The people are very carefull to have fine & fair weather for this business for rain falling during the the [sic] time that the weeds are exposd vastly Lessens the produce of Salts. The Kiln also said he should be built so that one of its longest sides fronts the wind which most generaly blows as when the wind blows in that direction it burns much faster & more even. When every thing said he succeeds well the Kelp is a valuable part of our estates. Two men can burn above a ton in a day which brings in £4=10 or £5 in hard money.

During the course of this conversation the Kiln was ready for laying what they call a floor. The people then ceasd to heap on fresh sea weed & in a short time the Kiln was [37] thoroughly ignited. The bottom was then coverd with red hot ashes above 18 inches deep. This said our host would be mere ashes worth little did not the operation they are now going to perform render it a solid body & by that means marketable. The men in the mean time took Each a pole of about 8 feet long headed with an iron Croockd like a hough[2] with this they briskly stirrd the ashes to & Fro till by degrees they came to a mass half vitrified & very much resembling soft dough. This they beat & pokd about with their poles for about half an hour. Then they let it settle. It soon was coverd with a hard crust almost resembling Lava which in another half hour was ready for another bed of ashes. Accordingly they began anew to Burn the sea weed & we left them myself at least totaly unable to account for the vitrifaction of the matter so suddenly being producd by merely stirring the mass about.

---

[1] Kelp was produced from ashes of burnt seaweed to make a kind of salt with soda and potash. The practice was well established in Scotland and by the 1800s was valued at £80,000 a year. Kelp was exported and used as raw material in a variety of chemical processes, especially in the manufacture of soap and glass. In the 1820s, however, the kelp industry declined due to cheap foreign imports of similar products.

[2] A *hough* is an old variant of hoe, an agricultural implement, one type is shaped a bit like a trough.

12

Last night being very fine we movd in the night & towards morning the tide being spent [38] came to an anchor on the Morven side opposite a small gentlemans house calld /*Drumling*/ Dramnen. The Master of it M^r M^c.Clean[1] having found out who we were very cordialy askd us ashore. We accepted his invitation & arrivd at his house where we met an english gentleman M^r. Leach[2] who no sooner saw us than he told us that about 9 leagues from us was an Island which he beleivd no one even in the highlands had seen on which were pillars like those of the Giants Causeway.[3] This was a great object to me who had wishd to have seen the Causeway itself would time have allowd. I therefore resolvd to proceed directly especialy as it was /*directly*/ just in the way to y Columb Kill. Accordingly having put up two days provision & my little tent we put off in the boat about 1 O'Clock for our intended voyage, having orderd the Ship to wait for us in Tobir more, a very fine harbour on the Mull side.[4]

[39] Staffa[5]

At 9 O'Clock after a tedious passage having had not a breath of wind we arrivd under the directon of M^r. McLeans Son & M^r. Leach. It was too dark to see any thing so we carried our tent & baggage near the only house upon the Island & began to Cook our suppers in order to be prepard for the Earliest dawn to enjoy that which from the conversation of the Gentlemen we had now raisd the highest expectations of.

Our tent was small it weighd altogether only 27 lb. We were 9 in number we might sleep in it but not without crowding. It was therfore resolvd that some might sleep among the Children in the house & 4 volunteers with Dr Solander at their head undertook the business. The house was smoky having no kind of vent for the smoak but the door. This was judgd a trifling inconvenience. Lice was the only fear so an enquiry was enterd into. The woman assurd the the [*sic*] gentlemen that no such vermin harbourd there. On the

---

[1] Sir Allan Maclean (d. 1783), the 6th baronet, was master of Drumnen (today it is called Drimnin), situated on the shores of the Sound of Mull (see *Historical and Genealogical Account of the Clan Maclean etc.*, pp. 207–9).

[2] Here Pennant has added in the 1776 edition a note on Mr Leach (as if from Banks): 'I cannot but express the obligations I have to this gentleman for his very kind intentions of informing me of this matchless curiosity; for I am informed that he pursued me in a boat for two miles, to acquaint me with what he had observed: but, unfortunately for me, we out-sailed his liberal intention' (*Tour in Scotland and the Hebrides*, p. 299).

[3] The Giant's Causeway is a famous basaltic promontory on the coast of County Antrim.

[4] Tobermory is a seaport on Mull. In 1588 the *Florida*, one of the ships belonging to the Spanish Armada, was blown up in the harbour. A town was built in 1787–8 by the British Fisheries Society with an excellent harbour which became a thriving seaport.

[5] Staffa is a grassy island, composed of volcanic tuff and basalt, in the county of Argyll, one mile (1.6 km) in length and half a mile (0.8 km) in breadth. The most famous cavern is Fingal's Cave. Banks was the first notable visitor; he would be followed by such celebrities as Sir Walter Scott (1810), Felix Mendelssohn (1829), Queen Victoria and Prince Albert (1836), Jules Verne (1839), David Livingstone (1864) and Robert Louis Stevenson (1870). In 1772 Banks found one family living there. Drawings of Staffa and Fingal's Cave are to be found in BL, Add MS 15510, ff. 20–32, 35.

In the second edition of his book *Tour in Scotland and the Hebrides*, Pennant included an excerpt from Banks's journal, lent to him for the purpose. 'The Account of Staffa communicated by Joseph Banks, Esq.' was subsequently printed in all editions of Pennant, in the 1790 edition from pp. 300–310 (discrepancies and additions by Pennant are here noted in footnotes). The following is the first of Pennant's notes (p. 310): 'As this account is copied from Mr. Banks's journal, I take the liberty of saying (what by this time that gentleman is well acquainted with) that *Staffa* is a genuine mass of *Basaltes*, or *Giant's Causeway*; but in most respects superior to the Irish in grandeur.'

strengh of that assurance our gentlemen [40] having Eat their suppers betook themselves to rest.¹

13.

The impatience which Every body felt to see the Wonders we had heard so largely describd prevented our mornings rest. Every one was up & in motion before the break of day & with the first light arrivd at the S W part of the Island the scene of the most remarkable pillars where we no sooner arrivd than we were struck with a scene of magnificence which exceeded our expectations tho formd as we thought upon the most sanguine foundations. The whole of that End of the Island /was/ supported by ranges of natural pillars the most /of which were/ above 50 feet high standing in natural Colonades according as the bays or points of Land formd themselves: upon a firm basis of Solid unformd rock. Above these the Stratum which reaches to the Soil or surface of the Island varied in thickness as the Island itself /was/ formd into hills or vallies each hill which hung over the Columns below forming an [41] ample pediment. Some of /which/ these /were/² above 60 feet in thickness form the base to the point formd by the Sloping of the hill in each side almost into the shape of those usd in Architecture.

Compard to this what are the Cathedrals or the palaces built by man. Mere models or playthings imitations as diminutive as /the/ his works /of man/ will always be when compard to those of nature. Where is now the boast of the Architect. Regularity, the only part in which he fancied himself to exceed his mistress nature is here found in her posession & here it has been for ages uncounted.³ Is not this the school where the art was originaly studied & what had been added to this by the whole grecian school. A Capital to ornament the Column /which/ of nature /had given them/ of which they could execute only a model & /that/ for that very capital they were obligd to a bush of Acanthus.

How amply does nature repay those who study her wonderfull works.

With our minds full of such reflections we proceeded along the shore treading upon another [42] Giants Causeway. Every stone being regularly formd into a certain number of sides & angles till in a short time we arrivd at the mouth of a Cave the most magnificent I suppose that has Ever been describd by traveler. In depth from the pitch of the arch to the bottom 250 feet its hight at the entrance 117=6 at the bottom 70 feet the whole supported by regular pillars ranging on each side. The bottom was water shoaling gradualy from three fathoms to 9 feet its breadth at the pitch of the Arch 53=7. at the farther end 20 feet.⁴

The mind can hardly form an Idea more magnificent than such a space supported on each side by ranges of Columns & roofd by the bottoms of those which have been broke off in order to form it between the angles of which a yellow Stalagmitick matter has exsuded which /serves/ to define the angles precisely & at the same time vary the Coulour with a great deal of Elegance & to render it still more agreable. The whole is lighted from without so that the farthest extremity is very plainly seen from [43] without

---

¹ Pennant omits this paragraph.
² Banks has deleted 'were', by mistake as it is certainly needed.
³ Pennant's version reads 'has been for ages undescribed' and adds a footnote: '*Staffa* is taken notice of by *Buchanan*, but in the slightest manner; and among the thousands who have navigated these seas, none have paid the least attention to its grand and striking characteristic, till this present year' (*Tour in Scotland and the Hebrides*, p. 301).
⁴ Banks spent some 12 hours systematically measuring the cave.

& the air within being agitated by the flux & reflux of the tides is perfectly dry & wholesome free intirely from the damp vapours with which natural cavern*s* in general abound.

We askd the name of it. Ouwa Eehn said our guide[1] the Cave of Fiuhn. What is Fuihn [*sic*] said we. Fuihn Mac Coul whoom the translator of Ossians works has Calld Fingal. How fortunate that in this cave we should meet with the remembrance of that cheif whose existence as well as that of the whole Epick poem is almost doubted in England.

Enough for the beauties of Staffa I shall now proceed to describe it & its productions more Philosophicaly.

The little Island of Staffa lies on the West Coast of Mul about 3 leagues NE from Jona[2] or Y Columb Kill its greatest length is about an english mile & its breadth about half a one. On the west side of this Island is a small bay where boats generaly land a little to the southward of which the first appearance of pillars are to be Observd. They are small & instead of being placd upright lie down on their sides each forming a segment [**44**] of a Circle. From thence you pass a small Cave above which the pillars now grown a little larger are inclining in all directions. In one place in particular a small mass of them very much resemble the ribs of a ship.[3] From hence having passd the Cave which if it is not low water you must do in a boat you come to the first ranges of pillars which are still not above half as large as a little beyond. Over against this place is a small Island Calld in Erse Boosha la[4] seperated from the main by a channel not many fathoms wide. This whole island is Composd of Pillars without any stratum above them. They are still small but by much the neatest formd of any about the place. The first division of the Island for at high water it is divided into two makes a Kind of a Cone the Pillars Converging together towards the Centre. On the other they are in general laid down flat & in the front next the main you see how beautifully they are packd together their ends coming out square with the Bank which they form. All these have their transverse sections exact[5] which is by no means the

---

[1] Pennant's version was: 'We asked the name of it, said our guide, the cave of Fhinn; what is Fhinn? said we. Fhinn Mac Coul, whom the translator of Ossian's works has called Fingal' (Pennant, *Tour in Scotland and the Hebrides*, p. 303).

[2] Iona is a small island in the Inner Hebrides, the site of one of the most important monasteries in the early Middle Ages. In 563 the monk Columba (521–97), went into exile there from his native Ireland with 12 companions, their aim being to convert the natives of the Hebrides, the Picts. He built two churches, an Augustinian monastery and a convent. The monks brought with them the ability to write and Iona became a renowned centre of learning. Around this time the island's famous high crosses were sculpted, perhaps what is known as the 'Celtic cross'. However, due to the Viking raids that began in 794 the monastery was abandoned in 849. The Benedictine abbey was built in 1203 and a Benedictine convent founded in 1208. The monastery itself flourished until the Reformation when buildings were demolished and all but three of the 360 carved crosses destroyed. In 1549 an inventory of 48 Scottish, 8 Norwegian and 4 Irish kings was recorded as being buried there. Iona Nunnery survives as a number of 12th–13th century ruins of the church and cloister. The following year Samuel Johnson visited many of the islands including Iona, apparently solely because of Boswell's curiosity. See Florence Marian McNeill's classic, *Iona. A History of the Island*, first published in 1920.

[3] Pennant adds a note here: 'The Giant's Causeway has its bending pillars; but I imagine them to be very different from these. Those I saw were erect, and ran along the face of a high cliff, bent strangely in their middle, as if unable, at their original formation, while in a soft state, to support the mass of incumbent earth that pressed on them.' (*Tour in Scotland and the Hebrides*, p. 304).

[4] Booshala Isle (*Am Buachaille*) is one of the Hebrides lying south of Staffa, entirely composed of basaltic pillars as can be seen in the illustrations in BL, Add MS 15510, ff. 32, 34, 37–9.

[5] Pennant inserts 'and their surfaces smooth' (*Tour in Scotland and the Hebrides* p. 304).

case with the larger ones [45] & in general they are smooth on all their surfaces when on the other hand the Large ones are crackd in all directions. I much question however if any one in the whole Island[1] is two feet in diameter.

The main Iland opposite Boosha la & farther towards the N W is supported by ranges of pillars pretty Erect & tho not tall as they are not uncoverd to the base of Large diameters. At their feet is an irregular pavement made by the upper sides of such as have been broken off which reaches as far under water as the Eye can reach. Here the forms of the Pillars are apparent. These are of three four five Six & Seven sides but the numbers of five & Six are by much the most prevalent. The largest I measurd was of Seven it was 4^{ft}. 5^{inch}. in diameter I shall give the measurement of its Sides & those of some other forms which I met with

N°. 1 4 sides diam. 1 ^{ft}. 5 ^{inch}			N°. 2. 5 sides diam. 2. 10		
S^{ide}	ft	inch.			
Side 1:	1.	5.	1:	1.	10.
2:	1.	1.	2:	1.	10.
3:	1.	6	3:	1.	5
4:	1.	1.	4:	1.	$7\frac{1}{2}$
			5:	1.	8.
[46]					
No. 3. 6 sides diam. 3: 6:			No. 4. 7. sides diam 4. 5.		
1:	0.	10.	1:	2.	10
2:	2.	2.	2:	2.	4
3:	2.	2	3:	1.	10
4:	1.	11.	4:	2.	0
5:	2.	2.	5:	1.	8.
6:	2.	9	6:	1.	6.
			7:	1.	3.

the surfaces of these large pillars in general is rough & uneven full of Cracks in all directions. The transverse fissures[2] in the upright ones are by no means regular but the perpendicular ones never fail to run in their true directions. The surfaces upon which we walkd were often flat having neither concavity nor convexity. The larger number however were concave tho some were very evidently convex. In some places the interstices /*between*/ within the perpendicular fissures were filld up with a yellow spar. In one place a vein passd in among the Mass of pillars Carrying here & there small threads of Spar there. Tho they were broke & Crackd through & through in all directions yet their Perpendicular fissures might easily be tracd from whence it is Easy to infer [47] that whatever the accident might have been that Causd the dislocation it happend after the formation of the pillars.

From hence proceeding along shore you soon arrive at Fingals Cave. Its dimensions tho I have before given I shall here again repeat in the form of a table.

---

[1] Pennant inserts 'of Bhuachaille' (ibid., p. 305).

[2] Pennant has 'figures', ibid., p. 304. In the following paragraph when Banks writes 'fissures' Pennant has changed this to 'figures'. Von Troil uses Banks's journal as well, not totally accurately (see *Letters on Iceland*, pp. 277–82).

		ft.	inch
Depth[1] of the Cave	from the rock without.	371.	6.
	from the pitch of the Arch	250.	0
Breadth of d°.	at the mouth	53.	7.
	at the farther End	20.	0.
hight of the Arch	at the mouth	107.	6[2]
	at the End	70.	0
hight	of an outside pillar	39.	6.
	of one at the NW. Corner	54	0.
Depth of Water	at the mouth	18	0.
	at the bottom	9.	0.

the Cave runs into the rock in the direction of NE by E by the Compass.

Proceeding farther to the NW you meet with the highest ranges of pillars the magnificent appearance of which is past all description here they are are [sic] bare to their very basis & [48] the Stratum below them is also visible. In a short time it rises many feet above the water & gives an opportunity of Examining its quality. Its surface rough & uneven has often large lumps of stone sticking in it as if half immersd. Itself when broken is Composd of a thousand heterogeneous parts which together have very much the appearance of the Surface of a Lava & the more so as many of the lumps in it appear to be of the very same stone of which the Pillars are formd. This whole Stratum lies in an inclind Position dipping gradualy down towards the S.E. As herabouts is the situation of the highest pillars I shall mention my measurements of them & the different Strata in this place premising that the measurements were made with a line held in the hand of a person who stood at the top of the Cliff & reaching to the bottom to the Lower end of which was tied a white mark which was observd by one who staid below for the purpose. When this mark was set off from the water the Person below noted it down & made a signal to him above who made then a mark in his rope when ever this mark [49] passd a notable place. The same signal was made & the name of the place noted down as before. The line being all hauld up & the distances between the marks made upon it measurd & noted down gave when compard with the book kept below the distances requird as for instance in the Cave. N°. 1 in the book below was Calld from the water to the foot of the first pillar. In the book above no 1. Gave 36. feet 8 inches the hight[3] of that ascent which was composd of Broken Pillars.

N°. 1. Pillar at the West Corner of Fingals Cave

	Ft	inches
1. from the water to the foot of the Pillar.	12:	10
2. hight of the Pillar	37.	3.
3. Stratum above the Pillar	66.	9.
N°. 2. Fingals Cave		
1. From the water to the foot of the Pillar.	36.	8.
2. hight of the pillar	39.	6.
3. from the top of the pillar to the top of the Arch	31.	4

[1] Pennant has 'Length', (*Tour in Scotland and the Hebrides*, p. 305).
[2] Pennant has '117 instead of 107 (ibid., p. 305); 107 is correct.
[3] 'highest' in Pennant, ibid., p. 306.

4. thickness of the Stratum above	34.	4
by adding together the three first measurements we get the hight of the arch from the water	107.[1]	6

[50] <u>No. 3.</u> Corner Pillar to the Westward of Fingals Cave

Stratum below the pillar of Lava-like matter	11.	0
Hight of Pillar	54.	0
Stratum above the Pillar	61.	6.

<u>No. 4</u> another pillar to the Westward

Stratum below the Pillar	17.	1.
Hight of the Pillar	50.	0.
Stratum above	51.	1.

<u>No. 5</u> another pillar farther to the Westward

Stratum below the Pillar	19.	8.
hight of the Pillar	55.	1.
Stratum above	54.	7.

The Stratum above the pillars which is here mentiond is uniformly the same consisting of numberless small pillars bending & inclining in all directions sometimes so irregularly that the stones can only be said to have an inclination to form into those shapes in others more regular but never breaking into disturbing the Stratum of Large pillars whose tops every where keep a uniform & regular line.

[51] Proceeding /now/ along shore round the North End of the Island you arrive at Oua na Scaroe or the Cormorants Cave. Here the Stratum under the Pillars is lifted up very high. The Pillars above it are considerably less than those at the N W end of the Island, but Still very considerable. Beyond is a bay which cuts deep into the Island rendering it in that place not more than a quarter of a mile. Over on the sides of this bay especialy beyond a little valley which almost cuts the Island into two are two stages of Pillars but small however having a stratum between them exactly the same as that above them formd of innumerable little pillars shaken out of their places & leaning in all directions.

Having passd this bay the pillars totaly cease. The rock is of a dark brown stone & no signs of regularity occur till you have passd round the SE end of the Island a space almost as large as that occupied by the Pillars which you meet again on the West side beginning to form themselves irregularly as if the Stratum had an inclination to that form & soon arrive at the Bending pillars [52] where I began.

The stone of which the Pillars is formd is a Course Kind of Basaltes very much resembling that of the Giants Causeway in Ireland tho none of them are near so neat as the specimens of the latter which I have seen at the British museum owing cheifly to the Colour which in ours is a dirty brown in the Irish a fine black. Indeed the whole production seems very much to resemble the Giants Causeway with which I should willingly compare it had I any account of the former before me.[2]

About 4 O'Clock our drawings & measurements of the Pillars were finishd & having resolvd to proceed to y Columb Kill that night we hasted to the tent in order to get our dinner. The Gentlemen who Slept in the house last night had during the morning become sensible that they were attended by some Guests whose Company they did not much

---

[1] Pennant has '117' instead of 107 (ibid., p. 305); 107 is correct.
[2] Here Pennant ends his extract from Banks's journal (ibid., p. 310).

approve. They therefore complaind to the woman of the house. With some peivishness the man who overheard answerd in Erse with a great deal of Sang Froid lice indeed if they have any lice [53] they certainly brought them here for I am sure there were none upon the Island when they came. Pleasd at his presence of mind we took leave having satisfied him for the Potatoes fish & milk which notwithstanding his poverty he has supplied us with during our stay with the utmost hospitality & which with the wild Pigeons & shags we shot had supplyd us with the greatest part of our diet.

At 5 we embarkd & before 8 arivd at Iona[1] or Y Columb Kill[2] famous for its religious foundations suppos'd to be the Source from whence Cristianity has flowd over the Island of Great Britain.

We were receivd here by a number of people who told us that they had heard of our Coming & proferd us every convenience the town could afford. But we soon found the difference between these & the simple people we had had to do with before. Few Strangers as these people had seen those few had Corrupted the hospitality of their countrey. One of the first questions askd us after we had agreed to accept their offers was how much we would give a question which had not been put to us since we came into the highlands till this time.[3]

[54] Y Columb Kil

As it is a much easier matter to deal with people for a favour before it is receivd than after we rejoicd that they had not arrivd at the next step of civilization, that of Bestowing and after the recipt requiring an enormous recompence. Our bargain was soon made. We were furnishd with an empty house, plenty of Clean straw & sour Curds & cream & a good Fire which we could well have dispens'd with as money could not purchase a chimney to let out the Smoak. We therefore put it out & having eat our sour Supper retird to rest.

14.

Tho we were up very early this morn the rain which fell in plenty would not allow us to pursue our enquiries. It was matter of rejoicing however that yesterday had been fine. About five it grew more moderate & we proceeded to the ruins of a nunnery[4] which stood near our Lodgings. Here was little worth observing every thing being in an Absolute state of Desolation the very chappel turnd into a Cow house in which no one monument was to be seen but one which had lately been dug by M[r]. Pennant[5] from under above 3 feet of [55] Cow dung /which/ on this was inscribd the name of a prioress. On one end of the stone was the figure of the Lady on the other that of the virgin & child the virgin having on her head an episcopal mitre.

---

[1] On Iona Banks put his artists to work. In BL, Add MS 15510 there is an illustration by Cleveley of the Great Church of Iona (f. 2), another one of a ruined ancient church (f. 4) and the church of 'Columb Kill' (ff. 5, 7, 8) as well as tombs in the church of Iona (ff. 10–12). There is a drawing of a woman on Iona (drawn 14 August 1772) in BL, Add MS 15510, f. 19.

[2] According to Martin, this meant the Isthmus of Columbus the clergyman, Colum being his proper name and Kil, meaning a church 'was added by the islanders by way of excellence' (*A Description of the Western Islands*, p. 286). Martin has a chapter on Iona (pp. 286–93). I am indebted to Andrew Cordier, archivist of the St Kilda Club, for information on the early Martin Martin.

[3] This was the first time they were asked to pay for their board and lodging.

[4] Columba had built the nunnery, endowed by the Kings of Scotland and the Isles (Martin, *A Description of the Western Islands*, p. 286).

[5] This is the first mention that Banks made of Pennant in his journal. See further on their relations in Appendix 2.

From hence we proceeded to the Great Church which like the other is an absolute ruin inhabited however by Cornish Choughs[1] Royston Crows[2] & Jackdaws. It is built in the form of a Cross pretty large & is on every side surrounded by chapels &c. both adjoining to & detachd from it. The Church yard is totaly overgrown with the largest plants of Petasites[3] I have ever seen which renders it impossible to Search after inscriptions in the summer time. There is however a hansome Cross dedicated to St Martin & a broken one to St John.

Our guide who boasted that he was descended both by Father & Mother from those who came over with St Columba carried us under the ample shade of the Petasites stopping us every here & there to inform us of the places where Kings & nobles had been interrd. Here said he is a King of France here one of Sweeden here 4 of England here [56] 8 of Norway & here 40 of Scotland. For /the truth of/ all these things however we were Obligd to confide in his Knowledge derivd as he told us from his grandmother as neither stone nor inscription gave us the least light.

Powerful Columba to have Kings so much revere his foundation as to bury in the open ground while saints & abbots only enjoyd the Cover of the church.

In a short time we arrivd at the Chapel of Oran a fellow Saint or as our Guide told us a brother of Columba who to forward the great work undertook in obedience to a vision of Columba to be buried alive in this place & was accordingly interrd. The next day he was dug up & found alive. No sooner was he uncoverd than he began to blaspheme crying out You are all deceivd hell is a trifle & the Devil a mere illusion invented to deceive you. Columba hearing this with great presence of mind, cry'd out Earth upon the head of Oran. He was instantly Obeyd & poor Oran buried again never more to arise till the Last trumpet shall awake him.[4]

[57] This story is told in almost the same words in Macphersons dissertation on the origin &c of the caledonians p. 375[5] but as the tradition is singular & I had it from the mans own mouth I could not avoid repeating it.

In orans Chapel[6] it was Easy to Observe that tho in the Early times Kings buried in the churchyard laterly Laymen of Less dignity got places in the church. Here were Knights in abundance but none very old. On the North side under an arch above the pitch of which is the remains of a crucifix a singular inscription may be seen upon a stone exactly resembling those Laid over the dead calld here lay stones. We read it thus

   Hic est crux Lacclenni Meic Fingone
   Et ejus Filii Johannis Abbatis de Hy
   facta anno Domini M CCCC LXXX IX[7]

---

[1] *Pyrrhocorax pyrrhocorax*: Cornish choughs.

[2] *Corvus cornix*: Royston crows, the hooded crow.

[3] *Petasite*: the common butterbur.

[4] Pennant tells this story in *Tour in Scotland and the Hebrides*, pp. 286–7.

[5] James MacPherson. See p. 65, n. 6.

[6] Oran's Chapel is the oldest surviving building on Iona, built as a mortuary chapel by Somerled (d. 1164). His son Ranald re-established the monastery as a Benedictine abbey and also founded the Augustinian convent (see McNeill, *Iona*, p. 44).

[7] Pennant read it thus: 'Hæc est crux *Lauchlani Mc.Fingon* et ejus filii *Johannis* Abbatis de HY. Facta an. Dom. M^{o++}CCCCLXXXIX' [Here is the cross to Lachlan McKinnon and his son John, Abbot of Iona, made in the year of our Lord 1489] (*Tour in Scotland and the Hebrides*, p. 250).

Posibly the Father was interrd here tho the stone does not declare it. As for John the abbot he lies under a pompous tomb of Black marble in the Church itself.

In passing through the Church yard we were very frequently shewn the burying places [58] of Particular families as MᶜNeals MᶜDonalds &c. who we were told bury there to this day.

Among the superstitions existing yet which seem to derive their origin from the most ancient times we Observd two singular ones.

In the way from Orans chapel to the great Church were 6 stones formd conicaly as if intended to beat or bruise corn or any thing which might be laid under them. These rested on a flat plate of Stone our Guide desird us to turn each of them round. When every one of us had separately fulfilld his directions he told us that Columba had placd those stones there & orderd that every stranger who came should turn them once round & at the same time predicted that whenever the stone on which they stood was worn through by this operation the world would be at an end. At present the stone is a good deal hollowed & one end quite worn through but some wise man willing to give the world a reprieve has movd the stones quite over to the other by which manoeuvre the age of the world is likely [59] to be prolongd 1500 or 2000 more than ever Columba intended. The other is the rubbing stone as it is Calld; a stone like a tombstone a little hollowd out which lies near the West door of the great Church. This is one of 4 one of which is said to be placd at each end of the Island. The use of them is that any mariner wanting a particular wind shall Come here & clean the stone which has the wind he wants in its power by which means he will certainly obtain it. Our stone had the power over the north wind & had been palpably cleand a very small time before we saw it as that wind however did not suit us we coverd it up piously with the dirt Which had probably been more piously taken off.

Having thus attempted to invalidate the power of the North wind & seen all that our guide could shew us we proceeded towards our boat. In the way we met many wild Pigeons & shot some. The rocks of which the Island is Composd we saw also. They are of Granate red & /white/ black & seem to have been the cheif material in the buildings.

[60] Oransay.[1]

By 12 we set sail intending in our voyage home to have visited Carnbrugh & Pladda[2] two Islands laying near Staffa whose appearance promisd a similar construction of rocks but the wind not coming fair prevented us. We passd however pretty near them but could not with our glasses perceive any pillars on the sides next us.

At 9 we arrivd at the ship having had a very bad passage for want of wind. We found her lying in Tobir more a prodigious fine harbour on the Mul side capable of containing in Safety a large Fleet.

---

[1] Oronsay is a small island in the Inner Hebrides, just over 2 sq. miles (5.2 km²) in size. A monastery was founded there in 563. Later in the 14th century the Augustinians erected the Oronsay Priory dedicated to St Columba, probably on the same site. Pennant's account mentions that he found ruins of an Augustinian priory there (*Tour in Scotland and the Hebrides*, pp. 270–74).

[2] These three islands all belong to the Treshnish Isles, an archipelago lying west of Mull in the Inner Hebrides. Their modern names are Cairn na Burgh Beag (the smaller) and Cairn na Burgh Mòr (the larger), together nicknamed the Carnburgs, obviously current when Banks was there. Fladda is the northernmost of the Treshnish islands.

15.

Our Freind Mr M{c}Clean having offerd to shew us sport in hunting Roebucks calld here Re if we would stay. We set out with him this morn to a small Island of his Calld Oransay situate in the mouth of Loch Sunart.[1] As the Deer were to be drove by hound & horn in order to be forcd by passes where we were to be station'd we took a Croud of all kind of noises French horns Chinese Gong &c &c. so we literaly made the woods ring but without success. Some Roes indeed were seen.

[61] Account of Islands

but not one shot so at night we returnd suppd with our polite Landlord & afterwards returning on board resolvd for sea immediately as we now had nothing to see between this place & St Kilda at least that we knew of.

As some things which I observd among these Islands relative to the people as well as other things were omitted in their proper places I shall take this opportunity of digressing a little that they may not be forgot.

The Soil of the Islands in general is very rough & craggy. Many of them are scarcely worth improvement except in the valleys which are very small. Ila[2] is by far the best the whole being as I before mentiond situate upon a bed of Marl. Mr. Campbels estate there is certainly most princely. The whole Island except a few Acres belongs to him. He has within himself a good & safe harbour & several anchoring places. Trade might flourish as several ships touch there even monthly. His mines are in a flourishing condition & promise much better than at Present. What might not a man make who could set down to improve such an Island [62] which literaly wants nothing but fencing to make it of ten times its present value.

A singular circumstance occurs in the Mines of that countrey which I do not remember to have met with any where else but I am told that it exists over the greatest part of the north of Scotland. Among the regular strata every now & then one intervenes exactly standing upon its Edge. These are of that hard Kind of Stone with which the Streets of London are now pav'd calld here Whyne.[3] These Strata proceed through a whole Countrey in a S & N direction Cutting through every species of Stone which lies in their road. One of them of 9 feet thick passes through Mr. Freebairns Lead mine[4] cutting the vein in two which is found again beyond it exactly as before. They are in general from 6 feet to 6 yards in breadth & from their extrordinary hardness often appear above ground in the form of a wall Especialy where they break into the Sea. This has causd the people to call them Whynne dykes.[5] [63] Very few of the Islanders make any winter provision for their cattle or sheep. These animals have in these Islands a resource which hunger would I should think scarcely drive ours to make use of Sea weed of different Kinds which they

---

[1] Loch Sunart is a sea loch.

[2] Islay.

[3] The 'whyn stone' or whinstone is a name for various very hard dark-coloured rocks or stones, as greenstone, basalt, chert or quartzose sandstone.

[4] Pennant wrote: 'am most hospitably received by Mr. *Freebairn* of *Freeport*, near *Port-askaig* ... Visit the mines, carried on under the directions of Mr. Freebairn, since the year 1763: the ore is of lead, much mixed with copper, which occasions expence and trouble in the separation: the veins rise to the surface, have been worked at intervals for ages, and probably in the time of the *Norwegian*s, a nation of miners' (*Tour in Scotland and the Hebrides*, pp. 248–9.) By the early 19th century whisky distillation had taken over from lead mining as a more profitable venture.

[5] There is an illustration of a whyn dyke in the Sound of Islay by Cleveley in BL, Add MS 15509, f. 31.

Eat plentifully & grow fat as I have been asssurd. Arundo Arenaria or Sea reed grass which grows upon the Sand hills near the Shore is a favourite food of their Cattle in winter & they reckon the lands that Produce it of Great value for wintering their Black cattle.

Black cattle & Kelp are the cheif produce from which they draw their returns. Corn they grow not sufficient for their own consumption. They give as a reason for it that their harvests are generaly wet. The true one I take to be the want of Fences which totaly prevents their growing winter Corn.

The better sort of People which we met with live much in the Stile of Farmers of 100 or more pounds a year rent. Their houses are hardly [64] so good nor have they such a variety to offer. In general they are so few having any more than leases of 19 years in general paying not a tenth part of the real or 100th. of the improvable value.

They receivd us every where with hospitality. We were so much aware of it that we did not bring a letter of recommendation nor had we ever occasion for one having more than once walkd up to a strange house with as much freedom as we would do to a publick house in England & met in it as cordial a reception as if a bill was to have been brought in.

Notwithstanding this we found it very difficult to procure any thing at all out of their way which we might want. They willingly gave whatever they had but did not wish to put themselves or their people out of their way to procure any thing. For instance we never could or did get specimens of the fish of the countrey.

In general you meet with a number of people of one name in Ila for instance there are [65] very few of any other name than Campbel. In this Case it is usual in speaking to any one not to Call him M^{r}. Campbel but to name only his place as Laggan, Bomore,[1] Killara &c.

In few houses is bread to be met with not one in a hundred. Instead of it the people eat dry Oat cakes like those usd in Wales or thin Barley Cakes like pancakes which latter I confess I myself preferr'd. They brew no ale but use spirit instead of it. Every man of any condition has in his house a still with which he distils Malt spirits cheifly from barley but sometimes from oats. This being all done with a turf fire acquires a strong Goût palatable enough to a highlander but odious to any other palate. My usual drink was milk which in this Countrey is very excellent better I think far than that of the Guernsey cattle. Themselves drink a good deal of it.

The inferior people live but very poorly. Their huts are poor to admiration. I have seen few indians live in so uncomfortable [a] manner nor Could I have thought that any thing but flies [66] could – induce – men to live in houses without chimneis which many houses are without. Chimnies indeed properly speaking are a rare commodity in general the remedy they apply to Smoak is no more than a hole in the top of their roofs.

Among all their poverty they seem however contented. They have still a Clannish attachment to their superiors & if they or any one who they look upon as above them undertakes to direct they obey with much more implicit obedience than englishmen will do shewing at the same time a decent respect which tho rather humble does not Produce any false shame. Every man answers with an Ease & freedom which an english man has

---

[1] Bowmore is now the capital of Islay, situated on the eastern shore of Loch Indaal. The first houses were built in 1768, thus practically brand-new when Banks saw them. There is 'A view of the Town of Bomore on the Island of Ila' by Cleveley in BL, Add MS 15509, f. 4.

little Idea of this /whether/ this /*I believe*/ proceeds from nature or Education /I/ is dificult to say but I firmly belive that it is the basis of that superiority which the Scotch in general enjoy over our nation.

   Education is here paid the strictest attention [67] to Even where a publick hardly exists a man of an income of fifty pounds a year who did not keep a private tutor in the house for his children would be thought very ill of. To this again they owe a great deal as Education under the eye of a parent must always excell that which can be bought of a master who feels no affection for the child he instructs.

   I should wish to be able to say a little about the Language of this countrey, but profess myself utterly unable; all I could Learn is, that it is Calld by them Galick: the name Erse, by which it is commonly known; they do not allow to have any signification in it. It is precisely the same as the Irish, & radicaly no doubt the same as the Welsh: but now differing as a dialect so far, that the languages could not be mutualy understood without some study; tho a little would probably suffice.

16.
Having spent a great deal more time in these Islands than we originaly intended & being yet desirous if Possible of Seeing St Kilda we resolvd to lose no time. So the weather being moderate got up our Anchor.

                             [68] Among the Islands.
& put to Sea very Early in the morn. At night we had many Islands in Sight Egg,[1] Canna,[2] Rum,[3] Tire ey,[4] Col,[5] Skie[6] &c &c. About Sun set saw the paps of Jura[7] which by the draughts appeared to be 24 L*eagu*es distant. They were 8° above the Horizon D^r. Lind who work'd the distance by a very ingenious proposition of his own founded on Knowing the hight of that which we had measur'd made the distance 54 Sea miles. Who was right I do not venture to determine.

---

[1] Eigg, one of four Small Isles in the Inner Hebrides, roughly 11.6 sq. miles (30 km²) in size. It is famous for its colourful history: it had a monastery but the monks were massacred in 617 by a Pictish queen, and in the 16th century all the inhabitants were massacred in The Massacre Cave. Sir Walter Scott was instrumental in getting the 395 victims given a Christian burial.

[2] Canna is the westernmost of the Small Isles island group in the Inner Hebrides. According to Pennant when he was travelling there, the island had a population of 220 of which all except four families were Roman Catholics. There was, however, no church, the minister and the Catholic priest residing on Egg (*Tour in Scotland and the Hebrides*, p. 315). Canna later came into the possession of John Lorne Campbell (1906–96), the Gaelic folklorist and scholar, who left it to the National Trust for Scotland, in 1981.

[3] Rùm, the largest of the Small Isles, is about 15.4 sq. miles (40 km²). By the late 18th century it had a population of 400, being tenant farmers and their families. At the beginning of the 19th century the inhabitants were forced to emigrate to America (the Highland Clearances) by its new owner (see Pennant, *Tour in Scotland and the Hebrides,* pp. 317–24).

[4] The Isle of Tiree is the most westerly isle of the Hebrides and the most fertile, 11.6 sq. miles (30 km²) in size. More than 20 forts from the Iron Age survive on Tiree. St Columba founded a monastery here. Crofting is still important and the kelp industry used to be significant (see Martin, *A Description of the Western Islands*, pp. 294–6).

[5] The Isle of Coll, 11.6 sq. miles (about 30 km²), now has under 200 inhabitants. See further Martin, *A Description of the Western Islands*, pp. 298–9.

[6] Skye is the largest and most northerly island of the Inner Hebrides. In 1755 the population was over 11,000 and increasing. Samuel Johnson and Boswell visited Skye in 1773, the year after Banks. Martin visited it in 1703 (*A Description of the Western Islands*, pp. 190–250).

[7] There is an illustration of the 'Paps of Jura' by Cleveley in BL, Add MS 15509, f. 37.

17.

Wind west Sailing between Skie & the outer Islands with a good deal of Sea. Saw upon the water a very Large Shark probably the Basking Shark of Pennant.[1] I saw it however so ill that I can found no opinion in the world upon my Experience.

18.

We were now off the But*t* of Lewis. The question to be determind was whether or no we should beat for St Kilda. It was determin in the negative the weather being dirty a great swell & foul wind so we turnd our heads towards Iceland. In a short

[69] to Iceland.

time sea sickness reignd among us as much as Ever. Those who had been the most at sea were hardly excepted. In short the motion of our small vessel was so quick & jerking that the experience we had got in Larger ships seemd of little Service to us.

19.

Dirty rough weather every body sick.

20.

Moderate & soon after calm. Not very agreable as all were now impatient to arrive at our next Land. In order to take some advantage however the boat was hoisted out. Tho many gulls & other birds had been seen in the morning few now appeard; the least Auk /Alca pica Linn./[2] was shot. It seemd to be a young bird & varied from linnæus's & Pennants descriptions a little cheifly in having no white bar across the wings. Three individuals of /Phyllodoce velella/[3] calld by our seamen by the [name] wind sailors or Sallee men were taken very large. I suppose them to have been driven from their proper stations which are about the tropicks by winds as I never remember to have heard or read of their having before been seen to the northward of the [70] mediterranean & our Latitude was now 59°:44′.

21.

Got an Easterly wind which put every one into spirits. Some Gulls still were seen which is not wonderfull as Ferro[4] was distant by account only 30 leagues: many Shearwaters /Procellaria Puffinus & Glacialis /[5] some Terns.

22.

Wind fair but so Strong that sickness again got footing among us.[6] Birds exactly as yesterday.

23.

Weather Birds &c. as yesterday. I never saw Gulls or Terns in the South Sea follow a ship so far or rather never saw them so far from the land as we have been. Possibly the Shoals

---

[1] Pennant saw this basking shark on the shore of Arran. It had been harpooned some days before his arrival. He went to have a look, finding it 'a monster', and a two-page description follows (*Tour in Scotland and the Hebrides* pp. 192–4).

[2] Banks inserts '*Alca pica* Linn.', now called a razorbill and the scientific term is *Alca torda*.

[3] *Phyllodoce velella*: Banks mentions seeing this creature several times in his *Endeavour* journal and Beaglehole calls it *Velella velella* in his notes, which is a hydrozoan, related to the jellyfish.

[4] The Faroe Islands.

[5] *Procellaria puffinus* is the shearwater and *procellaria glacialis* is the fulmar.

[6] By 1799 Banks had forgotten the recurrent bouts of seasickness and wrote to Henry Brougham (1778–1868): 'You ask me concerning the weather met with in the Sir Laurence when I visited the Western Islands & Iceland, it was to the best of my Remembrance uniformly fine, I do not recall one gale of wind' (21 July 1799, New York Public Library, Myers Collection, 2285).

of herrings &c. with which these seas abound teach them longer flights than their southern congeneres have any occasion for. At night an alarm of Land was given which provd false.

24.
Got the wind at N. to our no small discontent. Many birds Shearwaters Gulls & Terns. One of the Gulls came on board & setled on the deck. A water wagtail who had attended the ship for two days became so tird that he setled upon [71] the deck & ran about in search of food among the people.

In the Evening several flights of large dark brown birds passd the Ship flying in ranks as ducks generaly do in every other particular they resembled Shags.[1]

25.
Our circumstances much as yesterday till afternoon when Land was seen. The Thermometer stood at 44. a degree which felt rather colder than we should have chose.

26
The wind still blowing exactly off the Land. In the morn we were about three leagues from the Westernmost of the Gier-fugl-Skir or Penguin Rocks.[2] Calld by the Translator of Horrebow vulture Rocks by some strange mistake.[3] We saw three the outermost 6 or 7 leagues from the land appears most remarkable being a square column standing by itself in the water about as high as a ships main mast in appearance at this distance.

In the Evening a large white cloud which had been seen all day was shewn to us as something remarkable. We immediately Knew it to be the snowy top of some high Mountain the uncoverd part of which was still by its distance depressd
[72] *off* the coast of Iceland
below our horizon. This our Charts shewd plainly to be the Western Jocul[4] tho distant 25 Leagues at least.

27.
Wind blowing directly off the Land we stood on toward the Snowy mountain. Weather cold & raw Therm. 42.

28.
This morn we were very near the Land so that we plainly saw the shore which was flat & had many houses scatterd near the Beach. Round each of them for a small space the ground lookd green & pleasant but every where Else exceedingly black & barren. Behind many hills rose of a midling hight consisting cheifly of Long ridges.[5]

Many boats were fishing all round us, we doubted not that on shewing our colours some would come on board but notwithstanding that & all the signals we could make they seemd rather to avoid us. This Obligd us to hoist out a boat in order to Speak to

---

[1] A shag is a cormorant, especially the crested cormorant *Phalacrocorax graculus*, which in the breeding season has a crest of long curly plumes.

[2] *Geirfuglasker* are not penguin rocks but the rocks of the *Pinguinus impennis*, the Great Auk. The Latin term doubtless is to blame for this mistake. There are no penguins in Iceland.

[3] Chapter 50 in Horrebow's *Natural History of Iceland* is entitled 'Concerning the geir or vulture' (p. 68), and on the map Geirfuglasker is marked in as Vulture or Birds Island.

[4] Snæfellsjökull, the glacier and volcano on the end of the Snæfellsnes peninsula. In Jules Verne's *Journey to the Centre of the Earth*, entering the crater of the volcano was the starting point.

[5] Uno von Troil commented on the first sight of Iceland 'which though not pleasing, was uncommon and surprizing' (*Letters on Iceland*, p. 3).

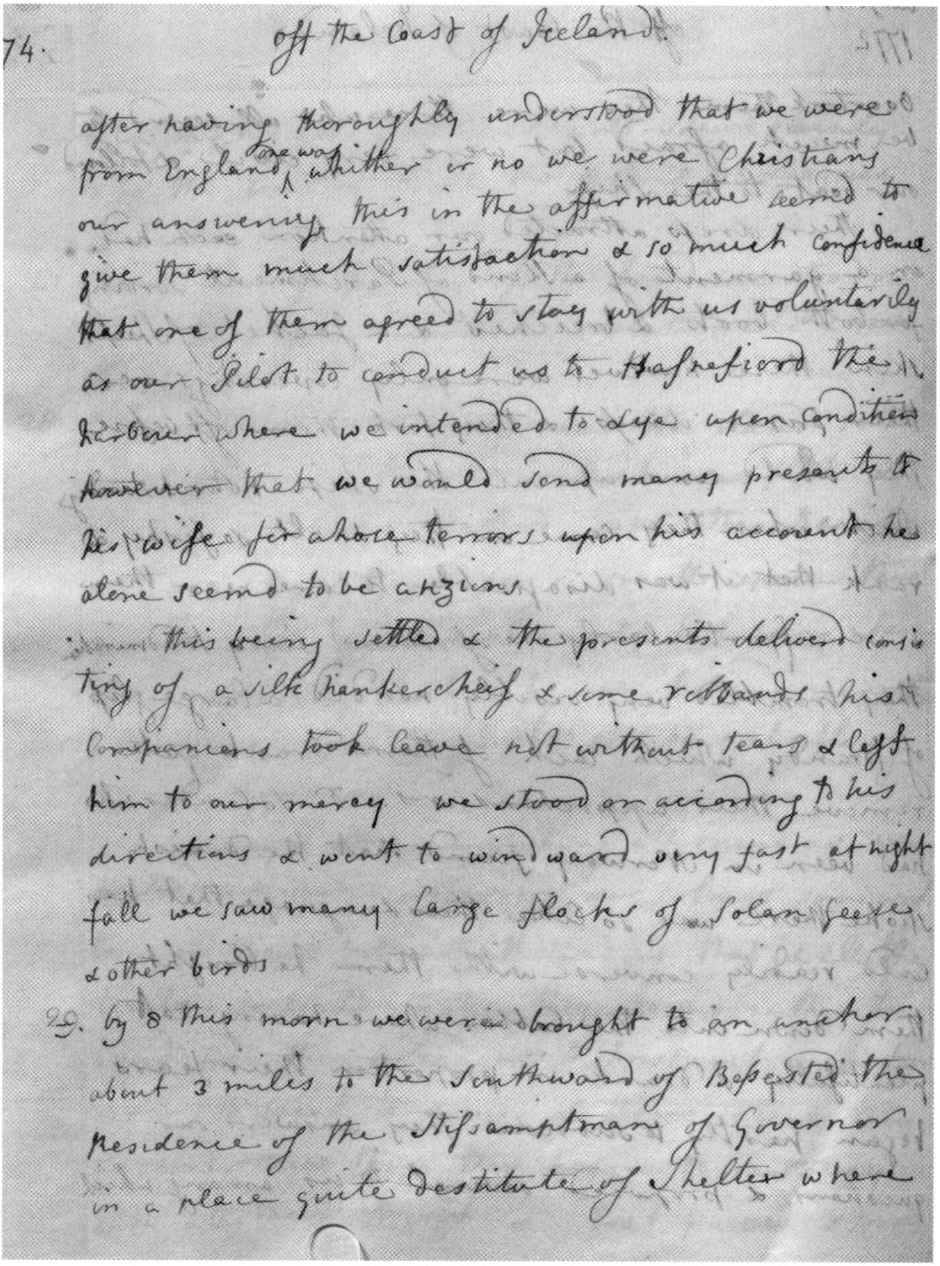

Figure 1. Page from Banks's Journal part I, for 28–29 August 1772. Courtesy of McGill University, Blacker-Wood Library, Montreal.

some of them which they no sooner saw than they began to row away with all their Streng*th*. Our boat pursued & soon

[73] off the Coast of Iceland

Overtook them. They were three who all seemd to be much afraid but were very civil & followd our boat to the Ship.

Their dress attracted our attention. Each had on a garment of a Kind of Parchment serving for both boots & breeches & a Jacket of Sheeps skin. These however were only coverings over their proper dress & they took them off before they would come up into the ship notwithstanding which when they came in they smelt so fishy & rank that it was disagreable to come near them & were (particularly one of them) Lousy to admiration. They trembled very visibly nor did a large glass of Brandy which each of them drank quite remove their apprehensions. D^r. Solander who had been in Norway found that the danish spoke there was so like their Language that he could readily converse with them.[1] He brought them down into the Cabbin where having eat plentifully & drank in proportion their fears began partly to subside. They answerd our questions & proposed several to us among which [74] after having thoroughly understood that we were from England /one was/ whither or no we were Christians.[2] Our answering this in the affirmative seemd to give them much satisfaction & so much confidence that one of them agreed to stay with us voluntarily as our Pilot to conduct us to Hafnefiord the harbour where we intended to Lye upon condition however that we would send many presents to his wife for whose terrors upon his account he alone seemd to be anxious.

This being settled & the presents deliverd consisting of a silk hankercheif & some ribbands[3] his Companions took leave not without tears & left him to our mercy. We stood on according to his directions & went to windward very fast. At night fall we saw many large flocks of Solan Geese & other birds.

29.

By 8 this morn we were brought to an anchor about 3 miles to the Southward of Bessested[4] the Residence of the Stifsamptman of [*sic* for 'or'] Governor in a place quite destitute of Shelter where

[75] off Bessested

we were told we were to lye till the Stifsamptman should give leave for us to be brought into a safer place. Dr Solander went in the boat to wait upon him & carried with him our pasport. About 12 he returnd having met with a most polite reception & assurances that we should have every assistance that was in his power to give. During his absence a multitude of Icelanders came on board none of whoom were so stinking & filthy as those we saw yesterday. In general they were clean & tidy well lookd people. Of them we bought muscles & fishing over the side caught great plenty of fine flounders.

---

[1] An Icelandic contemporary source says that the Icelanders had little difficulty in understanding the Swedish (Espólín, *Íslands Árbækur*, p. 3).

[2] The Icelanders thought the visitors were Turks (Espólín, *Íslands Árbækur*, p. 3). In 1627 Barbary corsairs descended on Iceland and carried 400 Icelanders away into slavery, as well as slaughtering others. This episode called *Tyrkjaránið* (the Turkish Raid) remained vivid in the memory of the Icelanders.

[3] Very much as if the Icelanders were 'noble savages'.

[4] Bessastaðir, then the residence of the royal governor, today the residence of the President of Iceland.

As soon as the Dr Returnd the Pilot[1] of the place who had been with him having now got his orders proceeded with us to the Harbour & by dinner time we were at an anchor in the Birth where we were to stay while upon the Island. It is calld Hafnefiord.[2] It is situated in the SW corner of the Island at the bottom of a Bay calld Faxa Fiordur[3] to all appearance an indifferent harbour as it is open to the N W wind but all the People concurrd in saying that wind never prevaild here.

The instant we had dind we Landed Eager to see

[76] Hafnefiord

the countrey & resolvd to make our first excursion a visit to the Stifsampman who Livd about 3 miles from where we lay.

By 4 we Landed upon a countrey rougher & more ragged than imagination can easily conceive. The rocks which were excessively hard rose up into peaks 8 or 10 yards perpendicular & sank again into small vallies or rather holes of a like depth. Near the Sea Shore a tolerable proportion of sweet but short grass was to be found but when we had advancd a quarter of a mile into the countrey nothing was to be seen but dryas[4] & a few mountain plants thinly scatterd among the Stones.

This singular appearance of the Rocks so different from /any/ what any of us had before seen was Evidently occasiond by the operations of fire. The hardness of the Stone its irregularity & above all the many holes formd in it by its unequal hardning after its fusion evidently provd it. We rejoicd in our situation fortunately chosen in a place where we might have an opportunity of Examining carefully one effect at least of a volcano.

[77] Our Guide[5] tho a sensible man on being askd how this part of the Island became so burnt answerd that he had heard that when the Norwegians first came to settle Iceland they found it preoccupied by certain irishmen[6] whoom it was impossible to dislodge by any other means than by burning the whole surface of this part of the Island which was accordingly done. Absurd as this story was it [was] additional proof that this singular disposition of Rocks was the effect of fire. As such we receivd it with pleasure.

In our way we met the Sysselman[7] an officer one of whoom presides over each division or district & whose power is similar to but rather greater than that of an English Justice of the Peace. He saluted us & Said he was heartily glad to find that we were Peaceable people.

At last we arrivd at the Stifsamptmans who receivd us with all possible politeness. With him was the amptman or deputy governor who vied with the Stifsamptman in shewing us every mark of civility. There was no house they told us where we could possibly live but

---

[1] This was Stefán Þórðarson (Stephen Tordenson), who was induced to remain as a pilot, see Roberts, p. 128.

[2] Hafnarfjörður. There is an illustration of the people of Iceland and some lava (which Banks correctly calls *hraun*) in BL, Add MS 15512, f. 1, drawn on 4 September 1772.

[3] Faxaflói, a large bay in western Iceland.

[4] *Dryas octopetala*, mountain avens, an Arctic-Alpine plant.

[5] Þorsteinn Jónsson, the farmer at Hvaleyri. See below, Document 24 and Plate 5. There is an illustration in colour by Cleveley in BL, Add MS 15511, f. 17 entitled 'View of Torsten's house at Hvaleyre', and one uncoloured (f. 18) drawn from a different angle. The following illustration is a coloured version of the same (f. 19).

[6] According to Ari 'the Learned' Þorgilsson in *The Book of Icelanders* (*Íslendingabók*), written in the the 12th century, Irish hermits or anchorites were living in Iceland when the first settlers arrived. They soon left not wishing to dwell with heathens 'leaving behind Irish books, bells and croziers'.

[7] Guðmundur Runólfsson (c. 1709–80) was the county magistrate. There are several sketches (ff. 20–22), made on 2 September, resulting in a coloured illustration BL, Add MS 15511, f. 23.

that in which the Danish [78] merchants resided during their stay [see Plates 2 and 3]. That was lockd up but could be opn'd by the concurrence of the Sysselman with them which they did not doubt so that on monday morn we might take possession.

In the mean time as the people here are very much inclind to the strictest principles of religion he advisd us not to take any step towards setling ourselves on the morrow which was sunday but to wait with Patience till monday when we should have the Doors of the Merchants house opned to us.[1]

Every thing was now settled amicably in the highest degree so we venturd to ask questions about the opinion of the People concerning us. To this the Ladies answerd very freely that they had been much alarmd. At first they thought that we were come with a hostile intention being well acquainted with the disputes now in agitation between Denmark & England. They thought that we were the Prelude of an English fleet sent to take possession of the Island.[2] That our being so well mannd had given a great sanction to that opinion. Some indeed said they were of opinion that we were come in search [79] of Some people who might have fled from Denmark in these troublesome times & were supposd to have conceald themselves in Iceland.

These scruples being Laid we began to ask concerning the State of the Island. We were sorry to hear that no volcanoes were now burning but proportionaly glad to hear that we might examine the effects of former ones in almost every corner of the place which was destind for our residence. After this we took leave & the pilot who had attended us here a sensible man whoom we all likd was orderd to attend us during our stay.

30.
It being sunday we resolvd to go to church in order to give the people a good impression in favour of us strangers so we went all dressd in our Best apparel.

The church was small but well filld. Candles were lighted upon the altar & [a] great deal of time was spent in singing the whole congregation joining in concert most unmusicaly. The bells were hung in the middle there being no steeple. During the time of the Clergyman (who was a dean)[3] praying at the altar he dressd himself in an Embroiderd dress exactly like the vestments of the Catholicks. He often sung by [80] himself which as he happened to have no voice & not the least idea of musick excited most ridiculous ideas in us bystanders.[4] We behavd with all moderation & decency & during the whole day not the least sign of either work or amusement was seen among our people, which as there were above 30 just Landed on a new countrey was rather extrodinary.

---

[1] There is a pencil sketch of 'Warehouses belonging to the King of Denmark' and a coloured painting by Cleveley, 'View of the Danish Storehouse where we live'd at Hafnefiord' (BL, Add MS 15511, ff. 12, 13). The latter is Plate 2.

[2] King Christian VII of Denmark and Queen Caroline Matilda, sister of George III, were experiencing great marital stress. The Queen's lover, Johan Friedrich Struensee, had been executed on 28 April that year in a particularly brutal manner.

[3] Guðlaugur Þorgeirsson (1711–89), clergyman at Garðar on Álftanes and dean (*prófastur*) of Kjalarnesþing. He is considered to have been intelligent and knowledgeable, a scholar, and was mentioned as a possible candidate for bishop at Hólar, albeit in 1754.

[4] All travellers to Iceland agreed that the Icelanders had no sense of music. See e.g. Niels Horrebow who described them 'as having no skill in musical modulations, they roar out in a very harsh and uncouth manner' (*The Natural History of Iceland*, p. 140).

After church we went to the Stifsamptman[1] & dind according to yesterdays invitation. He entertaind very genteely after the Danish manner. After Dinner he walkd us through his grounds & shewd us his Garden which was partly sunk underground & partly surrounded by immensely high walls of Sods & Stone. Here grew Cabbage of many Kinds Turnips & several other /kinds/ sorts of garden stuff in perfection.[2] Besides this he had a Kind of conservatory made with deals which according to the weather were lifted off or laid on as a shelter. Its utmost produce however was only Cabbage &c. a little better than the Garden could give. Below his house was his farm of about an acre of Land in which [81] were wheat Rye & Barley in all appearance growing very well. He told us that he did not hope for a crop of ripe corn for that either wind or frost always destroyd it about the time of its coming to perfection.

31.
As I had heard last night that the Sysselman made some little difficulties about opning the Houses I had little expectation of getting posession today so took my fishing rod & went to a place near the ship where a small brook ran into the bay & I had Seen many Trouts. I Soon caught a large dish & upon leaving off was agreably surprizd by the Sysselman who came down in order to give us possession of the Houses which he did with some ceremony. The furniture tho scarce worth twenty shillings we took an accurate schedule of & giving him a copy took possession of the[3] Keys. We had now 4 rooms in three houses a dining room in which some of us slept a drawing room proper in which the draughtsmen drew & slept a Kitchen & a loft where the Servants livd. The rest of the Houses were lodgements of goods now full which were Seald up with much ceremony but little wax. [82] Thus much for this day. At night we went on board well satisfied with its transactions

[September] 1.
The most of this day was employd in getting our furniture & bedding on shore. We receivd a hansome present of fish from the amptmans Lady.[4] As we were very particularly acquainted we found that the Stifsamptman had given very hansome orders in our favour. We were to be supplied with every thing at the Companys price. In the Evening we bought a sheep for one Rixdollar or 4/6. Also some turf for our cookery which was but indifferent. After this went out to botanise a little in order to find what our future prospect was to be.

2.
Slept ashore last night our Lodgings /ashore/ were not much less crouded than those on board but the convenience of not being obligd to spend time in passing to & from the ship

---

[1] Laurits Andreas Andersen Thodal (1719–1808), a Norwegian, was governor of Iceland from 1770 to 1784. He had the reputation of having been an exceptional governor and the first non-aristocratic one. Interested in agriculture, he brought with him Norwegian agricultural implements, his cabbage patch being a particular success (Ketilsson, *Stiftamtmenn, og amtmenn*, pp. 21–32). He had been a secretary to J. H. E. Bernstorff (1712–72), the Danish foreign minister, and went on diplomatic missions to Stockholm and St Petersburg, attaining the lowly rank of *justitsråd* (see Glossary).

[2] Thodal was well known for his gardening efforts, growing spinach, chervil, radishes, Savoy cabbage etc. Later that year on 13 September 1772 he wrote a report from Bessastaðir to Christian Martfelt, of the Kommercekollegiet and secretary of the Royal Danish Agricultural Society (Landhusholdnings Selskab), detailing the success of his efforts, harvesting eight barrels of turnips for example (RA, Etatsråd Martfelts Papirer VI, Kommercekollegiet 1735–1816, box 2124).

[3] Harold Carter read this as *two* (transcript in possession of The Banks Archive Project).

[4] Sigríður Magnúsdóttir (d. 1807), who was herself a daughter of an *amtmaður* or district governor (see further Glossary), thus she was high-ranking in Icelandic society, a lady as Banks underlines.

made them very usefull. Many people came in the morning bringing milk butter & berries of Empetrum nigrum[1] & vaccinium uliginosum[2] all which we bought & to encourage trade gave every one who came a small present 2 yards of ribband or a little tobacco. To do them justice no people could be more civil than they were or more thankfull for the small presents they receivd.

3.

Hauld the Seine[3] & caught above 50 prodigious fine Trouts at the mouth of the River in the Salt water. Myself with Dr Solander Botanizing we found but few plants from the lateness of the season. Many no doubt were gone out of Blossom. In general those that we found were such as grow upon high Lands in England as may be seen from the List in the appendix.[4] D^r. Lind whose medical abilities had been discoverd the very first day of his arrival, had a great Levy. He dispensd many & various medicines & after he had done treated the whole of his patients with an Electrical shock[5] which seemd much to surprize them but did not produce any of those humourous Effects which all of us expected. On receiving the shock Every one lookd as a fool who had receivd an unexpected Slap on the face. Nothing lively appeard no good prognostick of Bright parts in our new freinds

4.

The Sein hauld to day in the Same place as yesterday producd no fish. We botanisd again but scarce /Caught/ found a plant which we had not seen yesterday. In the Evening we receivd visits from [84] the Stifsamptman Amptman & Sysselman all of whoom came to ask if our Lodgings were quite convenient & ourselves quite satisfied. To both these questions we had the greatest reason to answer in the affirmative.

5.

M^r. Troil & myself wandering to day beyond a place Calld in the Chart Whaleire[6] fell accidentaly upon an old Stream of Lava which Seemd to be of immence extent as it occupied the whole countrey as far as the eye could reach filling every valley in its cource that occurrd either on one side or the other from its Edges. About half a mile towards its center the whole was composd of small hillocks the surfaces of which were pretty smooth in general but wrinkled exactly as metal after fusion when the Scoria began to harden upon it. These wrinkles being thrown into a thousand various appearances I suppose as the wind or other cause had affected the melted substance. Within this was a scene more easily to be conceivd than describd. The stream of the Lava had here been Strong & by breaking continualy flakes of its surface as soon as it became hard & carrying them along often upon their edges had accumulated upon its [85] surface Hillocks composd cheifly of Plates of Stone often of Large dimensions standing upon their Edges intolerable to

---

[1] *Empetrum nigrum*: the crow-berry.

[2] *Vaccinium gliginosum*: the bog bilberry.

[3] The *seine* is a fishing net designed to hang vertically in the water, the ends being drawn together to enclose the fish.

[4] Banks is probably referring to the check list of plants they had with them, which is in Solander's *Flora Islandica* in the NHM.

[5] This was not Banks's first use of electricity. He had two electrical machines, made by the famous London instrument-maker Jesse Ramsden (1735–1800), on board the *Endeavour*. In Madeira in 1768 he used it in revenge upon the governor of the island 'by an Electrical machine which we had on board; upon his [the governor] expressing a desire to see it we sent for it ashore, and shockd him full as much as he chose' (Beaglehole, *Endeavour Journal of Joseph Banks*, I, p. 160).

[6] Hvaleyri, on the Álftanes peninsula.

Figure 2. Poor man of Iceland. An unsigned pen and ink drawing, dated 4 September 1772. There is a description saying he is wearing a blue waistcoat and a light coat with metal buttons. The cap he is holding is blue. Add MS 15512, f. 23. Courtesy of the Trustees of the British Library.

walk upon & rougher to the Eye than any thing I have seen before. This Lasted near 2 miles beyond was a flat countrey coverd with smooth Lava such as before describd as far as the Eye could reach probably as far as the Edges of the next mountains about 10 miles off. Below was the sea into which this immence mass of fire had dischargd itself.

On our return home passing over the rough tract of countrey /we/ on which our houses were situate the analogy between the two compleatly convincd us that it also was a bed of Lava but of a much older date than that we had seen. It seemd probable also that it had been torn to peices by earthquakes attending may be that Later Eruption.

On Enquiring among the most sensible of the Icelanders they gave us the following account "Our traditions inform us that soon after the coming of the Norwegians to this Island all the South west part of it was on fire this place particularly & all Gulbringe Syssel.[1] Before that time we are told the [86] Geir fugla Schir[2] Rocks which lay off Reikaness[3] were joind with the Continent but then the intermediate Land fell in so deep that ships may safely pass between them. Those rocks which stand out of the water perpendicular higher than any ships mast are we know composd of the very same material as these Runs of Lava which we here call Hrauns & that these are runs of Lava is clear from comparing them with such as have hap*p*end in our times. The two you have seen are the nearest to this place but all over the countrey We speak of Every valley

[1] Gullbringusýsla.
[2] Geirfuglasker.
[3] Reykjanes.

is filld up with the same material for an extent of Countrey of above 20 danish 120 English miles in Leng*t*h. The Source of these streams of fire we cannot with certainty ascertain but suppose it to be in a hill calld Hellers Heide[1] a hill of no great hight distant from Reikaness about 100 miles. The time of this Eruption must have been in the 10th or 11th. Century but [as] it is not mentiond in our old Histories we are not certain about it."

How far this Legendary story may be true I will not venture to say. As to the whole having been done at one Eruption I confess I doubt but the [87] fact of that whole tract of countrey being overflowd as it were with Lava is undoubted true[2] I have learnd it from the concurrent testimony of many who are well acquainted with the countrey. I incline however to beleive that this Lava owes its origin to many different eruptions & possibly still a greater number of Craters. All of these must have been however in situations not very Elevated as there is no high hill in the whole neighbourhood.

6.

This day being sunday of course we abstain from business of all Kinds. The Stifsamptman & Amptman[3] with their familys came to visit us & dine. The Gentlemen wore Danish Dresses but the ladies all Icelandick the Cheif singularity of which consists in the ornament of the head which is a Cone of white Cloth about 18 inches high & bending a little forward round the bottom of which a silk hankercheif is tied which compleatly covers all their hair. For the rest it consisted of divers Jackets & peticoats differing indeed from ours but not very Strikingly except in the ornaments which were of silver & gold Fillagree & were worth from 50 to 80 pounds a [88] each dress they consisted of chains round the neck from one of which hung a medal plates of fillagree on the breasts small bobbs in rows below the Sleves bosses on the apron strings & a girdle which was generaly of Gold. Upon the whole the Dress tho certainly not very pleasing to an European Eye had some merit only that the hair being hid gave a nakedness to their faces very unbecoming [see Plate 6].

They seemd to admire our dinner which being servd up in courses appeard very different from any Danish entertainment that they had seen. The variety of wines also surprizd them but most of all the French horns which playd to them at their desire they having explain to us that musick was a laudable occupation even on a sunday. They staid with us till it was dark & then mounting their little Horses both men & women Gallopd away over the rough beds of Lava along their narrow paths with a nimbleness & fearlessness to us quite astonishing for as English horses could not we were confident have stood 3 steps upon such road English men would certainly have been much alarmd to have been hurried over it with such velocity.

Blacker-Wood MS B 36, 1772.

---

[1] Hellisheiði.

[2] This is not easy to decipher. Carter, in his journal transcript, wrote 'true', Rauschenberg 'in' ('Iceland Journals', p. 217). True makes more sense.

[3] Ólafur Stephensen (1731–1812) was at this point of his distinguished career the newly-appointed *amtmaður* or district governor (see Glossary) of the North and East of Iceland (1770–83). See Appendix 13.

Figure 3. Icelandic woman with two children. An unsigned pen and ink drawing of an Icelandic woman with a boy and girl. Banks described the Icelandic dress of the women in detail in his journal for 6 September. He found it 'not very pleasing to an European eye' and considered the headdress 'very unbecoming'. Add MS 15512, f. 3. Courtesy of the Trustees of the British Library.

Banks's Notes from 29 August–15 September 1772

'Transcript of notes in Banks's autograph found in a folder of miscellaneous sheets in the NHM during a visit on Friday April 6 1963. From a photostat sent to Harold B. Carter by ACT[1] the following week.'[2]

[29 August 1772]
... next year a fleet will follow us to take the Island we were told to be religious so resolve to go to church a man appointed to attend us[3] Insh [?][4] burnt out of the island
30.
went to church people religious bells in Church Candles lit much singing ridiculous[5] dine with Stissampt.[6] See hay Corn Garden Conservat*y* we have well all day no work
31.
Get houses trouts angling some difficulty with the Sisselman
   Mrs Amptmans Trouts
   So not done till night
[September] 1.
getting our things ashore Employd all day the Stissamptmans order has come we are to be supplyd with everything at the Companys price buy a sheep for botanise buy Turf Cookery
2.
people civil to excess many came give Something to Everyone ribbands Tobacco Silk handkerchiefs / <u>Troil goes to See Learned man/</u>[7]
   Sein many fish
3.
   Go up the river some rain see Aurdneys[8] house Dr Lind great Levy Electricity few plants Troil goes to see Learned man
   Sees spring
   Not 188° Vi.d T.

---

[1] A. C. Townsend, an employee of the of Natural History Museum in 1963.

[2] The Carter transcript is in the possession of the Sir Joseph Banks Archive Project. The location of the original documents in the Natural History Museum is currently unknown, thus it has not been possible to compare this transcript with the originals.

[3] This was Þorsteinn Jónsson: see Appendix 13.

[4] Banks probably wrote *Irish* as in his fair journal (see p. 83 above).

[5] In the fair copy Banks had chosen the word 'unmusicaly' (see p. 84 above).

[6] Stiftsamtman, the governor Thodal. See p. 85, n. 1.

[7] By a process of elimination this must be the Reverend Guðlaugur Þorgeirsson (see the Banks journal, p. 84, n. 3 above). In the fourteenth chapter of *Letters on Iceland* Von Troil wrote on Icelandic literature. He named two learned gentlemen he was 'intimately acquainted with': the rector Hálfdán Einarsson (1732–85) at Hólar and the county magistrate Bjarni Halldórsson (1703–73) who lived at Þingeyrarklaustur, both living in 1772 in the north of Iceland. He named others, but Þorgeirsson was the only one mentioned living in the vicinity of Hafnarfjörður in 1772 and certainly deserved to be called 'learned' (*Letters on Iceland*, pp. 171–3).

[8] The original manuscript might reveal what this should read.

4.
  Sein no fish Botanise no plants. Even*ing* Sysselman Ampt. Sissampt. Visit us
5.
  Vide see Lava
6.
  Sunday nothing done amptman Stissampt dine with us with ladies in Islandick dress play horns not irreligious
7.
  Visit Lava again it seems to occupy a large part which was sea comes over the hills covering a great space of Countrey from as the Sysselman told us the midle full of floating slags sides filling up every valley for many miles like the top of melted metal coverd with thin scoria wavd in all directions sometimes like a rope Shot many partridges upon it Danish ship[1] cam in
8.
9.
10.[2]
11.
  Strong wind & rain
12.
  Strong wind Cap Edlu rhun[3] 30 birds
13.
  Dine with Parson[4] family at home
14.
  hot baths sea[?][5]
15.
  Nasty weather but not strong

[*The following notes are written across the lower half of the page and at right angles to above.*]
  1.st water
  Ardnaness vog[6]
  2. d°
  Copa vog[7]
  Garde Rhun[8]
  Hedna Rhun[9] smooth

---

[1] The *Vrow Christiana* (see the Roberts journal, p. 130, n. 6 below).
[2] According to the Roberts journal there was not much happening (see p. 132 below).
[3] Kapelluhraun, lava field near Hafnarfjörður.
[4] See the Roberts journal, p. 132 below.
[5] According to Roberts, this day Banks and Solander went to Reykjavík (see p. 132 below).
[6] Arnarnesvogur.
[7] Kópavogur, an inlet near Hafnarfjörður, now a town between Hafnarfjörður and Reykjavík.
[8] Garðahraun, lava field near Hafnarfjörður.
[9] Probably *helluhraun* which is the Icelandic word for pa-hoe-hoe lava, the smooth lava.

Cap Edlu
Rhun
Kap odlu[1]

Graver vog[2]
Kolla fiord[3]
escan[4]

---

[1] Two different spellings for *Kapelluhraun,* lava field south of Hafnarfjörður.
[2] Grafarvogur, inlet north of Reykjavík.
[3] Kollafjörður, inlet near Reykjavík.
[4] Probably the *Esja*, a mountain north of Reykjavík.

THE ICELAND JOURNAL OF SIR JOSEPH BANKS
PART II: 17 SEPTEMBER–22 OCTOBER 1772

September 17. 1772[1]
<u>Iceland</u>

hight of part of Djawen[2] Clift by the barometer & Line[3]

top of the Clift	28 –	<u>937</u>
		1000
Bottom	29 –	<u>14</u>
		1000
		feet
gives	70 –	<u>84</u>
		100
		feet
hight by line	66 –	9

---

[1] This part of the journal covers the journey to and from Hekla, 17 September to 30 September, and the six days in the Orkneys from 16 October to 22 October. By early December Banks was back in London, having travelled down from Edinburgh, which he visited with Solander and Lind. This journal is not always easy to decipher. It is often a series of notes with little or no punctuation. Here capital letters and full stops have been added to mark sentences. There are no folio numbers so the dates, which are always present, have been given a line to themselves for purposes of clarification for the reader.

[2] Probably, as Rauschenberg suggests, *Almannagjá*, phonetically *gjáin* (chasm, cleft) and *djawen* sound alike ('Iceland Journals', p. 218). Banks had now arrived at Þingvellir where the Icelandic assembly the *Alþingi* convened every summer from 930 to 1798. Many significant events in Iceland's history have taken place at the site, e.g. the adoption of Christianity in the year 1000 and the foundation of the republic in 1944. It is geologically almost unique. The Þingvellir area is a part of the North Atlantic rift system at the point of the meeting of the North American tectonic plate and the Eurasian one, the Mid-Atlantic Ridge. Almannagjá is 4.8 miles (7.7 km) long, its greatest width measuring 70 yards (64 m). It marks the eastern boundary of the North American plate. Þingvellir and the Great Rift Valley of Eastern Africa are the only sites on Earth where the effects of two major plates drifting apart can be observed. It is now a national park and a designated UNESCO World Heritage Site. Thus Þingvellir is of great importance, historically, geologically and culturally.

[3] The first page of the journal is a series of notes until it begins with the date 18 September. The barometer was used to measure the height of mountains as well as changes of atmospheric pressure. Banks's barometer might have been the siphon barometer, which was developed as a portable instrument about 1770. The 'line' probably refers to a measuring line, rather than a sub-division of an inch.

Mem: brooks running in & out of Lava
Clags[1] at Geiser.

18.
9 o'clock. Mem. our provision of horses people & <u>Cloths</u> /*pass Amptmans house*/ pass Holm Rhuin[2] rough the sides very much fusd. Amptmans house.[3] Uneven countrey much burnd. Two Bays[4] or houses only. Sleep at the last Hederbae /*heydebay*/[5] in a very stin*t*[6] room. People hansome & civil, give us much milk & butter. Children survey us with much attention. By the side of Tingwalle watn[7] the largest lake in Iceland full of Beautifull Islands,[8] was there but wood.[9]

19.
At 7 turn out in order to procee*d* Tingwalle pass along the north side of the lake under hills formd of Tufa meet Tingwalle Rhuin[10] at the End of which is the first of the Djawen[11] a fissure in lava. See drawing. Highest side /*97*/ 107 feet 6 inches width 105 feet formd of many Lavas running over each other seldom one above 10 feet several not so many inches. Surface always spongy bottom compact except where it touches the surface of /*some*/ the lower one. Surfaces of different lavas visible under each other by their curld surfaces. Seem to have run all at one eruption in different flows. Some make arches having originaly flowd in hills.

Just here Lava runs into a low Countrey probably formerly part of Tingwalle watn for about two miles filling all up. Possibly some under stratum giving way there made this Chasm.[12]

---

[1] A *clag* is a sticky mass usually of mud or clay.

[2] Banks is writing phonetically – what he wants to write is *hraun* which is the Icelandic term for lava and also for lava field, for which he finally gets the correct spelling, see pp. 104 and 109. The lava field is *Hólmshraun*. I am indebted to Hallgrímur J. Ámundason for illuminating the geography in the journals.

[3] The house of district governor Ólafur Stephensen at Sviðholt on the Álftanes peninsula.

[4] Banks's use of the word 'bay' here means farm, phonetically spelt, the Icelandic term being *bæ* pronounced like the English verb to buy (the accusative case of the noun *bær*).

[5] Heiðarbær, a farm on the west side of the lake at Þingvellir, owned by the church at Þingvellir.

[6] Or possibly 'stin*ky*'.

[7] Þingvallavatn is indeed the largest lake in Iceland, 32.4 sq. miles (84 km²) in area. *Vatn* means lake or water in Icelandic, so *Þingvallavatn* is the lake at Þingvellir.

[8] There are two islands in the lake, Nesjaey and Sandey.

[9] According to Ari the Learned in *The Book of Icelanders*, written c. 1122–33, at the time of settlement all Iceland was wooded from the mountains to the coast. But in time the grazing of the sheep and the use of the timber for practical purposes eradicated the woods.

[10] Þingvallahraun. The Þingvellir lava field forms the bottom of the northern part of Lake Þingvallavatn. It originated in a major fissure eruption. The many single flows of this lava are easily visible in Almannagjá (see above p. 93, n. 2).

[11] Djawen – *gjá* or fissure. The major one is Almannagjá.

[12] Sketch diagram of strata of lava from Banks's Journal part II, for 19 September 1772. Courtesy of the Kent History and Library Centre, Maidstone.

hight of Lower wall opposite the measurement[1] of the
highest ———————————————— 36 – 5
highest part ———————————————— 45 – 4.
allow for slope of Earth ———————————————— 5 – 0

the nearer the top still the smaller parts it is crackd into fissures — to the Southward best. Some strata 30 feet thick. These thick ones not spongy on their lower edge vary[2] much in density.[3] One of twenty quite spongy. Holes in it oblogn[4] & much larger than in more superficial.

Larger strata flowing slower probably & cooler. All strata very irregular in thickness, sometimes forming wedges & /where/ their surfaces have formd hillocks /bottoms/. Opposite side to the *South*ward much higher. Cheif side also $\frac{1}{4}$ opposite Lodes very gently, crack in Consequence very deep. Whether different Lavas of one or of many Eruptions can not say. To the *North*ward very fine & vastly romantick & pretty the Djawen there becoming the bed of a river[5] which falls in Cascades through it. Have not time to draw beauties. Djaw*en* giá[6] reaches from Tingwalle watn to Armáns feild[7] about 6 miles Engli*sh*. 12 ^{H} Bar. 29 $\frac{122}{1000}$ Therm. 68[8]

Tingwalle Parson[9] lazy[10] ignorant as devil. Seat of justice long time [*illegible – crossed out*]. Here is a Room for the Court & a Timber house for the Amptman.[11] Many Walls of houses which in July are inhabited by the Bishop, Stifsamptman & the best people being coverd with wadmel.[12] The only one of high appeal Civil & criminal.[13] Women

---

[1] Banks is using feet and inches.
[2] Rauschenberg said 'very' ('Iceland Journals', p. 219).
[3] Could possibly be 'depth'.
[4] Banks no doubt meant to write 'oblong'.
[5] The Öxará river (River of the Axe) which flows into the lake.
[6] This word is superimposed. The major fissure bearing various names beginning with Almannagjá.
[7] The mountain Ármannsfell, 2,520 ft (768 m) in height.
[8] The 'H' denotes humidity, 'Bar' the barometer and 'Therm' the temperature by the thermometer. Trausti Jónsson of the Icelandic Meteorological Office has kindly deciphered these measurements; for an example, see page 107, n. 4.
[9] Magnús Sæmundsson (1718– 80). He had served as the clergyman at Þingvellir since 1745. He is noted as having been intelligent, a poet and excellent horseman. He drowned in the lake.
[10] The word 'lazy' is written between the lines.
[11] This wooden building was becoming dilapidated and in 1798 the officials left Þingvellir, refusing to continue their judgments in such a draughty hut.
[12] *Vaðmál* is Iceland's traditional woollen cloth, resembling twill. In medieval times it was used as currency.
[13] There were two courts at the Althing at the time, the court of the two *lögmenn* (the highest posts for native Icelanders, see Glossary) and the *Yfirréttur*, the highest domestic court, with the possibility of appeals being made to the Supreme Court of Justice in Copenhagen.

who have murderd their bastards drownd by the Law in a pool in the river under a Cascade. Such Examples very scarce.[1] In the youth of the parson 50 years old was one. She was /sewd/ tied up in a poke reaching to the middle of her legs. A rope came from her cross the river held by an executioner. She stood thus on the brink an hour then was pulld[2] in & pokd down with a pole till dead.[3]

Executions very scarce[4] murther only punishd by decollation.[5]

Very fine Chars in Tingwalle watn.[6]

Beyond Thingwalle the Rhuyn is Continued 3 miles then rises at once with 2 or 3 djawen [fissures] so the whole valley has sunk by some earth quake or softness of under stratum. The clergyman says that the whole bottom of the Lake is Lava & curld on its surface as the top of it generaly is so it also has been formd by the same accident as sunk the valley.[7]

Proceeded[8] over the Lava a long way to a ridge of hills formd of Tuffa one calld Dimen.[9] The inhabitants have a tradition that these hills have thrown out hot water & ruind this countrey, went along by such hills two miles at least. Came to /*Helleraa*/ a cave Laugervands hellra[10] where we were to have lodgd situate under the Ridge of Tufa mountains, coverd with angular stones of all sizes from my head to my fist which have been thrown from some Craters. All of them Lava of some former Eruption.

Many were immersd in Tufa sometimes forming Strata. Different Strata of the hill sometimes fine like sediment stone with Strata[11] sometimes coarse like Cinders. Three houses only today. At night arrivd at the best Baiy I have seen.[12] Master most glad to see us. The parson of /*Lauger*/ Thingwalle came with us to shew us the cave an Island curiosity. He rode ½ a mile in sight of Hecla before he told us which it was. His son came further at Laugerwatn. We saw a smoke & ask'd what it was. Tis a hot bath where your [*sic*] are to Sleep. Our Landlord very glad to see us had prepard 2 beds. In one I slept Cold &

---

[1] Eighteen women are known to have been drowned in this pool.

[2] Rauschenberg read 'hurled' ('Iceland Journals', p. 219).

[3] This is the only description extant of the process by which the condemned women were drowned.

[4] According to the first national census of 1703 there were seven executioners in Iceland. By the early 19th century nobody was willing to take on the task and condemned criminals were sent to Denmark or Norway for execution.

[5] That is to say the men.

[6] Chars are trout, *Salvelinus apinus*, in Icelandic known as the *murta*.

[7] The bedrock of the Þingvallavatn catchment consists mostly of postglacial lavas that are most extensive in the central part of the rift valley. The Þingvellir rift valley is surrounded by volcanoes, which illustrate the connection between rifting and volcanism. Postglacial lava flows that measure an estimated 7.2 cubic miles (30 km³) have flowed into the rift valley, and fissure eruptions within it have left their marks.

[8] Here and in a few other places Banks did not indent the new paragraph, but an indent has been added for the sake of clarity. The Hakluyt Society aims to produce an accurate text, but not a facsimile appearance.

[9] *Stóri-Dímon* (Big Dímon) and *Litli-Dímon* (Little Dímon) are two mountains in Rangárvallarsýsla.

[10] Laugarvatnshellar are man-made caves hollowed out of *móberg* or tuff, inhabited until 1921.

[11] Rauschenberg read 'water' ('Iceland Journals', p. 219).

[12] They have now arrived at the Laugarvatn farm. The farmer was probably Gunnar Högnason (b. 1724), a *lögréttumaður* (see Glossary) who lived in Laugardalur at least during the years 1751–81, when he probably died.

miserable without Sheets & scarce any blankets. Master most Civil gave us hot milk for Supper. Killd a sheep.

20.

At sun rise rose most beatiful sunrise Gilding the Countrey & hot baths smoking in sight round a little lake Laugarwatn.[1] Went down to the nearest & largest a basen of about 10 yards diam*eter* boiling most excessively sometimes a fathom high. The boiling part about 3 feet diam*eter*. Vast quantities of Air rise at each boiling more than /of/ water. The bading[2] place. The men prepare it for their sweathearts, but never baths together.

Therm*ometer* 210[3]

Buried in Earth 213[4] where was a spout of steam. Many such all round 3 springs Equaly hot the smaller.

Near the spouts of Steam accumulations of Vitriol or allum which by lying a short time became insipid.

Steam very plentifull smelt like rotten eggs not strong tho deposd a powder yellow like sulphur.

Very little Lebes[5] about it a Bolus[6] or clay under the sand of the lake. Bones boild in it brittle like Papins digesta.[7]

11 ^Humidity^. 11'. Bar*ometer* 29:190.   Ther*mometer*
                  Therm*ometer* 55.[8]

In winter says our Landlord it is beautifull. The Steam freezing forms a funnel round it. Sometimes it throws more water 4 times than now sometimes less.

Set out Iceland wood dificult says our Landlord to cut a whip stock.

   Value of land from him.

House & Land Lets for 9 R*ixdollars* per Ann*um*

Supports   10 Cows
            10 Horses
            400 Sheep

Sold Last year for 120 R*ixdollars*

Another valued at 160 where we lay.

            12 Cows
            18 Bullocks

---

[1] Laugarvatn is both the name of the farm and of a shallow warm lake fed by hot springs and easy to bathe in.
[2] i.e. bathing.
[3] Degrees Fahrenheit
[4] 213 degrees Fahrenheit in the ground.
[5] *Lebes* is defined by Banks in Document 12 (see p. 177 below). It is the hard residue left when the water from geysers has evaporated.
[6] *Bolus*: perhaps he means 'bole', a reddish soft variety of clay used as a pigment.
[7] *Papins digesta*: the Papin digester was a type of pressure cooker invented by the French physician Denis Papin (1647–1712?).
[8] The reading here would be the humidity 11 of 12 lines (the Leslie humidity scale). A hygrometer could be divided into 10 or 12 lines, this one probably 12. The barometer read 29 inches, which could be used to measure height as well as atmospheric pressure, and 190 would be the temperature of the heat in the sand, while the air temperature would be 55. Though, as there was another thermometer fastened to the barometer, in some cases Banks might be referring to the temperature at the position of the barometer itself.

>   8 Heiffers & calves
>   14 horses
>   82 sheep

Before the Pest[1] summerd 300 winterd 200.[2] Lands here are divided into estates which are never subdivided.

> 3 tenures   King Land
>   Church Land
>   Freehold

<u>Kings Land</u> Given by Landfodgt[3] to whoom he will. The Family who get it posess as long as they have an heir & can pay the receivd[4] rent very small & Tax 1 *Rixdollar*

<u>Church land</u> Given away by the Bishop & Amptman & held in the same manner.

<u>Freehold</u> as in other countries each Estate pays 1 *Rixdollar* per Ann*um*[5] to the King in Lieu of Landtax.

Land this morn more fertile several /*farms*/ houses /*much more frequent*/ upon the Flat countrey under hills of Tufa. Pass rapid river Broaraa[6] over bridge Gia 4 feet wide takes almost the whole river a large stone thrown into it covers it over. Pebbles laid on that. The Foundation of the Countrey is Lava seen here tho all is coverd many feet deep with mould.

Pass Hiutlidar Rhuin.[7]

Baiy with manure on the Ground the first we have seen.

At 6 arrivd at Mola[8] went on hill to see for[9] Geiser saw it spout took it for smoak

21.

At day break set out for Geiser arrivd in 20 minutes. [See Plate 7.] It is situate on uneven Ground under a hill of Lebes?[10] overturnd. Many hot springs some boiling mud some smoking only 3 spouting.

Geiser the large an immence Caldron funnel shapd /*breadth above*/

>   breadth   above – – – 56
>   below – – – 19

depth from level of uppermost edge of crater to water when we came – – – 13

to Edge of inner Crater ——   3.11

Crater formd of Lebes Edgd with it in waves beautifull nearly circular A Volcano of Water & stones.

---

[1] He is referring to a sheep distemper of 1761, when English rams were brought to Iceland to improve the wool. The distemper spread all over Iceland with a high mortality rate.

[2] He is referring to the sheep.

[3] The *landfögeti* (see Glossary).

[4] Rauschenberg read 'reservd' ('Iceland Journals', p. 220).

[5] Rauschenberg read 'tax' ('Iceland Journals', p. 220).

[6] Brúará is the second-largest clear water river in Iceland. In 1433 the unpopular Bishop Jón Gerreksson was placed in a sack and drowned in the river.

[7] Possibly Itiutlidar. Today Úthlíðarhraun.

[8] Múli, a farm in Biskupstungur, owned by Skálholt bishopric.

[9] Rauschenberg read 'her' ('Iceland Journals', p. 220).

[10] Banks's question-mark.

N°	Hour	min	Fath[1]	min[2]	sec[3]
1.	6.	42	5	0	20
~~2.~~	~~6.~~	~~47~~			
3.	6.	51	1	0	20
4.	7.	6	1	0	1
5.	7.	31	2	0	15
6.	7.	51	10	0	6
7.	8.	17	4	0	30
8.	8.	29	3	0	40
9.	8.	36	2	0	40

The water in the inner crater which had been rising by slow degrees ever since we began to observe now rose very considerably almost to the Edge.

10	9.	25	8	1	10

Partridge had been in to boil 7 minutes when this eruption happened it was drove into the breest left behind the legs done[4] enough. After it the inner crater was quite full afterwards the outer crater filld gently $\frac{2}{3}$.

11.	10.	16	3	1	0

12 Barom. 28.772.   Th. 50

Therm. 44.

12.   35

A noise[5] was heard 3 times like Guns under Ground the Earth in the neighbourhood shook & the Crater run over but soon sunk to /its/ just contain its water.

2.   8

Crater run over again & continued till

3.   15

when many repeated blows were heard like beating featherbeds. It overflowd much more than before but lasted not more than a minute.

4.43

Noises were heard like the Last but lowder followd by a great overflow

4.49

thumpd very hard & overflowd very much

12.	4. 51. 30.	4. 0			

This Last /overflow/ Eruption a most fine sight. Water cast not in a column but in many[6] Conical figures immence Column of smoak rose with it.

As we noted in the last observations no Ebullitions which did not throw water over the Edge of the Crater tho many such happned resolvd to note all of Every Kind as long as we could stay such happning only when the water was low in the Crater but after the 9th Eruption much Air coming up in the Smaller ones.

---

[1] A fathom was 6 ft or 1.8 m.
[2] Presumably minutes and seconds refer to the length of time the eruption lasted.
[3] The same headings apply to the remainder of the measurements.
[4] Rauschenberg read 'less one' ('Iceland Journals', p. 221).
[5] Rauschenberg read 'an exercise' (loc. cit.).
[6] Rauschenberg read 'more' (loc. cit.).

After they ceasd a continual boil was perceivable tho casting up no air Keeping the middle of the Basen calm no waves from wind forming there.

Instantly after the Great Eruption the water in the Crater sunk down down Lower than it had been when we came some feet. It soon began to boil up Emitting much air at quick periods some short time we continued in amaze whether it spouted or not I know not.

1.[1]	5 – 7	[blank]	18.	5 – 42		.10
2.	5 – 9$\frac{1}{2}$	[blank]	19	5 – 43$\frac{1}{2}$		.15
3.	5 – 10$\frac{1}{2}$	0.20	20.	5 – 47		1.10
4.	5 – 13$\frac{1}{2}$	0.15	21.	5 – 48$\frac{1}{2}$		.5
5.	5 – 14$\frac{3}{4}$	0.5	22.	5 – 49		.8
6.	5 – 17	0.5	23.	5 – 50$\frac{1}{2}$		.5
7.	5 – 18$\frac{3}{4}$	5	24.	5 – 51$\frac{1}{2}$		.40
8.	5 – 20$\frac{1}{2}$	6	25.	5 – 54		1.15
9.	5 – 21$\frac{1}{2}$	1.	26.	5 – 57$\frac{1}{2}$		.30
10.	5 – 23$\frac{1}{2}$	1.20	27.	5 – 59		1.0
11.	5 – 27$\frac{3}{4}$	.20	28.	6 – 10	15	1.0
12.	5 – 30$\frac{1}{4}$	.30	29.	6 – 19		5
13	5 – 31$\frac{3}{4}$	.45	30.	6 – 23		30
14.	5 – 33$\frac{1}{2}$	.45	31.	6 – 26		20
15.	5 – 35	.30	32.	6 – 29		10
16.	5 – 36	.10	33.	6 – 30	10	30
17.	5 – 38	1.20				

During all this time the inner Crater filld by imperceptible degrees but was not full when we went away. A little above & to the westward is another spring formd like it in miniature the basen[2] of that was flat bottomd & always full. In it were 3 or 4 holes out of 2 of which the water spouted never higher than a fathom always above a yard at very quick intervals.

Account of Eruptions for $\frac{1}{2}$ an hour

very curious the old lebes in the mitt[3] of the mountain. A plain mark of a old eruption of water.[4]

1.	3.45	1.
2.	3.47$\frac{1}{2}$	1.
3.	3.50$\frac{1}{4}$	1.15
4.	3.53$\frac{1}{2}$	45
5.	3.55	1.15
6.	3.57$\frac{3}{4}$	1.
7.	4.0	1.15
8.	4.3	1.15
9.	4.5$\frac{3}{4}$	1.
10.	4.8$\frac{1}{2}$	1.15
11.	4.11$\frac{1}{4}$	1.$\frac{1}{4}$
12.	4.14	1.

---

[1] In the original the statistics were written in a single column, but here they have been arranged in 2 columns for convenience.

[2] Rauschenberg read 'coun' ('Iceland Journals', p. 221).

[3] Middle.

[4] Several measurements are now crossed out.

Above that still to the westward is a hole with but little lebes about it quite dry giving but little Steam except when it spouts which hapned about 6 times only in the day. None of us were very near at any time my servant [it] was who said it went 5 fathom. I beleivd him from what I saw at a distance.

The Steam of these wells was a little sulphureous not so much so as Laugerness.[1]

Rainbow round our heads seen only by ourselves.

Our lodging house Mola[2] I cannot recommend to any body who may come to see. It is 20 minutes hard ride. The nearest /bay/ is calld Leugar Heukadæle[3] next better house. At half past 6 set out for Mola. Slept there.

22.

Set out for Scalholt[4] seat of Bishop.

Morn fine.

   H. 8.19

   Bar. 28-844.   Th. 45

   Therm. 42

Pass over flat boggy countrey near several small lakes. See many wild Swans. Stratum Lava where seen but seldom seems to have been coverd with ashes.

Dine at Hross haye.[5]

   H. 12-50

   Bar. 29=700.   Th. 51.

   Therm. 46.

Arrive at Scalholt

Church of Wood small but Cathedral like Shrine of St. Thorlaci.[6] /Not a shrine. The old shrine was loost and a new made with 3 plates from the old/ Recluse who has lived here. Vestments old of Popish times. Of present Crosses altar peices Candles. Bishop venerable sensible![7] School 28 boys from King [*blank*] Rixdollars each per Annum[8] their lodgings & school room appear/ance/ of little town all occupied by Bishops dependants. Supper 1 Rice milk 3 Roast mutton with burnt butter 2 fruit with currant sauce. Good bed but

---

[1] Rauschenberg read 'larger ones' ('Iceland Journals', p. 222). Laugarnes is now a suburb of Reykjavík, originally with hot springs, where the custom was for the women to take their washing. Banks had inspected the hot springs there earlier (see p. 132, n. 5 above).

[2] See p. 98, n. 8 above.

[3] Laug was a *hjáleiga* or leasehold from the church-farm Haukadalur in Biskupstungur, a subtenancy from the main farm.

[4] The bishopric of Skálholt was abolished in 1785, though Bishop Hannes Finnsson continued to live there until his death in 1796. Skálholt is now the seat of the suffragen bishop of southern Iceland.

[5] Hrosshagi, a farm in Biskupstungur, owned by the Skálholt bishopric.

[6] St Þorlákur Þórhallsson (1133–93) is the patron saint of Iceland. A deacon at 14, ordained at the age of 18, he then studied in Paris and possibly at Lincoln. He became bishop of Skálholt in 1178 and was declared a saint by the Althing in 1198. This, however, was not recognized by the Catholic church until 1984 when Pope John Paul II canonized him officially.

[7] The bishop in 1772 was Finnur Jónsson (1704–89). He had been bishop since 1754, was extremely learned and is the author of the monumental *Historia Ecclesiastica Islandiae* (1772–8), which has yet to be translated into Icelandic.

[8] Rauschenberg read 'of' ('Iceland Journals', p. 222). The schoolboys received a stipend of about 25 rixdollars.

short Eidder down. Room Bishops con dean[1] stupid! Other Gentlemens rooms no floors, beds of Eidder down.

23.

Elym aron[2] on Dunghil[3] in river? Rector[4] Conrector Pastor Emeritus[5] Congratulations Pas*tor* Em*eritus* drunk! Breakfast Coffee Roast mutton Cold cheese wine Brandy.

H 9.20
Bar. 28 - 938   Th. 43.
Visit Rector & Conrector
Income of Bishop
    „   „ Rector   60 Rthlr[6] and food
    „   „ Conrector   30 Rthlr and food
Troilius[7] buys mss of Rect*or* & Isl. Dict*ionary* of Conrect*or*[8] for which he asks only 1 Rx. All the town is dependand upon the Bishop who supplys the people with food every month. Observe heer the huers[9] by Skalholt

Mialkur huer[10]	heer the*y* have dye'd
Laugaras huer[11]	some cloth*er*s.
Drauga huer[12]	The Th. var*ie*s the most of
Hildar huer[13]	them 210

At 3 /*set out Pass fine valley Bishops property*/ at [*sic*] set out see pillers Iceland ferry boats

---

[1] The condean, second in command to the rector, was Páll Jakobsson (1733–1816). He was condean (or *konrektor*) 1758–81, then rector until 1784. He was an excellent Latinist, and wrote an Icelandic-Latin dictionary, now preserved in the National Library of Iceland (Lbs. 1344, 4to), as well as a Latin grammar and poetry.

[2] This is the way it looks. All three transcribers of this journal have deciphered these letters in the same way. Probably Banks is referring to a plant – *Elymus*. There are seven types to be found in Iceland, among them the *Elymus arenarius or Leymus arenarius* (blue lyme grass). In Bacstrom's 'Flora Islandica: Check List of Icelandic Plants' (in the NHM), prepared for Banks and Solander before they left for Iceland, the *Elymus arenarius* is listed, and marked that they found this plant near Hafnarfjörður. However, it is also to be found near Hekla (Babbington, 'The Flora of Iceland', p. 345; and Greipsson and Davy, 'Leymus arenarius').

[3] Both Carter and Rauschenberg read this word as 'Drenghil'. Banks's handwriting is difficult to decipher but the word looks more like 'Dunghil', which makes better sense. Thus Banks probably found blue lyme grass on a hill (holm) in the river.

[4] The rector of the Skálholt School was Bjarni Jónsson (1725– 98), a clergyman who was considered an outstanding teacher. See further Appendix 13. He composed an ode in Latin and Icelandic in honour of Banks. The Latin text is published in Hooker, *A Tour in Iceland*, II, pp. 271–94.

[5] The pastor emeritus was Halldór Finnsson (1736–1814), son of Bishop Finnur Jónsson who was the clergyman of Skálholt cathedral at the time. He had been teaching at the episcopal school since 1763 and calling him emeritus was a mark of respect. He was also considered a good preacher. I am indebted to the historian Guðlaugur R. Guðmundsson for information regarding the school at Skálholt in 1772.

[6] Rthlr is another abbreviation for rixdollar or *rigsdaler* in Danish. See 'Note on Currency, Weights and Measures'. According to an ordinance regarding the Latin school from 1743, the rector was to receive this amount in ready cash, as well as food and 'comfortable' lodging, bedding, light etc. (*Lovsamling for Island*, II, p. 441).

[7] Uno von Troil: see Appendix 1.

[8] Probably a copy of the dictionary mentioned in n. 1 above.

[9] *Hver*: hot spring

[10] Mjólkurhver – the milk hot-spring (doubtless milky in colour). This place-name is unknown in the region today.

[11] Laugaráshver – the hot spring of Laugarás, a farm in the vicinity, in the parish of Skálholt.

[12] Draugahver – the hot spring of the ghosts.

[13] Hildarhver – the hot spring of Hildur (a female name).

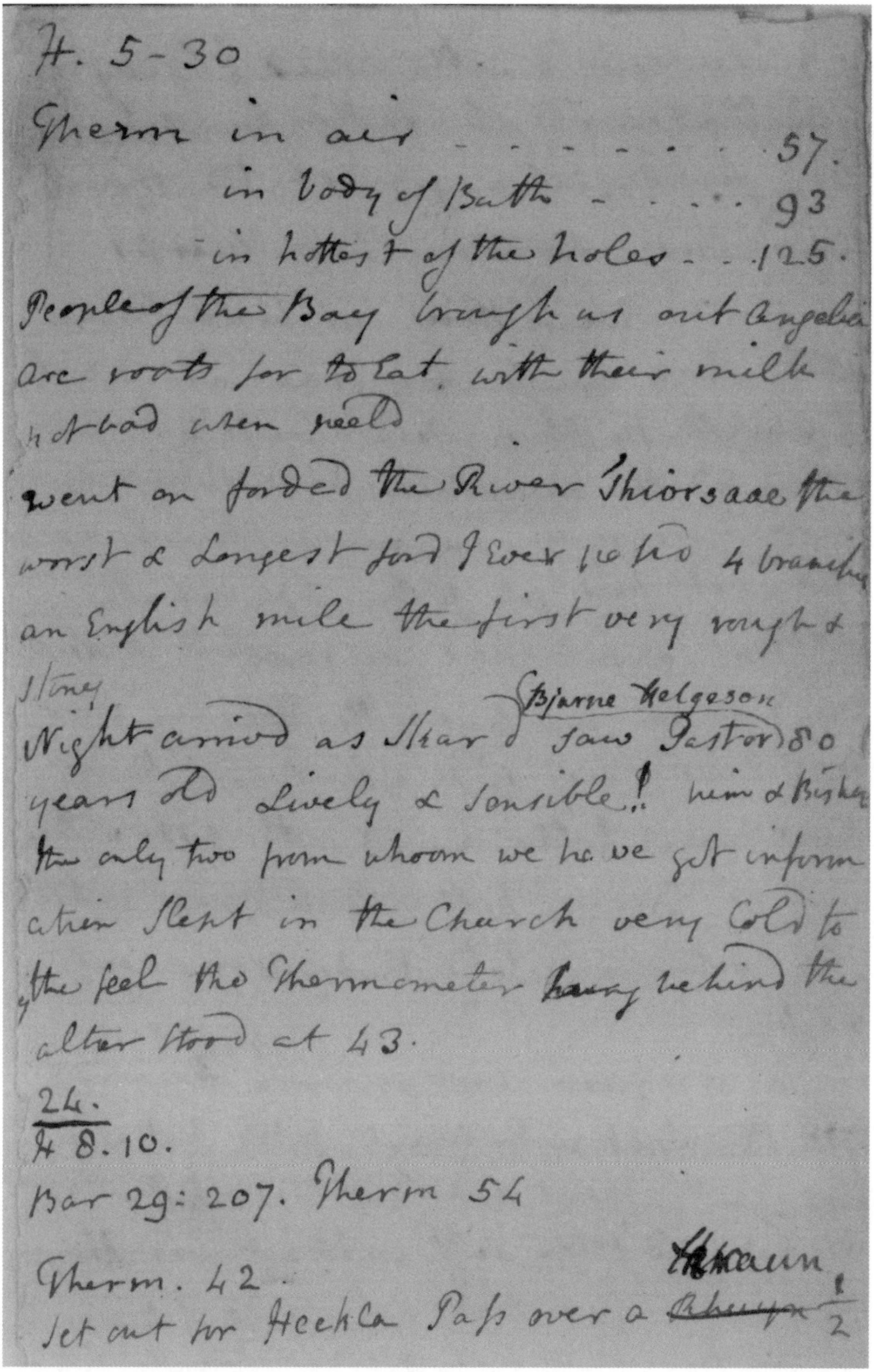

H. 5-30
Therm in air - - - - - - - 57.
　　　in body of Bath - - - - 93
　　　in hottest of the holes - - 125.
People of the Bay brought us out Angelica
are roots for to eat, with their milk
not bad when peeld
went on forded the River Thiorsaae the
worst & Longest ford I Ever met too 4 branches
an English mile the first very rough &
stony
　　　　　　　　　Bjarne Helgeson
Night arrivd at Skar & saw Pastor 80
years old Lively & Sensible! him & Bishop
the only two from whoom we have got inform
ation Slept in the Church very Cold to
the feel the Thermometer hanvg behind the
alter stood at 43.
24.
——
48.10.
Bar 29: 207. Therm 54

Therm. 42.
Set out for Heckla Pass over a ~~Abuyn~~ ½

Figure 4.　Page from Banks's Journal part II, for 23 September 1772. Courtesy of the Kent History and Library Centre, Maidstone.

us horses swim.[1] Beyond fine Countrey by river Bishops property. Hvita/*ra*/ae[2] the name of the river. Enter upon scene of desolation Countrey overspread with ashes & large lumps of very spongy Lava. Reika Meller[3] a small Bay[4] near that 100 years ago a small Crater is said to have opned & coverd all this countrey for several miles as it is. Saw a hillock as the Place. House felt the last Earthquakes of Heckla[5] so strong when situate Close by this hillock that it is removd ½ of a mile.

Pass by a large river upwards /Thiorsaae/[6] that we may Ford. Mountains near it seem small pillars like Staffas upper stratum under them Fine homogenious Tufa like Clay in Strata. Come to /*Kiosarholt*/ Thiorsaarholt[7] dry bath 4 holes in the Earth under a Turf Cover Steam very fine no smell.

H. 5-30[8]

Therm in air - - - - - - - - - - - - - - - - -57
     in body of Bath - - - - - - - - - -93
     in hottest of the holes - - - - -125

People of the Bay brough*t* us out Angelica arc[9] roots for to Eat with their milk not bad when peeld.

Went on forded the River Thiorsaae the worst & Longest ford I ever passd 4 branches an English mile the first very rough & stony.

Night arrivd at Skard.[10] Saw Pastor /Bjarne Helgeson/ 80 years old Lively & sensible![11] him & Bishop the only two from whoom we have got information. Slept in the Church very Cold to the feel the Thermometer hung behind the altar stood at 43.

24.
H 8.10.
Bar 29:207.   Therm 54
Therm. 42

Set out for Heckla Pass over a /*Rhuyn*/ Hraun[12] ½ a mile coverd with ashes flying like the desarts of Arabia. Countrey then fertile till approach the mountain when ashes & Lava in large lumps have spread [over] Every thing but here & there a spot where Grass very fine. Arrived at Næverholt[13] the Place where most who have attempted the mountain have lodgd. Dind on sheeps cream &c. in ruind church. As no one yet had got to the top but

---

[1] There were ferries across a few rivers in Iceland at the time but the horses always swam; a difficult operation, especially in the glacial rivers.

[2] Hvítá – the White River, a glacial river, one of the major rivers in southern Iceland.

[3] Probably Reykjavellir in Biskupstungur, a farm owned by Skálholt bishopric.

[4] Farm.

[5] That would have been the eruption which began in 1766 and lasted for two years.

[6] Þjórsá is the longest river in Iceland, a glacial river, 142.9 miles (230 km) long.

[7] Þjórsárholt, a farm in private ownership, in Gnúpverjahreppur.

[8] It is difficult to know what these figures mean, possibly subdivisions of a hygrometer scale.

[9] *Angelica archangelica L.,* which is very common in Iceland (*ætihvönn*), commonly known as Garden Angelica, Holy Ghost, Wild Celery, and Norwegian Angelica.

[10] Skarð was a parsonage in private ownership, on the eastern side of Þjórsá, in Rangárvallarsýsla.

[11] Bjarni Helgason (1692?–1773) had been the clergyman at Skarð since around 1762. He died the following year. Bishop Harboe, who visited Iceland in 1741–5 on the orders of the Danish government to report on the state of religion and education there, noted that Helgason was a little too worldly and fond of his drink.

[12] Banks now corrects the spelling, subsequently using *hraun* which means both lava and lavafield.

[13] Næfurholt, a farm with a church in Rangárvallarsýsla.

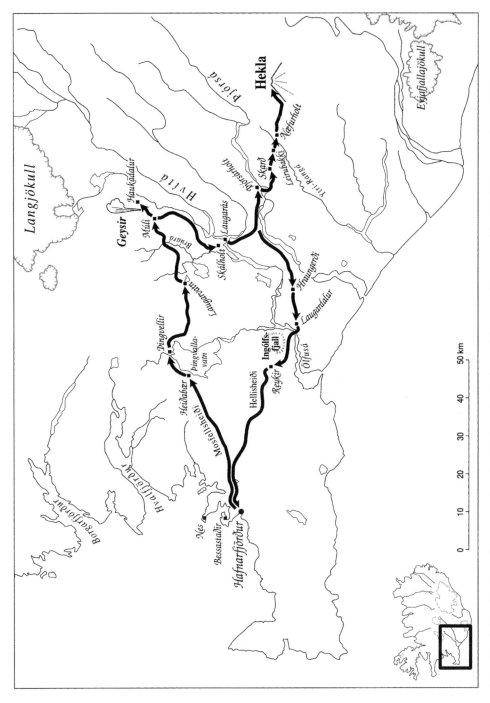

Map 3. Map showing the route taken by Sir Joseph Banks on his excursion to Hekla, 18–28 September 1772.

all returnd unsatisfied we resolvd to sleep nearer in our tent. The Evening was very fine we rode about an hour first over hills coverd with ashes then among Green hills Cross river with little water. Are told that it runs into Heckla.

Arrive at a green spot under a Hraun /Graufel/[1] of 1766 the Last in the ascent. Pitch tent very pleasant. Shoot 3 partriges put them in Dr Linds Kitchen. Self walk over Hraun see Crater get to it an opning of $\frac{1}{2}$ a mile round the West side taken away by Eruption Hraun lying as if it came out of it the Tufa & ashes of which /it/ the west side had been composd still floating upon it. The sides & bottom which remaind nothing but ashes Cinders & peices of Lava in all states cheifly burnt. Name of it Rød Øldur.[2] Cinders some black some red by water?[3] Scene of Ashes & desolation all round almost inconceivable.

Returning home found 2 hills of Tufa splitting the Lava. Slept under the tent 6 crowded but very warm. At one rose.

25.

Got our breakfasts

H. 3.30.   Bar. 28.800.   Th. 48.
            Th. 28.

At 4 set out on horseback passd ashes & cinders for two hours then again arrivd at side of Hraun. Left horses crossd it & came to its origin rough mass of Cinders with large blocks of Scoriæ scoriorum[4] often had standing upright in them. This appearance running up to an opning between two hills. A little higher observd Crater Edgd from whence we supposd water to have come. Ashes all about it red many masses standing upright. /East/ South side of crater taken away.

From hence ascend N Heckla Wind against us so strong we could hardly get on frost laying & cold very severe arrivd at the top of N Heckla. No one was ever higher of Gentlemen[5] tho Laugman[6] who wrote the description of Iceland built a house half way up to assist him. Fog & flying Clouds ourselves coverd with the Ice Cloaths like Buckram all hands a hearty draught of Brandy.

Ascended 1st peak here & there places where snow would not lie a little heat & Steam arising from them. On one of these rested to Observe Barometer.

H. 9.°15.   Bar. 24.838.   Th. 27
             Therm. 24.

---

[1] Gráfell, a mountain, 1,118.8 ft (341 m) high, situated in a lava field west of Hekla.

[2] Rauðöldur, about 1,640.4 ft (500 m) high, is a volcano with an exceptional crater.

[3] Banks's question mark.

[4] *Scoria* is an igneous rock of basaltic composition and containing numerous cavities caused by trapped gases.

[5] See Introduction, p. 19 above. Uno von Troil wrote wrongly that they had 'the pleasure of being the first who ever reached the summit of this celebrated volcano.' He went on to explain that the reason was in part superstition and in part the difficulty in ascending the volcano (*Letters on Iceland*, p. 5).

[6] Carter wrote 'Langman' (transcript); Rauschenberg 'Lauginan' ('Iceland Journals', p. 223). There is no one of either of these names. Eggert Ólafsson (1726– 68) was the first known man to climb the volcano in 1750 (with Bjarni Pálsson). He wrote a diary in Latin and a travel account. His title in Danish was *Vice Lavmand*. This must be what Banks is referring to. Though it is strange to say the least that Banks and his men (von Troil above and see Roberts later, p. 116 below) continued claiming to be the first men up Hekla.

Our Water all froze Dr Lind filld Wind machine[1] with warm water it rose to 1=6.[2] & then froze into spiculae[3] so we could not see longer.

Though*t* we were on the highest peak but saw now one above us all ran towards it in the valley between a chink with much warmth coming up. Dr Solander rests here with one icelander. The rest Hurried up to the top. Intensely cold on the very top of all a spot 3 yards broad out of which came heat that we could not set upon & much Steam[4]

    H 9=25.   Bar. 24.722.   Th.38
    Th. 24

Set out for the bottom very fast resolvd to pass the Snow before we stopd which we did all safe tho almost Froze.

Were told that till the Last Eruption the two Peaks were Clift[5] now Coverd with ashes. Came by a long hill of Tufa from whence the people say water Came the Last Eruption. Saw three Streams of Lava down its side From top of hill. Prospect wide if Clear were told that 4 seas might be seen. Saw S. & W. Countrey all round Coverd with Lava some extending to the sea S. Last Eruption only one Lava run upon an old one. Our old Lava was smooth till the Earthquakes of Last Eruption made it rough?[6]

Only one specimen of Sulphur was found very little Pumice.

Last Eruption broke out at once with an Earthquake S. Wind Carried quantities of Ashes to Holum[7] 180 miles.

Horses ran about till they dropd down fatigued. People who livd near the mountain lost their stock either Choakd with ashes or Starvd before they were removd to grass some livd a year lingering & when opend were full of Ashes.

At two O Clock arrivd at the tents a good deal fatigued.

    H. 2=29.   Bar. 28=920   Th. 52.

Weather fine & warm our other Therm*ome*ter broke in Coming down the mountain. Resolvd to proceed back to Scard[8] where we lay the night before last. See many Wild Geese upon the plain. Arrive.

    H. 7.30.   Bar. 29:392.   Th. 42.

Sleep in Parsons house. Good bed. Clean sheets. Rare in Iceland.

---

[1] Dr Lind's wind machine is, according to Rauschenberg, 'perhaps the best known of the eighteenth century devices which attempted to measure wind pressure rather than velocity' ('Iceland Journals', p. 223).

[2] The measurement regarding wind was in inches so this denotes 1.6 inches. 1 inch was a 'high wind', 2 inches 'a strong wind'.

[3] *Spicula* a sharp-pointed or acicular crystal or similar formation.

[4] Trausti Jónsson of the Icelandic Meteorological Office has kindly deciphered these measurements: Th 24 is in Fahrenheit or −4.4°C. Th 38 is probably the temperature in the barometer, which would be in Centigrade 3.3°C. Bar is the air pressure which indicates, taken at face value, 1,412 m. These are all very likely measurements.

[5] Meaning 'cleft'.

[6] Banks's own question mark.

[7] Hólar, the bishopric in Skagafjörður in northern Iceland. (There were two bishoprics in Iceland in 1772, Hólar and Skálholt.)

[8] See p. 104, n. 10.

26.[1]

H. 9=23.   Bar. 29=219:   Th. 41.

Morn Weather Clear a small smoak coming from the top of the highest Peak of Hecla. [See Plate 4.] Set out Pass Kioseraa[2] over the bad Ford. Very safe same road as we came. Hills like pillars stand on Stratum Clay like fine in Strata with round Pebbles in it. Come to bay where the man livd who got us Iceland Agate. He had got a bag full from a bog surrounded by hills. It lays in the Earth in loose pebbles.

Got of him also Quartz pebbles gatherd among the Gravel of Kioseraa. Began to Rain with West wind very uncomfortable. Pass large flat boggy Countrey here & there interspersd with bad[3] Spots of Hraun so suppose the foundation of the whole to be Hraun. Pass Mark Hraun[4] & Lerku Hraun[5] both very old both very old [*sic*]. Near our house Observe 2 holes in the Hraun circular appearing very much like Craters 50 yards in Diameter.

Arrivd in Hraun Garde[6] Sleep in Church.

27.

Morn Rainy Strong wind SE. so cannot travel.

H. 7.45   Bar. 28=278.   Th: 45

Breakfast upon hasty pudding made of fialle gras Lichen Islandicus[7] good & Glutinous certainly nourishing wholesome for the Breast.

At 12 weather more moderate set out. /*Pass under Ingulfs fell a mountain of Tufa very full of Stones which breaks in the Thaw sending down large pieces*/ Pass flat Countrey Hraun underneath?[8] Ølvesaa[9] swimming ferry pass under Ingulfs Fial[10] Mountain of Tufa discharging Large peices in the thaw under it large Flat like Lincolnshire fenns. See Orebakke.[11] Much rain & wind arrive at Rycombe[12] almost dark.

Sleep in the Church much hay very Comfortable.

28.[13]

Half past 5 fine morn rise go to see hot springs find an innumerable Quantity boiling hot whole hills of Lebes[14] one Kettle under Eastern hill boiling violently has spouted like

---

[1] On 26 September 1772, in Skálholt, Bishop Finnur Jónsson was writing a letter to Solander, telling him that he had been collecting shells for him (Duyker and Tingbrand, eds, *Daniel Solander*, pp. 293–5).

[2] Kjósará. There is no river of that name. He doubtless means Þjórsá; see his own correction on p. 104 above.

[3] Possibly 'bare'.

[4] Merkurhraun.

[5] There is no such place. He must be referring to Merkurhraun or Grenjahraun.

[6] Hraungerði parsonage, owned by the Skálholt bishopric.

[7] *Fjallagras*: *lichen islandicus*.

[8] Banks's own question mark.

[9] Ölfusá. The Ölfus River, a confluence of the glacial river Hvítá (the White River) and the clear water river Sog, running from the lake at Þingvellir.

[10] Ingólfsfjall. The mountain of Ingólfur, the first settler in Iceland, traditionally believed to have arrived in 874.

[11] Leirubakki, a church with a farm in private ownership, in Rangárvallarsýsla.

[12] No doubt Reykir, north of the town Hveragerði. Solander mentions two of the hot springs by name *Baðstofuhver* and *Fjallhver* (see Hermannsson, 'Banks and Iceland', p. 12).

[13] On 28 September Bjarni Jónsson, the rector of the Cathedral School at Skálholt, wrote to Solander. He was sending his luggage and an ode, which 'deserved no reward'. He is eager to hear about the Hekla trip. The people in Skálholt had noticed that Hekla was covered with clouds on the day of the ascent. Sends a letter regarding his son and has recommended him to Banks. He is prepared to send sheep to Hafnarfjörður for their voyage (Duyker and Tingbrand, eds, *Daniel Solander*, pp. 296–9).

[14] *Lebes*: see p. 97, n. 5 above.

Geiser 9 years ago /by the fall of the earth/. Another by the river went through its revolution in about an hour squirting the water[1] several fathoms forwards. The water then sinking almost out of Sight then filling by degrees then boiling then squirting &c. Smells of Sulphur some sublim'd over well innumerable smaller wells. Lebes has in its petrefactions of Leaves & Stalks.

Wind SE blowing hard & snowing.

Set out pass Hellers heide[2] a mountain of Tufa Coverd with running lava wonderfully supposd fountain of all our Hrauns. A little snow on it much Cold from thence over flat countrey. Join an old Road & at 5 Arrive at Hafnefiord.[3]

29.

Fine weather. Amptman[4] visits us Gives books stones Saddle &c.

30.

Dirty weather wind SE.

[*Here follow two blank half-pages, bottom left and top right, as if Banks intended to add something later.*]

[*Now back in the Orkneys*][5]

[October] 16.

See old lead works copper. Walk to Circles of Stone temples of Sun & moon Druidical, Places of Judgment bridges whether or no coæval with circles Tumuli.

17.

Go to Hoy[6] See Iron mine & Haematite[7] vapours min [*sic*] rise upwards hill of Hoy man of Hoy Tuyn,[8] Holm, Gia, Fortification on point like Heppah.[9]

18.

See fig. stones caves & cliffs. Burying places in the Links. People of Harra superstitious in relation to the Standing stones. Story of Girl that parish still has people who speak Danish /Vitrified/ burnt stuff near the Parsons with bones.

Burying places coffins of Slate 4-3 2.4 sides & top no bottom immence abundance of Tumuli commonly Coverd with Stones on the whole face of the bray[10] ashes or bones

---

[1] Rauschenberg read 'crater' ('Iceland Journals', p. 224).

[2] Hellisheiði.

[3] Hafnarfjörður.

[4] Ólafur Stephensen: see Appendix 13.

[5] According to von Troil they sailed on 9 October, arriving at Leith a few days later on 13 October. Subsequently they stayed visiting for 14 days in Scotland and then went on to London ('Själfbiografi', pp. 167–8). This, however, does not make much sense, i.e. going back to the Orkneys from Leith. Perhaps, when it came to writing his autobiography Banks had got his dates mixed up.

[6] Hoy is the second-largest of the Orkney Islands. It means High Island. The Old Man of Hoy, a detached pillar of rock 450 ft (137 m) high, is a famous landmark.

[7] Native sesquioxide of iron (Fe_2O_3), an abundant and widely distributed iron ore occurring in various forms (crystalline, massive or granular); in colour, red, reddish-brown, or blackish with a red streak.

[8] Tyhn might be a phonetic spelling of 'town'.

[9] That is, it looked like a Maori *pah* (according to Lysaght, 'Joseph Banks at Skara Brae and Stennis', p. 2). This was a kind of fortification found in New Zealand.

[10] Probably 'brae', Scottish for steep bank, especially beside a river.

sometimes both in one coffin beads 400 locket with Dr Ramsay[1] found among ashes bones in the same Coffin hair.
19.
Go to Kirkwall[2] see Church[3] 125' Town house Bishops d^o[4] Earls,[5] Castle, Get Freedom Town little manufacture 1600 people.
20.
See Burying places open one See
21.
Idle
22.
Idle tird resolve to go away fair or foul.

[1] Robert Ramsay (d. 1827) was an antiquary and professor of natural history at Edinburgh. There is an extant letter from him to Banks, dated 20 December 1772, where he sends him an account of the ancient burial-places in the Orkneys (NHM, DTC, 1, f. 39). Banks also became acquainted with George Low who wrote *A Tour through the Islands of Orkney and Schetland* in 1774, though it was not published until 1889 in Kirkwall (see p. xxvii).

[2] The council bestowed upon Banks the freedom of the town. Kirkwall is the largest town and capital of Orkney.

[3] The St Magnus Cathedral was built for the bishops of Orkney when the islands were ruled by the Norse Earls of Orkney. Its construction began in 1137. During Catholic times the Cathedral was presided over by the Bishop of Orkney, whose seat was in Kirkwall. Today it is a parish church of the Church of Scotland.

[4] The Bishop's Palace was built at the same time as the cathedral. King Haakon IV of Norway, wintering there after his defeat at the Battle of Largs between Norway and Scotland, died there in December 1263, spelling the end of Norse rule over the Outer Hebrides. The King was buried in St Magnus Cathedral until it was possible to remove his remains to Bergen. In 1468 Orkney and Shetland were pledged by Christian I of Denmark and Norway for the payment of the dowry of his daughter Margaret, betrothed to James III of Scotland; which still remains the case as the dowry has not yet been paid.

[5] The Earl's Palace in Kirkwall, near St Magnus Cathedral, was begun in 1607. It became the residence of the Bishops of Orkney during the 17th century and then fell into decay. It is now a ruin.

Banks's Notes from 28–31 October and 18–21 November 1772

[Transcript by Harold B. Carter[1]]

Oc*tobe*r
28.
Morn Fifeness abreast Countrey fertile & full of towns saild along the coast of fife many towns corn still out at night anchor off Leith
29.
Go ashore at Leith to Edinbourgh see new town bridge old town dine with the professors[2]
30.
See Physick garden[3] excellent houses of many storys property in the air
31.
bad w*e*ather but Society[4] so Learned & unpedantik resolve to stay sometime
Go out Sleep at Hopton House[5] see [*blank*] & [*blank*] gardens
Go on see Borrowstoness[6] coal &c.[7] Sleep at Dr Roebucks[8]
proceed see Canal locks & drawbridge. Carron works[9] their bellows
Mr Lewes[10]
dine at Carron hall with Mr Dundass[11] Sleep at Hopton House
proceed to visit L*or*d Hopetoun[12] at [*blank*]
sleep there & return to Edinburgh

---

[1] This is from a transcript, made by Harold B. Carter, in the possession of the Sir Joseph Banks Archive Project. The present location of the originals in the Natural History Museum is currently unknown, thus it has not been possible to compare this transcript with the originals. During his stay in Edinburgh Banks lodged with Mrs Thompson in Rider's Court off the High Street. Carter describes Banks's stay in *Sir Joseph Banks*, p. 114.

[2] According to Carter, these were the professors from the Old College of the University. He also dined with James Boswell, 'whose receptive ear must surely have ... gleaned arguments enough from the encounter to lay a foundation for the tour in the following year with Samuel Johnson to the Hebrides and West Highlands' (*Sir Joseph Banks*, p. 114).

[3] Now The Royal Botanic Garden in Edinburgh, it was established in 1670 as a physic garden.

[4] The Royal Society of Edinburgh was founded in 1783 so Banks must be alluding to some forerunner.

[5] Hopetoun House, a stately house on the outskirts of Edinburgh, the home of the Earls of Hopetoun. According to its website, it is 'Scotland's finest stately home'.

[6] Borrowstones is Borrowstounness, now known as Bo'ness, in 1772 a centre of coal mining with a harbour.

[7] According to Carter, Banks saw there the coal-mines and saltworks and 'almost certainly heard much about the Roebuck-James Watt partnership and their new steam engine' (*Sir Joseph Banks*, p. 114).

[8] Dr John Roebuck (1718–94), English inventor and industrialist. He is known for developing the industrial-scale manufacture of sulphuric acid. He was a fellow of the Royal Society and lived in Kinneil House.

[9] The Carron Company ironworks, established by Roebuck in 1760.

[10] Mr Lewes, unidentified.

[11] Henry Dundas, 1st Viscount Melville (1742–1811), Scottish lawyer and statesman and William Pitt's closest collaborator. In 1772 he was Solicitor-General for Scotland, to which he had been appointed in 1766. Harold Carter believed this to be their first meeting (*Sir Joseph Banks*, p. 114).

[12] John Hope, 2nd Earl of Hopetoun (1704–81).

<u>Mem</u> Castle hill & its prospect Holyrood house advocates Library Court of Justice parliament house Collidge its library[1]

Nov*embe*r
18.
  Set out for London Sleep at Lord Ellibanks[2]
19.
  See pillars at Dunbar[3] a small Island much worn by the sea top little pillarity Tower down much pillars divided by a vein of Stony matter many veins of Quarts intersect several network on the faces of some like Ludus Helmontii[4] [?] Stone red like sandstone Especially on the outside where the sea beats but hard: Lava ? Sleep at S*i*r J*no* Halls[5] at [*blank*]
see shot like Iron ore of no value
20.
  proceed to Alnwick through N. Berwick fortified
21.
  See Alnwick Castle[6] motley work Castle & walls of defense rampart leveld by Crown [?]
Statues aloft as if all was manned rooms beautiful but furnishd with Paper & stucco instead of Tapistreys pictures & Hangings how far are magnificent Ideas to be raisd by imitations?
in me not Library, Chappel, State bedchamber, town house &c. Gothick Gate d°
drawing room; Glass, & Copy picture, proceed through Morpeth to Newcastle

---

[1] i.e. on his return to Edinburgh Banks visited Holyrood House, the Law Courts, the Parliament, the Advocate's Library and the Old College Library.

[2] Patrick Murray, 5th Lord Elibank (1703–78), author and economist. He lived at Ballencrief near Haddington.

[3] Dunbar Island, one of the Islands in the Firth of Forth on the east coast of Scotland.

[4] A calcareous stone, the precise nature of which is not known, used by the ancients as a remedy in calculous affections.

[5] Sir John Hall, unidentified.

[6] Alnwick Castle is a castle dating from the 11th century and a stately home, residence of the Duke of Northumberland. It is the second-largest inhabited castle in the UK, after Windsor Castle.

Notes by Banks Regarding His Expedition to Iceland in 1772[1]

Mem.

~~land & sea breezes~~
Descrip of Harbour
Hight of Jocle[?][2]
Irish people
defenceless state
amptmans Lady like to miscarry for fear
people do not love to walk
Bishop of Scalholt 2000 Rx[3]
black bread supplyd to every body by the Danes[4]
Shortness of Hay
~~Corn destroyd by wind~~
no kind of fruits not gooseberries
Excellence of Empterum & Vacc ul. Rub. sax.[5]
~~Hillocks raid by Frost difficult to men[6]~~
~~Most Honest~~
people gatherd to oppose us where we took pilot
Womens
Mem. List of Plants
List of Diseases
Journal of Weather

people
well lookd but not hansome scarce so tall as us rather fair, timourous, no Idea of war
not sharp dealers, very honest vastly civil, Idle, inquisitive, fond of History
Hospitable, tolerable adress, religious seldom commit crimes, fond in matrimony
rather chaste or cold, not quick parts
Especialy men, fond of drink, take much snuff chew snuffbox too idle to walk
ride small distances

men civil to their women
women rather Laboriously Employd
men catch fish women bring them home to clean &c.
not bright or lively

---

[1] From a transcription made by Harold B. Carter in the possession of the Sir Joseph Banks Archive Project. The location of the original documents in the Natural History Museum is currently unknown, thus it has not been possible to compare the transcript with the originals. These are three separate sheets. The original notes were written in very short lines, some of which have been amalgamated here to assist understanding.

[2] This probably is supposed to mean glacier (*jökull*).

[3] The annual pay of the bishop was 2,000 rixdollars.

[4] As no corn was grown in Iceland for climatic reasons, one of the Danish merchants' main imports was rye for bread.

[5] *Empetrum nigrum* L. (crowberry) and *Vaccinium uliginosum* L. (bog bilberry) and *Rubus saxatilis* L. (stone bramble).

[6] Banks is referring to a *púfur*; these are small hillocks pushed up by frost which can be difficult to traverse.

[New page]

Climate

never very warm or very cold no Fires by inhabitants Land & Sea breezes S E winds
Snow on the Hills but never very thick or laying long in the valleys
vast high winds make the very stones fly about & houses shake destroy crops of Corn
carrying away vegetable mould 3 feet raising Clouds of ashes
Hillocks raisd by frost

[New page]

Productions

Horses, cows, sheep, no hogs, Foxes, few Rats or mice cats dogs
few fowls. Geese & Turkeys Ducks wild plenty Ptarmigans swans snipes Curlews sand
Larks Sula[1] sea fowl

Cod Haddock Holibut Flounder Salmon Trout Small Herrings Lance

Hay short no corn no potatoes why, turnips, Cabbage Lettice peas cauliflower
Iceland wood
no kinds of fruits but Vacc. myrt. – ulig[2]
Rub. sax. Empet. Excellent[3]

Countrey

very much burnt where we have been most valleys filld with Lava most or
all hills Tufa. flat ground rising in innumerable hillocks from cold runs of Lava Hrauns[4]
rivers large but cataracts
Harbours not apparently good plentyof Trouts Chars sea fish Hills lye-in ridges most
places burnt long runs of Lava

Food

Grant of Flower Lichen Island
Dulces & some more fruits

fish fresh fish heads stock fish
& butter whale fresh & pickled seal
Birds of all kinds

drink milk a little ale fond of Brandy syre[5]

NHM

---

[1] The *sula is* the gannet (*Sula bassanus*).
[2] See above, p. 113, n. 5.
[3] Banks is referring to the *Vaccinium uliginosum* (bog bilberry) mentioned on the previous page, but has used the fuller term *Vaccinium myrtillus ulig*. Actually there were probably also wild strawberries growing in Iceland at this time.
[4] *Hraun* (Icelandic): lava or lavafield.
[5] *Syre* (Icelandic: sýra) is a drink, sour whey.

THE ICELAND JOURNAL OF JAMES ROBERTS

A
Journal of a Voyage[1]
to the
Hebrides or Western Isles of Scotland
Iceland and the Orkneys
undertaken
By
Joseph Banks Esqr.
in the year
1772
By
James Roberts

---

[1] For the frontispiece to the Journal of James Roberts, see Plate 8.

[iii]
Preface

The cause of this Voyage was owing to M^r. Banks being disapointed of his second intended one round the world, after having engaged a number of Eminent Artists with other persons of Ingenuity Necessary in a Voyage of such magnitude and consequence.

The enterprising Genius of M^r. Banks is not confined to trifling remarks, or useless discoveries, his Philosophical researches were intended to improve the mind, and as far as possible, become a universal Benefit by observing and explaining the wonderful works of Nature in all her various Eliments and productions, which would find employment for the Chymist, Philosopher, Historian, and the Poet by opening new scenes drawn by the pencil of Truth to amaze, amuse, instruct and enlarge the understanding.

But none can tell the future by the present. The Ship M^r. Banks intended to have gone in which [iv] was call'd the *Resolution* after having been fitted up at a vast expence for the accommodation of himself and those Gentlemen who he had engaged to accompany him, was found incapable of such service.

This discovery was made in the following manner, every thing being ready many of Mr Banks people with the Seamen were on Board, when as is customary for the Benifit of exercise, the commanding officer ordered them upon deck to play, which they did, and in the pursuites of their rude sports running from one side of the Ship to the other, it was observed to roll and be agitated on the water in a very strange manner, this was particularly noticed by the Pilot who declared he would not take charge of the Vessel, as it was impossible it could ever perform the Voyage with its present Encumbrances.

This being communicated to M^r. Banks, changed his Intentions, and he was at a second trouble in taking away all those Stores which he at so great an expence had provided for such a long Voyage. [v] Yet having retained as before mentioned, such a number of Ingenius persons whom he was under an obligation to satisfy, he thought of a Voyage to Iceland, as a place well worthy [of] his observations, which he undertook in company with D^r. Solander, Dr Lind, M^r. Troil &c. &c.

In our way we visited several of the Western Islands of Scotland, perticularly Staffa, one of the most surprising places in the known world. Having examined and made observations on all that was worthy notice in Iceland and ascended to the top of Mount Heckla (being the first that ever did[1]) we returned by the Orkneys to Edenborough [Edinburgh] of which Voyage this as near as can be is a faithful Journal.

---

[1] Roberts, like Banks, believed wrongly that they were the first to ascend Hekla.

[1]

A Journal
To the Hebrides, or western Isles of Scotland,
Iceland, and the Orkneys, in the year 1772

1772 July[1]

This summer M^r. Banks having engaged the Brig *Sir Laurence* on purpose for a Voyage to the Hebrides, or Western Isles of Scotland, Iceland and Orkneys, for which he gave one Hundred pounds p^r Month. He took with him Dr Solander, Dr Lind, M^r: Troil and many other learned and Philosophical Gentlemen, it was my Fortune to be one of the number in that Expedition.

Saturday 11th

On the 11th of July the Brig (commanded by M^r. James Hunter) then Lying at Limehouse hole,[2] having taken the Draughtsmen and most of the party on Board, that night at 10 o'clock they got underway and at one on Sunday morning Anchored in the Gallions.[3]

[2] Sunday 12th

From the Gallions they proceeded to Grevesend, where they anchored at two o'clock in the afternoon. At twelve the same day M^r. Banks, D^r. Solander, M^r. Troil Count Lauraguais[4] M^r. Scott & myself set out from London, for Shuters Hill,[5] where we dined and in the Evening joined the rest of our company at Greavesend, we supped there and went on Board at half /past/ ten, by eleven hove up the Anchore and sailed for the Nore,[6] where we anchored at three o'clock on Monday morning.

Monday 13th

Sail'd for Dover, where M^r. Banks D^r. Solander and the count went on shore, we then with a Fair Wind sailed for the Downs where we Anchored at four o'clock in the afternoon.

Tuesday 14th

Nothing remarkable. Wind W.S.W.

Wednesday 15th

Wind West South West with hazy weather the first part of that day, the latter was more clear, we waited **[3]** for M^r. Banks and his company 'till half past six in the Evening, when the wind shifted to the North, we then hoisted a signal for the Gentlemen who came on Board immediately, all but Count Lauraguais, who staid on shore to take his passage for France – We soon afterward got underway with little wind at N. W.

---

[1] In the original manuscript the dates appear in the margin, but have been given as headings here to improve clarity.

[2] Limehouse Hole is situated on the north bank of the river Thames in the East End of London.

[3] Gallions Reach was a bend in the Thames between East Ham and Plumstead Marshes, below Woolwich.

[4] See the Banks journal, p. 50 above.

[5] Shooter's Hill was a hamlet near Greenwich and Woolwich.

[6] The Nore is an anchorage near the mouth of the Thames.

Thursday 16th

First part of the day clear with a calm. second part a fresh Breeze at W.N.W. distance from Beachy head[1] four Leagues.[2]

Friday 17th

Fresh Gales with Squals at W.N.W. Distance from Beachy head seven Leagues.

Saturday 18th

Wind and weather the same as yesterday; Distance from Beachy head twelve Leagues, in the afternoon saw the Isle of Wight.

Sunday 19th

A stiff Breeze at W.N.W. distance from the Isle of Wight five Leagues, in the afternoon most of the [4] Gentlemen Sea Sick.

Monday 20th

First part of the day rainy. second part fair; wind W.N.W. at half /past/ five this morning M^r. Banks, D^r. Solander, M^r. Miller and M^r. Bacstrom, went on shore at Cowes[3] to seek for Plants, they sent on Board Fresh provisions with Greens, which were very desirable refreshments. they returned on Board at half past one, and after dinner M^r. Banks, D^r. Solander, D^r. Lind, M^r. Gore and the rest of our people, went on shore at another little Town in the Isle of Wight, we went to the Castle Tavern to order Supper, then again to search for Plants. In our way we saw the Salt Pans, these Pans are to Extract Salt from the Sea water; on our return a poor man seemed to claim our assistance which we readily Granted. this was to help him home with a load of Straw which without our aid he could not have accomplished. We went to Supper at ten and got on Board at half past Eleven.

Tuesday 21st

A fresh gale at W.N.W. got underway Early this morning, past the Needles[4] at half past three, the [5] Sea run high, – in sight of Portland.[5]

Wednesday 22nd.

Little wind at East. – Portland in sight, at seven in the Evening distance from Start Point about nine Leagues, and at Eleven distance about four Leagues, at twelve in sight of Plymouth.[6]

Thursday 23rd.

Gentle Airs at N.E. at nine o'clock this morning hoisted out the cutter when M^r. Banks, D^r. Solander, D^r. Lind, M^r. Troil, M^r. Walden, and the two M^r. Millers with M^r. Briscoe and myself went on shore at Plymouth, the Ship being about three Leagues off; she came to an Anchor in Plymouth Sound about noon. We dined at the Prince George Inn, after dinner Captain Hunter, M^r. Gore, M^r. Clevely, M^r. Bacstrom and M^r. Scott came on shore; in the Evening some of the Gentlemen went to the Assembly,[7] and some to see the Dock Yard, at ten they all returnd to Supper.

---

[1] Beachy Head in Sussex is the highest headland on England's south coast rising to 531.5 ft (162 m).

[2] A league was a measure of distance, varying between countries, but in British marine use, 3 nautical miles.

[3] See the Banks journal, p. 51 above.

[4] See the Banks journal, p. 52 above.

[5] The Isle of Portland in Dorset, nearly midway between Portsmouth and Plymouth. It is really a peninsula rather than an island, being attached to the mainland by a ridge of shingle, Chesil Beach.

[6] See the Banks journal, pp. 52–3 above.

[7] The public assembly, which formed a regular feature of fashionable life in the 18th century, is described in 1751 as a 'stated and general meeting of the polite persons of both sexes, for the sake of conversation, gallantry, news and play' (*Oxford English Dictionary*).

Friday 24th[1]

This morning light Airs at S.W. towards the Evening moderate Breezes at W. and W.S.W. about seven this [6] Evening the Draughtsmen with M^r. Briscoe, M^r. Scott and myself got on Board. M^r. Banks, D^r. Solander, D^r. Lind, M^r. Troil, M^r. Gore, and M^r. Riddle went to the Play[2] and did not come on Board till one o'clock in the morning.

Saturday 25th

This morning we got underway with moderate breezes at S.S.W. at noon past the Eddystone Light-house,[3] a hollow Sea with Rain, most part of the day.

Sunday 26th

A fresh Gale this morning with showers of rain, M^r. Banks with M^r. Walden caught four Dog-fish, in the Evening the wind Westerly; at seven o'clock Fowey[4] in sight, distance about six Leagues.

Monday 27th

A fresh Breeze at N.W. Past Fowey at nine o'clock this morning, distance from Falmouth[5] about six leagues, some Fishermen here taking us to be Smuglers, came on Board in expectation of purchasing contraband goods, but were soon undeceived, at half past Eleven we were distance from the Lizard[6] seven Leagues which we past at four in the afternoon, at night [7] most of the people Sea-sick occasioned by the rolling and pitching of the Vessel.

Tuesday 28th

This morning a gentle Breeze at S.S.W. fine pleasant weather, in the afternoon wind at N.W. past mounts Bay[7] at six in the Evening, distance one League, at eight were about four Leagues from the Lands end which we past in the night.

Wednesday 29th

A fresh Breeze at S. this morning, and no Land in sight, in the afternoon the wind Blew excessive hard and the weather became very thick and hazy, at half past three we were going before the wind at the rate of ten knots, in a direct line on some Rocks call'd the <u>Smalls</u>,[8] on the coast of Wales, they were about two miles a head when we first saw them: our case in this situation was dreadful to think on, for had we run on them the Vessel must inevitably have been dashed to peices [*sic*], but Providence, the mariners guide in dangers, was now our friend, for the Sea running so very high and breaking [8] on them with such violence was the means of our discovering 'em; as soon as we perceived our danger we got our Tacks on Board, and by so doing weathered them, but in performing this task, it caused the Vessel to roll in so violent and agitated a manner, that the Table, and most of the Chests between decks broke loose, and were rolling from one side of the Ship to the other for two hours, at least, before we could get them in their places again, and to add to our distress, the Fire rolled from under the Copper,

---

[1] Roberts did not normally use superscript for ordinal numbers ending in 'th', whereas he did for 1st and 2nd, although he was not entirely consistent.

[2] Neither newspapers nor theatre bills in Plymouth are extant from this year.

[3] See the Banks journal, p. 53 above.

[4] Fowey is an old seaport in Cornwall.

[5] Falmouth is a market town and seaport in Cornwall.

[6] See the Banks journal, p. 54 above.

[7] The bay where St Michael's Mount is to be found.

[8] See the Banks journal, p. 54 above.

which caused a new alarm, for we were much afraid of the Ship taking fire, but thanks to Providence! this danger was got under by throwing plenty of water on it which soon extinguished it. There were some Hogs and Ducks loose upon Deck which we were obliged to throw down between decks for fear of their rolling overboard, and as the Sea run so very high and broke over us in such large quantities, we were obliged to have the Hatchways Batted down, which made it quite dark between decks, the Hogs were [9] grunting, and Ducks Quacking, which together with the rolling of the chests from one side of the Ship to the other, made the most confused noise that can be Imagined, and rendered it dangerous to be between decks for fear of getting our limbs broke, and those who were there, were fill'd with the utmost terror and consternation supposing every moment they were going to the Bottom.

The Ship continued rolling in this manner 'till seven o'clock in the Evening, when it cleared up and the weather became moderate to the great Joy and satisfaction of every one on Board.

There is a custom well known among Seamen, which is, that all those who have not past the Lands end, or any other noted head-land, must pay a fine of a Bottle of Rum or Brandy, otherways half a crown in money or be hoisted up to the Yard-arm and be Ducked three times in the Sea; we had a tolerable company of Fresh water Sailors on Board, none of whom had an Inclination to be Soused in Davy Jones's[1] Pickling-tub, so that the half [10] crowns were produced in plenty, to the amount of two pounds ten shillings, besides sixteen Quarts of Rum which was equally devided among the Seamen.

The latter part of the Night the wind began to blow again very hard and continued to do so 'till past twelve o'clock.

Thursday 30th

Moderate Breezes at S. and S.W. at four o'clock this morning Dublin in sight, distance about twelve Leagues, at eleven could see Dundram, bearing N.W. by N. at the distance of about ten Leagues, at half past one in sight of the Isle of Man, and at five in the afternoon the coast of Scotland in view, we also saw Ilay,[2] distance about twelve or fourteen Leagues, at six Entered the Straits of Galloway. Ireland on one side and Scotland on the other, both in sight. Some showers of Rain in the Evening.

Friday 31st.

This morning a Gentle Breeze at W. past several small Islands on the coast of Scotland, at three this afternoon a calm, distance from Islay five [11] Leagues, in the Evening moderate Breezes; at seven we made a Signal for a Pilot but none came.

Saturday Aug^t: 1st.

Moderate Breezes. – at Eight o'clock this morning came to an Anchor in Ilay Bay or Loughandall.[3] Soon after M^r. Banks, D^r. Solander, and some more of the Gentlemen landed at a small town called Beaumore,[4] and went in search of Plants, in the mean time we got our Tents on Shore and Pitched them close to the Town.

The Gentlemen returned and we dined in the Tents, the dinner was dressed by our own cook at a Public house just by, the Victuals tasted very much of smoke owing to the

---

[1] *Davy Jones* is a nautical slang term for the spirit of the sea.
[2] Today Islay.
[3] Loughandall is now called Loch Indaal, a sea loch on Islay.
[4] Bowmore: see the Banks journal, p. 57 above.

Peat, which is their common fuel having no coals; at night the Gentlemen went on Board, Mr. Briscoe & myself slept in the Tents.
Sunday 2nd.
This morning the Gentlemen came on shore, and Mr. Miller with the rest of the Draughtsmen went to Kirk, or church, which they report much resembles an old Barn, where they heard a Sermon in English, they likewise went to the Parish Church (where the [12] Sermon was preached in Erse[1]) to see the Sacrament Administered, and here it should be observed, that the Natives are Presbyterians, the Sacrament is Administered only once a year, and the method is this, as Mr. John Miller informed me.

Two of the Elders walk up the right side of the middle Aisle, and two on the left, to examine those who are Assembled to receive it; if they are properly prepared or not, if they are, one of the Elders takes some little bits of Diet Bread in the Palm of his hand which he holds out for each to take a bit, he then Delivers Wine in a half pint of which each takes a sip; but what appears to me rather extraordinary, is that they Administer the Sacrament even to Girls at the age of Fourteen, and the reason they give for this proceeding is that the Inhabitants of this Island marry so young, and they think it necessary that they should take the Sacrament before they enter into the matrimonial State.

The Gentlemen dined in the Tents and after [13] Dinner were visited by some Ladies & Gentlemen who drank Tea there, after which they mounted their horses and left us. Most people in this Country keep a horse, if they have means, and their horses usually carry double, and are so very hardy they tell us, that 'tis nothing uncommon to find them waiting for their owners Eight & Forty hours by the side of a Peat Stack without anything to eat or drink, yet notwithstanding this they generally are in very good condition.

The Gentlemen went on Board at half past Eleven, Starlight, we had some drizzling rain most part of the forenoon.
Monday 3rd.
This morning Mr. Banks, Dr Solander, Dr Lind and Mr. Troil came on shore and went to view some Lead Mines, and in their way left Mr Miller at Kilarew[2] to make Drawings of some old Tomb Stones in the Church Yard,[3] the Parish church[4] having formerly been there, but falling into ruins by the ravages of Time they have Built another for that [14b[5]] purpose at Beaumore though Killarew is, I think, far preferable.*

[14a][6] *M{r} Briscoe and me this day call'd at Killarew and visited two young women who had been at Beaumore to see our Tents, they were remarkably civil, and brought out for us, three different kinds of whisky; one of the young women wou'd but just taste it; but the other (her sister) was not a little fond of it. she quite forced it on us, and drank with us herself 'till we all alike became Intoxicated and knew not what we did, & then took

---

[1] Another name for the Irish Gaelic language, though in the 18th century the term was also used for the Gaelic of the Scottish Highlands.

[2] Kilarrow: see the Banks journal, p. 58, n. 3 above.

[3] All three artists made drawings of remains of ancient crosses, tombs and tombstones in Kilarrow churchyard, BL, Add MS 15509, ff. 9–21.

[4] The Kilarrow Parish Church or 'Round Church', as it is often called because of its shape, was built in 1767.

[5] The **b** added by an archivist; and see next note.

[6] Here Roberts has added a note, marked by an asterisk, and written on the left-hand page (numbered 14a by an archivist, though more correctly f. 13v).

leave, as I suppose, in the best manner we cou'd, intending to return to Beaumore together, but the strength of the Whisky had got the better of both our upper & lower understanding, for I remember I tumbled down several times, & I lost my companion. I got to Beaumore before night, but how long I was on my journey 'tis impossible for me to tell, Mr Briscoe I understood got about half way there & there 'tis supposed he fell down & cou'd not get up again, for as Mr Banks & the Gentlemen were returning from the Lead mines they found him fast asleep & his gun lying by him, it was with some difficulty they waked him, when at the same time a man with a cart was passing by, whom they would have hired to carry him to the Tents, but he was going another way so would not accept of the Job. soon after they perceived two men with a horse whom they agreed with, & put Mr Briscoe on the horse, one man got up before him & the other behind him & in this manner convey'd him safe to the Tents. The Gentlemen took care of his gun, we had shot a few small Birds & collected some specimens of different Plants which we put in a Bag that we took with us for the purpose, but the Birds, Plants & Bag we lost, & never saw 'em more. I guess we were both put to Sleep as soon as we arrived at the Tents, as we waked a little before midnight quite sober & well, and it was not a little Satisfaction to understand that some of the Gentlemen were not much better than ourselves, for when they were getting up to go on Board, some of them stagger'd and others were obliged to be led to the Boat. The Crew were careful and got them all safe on Board.

So much for Drunkenness, which if once it becomes habitual is the most dangerous and amongst the worst of the all vices.[1]

[14b continued] M^r. Banks in the morning told M^r. Briscoe and me to go a Botanizing, that is, in search of Plants which we did.

The Gentlemen returned from the Lead Mines about six o'clock in the Evening and dined at eight, went on Board late, it rained most part of the day.

Tuesday 4th

This morning the Gentlemen came on shore at Ten o'clock. M^r. Miller went again to Killarew to make Drawings, in his way there he went into a Weavers house and made a drawing of it, the Gentlemen dined in the Tents, and Supped likewise; after which they went on Board. Some small showers of rain to day.

Wednesday 5th

At Eleven o'clock this morning the Gentlemen came on shore, the weather was both rainy & hazy, wind S.W. Mr John Miller went to Killarew **[15]** again to finish his drawings. M^r. Banks, Dr Solander, Dr Lind and M^r. Troil set out for M^r. Freeburn's,[2] who lives in the Sound of Ilay where they intend to wait 'till the Brig arrives, in their way, they went a second time to view the Lead mines which are rented by Mr Freeburn.

To day we struck the Tents and got them on Board, and in the Evening, after dining at the Public house, we went on Board with Fair weather, being the first we have had since our coming to Ilay.

Thursday 6th

This morning wind S. and N.W. hove up the Anchor and came to Sail at Eight o'clock, in the Evening, distance from Ireland six Leagues, past the Mull of Chanach,[3] very high rocks.

[1] The note on the left-hand page ends here.
[2] Charles Freebairn.
[3] There is a drawing by J. F. Miller of the 'Mule of Chanach' on the Isle of Islay, BL, Add MS 15509 f. 32.

Friday 7th

Pleasant weather, early this morning we past Ardmore [Point], Bane of Yarag, Killa Point,[1] and Jura Point;[2] here Monsieur Thorot,[3] a famous French Admiral generally lay at an Anchor, and in Ilay he used to purchase horned cattle, at Eight o'clock [**16**] this morning we came to an Anchor in the Sound of Ilay almost opposite M[r]. Freeburn's house, distance about one mile and a half, at nine o'clock M[r]. Banks sent for M[r]. John Miller to come to M[r]. Freeburn's house which is called Freeport of which he made a Drawing.[4] To day we got our Tents on shore at Jura, the opposite side of the Sound to M[r]. Freeburns, and set them up to dry, the weather being so bad while we were at Ilay, we were obliged to bring them away wet; got them on Board again in the Evening dry, at six M[r]. Banks with the rest of the Gentlemen came on Board, at seven small showers of rain.

Saturday 8th

Fresh Breezes and cloudy weather. at Eight o'clock this morning M[r]. Banks, D[r]. Solander &c. with the Draughtsman M[r]. Briscoe and myself went in the cutter to the Island of Oransay,[5] about Eight or Nine Leagues from the Sound of Ilay. After we had left the Brig we had heavy Squalls of wind with hard showers of rain, the Sea ran very high and our Boat ship'd a great deal of water which made it very disagreeable, at half [**17**] past Ten we landed, and went to the principal house upon the Island, which belongs to one M[r]. M[c]Neil[6] who is Sole proprietor of the whole place, in this Island there is about Three hundred Inhabitants, Eighty cows besides Oxen and they annually export a great number of horned cattle; here is a Free School where the Natives learn the English Tongue. here is also the remains of an ancient Monastery which the Inhabitants suppose was founded by Columbus in the fifth century. our Draughtsman made a Drawing of it[7] and likewise of several of the Tombs[8] which are in it; we dined with a M[r]. M[c]Dougall[9] who behaved with much civility; and at seven in the Evening we left the Island with little wind and Fair weather, and got on Board the Brig at half past Ten at Night.

Sunday 9th

This morning at Three o'clock hove up the Anchor and got underway, in the forenoon M[r]. Banks, D[r]. Solander &c. with M[r]. Briscoe and myself left the Brig to go to an Island call'd [**18**] Scarba, in expectation of seeing a curiosity in the Gulf of Cariviacha,[10] where 'tis reported if the wind happens to blow fresh against a Spring Tide (which there always runs in whirlpools) it then Spouts the water up, at least, ten to twelve Fathoms, but we were disapointed, for this Phenomena did not happen while we were there, so we were obliged to take it upon trust, since we could not have Ocular demonstration. This Island is about nine English miles in circumference, and belongs to the Duke of Argyle and

---

[1] Could be Beinn Uraraidh and Keills.
[2] Probably just the end of the island of Jura. On Jura see the Banks journal, pp. 61–2 above.
[3] See the Banks journal, p. 62, n. 7 above.
[4] Illustrations of 'Mr. Freebairns house Freeport on the Isle of Ila': BL, Add MS 15509, ff. 2–3.
[5] See the Banks journal, p. 62 above.
[6] Banks calls him M[r]. MacNeal, but the correct spelling appears to be Mac-Neill.
[7] They are to be found in BL, Add MS 15509, ff. 38–41.
[8] Illustrations of the tombs in the McFee chapel at Oransay are to be found in BL, Add MS 15509, ff. 44, 49 and 51. More drawings from Oransay are to be found in this file.
[9] Banks mentions a Captain Macdougal, presumably the same person (see p. 62 above).
[10] The Gulf of Corryvreckan.

M[r]. Maclane.[1] We set up a small Tent as soon as we came on Shore, in which we dined on some cold provision that we brought with us; soon after dinner we had a visit from M[r]. Maclane's Son, who went on Board our Boat with us, and steer'd us toward his Fathers house, we landed about six; some some [sic] of the Gentlemen went with him to his **[19]** Fathers and others staid in the Boat Fishing; at nine we got on Board the Brig (accompanied by M[r]. Maclane) which came to an Anchor about noon off a small Island called Lunga.[2]

Monday 10th

Moderate Breezes at N.W. this morning hove up the Anchor and got underway, and in the afternoon came to an Anchor again in Lochdoun[3] off Mull; in the Evening M[r]. Banks, D[r]. Solander, M[r]. Briscoe and myself went in the Boat, shot some Sea Fowl & caught some Fish.

Tuesday 11th

This morning hove up the Anchor and came to sail with the wind at W.N.W. and pleasant weather, at twelve o'clock past Duart Castle[4] & at half after two past Autaurnist Castle,[5] in the Evening we came to an Anchor off a small town called Savary[6] in the Sound of Mull, it is on the main Land of Scotland. at ten at night got underway again.

Wednesday 12th

Came to an Anchor off Morven[7] about one o'clock to day, at two hove it up and got **[20]** underway again, M[r] Maclane[8] & M[r] Leach[9] came with us from Morven and went with M[r]. Banks, D[r]. Solander, D[r]. Lind, M[r]. Troil, M[r]. Gore, the two M[r]. Millers, M[r]. Clevely M[r]. Briscoe and myself to an Island called Staffa,[10] about Eleven Leagues from Morven in the Sound of Mull (where M[r]. Maclane & M[r]. Leach came on Board) we Landed at Staffa about ten o'clock at night and immediately went to the only house on the Island and Pitch'd our small Tent by it; we had for Supper some Fish, called Glass Eyes.[11] and Potatoes stew'd together over a Peat fire; at twelve o'clock some went to Sleep in the house upon Straw with the man and his wife and their four Children (which was all that inhabited the Island) the rest went to Sleep in the Tent.

Thursday 13th

This morning when the Gentlemen who lay in the house got up, they had very little reason to compliment their host for their accommodation, **[21]** indeed they complained bitterly of their repose having been very much interrupted by a Species of Familiar Vermin which I have heard call'd (by some people) Bosom Friends.[12] they however were found to be likewise backbiters and had stuck so close to them all night that it was impossible to obtain that happy refreshment which balmy Sleep bestows to the weary Traveller, indeed

---

[1] Banks calls him MacNeal 'the principal Gentleman in the Island' (see the Banks journal, p. 63 above).
[2] Lunga is one of the Treshnish Isles. It is not mentioned by Banks.
[3] Lochdoun is now Loch Don.
[4] Duart Castle still exists.
[5] Artaurinish Castle: see the Banks journal, p. 65 above.
[6] Savory is its present spelling. This place is not mentioned by Banks.
[7] See the Banks journal, p. 65 above.
[8] Probably Sir Allan Maclean (d. 1783).
[9] Mr Leach is unidentified.
[10] See the Banks journal, pp. 67–73 above.
[11] Apparently a type of pike.
[12] *Bosom friends*: probably lice, as Banks mentions (see pp. 72–3 above).

so fond were these little things of their new acquaintance that they would not forsake them when they quitted the apartment, but suffered themselves to be carried by the Gentlemen where ever they went and by continually irritating their Skin made them scratch and scrub so, that any person who was a stranger to the Gentlemens Country wou'd undoubtedly have pronounced them some of the true Caledonian breed, yet it must be confessed it was much against their inclination that they took away any of the property belonging [22] to the old Wife and her Bearns.[1] after Breakfast we took our Guns and went Shooting about the Island, which is about three miles in circumference.

Yet small as this Island is, it abounds with some of the most wonderful curiosities in Nature, some of which I shall endeavour to describe. Fingal's Cave is one of the greatest curiosities in Staffa, the Entrance of which is from the N.E. the cave is form'd of Natural Pillars which are very regular and very extensive on both sides. & arranged so as to support an Arch'd Vault, the Pillars are close together; at the end of the cave water of several feet depth is found according to the Ebbing and Flowing of the Tide, a great many pieces of Pillars compose the Bottom, or Floor, the cave is light enough withinside to see the Pillars to the furthermost end. you may walk upon the tops of the broken Pillars which rise above the water, but we went /in/ a Boat, the Pillars which are a vast many are call'd Basalts and [23] generally are of blackish grey colour, though some streaks of yellow may be observ'd, which seems to devide them from each other, and render them more distinct. these wonderful Pillars, which are seen in various directions in almost every part of the Island, are supposed to have been produced by firey Irruptions from the Bowels of the Earth, how many ages ago 'tis impossible to guess however this was; their appearance is the most magnificent and awful that fancy can imagine, for the whole Island is supported by a range of Pillars, there are more Caves in the Island, but that of Fingal is the most surprising, to the end of which, as I said before, we went in our Boat, the rocks, or broken Pillars, which hang over it withinside the roof, resemble very much the Leathern Buckets hung up in Gentlemens Halls. our Draughtsman made a Drawing of this cave and several other parts [24] of the Island.[2] from some of the rocky places which descend into the Sea, and are not less than thirty, or forty yards high, I have walked from the waters Edge to the top, almost in as regular a manner as I could up a pair of Stairs; the Pillars before mention'd are some five, and some six square, and are in various directions; some perpendicular, some Horizontal, some slanting and some bent in such a manner as to resemble very much the ribs of a Ship, and the whole together must impress a thinking mind with the power and wonderful greatness the immensity, and omnipotence of the Author of Nature.

The Island of Staffa is Inhabited only by one Family, who are stationed there to look after a few Cattle and Sheep, their Diet is consonant to the rudeness of their Situation, of Barley Oats and Potatoes they have a little, and Fish they have plenty, but when they are in need of [25] other necessaries, which their Situation cannot come at, they make a fire upon the highest part of the Island. which is a Signal to let the people in Mull know that their assistance is wanted.

In the afternoon we got several Specimens of these curious rocks and Pillars into the Boat, and at Six in the Evening left the Island and Steer'd for Columb-Kill, or Iona,[3]

---

[1] *Bairns* (Scottish): children.
[2] The Staffa drawings are in BL, Add MS 15510, ff. 20–32.
[3] See the Banks journal, pp. 73–5 above.

another Island about three Leagues distance and Landed there about nine o'clock at night; we shot some Sea Fowl on the Beach, which is a kind of white Sand mixt with small Shells and Stones, we brought our Boat to a Grapling[1] near an old Monastery and soon after went to Supper on Sour Milk, curds, Eggs. and what provision we had of our own; M^r. Banks & some of the Gentlemen slept in an old house which was well Littered down with Straw. M^r. Briscoe and myself staid with the Boats crew and covered ourselves with the sails. We had very hard showers of rain in the Night.

[26] Friday 14th

At six o'clock this morning went to view the Monastery, and an Antient [sic] Cathedral Church both supposed to be built by an /Irish/ Saint called Columba many centuries ago, in the Monastery we discovered a Tomb Stone on which was the following Inscription. <u>Sancta Maria ora pro me.</u> This was under the figure of the Blessed Virgin, <u>Hic Jacet Dona Anna Donaldi Tortete filia quondam Priorissa Iona qua, obiit anno M^o. D^o. XI May.</u> This Inscription on her left hand, and over the Virgins head the following
     Atam Altissiomo 2 mandamus[2]
This Tomb lay Buried under some Cow dung in the Monastery, here are still the Tombs of several Scottish, Irish, and Danish Sovereigns. S^t. Columba the founder of this Monastery and Church came into these parts to preach the Gospel, and became so Eminent for his Piety & wisdom, that many Crown'd heads consulted and advised [27] with /him/ after living some time on this Island and having converted the Inhabitants to Christianity he returnd to Ireland, from that time 'till some centuries past, it became a receptacle for Nuns. This Monastery was kept by one Anna Donaldi who was the Abbess.[3] Our Draughtsmen made several Drawings of this antient Monastery and Cathedral.

The Island is about Eighteen miles in circumference and appears to be the most Fertile of all the Scottish Islands I have seen. in the morning we got on board our Cutter and left the Island with moderate Breezes and pleasant weather, and got on Board the Brig at Tobir-more,[4] in the Sound of Mull about nine o'clock in the Evening. A calm Moon light Night.

Saturday 15th

Little wind at N by E. and pleasant weather. this morning M^r. Banks, D^r. Solander &c. went with M^r. Maclane and M^r. Leach in our Cutter to a small Island called Orancey,[5] [28] a Deer hunting, we saw several. but Shot none. at night we went in the Cutter to M^r. Maclane's house (which is on the main[land]) with a fresh Breeze. M^r. Banks thought it best, for fear of any accident, to discharge his Piece which he, holding it over the Boat's side with one hand, fired, but the recoil was so great that it fell out of his hand over board,

---

[1] Bring to a grappling means 'come to anchor'.

[2] The Iona Nunnery was Augustinian, founded c. 1200, and is one of the best preserved medieval nunneries in Britain. Roberts has made some mistakes while copying this inscription. This is the Anna MacLean stone. The restored inscription would read: *Sancta Maria ora pro me. Hic iacet Domina Anna Donaldi Terleti filia quondam Priorissa de Iona qur obiit anno MDXLIII eius animam altissimo commendamus.* Translation: 'Holy Mary, pray for me. Here lies lady Anna, daughter of Donald, son of Charles, sometime Prioress of Iona who died in the year 1543. We commend her soul to the All-Highest.' The Royal Commission on Ancient and Historical Monuments of Scotland has produced an inventory on Iona (1982). This stone is number 204. Pennant described the same stone (*A Tour in Scotland and the Hebrides*, p. 283).

[3] The nunnery is now in ruins and little is known about the nuns who lived here.

[4] Tobermory.

[5] Oransay.

and was of course lost, we supped at M^r. Maclane's and got on Board the Ship at twelve o clock, Moonlight.

Sunday 16th
Pleasant weather, hove up the Anchor about two o'clock this morning and got underway with little wind at N.W. in the afternoon lowered down the small Boat and pick'd up a Sheepskin Blown up like a Bladder, with about thirty fathoms of small line to it. we suppose it to have been some Fishermans Bouy, which he has lost, or perhaps, belong'd to some Smugglers.

[29] [Monday] 17th
Moderate Breezes at N.W. with some small showers of rain, two Sail in sight.

Tuesday 18th
Fresh Breezes at N.W. by N. and hazy weather, in the afternoon lost sight of the Land.

Wednesday 19th
This morning moderate Breezes at N.W. at noon, calm with drizzling rain, distance from Iceland 199 Leagues. at three in the afternoon a gentle Breeze sprung up at N. at six it changed to S.W.

Thursday 20th
Light Airs this morning at W. clear weather in the afternoon and calm.

Friday 21st.
Fresh Breezes at N.E. by N. and cloudy weather with some rain, the wind blew very hard in the Evening and continued all night.

Saturday 22nd
A hard Gale at E and N. by E with heavy showers of rain and cloudy weather, at five in the afternoon spoke a Snow[1] belonging to [30] the Danish company Service which Trade to Iceland for Fish &c. She was Bound to Copenhagen,[2] the Gale continued all night.

Sunday 23rd
This morning a hard Gale of wind at E.N.E. with some small Showers, in the afternoon moderate Breezes with cloudy weather, one Sail in sight.

Monday 24th
The weather still continues cloudy with a stiff Breeze.

Tuesday 25th
Fresh Breezes at N.W. by N. and cloudy weather, in the afternoon Iceland in sight.

Wednesday 26th
This morning little wind at N.W. and clear weather. in the afternoon the wind blew fresh. Land in sight with a deal of Snow upon the mountains.

Thursday 27th
Fresh Breezes at E.N.E. and cloudy weather, at four o'clock in the afternoon saw three [31] Flemish Fishing Vessels,[3] our Captain went on Board one of them to enquire how far we were from Bessastad,[4] the place we intended to Anchor at. they had been Fishing on the Banks adjacent to Iceland and had been about six months from Ostend, they had caught 36 Lasts or 72 Tons of Fish, our Captain return'd on Board with some fine Salt Cod which was a present from the master of the Vessel. The same Land in sight as was yesterday.

[1] A small sailing vessel resembling a brig.
[2] The Danish ships left late in the summer or early autumn.
[3] The Dutch had been fishing in Icelandic waters since the 16th century.
[4] Bessastaðir.

Friday 28th

Fresh Breezes at E. and clear weather. at noon we were about two or three Leagues of the Land, and saw a Boat with three men in it fishing. Captain Hunter and D^r. Solander went in our small Boat immediately towards them, but the poor men were so frightened when they saw them coming, that they haul'd in their fishing lines and rowed away as fast as possible, but our Boat soon overtook them, the Doctor ask'd why they went away ? And they replied, they Imagined England was at war [32] with Denmark and we was a Privateer, but being told to the contrary the Doctor with /much/ persuasion got them on Board of us, we bought their fish and gave them some Brandy, Meat, and Biscuit, which they received very greedily. After they had been on Board some time, two of them went on Shore in their Boat and the third; whose name was Stephen Tordenson,[1] staid to Pilot us into the Harbour called Hafnefiord.

Saturday 29th

Little wind at N.W. and pleasant weather, at nine o'clock this morning we came to an Anchor in Hafnefiord harbour, many of the Natives came on Board who all seem'd to be very poor people. Soon after we came to an Anchor we made ready our Fishing lines and caught a prodigious quantity of very fine Flounders, which this Harbour abounds with. in the afternoon M^r. Banks, Dr Solander, D^r. Lind and M^r. Troil walk'd to the Governor of the Islands house, and here it may not [33] be amiss to relate a whimsical mistake which happened, owing to two of M^r. Banks's servants, who were with him, in their Livery's, which is a Blue Coat turn'd up with Scarlet and Laced with Silver, a Scarlet & Silver Shoulderknot, a Scarlet Waistcoat lac'd with Silver, Scarlet Breeches with broad Silver kneebends and a Hat with broad Silver lace.

When M^r. Banks and D^r. Solander were within a quarter of a mile of the Governors house, they were met by a Gentleman who had heard at the Governors of M^r. Banks and Dr Solanders's arrival in Iceland, and seeing the two Servants Dressed so gay concluded they must be the Gentlemen, as they were behind, 'tis the custom in Iceland for the servants to go before, which without any other circumstance, is a sufficient apology for the mistake, the Gentleman Stop'd to enquire from the very persons themselves, which was M^r. Banks and which of the servants was Dr Solander, but [34] when he was inform'd they were M^r. Banks's Servants, he seem'd very much surprised as he thought from their appearance, they had very much the looks of Gentlemen. In the Evening M^r. Banks &c. Returnd on Board from the Governors.

Sunday 30th

Little wind at N.W. and pleasant weather, this morning we all went to church, which is about two or three miles from Hafnefiord, the people are all Lutherans, the Female part of the congregation appeared very devout and attentive to what their Pastor said, and some of them I observed to shed tears, the men seldom sit with the women, but as we were Strangers, out of respect, had that Indulgence.

After Devine Service M^r. Banks, D^r. Solander D^r. Lind and M^r. Troil went to Bessested where they dined and spent the Evening with the Stiftsampman,[2] or Governor, the rest return'd to dine on Board the Ship. After Dinner we walked into the Country a little way, which is [35] composed scarcely of anything but burnt rocks

---

[1] Stefán Þórðarson, the pilot. He is not mentioned by name by Banks.
[2] Thodal: see the Banks journal, p. 85, n. 1, above.

occasion'd by the Irruption of the Volcano's, which abound in this Island more than any other part in the known world, here are plenty of horned Cattle, which feed by the side of Rivers or Bogs, as Grass /is/ very rare to be met with here. there are very few Sheep owing, as 'tis said, to a distemper which happened among them about Eight years ago,[1] they have no Hogs, or Poultry,[2] their chief Food being Fish, Butter and Milk, after four o'clock they conclude the Sabbeth over, and spend the remainder of the day in Sports and plays, of different kinds, in which some of our people join'd. In the Evening we returned on Board, and at nine M[r]. Banks D[r] Solander &c. Return'd from the Governor's.

Monday 31st.
Calm pleasant weather, this morning M[r]. Banks went on shore to fish for Trout, some of the other Gentlemen amused themselves on Board fishing for Flounders which [36] they caught in abundance. M[r]. Banks return'd on Board to dinner, after which he went on shore again to Angle for Trout, several other Gentlemen went on shore, and at Eight o'clock in the Evening return'd on Board.

[Tuesday] Sept: 1st.
Little wind and cloudy weather, this morning got all the Bedding and other Necessaries on shore to some houses which the Governor Indulged us with, these houses are widely different from the generality of Iceland houses, being built with Boards and are much higher, they belong to the Danish Merchants, who come here in the Summer to Trade with the Natives for Fish and what other commodities they have to Traffic with. We shot a great many Ravens which the Natives were very well pleased with, as these Ravenous Birds destroy many of their Lambs, but we were given to understand that we must not Shoot any Hawks as the King of Denmark sends his Falconers annually to [37] catch them for the diversion of Hawking.[3] a little rain in the night.

Wednesday 2nd.
This morning little wind and cloudy weather, I went with M[r]. Banks and the Gardener to collect Plants. In the afternoon, had a little rain.

Thursday 3rd.
Little wind and cloudy weather with rain, to day we haul'd the Seine, near our houses and caught nineteen Brace of very fine Trout.

Friday 5th [sic] **[4th]**
Pleasant weather with a little wind. this morning M[r]. Banks made a present to the Governor, and Deputy Governor of Six hams and Six dozen of Porter each, in the afternoon they both paid him a visit.[4]

---

[1] See the Banks journal, p. 98, n. 1, above.

[2] The settlers had brought pigs in the 9th century but these died out in the 15th century (they were introduced anew in the early 20th century). Poultry was also brought by the settlers and though they have survived they were certainly rare by the end of the 18th century.

[3] As early as the 11th century Icelandic falcons were sought after by kings and in the late 13th century King Haakon of Norway presented one to Henry III. By the 17th century the Danish king had more or less gained control of the falcon industry. A royal falcon house was erected near Reykjavík in 1763. The last royal falcons were exported in 1806. Today in Iceland falcons are a protected species.

[4] See Document 4.

Saturday 5th

Little wind and pleasant weather. this morning Mr. Banks made the Sisselman[1] a present of two Hams and one dozen of Persico.[2] in the afternoon we haul'd the Seine and caught twenty Eight Brace of very fine Trout, a fine [38] moonlight Evening.

Sunday 6th

Little wind and cloudy weather, this morning I went in the cutter to Bessested to fetch the Stifts Ampman to Dinner, at half past three I return'd with him and his Daughter, a young Lady about Eight or ten years of age,[3] his Secretary, the Ampman and his wife, with two young Ladies, and two Boys.[4] in the Evening their Servants came with horses to fetch them, and about Eight o'clock they departed. A fine Evening.[5]

Monday 7th

Moderate Breezes and clear pleasant weather. This morning I went with Mr. Banks, Dr. Solander &c. to see the Lava from the late Eruption, while we were gone Mr. John & James Miller & Mr. Clevely went to the Sisselman's house and made a Drawing of it, at four this afternoon we returned, the Sisselman dined at our house. In the afternoon the Stifts Ampman sent his Secretary to inform us there was a Ship in sight, at Eight she came to an Anchor near our [39] Vessel. Mr. Banks immediately sent Mr. Walden (who is a Swede and understands the Danish language) on Board of her to enquire what she was and from whence she came, he was inform'd her name was the *Vrow Christiana*, that they made Iceland on the twenty fifth of August, that they came from Copenhagen and in their passage they fell in with an English Vessel Bound to Glasee,[6] at that time the wind blowed excessively hard and the Sea run very high, which overset the Vessel and wash'd all her hands overboard, the Danish Captain immediately hoisted out his Longboat and saved them all but the Captain and three Boys, the Vessel sunk soon after, the people that were saved were almost dead with Fatigue, but by the Humanity of the captain they soon recovered and meeting soon after with a Vessel Bound to Petersburg these people were put on Board much to their Satisfaction after being relieved from such Eminent danger.

A fine moonlight evening.

[40] Tuesday 8th

Little wind and hazy weather with drizzling rain. This morning the Captain of the Danish Ship with the Supercargo Breakfasted at our house, the Captain spoke tolerable English. The Icelanders are always extremely uneasy at the appearance of a Danish Ship,

---

[1] The *sýslumaður* (see Glossary) was Guðmundur Runólfsson (1709–80).

[2] Persico is a kind of cordial prepared by macerating the kernels of peaches, apricots etc. in spirit.

[3] Miss Klow was Thodal's stepdaughter but according to Icelandic sources she was 13 or 14 in 1770 when he was appointed to the office of governor. His wife died shortly after arriving in Iceland but he continued on as governor for 15 years. See further Document 4.

[4] Ólafur Stephensen (Stefánsson) (1731–1812), his wife Sigríður Magnúsdóttir (d. 1807), his sons Magnús (1762–1833) and Stefán (1767–1820), and daughters Þórunn (1764–86), who married bishop Hannes Finsson (1739–96), and Ragnheiður (1774–1826), who married Jónas Scheving (1770–1831), a *sýslumaður* (county magistrate). See further Appendix 13.

[5] From here the first part of Banks's journal ends.

[6] Glasgow, sometimes called Glasgee by Scots.

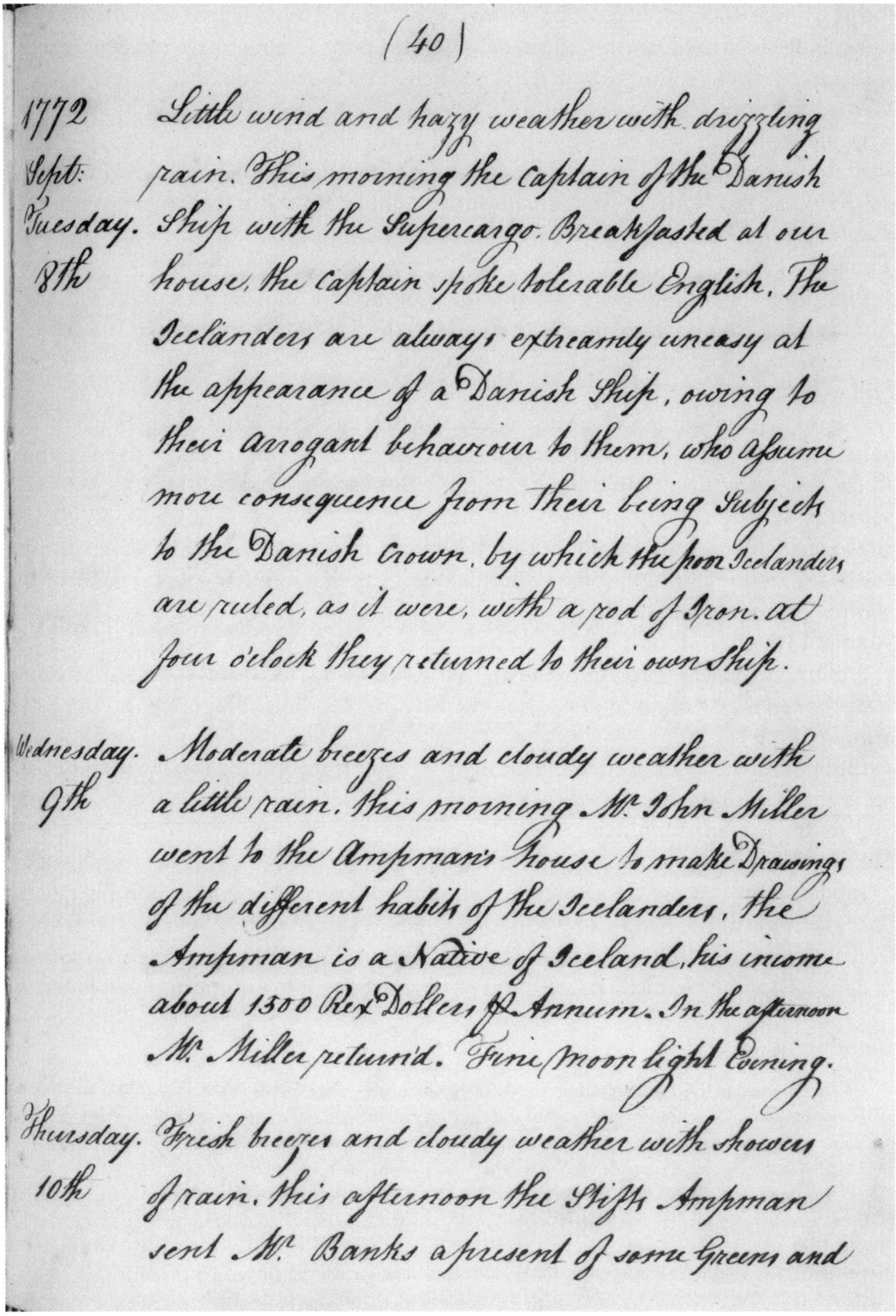

(40)

1772 Sept: Tuesday. 8th	Little wind and hazy weather with drizzling rain. This morning the Captain of the Danish Ship with the Supercargo. Breakfasted at our house, the Captain spoke tolerable English. The Icelanders are always extreamly uneasy at the appearance of a Danish Ship, owing to their Arrogant behaviour to them, who assume more consequence from their being Subjects to the Danish Crown, by which the poor Icelanders are ruled, as it were, with a rod of Iron. At four o'clock they returned to their own Ship.
Wednesday. 9th	Moderate breezes and cloudy weather with a little rain. This morning Mr John Miller went to the Ampman's house to make Drawings of the different habits of the Icelanders. The Ampman is a Native of Iceland, his income about 1500 Rex Dollers p. Annum. In the afternoon Mr Miller return'd. Fine moon light Evening.
Thursday. 10th	Fresh breezes and cloudy weather with showers of rain. This afternoon the Stift Ampman sent Mr Banks a present of some Greens and

Figure 5. Page from James Roberts's Journal for 8–10 September 1772. Courtesy of State Library of New South Wales, A 1594.

owing to their arrogant behaviour to them, who assume more consequence from their being Subjects to the Danish Crown, by which the poor Icelanders are ruled, as it were, with a rod of Iron. At four o'clock they returned to their own Ship.

Wednesday 9th

Moderate breezes and cloudy weather with a little rain. this morning M^r. John Miller went to the Ampman's house[1] to make Drawings of the different habits[2] of the Icelanders, the Ampman is a Native of Iceland, his income about 1500 Rix Dollars p^{er} Annum. In the afternoon M^r. Miller return'd. Fine Moonlight Evening.

Thursday 10th

Fresh breezes and cloudy weather with showers of rain. this afternoon the Stifts Ampman sent M^r. Banks a present of some Greens and [41] Turnips[3] which were very acceptable as indeed vegetables always are to Sailors.

Friday 11th

A hard Gale of wind and cloudy weather with heavy Showers, our habitation was overflowed with the rain which beat in so fast. Cap*tain* Hunter was obliged to let go three Anchors for fear of the Ship driving.

Saturday 12th

Moderate Breezes and cloudy weather with some showers of rain, while we were out to day in search of Plants &c, some of our people at the house purchased an Eagle of the Natives. A moonlight Evening.

Sunday 13th

Light Airs and clear pleasant weather, to day two Clergymen[4] and some Ladies dined at our house, they went away at seven in the Evening. Moonlight night.

Monday 14th

Little wind, the weather clear and pleasant. This morning. M^r. Banks, D^r. Solander &c. &c. went so see some hot Springs at a place called [42] Reikavik,[5] at nine in the Evening they return'd. A fine Moonlight night.

Tuesday 15th

Moderate Breezes and cloudy weather with some Showers of rain, this morning Doctor Solander went to the Ampman M^r. John and James Miller with Mr Clevely went there likewise to finish some Drawings, they return'd about dinner time; the Eagle was dress'd to day, some of the Gentlemen said it was so strong they could not touch it, but I was not

---

[1] Ólafur Stephensen lived at Sviðholt on the Álftanes peninsula, near Bessastaðir.

[2] He means dress. There are several drawings and paintings of the wife and daughter of Stephensen in Icelandic dress among the Banks illustrations in the British Library.

[3] On Governor Thodal's vegetable garden see the Banks journal, p. 85 above.

[4] These must be the Revd Guðlaugur Þorgeirsson (1711–89) of the Garðar parish (the closest to Hafnarfjörður), mentioned by Banks in his notes (p. 84, n. 3 above), a candidate for bishop at one time, and his son and assistant clergyman Ari Guðlaugsson (1740–1809), unmarried in 1772. The 'some ladies' were probably Þorgeirsson's wife Valgerður Þórðardóttir (d. 1775) and some of the five daughters of the family.

[5] At this time Reykjavík was just beginning to seem like a village because in the 1760s workshops for woollen manufacture had been erected there in an attempt to regenerate the Icelandic economy. The Reykjavík parish of 1786 had a population of 289 and the census of 1801 recorded the total number of inhabitants as 307. The hot springs or warm springs (certainly not geysers) were at a place called Laugarnes and were used for washing clothes until the middle of the 20th century. The main shopping street in Reykjavík today is Laugavegur – the road to the warm springs where women would trudge with their laundry.

of that opinion, as it might be Eat very well, for I made a hearty supper of it at night when it was minced, which M'r Douze[1] (our French Cook) had done very nicely.

Wednesday 16th

Moderate Breezes, cloudy weather and some showers of rain, to day the Danes got some of their Stores on shore & Lodged them in their Store houses.

Thursday 17th

At nine o'clock this morning came on a hard Gale of wind, the weather very cloudy with heavy showers of rain and some hail. This [43] Evening the Stifts Ampman & Ampman sent us horses that we might be ready Early in the morning to set out on our intended Journey to Mount Heckla.

Friday 18th

Fresh breezes and cloudy weather with some showers of rain, this morning at ten o'clock M'r. Banks, D'r Solander, D'r. Lind, M'r. Troil M'r. Clevely and myself set out on our journey to Mount Heckla, about one Hundred and eighty English miles from our habitation [see Plate 4]. Each of us was mounted on horseback and several horses were loaded with Provision and Liquor and a small Tent, we carried two Blankets each, being the whole of our Bedding, with these and some other trifling necessaries we set out, the Stifts Ampman sent two of his servants to be our guides, and some spare horses for fear of any accident, our Caravan consisted of nineteen horses in the whole. On the road to mount Heckla we saw a great many [44] Boiling wells, and having purchased Fish and Mutton of the Natives, we Boiled some in these wonderful wells, and it tasted equally as well as if it had been Boiled in a Pot over a Coal fire, and was done much sooner, but the Natives never dress anything in this water, and the reason they give is, that they Imagine it will taste of Sulphur.

The principal of these wells is called Geyser,[2] which we had the opportunity of seeing, it Boild with such rapidity that it threw up a Column of water near twenty fathom high in which we plainly saw Stones some of which were as large as a penny Roll [see Plate 7], we were given to understand that in Winter, and bad weather, its Boiling is greater and the water is thrown up much higher; it is to be understood that it does not keep continually Boiling, but only two or three times in an hour, and sometimes not above once in half a day, or a whole day, according to the weather, and this we were [45b[3]] informed likewise that the longer it is without boiling the higher it throws the water up when it does boil. M'r. Banks had shot a Ptarmigan*[4]

[45a] *Ptarmigan is a Bird somewhat like a Partridge, but larger, it is feathered down to the claws, they are remarkable for changing their color, which in the Summer is very like a Partridge, and not easily to be discover'd, as their color resembles very much the Lava, or rocks, amongst which they /are/ generally found in Iceland, in the Winter their feathers become as white as Snow, with which no doubt the country is covered. I suppose Nature ordered it so that they might not be so easily distinguish'd by their Enemies, they are well known in Norway, Denmark, Sweden &c. And in the Heighland of Scotland, and are in Iceland prodigiously numerous, they are very silly Birds and so tame that they will suffer

---

[1] In the passport his name is spelled as Douvez.

[2] Hermannson believed they had arrived at the Great Geysir on 13 September ('Banks and Iceland', p. 11). During the account of the journey to Hekla, Roberts ceased to write dates in the margin.

[3] The **b** added by an archivist.

[4] Here Roberts has added a note on the ptarmigan, marked by an asterisk and written on the left-hand page (numbered 45a by an archivist, though more correctly f. 44v).

you to come very near before they will rise, and then take but very short flights, they are delicious Eating and taste more like a Grouse than any Bird I know.[1]

[45b continued] but a little time before our arrival at the well. and in order to try an experiment made a string fast to the two legs with a stone to sink it, and then threw it into the well and let it remain just seven minutes before he attempted to draw it up, and when he did, there was only the two legs left which were made fast to the string, the Body being entirely consumed by the Boiling of the water. and the flesh on the legs was nearly gone, the little that was left was Boiled a deal too much, we had each a taste of it on account of the Novelty of its being Boiled in Geyser, about a fortnight before our arrival, the water was thrown with much force and volisity [sic] to a great distance, and falling on some Sheep, was the cause of their death, not owing to any bad quality in the water **[46]** itself (as what was dress'd in it was wholesome) but to the intenseness of its Boiling heat. We pitched our Tent at a proper distance from this wonderfully curious well, and staid there the greatest part of one day to watch its motions: about noon when the well had been a considerable time without Boiling we heard a dreadful noise within the Bowels of the Earth, like Thunder or the firing of cannon at a distance, we observed at the same time that the ground shook under us and did so for a considerable distance all round, and soon after the Column of Water before described was thrown up. This well is upon a rising ground, the mouth of it not unlike a coalpit but rather larger, a person from Denmark attempted to sound it, but after extending two hundred fathoms of line could find no Bottom, near to Geyser are several other lesser Wells, which Boil in the same manner. **[47]** The second night after we left Geyser, we stop'd at the Bishop of Skallholt's;[2] who behaved with great kindness and friendly civility, his house is not remarkable for its grandeur, being very low, built with Iceland Stone and wood, and covered with Turfs, a yard or more thick, like other Iceland Dwellings, they get all their wood from Denmark having none of their own, for Iceland does not produce any thing thicker than a Broomstick, Shrubs of Birch being all that is to be seen; yet it is certain large Trees have grown there formerly, but have been entirely destroyed by the devastations from the Volcano's, however this Birch is useful to them as they make charcoal of it. About half a days journey from Skallholt we came to a dry Bath,[3] it is covered with Stone and Turf over that, and has the appearance of a little hut, you have to go down, about four or five steps, there is not the least prospect of water, but a steam is perceivable to come into it from several holes **[48b[4]]** which is very hot, this dry Bath, as it is call'd, will contain three or four people. The Bishop of Skallholt is afflicted very much with the Rheumatism and visits this Bath at times, which he finds to afford him great relief, he stays in it about half an hour at a time.

On the twenty fourth in the Evening we arriv'd at the foot of Mount Heckla, and pitch'd our small Tent in which we lay, for there is no house within three or four miles of the Mountain*[5]

---

[1] The note on the left-hand page ends here.
[2] The bishop was Finnur Jónsson (1704–89). For an account of him, see the Banks journal, p. 101, n. 7 above.
[3] As the Banks journal makes clear, this was in Þjórsárholt.
[4] The **b** added by an archivist.
[5] Here Roberts has added a reflective note, marked by an asterisk and written on the left-hand page (numbered f. 48a by an archivist, though more correctly f. 47v.). (Obviously other sleeping arrangements were made for the two servants provided by Governor Thodal.)

[48a] *'T'is Natural to suppose that every one wish'd to make his situation as comfortable as possible in these dreary regions, our Tent would just hold us six, and that there might be no partiality, we concluded to draw lots with bits of sticks some shorter than others, which grasp'd in the hand and all made even at the top, which was long, and which was short, was all guess; and chance so ordered it that in this simple Lottery, my lot was to lay next to M^r. Banks; 'tis easy to suppose that these who lay nearest the sides of the Tent would have the coldest berth, which was the cause of our deciding it in this manner.

This Anecdote is mentioned to prove how Amicable our little Society was, and that no distinction was made in regard to Superiority in point of Fortune, here was no Master and Slave, but a willingness to assist each other put us all upon an equality, except in point of Learning and Philosophical knowledge.[1]

[48b continued] in the morning we got up very early as the weather was fine and prepared ourselves for our Journey to the top of Mount Heckla, the prospect of which was tremendous and the Idea of the combustibles which were within and none but Nature! itself could tell when they might burst forth with all their destroying Violence, would have Intimidated the hearts of the boldest upon Earth, but [49] these considerations at present, had, no weight with our Philosophical Adventurer's; when we had got about half way we found our clothes very damp with the fogs and clouds by which we were surrounded, and which the wind Blew about in various directions, there was some snow, and as we advanced higher we found the cold more Intense, so that our Clothes began to freeze and become stiff and hard, we had tied our handkerchiefs over our heads and tuck'd the hair up to prevent the wind and wet makeing it troublesome, yet accident left some of my hair loose, which after being damp was now froze in such a manner that it became as stiff and as hard as wire, when we had got near three parts of the way to the top the appearance of Snow was lost and we became Sensible of a change from cold [50] to warmth, which in a short time increased very much, and when we arriv'd at the top it was so hot that it was with difficulty we cou'd hold the stones in our hands which we had picked up as Specimens, our clothes which were before so much frozen, were now thaw'd and became so wet that they were very uncomfortable; as we ascended the Mountain the wind was so high that we were often obliged to lie down for fear of being Blown into some of the dreadful Craters by its fury, and for all this precaution I did not escape the danger but was absolutely thrown into one of them, from which with a good deal of difficulty I got out, the inside being nothing but cinders my feet slipt for want of a more solid foundation, we[2] had now gained, as we imagined the Summit of the mountain, at least our guide who was not very fond of the journey would [51] have persuaded us so, and the clouds, by which we were surrounded, hindered us from knowing to the contrary, when suddenly they dividing a higher Summit appear'd which we determin'd to ascend; here was no Snow at all to be seen but the Ashes and Cinders was very wet; here we experienced at one, and the same time, a great degree of heat and cold, the Mountain beneath being so hot and the Air above so cold. Having finished our observations we thought it safest to descend with all convenient speed. The Mountain is chiefly composed of Grit, Ashes, Cinders and burnt Stones with some Pumice which was very light and

---

[1] The note on the left-hand page ends here.
[2] Here 'x 67' has been superimposed, with a pencil and probably not by Roberts, though what it is supposed to indicate is a puzzle. The manuscript is only 64 pages long.

sometimes contain'd Sulphur. The situation of the mountain is in the southern part of the Island, about four miles from the Sea, and is divided into three points at top the highest of which is the middle, and by an observation which was [52] made, five thousand feet higher than the Sea. Though many have attempted to ascend this wonderful mountain, yet we have the satisfaction to say are the first who ever compleated that Intention, which perhaps was facilitated by the last Eruption which has considerably leveled the Mountain to what it was before. Some of the runs of Lava are fifty or sixty Miles in length, the last Eruption of Mount Heckla was in seventeen hundred and sixty six. it began on the fourth of April and continued to the seventh of September.[1]

The morning after we left Mount Heckla we observed a great smoke on the top of it, and the weather became rainy, so that we were very fortunate in ascending it while it was fair. In our journey to and from Mount Heckla we sometimes slept in our Tent, sometimes in their houses, sometimes in their out houses where they [53] keep their fish, and sometimes in the Church's [sic] which is common to the few who travel this country. On the 28th at six o'clock in the Evening we returned to Hafnefiord.

Thursday 29th

Moderate Breezes and cloudy weather with a little rain. Several of us in the Evening went to a large Pool of water, about a mile from our habitation, in order to shoot Swans which are here in great plenty, but we had no success as they often get together in the middle of the water, which was too far for our guns to reach them, and when we did shoot any we seldom could get them away, the best chance we had was when they wou'd fly to another lake and then we sometimes shot one or two as they flew over our heads, they are very coarse Eating.

Wednesday 30th

Moderate Breezes and cloudy weather, we hauld the Seine this morning and caught a great number of very small Herrings.

[54] Oct. 1st.[2]

Fresh Breezes most part of the day and clear weather.

Friday 2nd

Light Airs and pleasant weather, to day began to make preparations for quitting Iceland, got the Cow on board, and some things from the houses.

Saturday 3rd

Fresh Breezes this morning with a little rain, in the afternoon the wind blew very hard with heavy showers of rain. M^r. Banks went to the Governor and Deputy Governor, and made them several presents.

Sunday 4th

Moderate Breezes and cloudy weather with some showers of rain, to day M^r. Banks, D^r. Solander &c., went to Bessastad and dined with the Stifts Ampman, about eight they return'd.

Monday 5th

Fresh Breezes and cloudy weather with some showers of rain.

---

[1] This eruption of Hekla indeed commenced in April 1766 but did not cease until March 1768, the longest-lasting in historical times. The lava covered 0.3 cubic miles (1.3 km^3), also a record.

[2] From here we only have Roberts's record of what happened until 16 October.

Tuesday 6th

Little wind and cloudy weather with some rain. To day the Ampman came to Mͬ. Banks [**55**] and brought a musical instrument¹ and other things the productions of Iceland as presents, and Mͬ. Banks made various presents in return.

Wednesday 7th

Moderate Breezes and cloudy weather with drizzling rain; to day we got all our Bedding and other necessaries on Board, and in the Evening went on Board ourselves, at Night Mͬ. Hartman, Mͬ. Neyborough and Captain Paterson² supped on Board our Vessel.

Before I take leave of Iceland it may not be amiss to say something of the Island and the Inhabitants in general and first

The Island in general

Iceland is reckoned among the largest Islands in the known world, it is very little favored by Nature, consisting chiefly /of/ Volcano's or Burning Mountains, of which Heckla is the principal; Boiling Wells, of which Geyser (before described) is the most astonishing, burnt stones, Grit, [**56**] and sand, very little Grass and no wood sufficient to be call'd Trees, the chief being Shrubs of Birch, the Mountains continually covered with Snow and every prospect dreary and horrible to look at. Iceland contains about two thousand people,³ with plenty of Cattle and Sheep, and the Sea supplies them with a variety of excellent Fish. They are Subject to the King of Denmark, their Religion is that call'd Lutheran, their manners are Simple and they are remarkable for their Honesty, their dress is uniformly the same among all ranks, excepting the quality of the stuff it is made of, that in general use is call'd <u>Wadmal</u>, and is Black, the same with the women as the men.

Excepting the Governor's and some other principal people's, the houses in general are very bad, being made of wood which is brought by the Sea (from what parts can only be guessed at) they are raised upon Lava much in the same /manner/ as we make our Stone walls for Enclosures, [**57**] with moss stuffed between, the Roofs are covered with Sods laid over Rafters, or sometimes the ribs of Whales, the walls are not above⁴ three yards high, and the entrance much lower, and 'tis nothing uncommon to see Sheep and horses feeding on the tops of the houses.

The principal Employment of the Icelanders is Fishing, Fowling and tending their Cattle. Their trade is prohibited to the Danish Company, but they sometimes privately deal with the Dutch, who serve them with a better Commodity. Literature has been cultivated here for many Centuries back, and they have produced many Eminent Poets, and Historians, and the art of printing has been established there ever since 1531 when one John Mathiesson printed the first Book in Iceland.⁵ This short sketch will serve to prove, that under so many disadvantages from Nature, that no Clime is excluded from

---

¹ This must have been a *langspil*, most resembling a violin. There is a description and illustration of the instrument in Mackenzie's *Travels in Iceland*, p. 147.

² All unknown, probably foreign merchants trading with Iceland.

³ This is a gross understatement. Economic historians have calculated that in 1772 the population was 48,623.

⁴ Superimposed on *above* is '*x 74*' written in pencil and probably not by Roberts, like the mark mentioned on p. 135, n 2. What this means is unclear.

⁵ It is strange that Roberts should choose to mention him. Jón Matthíasson (d. 1567) was a Swedish printer who came to Iceland with the last Catholic bishop of Iceland Jón Arason. This bishop brought the first printing press to Iceland which was first set up in his see Hólar, the bishopric in northern Iceland. This suggests that Roberts may have accompanied those going north to Hólar to collect books for Banks.

the use of those rational Faculties with which **[58]** the almighty has endowed the human race, and at the same time that a contented mind constitutes real happiness, which does not depend on Titles, Wealth, Grandeur and Luxurious Living.

Thursday 8th

This morning at four o'clock hove up the Anchor and came to sail with moderate breezes and pleasant weather, the *Vrow Christiana*[1] Saluted us with her three Guns, about Seven o'clock in the Evening we lost sight of the Land.

Friday 9th

This morning fresh Breezes and cloudy weather with a little rain.

Saturday 10th

Fresh Breezes and cloudy weather this morning, in the afternoon the wind increased with rain.

Sunday 11th

To day we had a hard Gale of Wind at S.W. and foggy weather with some Showers of rain.

Monday 12th

The wind continues to Blow very hard at S.W. and the weather cloudy with a little rain.

[59] [Tuesday] Oct. 13th.

Hard wind and foggy with heavy rain, in the Evening the wind was variable.

Wednesday 14th

Hard Gales this morning at N.W. by W. and cloudy weather, the wind more moderate in the afternoon.

Thursday 15th

Fresh Breezes at W. and cloudy weather at five o'clock this afternoon we saw Land on the Starboard Bow, distance about three Leagues.

Friday 16th

A fresh Breeze to day at S. and foggy weather, at nine o'clock this morning saw the Orkneys, and at three in the afternoon came to an Anchor at Stromness[2] in the Island of Pomona.[3]

Saturday 17th

Moderate Breezes with a little rain, this morning M^r. Banks D^r. Solander &c. &c. &c. went on Shore at Stromness, and dined at the Inn, we return'd on Board at half past Eight in the Evening.

[60] Sunday 18th

Little wind and pleasant weather, went this morning with M^r. Miller, M^r. Clevely and M^r. Walden to a place called Stenhouse[4] to assist them in measuring some

---

[1] This ship is mentioned earlier, see p. 130 above.

[2] Stromness is the second largest town in the Orkneys, a seaport with an excellent harbour accessible to ships of large burthen. Kelp was manufactured there at the time.

[3] Pomona is usually now called Mainland Island in Orkney, the largest of the Orkneys. There is a sketch in pencil of the island in BL, Add MS 15511, f. 1, and a watercolour, 'View of the Town of Stromness in the Island Pomona' by Cleveley (f. 2).

[4] Doubtless Stenness. The series of drawings of 'Stennis' is much more comprehensive than that of Skail (the Walden survey), suggesting some have been lost. They are in BL, Add MS 15511: 'A Plan of the Circle of Loda in the Parish of Stenhouse in the Island of Pomona with the country adjacent taken from an actual Survey by Fred. Herm. Walden.', f. 3; 'View of the Circle of Loda', a watercolour drawing, signed John Frederick Miller, 1775, based on his pen and ink sketch of the stones of Stennis, 1772, ff. 7, 8; 'View of a semicircle of Stones on the

Stones[1] which stand a little distance from each other in the form of a half moon, some of them were above twenty feet high, six feet wide and about a foot and a half thick, returned to Stromness in the afternoon. This Island is the principal among the Orkneys. it contains ten Parish Church's and is 16 by 27 miles,[2] the Orkneys, and most part of Sheatland belong to Sir Laurence Dundass,[3] the Harbour of Stromness is a very good one and will contain a great number of Shipping, and altho' so late in the Season there was no less than twenty two Trading Vessels lay here.

Monday 19th
This morning moderate Breezes with cloudy weather, fresh Breezes and foggy in the afternoon, to day M[r]. Banks and D[r]. Solander went in search of Plants and M[r]. Briscoe and myself [61] went a Shooting. M[r]. Banks &c. return'd about seven dined at the Inn and then went on Board.

Tuesday 20th
Fresh Breezes and foggy weather. M[r]. Banks &c. went to a place call'd Kirkwall[4] in search of curiosities, return'd in the Evening.

Wednesday 21st.
Moderate Breezes and Hazy weather, at four o'clock this morning M[r]. Banks, D[r]. Solander, with the rest of the Gentlemen, and Servants, and many of the Natives with Spades went to a place call'd Sandwick, where we open'd two Antient Tombs, or Tumuli, in each of them was found the Bones of a man, and woman, the form of their Interment was somewhat Singular; they were laid in a very coarse mat which was entirely rotten, the Bones of the woman were laid at the man's feet, the Tomb was form'd of Flag-stones, one on each side, one on each end, and one on the top, the other was the same with the addition of one at the Bottom. I measured one of the Tombs, it was four [62] feet Eight Inches long, two feet Eight Inches broad, and two feet four Inches in depth, the other was nearly the same; the man was laid with his knees almost up to his Chin, which perhaps was the custom of the times, for without his being Buried in this manner; the place could not have contain'd him, as he must have been about seven feet high in proportion to his thigh bone, which measured nineteen Inches, the Draughtsmen made Drawings of both the Tombs. In the Evening our party (which consisted of about thirty hands) return'd to Stromness we had with us our two French-horns and Saunders with his Violin,[5] so that we spent the day very agreeably.

Banks of Stenhouse Lake in the Island of Pomona'. John Cleveley jun. delin. 1772, f. 10; 'A Plan of the Links or Sandhills near Skail on the Island Pomona taken from an actual Survey by Fred. Herm. Walden', f. 11. See further Lysaght, 'Joseph Banks at Skara Brae and Stennis'.

[1] These were the stones of Stenness, a Neolithic site thought to be about 3,000 years old, 'one of the most extensive and complete Druidical relics in the country, consisting of a circle, nearly entire, of massive and lofty columns, beyond which are a semicircle, with several single stones irregularly placed, and numerous cairns' (Lewis, *A Topographical Dictionary of Scotland,* I, p. 432). There used to be slate quarries there.

[2] In fact, 14 miles by 19 miles, or 30.6 km by 22.5 km (Lewis, *A Topographical Dictionary of Scotland*, II, p. 380).

[3] Sir Lawrence Dundas (1710?–81) of Upleatham, York, 1st baronet (so created 1762). He was a Scottish landowner and politician.

[4] Kirkwall is the capital of Orkney: see the Banks journal, p. 110 above.

[5] Saunders is unidentified, probably a member of the crew.

Thursday 22nd.

Moderate Breezes and Hazy weather, to day M^r. Briscoe, the Gardener and myself went a shooting and had a very good Success, we kill'd a great number of Sea Fowl and Starlings. In the Evening bought several pair of white worsted Stockings which are here very fine and very cheap.

[63] [Friday] Oct: 23rd.

Moderate Breezes and cloudy weather; went on shore as usual and returnd on Board about Eight o'clock in the Evening.

Saturday 24th

This morning moderate Breezes with hazy weather. Fresh Breezes in the afternoon with heavy rain, at noon hove up the Anchor and turn'd the Vessel to windward out of the Harbour, at two came to an Anchor round the point, call'd the back of the Holms.[1]

Sunday 25th

To day we had little wind with hazy weather and drizzling rain.

Monday 26th

Hove up the Anchor at Six o'clock this morning and came to Sail with a fresh Breeze at N.E. and cloudy weather; about nine we cleared the Orkneys.

Tuesday 27th

Fresh Breezes this morning and cloudy weather, two Sail in Sight, moderate Breezes in the afternoon and clear weather, in the Evening the wind shifted to W.N.W.

[64] October 28th.

Little wind and clear pleasant weather, Entered the River Forth at Eight o'clock this morning, and at Six in the Evening came to an anchor in Leith Roads, at ten o'clock at night, excessive hard rain.

Thursday 29th

This morning M^r. Banks D^r. Solander D^r. Lind M^r. Troil and myself went on shore at Leith about two miles from where our Ship lay, from thence we walk'd to Edinburgh (which is likewise about two miles) and took a Lodging at M^{rs}. Thompson's in Riders Court.

Thus ended a Voyage to Iceland and other Islands the most Romantic and diversified with the wonders of Nature of any in the whole world, in the short space of three months and Eighteen days.

---

[1] There are numerous small holms in the Loch of Stenness, the largest lake on Mainland Island.

CALENDAR OF LETTERS AND DOCUMENTS

1.	From Andreas Holt to Claus Heide	24 April 1772	p. 152
2.	From Claus Heide	25 June 1772	p. 153
3.	Banks's Iceland Passport issued by Baron Diede von Fürstenstein	2 July 1772	p. 155
4.	To Lauritz Andreas Thodal	4 September 1772	p. 158
5.	From Bjarni Jónsson	22 September 1772	p. 159
6.	From Bjarni Jónsson	28 September 1772	p. 160
7.	From Ólafur Stephensen	2 October 1772	p. 162
8.	From Bjarni Pálsson	4 October 1772	p. 164
9.	From Teitur Jónsson	27 December 1772	p. 166
10.	To Thomas Falconer	12 January 1773	p. 168
11.	From Thomas Falconer	16 January 1773	p. 174
12.	To Thomas Falconer	2 April 1773	p. 176
13.	From Uno von Troil	14 April 1773	p. 179
14.	From Thomas Falconer	17 April 1773	p. 181
15.	From James Lind	26 May 1773	p. 182
16.	From Ólafur Stephensen	24 June 1773	p. 183
17.	From Uno von Troil	30 October 1773	p. 184
18.	From Thomas Falconer	8 December 1773	p. 185
19.	From Pieter Boddaert	20 April 1774	p. 187
20.	From Bjarni Jónsson	12 August 1775	p. 188
21.	From Ólafur Stephensen	25 August 1775	p. 191
22.	From Uno von Troil (with list of plants)	25 March 1776	p. 193
23.	From Bjarni Jónsson	12 August 1776	p. 197
24.	From Þorsteinn Jónsson	12 August 1777	p. 198
25.	From Ólafur Stephensen	20 August 1777	p. 198

26.	From Andreas Holt	30 November 1779	p. 200
27.	To Thomas Pennant	4 May 1783	p. 202
28.	From Jonas Dryander	24 October 1784	p. 203
29.	From Peter Anker	28 September 1785	p. 204
30.	From Peter Anker	2 December 1785	p. 206
31.	From Thomas Bugge	1 July 1786	p. 207
32.	From Uno von Troil	30 August 1786	p. 208
33.	From James Home	9 March 1789	p. 210
34.	From Dr William Wright with Catalogue of Plants	25 December 1789	p. 211
35.	To Grímur Thorkelin	9 April 1791	p. 215
36.	From Grímur Thorkelin	31 January 1792	p. 215
37.	From Uno von Troil	16 February 1799	p. 217
38.	Memorandum from Banks to Henry Dundas: Remarks concerning Iceland	30 January 1801	p. 218
39.	From Jörgen Jörgensen	21 October 1806	p. 224
40.	From Magnús Stephensen	17 October 1807	p. 226
41.	To Dr William Wright	21 November 1807	p. 229
42.	From Perrot Fenton	21 November 1807	p. 230
43.	From Dr William Wright	27 November 1807	p. 231
44.	From Dr William Wright	27 November [1807]	p. 233
45.	Banks: Project for Conduct of S.J.B.	[Late November 1807]	p. 233
46.	From Lord Hawkesbury	29 November 1807	p. 236
47.	From Dr William Wright	29 November 1807	p. 236
48.	From James Frederick Denovan	1 December 1807	p. 238
49.	From Gunnlaugur Halldórsson	2 December 1807	p. 239
50.	Banks: Questions to Icelandic merchants	[9? December 1807]	p. 240
51.	Banks: Answers given by the merchants	[9? December 1807]	p. 243
52.	Banks: Note on Icelanders	9 December [1807]	p. 245
53.	From Bjarni Sívertsen, Hans Georg Bredal, and Westy Petræus	15 December 1807	p. 245
54.	From Hans Thielsen	17 December 1807	p. 248

55.	From John Christopher Preidel	[19 December 1807]	p. 249
56.	From Westy Petræus	22 December 1807	p. 250
57.	To Alexander McLeay	24 December 1807	p. 251
58.	From Westy Petræus	24 December 1807	p. 252
59.	From Alexander McLeay	28 December [1807]	p. 253
60.	To Georg Wolff	28 December 1807	p. 253
61.	To Mr Stephensen	[Late December 1807]	p. 254
62.	Banks: Project for the annexation of Iceland	[Late December 1807]	p. 256
63.	To Lord Hawkesbury	30 December 1807	p. 257
64.	From Bjarni Sívertsen	31 December 1807	p. 265
65.	To Bjarni Sívertsen	1 January 1808	p. 266
66.	To Viscount Castlereagh	2 January 1808	p. 267
67.	Banks: Note re Sívertsen	3 January 1808	p. 267
68.	From John Christopher Preidel	4 January 1808	p. 268
69.	Preidel: Petition to the Lords of the Treasury	4 January 1808	p. 268
70.	From John Christopher Preidel	5 January 1808	p. 269
71.	From Adser Christian Knudsen	16 January 1808	p. 270
72.	From Westy Petræus, Bjarni Sívertsen and Hans Georg Bredal	5 February 1808	p. 271
73.	From Hans Thielsen	22 February 1808	p. 274
74.	To Lord Hawkesbury	24 February 1808	p. 275
75.	From Alexander McLeay	25 February 1808	p. 276
76.	To [Edward Cooke]	26 February 1808	p. 277
77.	From Edward Cooke	26 February 1808	p. 277
78.	From Bjarni Sívertsen	26 February 1808	p. 278
79.	From Bjarni Sívertsen	29 February 1808	p. 278
80.	To Edward Cooke	3 March 1808	p. 279
81.	From Boulton & Baker	7 April 1808	p. 280
82.	To Edward Cooke	7 April 1808	p. 281
83.	To Boulton & Baker	7 April 1808	p. 282
84.	From Edward Cooke	13 April [1808]	p. 283

85.	From Bjarni Sívertsen	18 April 1808	p. 283
86.	Banks: Note on licences	[April 1808?]	p. 285
87.	From John Christopher Preidel	20 April 1808	p. 285
88.	From Boulton & Baker	29 April 1808	p. 287
89.	From D. B. Baker	[29 April 1808]	p. 288
90.	Banks: Note on condemned ships	[29 April 1808]	p. 289
91.	From Boulton & Baker	16 May 1808	p. 289
92.	To Boulton & Baker	16 May [1808]	p. 290
93.	From Boulton & Baker	20 May 1808	p. 290
94.	To Boulton & Baker	20 May 1808	p. 291
95.	From Jörgen Jörgensen	27 May 1808	p. 291
96.	From Jörgen Jörgensen: Enclosure	27 May 1808	p. 292
97.	To Earl Bathurst	27 May 1808	p. 294
98.	From Hans Georg Bredal	31 May 1808	p. 297
99.	From Kjartan Ísfjörð	23 August 1808	p. 298
100.	William Wellesley-Pole to William Fawkener	7 January 1809	p. 300
101.	Markús Magnússon to Jörgen Jörgensen	[18 January 1809]	p. 301
102.	From Boulton & Baker	8 February 1809	p. 302
103.	Petition from Boulton & Baker to the King in Council	20 February 1809	p. 304
104.	From Boulton & Baker	22 February 1809	p. 304
105.	From Boulton & Baker	25 February 1809	p. 306
106.	From Bjarni Sívertsen	28 February 1809	p. 307
107.	From Boulton & Baker	27 March 1809	p. 308
108.	From Boulton & Baker	30 March 1809	p. 309
109.	Petition from Boulton & Baker to the King in Council	30 March 1809	p. 310
110.	From Jörgen Jörgensen	14 April 1809	p. 311
111.	Jörgensen: Memorandum on Iceland	[April–May 1809]	p. 311
112.	Banks: Notes on the voyage of the *Clarence*	[15? April 1809]	p. 318
113.	To Earl Bathurst	16 April 1809	p. 320

114.	From Samuel Phelps	21 April 1809	p. 322
115.	From Samuel Phelps	22 April 1809	p. 323
116.	From George Steuart Mackenzie	20 May 1809	p. 323
117.	To George Steuart Mackenzie	[20 May 1809]	p. 325
118.	To Ólafur Stephensen	28 May 1809	p. 325
119.	From John Barrow	2 June 1809	p. 326
120.	From Charles Konig	24 June 1809	p. 327
121.	From Ólafur Stephensen	10 August 1809	p. 329
122.	From Magnús Stephensen	25 August 1809	p. 330
123.	To William Jackson Hooker	1 October 1809	p. 334
124.	Memorandum from Count Trampe to Earl Bathurst	6 November 1809	p. 335
125.	Banks: Notes on Trampe memorandum to Lord Wellesley	[November 1809]	p. 357
126.	From Lorenz Holtermann	6 November 1809	p. 358
127.	From Boulton & Baker to Earl Bathurst	23 November 1809	p. 359
128.	From Jens Andreas Wulff	[November 1809]	p. 360
129.	From Jörgen Jörgensen	24 November 1809	p. 360
130.	From Bjarni Sívertsen	25 November 1809	p. 364
131.	From Samuel Phelps	29 November 1809	p. 364
132.	From Boulton & Baker	5 December 1809	p. 366
133.	From Jens Andreas Wulff	5 December 1809	p. 367
134.	To Lord Liverpool	11 December 1809	p. 368
135.	To [Culling Charles Smith], Foreign Office	13 December 1809	p. 371
136.	To Boulton & Baker	26 December [1809]	p. 372
137.	To Boulton & Baker	27 December [1809]	p. 373
138.	From Corbett, Borthwick & Co.	[undated, 1810]	p. 374
139.	To [Culling Charles Smith]	12 January 1810	p. 376
140.	Banks: Memorandum on Icelandic Revolution to Foreign Office [Marquess Wellesley]	12 January 1810	p. 377
141.	From Sir George Steuart Mackenzie	23 January 1810	p. 381
142.	To [Culling Charles Smith]	25 January [1810]	p. 382

143.	Banks: Note to Foreign Office	[1 February 1810]	p. 382
144.	Banks: Draft of Order in Council	[December 1809–February 1810]	p. 383
145.	From Kjartan Ísfjörð	10 February 1810	p. 385
146.	From Rooke & Horneman	10 February 1810	p. 387
147.	Statement from Henrik Christian Paus to Holger Peter Clausen	[spring? 1810]	p. 387
148.	Banks: Report on Jörgensen	2 April 1810	p. 388
149.	From Kjartan Ísfjörð	18 April 1810	p. 391
150.	To Count Trampe	23 April 1810	p. 391
151.	From Culling Charles Smith to [Marquess Wellesley]	23 April 1810	p. 393
152.	To [Culling Charles Smith]	24 April 1810	p. 393
153.	To [Edward Cooke]	18 May 1810	p. 394
154.	From George Steuart Mackenzie	20 May 1810	p. 395
155.	From Hans Frederik Horneman	30 May 1810	p. 396
156.	Banks: Observations on Jörgensen's narrative	9 June 1810	p. 397
157.	Banks: Note on Jörgensen's account	[9–15 June 1810]	p. 404
158.	Banks: Memorandum on Jörgensen's account	[June 1810]	p. 404
159.	To William Jackson Hooker	15 June [1810]	p. 405
160.	To William Jackson Hooker	16 June 1810	p. 407
161.	From William Jackson Hooker	20 June [1810]	p. 408
162.	To William Jackson Hooker	29 June 1810	p. 409
163.	From William Jackson Hooker	22 July 1810	p. 409
164.	To William Jackson Hooker	26 July [1810]	p. 411
165.	To Grímur Jónsson Thorkelín	26 July 1810	p. 411
166.	To [Culling Charles Smith], Foreign Office	26 July 1810	p. 412
167.	From William Jackson Hooker	27 July 1810	p. 413
168.	To [Culling Charles Smith]	28 July 1810	p. 415
169.	From Jens Andreas Wulff	30 July 1810	p. 416
170.	Corbett, Borthwick & Co.: Report on the *Rennthier*	3 September 1810	p. 417
171.	From Rooke & Horneman	6 September 1810	p. 418
172.	From John Christopher Preidel	11 September 1810	p. 420

173.	Petition from Hans Thielsen to the Lords of the Privy Council	September 1810	p. 422
174.	From Hans Thielsen to the Lords of the Board of Trade	11 October 1810	p. 423
175.	From Culling Charles Smith	15 October 1810	p. 424
176.	Banks: Memorandum on Mackenzie	[15? October 1810]	p. 424
177.	To Culling Charles Smith	[15] October 1810	p. 425
178.	Petition from Holger Peter Clausen to the Board of Trade	[29 November 1810]	p. 426
179.	Banks: Note on oil and fish	[Late November 1810]	p. 429
180.	From Holger Peter Clausen	5 December 1810	p. 429
181.	From Boulton & Baker	11 December 1810	p. 430
182.	From Jacob Nolsøe to Consul Wolff	12 December 181[0]	p. 431
183.	From L. Lobnitz and H. Hammershaimb to Captain Bohnitze	12 December 1810	p. 432
184.	From Hans Frederik Horneman	12 December 1810	p. 433
185.	From Hans Frederik Horneman	13 December 1810	p. 433
186.	From Corbett, Borthwick & Co. to the Lords of the Board of Trade	21 December 1810	p. 435
187.	From Holger Peter Clausen	24 December 1810	p. 435
188.	From Corbett, Borthwick & Co.	[25 December 1810]	p. 436
189.	From Corbett, Borthwick & Co.	1 January 1811	p. 440
190.	From Findlay, Bannatyne & Co.	28 January 1811	p. 441
191.	From Corbett, Borthwick & Co.	12 February 1811	p. 442
192.	From Georg Wolff to Jacob Nolsøe	27 February 1811	p. 443
193.	To William Jackson Hooker	1 April 1811	p. 445
194.	From Jacob Nolsøe to Georg Wolff	24 April 1811	p. 446
195.	From Hartvig Marcus Frisch, Friedrich Martini and Hans Jensen to Corbett, Borthwick & Co.	6 May 1811	p. 446
196.	From Henrik Henkel to Hans Frederik Horneman	13 May 1811	p. 448
197.	From Corbett, Borthwick & Co.	24 May 1811	p. 451
198.	From Findlay, Bannatyne & Co.	1 June 1811	p. 453
199.	From Corbett, Borthwick & Co.	10 June 1811	p. 453

200.	Petition from Corbett, Borthwick & Co. to the Lords of the Board of Trade	10 June 1811	p. 455
201.	From Geir Vídalín	9 July 1811	p. 456
202.	Banks: Draft of Instructions to the British Consul	[July 1811]	p. 457
203.	From John Parke	9 July 1811	p. 458
204.	From John Parke	13 July 1811	p. 460
205.	From Jacob Nolsøe to Georg Wolff	15 July 1811	p. 461
206.	From Corbett, Borthwick & Co.	19 July 1811	p. 462
207.	From Corbett, Borthwick & Co.	2 August 1811	p. 463
208.	From John Parke to the Marquis Wellesley	20 August 1811	p. 464
209.	From Corbett, Borthwick & Co.	21 August 1811	p. 466
210.	From Everth & Hilton	[August–September] 1811	p. 467
211.	From John Everth	3 September 1811	p. 467
212.	To William Jackson Hooker	2 October [1811]	p. 468
213.	From Corbett, Borthwick & Co.	5 October 1811	p. 469
214.	From Thomas Lack	15 October 1811	p. 471
215.	From Corbett, Borthwick & Co.	15 November 1811	p. 472
216.	From Captain William Henry Majendie	27 November 1811	p. 474
217.	To Captain William Henry Majendie	27? November 1811	p. 475
218.	Report of the *Gratierne*	2 December 1811	p. 475
219.	From Corbett, Borthwick & Co.	7 December 1811	p. 476
220.	Petition from Corbett, Borthwick & Co. to the Board of Trade	[7–11 December 1811]	p. 478
221.	To William Jackson Hooker	18 December 1811	p. 479
222.	From Holger Peter Clausen	20 December 1811	p. 479
223.	Petition from Holger Peter Clausen to the Privy Council	[20 December 1811]	p. 480
224.	From Holger Peter Clausen	26 December 1811	p. 480
225.	Statement from the Master of the *Freden*	26 December 1811	p. 481
226.	From Boulton & Baker	2 January 1812	p. 483
227.	Banks: Notes on Trade	4 January 1812	p. 483
228.	From Rooke & Horneman	16 January 1812	p. 484

229.	To William Jackson Hooker	10 February 1812	p. 485
230.	From Holger Peter Clausen	23 February 1812	p. 486
231.	From Kjartan Ísfjörð	28 February 1812	p. 488
232.	From Jacob Aall Jr	1 April 1812	p. 489
233.	To Jacob Aall Jr	[April 1812]	p. 490
234.	From Thomas Bugge	12 May 1812	p. 491
235.	From Count Christian Ditlev Reventlow to Count Johan C. T. Castenschiold	12 May 1812	p. 494
236.	To [Count Christian Ditlev Reventlow or Thomas Bugge]	May 1812	p. 495
237.	From Morten Wormskiold	5 June 1812	p. 495
238.	To William Jackson Hooker	1 July 1812	p. 498
239.	To William Jackson Hooker	13 July 1812	p. 499
240.	From Magnús Stephensen	8 August 1812	p. 499
241.	From Count Johan C. T. Castenschiold	22 August 1812	p. 505
242.	From Christian Ignatius Latrobe	4 September 1812	p. 506
243.	From Holger Peter Clausen	10 September 1812	p. 507
244.	From Christian Ignatius Latrobe	17 September 1812	p. 509
245.	To Christian Ignatius Latrobe	27 September [1812]	p. 510
246.	To [Count Johan C. T. Castenschiold]	[November 1812]	p. 511
247.	From Hartvig Marcus Frisch to Christian Latrobe	18 December 1812	p. 512
248.	Banks: Project for Displanting and Dispeopling of the Danish Colonies in Davies Strait	[1812]	p. 513
249.	To Magnús Stephensen	[Undated, December 1812]	p. 514
250.	From Corbett, Borthwick & Co.	29 January 1813	p. 516
251.	From Boulton & Baker	1 February 1813	p. 517
252.	List of Ships clearing for Iceland in 1812	13 February 1813	p. 517
253.	From Horne & Stackhouse	28 April 1813	p. 519
254.	From George Steuart Mackenzie	11 May 1813	p. 520
255.	From Christian Ignatius Latrobe	13 May 1813	p. 521
256.	From George Steuart Mackenzie	16 May 1813	p. 522
257.	From Corbett, Borthwick & Co.	22 May 1813	p. 523

258.	Banks: Memorandum, Notes on Iceland	[June 1813]	p. 526
259.	Banks: Draft re Memorandum	[June 1813]	p. 537
260.	From the Earl of Clancarty	6 June 1813	p. 539
261.	To the Earl of Clancarty	11 June 1813	p. 539
262.	From Jörgen Jörgensen	24 August 1813	p. 541
263.	From Hans Frederik Horneman	28 August 1813	p. 544
264.	From Jörgen Jörgensen	28 August 1813	p. 545
265.	From Hans Frederik Horneman	28 August 1813	p. 547
266.	From Jörgen Jörgensen	30 August 1813	p. 548
267.	From Hans Frederik Horneman	31 August 1813	p. 549
268.	From Jörgen Jörgensen	31 August 1813	p. 550
269.	To William Jackson Hooker	2 September 1813	p. 551
270.	From Jörgen Jörgensen	6 September 1813	p. 552
271.	From Hans Frederik Horneman	7 September 1813	p. 554
272.	From Jörgen Jörgensen	9 September 1813	p. 555
273.	From Jörgen Jörgensen	10 September [1813]	p. 556
274.	From Hans Frederik Horneman	21 September 1813	p. 557
275.	From Hans Frederik Horneman	30 September 1813	p. 560
276.	From Jörgen Jörgensen	6 October 1813	p. 562
277.	From Hans Frederik Horneman	26 October 1813	p. 563
278.	From Corbett, Borthwick & Co.	9 November 1813	p. 565
279.	From Holger Peter Clausen	12 November 1813	p. 567
280.	From Holger Peter Clausen	12 November 1813	p. 568
281.	From William Horne to John Thomas Stanley	[Early January 1814]	p. 570
282.	From John Thomas Stanley	13 January 1814	p. 572
283.	From Hans Frederik Horneman	25 January 1814	p. 573
284.	To Horne & Stackhouse	25 April 1814	p. 575
285.	From Holger Peter Clausen	28 June 1814	p. 575
286.	From Corbett, Borthwick & Co.	18 March 1815	p. 576
287.	Petition from Corbett, Borthwick & Co. to the Lords of the Board of Trade	18 March 1815	p. 577

288.	Circular from Rasmus Christian Rask	21 September 1815	p. 578
289.	From Rasmus Christian Rask	25 September 1815	p. 579
290.	From George Steuart Mackenzie	7 October 1815	p. 580
291.	From George Steuart Mackenzie	15 October 1815	p. 581
292.	Royal Ordinance of Frederik VI on liberty of trade to Iceland	11 September 1816	p. 583
293.	From Count Eduardo Romeo von Vargas-Bedemar	13 April 1817	p. 586
294.	From King Frederik VI	17 September 1817	p. 587
295.	From John Ross	25 July 1818	p. 589
296.	From Count Niels Rosenkrantz	9 September 1819	p. 590
297.	Ólafur Ólafsson: Poem in celebration of Banks	[1820]	p. 590

THE ICELAND CORRESPONDENCE AND DOCUMENTS OF SIR JOSEPH BANKS, 1772–1820

1. Andreas Holt to Mr. [Claus] Heide,[1] **24 April 1772**

Copenhagen

Sir!
At the same Time I have the Pleasure of saluting You and paying my Compliments for your past Favours, I also send you with Mr. Wolff[2] a Natural Curiosity from Iceland, consisting of two Samples of Wood, which in large Blocks is found pretty abundantly in the middle of Rocks and Mountains, especially near the Sea-Shores of the western Parts of that Island.[3]

This curious Phenomenon is without all Doubt produced by the Means of some extraordinary Revolution of the Sea, and I guess, I dare say with the greatest Probability, that this Wood is from /any/ some Parts of America, as the Quarter of Iceland, where it is most plentifully found, faces the new World, and I do not find, that any European Wood Bears Resemblance to it.

All original Mountains in Iceland are, as the Norwegian and German Miners term them, Flöte=Bjërge,[4] for I have no where observed any Gang=Bjerg,[5] and the Beds are distinctly separated, consisting of different kinds of Stones both in Regard to Colour and Texture. Below such Beds the mentioned Wood lays buried, but never many Fathoms above the Sea.

The westerly Quarter of Iceland abounds with many Firths, being divided by some Tracts of Land. The quantity of the /said/ subterranean Wood is so prodigious, that the

---

[1] After having decided to go to Iceland, Banks consulted Claus Heide, a Danish merchant resident in London, seeking information about what might be of interest to see in the island. Heide had recently received this letter from Andreas Holt, who visited Iceland in 1770 as the chairman of the Royal Commission of 1770 (*Landsnefndin fyrri*), which travelled around Iceland that summer investigating the economic situation.

[2] Georg Wolff, a Norwegian merchant in London. See Appendix 13.

[3] This is what is called in Icelandic *surtarbrandur*, a variety of lignite (brown coal), usually black. Mainly formed of tree trunks and branches, it is most frequently found in the tertiary formations in Iceland. The structures of the compressed and coalified trees are often preserved in *surtarbrandur*. All geological information in the notes has been supplied by Professor Leifur Símonarson of the University of Iceland.

[4] 'Flöte=Bjærge': the Tertiary basalt formations in Iceland are mainly composed of basaltic lava flows accumulated in flood eruptions. They were originally more or less horizontal, but during the Ice Age the flat lava pile was dissected mainly by glaciers and rivers and deep valleys were formed, usually separated by flat-topped mountains.

[5] 'Gang=Bjerg': dyke in English, in Icelandic *gangur*. It is a sheet-like body of volcanic rocks, which cuts across the bedding or structural planes of the host rock, frequently found in the Icelandic Tertiary formations.

Bed of it, which is seen in /near/ the Firth a, will sometimes be found near the Firth b, tho' parted at the Distance of some Leagues by the mountainous Tract c, Fig: A.
[*Figure drawn in left-hand margin:*[1]]

I shall add no more than to assure that I am always with due Respect and best Wishes,
Sir,
Your
most obedient humble Servant
[*Signed*] A. Holt

BL, Add MS 8094, ff. 31–32. Autograph letter.[2]

2. Claus Heide to Banks, 25 June 1772

Well Close Sq[r]:

Sir

I caled[3] to Day betwixt one & Two of the Clock at your House but had not the Pleasure to find you at Home – my Erand was to talk a Little more colesely [*sic*] of what was Proposed yesterday in so Short a discourse,[4] and having now Maturely considered the affair, which realy is of Concequence to you & the Doctor [Solander], whose design must be Laid & Executed in a Safe manner. I can not but think it the best way to pursue my first advice Viz[t] to Write to Copenhagen, before Setting out, and in 3 Weeks time

---

[1] Sketch diagram in letter from Andreas Holt to Claus Heide, dated 24 April 1772. BL, Add MS 8094, ff. 31–32. Courtesy of the Trustees of the British Library.

[2] Note in Banks's autograph, verso:

Copenhagen – 1772	from Andr. Holt
Andreas Holt –	to M[r]. Heide
Dated – 24 April	about Sorte brandt
rec*eiv*ed – 15 May	
answered – [*blank*]	

[3] His English is poor. There are numerous spelling mistakes in this letter.

[4] Doubtless concerning the feasibility of Banks leading an expedition to Iceland. Banks had withdrawn from the *Resolution* expedition to the South Seas at the end of May 1772. See Introduction, pp. 6–7 above.

Proper Pasports[1] &c. – may be had according to all Probability, nor can you be ready in a Less time. I have not been able to meet with any Person or Procured better Information than before which is Chiefly out of books[2] but if I remember right was told Some time ago that forreign Ships must not fish ab*ou*t Ice Land at a Less Distance than 4 Danish miles[3] or ab*ou*t 20 English, neither Trade with the Inhabitants nor come into Harbours without great Necessity, the Ship & goods being otherwise Liable to be Confiscated.[4] There was a few Dutchmen[5] taken a good many years ago which proved good Prize to some mann of Warr Cruisers fitted out for that purpose but have not heard Lately /*heard*/ any Such things done. I must therefore earnestly beg your serious consideration, to what seems to me to /be/ best, which is that I shall Write to morrow night to one of my particular Friends the most Leading Director of the Iceland Comp*an*y at Copenhagen[6] /for a License[7] of theirs of higher Powers/ and in Setting your affair in the most advantageous Light Possible, my very good Corespondent Andreas Holt[8] who was likewise known to Doctor Solander when here in England 1766, returned the Latter end of Last year from Iceland, where he had been Sent about Some thing of the Company's Trade, is a very Proper Person to Consult by me in Writing, the most Proper places for you to go to Vieuw Mount Heckla,[9] or what place is burning at present to give the names of People of the greatest note, likewise Letters of recomendation, and also to Supply you with money or other things needfull for which I Promise to be accountable, without Such Certainties to proceed upon you run too great a risque, for if as you proposed yesterday, the needfull things Should be Sent to Iceland from Copenhagen you have not Concluded what place most proper to go to, nor is there Any posts in that Country whose Extent is about as great as England & Scotland put together,[10] and besides we don't know wither or not all the Company's Ships may by this time /be/ gone for Iceland – I should be glad to know to morrow morning ab*ou*t your aprobation or the Contrary to my proposal, and altho imperfect in Words, yet belive my meaning may

---

[1] See following document.

[2] The list of books that Banks took with him to Iceland is in Appendix 4.

[3] This is not an exact equivalent.

[4] Heide is referring to the ordinance regarding the Iceland trade 'Forordning ang. den islandske Handel og Skibsfart' of 13 June 1787 (see Appendix 8), though there is no mention of the 4 Danish miles.

[5] The Dutch carried on an illicit trade in Iceland throughout the 17th and 18th centuries.

[6] During the period 1764–74 the trade monopoly was in the hands of the General Trading Company (Almenna verzlunarfélagið), one of the largest trading companies in 18th-century Copenhagen with interests in the Baltic, Mediterranean and the North Atlantic and a triangular trade with Africa and St Croix (Gunnarsson, *Monopoly Trade*, p. 120). An important shareholder was the prominent merchant Niels Ryberg (1725–1804), who is probably the gentleman referred to. In 1767 Ryberg had founded a transit depot in Tórshavn, in the Faroe Islands, Ryberg's Handel. In fact it was used for smuggling with goods coming from America and Bergen and continuing to Scotland and Ireland. This ended in 1788 (West, *Faroe*, pp. 45–8). Heide was Ryberg's major contact and collaborator in London. See further Rasch, *Niels Ryberg*.

[7] All ships sailing to Iceland needed licences or *söpasser* as they were called. See Copenhagen, RA, Rentekammer [hereafter Rtk] 373.121, 'Protokol over udstedte islandske, algeirske, færöiske og grönlandske söpas 1787–1841'.

[8] Andreas Holt (see Appendix 13), having been in Iceland in 1770, would thus have been an ideal person to give Banks information on the current state of the island.

[9] In the Middle Ages Hekla was believed to be the entrance to Hell. The volcano had last erupted in 1766.

[10] Iceland is about 39,768.5 sq. miles in area, Scotland 30,394 and England 50,328.4 (about 103,000, 78,720 and 130,350 km^2, respectively). Thus, this is a gross exaggeration.

be understood by you & the Doctor for whome I have a due Esteem, and on all occasions am,

> Sir, your most humble,
> Servant
> [*Signed*] Claus Heide

BL, Add MS 8094, ff. 29–30. Autograph letter.[1]
Hermannsson, 'Banks and Iceland', pp. 5–6.

3. Passport Issued to Banks by Baron Diede von Fürstenstein, 2 July 1772

Passport issued by Baron Diede de Fürstenstein, Envoy Extraordinary of His Danish Majesty to the British Court to a party of learned gentlemen, who are going to visit the Islands of Iceland and the Faroes, to make observations there in astronomy, botany and other aspects of natural history.[2]

> 2 July 1772
>
> London

William Christopher Diede, a Freeman of the Holy Roman Empire, Baron of Fürstenstein, a Gold-decorated [*auratus*] Knight of the Danish Order, a Gentleman-in-Waiting of the first rank of the King of Denmark and Norway, and now Ambassador Extraordinary of this most sacred Majesty to the British Royal Court:
To one and all who read this letter of mine I make known and witnessed, that I, by virtue of the authority delegated to me in this Royal Court have granted protection with my safe-conduct (laisse-passer) and commendation the very distinguished men Joseph Banks, an English Esquire and Lord of the Manor of Revesby, and Daniel Solander, a Swedish doctor of medicine and of the Laws, being, because of their recent voyage to the Antarctic Pole,[3] already very famous everywhere, but especially in learned society, and now considering also a voyage towards the Arctic regions, including the coast of Iceland, for the purpose of observing Mount Hekla and the Faroe Islands. Accordingly, I demand of each and all officers of every nation, especially of Denmark, in charge of fleets, ships, harbours and forts, magistrates of counties and towns, and reverend functionaries of the church, that, with the kindness with which it is right to approach each one of them in accordance to his rank and condition, they should grant (I ask) these men and also their new companions on this new voyage, whom I shall enumerate by name afterwards, together with the twelve servants of all these who are voyaging and with their provisions,

---

[1] Note in Banks's autograph, verso: 'Mr. Heide.'

[2] Roy Rauschenberg translated most of the passport in 'Iceland Journals', pp. 217–18. This is a new translation by Charles Burnett. This first part is in French and is not included in Rauschenberg's transcription. What follows is in Latin. Someone (probably a Dane, judging by the handwriting) has translated the occupations of Banks's party into Danish in Gothic handwriting (e.g. *Skriferer*: secretaries, *Kammertienere*: valets, *UrteGaards=Dreng*: kitchen gardener).

Banks was well on his way to Iceland before the formalities had been concluded, leaving Gravesend on 12 July. On 11 July the Danish Chancellery (*Kancelli*) in Copenhagen had written to Baron Diede that, as the Iceland ships had left, they were sending the originals of orders and recommendations from the King of Denmark to his officials in Iceland, to England, hoping that the ambassador could send them on to Banks before the expedition left the Scottish Isles (RA, Departmentet for de Udenlandske Anliggender [hereafter DfdUA] 892.

[3] This is obviously a misunderstanding.

not only a safe passage by land and sea, sojourn, travel and return, but also that they should attend them with all good will and where necessary assist them as obligingly as possible. I promise in this letter, by equal obligations of humanity and friendship, that I in turn shall help those who, having been similarly recommended by these same officials and magistrates, cross my path. I have ordered this letter, written by my own right hand, to be confirmed by my personal seal, in order to provide surety for all this.

Issued in London, the 2nd day of July, 1772
William Christopher Diede of Fürstenstein.

List of all the companions[1] of the nobleman Joseph Banks, Esquire, at whose expense the ship called 'Lawrence' has been built and is to be navigated by the Shipmaster, John Hunter.[2]

Daniel Solander, Doctor of Medicine and of the Laws and member of the learned societies of Uppsala, London and Paris.[3]
John Gore, Esquire,
John Riddel, Esquire,
James Lind, Doctor of Medicine, of Edinburgh.
Uno von Troil, Doctor of Philosophy.
Frederick Hermann Walden.
and Sigismund Bacström, } Secretaries. *Skrifere.*
John Frederick Miller
James Miller
John Clevely } Artists. *Teignere.*
James Hay, Astronomer.

* * * * * *

The servants of various kinds of the aforementioned
Alexander Scott,
Peter Briscoe, } Valets. *Kammertienere.*
James Roberts
John Asquith,
Peter Sidserf,
Alexander Samarang
Nicholas Young } Servants, *Laquayer.*
John Marchant.
Robert Holbrook.
John Taylor.

James Donaldson.[4] Kitchen-gardener,[5] *Urtegaards=Dreng.*
Antoine Douvez. Cook. *Kokken.*
William Christopher Diede of Fürstenstein

---

[1] For short biographies of Banks's entourage, see Appendix 1.

[2] The name should be James Hunter.

[3] The learned societies: Kungliga Vetenskaps Societeten in Uppsala, The Royal Society in London and Académie des Sciences in Paris. Rauschenberg believed the Latin word for Paris possibly to refer to Lund in Sweden ('Lund?') ('Iceland Journals', p. 218), but *Lutetiensis*, as it says in the Latin passport, is the genitive of *Lutetia* – Paris.

[4] James Moreland was the gardener, see Banks's journal, p. 48.

[5] In her fair copy Sarah Sophia Banks has added: '(provider of vegetables)'.

Canberra, National Library of Australia, Banks Papers, MS 9/118.[1] Possibly original.
McGill University, Blacker-Wood Library, MS B 36, 1772, ff. 89–93 (appendix to The Banks Iceland Journal). Copy.
Kent History & Library Centre, U1590/S1/2. This is a fair copy made by Sarah Sophia Banks, titled an appendix (as above), 5 pages after f. 84.[2]

French original:
Passeport
donné par Le Baron Diede de Fürstenstein, L'Envoyé Extr*ordinair*e de SA MAJESTÉ DANOISE à la Cour Britanique, à une Association de Scavans, qui vont faire un Tour aux Isles d'Islande & de Ferroe, pour y faire des Observations Astronomiques, Botaniques, & autres relatives à L'histoire Naturelle.
Le 2 Juillet 1772

Latin original:
Guielhelmus Christophorus Diede, Sacri Romani Imperii Liber Baro in Fürstenstein, Ordinis Dannebrogici Eques auratus, Unus e Cubiculariis primi Ordinis REGIS DANIÆ & NORWEGIÆ, & jam dictæ Sacratissimæ Majestatis nunc Temporis ad Aulam Brittannicam Ablegatus Extraordinarius, Omnibus, & Singulis hasce meas Litteras inspecturis Notum Testatumque facio, Me, vi delegati mihi in hac Aulâ Ministerii, Clarissimos Viros, Josephum Banks, Armigerum Anglum, & in Reversby [*sic*] Dominum, & Danielem Solandrum, Suevum, Medicæ ut & Legum Doctorem, novissimâ versus Polum Antarcticum Navigatione jam ubique, præsertim in republica Litterariâ, celebratissimos, nunc autem etiam versus Arcticum Vela Facere &, inter alia Littora, Islandica, montis Heclæ conspiciendi gratiâ, ut et Ferroensium insularum, visitare meditantes, Salvo Conductu & Commendatione Meis muniisse. Proinde ab Omnibus & Singulis, Cujuscunque Gentis, præsertim Danicae, Classium, Navium, Portuum, & Fortalitiorum Præfectis, Dictionum & Oppidorum Magistratibus, necnon Reverendis ecclesiastico Munere fungentibus, eâ Comitate quâ Eorum quemque, pro Status & Conditionis ratione, adire par est, id rogo: Ut Ipsis, & porro Eorum, posthæc nominatim enumerandis, novis novi itineris sociis, cum Duodecim omnium horum itinerantium Famulis, & cum Sarcinis, non solum, terrâ Marique, tutum Iter, Moram, Transitum, Reditumque concedant, sed & Eos omni Favore, & ubi opus fuerit, auxilio, quam amicissime prosequantur.

Paribus Humanitatis & Amicitiæ Officiis Me vicissim iis præsto fore, qui, ab iisdem Præfectis & Magistratibus similiter commendati, mihi obvii fuerint, hisce Litteris spondeo. Quas in horum Omnium fidem, Manu meâ propriâ subscriptas, simul sigillo meo Gentilitio firmari jussi.
    Dabantur Londini, Die 2da Julii 1772
    Guilhelmus Christophorus Diede
        in Fürstenstein

    Designatio
omnium Comitum Generosi Josephi Banks, Armigeri, cujus Sumptibus Navis, Laurentius dicta, & per Navarcham Johannem Hunter vehenda, instructa est.

---

[1] Verso, note: 'Passport Given by Baron Diede of | Furstenstein to Sir Joseph Banks & others | and observations on Iceland.' These observations are a 'Journal of the Barometer, Thermometer, and direction of the winds at Havnefiord in Iceland, & at Hawkill near Edinburgh, at eight o'clock in the morning for the month of September, 1772'; 'Account of the Barometer on a Journey to and from Mount Hecla and its difference with on*e* kept at Havnefiord [*blank*] feet above the Sea at high water'. There is also a list of the latitude of various places visited in the Western Isles and one page in Danish describing the route from Skálholt to Hekla and back.

[2] Sarah Sophia wrote as a note : 'I fear in copying this Passport there will be unavoidably several mistakes.'

Daniel Solander Medicinae itidemque Legum Doctor, & Membrum Societatum eruditarum Upsaliensis, Londinensis, & Lutetiensis.
Johannes Gore, Armiger,
Johannes Riddel, Armiger,
Jacobus Lind, Medicinæ Doctor, Edenburgensis.
Unno de Troil, Doctor Philosophiæ.
Fridericus Hermannus Walden ⎫
Sigismundus Backstroem ⎭ ab Epistolis. *Skrifere.*
Johannes Fridericus Miller ⎫
Jacobus Miller ⎬ Delineatores, *Teignere.*
Johannes Clevely ⎭
Jacobus Hay, Astronomus.

* * * * * *

Omnium supradictorum Famuli varii generis.

Alexander Scott, ⎫
Petrus Briscoe, ⎭ Vestimentorum Custodes. *Kammertienere.*
Jacobus Roberts, ⎫
Johannes Asquith, ⎪
Petrus Sidserf, ⎪
Alexander Samarang ⎬ Asseclæ qui sunt a pedibus, *Laquayer.*
Nicolaus Young ⎪
Johannes Marchant. ⎪
Robert Holbrook. ⎪
Johannes Taylor. ⎭
Jacobus Donaldson. Olitor, *ÚrteGaards=Dreng.*
Antonius Douvez. Coquus. *Kokken.*

Guilhelmus Christophorus
Diede in Fürstenstein   (Seal)

4. Invitation from Banks to Lauritz Andreas Thodal, 4 September 1772.[1]

Mr. Banks presents his compliments to Governor Thodal[2] and requests the honour of the company of the governor and both young ladies[3] at dinner in the Hafnarfjörður merchant house[4] next Sunday.[5]

---

[1] Dated from the Journal of James Roberts. See Friday 4 September 1772, p. 129 above, where he wrote: 'This morning Mr. Banks made a present to the Governor, and Deputy Governor of Six hams and Six dozen of Porter each, in the afternoon they both paid him a visit.' I am indebted to Professor Harald Gustafsson of the University of Lund for help with this letter.

[2] See further on Thodal in the Banks journal, p. 85, n. 1 above.

[3] Shortly after his arrival in Iceland Thodal had lost his wife Anna Helene Klow (née Valentinsen), the widow of a Norwegian *stiftamtsskriver* (secretary to the governor) in Bergen named Klow (d. 1768). Thodal and Widow Klow had married in the spring of 1770 shortly before setting off to Iceland. At this point Thodal had a stepdaughter and stepson living with him. Miss Klow would then have been about 15–16 years of age. In the original Swedish of this letter they are spoken of as *Fröknarnes,* young ladies. Presumably this might be based on a misunderstanding, two stepchildren instead of two young ladies. She eventually fell in love with Carl Pontoppidan, at the time a merchant engaged in the Iceland trade, but Thodal opposed the match and she died aged 30.

[4] Banks and his entourage had established their quarters on land and this warehouse had been opened up especially for him for this purpose. See Plates 2 and 3.   [5] 6 September 1772.

Mr. Banks also bids the governor accept a little English porter[1] and a few English hams.

NHM, Solander MS, *Plantae Islandicae et notulae itinerarie etc.* Autograph letter by Solander. Translated from the Swedish.

Swedish original:
Herr Banks presenterar Sin vördsamme hälsning till H' Stiftsamtman Thodal och beder at få den äran af H' Stiftsmans och bägge Fröknarnes Sällskap at spisa middag i Hafnefjords kjöpmans hus, nästkommande Söndag.

H' Banks också ber H' Stiftsamtman hålla till godo litet ängelsk Porter och få ängelska Skinkor.

5. Bjarni Jónsson to Banks, 22 September 1772

Skálholt

Song of Joy (*Glededans*)

A song of joy[2] performed by the Muses in Skálholt on the occasion of the visit of the venerable Sir Joseph Banks who travelled from England to Iceland to study and observe the natural wonders of this poor Island accompanied by a very illustrious and learned party of one naturalist, that is to say the famous and scholarly Doctor D. Solander, one astronomer, one antiquarian, three artists, two secretaries, one captain of the navy and one lieutenant and offered with honour and due respect to the English gentlemen and his learned followers.

Skálholt, 1772 anno Domini 22 September by Bjarni Jónsson Rector of the Skálholt School.

Maidstone, Kent History and Library Centre, Banks MS U951 Z32/19, ff. 21–25.[3] Translated from the Icelandic.

Icelandic original (also extant in a Latin version):
Glededans slegen Af Saunggidiunum ad Skálhollte i Tilkomu Tignarlegs Manns Hr: Josephs Banks Vopnbera hvor ed reiste frá Englandi til Islands til ad ransaka og adgiæta markverda hluti Þessa fátæka Eylands í Ríki Nátturunnar ásamt med yfred frídu og velmentudu Foruneiti sem samanstód Af einum Natturu Speking, nefnilega þeim Nafnfræga og hálærda manni Doctor D: Solander, Einum Stjörnu Meistara, einum Fornfrædameistara, þremur Málurum, Tveimur Skrifurum, einum Skips Capteine og einum Lieutenant Til heidurs og Skilldugrar Virðingar Fyrir þann Enska herra og hans lærdu Filgiara framborenn.

Að Skalhollte, árum epter Gudsburd 1772 þann 22 Septembris af Bjarna Jónssyni Skólameistara ad Skalhollti.

[1] A strong beer.
[2] This is the 'Tripudium', 10 pages of Latin text with notes, 25 verses in all. It is called in Danish a *Gratulations Vers* and was printed in Hrappsey, Iceland, in 1773 in the first edition of the Danish-language journal *Islandske Maaneds-Tidender*, pp. 2–6, in both Icelandic and Latin. It was also printed in Latin in Hooker, *A Tour in Iceland*, II, pp. 279–92 with notes in Latin.
[3] Note in Banks's autograph, verso: 'Carmen gratulatorium | Rectoris' [Trans.: Congratulatory poem from the rector]. (I am indebted to Dr. Þórunn Sigurðardóttir for help with this document.)

6. Bjarni Jónsson to Banks, 28 September 1772

Skálholt

Most celebrated man!

Since you have visited me with your learned companions, I offer you the greatest thanks. Now I send to you, most cultured man, a Congratulory Song,[1] a little more correct and larger than before. Indeed this is a very meagre little present, and falls far below your merit. But nevertheless, may you accept it in a generous spirit, and interpret this [gift] favourably, reflecting rather upon the spirit of the giver than on the meanness of the gift; the more evident tokens of your refinement I have, the more certain is the hope I cherish in this matter. I would like you to allow it to be published when, with God your guide, you return to England.[2] Also, an English translation could be added, if it pleases, so that your good actions towards our people may shine out more brilliantly, and in this way be spread abroad among the multitude.

I would like to know how your journey to Hekla turned out; I hope that it met your expectations.

Allow me, most cultured man, to explain to you the situation of my family in a few words. I have now held the office of Rector in the school at Skalholt for 20 years, being almost fifty years old. From my wife (whom you saw with my three daughters in my Museum), I have had 14 children, of which five only have survived, two sons and three daughters, of which the youngest is now three years old.[3]

One of my sons, Paul Bjarnsen,[4] has spent almost three years studying the liberal arts in the Copenhagen Academy. He has a reasonably sharp intelligence and one naturally suited to letters, but he is among those whose progress is hindered by straightened [straitened] circumstances at home. Up to now I have borne all the costs of his studies, as far as I could. But since the purse has now been completely emptied, I have to call him back home, although against my will. For, my annual salary of 60 Imperials[5] is scarcely sufficient for my domestic needs, let alone expenses of this magnitude. Since I am not able any more to support his weakness from my straightened means, I flee to you, most

---

[1] It is mentioned in the first copy of the *Islandske Maaneds-Tidender* that Bjarni Jónsson had composed *nogle* meaning 'some' odes (*Gratulations vers*). Among the Banks papers in The Kent History and Library Centre, U951 Z32, there are several odes (most published by Hooker, *A Tour in Iceland*, vol. II): This includes 'Pro felici in Islandiam itinere et in Patriam reditu Magnatum Britannorum, Anno MDCCLXXII. Votum' (*Votum* has been pencilled in and is in Banks's autograph), and 'Poeses Islandorum' by P. Jacobæus (Páll Jakobsson, see the Banks journal, p. 102, n. 1 above), the condean (*konrector*) of the Skálholt school (Z32/21, ff. 1–4, not published in Hooker).

[2] This must be the Hekla ode to Banks and Solander: 'Heklæ Vale Anglis Heroibus', Banks MS U951 Z32/20 (published with notes in Hooker, *A Tour in Iceland*, II, pp. 273–6).

[3] His wife was Helga Sigurðardóttir and his three surviving daughters were: Ingibjörg (b. 1761), married to the Revd Þórhallur Magnússon; Ragnhildur (1759–1819), married to the Revd Illugi Hannesson, and Helga, who died unmarried in 1800. The sons were Páll, mentioned in the letter, and the Revd Helgi Jónsson (1757–1816).

[4] Páll Bjarnason, his eldest son (b. 1750), simply called *stúdent* in Icelandic sources which means he had completed the entrance examination for university. Little did Bjarni Jónsson know when he wrote this letter that his son would die of measles in Copenhagen in 1772 (buried 5 November) while pursuing further studies.

[5] An *imperial* was never a type of currency in the Danish state (though it was in Milan in medieval times and in 18th-century Russia). The rector probably had no idea of the Latin term for the *rigsdaler* (rixdollar) and found this a fitting substitute. The salary of the rector at this time was 60 rixdollars, plus bedding, candles, board

cultured man, asking you most humbly, whether you would consider this poor, virtuous young man as your own son, and help his lack of means both materially and with good counsel in such a way that he can successfully finish weaving the web of his studies that he has begun.

I confess that this request is bold, since I have deserved no good thing from you, nor could I deserve it. But the hard thrusts of poverty order one to try everything, and the reputation that has reached my ears of your singular kindness and generosity, and especially the opinion which I formed of you at our first meeting, persuade me to exceed, on this one occasion, the bounds of shame, with this impertinent request; for which, I pray, most cultured man, that you will forgive me.

If you provide any room for my prayers, and deign to give my son your favour and patronage, you are doing something worthy of your name, and most pleasing for me.

But I have nothing to give in return; nothing by which I can declare my heartfelt thankfulness. But if you would like to make use of my labour, either in procuring histories of Iceland or in translating them into Latin, you will find me ready and prepared to take up this kind of duty. Nor will you find my son unuseful for you in this office. But I fear lest I am making you sick and tired by this untimely prolixity; for this I beg your pardon over and over again.

May God return you safe and sound to your home, and favour all your glorious endeavours!

Farewell, most cultured man, and, mindful of us, may you live happily for a long time.

With greatest esteem
[Signed] Bjarni Jónsson

Maidstone, Kent History and Library Centre, Banks MS U951 Z32/25, ff. 37–38.[1]
Part of this letter is published in Latin in Hooker, *A Tour in Iceland*, II, pp. 277–8.
Translated from the Latin.

Latin original:
Vir Celeberrime!
Quod me cum erudito tuo comitatu invisisti, grates ago quam maximas. Mitto tibi jam, Vir humanissime, Carmen Gratulatorium, paullo correctius auctiusque quam antea. Nimis quidem exiguum hoc est munusculum, longeque Tuam infra dignitatem positum, sed velis nihilominus benigna id suscipere fronte, inque meliorem partem interpretari, animum potius datoris quam doni vilitatem respiciens; qua de re eo certiorem spem foveo, quo evidentiora humanitatis Tuæ habeo indicia. Velim id typis vulgari permittas, in Angliam, Deo duce, cum redieris. Adjici etiam posset Versio Anglica, si ita visum fuerit, ut vestra in gentem nostram merita eo clariora evadant, atque hoc pacto in vulgus emanent. Iter vestrum ad Heklam quomodo cesserit scire gestio. Utinam bene et ex animi sententia! Permitte mihi, Humanissime vir, ut rerum mearum statum paucis Tibi expediam. Rectoris munere in Schola Schalholtensi per annos xx jam functus sam, fere quinquagenarius. Ex uxore (quam cum tribus filiabus in Museo meo vidisti) 14 liberos suscepi, ex quibus 5 tantum superstites sunt, filii scilicet duo, tresque filiæ, quarum natu minima nunc trimula.

Filiorum alter nomine Paulus Biarnsen elapsum proxime triennium in Academia Havniensi bonis artibus operam impendet. Habet is ingenium satis acre et ad literas aptum natum, sed in eorum est numero, quorum progressibus obstat res angusta domi. Sum[p]tum quidem omnem in studia ejus,

---

[1] Note in Banks's autograph, verso: 'Epistola Rectoris | Scalholtiæ.' [Trans: Letter from the rector of Skálholt School.]

pro virium modulo, hactenus feci. Sed exenterato jam prorsus marsupio, domum eum, quantumvis invitus, necesse habeo revocare. Annuum enim 60 imperialium salarium vix domesticis usibus, nedum tantis impensis sufficit. Cumque tenuitatem ejus ex meis angustiis ulterius sustentare nequeam, ad te confugio, Vir humanissime, humillime rogans: velis pauperem hunc et probum adolescentem pro Tuo agnoscere, ejusque inopiam re et consilio ita adjuvare, ut inceptam studiorum telam feliciter pertaxere queat.

Fateor equidem temerariam hanc esse petitionem, cum nihil boni a Te promeruerim, nec promereri possim. Sed durum paupertatis telum omnia jubet experiri, eaque quæ de singulari tua humanitate et munificentia ad meas aures pervenit fama, præsertim vero, quam de Te ad primum aspectum concepi opinio, suadent, ut importuno hoc petito verecundiæ limites semel excedam, quam Tu mihi, Humanissime Vir, culpam, precor, ignoscas.

Si his meis precibus locum dederis, filiumque meum favoris tui et patrocinii participem reddere dignatus fueris, rem facis tuo nomine dignam, mihi vero gratissimam.

Sed nihil habeo, quo vicem reddam, nihil, quo animi gratitudinem declarem. Quod si forte opera mea uti placeat, vel in conquirendis Historiis Islandicis vel in Latinam linguam transferendis, ad id officii genus prom[p]tum me et paratum invenies. Nec filius meus hac in parte tibi erit inutilis.

Sed vereor, ne intempestiva hac prolixitate tædium Tibi et nauseam attulerim, quam Tu mihi veniam des, etiam atque etiam rogo.

Deus Te salvum et sospitem patriæ reddat, omnesque tuos gloriosos conatus secundet. Vale, Vir Humanissime, nostrique memor felix diu vive.

Tui Nominis observantissimus Cultor
Biarnus Johneus
Skalholti iv Calend; Octobris

7. Ólafur Stephensen to Banks, 2 October 1772

Sviðholt[1]

List
Of the Icelandic minerals delivered to Mr. Banks in 1772

No. 1. Black agate (*hrafntinna*[2] in Icelandic) found on a mountain opposite the [former] monastery at Reynistaður, within Hegranes County[3] in a place covered with stones and soil.[4]

No. 2. White agate[5] (*Glerhallur* in Icelandic) found in Glerhallavík at the foot of the mountain Tindastóll in the same county. The sea casts this mineral upon land and it can be found in other places by the coast, though Glerhallavík offers the best examples.

No. 3. Another kind of the same white agate, with yellowish markings,[6] found on the beach of Hvalfjörður near the coast of Borgarfjarðarsýsla.

---

[1] Sviðholt is on the Álftanes peninsula south of Reykjavík.

[2] Black agate: Obsidian, a black, wholly glassy acid, volcanic rock (rhyolite). Stephensen uses Roman letters for the Icelandic words, otherwise Gothic script.

[3] Today Hegranes County is Skagafjarðarsýsla in the north of Iceland.

[4] In Danish, *jordsvær* is a term for soil or peat.

[5] White agate: *Glerhallur* is an Icelandic term for chalcedony, a fine crystalline and banded variety of silica, consisting essentially of fibrous or ultrafine quartz.

[6] The colour indicates that this could be opal, an amorphous variety of silica.

No. 4. A type of crystal found in Hvammsvík in Kjósarsýsla. The other kind is a former crystal (the six-sided one)[1] found in Múlasýsla, in Eastern Iceland.

No. 5. A reddish stone, which is a compound of flint[2] and various other minerals, found in Akrafjall in Borgarfjarðarsýsla.

No. 6. A white stone, appears to be of plaster of Paris,[3] found in Saurbæ parish in Borgarfjarðarsýsla.

No. 7. A small whitish stone with markings on all corners, found on Skarðsheiði[4] (a so-called mountain in Borgarfjarðarsýsla).

No. 8. Two other whitish stones, one of them is hollowed through with small excrescences or crystallizations internally, found on the mountain (called Esja) in Kjósarsýsla.

No. 9. Conches,[5] also called in Icelandic Igulkier[6] and Marbendils-Smÿde,[7] found on the coast of Borgarfjarðarsýsla and Gullbringusýsla.

No. 10. *Surtarbrandur* (*Lignum fossile*) which has broken off from a mountain, above the valleys of Skagafjörður in the North-Western part of Iceland and is believed to form a stratum through the whole mountain.[8]

[*Signed*] O Stephensen

Wisconsin, Banks Papers, ff. 1–2. Autograph letter. Translated from the Danish.[9]

Original in Danish Gothic script:
Notice
Over de til herr Banks leverede Steen arter i Island 1772

No. 1 Sort agat (:paa Islandsk hrafntinna:) taget paa et field over for Reineness stadar Kloster, inden Hægranes Syssel paa et Stæd belagt med Steene og Jordsvær imellem.

No. 2 hviid Agat (:paa Islandsk: Glerhall:) taget i Glerhallavÿk under fieldet Tindastöl inden samme Syssel. Denne Steenart kaster Söen paa Land, findes ellers paa fleere Stæder ved Söekanter, men ingenstæds, saa goede som i Glerhallavÿk.

No. 3 En anden Sort af samme hvid Agad, med gulagtig Tegning, tagen ved Hvalfiords Stranden udi Borgerfiords Syssel.

---

[1] Apparently rock-crystal or crystal-clear quartz, commonly found as amygdules in the Icelandic Tertiary formations.

[2] In Icelandic, *tinna* or *eldtinna*. This is probably jasper (Icelandic *jaspis*) a fine-grained variety of silica. Flint has not been found in Iceland.

[3] *Gypsum* is a better word (in English). This is probably some amygdule from the Tertiary formations.

[4] *Heiði* in Icelandic actually means plateau, but the *Skarðsheiði* is definitely a mountain.

[5] Molluscan shells (mainly gastroped shells), *Konkylier* in Danish and *kuðungur* in Icelandic.

[6] *Ígulker*: the Icelandic term for sea urchin.

[7] *Marbendill* (in Icelandic) is a fabulous creature in Icelandic folklore, the upper part from the hip is a man, combined with a seal as the lower part. *Smyde* (in Danish?) might refer to this shape (viz. *smíða* in Icelandic: to build or form.)

[8] On this document there are some notes pencilled in mentioning *Jaspis lutescens, Aggregat ex albo & nilso* (?) and *Lebeetordes* from Borgarfjörður-Syssel. It is of interest to compare this list with the lengthy appendix compiled by the mineralogist Sir George Steuart Mackenzie, who visited the island in 1810 (*Travels in Iceland*, Appendix III, pp. 435–56).

[9] Note in Banks's autograph, verso: 'account of Minerals | sent by Amptman | Stephenson'.

No. 4 En Art Christall tagen i Hvamsvÿk inden Kiose Syssel. Den anden Sort, forhenværende Christall (:Sexkantet:) er funden udi Mule Syssel, som ligger i Islands Österdeel.

No. 5 En Rödagtig Steen, som er Blanding af Flintesteen, og adskillige andre Steenarter, tagen fra Akrafialle inden Borgerfiords Syssel.

No. 6 En hviid Steen, seer ud til at være af Gipsart, tagen udi Saurbæ Kirkesogn inden Borgerfiords Syssel.

No. 7 En hviidagtig Liden Steen med Tegning paa alle Hiörner, er tagen paa Skardsheide (: et saa kaldet Field i BorgerfiordsSyssel).

No. 8 2 andre hviidagtige Steene, hvoraf den eene er igiennem huldat med smaae geræxter indvændig, Christalliserede fundne paa et Field, (:Esian kaldet:) inden Kiose Syssel.

No. 9 Cochylierne, samt de paa Islandsk kaldede Igulkier og Marbendils-Smÿde, ere tagne ved Strandbrædden udi Borgarfiords Kiose og Guldbringe Syssel.

No. 10 Surtarbranden (:Lignum fossile:) er brækket ud af et Field, som ligger oven for Skagefiordsdalene i Landets norderdeel og formeentes at gaae som et Stratum igiennem heele Bierget.

8. Bjarni Pálsson to Banks and Solander, 4 October 1772

Nes in Seltjarnarnes

Most famous, learned, and experienced men!
Most skilled observers!

I know that you, famous, learned men, who far exceed my praise (but not my admiration), have undertaken quite a difficult and expensive journey to this island and have spared no efforts, no expenses, being guided only by that spirit and burning with that desire to see more intimately the abundant works of Jehovah, to investigate them and then (no doubt) communicate them to the learned world – a rare, but a precious, example for all those, as I think, who are not completely lacking a taste for knowledge!

Therefore and further,[1] just as nature and kindness oblige me to help such a plan as far as is possible for my little strength, I was informed on the evening before by our chemist, B. Jónsson,[2] that some of you planned to make a trip here yesterday in order to visit our office and the small natural or 'curious' collections (if they deserve to be addressed by that name). I do not doubt at all that he told me the truth, but the very stormy weather that occurred was certainly sufficient to deter you from your plan and to keep you away.

I confess that my small collection is disordered and untidy, without classification etc.

Meanwhile, however, if I have by chance one or two specimens which please you, I have no doubt that we will agree to share them. Likewise, my bookshelf is, indeed, meagre, but is nevertheless such that it should not be thought completely unworthy of the sight of learned men. Only, I would add without boasting that certain things can be found in it

---

[1] There appears to be a breakdown in the Latin syntax here.
[2] Björn Jónsson (1738–98) was the first apothecary to be appointed in Iceland, actually that same year in May. Until then the state physician had also been the apothecary. He was trained in Denmark and had been Pálsson's assistant.

that will perhaps satisfy your desire or can placate a spirit desirous of curiosities – things that I would not at all deny you the sight of. Thus I have explained the whole matter and aim of this unpolished letter in a few words, which you should accept in a good spirit.

Do come then, if you like, not to a rich feast,[1] but at least for the aim that I have already described.

I remain, as far as I can be, the servant and devoted client of your famous and most worthy names,

[*Signed*] Bjarni Pálsson, state physician of Iceland.[2]
Written at my home, which is Nes in Seltjarnarnes[3]
on the fourth day of October, 1772.

Tell me through a messenger whether you should be expected and when. Farewell!

Kent History and Library Centre, Banks MS, U951 Z32/24, ff. 35–36.[4]
Another translation is to be found in Duyker and Tingbrand, *Solander*, pp. 300–302. Translated from the Latin.

Latin original:
Viri Celeberrimi, Doctissimi, e[x]perientissimi!
Observatores solertissimi!

Notum quidem mihi est vos, viros eximios, doctissimos et mea laude (non vero admiratione) longe majores, iter satis arduum et sumptuosum hanc in insulam suscepisse et nullis laboribus, nullis impensis pepercisse, eo tantum animo actos & desiderio flagrantes, ut amplissima Jehovæ opera intimius perspiceretis, rimaretis et orbi literato deinceps (sine dubio) communicaretis! Raro sane sed omnibus, uti puto, qui omni gustu scientiarum non prorsus destituti sunt, charo exemplo!

Ergo, quemadmodum tale institutum quoad per exiguas vires possibile est adjuvare natura me et philanthropia obligant et insuper. Pridie vespera per Pharm[a]copolam nostrum B. Jonæ certior redditus sum quosdam vestrum iter huc facere heri instituisse ut et officinam nostram et exiguas collectiones naturales vel curiosas (si eo nomine salutari merent) inspiceretis, quod verum ipsum retulisse nullus dubito, sed incidens tempestas valde procellosa certe sufficiens erat vos a proposito deterrere et retinere!

Fateor quidem meam collectionem exiguam, tumultuariam, incomptamque esse, sine classificatione etc.

Interim tamen si forte unum aut alterum exemplum penes me sit quod vobis arrideat, nullus dubito fore ut de communicatione inter nos conveniat. Pariter, macilenta quidem mea est suppellex libraria, talis tamen ut visui doctorum virorum non prorsus indigna sentienda sit, & tantum sine jactatione addere sustineo, quod in ea quædam reperiantur quæ forte desiderio vestri satisfaciant vel animum

---

[1] The feast proved memorable: see Introduction, p. 19 above.

[2] Pálsson wrote in Latin *medicus ord*, i.e. *medicus ordinarius* which was his official title as state physician. He was the first state physician (*landphysicus*) in Iceland (1760–79).

[3] Seltjarnarnes, a peninsula just outside Reykjavík. The 18th-century building still exists and is now a museum, *Nesstofa*.

[4] This letter was addressed in German in the following manner:
Zu den
Vohlgebohrnen, Hohgelehr–
ten Herren Observateurs
Aus England
[Trans.: To the esteemed, very learned gentlemen naturalists from England.]
Note in Banks's autograph, verso: 'Letter from the | Landphisicus'.

curiositatum avidum pacare queant, quæ similiter visui vestro minime subtraham. Atque sic rem totam et scopum hujus impoliti scripti paucis explicavi quod æqui bonique consulere velitis!

Adeste ergo, si lubet, minime ad convivium opiparum sed saltim scopo jam depicto!

Ego, quoad per me possibile est, permaneo Vestri nominum eximiorum æstimatissimorum servus et cliens
addictus
Biarno Pauli
Medicus ord: Islandiæ
Domi meæ quæ Nese est
Seltiórnensium
Die IV viiibr MDCCLXXII

9. Teitur Jónsson to Banks and Solander, 27 December 1772

Skálholt

To the famous, most noble, most reverend, most liberal and most wise men, very many greetings!

Here in Skalholt your conversation was pleasing to all, so you remain celebrated on everyone's lips, indeed praised and remembered by our people. But after your departure the news came to me that you had travelled round the whole world within a period of three years. When I heard of this immense journey, I was filled with admiration, indeed, everyone was drawn to admire you, pondering in their minds [the fact] that you are no less supported by unbelievable strength of spirit, than endowed with the strongest possible bodily constitution. That Jacques Le Maire[1] and Willem Schouten[2] circumnavigated the world has been recorded; your undertaking and attempt are worthy of the veneration of all people. Moreover, your generosity, being almost king-like, deservedly holds everybody's attention. Such rare munificence is recorded in the *exempla* of Bishop Rev*erend* John Widalin,[3] whose sermons on the gospel throughout the year were bought by you, most noble Troil, at Skalholt. He was once criticized on the grounds that 'parsimony is the best [source of] revenue',[4] and that 'it is too late when [the resources] have run dry';[5] the careful man needs to be parsimonious. To this, the Bishop replied: Death is always imminent, it can never be far away. There is added that steadfast and natural character and

---

[1] See Le Maire, Jacob, *Australian Navigations Discoverd by Jacob Le Maire in the Years 1615,1616 and 1617 ... etc.* Le Maire (c. 1585–1616) was an Amsterdam merchant wishing to trade 'in strange and far distant parts'. With him sailed Willem Cornelisz Schouten of Hoorn (c. 1567–1625), a famous navigator, having often sailed to the East Indies as skipper, pilot and merchant. Their aim was to search 'in the most southerly and unknown part of the earth, to look for a thoroughfare south of the Strait Magellani' ('To the Reader', pp. 165–6).

[2] Schouten's account: *Journal ou description du merveilleux voyage*, was often reprinted and translated into Latin and German during the years 1618–20. The geographer Alexander Dalrymple (1737–1808) included the voyage of 1616 in *A Collection of Voyages Chiefly in the Southern Atlantick Ocean*.

[3] Jón Vídalín (1666–1720), bishop of Skálholt in southern Iceland 1698–1720, was very learned and perhaps the best Latin poet of his times. He is best known for his *Húspostilla* (12 editions from 1718 to 1838), the most popular collection of sermons during this period, read in all homes.

[4] A quotation from Cicero, *Republica*, IV, vii: *optimum autem et in privatis familiis et in re publica vectigal duco esse parsimoniam*.

[5] A quotation from Seneca, *Epistle*, I, v: *sera parsimonia in fundo*.

endowment, that gentle and pleasing seriousness, that hoary weightiness of someone who had scarcely entered upon his adult years, that demanding law of self-control, without which each man in vain may direct his mind to great things, and that extraordinary modesty which, as he says, 'must not be scorned by great men, and is valued by the gods'.[1] Therefore, since you adorn your excellent selves with a refined character and remarkable studies, it is proper to add my support and, together with other extraordinary praises, adorn what remains with the words of the Poet,[2] Book 1, *Odes*, No. III:

> May the goddess who rules over Cyprus,
> May Helen's brothers,
> The gleaming stars and the Father of the winds,
> Confining all but Iapyx,[3] guide thee so!

I greet you illustrious, excellent men! What do I say? I embrace you with the salutation of the Poet and that of the Apostle. I will not stop commending you and all your deeds to the greatest and best God by whose nod and judgement all things are ruled. May His very self favour your journey, lead you prosperously through the world and roll out [your journey] to its end. Finally, having been restored to your homelands,[4] may He carry you up to the highest honours in the celestial court. May you think it worthy to accept these small things in a good spirit!

Teitur Jónsson, most dedicated to your nobility

[P. S.] I am very happy for you, most noble men, that the fire-breathing and fire-vomiting cruel maiden (Hekla) has sent you away unharmed, in peace.

Kent History and Library Centre, Banks MS, U951 Z32/27, ff. 41–42.[5] Another translation may be found in Duyker and Tingbrand, *Solander*, pp. 306–8.
Published in Hooker, *A Tour in Iceland*, II, pp. 292–4.
Translated from the Latin.

Latin original:

Viris

Illustrissimis, Nobilissimis, summe reverendis, amplissimis et consultissimis!

Plurima Salus!

Vestra hic in Skalholtia conversatio omnibus erat grata, quare in omnium ore versamini in celebritate, imo indigenarum laude et memoria. Verum post vestrum discessum res novæ mihi sunt relatæ, quod totum orbem intra triennium emensi fueritis, quo vasto itinere audito, miratio mihi facta est, immo omnes in admirationem vestri trahit, suo animo apud se perpendentes, quod non minus incredibili animi robore septi, quam firmissima corporum complexione præditi sitis. Quod Jacobus de Lamaire et Wilhelmus Soutensis terrarum orbem peragraverint, memoriæ est proditum: vestrum susceptum

---

[1] A quotation from Tacitus, *Annals*, 15, viii.

[2] Horace or Quintus Horatius Flaccus (65–8 BC), Roman poet. See e.g. Alexander, *The Complete Odes and Satires of Horace*, p. 7.

[3] According to the *Oxford Latin Dictionary*: The 'WNW wind which favours the crossing from Italy to Greece'.

[4] 'Rediti' appears to be a 'nominativus pendens', unrelated syntactically to the main clause of the sentence.

[5] Note in Banks's autograph, verso: 'Epistola | Pasteris Emeriti' [Trans.: Letter from the Pastor Emeritus].

molimenque omnium est cultu dignum. Præterea liberalitas vestra pene regifica omnium merito retinet animos; qualis munificentia rara est in exemplis, Præsulis Magistri JOH[A]NNIS WIDALINI excepta, cujus super anniversaria Evangelia orationes incolæ habent, a Te, Nobilissime Troili, Schalholti em[p]tas. Huic olim objectum, parsimoniam optimum esse vectigal, seram esse in fundo, opus esse cauto. Ad quod Præsul: Mors semper impendet, nunquam longe abesse potest. Accedit stabilis ille atque ingenuus mos et decor, ac inoffensa gravitas, nec sine jucunditate senile illud iam vixdum viriles annos ingressi pondus, illaque exacta frugalitatis lex, sine qua frustra aliquis mentem applicet magnis, modestiaque inusitata, quæ neque summis, ut ille ait, mortalium spernenda est, atque a Diis æstimatur. Igitur cum vos ipsos loco ornatissimos lepidis exornetis moribus, singularibusque studiis, tum meum adjicere decet suffragium, et una cum aliis laudibus exquisitis ornare quod reliquum est, Poetæ verbis Lib: 1 Carmin: Ode iii:

> Sic te Diva potens Cypri
> Sic fratres Helenæ, lucida sidera
> Ventorumque regat pater
> Obstrictis aliis præter Iapyga!

Vos illustres optimos, saluto! Quid dico? poetæ et ipsa Apostolica salutatione amplector. Vos Deo opt[imo] Max[imo], cuius nutu et arbitrio omnia reguntur, omniaque vestra commendare non desistam. Ipsissimus vestrum per orbem iter fortunet, et ad umbilicum ducat! Postremo patriis redditi terris, ad summas in cælesti curia evehat dignitates. Hæc pauca boni æquique consulere dignemini.
Vestræ nobilitatis
addictissimus
Theodorus Johannis P. Em.

Skalholti Isl[andiæ]
d[ie] 27 Dec[embris] 1772
[P.S.] In hoc vobis, viri nobilissimi, gratulor, quod igniflua, ignivoma, et crudelis virgo (Hecla) cum pace incolumes demiserit.

10. Banks to Thomas Falconer, 12 January 1773

New Burlington Street

Dear Sir

I know not whether I ever thankd you for the usefull & instructive letter[1] which you sent me during the time of my preparation for my intended voyage to the South Sea if during the hurry & confusion occasiond by my disapointment it was forgot I trust your candour has forgiven already an omission which I now ask pardon for.[2]

The only return I can attempt to make you is to communicate some account of what my last short voyage has offerd to my observation. The course I steerd was thro the Western Islands to Iceland from whence after having remaind 8 weeks I returnd by the orkneys to Edinburgh & from thence by Land to London.

---

[1] The letter referred to from Falconer to Banks is dated 16 April 1768, Kew, Archives of the Royal Botanic Gardens, Sir Joseph Banks Papers [hereafter JBK] I, f. 22; NHM, DTC 1, ff. 20–22.

[2] This letter is the first account of his voyage to the Western Isles which he visited before going on to Iceland. The reader is referred to the notes accompanying Banks's Iceland journal (pp. 45–79 above). The notes accompanying this letter are on points not raised in the journal. Document 12 completes the account. Capital letters and full stops have been added in this letter.

As the subject matter of the whole would be too long for one letter I shall divide it, in this I shall confine myself to what happend before we left the western Isles if that proves amusing to you you shall in the course of a few weeks receive the rest.

Disappointed in the execution of my favorite project of visiting the Southern Hemisphere but not destitute of hopes that the next year might revive it I did not chuse to discharge the draughtsmen that had been retaind for that purpose so with the concurrence of the gentlemen who had intended to accompany me it was resolvd that some short voyage should be undertaken tho the season was far advancd in order to shew our freinds that our disappointment had not been owing to any tardiness of our own or disinclination to undertake voyages. Four months was all we had left nothing within the compass of so short a time appeard an object of curiosity equal to Iceland a countrey which had been peopled by Europeans 800 years ago who for some ages signalisd themselves both in Learning & adventure but had now been almost unvisited by travelers for many ages. accordingly it was determind that that should be our route & having hird a Brig of 150 tons we saild from London on the 12th of July.

After a passage rather tedious we arrivd at Ila at the latter end of the month from hence we proceeded to Jura thence to Oransay. Ila is the most fertile of the western Isles that I have seen tho probably Bute[1] & some of those that lye nearer the main*land* are more so. The general soil is what is there calld whyn stone[2] such as london is pavd with but in the middle of the Island is a patch as it were of Lime in which are lead mines which are workd to some advantage also the mineral of Emery[3] & some Copper tho not iron workd. The Island in general abounds with a Kind of stone marl[4] of which the inhabitants have but just discoverd the use but from which they will probably in a very short time reap much benefit as it in a very short time reduces Ling[5] land to a good turf by merely being spread over it.

The misfortune of this & I beleive most of the Western Isles seems to be the thinness of population in general large parcels of Land are leasd out to people who are not capable of improving or even cultivating a tenth part of them nor will people be content without this extra quantity of Land which is usd for breeding of Cattle so that of necessity $\frac{9}{10}$ of the Countrey remains in much the same state as they were left in after the creation.

Jura is separated from Isla by a straight of less than an English mile in breadth here Thurot[6] laid his little fleet trusting that by the rapidity of the tides he might go out at any time in a contrary direction from that in which an Enemy might enter to pursue him. The Scotch speak of him with great affection he seems to have behavd very civily to them & to have signified that he had particular orders from his master so to do.

The Island itself is much less fertile than Ila owing cheifly to the swamps with which [a] great part of it is coverd.

---

[1] Bute is not mentioned in the journal. It is an island in the Firth of Clyde.

[2] See the Banks journal, p. 169 above.

[3] Banks does not mention emery in his journal. It is a greyish-black variety of corundum which can be crushed and used as an abrasive for polishing tough surfaces (hence emery paper, emery boards). Banks may have been mistaken as there are no emery deposits on Islay or Jura.

[4] See the Banks journal, p. 169 above.

[5] Ling is heather, the word being derived from Old Norse, in modern Icelandic *lyng*.

[6] See the Banks journal, p. 169 above.

Oransay a small Island seperated from Colonsay by a narrow Channel dry at half Ebb is low & rather fertile. Here are the remains of what is thought to be the second place of Religious worship which was founded in these Kingdoms the Ruin is large & in good preservation but from the ill contrivance of the different apartments plainly appears to have been built at different times. At the west end is a beautiful cross in good preservation with several characters upon it which we were not able to read.

From Oransay we proceeded along the shore of Jura to the northernmost end of that Island where in a straight between that Island & Scarba both the reports of the living & printed accounts had fixd the site of a whirlpool almost as extraordinary & as terrible as that of Mael Strom on the coast of Norway.[1]

Willing to examine so extraordinary a Phoenomenon with all possible care & attention we Landed on the little Island of Scarba & pitching a tent waited with patience for the turn of the tide which was to exhibit such wonders but in vain tho [the] tide turnd & ran with more than usual rapidity indeed [it] did not produce a whirl sufficient to have endangerd the smallest boat. The tides indeed were not high which made the waters less turbulent than they are at the time of spring tides but upon enquiry at the gentlemans house who lives nearest to it he told me that it was not at all terrible except at the concurrence of a spring tide & a westerly wind. In his time however only one boat had been lost in it & that a very small one managd by only two men.

From Scarba then a very small & unfruitful Island we took our departure & passing among the Luing[2] or Slate Isles found ourselves among a variety of tides that made it difficult to determine what course to steer. At this rather critical conjuncture it was I beleive that we passd the ship of our freind Mr. Pennant[3] which I had all along carefully lookd out for not without hopes if I should be fortunate enough to meet him of tempting him to enlarge his plan & wishing much to assure him at all events that as I had lookd upon myself while among these Islands as treading upon ground which by prior right he has taken possession of I should communicate to him every observation I could possibly make whenever he thought fit to publish any account of them.[4]

From hence we proceeded through the straights or as it is Calld the Sound of Mull passing between that Island & the Countrey of Morven the suppsd residence of Fingall both sides affording a variety of romantick & beautifull prospects especially that of morven. Here as we were obligd to stop every tide we landed often & in one of these excursions I saw the method of making Kelp an article which raises a great revenue to this countrey tho a few years ago totaly neglected or let out at a small price to people who came every year from Ireland for that /on/ purpose the process as it appeard rather curious to me I shall here describe.[5]

The crop which consists I believe of every Kind of sea weed except the sea Thongs is reapd once only in three years (for so long it takes to grow again to maturity after

---

[1] See the Banks journal, p. 63, n. 3 above.

[2] Luing is one of the Slate Islands, not mentioned by Banks in his journal.

[3] Thomas Pennant, zoologist and traveller. See Appendix 2.

[4] As Banks makes clear, Pennant was touring Scotland and the Hebrides at the same time. Banks permitted Pennant to use his journal for his edition of *Tour in Scotland and the Hebrides*, in which Banks's description of Fingal's Cave won praise from such readers as the author Horace Walpole (1717–97) as the only passage worth reading in the whole book. Later, relations between Banks and Pennant became strained. See further Appendix 2.

[5] See the Banks journal, p. 66 above.

being cut down) it is tedded¹ & made just as hay is & if it receives much rain during that time becomes totaly unfit for its purpose. When it is about two thirds dry that is that it will take fire /& burn/ with the addition at first of some hay or dry stuff it is ready to burn which is done in [a] Kind of trough built up of turf about 12 or 14 feet in leng*t*h & 4 or 5 in breadth. When the fire is once well lighted the weed will burn alone the people then attend constantly supplying it with plenty of weed till what they call a floor is formd that is till the bottom of the Kiln is coverd with 16 or 18 inches of ashes which then are as light & powdery as wood ashes falling from the fire. the moment that the weed is intirely consumd the attendants begin with instruments almost like hoes to stir about the ashes which by this kind of motion are reducd from a powder into a state maybe of partial fusion in which they resemble much in consequence the dough from which bread is made & from which when they cool into the hard stonelike form in which we receive them so hard that in using them we are obligd to apply large hammers to break them. In this manner they go on raising floor upon floor till the Kiln is full when they cover it with turf & let it remain till the Vessels come to fetch it away. In the stirring alone consists the whole mystery & dificulty of making the Kelp for the ashes unless reducd into that stony consistence are of little Value. I confess I was much surprisd at the effect of that operation which tho it may be well known to the chymists I was before totaly ignorant of you would Oblige me much by sending me your opinion of it.

The morning after this Highland lecture upon Chymistry it was our fortune to anchor immediately in the famous Harbour of Tobir more [Tobermory] near a gentlemans house calld Drumlin. M^r M'lean² the owner of it according to the Hospitable customs of the countrey came off to breakfast with us & invited us ashore which invitation we most readily accepted. At his house we found his son & an English Gentleman M^r. [*blank*]³ who told us that on a small Island not very distant from thence was to be seen an appearance much resembling the Giants causeway in Ireland he had he said been looking out carefully for M^r. Pennant in order to have given him the same information but not having been fortunate enough to see him rejoicd much in the opportunity he had of communicating it to us. As I still lamented the not having had an opportunity of visiting the giants causeway which I had at first intended this was an opportunity not to be lost I instantly orderd the ship to wait my return in the Harbour of Topir more & set out in the boat accompanied by both the gentlemen by 10 o'Clock in the morn. As the distance was full 40 miles & we had not a breath of Wind we were obligd to work hard at our oars nor did we arrive at the place of our destination till 10 at night when it was to*o* dark for us to Judge at all whether or not our curiosity was likely to meet with gratification equal to the pains we had been at.

The small Island of <u>Staffa</u>⁴ on which we were is situated on the West side of Mul*l* about 4 miles from the shore opposite an inlet calld <u>Loch na Gaul</u> & about 9 miles from the celebrated Island of Y Columb Kil or Iona. It is about an english mile in leng*t*h & half as broad inhabited by a few young Cattle Some Sheep & one family consisting of seven

---

[1] To *ted* means to turn over and spread out to dry.

[2] Sir Allan Maclean (d. 1783) was master of Drimnin at this time: see the Banks journal, p. 67 above.

[3] There is a blank in the original but according to Banks's journal (p. 67 above) the gentleman in question was a Mr Leach. Dawson, *The Banks Letters*, also gives the name as Leach.

[4] On Staffa see the Banks journal, pp. 67–72 above.

people living in a miserable highland house without a chimney according to the custom of the countrey where a stranger might well imagine smoak to be one of the necessaries of life as he must find much more of that than of any other Kind of provision in every house into which curiosity or necessity may induce him to enter.

In this house & a small tent which we had carried with us we slept not very comfortably you may imagine so at the earliest dawn of the succeeding day we were up & prepard to prosecute our undertaking.

We had been taught to expect something resembling the Giants causeway in Ireland but had not the least Idea that what we were to see was at all to rival that noted curiosity Either in the extent of the whole or the size of the component parts. Many such things were before known to exist in the Islands of Canna Sky*e* & Mul*l* but they seemd from the descriptions I had heard or read of them to be little more than Efforts of nature towards that regularity which in the giants causeway she had compleated.

Think then how agreable was our surprize when on walking down to the sea beach we were at once presented with the view of the ranges of Pillars near a mile in Extent some of which were 57 feet high supporting the Mass of the Island above them which as the ground of it swelld into hills or sunk into valleys seemd monstrous Pediments /*which*/ vast as the Columns were seeming to Lean heavy upon them & Claim all their streng*t*h to support.

Below these to the water was that kind of Pavement from whence the Giants causeway has got its name viz the tops of the Joints of the Pillars which have been broken off & remain shewing their angular figures which are separated from each other only by small cracks. These were Exactly as those of the Giants causeway formd of a hard black stone (possibly the Basaltes of the ancients) their Surfaces concave & convex & of all figures from triangular to heptangular tho pentagons & hexagons were far the most frequent. The thickest we measurd was $4\frac{1}{2}$ feet in diameter.

At the north End of this range of pillars was a very singular sight the pillars themselves there instead of standing upright lay on their sides & were bent in all possible directions yet remain whole. The joints indeed were visible in them but adheard nearly as fast to Each other as any other part of the pillar.

But curious as this was or striking as that vast assemblage of Columns above describd the beauty of a third object made it almost the most desireable sight of the three. I mean a Cave which had been hollowd out by the Sea among these pillars from the first breaking of the causeway that is from a to b in the sketch which I send[1] the depth is 371 feet from the point which is coverd by the arch that is from c 251 the Arch itself is from the water 117 the Sides consist of Pillars the roof also is fretted with the bottoms of those which are broken off. Underneath is water on which a boat may go quite to the farther end by means of which light is reflected so strong that at the farther extremity we could easily read a common printed book.

A cave of these dimensions has always been lookd upon as an extraordinary sight even without ornament & the aid of a few stalactites has made people boast of it as a beautiful wonder: what then must we esteem this cave to be ornamented as it is by such a range of columns, & fretted as the roof is & illuminated by its own light. I confess I cannot

---

[1] This sketch does not seem to have survived.

conceive that the grotto of Antiparos,[1] the famous cave in the neighborhood of Naples[2] can bear a comparison with it, nor can I say that I have in my life receivd so much pleasure from the contemplation of any artificial arrangement of Pillars as I had from this one.

We had nobody with us who had seen the Giants causeway but I have since spoken with several who all agree in the great superiority of this Island over it in curiosity as well as in beauty & extent. The bending pillars of which you receive sketches are not those that I have heard of these I take to be a great illustration of the subject.

After having spent just 12 hours upon this Island the whole time in a state of satisfaction which none but a traveler can feel we embarkd for Iona[3] or Y Columb Kil where we arrivd at night & /having/ spent part of the morning in viewing that Island remarkable only as the first Seat of Christianity. The buildings are extensive & rather pretty considering the time in which they were erected many people of high note have been buried here but cheifly without stones. We saw at least none whose antiquity claimd any degree of respect tho we were told that some were there in the ancient Irish character but did not meet with any such.

On each of the 4 sides of this Island which answer the Cardinal points is a stone in which seamen place great faith beleiving that if they clean carefully any one of them a wind will arise from its respective quarter when we were there the stone on the North side was nicely Swept & a northerly wind arising fannd us gently away to our ship where we arrivd at night.

The harbour of Topir more[4] in which we lay is the best in this part of the highlands capable of containing in safety an immense number of ships here tradition tells us that one of the great Spanish armada[5] was lost flying from our victorious fleet the name of her having been Spanish makes the people here dream of Dollars by bushels & wish much that they had a diving bell to search for them tho in all probability being an armd ship she was not chargd with that Kind of Lading.

The next morning we Landed on a small Islet near the main*land* of Scotland in order to hunt the roe but we were not fortunate enough to see any tho the scotchmen saw seven & my servants some.

The Season being now far advancd we resolvd to spend no more time among these Islands but to push directly for Iceland after having seen St Kilda. *A*ccordingly we saild passing in sight of Canna Rum Egg Sky*e* the Uists Harris & Lewis & many more arrivd at the but*t* of Lewis here we had the wind S W with thick weather a fair wind for Iceland but a foul one for St. Kilda so thinking it absolutely certain that M^r. Pennant who set out intirely to visit these Islands must have been upon it we resolvd for Iceland where my next shall pilot you if you signify that you have had any entertainment from this.

Beleive me your obligd & Faithfull
J. Banks

---

[1] Antiparos, one of the Cyclades in the Aegean Sea. A famous stalactite cavern is to be found there.
[2] Probably the Blue Grotto on the island of Capri in the Bay of Naples.
[3] See the Banks journal, pp. 73–5 above.
[4] See the Banks journal, p. 67 above.
[5] The *Florida* was sunk near Tobermory, little was salvaged except guns of brass and iron.

Banks Archive Project, transcript by H. B. Carter from the Hawley original.
BL, Add MS 56301 (7), ff. 10–19, modern copy by Warren R. Dawson.
Original letter in Hawley Collection, whereabouts unknown.[1]
Chambers, *The Letters of Sir Joseph Banks*, pp. 30–39.

11.   Thomas Falconer to Banks, 16 January 1773

Chester

Dear Sir

Never was I more agreeably surprised than at the receipt of your letter yesterday. I could not have expected so large a communication of knowledge on so short an acquaintance tho' none can receive it with more gratitude than myself. Your liberality of sentiment with respect to my kinsman Mr Pennant,[2] shews the true spirit of a scholar, & I have ventured to give him an account of what you have said. He himself was disappointed of his voyage to S^t. Kilda, but the loss is the less to be regretted as so long an account of that small island is already published.[3] Indeed the remarks of an English traveller would be more consonant to an Englishman; for we all judge by comparison, & a Scotch traveller would not consider a place as barren, which we should regard as worth nothing. For this reason Martin & others have given us undesignedly a false idea of the western isles & your remarks will set the world right in many particulars wherein they were ill instructed before. Your precision of measures, & the advantage of able artists are a great point; for when we first employ description we form an opinion /thru/ through the medium of another mans understanding who generally compares it with something else he has seen. Even the characteristical difference of nations must be attended to. The French use the terms magnifique & superbe to the close & dirty lanes of Paris, & the Italians are more hyperbolical. What an assistance is it then to truth to have the objects delineated by one common measure which, speaks univerally to all mankind. Some things indeed are incapable of this illustration, & there we can only describe the facts, as you have done in the account of the whirlpool of Scarba,[4] which you have so well confuted. Martin who describes it as I suppose he heard it (for it does not appear he ever saw it) contradicts himself in the next page; for an English Ship he owns passed through it safe, tho the men in a fright had described her with all her sails up. As for your description of making Kelp I who am no chymist can only admire it, but I shall inclose a solution of your difficulty by

---

[1] According to Warren R. Dawson, this letter was 'formerly in the possession of Dawson Turner, bound in a volume with the original drawings etc. D.T. Manuscripts Sale, 1859, Lot 26'. Dawson cited 'Hawley Coll.', a manuscript collection of Banks papers in the possession of Major Sir David Henry Hawley, 7th baronet of Mareham-Le-Fen, Lincolnshire (Dawson, *The Banks Letters*, pp. xxxviii, 318). A typed transcript, made by Harold Carter, is in the possession of the Banks Archive Project. A great many Hawley Papers are preserved in the Lincolnshire Archives, but this letter is not among them. See p. 40, n. 6 above.

[2] In 1759 Pennant had married Falconer's sister Elisabeth (d. 1764).

[3] Martin Martin (c. 1665–1719) was born in Skye. He accompanied the minister of Harris, John Campbell, on his annual visit to St Kilda in the summer of 1697, and wrote the first detailed accounts of life there. See his following books on St Kilda (two owned by Banks): *A Voyage to St Kilda*, London, 1698; *A Late Voyage to St Kilda*, London, 1698; and *A Description of the Western Isles of Scotland*, London, 1703.

[4] A lengthy account: see the Banks journal, pp. 63–4 above.

my worthy fri*en*d Dr Haggarth,[1] who understands it very well. What surprized me most was to find they did not use the sea Thongs for their purpose. Dr Merrett[2] in his notes on Neri p. 263 says that it was then chiefly made of that herb; from whence Bauhin[3] calls it the alga angustifolia vitrariorum, & Matthiolus[4] the alga vulgaris Venetorum. These old accounts are often mistaken. Ray[5] Syn: 43 says nothing of this property of the sea thongs. It is plain that Merret never saw Kelp made & the author he commends does not explain it. Yet from some means or other he has collected most of the facts you mention, tho indigested. Sea plants in general give the same salt, which (a little improperly at first sight) is called the <u>fossil</u> Alkali, & hence I do not wonder that so many sea weeds are collected for the Kelp. The concreted substance used to be called Rochetta from its stonelike substance, & was brought through Venice from Levant. When powdered it was called Polverine, which being liable to adulteration was reckoned the worse. Merret says p. 254, that being cut down dried & burnt it grew into a hard mass or stone, from whence it is plain he was unaquainted with the most necessary part of the process. I am extremely glad you met with the information about the cave in Staffa & am particularly obliged for the honour you have done me in sending me these drawings.[6]

The obliquity of the columns is a beautifull circumstance & may tend to illustrate the point. I myself am unable to solve it, but it is certain that fossil substances will form themselves exactly like salts. I have seen the artificial crystallization so exactly resemble a congerie of native chrystals, that it was difficult to distinguish one from the other. The tendency of some is to form round a Nucleus, of others upright, but in all these cases some accidental shock may divert them from their natural course. As I take it /pillars of the/ Giant's causeway & the Ile of Staffa have been formed under the earth but by time & accidents are now laid open. I suppose then that some body of earth had given way under them it may account for their oblique position. I would rather suppose this than to imagine some attraction to one particular sort of matter had directed their position for

---

[1] Dawson has written 'Haggarth' quite clearly, though no one can be identified with that name. This is doubtless Dr John Haygarth (1740–1827), FRS, physician, philanthropist. He had studied in Edinburgh, Leiden, London and Paris and was appointed physician to Chester Infirmary in 1766. William Falconer was a colleague at the Chester Infirmary. Two letters between Haygarth and Banks are extant and he had a close relationship with the Falconer family. See Booth, *John Haygarth, FRS (1740–1827)*.

[2] Christopher Merrett (1614–95), a naturalist, was a founding member of the Royal Society. Merrett translated Antonio Neri's *L'Arte vetraria*, 1612, a pioneering work, for the Royal Society, as *The Art of Glass ... With Some Observations on the Author*. Merrett is considered to have had a considerable influence on glass-making in England and other European countries.

[3] Gaspard Bauhin (1560–1624), anatomist, botanist and prolific author. Bauhin's major botanical work was *De plantis a divis sanctisve nomen habentibus ... Ioanni Bauhini*.

[4] Pietro Andreas Mattiolus (1500–77) of Siena, a physician. He wrote a book about herbs, translated into German by Handschius, Prague, 1563, and his collected works *Opera quae extant omnia* were edited by Bauhin (Caspar Bauhinus) in Frankfurt, 1598.

[5] John Ray (1627–1705) was one of the most eminent naturalists of his time, besides being an influential philosopher and theologian. Often referred to as 'the father of natural history' in Britain, Ray was elected a Fellow of the Royal Society in 1667. The work of Jean and Gaspard Bauhin provided him with the immediate model for his *Catalogue of Cambridge Plants* (*Catalogus Plantarum circa Cantabrigiam*). The *Syn* is *Synopsis Methodica Stirpium Britannicarum*, London, 1724 and often reprinted, rather than his *Synopsis Methodica Avium et Piscium*, London, 1713. Like Linnæus, Ray searched for the 'natural system'. Unlike Linnaeus, whose plant classification was based entirely on floral reproductive organs, Ray classified plants by overall morphology.

[6] The drawings from Staffa are to be found in BL, Add MS 15509.

the criterion of Basaltes is the perpendicular position of the columns; & as this is a singular substance I would rather account for it from an accident than disturb the established order. It is true that the more we see, the more we disturb the systems already formed. My time Dear Sir will permit me to write but little more, or I should have presented you with another sheet. To say the truth I was desirous of shewing my gratitude as early as possible, & if you will be kind as at some leisure hour to gratify me with the like remarks on Iceland (one of those parts of the world, hardly known but by name) you will increase an obligation which even now is greater than I can repay. I was highly entertained at Oxford with a sight of some curiosities you sent from Otaheita & new Zealand,[1] & from the clearness of your narrative could have wished you you could have found time to have drawn up your own travels, without having recourse to a second hand. I shall add no more but my sincerest wishes for your wellfare, & regretting it is not in my power to assist you in the liberal pursuit of science. I am with truest

regard your most obedient friend & Servant
[*Signed*] Tho[s] Falconer

Kew, Banks Correspondence, I, JBK/1/2, ff. 32–33. Autograph letter.
NHM, DTC 1, ff. 43–45. Contemporary copy.

12. Banks to Thomas Falconer, 2 April 1773

New Burlington Street [London]

Dear Sir

I must beg your pardon & more particularly D[r]. Haggarths[2] for the imperfection of my account of Kelp burning[3] but hope you will both attribute it to the known dificulty of drawing a description of any fact in which material circumstances only are inserted & those of little import left out a thing indeed impossible for any man to perform perfectly till he knows the system to which his description may be applyd I am inclined to beleive that D[r]. Haggarth is mistaken in his Solution because I omitted to relate in my description that after the ashes were sufficiently stirrd into their doughy consistence they still remaind glowing but hot enough to give fire to an additional quantity of Sea weed which was thrown upon them almost instantly in which state of heat I apprehend it is impossible they should imbibe a sufficient quantity of atmospherical water to produce the requird Effect. But to our Business

On the 18th of August we left the Butt of Lewis resolvd to steer immediately for Iceland & on the 29th came to an anchor at a place calld Besested[4] lying at the Bottom of a deep bay on the S.W. Side of the Island here as we found a tolerable harbour we resolvd to remain we went ashore therefore & to our great Joy found that the sides of the Harbour were constituted of the Surface of an ancient flow of Lava.[5]

[1] From the *Endeavour* voyage, Otaheite being Tahiti.
[2] See previous document, p. 175, n. 1. Dawson always writes 'Haggarth'. Capital letters and full stops have been added to this letter.
[3] See Document 10.
[4] Bessastaðir was the residence of the governor of Iceland.
[5] He had arrived in Hafnarfjörður which lies on a bed of lava.

The Governor Stifsamptman[1] as he is calld receivd us with great politeness & allotted us certain houses belonging to the Danish merchants who were returnd to Copenhagen for our residence here. We soon fixd ourselves not so conveniently as in English houses but more so than we could have done in the Ship. To give a detail of our daily proceedings from this time would be tedious & unentertaining I shall therefore proceed as well as I can to give as good an account as I am able of the Countrey & its inhabitants.

The Island of Island [*sic*] contains an area I beleive Larger than that of Ireland:[2] as we were only 6 weeks ashore upon it it cannot be expected that we could examine so large a space /*with any degree of accuracy*/ indeed we traveld 13 days on horseback & merely reachd Hecla as the extremity of our Jaunt. Wherever we traveld Lava & that stone which is Calld by *Sir William* Hamilton[3] Tufta[4] were the only strata. So much has this Island sufferd by Fire that we were told that since it has been inhabited the whole S.W. promontery & near or Quite a hundred miles inland from it was on fire at once. Many are the mountains which have occasionaly vomited up fire but in this land of eruption it has often hap*pen*ned that the level ground has opend itself into a Crater & thrown out water, stones Lava &c over all the adjacent countrey.

We seldom traveld half a day without [encountering] hot baths whose waters were generaly of a boiling heat these were constantly surrounded by Quantities of that Stone which by the mineralists is calld Lebes from its similitude to that formd on the bottoms of Kettles &c. Some of them boild so violently that in the center of them the water was raisd by the agitation of the heat 6 or 7 feet high almost constantly. Others from the impulse of a power which I confess myself unable to trace spouted up their water periodicaly to an immense hight particularly that calld Geiser situated about [*blank*] miles N. from Scalholt[5] of which you receive some drawings merits some description .

This wonderful volcano of water[6] if I may use that Expression which has continued its eruptions ever since the Island was inhabited about 800 years rises amidst hundreds of smaller wells from a basin which it has formd for itself of the Lebes which it deposits 57. feet in diameter formd like a funnel which gradualy slopes from the sides to a hole in the center 19 feet in diameter out of which the water rises.

We arrivd there in the morning at 6 O'Clock about half an hour after the eruption which generaly happens morning & evening was over. The Crater was then Empty & we had an opportunity of examining the Central hole; which for about 10 feet down was also free from water; but this was not quite safe for every now & then a burst would come up throwing the water some fathoms high without giving us the least warning this Continued till half past 8 the water rising gradualy all the time till it overflowed the edge

---

[1] *Stiftamtmaður*: governor (see Glossary). This was Thodal, see p. 85, n. 1 above.
[2] This is correct; Iceland is 39,768.5 sq. miles (103,000 km²), Ireland 32,595.1 sq. miles (84,421 km²).
[3] Sir William Hamilton (1730–1803). He is best known for his diplomatic career as British ambassador to Naples and his wife Emma's affair with Nelson. His scientific reputation rests on his hobbies – the study of volcanism and the collection of antiquities. His stay in Naples enabled him to study Vesuvius, Etna and the Lipari Islands. Elected a fellow of the Royal Society in 1766, he published his studies on volcanoes in the form of numerous letters in the *Philosophical Transactions* (1767–95).
[4] Banks apparently meant tuff or tufa, rock formed by the consolidation of volcanic ashes.
[5] Skálholt is the bishopric in southern Iceland.
[6] Geysir, the hot spring: see the Banks journal, pp. 98–101 above.

of the central hole. *From* this time till half after ten only one eruption hapend & that a small one. The Crater however was filld two thirds at half past 12 three strokes were heard underground like the firing of distant Cannon. The water in the Crater instantly rose so high as to overflow the Edges & the ground all around it was Shaken very perceptibly. At a quarter after three many distinct blows of the same kind as the last were heard the earth Shook much & the crater overflowd more than before. At 43 minutes after 4 louder noises were heard which shook the ground above a mile from the Crater & much water ran over. At 49 minutes after 4 still louder noises were heard than before & at 51. minutes after 4 the great eruption hapend a Column of water not solid but Consisting of many pointed Cones at once flew up to the hight of 92 feet above the edge of the crater & Continued by Jerks rising & falling 4 minutes. Its thickness we guessed to be about 26 feet but its hight was accurately measurd by Dr Lind who had prepard a base & /*stood*/ with a quadrant measurd its hight as accurately I beleive as that operation will admit of.

How this effect is producd is certainly a mystery which nature has Lockd up from the eyes of curious mortals with great care guarded as she here is by fiery dragons it is impossible probably for the art of man to gain an opportunity of inspecting her operations. Analogy then is the only resource & that tells us that as water when expanded by fire into the form of steam occupies 14000 times the space that it did while only water such an Expansion so much more violent than that of the air contain among the grains of Gunpowder seems & probably is sufficient to produce even the effects above describd this goes however no farther than a general guess at the Gross cause Experiment may in time assist me to account for the particularities of this Phaenomenon we are now Employd in trying some of the results of which if they are successful I may one day send to you.

You should have receivd this Letter sooner had not my absence in Holland[1] through which I have made a Short tour prevented me. I wish I could send you any thing from that Countrey worth Observation Holland is a great Hydraulick machine which must continualy work in order to discharge its water. Dutchmen Lazier than any other race of mankind have from that very principal invented more methods of Easing themselves from the curse of Labor than any other nation hence it proceeds that their Engines Mills &c are better & more numerous than those of any other nation.

M^r [Thomas] Pennant is in town he has had my Journal & the drawings which I made in the western Isles.[2] I lookd upon them as I told you as his right I while in that Countrey Lookd for him with assiduity conceiving myself as no more than a poacher who might get leave of the Lord to shoot upon the mannor but in return owd at least the offer of whatever he might Kill.

Adieu dear Sir I snatch an opportunity of sending this in the mean time assure yourself that you shall hear from me again the very first Leisure I have

Beleive me Your Affectionate
& very H*um*ble Servant
Jos: Banks

---

[1] Banks visited Holland from 12 February 1773 to 22 March 1773 and kept a journal throughout his visit. This journal has now been published by Kees van Strien: 'Joseph Banks, Journal of a Tour in Holland, 1773'. See further Carter, *Banks. The Sources,* p. 54.

[2] See Appendix 2.

P.S. we are employd in fitting out an expedition in order to penetrate as near to the North Pole[1] as Possible it consists of two Boom Ketches chose as the strongest species of Ships therefore the best to Cope with the Ice. *T*hey will sail before the middle of the next month commanded by a good Freind of mine Capt*ai*n Phipps[2] your opinion of the Frigid Zone cannot but be usefull to him & very agreable to me at this Juncture.

Banks Archive Project, transcript by H. B. Carter from the Hawley original.
BL, Add MS 56301 (7), ff. 20–24, contemporary copy by Dawson Turner.
Original letter in Hawley Collection, whereabouts unknown.
Chambers, *The Letters of Sir Joseph Banks*, pp. 46–9.

13. Uno von Troil to Banks, 14 April 1773.

Stockholm

Dear Sir!

Nothing can be more difficult to express then the sweets feelings we have by remembering a friend and then I hope You vill not find a description of my mind in reading this letter. No Sir! without others reasons I know not Your language enough[3] to rait[4] any thing pretty, and then I beg You will let me be excus'd, if I omit it, so well as great many compliments for all Your goodness and friendship for me, under that time I was happy enough to live with You. I[5] can't nevertheless forbear to assure You, that as I look upon that period of life, hvat[6] I have pass'd away with You,[7] as most usefull and entertaining, so can I never forget a friend, I have to thank you for it. Can I never find any opportunity to show You my thankfulness, so are You sure, I hope, that I keep it in the most gratefull heart.

My voyage[8] has, notwithstanding the bad season, been very agreable, and I have seen great many curiosities in the natural way, after that I left London. How many questions about Sir Banks' have I not answerd, and how much am I not obligd to You, for all politeness I have got from unknown people, only for that I was a friend of Yours? I speck not alone about Mr. and Mrs. Loten,[9] she the most amiable woman I know, but about all the compagnies I have frequented in Holland, Germany and Danemark. Every body has

---

[1] On the Phipps voyage see above, p. 26, n. 1 above.

[2] In the book on his voyage, Phipps wrote: 'To Mr Banks I was indebted for very full instructions in the branch of natural history ... which I acknowledge with particular satisfaction, as instances of a very long friendship which I am happy in an opportunity of mentioning' (*Voyage towards the North Pole* pp. 12–13).

[3] His poor spelling and language in this letter make that quite clear.

[4] Von Troil is trying to write 'write' (phonetically correct).

[5] Von Troil sometimes uses a lower case 'i' instead of a capital 'I' when talking about himself.

[6] Von Troil is trying to write 'what'.

[7] He is here referring to the expedition to Iceland in 1772.

[8] Von Troil's return to Sweden.

[9] John Gideon Loten (1710–89), FRS, was the Dutch governor of Ceylon, whom Banks had met in Utrecht in March 1773. Loten brought drawings of zoological specimens from Ceylon which Sydney Parkinson, a natural history artist employed by Banks, copied (Carter, *Sir Joseph Banks*, pp. 42, 46–7, 117). According to Lysaght, 'the originals, signed and dated, are in the Zoology Library of the British Museum' (*Banks in Newfoundland*, p. 103). The Natural History Museum has 40 watercolour drawings, mostly painted on vellum, dated 1767. The recent biography by Raat, *The Life of Governor Joan Gideon Loten*, details Banks's relations with Loten.

been curios to know any thing about You, and no men have ask'd more, than our Great Gustaf,[1] hwo [sic] under the greatest business of political cares has taken a pleasure to hear about You and Your discoveries.

I hope Professor Murray[2] already has given You notice about the drawings he has got from me.[3] I am not sure, but I believe You will hardly have any Germans who make it better and keeper than the English's engravers.[4] How will it be with Your voyage to the North pole?[5] I have from the newspapers seen that You have been in Holland,[6] and that You not have put out of Your mind that tour.

How sun will Your voyage be ready? and how goes it with your southsee collections. I hope somes of your plants are already engrav'd.

As we know very little about the eruptions in Island, wil You not be displeasd, to have a cataloge of all the eruptions, who have been observ'd, by the Islanders. I have found them myself in a historie we have in the Danish language about the fire spouting mountains in Island.[7]

I am afraid for Your friends in Otaheiti,[8] after that I have seen a book with titel: Observations of the present state of the waste lands of Great Britain, on occasion of the establishment of a new colony on the Ohio.[9] How is it? I know You will, and I wish they may be in their happy situation, and it is not possible if Your or any other nation will send colonies there.

What for plants will You have by Björnasbo?[10] I vill be a commissionair to You in sending them, but I hope once to have the pleasure to see You self these for collecting preferables copies. In the mean while I hope to be conserv'd in Your favour and friendship. – Live so well as I wish, I am sure You will be happy, and let me now and then know Your situation. It will be interesting for all Your unknown friends, and the most sensible pleasure, for

Your most obedient humble servant
[*Signed*] Uno von Troil

---

[1] King Gustavus III of Sweden (1746–92). He reigned from 1771 until his assassination in 1792.

[2] Johan Anders Murray (1740–91), a student of Linnaeus, a Swedish-German physician and botanist, later director of the botanical garden at the University of Göttingen, with whom Banks frequently corresponded in the 1780s. See: Dawson *The Banks Letters*, pp. 626–8.

[3] One of these was doubtless a drawing of the Great Geysir, which was printed on p. 244 in von Troil's *Letters on Iceland*.

[4] In 1773, 'there still remained the problem of engraving for reproduction and engravers to be found' (Carter, *Sir Joseph Banks*, p. 115). These refer to engraved copies of the *Endeavour* flora.

[5] In 1773 Banks was planning the Arctic voyage with his old friend Constantine Phipps (see Introduction, p. 26 above).

[6] In February–March 1773, Banks went to Holland with Charles Greville, nephew of Sir William Hamilton. There he met Dutch Greenland captains, gathering useful information about the ice between Spitzbergen and Greenland for Phipps's proposed polar voyage (Carter, *Sir Joseph Banks,* pp. 115–19).

[7] This is doubtless *Fuldstændige Efterretninger om de udi Island Ildsprudende Bierge, deres Beliggende, og de Virkninger, som ved Jord=Brandene paa adskillige Tider ere foraarsagede*, Copenhagen, 1757, an 88-page pamphlet dedicated to Otto Thott (1703–85), a renowned minister of state and a great bibliophile, by H. Jacobæus [Halldór Jakobsson (1735–1810)].

[8] Tahiti, where Banks and Cook observed the transit of Venus.

[9] Arthur Young's book *Observations on the present state of the waste lands of Great Britain. Published on occasion of the establishment of a new colony on the Ohio*, published in 1773.

[10] Unidentified.

P.S. How is it with thy dear Dr. Lind?[1] I should be very glad to hear him with his Bishop well.[2]

BL, Add MS 8094, ff. 36–37. Autograph letter.[3]

14. Thomas Falconer to Banks, 17 April 1773

Chester

Dear Sir

I can only return my sincerest thanks for another most entertaining letter. Iceland was a part of the world I was totally unacquainted with; & tho' no desirable situation to live in, yet more curious than the more cultivated Countries. From whatever cause Volcanos proceed they must discover the contents of the soil by the nature of the lava, & I should fancy would prove what minerals are pretty abundant. The trembling of the earth from the water is another proof to me that that phaenomenon arises from different causes. The effect of steam in a common fire engine shews its vast powers, & will account for a local earthquake, but we cannot recur to the same cause for one earthquake which extended from Iceland to Morocco, & from Boston to Turkey, as that did which we call the Lisbon earthquake.[4] The periodical rising of the steam at Geiser hath so puzzled my little philosophy, that, after several attempts, I have given up the point & shall wait impatiently for the further account, which you have been so kind to promise. The common doctrine of periodical springs, (viz. on the principle of a syphon,) & that less known & less probable solution, which Addison[5] stole from Berniers account of Indostan[6] (viz. the melting of the snows) can have no place here. The quantity of water must, I think, contradict both. You enquired, I doubt not, whether it was the same all the year round, & whether the water was impregnated with any sulphur or salts. The heat is probably uniform, & the efflux must then arise from a periodical flow of water from some higher grounds into the caverns under the plain. Had this been the only cavity in the neighbourhood, the natural philosophers would have laid hold on that circumstance to account for the appearance of the spout but, as there are so many wells in the environs, the varied appearance of this

---

[1] Dr James Lind (1736–1812), a member of Banks's Iceland expedition. See Appendix I.

[2] This is in all probability Dr John Douglas (1721–1807), FRS, Bishop of Salisbury, who whilst Canon of Windsor during the years 1776–81 had been involved in preparing Captain Cook's journals for publication. In 1807, Lind was a member of the funeral procession of the late bishop, which took place from St George's Chapel Windsor on 25 May 1807: *Gentleman's Magazine*, vol. 77 (1807), pp. 475–6, 583. (I am indebted to Dr Christopher Goulding of Newcastle-on-Tyne for this information.)

[3] Note in Banks's autograph, verso: 'Mr Troille | Received May the 7th. 1773 | [Answered] – November 29 – [1773]'.

[4] The Lisbon earthquake on 1 November 1755, All Soul's Day, was a disaster that had a lasting effect on Europeans. A great fire broke out, followed by three large tsunamis. Much of Lisbon, one of the most beautiful cities in Europe at the time as well as being one of the largest, was destroyed. It is estimated that about half the population of 275,000 perished. There was also loss of life in Spain and Morocco. Tsunamis reached as far as England.

[5] Joseph Addison (1672–1719), English essayist.

[6] François Bernier (1620–88), French physician, traveller and philosopher, was the author of many works. In this instance reference is being made to *The History of the late Revolution of the Empire of the Great Mogul ...* . This had been 'English'd out of French'.

/*single spout*/ one is much more extraordinary. As for the others, it is no more than what Aetna & Vesuvius furnish; a collection of water heated & thrown up, but the cause of the Geiser spout must be totally independant of all the rest, or they would be affected in some degree in the same manner. The bed is probably deeper, or else we cannot account for the perpendicular form of the spouts. If the bed is deeper, the heat is probably more violent, & consequently the force of the expansion is augmented. But why should I tire you with words, when the result will only be a confession of ignorance. Your kindness to one so remote as I am from all the means of information, or knowledge is what I can never forget. I am much pleased to find my kinsman[1] will join your observations to his own. Nothing is more liberal in men of science than such an union. I hope your journey to Iceland will not be lost to the world. No description hitherto has been published worth reading[2] & yet no Island deserves it better. I am much pleased the intended voyage to the North pole is under the direction of so able an Officer as your friend Captain Phipps. My knowledge of the frigid Zone is too imperfect to afford any usefull observations; but your request shall be as a command, & I will endeavour to throw some thoughts on paper. I have yet forgotten to make any apology for not answering your kind letter sooner. The truth is, I am but just returned from some short excursions, having lately spent a month with your acquaintance Dr Bagot, the Canon of Christ Church.[3] I must now conclude, having only good wishes to offer, & am, with sincerest gratitude, your most obedient friend & Servant

[*Signed*] Thos. Falconer

Kew, Banks Correspondence I, JBK/1/2 f. 34. Autograph letter.
NHM, DTC 1, ff. 49–50. Contemporary copy.[4]

15. James Lind to Banks, 26 May 1773

Edinburgh

Dear Sir

Since my arrival here on Saturday the 15th current, I have been so much hurried in geting a house &c. that I have not /had/ time till now to thank you for the civilities, and friendship I lately received from you – Civilities which are so deep impressed, and of such a nature, that if ever I should forget them, I ought to be looked on as the most forgetfull, as well as ungratefull of all humankind – Your generosity, opposed to the ungenerous treatment of Government,[5] first made me severly feel my disapointments, and at the same time shewed me under what great obligations you had laid me, Obligations, which I shall

---

[1] Thomas Pennant.

[2] None in English, perhaps.

[3] Lewis Bagot (1740–1802) bishop of Bristol, Norwich and St Asaph. Matriculated at Christ Church in 1757. In 1771 Bagot was appointed a canon to Christ Church.

[4] Note in Banks's autograph, verso: 'Mr. Falconer | Received April the 21 [17]73 | [Answered] November 30 – [1773].'

[5] Daniel Solander had written to Lind in December 1771, entreating him to come with them as astronomer: 'What great thing could you not do, preferable to any body else in the creation.' Banks was not acquainted with Lind at the time but sent via Solander 'his warmest wishes' of having Lind 'as a fellow Traveller' (Beaglehole, *Journals of Captain Cook*, II, pp. 901–3). See further on Lind in Appendix I.

most gratefully repay, as soon as ever I shall have it in my power so to do: in the mean time please to accept of my best thanks, and I beg Mr Banks to command my services, poor as they may be, when, and where, he pleases; He may depend on having my best wishes, and endeavours to serve him. I beg to be remembered to all friends, and to Dr Solander,[1] the Bishop[2] desires to join me in complements to you and him, for the particulars of all here and of my present state of confusion, I refer you to Dr Gawn,[3] as also for the account of Penante's generosity to the people of Aron[4] when on his late tour through the Western Islands of Scotland – excuse this scrawl[5] – I am with Respect and regard

Dear Sir
Your devoted
and much obliged
humble Servant
[*Signed*] James Lind.

PS. as soon as I get settld I shall set about the working modle of Geyser[6] and the portable Sketchers. I beg to know if Capt Phipps is saild.[7]

BL, Add MS 33977, f. 24. Autograph letter.[8]
Chambers, *The Indian and Pacific Correspondence of Sir Joseph Banks*, I, pp. 147–8.

16. Ólafur Stephensen to Banks, 24[9] June 1773

Sviðholt in Iceland

Greatly esteemed and most Honourable Mr. Banks!
It is my duty to send you my most humble thanks for all the goodness your Excellency, was pleased to show me during your visit here in this country, to which I am greatly obliged, though it is unmerited. I am glad to hear that you had a safe journey home as I have learned from Mr Uno von Troil in a letter from Copenhagen this year.

Since your departure from this country I have done my best to fulfil your Excellency's commission to collect and subsequently charge our best copyists to transcribe the antiquities and sagas,[10] as many as I have been able to obtain according to your own

---

[1] Daniel Solander, the Swedish botanist. See Appendix 1.

[2] See previous letter, p. 182, n. 3.

[3] No Dr Gawn has been identified.

[4] Arran, see Pennant, *Tour in Scotland and the Hebrides*, pp. 196–203. It is not clear what is meant by Pennant's 'generosity'.

[5] Over the word *scrawl* is written '112'.

[6] It was Lind who measured the height of the Great Geysir at Haukadalur with a quadrant.

[7] Phipps sailed at the beginning of June (*Voyage towards the North Pole*, p. 21).

[8] Note in Banks's autograph verso: 'Dr Lind | Recei*v*ed June 1st 177[3 *page cut*] | [Answered] Feb*ruary* 16 1774'. There are in fact two endorsements, but giving the same information.

[9] This number is illegible. All that can be made out for certain is a number and then ^de, which e.g. rules out 1st. In Document 21 Ólafur Stephensen mentions having sent Banks a letter dated 24 June 1773, doubtless this one.

[10] The Danish word used here is *historier*.

specifications and the Torfæus Register.[1] Those that are ready have been sent by ship to Justitsraad[2] Andreas Holt,[3] a deputy in the Board of Trade in Copenhagen,[4] with the request that he will convey them at a safe opportunity to London.

I shall neither forget nor neglect continuing with this task, as well as in everything that I am able to do, I will always with the greatest devotion have the honour to be your Excellency's most humble servant.

[*Signed*] O.Stephensen

BL, Add MS 8094, ff. 34–35. Autograph letter. Translated from the Danish.[5]

Danish original in Gothic script:
Höyvelbaarene allerhöystærede herr Banks!
Min Pligt tilholdes mig at aflægge herved min ydmigste Takksigelse for ald den goedhed Deres Höyvelbaarenhed uforskyldt behagede at udvise moed Mig den tid De opholdt Dem her i Landet, hvorimoed er höylig forbunden. Det fornöyer mig at hiemreysen gik Lykkelig hvorom tilligemed en Heelsen fra Herr Uno von Troil indeværende Aar er bleven fra Kiöbenhavn Underrettet.

Siden Deres Afreyse fra Landet har jeg efter Deres Höyvelbaarenheds Mig givne Commission giort mig ald Umage for at samle og siden Lade udskrive af de beste Skrivere her er at faae i Landet Antiqvitæter og Historier, saa mange jeg har kundet overkomme, efter Deres mig tilskikkede Specification og Torfæi Register, hvoraf med Skibene sender til Justitz Raad Andreas Holt, Committeret i Oeconomie og Commerce-Collegio i Kiöbenhavn hvis for denne gang har faaet færdigt, hvilket beder han vil befordre ved sikker Leylighed til London.

Jeg skal ey forglemme eller effterlade at continuere hermed fremdeeles, og i alt hvis Jeg maatte formaae erviise, at jeg med allerstörste Hengivenhed vil altid have den Ære at forblive/Deres Höyvelbaarenheds/ydmige tiener/O Stephensen

17. Uno von Troil to Banks, 30 October 1773

Stockholm

Dear Sir.
It is for me the greatest pleasure in the world, when through my friends in England I can obtain any notice, about You and Your prosperity, allthough I should be much happier if Your business any time should permit You to honour me with few words from Your proper hand. I have to thank You for the happiest part of my life,[6] and when I remember the past time, it is always with the sweetest memory of Your kindness and friendship. Be assur'd Sir! that no man in the mortality, can take more part in every thing what arrive to You, and I hope to remember a friend, when I thank [*sic*] on You. By Mr. Asp[7] I am inform'd

---

[1] As Hermannsson pointed out this is the list of sagas made by Torfæus (Þormóður Torfason (1636–1719), the Latinized version of his name being Thormodus Torfæus), historian and antiquarian, in his *Series dynastarum et regum Daniae*, Copenhagen, 1702, IV, pp. 3–10, dedicated to King Frederik IV ('Banks and Iceland', p. 16n.).
[2] *Justitsraad*: an obsolete Danish minor title (see Glossary).
[3] Andreas Holt: see Appendix 13.
[4] *The Kommercekollegiet*, the Danish Board of Trade.
[5] Note in Banks's autograph, verso: 'Amptman Stephensen | Rec*eiv*ed Sept*embe*r 1 1773'.
[6] The expedition to Iceland.
[7] Per Olof von Asp (1745–1808), Swedish diplomat in London from 1770 and later chargé d'affaires 1775–6.

that You the past summer have been in the country, and thus have You not made any voyage to the north, as I believe You propos'd to You [sic] in the spring.[1] By the newspapers have I seen that somes ships from Your contry have been there[2] without to make any news discoveries. I believe likewise it will be greater pleasure for You to give Yours Otahiteans[3] a second visit, and it should be a very great temptation for me, allthough a clergyman, to be one with of the party, if You should propose it to me. Now I suppose You are to work with Your collections from the south see. How have You been contented with Mr Murray in Gottingen[4] and the German engraver? Are not any of the Islands drawings engrav'd, and thank You not to publish any thing about that voyage! One of my countryman have drawings from Vesuvius, where I have seen pillars, if not so regular as by Staffa, yet curioses by a Volcan, and as he told me of basaltes. I hope the Staffa drawings are engrav'd, and if You not take it ill, I should wish to have a little copy from the two drawings of Geyser, with the first opportunity.[5]

I hope You have gott the medaillon of Sir Charles Linne,[6] I have send to You for somes weeks ago /with one Captain Grayson/.[7] He is very resembling.

Now the Young Mr Lindegren[8] have a pot with åker berrys,[9] to send to You, with the first opportunity. The next year I hope to send You more. How is it with Dr. Lind? I should be very happy if he is weal.

Favour with Your remembrance,
Your most obedient humble servent
[*Signed*] Uno von Troil

BL, Add MS 8094, ff. 38–39. Autograph letter.[10]

18. Thomas Falconer to Banks, 8 December 1773

Chester

Dear Sir

Your obliging letter ought to have been at least acknowledged sooner. I designed it but was prevented by company. Your great kindness to so distant a friend as myself will ever

---

[1] See Document 13.

[2] This is a reference to Phipps, see p. 26 above.

[3] Tahitians.

[4] On Johan Anders Murray see Document 13, p. 180, n. 2.

[5] An engraving of the Great Geysir was printed in von Troil's *Letters on Iceland*, inserted between pp. 244–5.

[6] Linnaeus.

[7] Unidentified. (Could either be a merchant or naval ship's captain.)

[8] Carl Lindegren and his brother Anders were prominent Swedish merchants in London (the firm of Anders & Carl Lindegren). Carl was a Trustee of the Swedish Church in London and 'effectively Solander's executor after his death' (Duyker and Tingbrand, *Solander*, pp. 273n, 360n). Von Troil mentions in his autobiography that Carl Lindegren was especially helpful during his stay in England, helping to finance his voyage to Iceland after Banks's invitation ('Själfbiografi', pp. 166–7).

[9] The generic name for the åkerberry is *Rubus arcticus*. Found in Northern Sweden, it is related to the blackberry.

[10] Banks wrote verso: 'Mr Troil | Rec*eiv*ed Nov*ember* 20 1773 | [Answered November] 29 [1773]'.

/demand/ my gratitude, & thanks are all I can return for it. Captain Phipps voyage[1] has not quite answered unless in deterring others from a like attempt. I tremble for the Southern emigrants, & tho I admire your spirit so much am glad you have not run the hazard of so dangerous a tryal. Yet after all I cannot but think there is an open sea near the Pole. The vast soundings in those latitudes do not indicate narrow seas. Should the experiment be repeated would not Iceland be a better rendezvous than Spitzbergen? But you know those parts so well I shall not expose my ignorance. When the whole is published I hope you will arrange the few plants got from thence, at least to render the work as usefull as possible. One other scheme of discovery yet remains in a very different climate & where the danger would scarcely be much less than in sailing round the Globe. I mean the navigation of the river Senegal[2] as high as one could go. The great Niger[3] in Africa is thought to be the source of those two great streams of Gambia & Senegal. So large a body of water must naturally furnish ample materials for natural History. The Climate as far as the Gum forests is hot & unwholesome, but the natives come in large numbers down to these forests from very inland parts & bring down as I have heard, carving & other works of art which shew that civility is far advanced in some of those regions. If the ground rises the air will be cooler, & provided Europeans can submit to feed like the natives, a healthy European is as likely to subsist on the river Senegal as in the East Indies. There is a new field for enquiry yet unexplored, but I have not time to explain particulars.

I have engaged some time at the request of my friends in antient Geography, & find it a larger & more troublesome plan that I expected, & too great for an invalid constitution. Not merely the longitude & latitude of places, but my enquiries must at times lead me into both natural & civil history. I am however but a novice in the work & proceed very slowly from various engagements, It is indeed an attempt which I do not yet chuse should be known to the world tho' as it is so likely to prove abortive. Your last kind letter on Iceland was an account of the boiling fountain of Geiser[4] if you have leisure to continue the subject you will ever find me gratefull for [your] communications,

I am Dear Sir, with truest [tear]
your most obedient & devoted friend [tear]
[Signed] Servant Thos Falcon[er]

Kew, Banks Correspondence I, JBK/1/2 f. 42.[5] Autograph letter.

---

[1] See Document 12.

[2] Banks was very interested in the exploration of Africa, founding the Association for Promoting the Discovery of the Interior Parts of Africa on 9 January 1788 with eight others. During the *Endeavour* voyage in 1771, Banks set foot on African soil.

[3] The legendary city of Timbuktu was known to be on the River Niger. Banks was to send Mungo Park (1771–1806) on a mission to discover the course of the Niger, which he reached in 1796. See further Middleton, 'Banks and African Exploration', pp. 171–6.

[4] See Document 12.

[5] Note in Banks's autograph, verso: 'Mr Falconer | Received December 10 1773'.

19. Peter Boddaert to Banks, 20 April 1774

Utrecht

Sir![1]

Having learned that you are back from your voyage to the Pole[2] [see Plate 16] I cannot fail to write to you to express again the great pleasure which I have had in making the acquaintance of a man of your merits and your wisdom (knowledge), I congratulate you Sir on your return and I hope that you will publish also your voyage to the Arctic, I read with the greatest pleasure your voyage to New Zealand,[3] I find human nature everywhere, virtues and vices; I hope Sir, that you will once more enrich the world of science with your discoveries in natural history, with the exception of your Sitodium altile,[4] you make no mention of plants, although it seems you have gathered many.

After your departure from here, our public garden has received a Lobelia[5] which appears to me to be of a new species although Mons. Scopoli[6] and Mr. Linneaus named it an Hyoscyamus.[7] If you wish to have a plant or seeds, I would be pleased to send you some. Mr. Burman[8] has sent me many seeds of plants sent by Mr. Pallas[9] from Tartar Russia.

Dare I remind you Sir to send me a Dionaea Muscipula,[10] you would do me a great favour and if there is anything here that I could be of use to you I beg you Sir to let me know, I will always consider it an honour to give pleasure to a man to whom the scientific

---

[1] Boddaert's French is imperfect, though this translation fails to give that impression. The underlining is the writer's.

[2] Boddaert was under the impression that Banks had accompanied Constantine Phipps on his Arctic voyage. This is not really surprising as it was believed by many that Banks had again voyaged north. For example an item in *Lloyd's Evening Post* reported: 'some English Gentlemen (among whom is Mr. Bankes) ... are determined to sail this year towards the Pole, from whence they intend to attempt to sail westward, towards America, and particularly to try to gain the coasts of California' (19–22 March 1773, p. 279).

[3] He is referring to the *Endeavour* voyage, when the entire coastline of New Zealand was mapped, including the discovery that it was two islands. Banks's journal of the voyage was first published in an incomplete edition by Sir Joseph Hooker in 1896, subsequently a complete edition by J. C. Beaglehole was published in 1962 (Carter, *Banks. The Sources*, pp. 53–4).

[4] *Sitodium altile*: the breadfruit, also known as *Artiocarpus altilis*. Banks was extremely interested in transplanting breadfruit from Tahiti to the Caribbean and persuaded the government to fund a voyage. It was Banks who decided that Captain Bligh would command that ill-fated voyage of the *Bounty* in 1787–9.

[5] Lobelia (*Lobelia inflata*) is named after the French botanist Matthias de Lobel, who died in London in 1616.

[6] Two letters are extant from Giovanni Antonio Scopoli (1723–88), a professor of botany and chemistry at the University of Pavia, to Banks (Dawson, *The Banks Letters*, p. 738). Scopoli adopted the Linnean system, applying it to the classification of plants and animals living in northern Italy.

[7] *Hyoscyamus*: henbane.

[8] Nicolaas Laurens Burman (1734–93) was a Dutch botanist. He became professor of botany at the University of Amsterdam. A Linnean, his works include *Specimen botanicum de geraniis*, 1759, and *Flora Indica*, 1768, which was eventually completed by J. G. König (see Introduction, p. 8 above). He had written to Banks from Amsterdam in 1772 wishing him a prosperous journey to Iceland (Dawson, *The Banks Letters*, p. 188). Banks met Burman in Holland in 1773.

[9] Peter Simon Pallas (1741–1811), FRS, naturalist and traveller. He sent Banks many letters and specimens from St Petersburg where he enjoyed a privileged position under the auspices of Empress Catherine II (Dawson, *The Banks Letters*, pp. 644–6). Banks sent a collection of plants from Kew to the Empress (Carter, 'Sir Joseph Banks and Catherine II of Russia 1795').

[10] *Dionaea Muscipula*: the Venus flytrap, a carnivorous bog plant. It is only found in the savannas of the Carolinas in the United States. Banks did send him one: undated letter from Boddaert in Utrecht, BL, Add MS 8094, f. 191.

world of Europe owes so much, if you would honour me with your reply to this please be good enough to send it with Mr. John Berendz,[1] a merchant in London.

Mr. Burman has entrusted me to pass on his compliments to you, he would with pleasure write. But he has not the time. I beg you Sir, to accept my respects and to believe me sincerely to be

your most humble Servant
[*Signed*] P Boddaert

BL, Add MS 8094, f. 50. Autograph letter. Translated from the French.

French original:
Monsieur!
Ayant entendait que vous chez de retour de votre voyage au Pole je ne pus Manquer de vous ecrire pour vous temoigner encor la plaisir, que j'ai eu d'avoir fait connaissance avec un homme de votre merite et des vos lumieres, je vous felicite Monsieur de vote retour et j'espere, que vous donnerez aussi votre voyage arctique, je lis avec le plus grand plaisir votre voyage aux Isles de la Nouvelle Zélande ; J'y trouve pas tout la nature humaine, des Vertus et des vices; jespere Monsieur, que vous enricherait encor le Monde Scavant de vos decouvertes en histoire Naturelle car outre votre Sitodium altile, vous ne faites aucune mention des plantes, quoique j'esperait, que vous en avez receuillé beaucoup.

Apres Votre depart d'ici, notre Jardin public, a recu un Lobelia, qui me parait un genre nouveau quoique Mons. Scopoli et M. Linné la nomment un Hyoscyamus Si vous desideré la plante ou la Semence, je me ferait un plaisir, de vous la faire avoir. Monsieur Burman m'a envoyé beaucoup des Semences des plantes, que M. Pallas a envoyé de la Russie Tartarique.

Oserai je Vous faire le souvenir Monsieur de m'envoyer une Dionea Muscipula, vous m'obligererez beaucoup; et si il y a quelque chose ici, dont je vous puisse etre utile je vous prie Monsieur, de me le faire Scavoir, je me le compterai toujours un honneur de faire plaisir, a un homme, de qui l'Europe Scavante a tant d'obligations; si vous voudrez m'honorer de votre reponse a celle ci, ayez la bonté, de la remettre a M. John Berendz Marchand a Londres.

Monsieur Burman M'a chargé de vous faire Ses compliments, il ecrirait Volontiers, Mais Son temps ne la pas permis, Je vous prie Monsieur de recevoir mes respectueux Compliments et de me croire avec Sincerité

Monsieur
votre tres humble Serviteur
P Boddaert

20. Bjarni Jónsson to Banks, 12 August 1775

Skálholt

To the most noble
Sir Joseph Banks
Bjarni Jónsson
sends many greetings.

The letter you sent me four years ago as you prepared to return from Iceland to England filled me with great joy, inasmuch as it is sprinkled with much of the wit of

---

[1] The name may have been misspelled. Unidentified.

your culture. I received a gift with this letter, most suited to your generosity, but as to me, wholly undeserving; namely a silver clock and 24 bottles of red wine, both [items] most pleasing to me, for which now at last I send you most sincere thanks. I still live, most cultured man, with my wife and four children,[1] in the same place in which you visited me, occupying the position of schoolmaster. My son,[2] whom I recommended most strongly to your humanity that time while living in Copenhagen, died after an interval of a few weeks, overcome by a malignant fever. I still mourn his premature death, and shall mourn it as long as I live. To such an extent is there nothing in human affairs lasting or stable!

From this most remote corner of the globe I have nothing to write to you worthy of your notice apart from the fact that our ancient monuments, by the attention and auspices of learned men, some within, some outside our country, are beginning to see the light of day. Among those which have been printed in Iceland, the annals commonly called the Annals of Skardsa are particularly worthy of mention, for they include the deeds not only of our own people but also of many other nations throughout Europe.[3] In the process of being printed now is the Saga of the Laxdalians,[4] an ancient work, and of the highest quality in that genre, with a Latin translation so that it may also serve the needs of foreigners. With our Hekla nothing remarkable is happening, for, since your departure it has remained for the most part peaceful, except for the fact that from time to time it emits smoke, which is a most certain sign that the fire below is not yet absolutely asleep or extinct. Recently I wrote a historico-critical treatise on the papist feast-days commonly known as Rogation Days,[5] once common throughout the whole of Europe, and usually celebrated with the greatest superstition. At the same time I wrote in the vernacular a certain poem called *The Charms of London*,[6] which I have decided, when the opportunity arises, to send to you, most cultured man. For you do not despise our things, however slight and meagre in weight. I am very happy that you have successfully returned home, your maritime labours happily at an end – these labours which are the cause of the great

---

[1] See Document 6, p. 160, n. 3 above.

[2] Páll Bjarnason: see p. 160 above.

[3] *Skarðsárannáll* was written by the Revd Björn Jónsson of Skarðsá (1574–1655). He had no formal education but was considered exceedingly learned. Björn Jónsson was in the employment of Bishop Þorlákur Skúlason (1597–1656). This annal, which spans the years 1400–1640, is his main work, and one of the major sources for the history of the Reformation in Iceland. Especially famous is his contemporary account, the oldest extant, of the 'Turkish Raid' of 1627 (see p. 82, n. 2 above). As Bjarni Jónsson writes, this annal is also noteworthy for its accounts of events in Europe. For example, the death of Queen Anne, wife of James I, is duly recorded for the year 1619 – but then she was, of course, a Danish princess, daughter of King Frederik II.

[4] *Historia Laxdælensium* in the Latin text of the letter. The first Latin translation of this saga *Laxdæla-saga sive Historia de rebus gestis Laxdölensium. Etc.* was not published until 1826. Þorleifur Repp was the translator (see Wawn, *The Anglo Man*). An Icelandic/Danish/Latin text was in the pipeline at the time that Bjarni Jónsson was writing to Banks, but the work proved to be much more time-consuming and costly than was originally envisaged (Helgason, *Hrappseyjarprentsmiðja 1773–1794*, p. 61).

[5] *Tractatus Historico-Criticus de Feriis Papasticis Vulgo Gagn-Dagar*, printed in Copenhagen in 1784. This work is mentioned as being ready for publication by von Troil (*Letters on Iceland*, p. 173).

[6] It is well known that the letter-writer's namesake Bjarni *skáldi* (the poet) Jónsson (c. 1575–80–c. 1655–60) wrote a poem about London, preserved in many manuscripts. The poem written by this Bjarni Jónsson is, however, unknown, though there is always the possibility that it will turn up in future research by the staff of The Árni Magnússon Institute for Icelandic Studies (my thanks to Dr. Margrét Eggertsdóttir who provided this information).

splendour of your name, not only among your own people but also everywhere throughout the world, especially when you discovered hitherto unknown lands.[1]

I hear that an account of your voyage has now begun to be published,[2] and indeed you do well, most cultured man, in that you do not grudge the people the fruit of your labours and of your learning. Would that I might also see this exceptional product of your genius! I have not received the silver chain which I had sent to Dr. Solander for gilding, which he could have easily forgotten to do, because of the occurrence of so many important affairs; nor is such a small matter worthy of being mentioned. Although you have given me free access to your kindness by letter, nevertheless it does not seem proper to use this favour on this occasion, especially since I can in no way repay in kind. Farewell, most cultured man! May you live long and happily, mindful of us, as the enrichment and embellishment of your country and the learned world!

I issued this at Skálholt on 12th August, 1775.
If you honour me with a letter, you can send it to Copenhagen, to the student Johannes Finnæus[3] who lives there. My wife again sends you very many greetings and we bid you once again farewell!

BL, Add MS 8094, ff. 87–88. Autograph letter. Translated from the Latin.[4]
Hermannsson, 'Banks and Iceland', pp. 20–22 (in Latin only).

Latin original:

Generossisimo Domino JOSEPHO BANKS
Salutem plurimam dicit
BIARNUS JOHNÆUS

Literæ, quas ad me dedisti anno abhinc retrorsum quarto, ad reditum ex Islandia in Angliam accinctus, magno me perfuderunt gaudio, utpote multo humanitatis sale conspersæ.

Accepi cum epistolio donum, Tua quidem munificentia dignum, sed quantum ad me, prorsus immeritum; horologium nempe argenteum et 24 lagenas vini rubri, utrumque mihi acceptissimum, pro quo nunc tandem grates tibi solvo meritissimas. Vivo adhuc, Vir humanissime, cum uxore et quatuor liberis, eodem, quocum me invisisti, ludi magistri officio fungens. Filius meus, quem Tuæ Humanitati id temporis Hafniæ degentem impensissime commendabam, paucis interjectis hebdomadibus, febri maligna correptus, supremum obiit diem, cujus ego præmaturum fatum adhuc doleo, et quoad vixero, dolebo. Adeo nihil est in rebus humanis satis firmum et stabile!

Ex remotissimo hoc orbis angulo quod tibi scribam, nihil habeo Tuis oculis dignum præterquam quod antiqua nostra monumenta eruditorum cura et auspiciis, partim intra, partim extra patriam,

---

[1] Bjarni Jónsson is of course referring to the *Endeavour* voyage.

[2] He is probably referring to the first excerpts of what would later become Uno von Troil's *Letters on Iceland*, which were first published in the Uppsala newspapers in 1773.

[3] Actually Hannes Finnsson. Johannes Finnæus would be Jón Finnsson, but *not* the son of the same name of Bishop Finnur Jónsson, who never went abroad to study, being of weak constitution and limited academic ability. Von Troil in his *Letters on Iceland* speaks of 'the learned Mr. John Finsson ... lately ... appointed his father's assistant and provost' (p. 173), but in the Icelandic translation of Haraldur Sigurðsson this has been corrected to *Hannes Finnsson*, another son of the bishop, rightly considered a great man of letters. There is no Johannes Finnæus in the list of Icelandic students registered at the University of Copenhagen (Jónsson, *Íslenzkir Hafnarstúdentar*).

[4] Note in Banks's autograph, verso: 'Biarne Jonsson | [Received] Oct*ober* 25 – [17]75 | [Answered] April 15 – [17]77'.

publicam lucem aspicere incipiunt. Inter illa, quæ domi prodierunt, annales vulgo Skardsaenses dicti præsertim nominari merentur, non enim nostra tantum, sed et permulta aliena apud Europæas nationes gesta continent. Sudat nunc sub prelo Historia Laxdælensium, opus antiquum et in eo genere optimæ notæ, cum versione latina, ut exterorum quoque usibus possit inservire. De Hekla nostra nihil jam memorabile occurrit, a discessu enim vestro quieta maximam partem permansit, nisi quod fumum subinde eructet, quod certissimo indicio est, nondum prorsus sopitam aut exstinctam esse subtus flammam. Scripsi nuperrime Tractatum historico-criticum de feriis papisticis vulgo Gangdage, per universam olim Europam usitatis, summaque cum superstitione celebrari solitis. Scripsi itidem vernaculo sermone carmen quoddam, Amœnitates Londinenses dictum, quod Tibi, Vir humanissime, data occasione mittere decrevi, neque enim Tu nostra spernis, quantumvis levia et exigui ponderis. Gratulor Tibi faustum in patriam reditum, post exantlatos feliciter labores maritimos, qui Tibi non tantum apud gentem Tuam, sed et passim alibi terrarum, magnum pariunt nominis splendorem, præsertim cum ignotas hactenus detexeris oras. Audio itinerarium Tuum jam edi cœpisse, et bene quidem facis, Vir humanissime, quod laborum Tuorum et eruditionis fructum publico non invideas. Utinam quoque mihi eximium hunc ingenii Tui fœtum videre contingat! Catenam argenteam, quam Domino Solandro inaurandam commiseram, nondum recepi, quod ob tot tantique momenti incidentia negotia oblivioni facile dari potuit, nec tantilla res digna est, ut memoretur. Quamvis liberum mihi aditum per literas ad me missas ad humanitatem Tuam mihi dederis, ea tamen venia hac vice uti non placet, præsertim cum vicem nullo modo reddere queam. Vale, Vir humanissime, nostrique memor felix diu vive in patriæ Tuæ orbisque literati utilitatem et ornamentum.

    Dabam Skalholtiæ
    pridie Idus Aug[usti] 1775

Si quo me epistolio dignaris, Hafniam id mitti potest ad studiosum Johannem Finnæum ibi degentem. Te plurimum salutat uxor mea, iterum iterumque vale.

21. Ólafur Stephensen to Banks, 25 August 1775

Sviðholt

Greatly esteemed and Most honourable Mr Banks

I permitted myself to write two letters to your Excellency on June 24th and August 6th 1773,[1] the latter was accompanied by 24 Icelandic sagas[2] of the best quality and a *Lignum fossile*[3] with its Strata and a ruler made of the same material. Since then I have not been so fortunate as to obtain any news of these letters nor whether your Excellency received the Enclosures. Thus it occurred to me that your Excellency and Dr. Solander must have gone on a journey to southern lands, which news from Copenhagen has since confirmed.[4] Therefore I sent no letters to London last year. But now the ships have arrived and brought the news that the honourable gentlemen are now returned safely to their fatherland, a fact I congratulate you on with all my heart, and I cannot refrain from the opportunity and the honour of sending 18 transcriptions of Icelandic sagas as well as some pieces of our Icelandic spar,[5] as it is found here. Though I hardly think it is the right

---

[1] The former letter is presumably Document 16, while the latter is apparently not extant.

[2] The Danish word used is *historier*. These books are probably among those later deposited in the British Library. See Introduction, p. 21 above.

[3] *Lignum fossile* is a variety of lignite (brown coal) called *surtarbrandur* in Icelandic.

[4] This rumour, however, was untrue.

[5] Spar is a crystalline, easily cleavable and non-lustrous mineral.

kind or will be of any use, I also enclose a second piece of *surtarbrand*, the largest I was able to obtain for you. I wish I could show by my deeds how much I wish to do all that is in my power to serve my honourable gentleman, which I consider an honour and great duty. I am very sorry to report that one of the pretty English tea-boards[1] that your Excellency so graciously gave me and I would have liked to keep in memory of you and conserved as well as possible, but on account of my servants' carelessness has been damaged, and I tried having it repaired in Copenhagen last year but nobody there was able to do the work. Dare I ask that it could be repaired in London, and returned to me next year. In that case I would again be in your Excellency's great debt and such goodness would deserve in return anything I could do. Thus I dare take the liberty to enclose the tea-board and I ask you humbly to pardon my letter.

I remain with my highest esteem,
Your Excellency's humble servant,
[*Signed*] O. Stephensen.

BL, Add MS 8094, ff.122–123. Autograph letter. Translated from the Danish.[2]

Danish original in Gothic script:
Höyvelbaarne
Allerhöjstærede Herr Banks
Ved 2de Skrivelser af 24de Junii og 6te Aug^{sti} 1773 gav Jeg mig den ære at opvarte Deres Höyvelbaarenhed, det sidste fulgte 24 Islandske Historier af det beste slags. 1 stk. Lignum fossile med sine Strata og en lineal giört af samme Materie; siden da Jeg har ikke været saa lykkelig at erholde nogen Efterretning om disse mine Skrivelser og medfulgte var Deres Höyvelbaarenhed indhændiget; saa gior Jeg mig den Tanke at Deres Höyvelbaarenhed og Herr Doctor Solander maatte være reiste til de Sydelige lande, hvilket og Efterretningerne fra Kiöbenhavn have bestyrket. Hvorfor Jeg og ikke heller afsendte noget brev til London forrige Aar. Men da Jeg nu med Skibene har faaet underretning om at Höyvelbaarne Herre var kommen lykkelig og vel hiem til Fæderne Landet igien, hvormed hiertelig Gratulerer; saa maae Jeg ingenlunde efterlade den Ære som Jeg tager mig ved at tilsende dem hermed 18 Stk. Afskrevne Islandske Historier samt endel af Vor Isl: Spat. saadan som det falder her. Hvilket Jeg troer næppe vil blive af det rette slags eller komme til nytte. 2de Stk. af Surterbrand lader Jeg og hos fölge, saa store som Jeg har været i Stand til at kunne skaffe dem. Jeg ville önske Jeg kunne viise i gierningen hvor stor Villie Jeg har til at kunde i alt hvis mig maatte være mueligt være Höjvelbaaren Herre til tieniste, hvilket Jeg anseer for min Ære og store Skyldighed. Det giör mig meget ont at eet af de smukke Engelske tee bredt, som Deres Höyvelbaarenhed var saa Gratieux at forære mig, og som Jeg ville til en amindelse have eyet og Conserveret paa beste maade om formedelst mine Folkes uforsigtighed kommet til Skade, hvilket Ville have ladet giöre til enten i Kiöbenhavn næst afvigte Aar, men ingen har samme stede været i Stand til at paatage sig arbeidet. Maatte Jeg nu giöre mig saa dristig at bede det maatte vorde Reparet i London, og mig til næste Aar tilbage sendes. Ville Jeg finde mig Deres Höyvelbaarenhed meget obligeret igien, og saadan store Godhed muelig ste maade forskylde, til denne Ende Vover Jeg at lade thee bredtet medfölge, hvilken Frihed, saavel som denne min Skrivelse beder Ydmigst maatte pardoneres som henlaver med allerstörste Höyagtelse/Deres Höyvelbaarenheds/ydmige tiener/O Stephensen.

---

[1] Probably a tea-tray. Banks bought a set of Queen's Ware (named after Queen Charlotte, consort of George III in 1765) for his Iceland voyage.

[2] Note in Banks's autograph, verso: 'Mr Stephenson | [Received] Feb*ruary* [*blank*] [17]76.'.

22. Uno von Troil to Banks, 25 March 1776

St[ockholm]

Dear Sir.

If not Your letters of the first of Januari,[1] had assur'd me of Your favour and remembrance, I should certainly have been afraid, that You look'd uppon me as the most ungratefull man in the world, when in so long a course of time, I not have answered Your very obliging letters, that I reciv'd with Mr. Wicksell.[2] I can not give somes valuables reasons,[3] why I have been so neglectfull, and I believe it hardly necessary, when I am sure that You now are going to pardon my fault.

For the drawings of Staffa,[4] and Your Southern Voyage,[5] I must humbly thank You, and wish, that in any thing, I could show my acknowledgement. Possibly I will in short time publish somes letters about our voyage to Island, and I suppose You will permitt me, ther to make use of them.[6]

The Labrador stone[7] You have sent with Mr. Wicksell I have shown to many littologists[8] heer, and all of them look uppon him as a very curiouse one. I am not sure, that I in English can give You their sentiments about him, an for that I have sent them in Swedish to Dr. Solander, who can give You notice about it.

No body could be more satisfied than I if it for me should be possible to procure You a compleat collection of all Swedish plants, who are not to be found in England. I hope You in short time will have the most of them, then Dr. Gahn[9] has told me, that he has sent You a collection of plants from Lappland, and heer I will give You a lot of plants, that a botanist from Switzerland has collected, and I shall send You as soon the sea is open. For the Schoenus ferrugineus, Lin.[10] I vill self collect many and fines specimens. How happy, if I once more could accompany my dear Mr. Banks in collecting plants, and that in my mothers estate![11] I should look uppon me as the happiest man in the vorld if I once could procure a man somes pleasures in Sveden, who has shown me so great deal of favour and friendship in your happy country and in our pleasant tour to Iceland. Likevise I shall send You a litle collection of Svedish minerals as a present from

---

[1] Not extant.

[2] Unidentified.

[3] Von Troil is wrongly adding an 's' to adjectives when they accompany a plural noun, as would be done in most European languages other than English.

[4] The Staffa drawings are preserved in BL, Add MS 15509.

[5] Nothing was published at this time, so these would be the engravings from the *Endeavour* voyage.

[6] On von Troil's book see Introduction, p. 22 above. There are no drawings of Staffa in this book.

[7] Probably a stone picked up by Banks on the Newfoundland and Labrador expedition of 1766. Banks had a Labrador collection consisting of stones he found while there (Lysaght, *Banks in Newfoundland*, p. 124).

[8] A lithologist is one who is versed in mineralogy which treats of the nature and composition of stones and rocks.

[9] Dr Henrik Gahn (1747–1816), was a Swedish physician who had studied botany at Uppsala University under Linnaeus. He subsequently turned to medicine, becoming one of Sweden's most celebrated doctors. During the years 1770–73 he visited Germany, Holland and England. Von Troil met Henrik Gahn during his European tour and again when in London in 1772.

[10] Brown bog-rush.

[11] Uno von Troil's mother was Anna Elisabet Angerstein.

Mr. Bergman Professor of the Chymie at Upsala.[1] He promises if You please to send You more, that You with the time may have it compleat, and if You not vil*l* have it for Your own use, You may serve some of Your friends with it. I suppose that You have got letters from Sir Charles Linneus. I told him that You vish'd to continue Your botanical correspondence with him, and he could no other than regarde it as a honour to him; but You must pardon him if he not so often as I wish writes to You, then he begins to grow old, and has not a litle deal of caprices, but I think alvays about him, as the Englishman told the Frenchmen about Your great Marlborough: <u>I have forgot the faults of so great a man</u>. Mr Kallström[2] lives in Stockholm and has the direction of the kings garden.[3] He was very much surprised to hear, that You makes him the honour to remember him, and will he certainly in short time have the honour to write to You. Somes notices about the Otaheite man[4] who has been in England should be very agreables to me; as I know what part You take in all thing concerning that country, I am sure, no body can give me more interesting news than You, and no body can be more glad thereof than I. Lastly I hopes You not vill take it ill Sir if I beg You to send me with the first occasion two Cylinders after the following measure.[5]

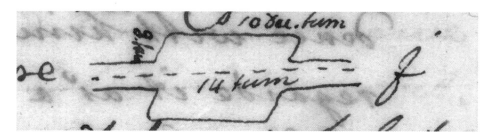

I should send money, but I do not know, what the cost, and I hope You will be good to lay out money for them; but for all You must let me know when You send them, then the*y* are contrabands heer. I should wish soon to see somes vords from You, to hear that You are well, any thing about your Southern Collection and any thing else. Be sure that is it the greatest pleasure, for a man that heartly loves and respects You, and who never can desist to be

Your most obedient humble servant
[*Signed*] Uno v. Troil

---

[1] Torbern Olof Bergman (1735–84), naturalist and chemist. Many of von Troil's *Letters on Iceland* are addressed to Professor Bergman and an essay entitled 'Professor Bergmans's Curious Observations and Chemical Examination of the Lava and other Substances produced on the Island' is included as an appendix to that volume.

[2] Andreas Kallström (1733?–1812), botanist. He had been a gardener in London and on 29 March 1776 this Swede sent Banks a letter telling him he was now Director of the King's Gardens and was about to draw up a catalogue of his plants, which he would send to him. Banks could then request what plants he required: BL, Add MS 8094, ff. 90–91.

[3] The name of this garden is Kungsträdgården (the King's Garden), formerly the King's formal garden and now a park in central Stockholm. Dating from the Middle Ages, in the 18th century it was a Baroque pleasure garden.

[4] Omai: see Introduction, p. 26 above.

[5] Sketch diagram in a letter from Uno von Troil, dated 25 March 1776. BL, Add MS 8094, f. 125. Courtesy of the Trustees of the British Library.

P.S. I had like to have for got to tell You, that I am to be marryed.[1] As a clergyman I can not always have a Miss Jorun,[2] and for that I must take a wife. I am sure you wish to hear me happy.

BL, Add MS 8094, ff. 124–125. Followed by list of plants, f. 126. Autograph letter.[3]

Genus	Species	English name
Circæa	*Alpine*	Alpine enchanter's nightshade
Veronica	*Maritime*	Our Lady's faith
Veronica	*Verna*	Spring speedwell
Eriophorum	*alpinum*	Alpine cotton-grass
Agrostis	*Arundinacea*	[Reedy bent grass]
Poa	*Nemoralis*	Wood meadow-grass
Festuca	*Rubra*	Red fescue
Festuca	*Elatior*	Tall fescue
Avena	*Pratensis*	Meadow oat-grass
Arundo	*Epigejos*	Wood small-reed
Asperula	*Tinctoria*	Dyer's woodruff
Galium	*Boreale*	Northern bedstraw
Tillæa	*Aquatic*	Water pygmyweed
Myosotis	*Lappula*	European stickseed
Anchusa	*Officinalis*	Common bugloss
Pulmonaria	*Officinalis*	Lungwort
Andresace	*septentrionalis*	Northern fairy candelabra
Lonicera	*Xylosteum*	Fly honeysuckle
Ribes	*uva crispa*	Gooseberry
Asclepias	*vincetoxicum*	Swallow-wort
Selinum	*Palustre*	Marsh parsley
Athamanta	*Libanotis*	[Parsley spignel]
Laserpitium	*Latifolium*	Laserwort
Allium	*scorodoprasum*	Sand leek
Allium	*schoenoprasum*	Chives
Fritillaria	*Meleagris*	Snake's head fritillary
Ornithogalum	*minimum*	[Small star of Bethlehem]
Convallaria	*Bif…?* [*Illegible*]	
Daphne	*Mezereum*	Mezereon
Elatine	*hydropiper*	Eight-stamened waterwort
Ledum	*Palustre*	Marsh labrador tea
Arbutus	*uva-ursi*	Blackberry
Pyrola	*Umbellate*	Prince's pine
Pyrola	*Uniflora*	Single delight
Sedum	*sexangulare*	Tasteless stonecrop
Sedum	*Annuum*	Annual stonecrop

[1] He married Magdalena Elisabet Tersmeden in 1776.

[2] Jórunn was not an uncommon name in Iceland at the time. The most likely candidate is Jórunn Ásmundsdóttir (b. 1749) who married in 1775 Þórarinn Þorsteinsson (son of Þorsteinn in Nýjabæ, Banks's guide while in Iceland), later becoming the mistress at Hvaleyri in 1801.

[3] Note in Banks's autograph, f. 126 verso: 'Mr. Troille | [Received] April 15 [17]76 | [Answered] Oct*ober* 6 [17]78.'

Prunus	*Domestica*	Plum
Mespilus	*cotoneaster*	Common cotoneaster
Rosa	*Villosa*	Apple rose
Rubus	*Arcticus*	Arctic raspberry
Rubus	*chamaemorus*	Cloudberry
Potentilla	*Verna*	Spring cinquefoil
Aconitum	*Napellus*	Monkshood
Anemone	*Hepatica*	Common hepatica
Anemone	*Vernalis*	[Spring pasque flower]
Ranunculus	*Repens*	Creeping buttercup
Ranunculus	*polyanthus*	[Goldilocks buttercup]
Ranunculus	*Arvensis*	Corn buttercup
Scutellaria	*Hastifolia*	Spear-leaved skullcap
Melampyrum	*Nemorum*	Wood cow-wheat
pedicularis	*sceptrum carolin*	Moor-king
Linnæa	*Borealis*	Twinflower
Fumaria	*Bulbosa*	[Fumitory]
Orobus	*Vernus*	Herb linn
Orobus	*Niger*	Black pea
Vicia	*Sepium*	Bush vetch
Trifolium	*Hybridum*	Alsike clover
Trifolium	*Montanum*	[county flower of Oslo]
Trifolium	*spadiceum*	[Large brown clover]
Scorzonera	*Humilis*	Viper's-grass
Hieracium	*Dubuium*	[Mouse-ear hawkweed]
Hieracium	*Auricular*	European hawkweed
Hieracium	*praemorsum*	Leafless hawk´s beard
Carduus	*Crispus*	Welted thistle
Inula	*Salicina*	Irish fleabane
Centaurea	*Jacea*	Brown knapweed
Viola	*Mirabilis*	Wonder violet
Carex	*Uliginosa*	[Star sedge]
	Digitalis	[Slender woodland sedge]
	Montana	[Soft-leaved sedge]
	[*illegible*]	
	[*illegible*]	
	[*illegible*]	
	[*illegible*]	
	Sylvestris	[Hook sedge]
Salix	*Fusca*	Brown willow
Salix	*Cinerea*	Grey willow
Juniperus	*Communis*	Common juniper
Osmunda	*struthiopteris*	Ostrich fern
Polypodium	*Filix fæmina*	[Female polypody, a fern]
Lycopodium	*complanatum*	Ground cedar

BL, Add MS 8094, f. 126.

23. Bjarni Jónsson to Banks, 12 August 1776

Skálholt

To the most noble
Sir Joseph Banks
Bjarni Jónsson
sends many greetings.

I am surprised, most cultured man, that I have not received any letter from you, especially since I have been reliably informed that a letter sent by me almost a year ago has come into your possession; but perhaps other matters have deprived you of the opportunity of writing for a while. Hence I send you this letter again, so that you may know that I still live with my wife and children,[1] mindful of your kindness. I have nothing to write which might be worthy of your attention. Hekla, for the most part, shows a peaceful aspect, except for the fact that a smoky cloud arises from her, which is a clear sign that the flame below is not yet completely asleep.[2] That silver chain which Dr Solander received for gilding, I would wish to be sent back to me, when the opportunity presents itself. But if by chance it is lost, there is nothing more I can say about the subject. For I regard it as completely unfitting to impose on you more often concerning such a small matter. Would that I might see the account of your journey, most learned man! But there must be many reasons which prevent this from happening. Farewell, most cultured man! May you live long and happily, mindful of us!

I wrote this at Skálholt in Iceland on 12th August, 1776.

BL, Add MS 8094, f. 89. Translated from the Latin.

Latin original:

Viro Generosissimo
Domino Josepho Banks
Salutem dicit plurimam
Biarnus Jonæus

Miror, Vir humanissime, quod nullas â te acceperim literas, præsertim cum epistolium a me anno proxime elapso missum, ad Te pervenisse certo mihi sit relatum. Sed alia forte obstacula scribendi occasionem Tibi præclusisse longius, quare has iterum ad Te literas mitto, ut scias me adhuc cum uxore et liberis vivere Tuæque humanitatis memorem esse. Qvod ad Te memoratu dignum scribam, nihil habeo. Hekla ut plurimum se quietam præbet, nisi quod fumida nubes subinde ascendens, flammam subtus nondum esse prorsus sopitam, evidenti sit argumento. Catenam illam argenteam, quam D. Solander secum accepit inaurandam, data occasione mihi remitti vellem; quæ si forte perdita est, non est quod amplius de ea quidquam loquar. De tantilla enim re Humanitatem Tuam sæpius interpellare, indignum prorsus judico. Itinerarium vestrum, Viri eruditissimi, utinam mihi videre contingat. Sed quominus id fiat, multa profecto obstant. Vale, Vir humanissime, et vive diu felix, nostrique Memor. Scripsi Skalholti in Islandia pridie Idus Augusti Anno MDCCLXXVI.

---

[1] His wife Helga Sigurðardóttir died in 1785, aged 57. Four of their children were alive when this was written (see p. 160, n. 3 above).

[2] The main purpose of Banks's expedition to Iceland in 1772 had been to see Hekla.

24. Þorsteinn Jónsson to Banks, 12 August 1777

Hvaleyri [Iceland]

From the excellent gentleman in London England Sir Joseph Banks I the undersigned have received as a gift one gilded silver chain for which I give my humble thanks and sign a receipt to the amtman[1] Ólafur Stephensen for sending it promptly to me.

[*Signed*] Þorsteinn Jonsson[2]

BL, Add MS 8094, f. 264. Autograph letter. Translated from the Icelandic.

Icelandic original:
Frá þeim háa herra í London í Einglande herra Josep Banks hefe eg under skrifaður meðtekið i Skienk eina gillta Silfur feste hvöria eg i undergiefna audmiuklegast þacka og kvittera hr^a Amtmanninn Olaf Stepansson firer hennar skilvislega afhending til min.

25. Ólafur Stephensen to Banks, 20 August 1777

Sviðholt [Álftanes]

Most highly esteemed and most excellent Gentleman

I humbly thank you for your Excellency's highly esteemed letter of April 15 last as well as the generous present of a silver writing-case, two tea-boards and two riding whips, which reached me safely in July last with the last ship to arrive in this country.

I do not know how I can show my highly esteemed gentleman some service. I only wish that I could be commanded to send from here something which would please your Excellency. I thus permit myself to continue the commission you gave me and will continue to send to England the Sagas and antiquities I have been able to collect this year and will continue to do so in order to complete your Excellency's collection.

I enclose a list of 35 Sagas[3] I now send as well as annals and antiquities, along with directions how to use the enclosed Calendarium perpetuum[4] which an Icelandic clergyman, who has never left the island, only by reading and using his own thoughts here in this country has invented and carved himself in wood, which I believe you will not be averse to accepting, because it is a rarity with regards to its place of origin (that is to say Iceland) to have among other rarities that my esteemed Gentleman collects, the Calendarium is set up for New Years Day 1778. I am certain that your Excellency as well as Doctor Solander and Dr Lind will find this invention acceptable.

---

[1] *Amtmaður*: district governor (see Glossary).

[2] Þorsteinn Jónsson, a farmer at Hvaleyri near Hafnarfjörður in Southern Iceland, was Banks's guide while in Iceland. This letter has a seal attached to it. His farm Hvaleyri is depicted in two of the illustrations from the expedition BL, Add MS 15511, ff. 17–19. See Plate 5.

[3] Missing.

[4] A *calendarium perpetuum* was, as the name suggests, a perpetual calendar, probably constructed out of wooden blocks that could be changed at will. Stephensen did indeed enclose a detailed description, in Danish Gothic script, preserved in BL, Add MS 8094, f. 263.

Among the books there is an old Icelandic law-book[1] written on parchment more than five hundred years ago, it is supposed to be the oldest in this country and is considered a rarity, as well as a papist book of psalms on vellum,[2] the contents of which are also of interest because of their age.[3] I permit myself to send a gold ring from papist times, used here, to show you the taste of our ancient ancestors, how the workmanship of goldsmiths was at the time and that they used wool instead of folium under their glass or crystals.[4]

If I could think of something that might please my greatly esteemed Gentleman then I assure you that nothing could please me more than to be able to show my sincerest thanks to my benefactor, if he would be so kind as to give me orders to send from here this and that which he may have forgotten to take with him when he honoured this country with his visit.

After permission has been given I ask with great modesty to have sent to me next year 7 bottles of wine of the same kind as the one enclosed, which was a gift from you. One of the six has broken so I only have five left including the one I am enclosing but I would like to have 12 in all, pardon me Esteemed Sir for taking this liberty. I commend myself to my esteemed gentleman's favour and grace and with the deepest respect for you and Doctor Solander I have the honour to remain

your Excellency's
humble servant
[Signed] O: Stephensen

[P.S.] The books were in two boxes and enclosed on the reverse is a humble letter of thanks from Þorsteinn from Hvaleyri[5] for the silver chain you sent him. Letters from England can reach Iceland the same year, if they are in Copenhagen by late April until the middle of May.

BL, Add MS 8094, ff. 256–257. Autograph letter. Translated from the Danish.[6]

Original Danish in Gothic script:
Höyvelbaarne Aller höystærede höy gunstige Herre!
For Deres Höyvelbaarenheds höystærede Skrivelse af 15de April sidstleden har Jeg den Ære ydmigst at takke saavelsom den der hos fulgte store Present af en Solver Skrivlade, tvende Tee=bretter og tvende Riidepisker, som alt er kommet mig rigtig til haande sidst i afvigte Julii maaned med det til Landet komne seeneste Skib. Jeg veed ikke hvor med Jeg skulle kunne viise nogen slags igien tieneste til min höy gunstige Herre, Jeg önskede alleene mig maatte befales noget at sende her fra som matte være Deres Höyvelbaarenhed til Velgefalt. Jeg tager mig, ellers den Frihed effter givne Commission

---

[1] BL, Add MS 4873, f. 60. The catalogue entry is as follows: 'Jónsbók: the code of laws promulgated in Iceland by Jón Einarsson in 1280, at the instigation of King Magnús Lagabætir; 15th cent.' *Jónsbók* was printed at Hólar in 1578 under the title of *Lögbók Íslendinga*. For centuries it was the book most read by Icelanders and the book used to teach people to read. Many vellum manuscripts were made, the most famous being *Skarðsbók Jónsbókar* from 1363, now preserved at The Árni Magnússon Institute of Icelandic Studies in Reykjavík.

[2] Vellum is fine parchment.

[3] BL, Add MS 4895, f. 96. The vellum is from the 15th century.

[4] Banks bequeathed much to the British Museum. However, this ring is not preserved there (letter from Victoria Smithson, dated 1 March 2012).

[5] See previous document.

[6] Note in Banks's autograph, verso: '[Received] Amtman Stephensen Dec^r [17]77.'

at Continuere med Oversendelse til Engeland af de Historier og Antiqviteter Jeg har kundet indsamle for dette Aar, hvor med skal fremdeles Continuere, for at faae Deres Höyvelbaarenheds Samling completeret.

Over de 35 Historier som nu sendes fölger en Liste saa vel som Annaler og Antiqviteter, samt en Forskrifft hvorleedes det med fölgende Calendarium perpetuum som en Islandsk Præst, der dog ikke har reist ud af Landet, alleene af Læsning og af egen Tanker har paa fundet og selv forarbeidet i træet, som Jeg har troet ikke skulle være ubehageligt for Raritetens Skyld og i henseende til Stedet hvor det er fra (nemel. Island) at have i bland andet rart som min gunstige herre samler, Calendarium er opsat i den Orden det skal staae i Nytaarsdagen 1778. Jeg er viss paa at Deres Höyvelbaarenhed saa velsom herr Doctor Solander og hr Doctor Lind skal finde Dem meget ret i Inventionen.

I bland bögerne findes den Islandske gamle Lovbog skrevet paa Membrana for meer end 500 Aar siden den er den aller ældste som her skal findes i heele Landet og anseer for een Raritet, samt en Papisk Psalme bog paa Membrana ligeleedes meget gammel artig for Indeholdens Skyld. Een Guldring fra de Papiske tiider ligesom den har her været brugelig tager mig og den frihed at sende for at lade see hvad Smag vore gamle Forfædre have havt, hvorleedes Guldsmeds Arbeidet her været i De Tiider, og at de have brugt ulden tog til Folium under Deres glas eller Christaller.

Kunne Jeg optænke noget som maatte være min höy gunstige Herre til Fornöyelse saa forsikrer Jeg helligt at intet skulde være mig til störrre fornöyelse end kunde viise min allerskyldigste takknemmelighed mod min store Velgiörer beder Jeg maatte faae Ordre at oversende et eller andet her fra som maatte være glemt at tage med da De giörde Landet den Ære at være her.

Effter givne Tilladelse udbeder Jeg mig men med megen Undseelse at maatte faae til næste Aar 7 Stk. Vinflasker af samme slags og medfölgende, som er en Present fra Dem hvor af 1 af 6 er gaaen i Stykker saa Jeg har kun 5 igien den nu sendte i beregnet men Jeg ville gierne have 12 stk i alt, pardoner min höy gunstige Herre denne min frihed. Jeg recommenderer mig i min höye Herres Gunst og Grace og med allerstörste Höyagtning for dem og Herr Doctor Solander have den ære at forblive

Deres Höyvelbaarenheds
ydmigste tiener
O: Stephensen

[P.S.] Bögerne ere i 2de Casser og her hos fölger en Revers Indeholdende underdanig takksigelse fra Thorsten paa Hvalöre for den tilsendte Sölv keede. Breve fra Engeland kand komme hid til Island samme Aar, naar de sidst i April og ind til Medium Maii kand være i Kiöbenhavn

26. Andreas Holt to Banks, 30 November 1779

Copenhagen

Sir!

The inclosed Letter[1] came pretty late to hands with a Ship from Iceland, and, tho' I[2] till now, have made all possible Inquiries for a returning Vessel to London, in order to forward a Box with Books and Papers, which along with the above Letter is directed to my Care, it has nevertheless not been in my Power to get any such Opportunity.

Next Spring I shall however watch the first Occasion for sending the said Box, and it shall in the mean while, as it is very well sealed up, stand quite safe in my Book=Room.

---

[1] Probably the letter from Ólafur Stephensen dated 20 August 1777, Document 25.
[2] He always uses a J rather than an I, as was common at the time.

For the generous Promise of sending and favouring me with an Exemplar[1] of Your so much celebrated and here with the utmost Longing expected Work,[2] I pay You, Sir, my most humble and respectful Thanks: I will not only always glory in the possessing of such a splendid Work, especially as it comes from Your own very worthy Hands, but I must intreat you for the future to lay now and then any Injunction on me, by which Means I may in some Measure prove capable of retaliating Your Generosity, and becoming Serviceable to Your good Commands.

My grateful Sentiments on the Head of your kind Intention should long since have been laid before You, but I did the whole last Summer expect the now sending Letter from Iceland and I thought in forwarding it, at the same Time to utter something of my sincere Gratitude; but one Month did elapse after an other: please therefore now to accept of the most ample acknowledgement.

When I do next Spring send your mentioned Box, I will beforehands advise /it/ by the Post, and thinking, it may afford some Pleasure, I intend at the same time /to/ forward a Description of the new, within a Couple of Years elegantly laid out and beautifully adorned Royal Botanic Garden in this Capital.[3]

The Publick here could not but expect the best from the for this Purpose nominated Commission, consisting of two Persons, that were to execute the Institution; but these Gentlemen have even surpassed the Expectation, both in respect to the Time, and fine Disposition of the whole. – The first of the Commissaries is one Mr Holm Counselor of Conferences and /General/ Director of the Posts;[4] the second is Doctor Rothbol-Friis;[5] the Knowledge, Genius, and true Taste of the former in the botanick-Way is very notable, and the Parts of the latter is celebrated in His Capacity as Professor of Physick and Botany in the University of Copenhagen, being also one of the eldest Disciples of the late Linnæus. –

Having proposed the Method of extirpating the fatal Distemper among the Sheep in Iceland, it is a real Joy to me to hear News about its' prosperous Event, and as I am sensible, Sir, You love the poor Icelanders, it will equally prove joyful to your good Heart to know, that one small Part of /the/ Northern Quarter is now only exposed to such Calamity, and the Sheep therein are at the later End of this Year to be killed. –

---

[1] *exemplar*: Swedish for copy.

[2] He is referring to the proposed publication of the *Endeavour* illustrations and engravings, which were never published. Banks was hoping to do so, but the death of Solander in May 1782 put paid to the plans. Banks was in the habit of sending scientific volumes as gifts (see Carter, *Sir Joseph Banks*, p. 174).

[3] The Botanisk Have (now 25 acres, or 10.1 ha), opposite the Rosenborg Palace in Copenhagen, was founded in 1600.

[4] Johan Theodor Holmskiold (1731–93), Danish botanist and doctor, was a protégé of Rottbøll (see following note) and went on botanical expeditions to Germany, Holland and France at his expense. He became a doctor and professor of natural history at the Sorø Academy where he founded a botanical garden. In 1767 he became the general director of the Posts. He later held high positions in the Danish court and in 1781 he supervised the foundation of a botanical garden for the University of Copenhagen. 'Counselor of Conferences' is a translation of *conferensråd*, a high-ranking Danish title, now obsolete, with which he was honoured in 1774 (see Glossary).

[5] Christen Friis Rottbøll (1727–97), botanist and classical scholar, was a student of Linnaeus. In 1770 he had published a work on plants to be found in Greenland and Iceland 'Afhandling om en Deel ...', based on J. G. König's visit (see p. 8 above).

The Land Physicus Biarne Paulsen is no more,[1] and the Bishop of Holum[2] did also last Autumn leave this World. –

Mr. Thodal,[3] the Governour, remains still in Iceland, tho' he might two Years since have returned; but this steady and brave Norwegian will continue there, till several Matters are well established and settled. –

with a perfect and true respect am I always

Sir
Your most obedient Servant
[Signed] A. Holt

BL, Add MS 8094, ff. 219–220. Autograph letter.[4]

27. Banks to Thomas Pennant, 4 May 1783

Soho Square

Dear Sir

On my return to town after the Easter holidays, I receiv'd a message from you by Mr. Dryander,[5] informing me that you had purchas'd from Miller,[6] whom I took with me as an articled Draughtsman to Iceland, certain drawings of that Country and enquiring whether I had any objection to your publishing some of them;[7] in answer to which, I must inform you, without entering into the Question of the propriety of buying things circumstanc'd as these were, that I have always consider'd them as stolen from me in a most unhansome as well as illegal manner, & have held myself ready to Prosecute Miller, if he should publish them in any shape whatever: so circumstanc'd, I trust that you will excuse me for refusing my consent to their publication; as I must consider such a measure a material injury to the journal of my Icelandic voyage, Obtain'd at no small expence, & probably if leisure allows me hereafter to print it,[8] to be attended /at/ with no small emolument.

Your Faithfull Servant
[Unsigned]

NHM, DTC 3, f. 37. Contemporary copy.
Chambers, *The Scientific Correspondence of Sir Joseph Banks*, II, p. 81.

[1] Bjarni Pálsson, the state physician, with whom Banks dined in Iceland, died in 1779. See p. 19 above.

[2] Gísli Magnússon, bishop of Hólar, the bishopric of northern Iceland, died in 1779. There is no record of Banks meeting him in Iceland, though Banks bought books from Hólar.

[3] Lauritz Andreas Thodal was governor in 1772 when Banks visited Iceland and remained in his post until 1785. See p. 85, n. 1 above.

[4] Note in Banks's autograph, left margin of last page of letter: 'Mr Holt | [Received] December 15 [17]79 | [Answered] – 31 [17]80'.

[5] Jonas Dryander (1784–1810) was botanical curator and librarian to Banks 1777–1810.

[6] This would be John Frederick Miller (fl. 1772–85). His brother James had died in 1782. The brothers were Banks's draughtsmen on the Iceland expedition.

[7] These drawings are preserved in BL, Add MSS 15509–15512.

[8] Banks never found the leisure to do so, and seems never seriously to have intended publication of this, or indeed, any of his journals. However, he was generous in making his material available to others. Pennant has already been mentioned and both von Troil and Hooker made use of Banks's Iceland material. Von Troil, for example, devoted letter XXII in *Letters on Iceland* to 'the Pillars of Basalt; to which is subjoined Mr. Banks's curious account of the island of Staffa' (pp. 266–93).

28. Jonas Dryander to Banks, 24 October 1784

London

Dear Sir

The day before /yesterday/ I at last received the parcel which Prof. Bugge[1] wrote to you long ago about. The first pamphlet I have looked over is the account of the eruption in Iceland last year;[2] it is in Danish; for which reason I look upon it as my province to give some account of this pamphlet, which I believe D^r. Blagden[3] would find difficult to understand, as he has not yet begun to read Danish books. The scene of this terrible devastation lays to the westward of that part of /the/ Island which you have visited, and may be about 80 (english) miles from Skalholt. The eruptions of Vesuvius are mere trifles compared to this Skaptar Jökull or Sidu Jökull.[4] It began the 1st of June with Earthquakes which were felt all round; and three different columns of smoke and fire were seen, which higher up in the air joined into one, which rose so high that it was seen at the distance of more than 170 miles. (I mean always English miles, reckoned 5 to one Islandick, which, according to the scale on Troil's map,[5] seems to be about the proportion of them); The atmosphere for a great distance was so filled with ashes, cinders, and pumice stones, as to darken the country, to the degree that people at noon could neither see to read nor write: these hot ashes falling down burnt the grass and destroyed all the cultivated parts. The 12th of June the lava began to run in Skaptargliufur (Gliufur is the same what the Dutch at the Cape call Kloof, a narrow passage between two ridges of mountains[6]) this Gliufur, which was from 600 to 1200 feet deep, from 120 to 600 feet broad and 20 miles long, was soon filled with lava quite to the brim. From thence the lava broke out on the inhabited plains, where it spread itself over a great extent of country, and continued running till the 12th of August.[7] The stream of lava on coming out on the plains, was exactly observed to have 420 feet perpendicular height: but, on widening, it lost in thickness, so that, where it stopt, its perpendicular height is only from 90 to 120 feet. The lava here covered churches, houses, small mountains, every thing which was in its way.

---

[1] Thomas Bugge (1740–1815), FRS, Danish astronomer. The earliest extant letter between Bugge and Banks is from 1782. The parcel probably contained a marine chronometer by Arnold which Bugge was sending for repair (Dawson, *The Banks Letters*, pp. 183–4); and see Document 31 and Appendix 13.

[2] This pamphlet, is *Om Jordbranden paa Island i Aaret 1783* ['On the Subterranean Fire in Iceland in the year 1783'], by the Icelander S. M. Holm [Sæmundur Magnússon Hólm (1749–1821)], at the time a student of theology at the University of Copenhagen. The 76-page pamphlet, with an extremely interesting map showing the flows of the lava, was published in Copenhagen in early 1784, one of the first accounts to appear. The same year it was translated into German. A year later Magnús Stephensen published his *Kort Beskrivelse over den nye Vulcans Ildspudning i Vester=Skaptefields=Syssel paa Island i Aaret, 1783* [Short account of the new volcanic eruption in Vestur-Skaftafellssýsla in the year 1783], published in Copenhagen. Hooker used Stephensen's account in his *A Tour in Iceland*, vol. II, appendix C. See further Rafnsson, 'Um eldritin 1783–1788', pp. 244, 251–4. The eruption of the volcano Laki was a major natural catastrophe, see Introduction, p. 17, n. 5 above.

Dryander's translation is excellent. However, the map he was using to determine distances was inaccurate. Regulations in 1683 stipulated that the Danish mile, used in Iceland, should correspond to 7.5 km. He bases his measurements of feet on the *faðmur*. There were 4,000 of them in a Danish mile.

[3] Sir Charles Blagden (1748–1820), FRS, physician, was Secretary of the Royal Society 1784–97.

[4] *jökull*: glacier (Icelandic).

[5] The map in Uno von Troil's *Letters on Iceland* was first published in England in 1780 as the frontispiece.

[6] *gljúfur*: ravine (Icelandic).

[7] See Holm, *Om Jordbranden*, p. 16.

Another stream of lava broke in the beginning of August thro' Hverfisfliots gliufur[1] to the eastward of the former; but this stream did not proceed far into the plains nor widen much, so that it did comparatively but little damage: it stopt the 10th of August. The vast extent of waste land between these two Gliufurs is entirely covered with lava, which continued extending itself into the desert country to the Northward, and was still running on the 1st October, when the last accounts came from Iceland last year. The circumference of the Fire Sea, as they call it, or the first extent of lava which I have mentioned, is full 150 miles; the second is 75 miles long and 35 miles broad north of the last mentioned Gliufur. The extent of the lava in the desert cannot yet be ascertained. Three large rivers and 8 small ones have been dried up: 17 inhabited places are destroyed. The sum of the inhabitants of all these places was 220. The effort of the stream of lava breaking out of Skaptargliufur into the plain was very violent, attended with earthquakes, and throwing up large rocks into the air. Rocks as big as the largest /*Hvales*/ Whales[2] were seen swimming in the lava. The ashes were carried with the wind as far as Faro, which is at the distance of 100 miles from Iceland.[3] These are the principal particulars I have collected in skimming over this pamphlet.

I have to day hardly any Journals to send. The Journals with the last French mail are not yet come; I don't know for what reason. The inclosed letters from Broussonet[4] came last Friday, so that a French mail certainly then came. To fill up the parcel a little, I send Simmons's last journal.[5]

If you do not come to town on the 4th of November I beg to know your determination by the end of this week, that I may in that case send you a parcel next Sunday.

I am with great respect
Dear Sir
Your most obedient and
most humble Servant
Jon. Dryander.

NHM, DTC 4, ff. 80–82. Contemporary copy.

29. Peter Anker to Banks, 28 September 1785

Bennett Street, St. James'

Sir!

I am favored with your obliging answer to my Note,[6] which I took the liberty to write not knowing, that you was out of Town for more than some Days.[7]

---

[1] Hverfisfljótsgljúfur.

[2] Whale in Danish is *hval* (*hvalur* in Icelandic). See e.g. *hvalfiske* in Holm, *Om Jordbranden*, p. 15.

[3] The effects of the eruption were felt all over Europe. Holm mentions the Faroe Islands, Norway and his own experiences in Denmark (*Om Jordbranden*, pp. 52–6). An interesting article describing foreign reactions to the eruption is Steinþórsson, 'Annus Mirabilis. 1783 í erlendum heimildum'.

[4] Pierre Marie Auguste Broussonet (1761–1807), FRS, French botanist.

[5] Samuel Foart Simmons (1750–1813), FRS, physician. In 1781 he became editor of the *London Medical Journal*, a new magazine which was continued under the name *Medical Facts and Observations* and duly catalogued by Dryander.

[6] This note does not appear to be extant.

[7] Banks spent September 1785 at his residences in Overton and Revesby Abbey.

The reason why I made free to request the favor of an Interview was on account of some Matters concerning Iseland,[1] the state of which you have a local knowledge of. I have been encouraged to take into consideration, if the Iseland Fisheries were lay'd open on the Plan of Newfoundland it would be an advantage to the Island and at the same time beneficial to the Crown of Denmark.[2]

The fisheries at present are but indifferently attended to and the State of the Inhabitants starving. I am also partly prevail'd on to make a Representation to his Majesty[3] to permit the fisheries to be made free, and as I think that this Country might more than any other be of Service to the Island, to allow none but the English the liberty of fishing and preparing the Fish in the Island for foreign Markets, for which should be pay'd to the Crown of Denmark $2\frac{1}{2}$ per Cent Cognizance of all In & Exports.

This is the Plan I wished to have been favored with your Opinion on, and if it is not to be too Troublesome permit me to request the same. As I should wish to know, if you think it would encourage immediate Expeditions to be made from this Country, which is the Opinion of those, who have prevail'd on me to take the affair in hand.

I should think the East Coast of this Country would be glad of it,[4] as Newfoundland is rather too far off and inconvenient to make Expeditions to.

Excuse the liberty I take in addressing you on this Subject, and permit me to remain

Sir
Your most obedient
and very humble Serv*ant*
[*Signed*] P: Anker

BL, Add MS 33978, ff. 36–37. Autograph letter.[5]

---

[1] In Danish Iceland is called *Island* (*Ísland* in Icelandic) but Isaland or Iseland was much in use.

[2] Earlier that year Peter Anker had been ordered by his government to inquire whether the British government would permit the exportation of Scottish sheep to Iceland to build up the greatly depleted stocks in the wake of the Laki eruption (see Introduction, p. 17, n. 5 above). This was not granted but led to plans being hatched by a friend of his, the Hon. John Cochrane, a younger son of the 8th Earl of Dundonald, concerning a British annexation of Iceland, the advantages being the sulphur mines, the fisheries and the possibility of using Iceland as a penal colony. See Agnarsdóttir, 'Scottish Plans for the Annexation of Iceland 1785–1813', pp. 82–91.

[3] In the documents of the Danish Board of Trade (Kommercekollegiet) in the Danish National Archives, among the papers of the minister of finance Count Ernst Schimmelmann (1747–1831), there is a document from the Hon. John Cochrane entitled 'Précis de Mr Cochran sur un Arrangement à faire pour la Pêche en Islande, entre le Danemark et l'Angleterre' (RA, Kommercekollegiet [hereafter KK], Schimmelmannske papirer, box 2155) or, as Cochrane called it, 'Thoughts upon the Iceland Fisheries'. It is undated but is doubtless connected to this letter. Cochrane writes in English that Anker 'is as desired to make some enquiries what could be done in this country [Great Britain] to assist Iceland in its present distress ...'. Anker had consulted his old friend John Cochrane, whose opinion was simply that Denmark should open all ports in Iceland to the British nation to carry on the Iceland fisheries and the Iceland trade. He compared the advantages inherent in the Iceland fisheries to Britain favourably with those of Newfoundland. See further in following letter.

[4] During the 15th century an abundant fishery was carried on off the coast of Iceland from East Anglian ports. See further Introduction, p. 11 above; and Agnarsdóttir, 'Iceland's "English Century"'.

[5] Note in Banks's autograph, on left margin of first page: 'Mr. Anker | [Received] Sept*ember* 30 [17]85 | [Answered] Oct*ober* 3 – [1785]'.

30. Peter Anker to Banks, 2 December 1785

Bennet Street, St. James'

Hon^d. Sir!

I have the honor to return by the Bearer le Traité General des Pesches,[1] you was so obliging to allow me the perusal of.

I have survey'd every Circumstance relative to the Newfoundland Fisheries, and as far as I am able to judge I think there is no doubt; but my plan of laying the Iceland Fisheries open to British Subjects may be attended with mutual Advantage to both Kingdoms.[2]

The want of Wood in Iceland for constructing Stages and other necessary Buildings may at an easy Rate be supply'd with from the Northern parts of Norway, where the Situation will not admit of an Exportation of that Commodity in Competition with the Exports of the Southern Provinces of that Kingdom.[3]

I am not well enough acquainted with the nature of the Iceland Fisheries to lay down a plan of Establishment, which might coincide with the Views of Great Britain. This I shall request you will be so obliging to point out to me, and in what manner you think the Crown of Denmark might receive an Acknowledgment for granting Great Britain an exclusive liberty of fishing with the free use of the Island whether a stipulated Contribution on the Burden of each Ship ought to be pay'd in preference to certain per Cent on the In & Exports. Or, if a Tax on the Establishments in the Island should be more advisable, as the Expence attending the Police & Government must be considered.

I have communicated the Plan to Mr. Nepean,[4] who has promised to consult with you on the Subject, and I shall request the favor to be directed by your Opinion.

---

[1] H. L. Du Hamel du Monceau and M. de la Marre, *Traité Général des Pesches*. Du Hamel du Monceau, FRS, was *Inspecteur général de la Marine*. The book is listed in Dryander, *Catalogus bibliothecae ... Joseph Banks*, II, p. 564. Amongst other things, it discusses the Icelandic method of fishing with fish-hooks (*hameçons*).

[2] In the National Library of Iceland there are several undated documents from Cochrane (Lbs. 424 fol.) which shed light on the background here. Cochrane had described the miserable conditions of the Icelanders. However, 'the young Prince [subsequently Frederik VI] ... a true father to his People' was interested in helping his subjects. Anker 'was desired to make some enquiries what could be done in this Country to assist Iceland in its present distress'. In Cochrane's opinion, 'to relieve the Icelanders from their present distress, Denmark has only to open the Ports of that Island to the British Nation' which would participate in the fisheries and in the Iceland trade. British merchants would supply the Icelanders with all necessaries in exchange for fish. There were mutual benefits for Denmark and Britain. Schimmelmann, the minister of finance, asked C. U. D. Eggers (1722–98), secretary to the Royal Commission of 1785, to examine the current situation of Iceland, to weigh the pros and cons. Eggers's conclusion, after considering the case carefully, was that a society of British merchants should be permitted complete freedom to fish in Iceland, with the same rights and privileges as the Icelanders, on condition they paid an annual fee for the privilege. He suggested the sum of 6,000 rixdollars per annum for the first 20 years (RA, KK, Schimmelmannske papirer, box 2155, 'Reflexions sur le Privet de Mr Cochrane', 4 April 1786).

[3] Cochrane had mentioned in his memorandum that 'great quantities of Timber, particularly Pine will be wanted for Buildings and for Stages to cure the Fish &c and as all the Timber thus used will come from Norway, will constitute a new and beneficial Trade to the Danish Nation'.

[4] Sir Evan Nepean (1752–1822), FRS, politician and colonial administrator, had begun his career in the Navy in 1776. He was purser of the *Falcon* sloop-of-war on the coast of North America, which would explain his being consulted. In 1782–94 he was the first to be appointed Under-Secretary of State in the Home Office. After holding several other government offices, he became governor of Bombay 1812–19. He frequently corresponded with Banks and they eventually developed a close working relationship. Nepean worked with Banks on the colonization of Botany Bay and the Francis Masson expedition to the Cape of Good Hope in 1786.

I have the honor to remain

> Hon*oure*d Sir,
> Your most respectfull &
> most obed*ient* very humble Servant
> [*Signed*] P: Anker

BL, Add MS 33978, ff. 42–43.¹ Autograph letter.

31. Thomas Bugge to Banks 1 July 1786

Copenhagen

Sir

By those lines I have the honour to introduce to your acquaintance a worthy friend of mine, Professor Torkelin,² a very able and skilful antiquarian, who has already³ published some valuable works in that branche of Knowledge. I recommend him to Your good offices. He brings three copys of a new map of Norway;⁴ one for the Royal Society, the other with a geographical description for You and the third for General Roi.⁵ Though /this Kingdom/ has not yet intirely been surveyed, the map is constructed upon the best astronomical and geographical observations, there ware [*sic*] to be got, and in great manny points it is to be depended upon.

I have not been satisfeyed by the Chronometer of Mr. Arnold;⁶ its rate of going has been very irregular, and finally it was quite stopped; because the gold spring and a tooth of the horizontal /or/ Cylinder wheel was broken. Mr Torkelin is bringing it back, and I hope, that Mr. Arnold will put it in order. I look upon this Accident not as a fault of Mr. Arnold, but as a unknowable fault in the materials.

By Dr: Muller⁷ I have received Your last favours of 31 March,⁸ and I am infinitely obliged to You for Your indeavours to promote on my Election /of the/ as a membre [*sic*] of the Royal Society.⁹ I am very glad that my paper on the nodus of Saturn will be

---

¹ Note in Banks's autograph, verso: 'Anker | [Received] 4 [December] [17]85'.

² Grímur Jónsson Thorkelín (1752–1829), Icelandic antiquarian and later Keeper of the Royal Archives. See Appendix 13.

³ As was the Danish custom, Bugge puts an umlaut over the y everywhere in this letter, but they have been removed.

⁴ This is the map of 1785 'Kartet over det sydlige Norge', considered the best map of that region of the 18th century, giving a great deal of new information and drawn by the Danish cartographer Christian Jochum Pontoppidan (1739–1807). In addition to the map itself, a book on it, *Geographisk Oplysning til Cartet over det sydlige Norge i trende Afdeelinger*, was published in 1785. The copy in the British Library is Banks's own. Later, Pontoppidan published a map of northern Norway in 1795.

⁵ William Roy (1726–90), army officer, surveyor and founder of the Ordnance Survey. He became a fellow of the Royal Society in 1767 and was a friend of Banks.

⁶ John Arnold, see the Banks journal, p. 52, n. 7 above.

⁷ Dr Muller is unidentified. It is a common name.

⁸ Banks and Bugge frequently corresponded. Banks had sent Bugge a letter dated 1 March 1786. In Det Kongelige Bibliotek (Royal Library), Copenhagen, many letters from Banks to Bugge are preserved, namely in Ny Kgl. Saml. 1304¹, ² and ³, 2⁰. They are scientific letters from 1782–1800.

⁹ Bugge became a member in 1788. Banks had become a member of Det Kongelige Danske Videnskabernes Selskab, the Danish equivalent of the Royal Society, in 1786.

acceptable;[1] and while it can not be read before November, I shall once more look it over, and send it early in Septembre. Your letter to the Royal Society is received safe.

A valuable posthumous work of our deceased natural philosopher O. F. Muller[2] is brought in order by his worthy disciple Dr. Otho Fabricius,[3] chaplain at the foundling hospital, and it will very soon be published.

I have had the pleasure to see Sir John Sinclair[4] several times at my house, and I am much pleased with this polite Gentleman. He was detained 10 days in the voyage by contrary winds, and by this reason he only has spent 8 days at Copenhagen, and he is not to be prevailed upon to stay longer. To morrow he is setting out to see the Swedish camp at Londscrona:[5] from hence he goes /trough/ through Car/skrona[6] to Stockholm and so farther to Petersburg. We parted very hearty friends, and he has gained the esteem of evry one, who has had the pleasure of his interresting conversation.

Beleive me to be
Sir
Your
most obedient
and humble Servant
[*Signed*] Th: Bugge

BL, Add MS 8096, ff 406–407. Autograph letter.[7]
Chambers, *The Scientific Correspondence of Sir Joseph Banks*, III, pp. 173–4.

32. Uno von Troil to Banks, 30 August 1786

Linköping

Dear Sir!
I very sincerily owe, that You have a great reason to think me guilty of ingratitude; but, my Dear Sir, give me leave to assure You, that want of any news worth Your attention, as

---

[1] Bugge's 'Determination of the Heliocentric Longitude of the Descending Node of Saturn', communicated by Banks, was published in the *Philosophical Transactions of the Royal Society*, vol. 77, 1787, pp. 37–43. Three further articles by Bugge appeared in the *Phil. Trans* and a paper titled 'Observations of the transit of Mercury over the Sun on 9 November 1802' was read to the Society on 10 February 1803, but never published.

[2] Otto Frederik Müller (1730–84), botanist and zoologist. In 1776 he had become a member of the Royal Danish Academy of Sciences. According to the *Dictionary of Scientific Biography* he was the 'foremost representative of the Linnean period in Danish natural history' (*DSB*, IX, p. 575). His principal works were *Animalcula Infusoria Fluviatilia et Marina ...*, and *Entomstraca seu Insecta Testacea, quae in Aquis Daniæ et Norvegiæ reperit etc*.

[3] Otto Fabricius (1744–1822), clergyman, zoologist and missionary in Greenland. His major work was *Fauna Groenlandica*, published in 1780.

[4] Sir John Sinclair (1754–1835) became MP for Caithness in 1780 and was the first president of the Board of Agriculture. After losing his first wife, he abandoned public life for a time and in 1786 travelled for seven months through the north of Europe, visiting the courts of most of the northern states, including that of Empress Catherine of Russia. It will have been at this point in his life that he met Bugge.

[5] Landskrona is a Swedish town in Skåne. It has a good harbour and strong fortifications.

[6] Karlskrona is spread out on 30 islands in the Blekinge archipelago. It was one of the largest cities in Sweden at the time, the base for the Royal Swedish Navy and the shipyards.

[7] Note in Banks's autograph, verso: 'Professor Bugge | [Received] Aug*u*st 18 [17]86 | [Answered] Febr*uary* 13 [1787]'.

living in a distant country, has been the principal cause of my long silence. I therefore hope You vill excuse my negligence, and flatter myself of Your being persuaded that, as long as I live, I never shall cease to acknowledge the uncommon kindness and humanity with wich You received a stranger, who had no other right to Your favour, than what was founded upon Your own Generosity. I know very well Your noble character. Your happiness and contentment, consist more in doing good to others, than in their returning of thanks; but as to me, I can not be easy or content with myself, without telling and assuring You, that, neither distance of places, nor vicissitudes of time, shall ever be able to produce the least alteration in my remembrance of Your undeserv'd kindness, and in my sincere friendship and perfect regard for Your person.

Nothing can make me more happy, than the information I get now and then from travellers of Your being well, and nothing affords me greater pleasure, than to perceive Your always continuing with the same zeal and assiduity to instruct Your fellows-creatures, by throwing more light upon usefull arts and sciences. But when recalling to my mind Your generous behaviour and politeness to me, I can not help being sorry, that it never has been in my power to show You any real proof of my gratitude. If the examination of nature had been my task, I might perhaps some time or other have been able to amuse a friend of Your taste by some discovery in that way; but all my occupations being of so very different nature, I have nothing, that I can suppose more interesting to offer You, than a portrait of the great favourite of nature &c, Archiater von Linné.[1] The original picture, made some years before his death, by one Mr. Roslin,[2] a famous Svedish painter at Paris, and who made it when visiting his northern country, was given to the Royal Academie of Sciences; and this copy, wich I now send You, is made by Professor Pasch,[3] our best painter at this time in Sweden. I hope You will accept of it as a present from a friend, thought [*sic*] it is of us other value [no other value to us?], than what depends upon Your opinion, as an admirer of that great man, who is represented by it. I should be very happy if You think it worth a room in Your precious collection, and if it could keep You in remembrance of Your very affectionate and infinitely obliged friend Mr. Troil.

The death of Dr. Solander[4] was a great loss to all the learned world, and Your tears the most honourable flowers on his tomb. I have read Your letter to Mr. Alströmer[5] on that subject over and over again, and I partake very sincerely in the regret of a man of Such rare merit, and who procured me the favour of Your invaluable acquaintance and friendship.

---

[1] *Archiatar* is a title used for distinguished physicians since ancient times. In 1747 Linnaeus received this title from the King of Sweden when he became 'physician in ordinary' to the Royal family (Duyker and Tingbrand, *Solander*, p. 434).

[2] Alexander Roslin (1718–93), Swedish portrait painter. He painted Linnæus in 1775. From 1750 he worked mainly in Paris, where he died.

[3] Lorens Pasch (1733–1805), Swedish portrait painter, sometimes called the Younger as his father Lorentz/Lorens Pasch (1702–66) bore the same name. He painted von Troil.

[4] Solander died in May 1782.

[5] Johan Alströmer (1742–86), FRS, natural philosopher. He was a close friend of Solander and a pupil of Linneaus. Banks donated to Alströmer an important South Seas ethnographical collection, the bulk of which was collected by Solander and himself during Cook's first voyage. It is now preserved in the Alströmer Collection in the Ethnographical Museum of Sweden and considered to be 'the oldest South Seas collection of ethnographica extant'. Alströmer became President of the Royal Swedish Academy of Sciences. He visited England in 1777–8 when he became acquainted with Banks and Solander and was elected an FRS before departing. There are several letters from Alströmer to Banks in BL, Add MSS 8095 and 8096. See further Rydén, *The Banks Collection*, quotation p. 8; and Jonsell, 'The Swedish Connection', p. 25.

I have been told that You are married,¹ and hope You, wen in that state, are so happy as You deserve. If you ever favour me with a letter, I hope You will inform me of any lucky change that may have happened to You as well in the litterary as in Your particular way of life; and being convinced of Your kind partaking in my wellfare, I think it my duty, to let You know, that fortune has been more favouring to me, than I ever expected, and that I am indebted to You for it, as an instrument of providence to open the field. The voyage to Island made me known to the king.² His particular grace has ever since been effectual. I am now Bishop.³ My Bishoprick is the first in rank and one of the best in the kingdom. The Arch-Bishop being dead, I have been the speaker, or the first commissioner for the Clergy, during the last diet. I am married: I have a very good wife and six children. I find myself very happy, spending the greatest part of my time in the fulfilling of my duties, and the rest with my books.

I recommend myself to Your remembrance, and wish very warmly for an opportunity to convince You, of the true estime and most sincere friendship with wich I always remain Your

most obedient humble servant
[*Signed*] Uno von Troil

BL, Add MS 8096, ff. 371–372. Autograph letter.⁴

33. James Home to Banks, 9 March 1789

Edinburgh

Sir

Perhaps I might now use the freedom of writting to you as having formerly had the honour of your acquaintance; but I shall rather claim that Liberty as a mark of that respect due to you as the great patron of Science, as peculiarly eminent for your Knowledge of Natural History, & as the great encourager of it in others. – Besides, the subject upon which I have now the honour of addressing you naturally points you out as the most proper person to whom I should apply for advice & information; since almost the only Knowledge that the world has of Iceland has been obtained from your Zeal for Science.

Mr. Stanley⁵ whom a generous design⁶ for Knowledge has tempted to visit that remote country this summer, has done me the honour to ask me to accompany him.⁷ The

---

¹ Banks married Dorothea Hugessen on 23 March 1779 at St Andrew's Church, Holborn.

² King Gustavus III of Sweden.

³ Uno von Troil became bishop of Linköping in 1780 and eventually archbishop of Sweden in 1786.

⁴ Note in Banks's autograph, right-hand margin on last page: 'B*isho*p Troille | [Received] Oct*ober* 20 [17]86 | [Answered] Nov*ember* 28 – [1786]'.

⁵ Sir John Thomas Stanley (1766–1850), travelled to Iceland in 1789. The spelling is 'Stanly' in DTC copy. See Appendix 13 and Introduction, p. 24 above.

⁶ In DTC copy, *design* is misread as *desire*.

⁷ James Home (1760–1844), became professor of Materia Medica at Edinburgh University in 1798. In fact he did not accompany Stanley to Iceland. But as Stanley wrote he had 'before the month of May to abandon the prospects of having any of these gentlemen [Home and others] associated with me in my Enterprize, and I had to select new Companions'. Among those he asked was Grímur Thorkelín (West, 'Stanley's introduction to the journals', *Journals of the Stanley Expedition*, I, pp. xv–xvi). Thorkelín (see Appendix 13) was well known to Banks so there is a connection there. Later Banks and Stanley were to become well acquainted, see Introduction, p. 24 above.

principal object that we propose to ourselves in this Expedition is the Natural History of that country. From your intimate knowledge of that branch of Science & your acquaintance with Iceland we are induced to apply to you for your advice in conducting this Expedition to the greatest advantage. Any information which you may be pleased to give us we shall be much obliged to you for & we beg leave to offer our feeble services for promoting the advancement of Natural History.[1]

I hope you will excuse the liberty I have now taken. It was an homage I thought due to the Head of Science in Great Britain.

I have the honour to be, Your most humble Servant
[*Signed*] James Home.

Kew, Banks Correspondence I, JBK/1/4 f. 337.
NHM, DTC 6, ff. 140–141. Contemporary copy.

34. Dr William Wright to Banks, 25 December 1789

Edinburgh

Dear Sir

I was duly favoured with yours, and thank you for the Contents. I have since that got a Copy of the Hortus Kewensis,[2] and have seen my Name several times as a donor to the Kings Garden.[3] There are a prodigious Number on the whole, and I find there are many in Jamaica not yet got home.[4] Were I again to pass a few years more beyond the Atlantic I would add greatly to this magnificent collection.

I got a parcel of seeds from Jamaica by the last shipping, but not gathered by a person of Science. Some had only the Country names, but all of them I know by sight and have ascertained them pretty nearly, and will try to get them assorted this day to go by Mr Stanly[5] who will have the hon*ou*r of delivering the parcel with a Card of Introduction. In this same parcel you will find a few Iceland seeds amongst them more Koenigia Islandica.[6] After serving the Royal Garden you will no doubt give a few to Dr Pitcairn[7] with my respectfull complime*n*ts.

I inclose you a list of Plants gathered by my nephew.[8] Mr Stanly has two Specimens of Each and I believe to present you with one or of any you want. Should that not happen

---

[1] The journals of this expedition were first published by John F. West. Later they were translated into Icelandic by Steindór Steindórsson.

[2] The first edition of the *Hortus Kewensis* was published in three volumes in 1789 under the name of William Aiton, gardener to George III at the Royal Botanical Gardens at Kew, and dedicated to the king. See further Carter, *Sir Joseph Banks*, pp. 250–52.

[3] The Royal Botanic Gardens at Kew.

[4] Dr William Wright (1735–1819), FRS, had served in a government post in Jamaica in 1760–64, sending live plants to Kew (West, 'Introduction', *Journals of the Stanley Expedition*, I, p. x). See further on Wright in Appendix 13.

[5] John Thomas Stanley embarked on an expedition to Iceland in 1789.

[6] Iceland purslane, named after Johan Gerhard König, see Introduction, p. 8 above.

[7] Dr David Pitcairn (1749–1803), physician.

[8] This nephew was James Wright (1770–94), who travelled with Stanley to Iceland and wrote one of the journals of the expedition. After the death of his father of the same name, his uncle provided for him and he studied medicine from 1785 to 1788 at Edinburgh University. On the relationship between James and William Wright, Banks and Stanley see *Memoir of the Late William Wright, M.D.*, pp. 77–80. The List of Plants follows.

you will advise me, and Specimens of those left here will be forwarded by first opportunity.

The inclosed Plant (supposed to be new) is not given to Mr Stanly. Of course you will not mention it to that gentleman. Mr Wright and Mr Stanly have parted on the best of terms which was not the case with others of their shipmates. There were faults on both sides.

From some untoward accidents & particularly the ignorance and knavery of Lieu^t Pyrie[1] Mr Stanly was unable to see little more of Iceland than you did some years ago.[2] As M^r W*right* has been so busy copying his Journal[3] I have scarcely dipt into it, and refer you in the mean time to the Principal who is possessed of the Various Journals kept on board. by astronomer & draughtsman Two young Gentlemen;[4] The master Crawford[5] and Mr Wright.

I have given my young man rather a superior medical and Philosphical Education to most others. I have him still going on at College and private Masters. I am glad to see he is fond of Botany and Natural History. He is desirous of getting out to India, but I have no Interest with our almighty advocate.[6] If you could procure him an appointment as assistant Surgeon in the Companies[7] Service I would send him up and both of us exceedingly thankfull.

I hope you have by this time received the copies of my Paper. I have sent copies to Dr Pitcairn & Pulteny[8] as well as to Dr Garthshore[9] and Mr Home.[10] You will therefore dispose of the others as you think proper I could wish a Set to Dr Smith at Chelsea.[11] Mr Dryander[12] if he was in it. I have also sent copies to Professor Murray[13] and Jacuin[14] and

---

[1] Lieutenant Pyrie, or rather Pierie, was the commander of the ship, a retired naval officer with 30 years' service (West, 'Introduction', *Journals of the Stanley Expedition*, I, p. viii).

[2] Stanley arrived on 5 July 1789, leaving at the beginning of September. Thus he spent some two weeks more in Iceland than Banks and he travelled further afield.

[3] James Wright's journal was published as the first volume of West's *Journals of the Stanley Expedition*.

[4] These were Isaac S. Benners and John Baine who both wrote journals included in West's *Journals of the Stanley Expedition*. See further Introduction, p. 24 above.

[5] John F. Crawford, part owner of the ship used by the expedition (West, 'Introduction', *Journals of the Stanley Expedition*, I, viii).

[6] Henry Dundas (1742–1811), a formidable Scottish politician, was Pitt's closest collaborator. At this time he was the Secretary for War.

[7] It was due to Stanley's influence that James Wright was appointed assistant surgeon to the Madras Medical Service in 1791, where he distinguished himself. He was thrown from his horse and died on 8 April 1794 at Attore.

[8] Dr Richard Pulteny (1730–1801), FRS, physician and botanist.

[9] Dr Maxwell Garthshore (1732–1812), FRS, an eminent Scottish physician, who moved to London in 1764 where he practised until his death. He wrote many medical papers in *The Philosophical Transactions*.

[10] Sir Everard Home (1756–1832), FRS, surgeon, was later first president of the Royal College of Surgeons.

[11] Sir James Edward Smith (1759–1828), FRS, founder and first president of the Linnean Society.

[12] Jonas Dryander: see Appendix 13.

[13] Johan Anders Murray (1740–91), Swedish botanist in Göttingen. See p. 180, n. 2.

[14] Nicolaus Joseph, Freiherr von Jacquin (1727–1817), FRS, naturalist, was appointed Professor of Botany and Chemistry at the University of Vienna and the Director of the Botanical Gardens of the university in 1768. He subsequently became rector of the University of Vienna in 1809. Banks bought a herbarium from him in May 1777 (Carter, *Sir Joseph Banks*, p. 173). His son, Joseph Franz Freiherr von Jacquin (1766–1839), was also a naturalist, botanist and chemist, and visited Banks in 1789. He also corresponded extensively with Banks (Dawson, *The Banks Letters*, pp. 445–7).

if you desire more I have two to spare. I have the Hon*ou*r to be with great Esteem and obligation Dear Sir

[*Signed*] your most obed*ien*t Serv*an*t Willm Wright

NHM, Botany Library, MSS WRI I.[1] Autograph letter.

Catalogue of Plants gathered in Iceland and The Faroe Islands By Ja^s Wright. A. M. Student of Physick Edinburgh.[2]

II[3]	*Veronica alpina fruticul^a*. [Alpine speedwell]
III	*Poa alpina Viviparum* [Alpine meadow-grass] *Eriophorum alpinum* [Alpine cottongrass] *Kœnigia Islandica* [Iceland purslane]
IV	*Cornus Suecia*– Faroe[4] [*suecica*: Dwarf cornel]

V	*Gentiana 5folia*
	  Nivalis [Snow gentian]
	  Amarylla [*Gentianella amarella*: Autumn gentian]
	  Campestris [*Genitanella campestris*: Field gentian]
	  Nov Spec ?[5] [new species ?]

VI	*Juncus triglumis 3fidus* [Three-flowered rush] *Anthericum calyculatum* [*Tofieldia calyculata*: Tofields's asphodel]
VIII	*Vaccinium uliginosum* [Bog bilberry] *Epilobium latifolium* [*Chamerion latifolium*: Dwarf fireweed / River beauty Willowherb] *Polygonum Viviparum* [*Persicaria vivipara*: Alpine bistort]

X	*Arbutus alpina*
	Pyrola rotundifolia [Round-leaved wintergreen]
	  Secunda [*Orthilia secunda*: Serrated wintergreen]
	Saxifraga Hirculus [Marsh saxifrage]
	  oppositifolia [Purple saxifrage]
	  Cespitosa [Tufted saxifrage]
	  Hypnoides [Mossy saxifrage]
	  Stellaris [Starry saxifrage]
	Silene acaulis [Moss campion]
	Arenaria ciliata [Fringed sandwort]

[1] Note in Banks's autograph: 'Dr Wright | [Received] De*cembe*r 29 [17]89.'

[2] David Hibberd very kindly worked through this list and provided the information, including the English names for the plants mainly as used by the Botanical Society of the British Isles in their plant-mapping schemes. The Latin names have been rendered into italics as is the convention. Presently accepted names which differ from Wright's are bracketed with the English names.

[3] The Roman numerals refer to Linneaus's *Systema Sexuale*, his classes and orders of plants, used for identification. There were several orders within each of the 24 classes.

[4] The Stanley expedition visited the Faroe Islands on their way to Iceland.

[5] This is what James Wright writes without any further explanation.

	Sedum Villosum [Hairy stonecrop]
	Lychnis alpina [*Viscaria alpina*: Alpine catchfly]
XII	*Dryas octopetala* [Mountain avens]
	Potentilla cineria [Grey cinquefoil]
	Rubus saxatilis [Stone bramble]
	Geum Rivale [Water avens]
XIII	*Comarum palustre* [Marsh cinquefoil]
	Thalictrum alpinum [Alpine meadow-rue]
	Ranunculus glacialis. Faroe [Glacier buttercup]
	reptans [Creeping Spearwort; Creeping buttercup]
XIV	*Bartsia alpina* [Alpine bartsia]
XV	*Cochlearia Danica* [*officinalis* – Common scurvygrass] [Danish scurveygrass]
XIX	*Erigeron Uniflorus* [One-flowered fleabane]
	Hieracium alpinum [Alpine hawkweed]
XXI	*Betula nana* [Dwarf birch]
XXII	*Salix herbacea* [Dwarf willow]
	reticulata [Net-leaved willow]
	arenaria [*Salix repens ssp. argentea*: Creeping willow]
	Myrsinites [Whortle-leaved willow]
	Lapponum[1]
	Myrtiloidus[2]
	Caprea [Goat willow /Pussy willow]
XXIII	*Holcus odoratus* [*Anthoxanthum nitens*: Sweet grass/Bison grass]
XXIV	*Bryum*[3] *hyperborum* [*Bryum hyperboreum*]
	scoparium [*Dicranodon scoparium*]
	tinctorium [*Eucalypta vulgaris*]
	Viridulum [*Fissidens exilis*]
	Heteromalum [*heteromallum*: *Dicranodontium heteromallum*]
	Hypnoidus [*Bryum hypnoides*]
	Lichen rangiferinus [*Cladonia rangiferina*: Reindeer lichen]
	Subulatus [*Cladonia subulata*: Antlered powderhorn]
	Borealis Islandicus [*Cetraria islandicus*: Iceland moss]

Description of a Plant on the top of a high mountain in Iceland from field notes J W[4]
   Planta humulis fruticulosa
    Cal 5 partilus persistens

---

[1] See Babington, 'Flora of Iceland', p. 328.

[2] Rather *Salix Myrtilloides*: see Babington, 'Flora of Iceland', p. 329.

[3] There are very few common names applied to species of *Bryum* and mosses in general.

[4] Dr Hörður Kristinsson, an Icelandic botanist, suggests this plant might be *sauðamergur* (Trailing azalea in English and *Loiseleuria procumbens* in Latin).

Cor: Petula 5 ovata patintia
Stam 5 convergentes, petalis breviora
Stylus eructus brevis, Stigma simplex
Germen ovatum, uniloculare, calyce
Serilectum – Semma minima numerosa
Quæ in matuitione non Vidi

NHM, Botany Library, MSS. WRI. Autograph letter.

35. Banks to Grímur Thorkelín, 9 April 1791

Soho Square

Sir Jos: Banks presents his Comp*lime*nts to M^r. Thorkelyn[1] as begs leave to trouble him with some Copies of a Book he has Lately Printed[2] to give to his Friends which will not be sold requesting he will be so good as to deliver them at Copenhagen to the Gentlemen whose names are written upon the Covers he begs Mr. Thorkelin's acceptance of one of them which is on the top of the Parcel & with best wishes for his Prosperous voyage & Safe arrival requests to be indulged in his future Correspondence.[3]

Edinburgh, University Library, La.III.379.f.24. Autograph letter.

36. Grímur Thorkelín to Banks, 31 January 1792

Copenhagen

Dear Sir Joseph!
There is nothing I more ardently wish, than to convince both yourself and the world, with how much gratitude I retain the remembrance of your goodness towards me during my residence in England. The share in deed which I enjoyed in your protection is by no means secrete to the world; it has occasioned Mr Snedorff,[4] Professor of Modern History and Statistic in this university to apply to me for an introduction to you.

---

[1] Grímur Jónsson Thorkelín spent the years 1786–90 in England.
[2] This would be *Icones selectae plantarum quas in Japonica collegit et delineavit Engelbertus Kaempfer* (popularly known as the *Icones Kaempferianae*), published by Banks in 1791. This included a series of engraved plates produced from drawings by Engelbert Kaempfer (1651–1716), the German naturalist and traveller in the Far East. See Jonsell, 'The Swedish Connection', p. 25. Banks edited this volume.
[3] The letter is addressed to 'Mr. Thorkelin, Charlotte Street No. 38 Portland Place' (and on the envelope, in Banks's autograph, 'with a Large Parcel of Books').
[4] Frederik Sneedorff (1760–92), Danish historian who had been born in Penrith, Cumberland. He was designated 'professor extraordinarius' in history in 1787 after giving a series of public lectures, and received a university appointment in history and statistics at Copenhagen in 1788. In 1791 he obtained leave to travel to Germany, Switzerland, France and England and it is in this context that the letter is written. He was not disappointed in England (though France in the throes of revolution had not been to his taste) and visited Oxford, Birmingham, Liverpool and his birthplace in Cumberland, where he ironically died in a carriage accident near Penrith.

He is a person universally esteemed for his learning and probity, and I am sure, that, though you have bestowed numberless good offices upon many, you never conferred one that was better placed than upon him. Permit me therefore to entreat you, to favour Mr Snedorff with your protection as a fresh proof of your friendship towards the Northern Muses. They most gratefully acknowledge your presents, which you intrusted to my care at my departure from London. I hope you have allready received the thanks due to your liberality from Messrs Schumacher,[1] Vahl,[2] Suhm[3] and others. His Excellency the Privy counsellor Holmskiold[4] desires, You will accept of his best thanks and compliments, till he shall himself be enabled to write at the finishing of his own work on the Mosses,[5] which is not far being ready for publication, if the impression of thirty Copyes intended as presents to the most illustrious Patrons of Botany & Natural History may be stiled as such. The Copy of your Publication of Plants, with which You favoured me, I have given to His Excellency the Marshall Bulow,[6] who admires and follows your Exemple, in so much, that he has at his sole expence sent a gentleman to Africa in order to study the Kingdom of Nature in that Country: and taken upon himself a great share of the expences which defray a like mission of a gentleman to Iceland. When ever those two Messengers shall return, I am confident, the Marshall will take the earliest opportunity of communicating to you the results of the hoped for discoveries.[7]

I am with the most sincere respect for You and my Lady Banks

Dear Sir Joseph,
Your
most obedient
very humble
and very much obliged Serv*an*t
[*Signed*] G Thorkelin

BL, Add MS 8098, ff. 99–100. Autograph letter.[8]

---

[1] Hans Christian Friedrich Schumacher (1757–1830), Danish physician and botanist.

[2] Martin Vahl (1749–1804), Danish botanist.

[3] Peter Frederik Suhm (1728–98) was a prominent and prolific Danish historian of Nordic history and man of letters. According to Dryander's catalogue of the Banks Library, Banks owned no books by him and there is no extant correspondence. Both, however, shared an interest in Iceland and Suhm supported in a princely manner the publication of numerous Icelandic manuscripts during the years 1774–87.

[4] Johan Theodor Holmskiold. He became politically involved during the Guldberg period, gaining high office and even after the coup of 1784 he remained in favour. He became a director of the new botanical gardens at Charlottenborg, and eventually a privy councillor (*Gehejmeråd*) in 1784. See also p. 201, n. 4 above.

[5] This is *Beata ruris otia fungis Danici a Th. H. impensa*, a magnificent work by Theodoro Holmskjold, as he is called on the title page, published in Copenhagen in 1799 in two volumes. Actually the subject matter was mushrooms not mosses. Banks's copy is in the British Library, as is the King's.

[6] Probably Johan Bülow (1751–1828), marshal of the Royal Court (*hofmarskal*), serving Prince Frederik for more than 20 years before falling out of favour in 1793. He then became a patron of science, subsidizing various scientists including Thorkelín and another famous Icelander, Finnur Magnússon (1781–1847), the Keeper of the Royal Archives.

[7] No correspondence is extant between the two.

[8] Note in Banks's autograph, verso: 'Mr Thorkelin | [Received] Feb*ruary* 18 [17]92'.

37. Uno von Troil to Banks, 16 February 1799

Upp*sala*

Sir.

I cannot enough reproach myself, for not having, in so long a time, answered to the letter, Mr. Silfuerhjelm[1] brought me from You, with the Drawings of Geyser. I hope, You will not look upon it as oblivion of a friend, whom I sincerely respect and love, and whose favour I can newer [*sic*] enough acknowledge.

Nothing more delightfull to me, than both the letter and the present, when the both, put me in mind of the happy time, in which I dayly enjoyed marks of Your friendship, and by Your generosity and in Your entertaining and instructive Company had the satisfaction of seeing this wonder of nature. Sir! As soon as my remembrance of this shall be no less gratefull than lasting, so much it flatters me to find, I am still in Your memory.

The drawings are excellent, and I am glad that Mr. Stanley has found my account of the place, extracted from Your observations, agreeing with those of his own. From the account he has given in the Edinburgh Transactions,[2] for wich I also thank You, it shows that many alterations have been made there, since the Year 1772. My situation in life, hinders me from ever more visiting that land, which might indeed be interesting, bot [*sic*] would be less diverting, when I could not do it under Your direction and in Your Company, which would give a double relish to the pleasure of being there. I embrace this opportunity, to deliver the wishes of the Royal Society of Sciences at Upsala, to be honoured by Your accepting a place therein; and I hope You will regard it, as a mark of the general esteem among the learned, which Your enterprises for promoting natural History, Your many Discoveries, and your Protection for learning in general have procured You.[3]

Mr. Forner[4] is well, and will, no doubt, some time mention Nova Literaria[5] from this place. I know nothing interesting to You, unless it be the monument of Porphyr of Sir Charles a Linné,[6] which by subscription will in a few weeks be erected in the Cathedral Church here.

Mr. Paykull[7] has published 2 parts of his Fauna Svecica, and will very soon give the Publick its continuation. Mr. Thunberg will also this year publish his Flora Capensis.[8]

---

[1] Baron Göran Ulrik Silfverhjelm (1762–1819), chargé d'affaires at the Swedish Embassy in London.

[2] Dr Joseph Black (1728–99), professor of medicine. See Introduction, p. 24, n. 1 above.

[3] Banks had become a Foreign Member of the Royal Society of Sciences in Uppsala on 15 December 1798 (Carter, *Sir Joseph Banks*, p. 586).

[4] Unidentified.

[5] *Nova Literaria* (Latin): 'what has been written recently'.

[6] Linnaeus died in 1778 and was buried in Uppsala Cathedral. The memorial, made by the sculptor Johan Tobias Sergel (1740–1814), was not completed until 1803.

[7] Baron Gustaf von Paykull (1757–1826), zoologist and writer. The three volumes of his *Fauna svecica. Insecta* were published in Uppsala in 1798–1800.

[8] Professor Carl Peter Thunberg (1743–1828), FRS, naturalist, oriental traveller and professor of botany at the University of Uppsala. He was a student of Linnaeus and has been called 'the father of South African botany' and the 'Japanese Linnaeus', viz. his work the *Flora Japonica* (1784). Von Troil is referring to Thunberg's work *Prodromus plantarum Capensium* (1794–1800), a summary of his findings in the Cape Colony. His *Flora Capensis* was not actually published until 1807–23, completed with the help of the German botanist J. A. Schultes.

I shall be happy in the continuance of Your favour and friendship, assuring You of the respect, wherewith I have the honour to be,

Sir,
Your
most obedient humble
servant [*Signed*] Uno von Troil

BL, Add MS 8098, ff. 488–489. Autograph letter.[1]

38. Memorandum from Banks to Henry Dundas[?][2] Remarks concerning Iceland, 30 January 1801[3]

The English who frequented the Seas that surround Iceland for the sake of fishing, in very early times, do not seem to have been prevented from visiting the Ports[4] of that Country till the Danes began to assert their Right of excluding all other Nations from trading with the Icelanders in 1465.[5] Licences were however granted to those Subjects of England and possibly of other Nations who chose to accept them, so that the Fishery does not seem to have experienced any material Interruption.

From the middle of the 16th Century when the Fishery of Newfoundland began to attract the attention of England, that of Iceland[6] appears to have gradually decayed and in 1615, England had 120 Ships and Barks only, employed in the Iceland Fishery, and 250 in that of Newfoundland. –

About the Year 1730, a Monopoly of the Iceland Trade was granted to an exclusive Company at Copenhagen.[7] The English had probably quite abandoned those seas when

---

[1] Note in Banks's autograph, verso: 'Von Troille | [Received] March 12 – 99'; also, on margin of first page: '[Answered] April 30 – [1799]'.

[2] Banks definitely wrote this memorandum for a member of the government, probably Henry Dundas, the Secretary for War and Pitt's closest collaborator. A Scottish aristocrat, the Hon. John Cochrane, a younger son of the Earl of Dundonald (see p. 205, n. 2), had sent many memorials to the British government urging the annexation of Iceland by offering the Danes another territory (Crab Island in the West Indies or the Duchy of Saxe-Lauenburg) in exchange. In January 1801 he seized the opportunity offered by the anti-British League of Armed Neutrality, to which Denmark belonged, and proposed that the state of hostilities would facilitate the acquisition of Iceland by making a military conquest feasible. 'A few troops and ships of war' would be enough and Cochrane believed that the island would surrender at once without loss of blood. Though Dundas and Pitt had taken little notice of Cochrane's former proposals, they appear to have done so now, Banks's memorial being the very proof of this. (On Cochrane see Agnarsdóttir, 'Scottish Plans for the Annexation of Iceland', pp. 82–91.)

For the origins of the British government's sudden interest in Iceland, see Introduction, p. 30 above. The government sought Banks's advice as he was virtually the only Englishman of authority to have any first-hand knowledge of the island. This memorandum reflects many of the points made by Cochrane in his memorials, suggesting that Banks had a copy of Cochrane's proposals to read and to comment on.

[3] The date is in pencil, written by a clerk.

[4] 'Parts' in DTC copy.

[5] Banks's information is correct, though the trading permits were rescinded the following year.

[6] Spelt *Island* in the draft.

[7] In 1733 a privileged company in Copenhagen, Det islandsk-finmarske Kompagni (The Iceland-Finmark Company), was granted the monopoly of the Iceland trade. On the Iceland trade see Introduction, pp. 12, 14–15 above.

this took place. Since that period all Foreigners have been strictly forbid to visit the Ports of that Country, and in truth it does not appear that any Nation has thought it expedient to make the attempt, tho' from the numberless Harbours with which the Coast of the whole Island abounds, the risk of detection could never have been considerable,[1] and yet, complete as this Monopoly has been, the Company who are said to have paid to the Danish Government no more than 4000 Rix Dollars a Year for their Privilege became bankrupt some Years ago.[2]

The decided preference which all Nations have given to the Newfoundland Fishery is probably owing to the total want of Wood in Iceland, and the great abundance of that Article in Newfoundland, which enables the Fishermen to erect Stages at cheap rate, elevated considerably above the Water, and thus to cure fish more quickly than can be done on the Beaches of Iceland, and make it into an Article of more merchantable[3] quality than the Iceland cured Fish appear to have been.[4]

Whether this or any other circumstance was the real cause of the decay of the Iceland and the Advancement of the Newfoundland Fishery, the whole appears to have been the effect of natural causes, and not to have arisen from Political Interference, we must not therefore expect that even the Capital and the Industry of English Merchants can raise in Iceland a rival to the Newfoundland Fishery. The Northern parts of America are now, and are likely to continue to be the great center [sic] of Export for the Supply of cured Fish to the Markets of the Catholick Countries.

If the Ports of Iceland were open to our Fishermen, great quantities of Fish might certainly be procured from thence for the English Market, and by being brought home half salted, might be sold at a very cheap rate. But salted Fish, though it may be obtained at a very cheap rate from the Northern parts of this Island, has never been a favorite food of the English, and it is not to be supposed that the present exorbitant Price of the necessary Articles of Food can continue long.

Next to the Article of Fish, the Sulphur Mines in Iceland appear to be an object of importance. The Danes tried to work them several Years ago, and by their own account, seem to have derived a Profit sufficient to encourage an Adventurer.[5] On the other hand it does not appear that they have brought to Market any quantity of Sulphur to rival the produce of Italy in that article. In truth, Sulphur is so cheap a commodity that as a Raw Material it seems of little consequence whether England purchases it of an independent State or of one of her own Dependencies, and it does not appear probable that it would,

---

[1] The Danish authorities, however, always had difficulty enforcing the monopoly trade in all parts of the country and recent research concludes that a good deal of illicit trade took place between Icelanders and English, Dutch and French fishermen.

[2] The above-mentioned company lost its licence in 1743, and was succeeded by Hörkræmmerlaget (The Grocers' Company) until 1759. From 1764 to 1774 the trade monopoly was in the hands of Det kongelige octroierede almindelige Handels-Kompagni i Köbenhavn (The Royal Chartered General Trading Company of Copenhagen). The King then formally took over the trade, until it went bankrupt in 1787. Eventually the 'Free Trade Charter' (fríverslun) took effect, granting freedom of trade to most subjects of the Dano-Norwegian kingdom (but strictly prohibiting foreigners from participating in the Iceland trade).

[3] DTC copy has 'marketable'.

[4] This takes up the theme of the Anker letters (Documents 29 and 30).

[5] Sulphur was first traded in the 13th century. The sulphur trade's heyday was during the 16th century when it became a royal monopoly. By the end of that century the mines were almost exhausted.

in any case, be thought adviseable to transfer the manufacture of vitriolic acid from the Mother Country to any Dependency whatever.

As the Writer of this visited Iceland in the Year 1772 he will no doubt be expected to give some account of that Country, as he saw it at that time. – It is nearly as large as Ireland,[1] but so thinly is it peopled, that the space of /the/ Country which forms in England a Parish is there a Farm, and instead of a Village a single house only is to be found upon it. – No Town of any kind appears to exist in all Iceland[2] the only aggregate of Houses met with are a few that surround the Bishop's Palace at Skalholt[3] and these were entirely inhabited by Priests, Deacons, Clerks &c who attended the duty of the Cathedral Church, and were maintained by monthly rations of food supplied to them by the Bishop.[4]

The Climate is not so severe as the latitude would lead us to suppose.[5] The Southern and Western Harbours are never frozen in winter; but on the other hand, the summer is not warm enough to ripen any of the ordinary fruits of Europe, at least none of them; not even Gooseberries or Currants were found there. – The Gales of Wind in the Autumn are tremendous, but the Harbours are so good, that Ships with proper care need not be under any apprehension of suffering by them. The great scourge of the Country is the frequent eruption of Volcanoes. It is but a few Years since, a Mountain, not before known to be a Volcano,[6] burst forth at once with a violence which wholly destroyed a tract of Country about as large as an English county; 25 Miles on one side of it, were covered several inches deep by ashes and cinders and for a space round it, not exactly defined, a shower of filaments of Glass almost as fine as the web of a Spider, covered all the land. – Wherever this happened, all animals that grazed upon the pastures so tainted, soon died, and it is said to be several Years before such land becomes again wholesome for Cattle or Sheep.

Tho' Trees of many kinds are found upon the Island, it is seldom that any one of them reaches the size of a common whip. – No Corn of any Kind can be grown there.[7] In 1772 the Governor[8] sowed Wheat Rye and Barley on a small plot, which the Writer of this saw on the 30 August, flourishing and then in full ear, but was told that it certainly would be destroyed either by frost or by wind. – The latter happened within a few days. – The straw was beaten down flat upon the ground by a gale of wind, and every grain detached from the Ear, and dispersed over the Field. – Some grains however were picked up and these appeared nearly ripe and fair.

Turnips, Cabbages, Lettuce, Pease, Cauliflowers, and some other Kinds of Garden Stuff prospered well in the Governor's Garden, under the shelter of banks of earth raised for

---

[1] Iceland is c. 39,768.5 sq. miles (103,000 km²), while Ireland is c. 32,595.1 sq. miles (84,421 km²).

[2] Reykjavík was founded in 1786 as a privileged or market town (*kaupstaður*); in 1801 there was a population of 307.

[3] See the description by James Roberts in his journal (p. 134 above).

[4] This is correct. There was also a school there with 24 students, but Banks arrived during summer so he would not have met them. A similar conglomerate was at the bishopric of Hólar in northern Iceland.

[5] As a result of the Irminger Stream of the North Atlantic Drift.

[6] The eruption of Laki 1783–4 (see Introduction, p. 17, n. 5 above).

[7] From the time of settlement in the 9th century until the 16th century, grain, chiefly barley, was grown in the most fertile parts of Iceland. The Little Ice Age, with its deteriorating climate, is blamed for its demise.

[8] Thodal, governor of Iceland 1770–85. Banks visited him in 1772 and describes the visit in his journal (see p. 85 above).

their protection. Potatoes were not known in the Island,¹ but there can be little doubt of their thriving whenever they shall be planted there.

Fuel is very scarce in all Parts. In the Southern, the most populous by far, Turf alone can be procured. In the North East the Sea throws ashore every Winter some Drift Wood, and in the North West there are some Beds of an inferior kind of Pit-Coal. It does not however appear, cold as the Climate is, that the Inhabitants ever use Fire for the purpose of warming their appartments. – They live in small rooms, the whole Family together, and sleep, the unmarried ones at least, several in a Bed their heads being placed alternately at the head and the Feet of the Bed. –

The greater part of the Country is covered with Ashes or with Lava – there is however a proportion of Pasture and of Meadow. – The Meadows produce very short hay and are mown with difficulty, for the Frost every Winter heaves up the Earth in Ridges, and prevents the surface from being level.

Cattle and Sheep are in Plenty, but no Hogs or Poultry are kept.² Fish is a chief Article of Food – instead of Meat. Flour of the seed of a Grass that grows on the Sea Banks is used, but the produce of this is very inadequate to the consumption of the Country. The Danes supply the Deficiency with Biscuit and Meal, and poor People occasionally use a kind of Liver wort³ found on the Mountains, that thickens their Milk when boil'd in it.

No kind of Carriages are at present in use; tho' Horses are in tolerable abundance. – It is said that Carts were formerly used,⁴ but no appearance of Roads across the Hrauns or Beds of Lava capable of receiving a Cart, were observed – indeed the Tracks in which the Horses travel over these Craggy Places are narrow, that none but those bred in the Country ought in prudence to be trusted upon them.

The Population of Iceland, small as it is, is gradually diminishing. In 1703 there were 50,444 Souls.⁵ In 1769, 46,201⁶ only – of these, 34,216 lived on the South side and 11,985 only on the North. The People are Mild, inoffensive, and very timid. – No Arms of any kind, either Cannons or Muskets or Soldiers of any description, were seen among them – not even at the Seat of Government:⁷ nor was there a Flag hoisted any where, to which a Foreign Ship might pay the compliment of a Salute. Even Fowling Pieces were scarce and little used, as was evident from the Quantity of Game (Ptarmigons) that was met with, and from these Birds being quite tame.⁸

---

¹ This is not correct, though the cultivation of potatoes in Iceland was admittedly in its infancy. The first potatoes in Iceland were grown by the German Baron Hastfer at Bessastaðir in 1758. The Icelandic pioneer was the Revd Björn Halldórsson (1724–94), who planted potatoes at Sauðlauksdalur in western Iceland in 1759. In 1761 the Danish government ordered royal officials in Iceland and the wealthiest farmers to cultivate vegetable patches. Around 1770 there were about 170 vegetable gardens in Gullbringusýsla alone and they were steadily increasing.

² Hogs were reintroduced in the 20th century, while poultry was rare though not unknown.

³ Called *fjallagras* and still gathered every year in Iceland.

⁴ There is no historical evidence for this statement.

⁵ This is almost correct. The latest research on the census of 1703 (the first national census of modern times) gives a total population of 50,358.

⁶ 46,207 in DTC copy. This shows the reliability of the original as compared to the DTC copy as the population figures of 1769 are exactly 46,201 as Banks wrote.

⁷ The Danish government had prohibited the use of weapons in Iceland while strengthening its hold on the country during the decades following the Reformation in 1550.

⁸ Here the DTC copy adds '& fearless'.

Even the richer People were ill supplied with the necessaries of life furnished by the Danish Company. Wood for building Houses and Boats; – Biscuit, Meal, Brandy and Hats, were the chief Articles of their consumption. In return they gave Fish, Salt Meat, Hides, Tallow, Wool, Salt Butter, Feathers and Eider Down, and a few Furs, chiefly of White Fox – all Ranks appeared unhappy and would as the Writer believes be much rejoiced in a change of Masters that promised them any portion of Liberty – the bettermost People[1] showed a predilection for England, and privately solicited the Writer to propose to his Government to purchase their Island from Denmark, and at the same time asserted that the whole of the Revenue derived by the Danish Nation from them, did not exceed 40,000[2] Rix Dollars a Year.[3]

They proposed to him to purchase a Farm, in hopes as they said, it would induce him to promote this Business. He enquired into Particulars and found one that maintained 10 Cows, 10 Horses and 400 Sheep. – The annual Rent of this was Nine Rix Dollars, and the Purchase 120. – Another farm maintained 12 Cows, 18 Bullocks, 8 Heiffers or Calves, 14 Horses and 82 Sheep – the Price of this was 120[4] Rix Dollars.

If it should be thought expedient to seize upon Iceland, either as an object of Exchange in the case of Peace,[5] or with intention to unite[6] it permanently to the Crown of the United Kingdom, the Writer has no doubt that 500 Men,[7] with a very few Guns, to be mounted on Horses when the Troops arrive, would subdue the Island without striking a blow. – There are only four Persons of Authority in it. The Stiftsamptman or Governor. The Amptman or Lieutenant Governor and two Bishops. – The two first resided in 1772, within a few Miles of each other,[8] and one of the Bishops[9] lives at no very great distance from them. The Attack therefore ought to be made in that Quarter, where by securing these three Persons, the Country would in all probability be at once in the possession of the Invaders. – It must be remembered however, that every part of the Coast abounds with safe Harbours, in which an Enemy might land any Force he could transport thither; but if the English are once settled in quiet Possession, a Hostile Army will find much difficulty in marching through the Country, where neither Bread nor Quarters can be obtained.

In conquering Iceland and Ferroe,[10] the United Kingdom would annex to itself the Dominion of all the respectable Islands in Northern Europe, a proud Pre-eminence for the British Isles to obtain. – She would emancipate from an Egyptian Bondage a Population,

---

[1] 'the bettermost people' was added by Banks in the draft.

[2] Both the DTC copy and the draft have 4,000 rixdollars, which is wrong.

[3] C. U. D. Eggers, secretary of the Royal Commission of 1785 studying the situation in Iceland, wrote in 1786 that the amount was 60,000 rixdollars.

[4] It should be 160: see the Banks journal, p. 97 above. This was information Banks received in 1772. The draft also has 120 – an easy mistake to make with 120 directly in the line above.

[5] This may show Banks's familiarity with Cochrane's proposals for exchanging Iceland for either Crab Island in the Caribbean or the Duchy of Saxe-Lauenburg. See above, p. 218, n. 2.

[6] DTC copy has 'annex'.

[7] In his memorandum Cochrane had suggested an army of 1,000 men.

[8] The governor Thodal lived at Bessastaðir and the *amtman* (district governor Ólafur Stephensen) at Sviðholt on the same peninsula.

[9] The bishop of Southern, Western and Eastern Iceland resided in Skálholt, in southern Iceland, about 62.2 miles (100 km) east of Bessastaðir.

[10] The Faroe Islands. According to Banks's passport he had planned to stop in the Faroe Islands but did not do so. The season was advancing.

consisting entirely of Fishermen, and consequently of Seamen, that would rapidly increase under her mild Government, and when allowed to use Shipping would furnish in a short time a supply of Seamen to the British Navy, of no inconsiderable importance; – she would open an increasing Market for her Manufactures in proportion as the Population of Iceland would in future increase. – She would possess herself of the Iceland Cod Fishery and would no doubt, by the force of British Capital recover some portion of its former importance – she would not only prevent the Faroe Islands from being the harbour of Smugglers, but would absolutely command the Herring Fishery, at the Point where the Shoals divide, and where of course the Fish is known to be in the best condition, and of the greatest value, and she would strike a blow,[1] at a very small expence, both of Men and of Money, that by depriving Denmark of a part of the Ancient Hereditary Dominions of the Danish Crown, would probably produce a greater effect on the Public Opinion of Europe, than would be done by depriving her of the whole of her Colonies, both in the East and the West Indies.[2] –

These are the advantages which occur to the Writer. – On the other hand it is obvious that the Island does not a*t* present nor is likely for some time to come, to produce any Revenue whatever.[3] – That not only the Troops employed on the Expedition, but the very people whom they are employed to conquer, must be supplied with all kinds of Vegetable Food, by the Nation to which they shall be in Dependency. – That no material advantage can soon be derived, either from the Fishery or from the Mines, and that /even/ the supply of Seamen which may in due time be expected is a remote object, when these and such other Objections as will naturally occur to people better versed in the political Interests of the United Kingdoms, than the Writer pretends to be, are balanced against these Advantages that appear to His Majesty's Ministers to be real, then alone can the propriety or impropriety of the Measure be finally decided upon. – [4]

---

[1] In the draft Banks had crossed out 'of no considerable importance in the Eyes of Europe'. On second thoughts that was obviously not going to help his argument which had as its aim the annexation of Iceland.

[2] This was precisely what the British government had done in January 1801 (and was to be repeated again at the fresh outbreak of hostilities in 1807).

[3] Cochrane had described the advantages for Britain if it took possession of Iceland in glowing terms, but as can be seen here Banks did not agree, especially as regards the potential revenues from the fisheries and the mining of sulphur.

[4] The Pitt administration did not take the advice proffered by Cochrane and Banks. Banks's memorandum considers the advantages of a British annexation in the light of a humanitarian gesture and a question of enhanced status and prestige. The British statesmen were perhaps more interested in economic benefits, which Banks emphasizes could only be long-term. There was clearly no hesitation regarding the seizure of the economically lucrative Danish colonies in the Caribbean and India, but what was the point of saddling England with the supposedly unprofitable Iceland in 1801? Another factor may have been that British fishing interests did not want competition from Icelandic fish products, as became clear during the Napoleonic Wars.

The timing of this affair is also important. Banks's memorandum would have reached the government at a critical moment. It was written at the very end of January. At the beginning of February Pitt resigned over the question of Catholic Emancipation and Addington was invited to form a new administration. When that government was finally completed on 14 March the threat of the Armed Neutrals was almost over. Parker and Nelson finally received their instructions on 15 March to proceed to Copenhagen, where the Danes were decisively beaten on 2 April and forced to agree to an armistice. Tsar Paul, the actual leader of the League, was murdered in March and eventually the League of Armed Neutrality was dissolved. England and Denmark were at peace and the occupied Danish colonies returned. Thus there was no longer an acceptable reason for contemplating the annexation of Iceland.

To the Writer of this, however, the Conquest of Iceland appears to be a wise Measure; it will subject Denmark to a considerable political humiliation in the Eyes of Europe, without at all diminishing her real Resources as an independent Nation; and though it will not immediately, it will in due time extend the Commerce, add to the Revenue, and increase the Nautical Strength of the United Kingdom.[1]

BL, Add MS 38356, ff. 39–48.[2]
NHM, DTC 12, ff. 157–166. Contemporary copy.
Hermannsson, 'Banks and Iceland', pp. 25–30 (with only one note).
The draft is in the Royal Geographical Society of South Australia, Banks, Sir Joseph, Notes on Iceland, MS 6c.

39. Jörgen Jörgensen to Banks, 21 October 1806[3]

Copenhagen

Sir!
The protection you have always afforded to such persons, who have been engaged in long voyages, especially in the South seas, has made me presumptious enough to trouble you with this letter; and I[4] hope by the generosity and humanity I lately discovered in you, when I represented the calamitous situation of the othaheiteans,[5] I had brought with me to London, that I may by your help be able to complete a piece of work, which I otherwise will have great difficulties in doing. When I wrote to you, Sir, in London concerning the otaheiteans I subscribed myself John Johnson; but my danish name is Jurgen Jurgensen,[6] which is the same name allowing for the difference in the pronounciation. I have now finished a work: The Transactions of His Majesty's Armed Tender *Lady Nelson* from the

---

[1] Note verso, written by an amanuensis: 'Remarks on Iceland | January 30 1801.'

[2] There are two amanuenses writing this lengthy memorandum.

[3] This is the second letter to Banks from Jörgensen – the first is not extant, but in this letter he introduced himself. As theories have been advanced that Banks and Jörgensen together planned the annexation of Iceland during the Napoleonic Wars it is interesting to note how these two unlikely accomplices made each other's acquaintance. In his 'Account of the Iceland Revolution' Jörgensen described his first efforts at corresponding with Banks thus: 'It was sometime after my arrival in Denmark, I transmitted a letter to Sir Joseph Banks, in which I begged his assistance in publishing a work, that was compiled of the different journals I had wrote during my stay in the British service. ... I shortly after received an answer from Sir Joseph Banks, wherein he expressed his willingness to promote the publication of my work, provided it met his approbation. I had several offers from booksellers, who had connections in Germany, ... [but] I thought it a much greater honor to have a work coming from my hands, which had been approved of by the Mæcenas of the age; ...'. He entrusted his papers to an Englishman travelling to England but later learned that the Englishman had been forced to flee leaving his journals behind. Jörgensen wrote that he was 'rather astonished, that I did not receive a letter from Sir Joseph Banks, expressing wether he had received my papers, or not' (BL, Eg. MS 2067, ff. 197–200). See also Plate 9.

[4] Throughout this letter Jörgensen writes i instead of I.

[5] i.e. the Tahitians. Jörgensen had been first mate on the whaler *Alexander*, which had anchored in Otaheite (Tahiti) in search of supplies. On leaving they took with them 'a chief of Otaheite with a young companion of his'. On arrival in London in June 1806 Jörgensen introduced them to Sir Joseph Banks, who took them under his care entrusting them to the Revd Joseph Hardcastle. Both died in little more than a year (Hogan, *The Convict King*, pp. 61–2).

[6] Actually Jürgensen is the correct surname. He is, however, much better known under the name Jörgen Jörgensen or Jorgen Jorgensen, though Danes now spell his name Jørgen Jürgensen.

twelfth of april 1803 until the 25th of april 1804, /*with my*/ Two voyages performed in the *Contest*, from port Jackson to Bass' Straits, and new Zeeland – A voyage in the Ship *Alexander* from port Jackson to England – The Transactions of the *Lady Nelson*, will give an account of every particular circumstance, relating to the setlements lately established in the Derwent river, in Port Dalrymple and Hunter's river.¹ A discription of the sliding keels, with rules for working them, which are grounded on three years experience.

In the voyages of the *Contest* will be given an account of the first attempt made to establish a new Setlement at Port Dalrymple; and of a voyage made to New Zeeland, where four extensive harbors /were/ discovered by me. – The voyages by this vessel were performed /*by me*/ between the 25th April 1804 and the 26th February 1805 – The voyage of the *Alexander* performed since that time, will give an account of the present situation of port Jackson, New Zeeland, Othaheite and St Catharine –, with a large reef discovered on our passage to othaheite, and allready known to you Sir.

My work is completely finished and ready for the printer. I do not know any person in England, /*there*/ who would interest himself in my favor, except you Sir should be good enough, just to recommend my manuscript, to any honest person you might choose to employ. If you should not be offended, /*of*/ by me asking you such a favor; I shall immediately on receiving an answer to this letter send my whole work, over for your inspection; and if you will employ a printer, or a proper person to publish it, I shall willingly pay all expences; or if the editor should wish to publish it on his own account, he is wellcome to do it. –

I would willingly let it be printed in Copenhagen, but several difficulties are in the way: first it will be nearly four months before it can be finished; second; it can not be printed neatly like in England; and lastly I have been prevented by the commission for to inspect into all books before they are printed, ~~to~~ in making use of several expressions in the introduction; of fear that they may be offensive to a certain diabolical government in Europe;² So I have taken my book, and declared if I have no liberty to write, what I pleases, my book shall be published in England, and if that is not possible, it shall /*be*/ not be published at all.³ –

I mean therefore to beg you to be good enough to take my manuscript, and give it to a person, and to let it be published in what form and what manner you pleases yourself; if any thing should be incorrect I should be glad if you would exactly leave out, what you should think is not proper. It is undoubte*d*ly a great presumption of me to trouble such a gentleman as Sir Joseph Banks, with such trifles as these; but when I consider your

---

¹ Jörgensen's role in the exploration of Tasmania is generally acknowledged, see e.g. *The Australian Encyclopedia*. He was later transported for life to New South Wales for robbery and died in Hobart in 1841.

² Presumably France.

³ This did not turn out to be the case. There is no record of this book being published in England; however *Efterretning om Engelændernes og Nordamerikanernes Fart og Handel paa Sydhavet* (Observations on English and North-American Voyages and Trade in the South Seas) was published in Copenhagen the following year. This was Jörgensen's first published work, a 39-page pamphlet. According to Jörgensen it was an extract from a longer work which he had written in English, as described in this letter. He hoped it would be of use to his *Fædreneland* (fatherland) (p. 3). In 1996 it was translated into English by Lena Nielson Knight, edited by Rhys Richards and published in Wellington, New Zealand: *Jorgen Jorgenson's Observations on Pacific Trade; and Sealing and Whaling in Australian and New Zealand Waters Before 1805*. Jörgensen was the first Dane known to have circumnavigated the globe.

wellknown Character as the promoter of all what is useful to your country, /if/ I flatter myself, that you will at least not be offended with me writing to you.

I do not know, whether you are acquainted with the new invented metal thermometers, invented in Copenhagen, by a older brother of mine, named Urban Jurgensen Author of a work lately published in french: Rules for the exact measuring of time by watches.[1] In case you are not acquainted with these thermometers, or have not got one in possession, it will be /a/ great pleasure to me, if you will signify the same to me in a letter, so I may have the honor of sending you one to London.

I should wish for certain reasons, that my name might be unknown, and this letter not made public, or come into any other persons hands, than your own. I hope you will excuse my rough way of writing as a seaman is not qualified to write so well, as any other person long used to converse and correspond with men of letters. However whether you should think it worth your while, to answer my letter, or concern yourself with my work, I shall constantly remain Sir!

with the greatest respect
Your hum*b*le ob*edient* Servant
[*Signed*] Jurgen Jurgensen

[P.S.] Letters addressed to me – Mr Jurgen Jurgensen directed to /my father/ Mr. Jurgen Jurgensen, Watch maker to His Majesty the King of Dannemark

SL, Banks Papers, I 2:12. Autograph letter.[2]

40.   Magnús Stephensen to Banks, 17 October 1807

Copenhagen

To his Excellency the right honourable Sir Joseph Banks

'Tis in a gratefull remembrance, that the frozen tho' heartburnt Island, Iceland, preserves the thought on the honnour you once did it, dearest Sir, when, upon your travel round the world, together with Dr. Solander and Mr. Uno von Troil in the year 1772,[3] you not only came there to Iceland, but took pleasure in staying there some time; perlustrating our famous Hecla and the boiling spring, Geyser, in the mean time spreading your liberalities among the poor inhabitants, and teaching every Icelanders heart to bless and revere the illustrious friend to humanity, Sir Joseph Banks. You often honoured my father's house with your presence, and even bestowed upon him your precious and munificent friendship, which he on his side strived to merit by alleviating your toilsome travels,

---

[1] Urban (Danish spelling of name) was Jörgensen's older brother (b. in 1777). He learned watchmaking in Switzerland and was one of the foremost watchmakers of his day. See Chapuis, *Urbain Jurgensen et ses continuateurs*. There is a family museum in Bienne (Biel). The work referred to is *Regler for Tidensnøjagtige Afmaalning ved Uhre* (Rules for the Exact Measuring of Time by Watches), published in 1804 at the King's expense.

[2] Note in Banks's autograph, left-hand margin of last page: 'Mr Jorgenson | [Received] Nove*mbe*r 7 [1806] | [Answered] – 17 [1806]'.

[3] With Solander but *not* von Troil on the *Endeavour* voyage 1768–71 and the expedition to Iceland was of course a separate voyage.

providing the means to further them, and procuring a chosen collection of the best manuscripts, Sagas[1] and historical monuments of the country.

In recalling to your memory these data, I hope, Dear Sir, you will favour with a kind thought the Manes[2] of your former friend and my father, Olav Stephenson, then bailiff, afterwards grand bailiff, of the Island, and living at Svedholt near Bessested.[3]

Often had I too the good luck of seeing you, being his eldest son, then a 10 years' boy. I am now his danish Majestys counsellor of Justice, and President of a new organized High court of justice for the whole Island. Ignorant of any misunderstanding between England and Denmark, did I last autumn, in private affairs undertake a voyage to Copenhagen; but under the Naze in Norway I was made prisoner by an English ship of war, the 19th of Septr. last, and carried to Leith in Scotland, where also the merchant ship, in which I left Iceland, *De tvende Söstre,* was laid under Embargo.[4] Two merchants of Iceland, owners of this named brig and her cargo, are still there detained with pregnant wifes and a number of children and servants, in a helpless and moneyless situation.[5] They are my friends, and men that have deserved well of my paternal Island, by having, under the many plagues of later years, with no small sacrifice, saved some thousands of inhabitants.

The case is now, that not only these honest men must be ruined by losing their all, with this and other in England and Scotland detained ships and cargoes, but what is still worse, the navigation to Iceland will be at an end, and I shall, once returning to my dear country, find it transformed into an uninhabited desert, and the people starved. For it is altogether unavoidable for this Island to escape hunger, if it is only to hold out a single winter without being supplied with provisions, corn, fishing lines, and iron; being things which the benevolent Danish government every year with considerable loss sent over, in order to keep alive the population. The King is no longer parttaker in the Iceland trade. It was ceded in 1788 to private merchants,[6] and all hoped for a happier period; but so many evils have of late assailed this poor island, viz. scarcity of grass, decline of fishery, laying to of the Greenland-ice, and eruptions of Vulcanoes, that the cattle and sheep are much brought down, and the export nothing more than a small quantity of stokfish. Now, if the war is to take effect, in its full rigour, against these already so unhappy Icelanders, there is no means of preventing a population of 47.000 honest souls from perishing most miserably thro' famine.

---

[1] The Icelandic Sagas were written down in the 13th and 14th centuries based on oral tradition.

[2] *Manes* means 'shade of departed person, as object of reverence' according to *The Concise Oxford Dictionary*. This surprising choice of words led Banks to believe that Ólafur Stephensen (d. 1812) was deceased, as witness Document 61 and then Document 118, when he rejoices at the news that his old friend is 'still alive'. For a portrait of Magnús Stephensen, see Plate 10.

[3] Sviðholt on the Álftanes peninsula near Bessastaðir.

[4] *De Tvende Söstre* had been captured by the *Peacock*, under Captain Peake, off the Naze of Norway. All papers regarding the capture are to be found in TNA, High Court of Admiralty [hereafter HCA] 32/1197. Stephensen kept a journal describing his voyage from Iceland on 1 September until 10 October 1807. All the men had been taken on board the warship, the women and children being left on the ship, which was manned by British sailors. Bjarni Sivertsen and Stephensen were invited to take their meals with the captain in his cabin. See Stephensen, *Ferðadagbækur* and the translation into English by Sigurður L. Pálsson.

[5] The merchants were Bjarni Sivertsen and Westy Petræus. See Appendix 13.

[6] See Introduction, p. 15 above.

That England, which was ever renowned for liberality and enlightened humanity, will not, I hope, stoop to employ its vast force against the defenceless & already totally impoverish'd Iceland. – It will not put a stop to its insignificant trade, which would be to pronounce the sentence of death over its inhabitants. It would be a blemish to the magnanimous British character, to side with the raging elements and other destroying plagues, in order to exterminate all Icelanders, by continuing to detain their few merchant vessels. Even in the capitulation of Copenhagen the sanctity of private property is promised;[1] and here all is private.

Being, as I before said, brought to Leith, I there got permission from the admiral to go on board an English ship of war, and arrived thus in Copenhagen on the 10th hujus.[2] But, learning that there is yet no peace made, and that even Ships from Iceland here detained by the English fleet, are daily sent over to England,[3] where the Iceland merchants and passengers live in great distress and want, and foreseeing with infinite horror the utter ruin of my country, brought on in this very first year of the warfare, by the ceasing of its navigation, and my dear beloved family threatened with the most dreadful of Deaths, I took the resolution,[4] prompted by the many proofs of magnanimity, which you, dearest Sir, did give not only Iceland but the whole world, and by the happy circumstance of your good will for Iceland and for my family in specie,[5] to recur most humbly and confidently to your bosom, for a speedy and earnest intervention for the poor Iceland, that it might be exempted, thro' the precious favour of the English Government, from all hostilities, even if war goes on, and that the detained Iceland ships might be put in liberty to pursue their trade to and from Iceland, with or without English passports, as long as the war does exist.[6]

But if this magnanimous favour is too great to be hoped for, which I don't presume, I beg you, Sir, most respectfully and most tenderly, to employ your influence in the best

---

[1] The capitulation of Copenhagen was signed on 7 September 1807. According to the 7th article it was stipulated that all property, public or private, excepting those ships and warships belonging to His Danish Majesty, would be respected (copy in the RA, DfdUA, England 1807–1808, box 1920).

[2] Of this month, i.e. October. The admiral was Rear Admiral James Vashon (1742–1827). He had entered the navy in 1755 and had served widely, on the Newfoundland station, in the West Indies, in America during the American War of Independence and in the Channel fleet. In 1804 he was promoted to rear admiral and commanded the ships at Leith and on the coast of Scotland until 1808, when he was made a vice admiral. He eventually became an admiral in 1814. He now permitted 'the Danish Judge Mr Marcus Stephenson' to go to Copenhagen on the *Peacock* (TNA, Admiralty [hereafter ADM] 99/181, from Lieut. Flinn, 29 September 1807). The Danish consul George Home had written a letter to Vashon: 'The Bearer Mr Stephensen ... is desirous to have the liberty of Admiral Vashon to go to Copenhagen ... [He] is one of the Chief Men in Office at Iceland.' On the letter has been written: 'The Admiral has no objection' (Reykjavík, Landsbókasafn Íslands-Háskólabókasafn, Department of Manuscripts [hereafter Lbs.], JS 93 fol.). In his journal Stephensen describes his meeting with Vashon and the voyage to Copenhagen.

[3] At least 18 ships of the 41 sailing from Copenhagen to Iceland in 1807 were captured. See Agnarsdóttir, 'Great Britain and Iceland', chap. III.

[4] On arriving in Copenhagen Stephensen had been appalled by the state of his 'beloved Copenhagen' writing in his journal: 'Here there is hardly any government'. The Crown Prince had stationed himself in Rendsborg in Schleswig-Holstein (now Rendsburg in Germany), so Stephensen realized he had to take matters into his own hands and chose to approach Banks. In Copenhagen there were several high-placed Icelanders who may have encouraged him to do so, e.g. Grímur Thorkelín the Keeper of the Royal Archives, Banks's friend. See Stephensen, *Ferðadagbækur*, pp. 41–2.

[5] *in specie*: in particular.

[6] As will be seen in the following letters, Stephensen got his wish.

way, for obtaining a permission, that the 4 merchant ships belonging to the merchants Westi Petræus, Biarne Sivertsen and Adser Knudsen,[1] might freely trade to and from Iceland, and consequently be relieved from the embargo in Leith, and notice thereof be given them, thro' the Danish Consul in that place, Geo. Home.[2]

I lastly harbour the flattering hope, to be honoured thro' the post, via Gothenborg,[3] or other sure occasion, with your precious answer, that I may know if this letter is come at hand, what part Your excellent heart does take in the fate of my dear Island, what success may be expected, & finally that you pardon the freedom, with which only the most imminent danger inspired me, of thus troubling your venerable age, to which it will add a new, and never perishing splendour, to have saved the whole little good but unhappy nation of Icelanders: let my knowledge of your noble heart and of your friendship for my family be my excuse.

May heaven preserve you, dearest Sir, still many years, as an example of humanity and a true pride for Science to the good of many thousands and of my poor Iceland, this is the wish of him[4]

who is
Sir
Your Excellency's
most humble and most
devoted servant
Magnus Stephensen

NHM, DTC 17, ff. 59–63. Contemporary copy.
Hermannsson, 'Banks and Iceland', pp. 33–5.

41. Banks to Dr William Wright, 21 November [1807]

Soho Square

My dear Doctor

I have heard[5] with /some/ Pain that some Iceland Ships have been sent in by our Cruisers and are now lying at Leith. The hospitable Reception I met with in Iceland made too

---

[1] He is referring to the ships *De Tvende Söstre* (Sívertsen), *Seyen* (Petræus), *Johanne Charlotte* (Sívertsen) and *De Fem Brödre* (Knudsen).

[2] George Home was the Danish consul in Leith, an Englishman who understood 'very little except English' according to Stephensen. 'We asked him to assist us to change money, etc., but nothing was to be had from him now, not a farthing ... he is said to be a man of little influence whose word should not be trusted' wrote Stephensen in his journal (*Ferðadagbækur*, pp. 27–8). However, this was an unfair assessment. See Document 56, p. 251, n. 1 below.

[3] Gothenburg was at the time a very important entrepôt port used for British exports, subsequently sent into the Baltic as neutral goods on neutral ships (Christie, *Wars and Revolutions,* p. 307).

[4] Stephensen was writing to the enemy. The King of Denmark had forbidden all communication with the English, so technically Stephensen's action was traitorous. This coupled with his suspect behaviour in the Icelandic Revolution (see Introduction, p. 32) forced him to write a formal apology to the Danish government at the end of the war (to Kaas, the Minister of Justice, dated 19 September 1815, eventually published in 1882, 'Varnarrit', *Ísafold*, pp. 5–8, 13–15).

[5] From the previous letter from Magnús Stephensen (Document 40).

much impression on me to allow me to be indifferent about any thing in which Icelanders are concernd.¹

I have therefore Orderd a Permission from Government for two Icelanders to Come here for the Purpose of Stating their Case & Solliciting government if I can possibly induce them so to do to include my innocent & virtuous Friends in the exception made in the Capitulation of the Trade of Zeeland² for the Private property of individuals.

Will you my dear Dr. take the trouble Either in Person or by a Trusty & well informd friend to inform these Good People of the Interest I take in their Case and advise them to Send up one or two of their members to London. The Persons I most wish to see are the Two merchant Proprietors of the *Tvende Söstre*³ or one of them. If Either of them can speak any Language besides Icelandic & Danish it will be sufficient for him to come alone. If neither of them can then let one of them come & bring with him some Icelander capable of Speaking French English or German. The Sooner /*I see these Per*/ this is done the better it will be if they should want money to bring them up I will thankfully repay any person who will do me the favor to provide them with a proper supply.

SL, Banks Papers, I 1:4. Draft in Banks's autograph.⁴

42. Perrot Fenton to Banks, 21 November 1807

Doctors' Commons [London]

Sir

In answer to your Letter of this date desiring me to transmit, for the information of the Lords of His Majestys Most Honourable Privy Council, a statement of the number of Iceland Ships which are now detained in the Port of Leith – I have to state that I have not any Information which is not contained in the List of Ships which I had the Honor of transmitting on the 18th *October*⁵ by which it will appear there are two Iceland ships at Leith viz: the '*Two Sisters*' (Strengun⁶) and the '*Conferentz Raad Pretorius*'⁷ but the

---

¹ An undated note preserved in the Sutro Library (Banks Papers I, 1:36) voices similar sentiments. Banks wrote the following: 'I could not see the Countrymen who had Receivd me with hospitality & kindness when [I] visited their Island as a Stranger become the innocent victims of the implacable temper of the Crown Prince of Denmark with[out] feeling a deep interest in their Favor.'

² Copenhagen is on the island of Zealand.

³ Banks is referring to Westy Petræus and Bjarni Sivertsen. *Tvende Söstre* means *Two Sisters*, often used in British sources regarding this ship.

⁴ Notes in Banks's autograph, front page: 'Dr Wright'; verso: 'The Two Sisters B. Stangen master Fish Oil Wool Skins. ['Stangen' must be a bastardization of Sivertsen.] | Conferentz Raad Prosarius [Prætorius] Warsberg [the master was S. Karsberg]'.

⁵ According to the Marshal's List, dated 31 December 1807, a third has been added, the *Rodefiord*, J. Christensen master (TNA, Public Record Office [hereafter PRO] 30/42/14/8).

⁶ Or Stangun; this is probably a bastardization of Sivertsen. In the Marshal's List the same spelling appears.

⁷ The *Conferentz raad Prætorius* was a ship owned by Norwegian merchants. The supercargo was Hans Georg Bredal (see Appendix 13). The ship was captured and brought to Leith.

Documents of those Ships not having yet been transmitted I have not the means of ascertaining the Names of the Danish Owners – I have by this Evening's Post, written to Mess.rs Ramsay Williamson & C.o of Leith[1] on the subject and immediately on the receipt of their Answer it shall be communicated by

Sir
your most obedient
humble Servant
P. Fenton Dep*u*ty M*a*rsh*a*ll

SL, Banks Papers, I 1:3. Autograph letter.[2]

43. Dr William Wright to Banks, 27 November 1807

Edinburgh

Dear Sir

Your letter of the 21.st Current came to hand in course of Post.[3] I lost no time in making the necessary inquiry about the Owners and people of the captured Icelandic vessels – I first waited on Rear Adm*i*ral Vashon[4] at Lieth [*sic*]; He received me kindly, and read your letter to me with attention. He is warmly interrested for the unfortunate sufferers: and has written to Government in their behalf – He applied to the board of customs to give up the private property of the owner & Masters of those Vessels as body & bed cloaths and some household furniture.[5] The former has been complied with, but the furniture could not be restored with*ou*t an order from the Treasury. The Admirals kindness and humanity to those good people does him much hon*ou*r.[6] He tells me that the Iceland Vessels if condemned is a Droit of the Admiralty.[7] This is a fortunate circumstance; I hope there will be no difficulty with Government to include the Iceland

---

[1] A merchant firm in Scotland. The vessel, *The Two Sisters* (*De Tvende Söstre*) and the ship's cargo were in the possession of this firm. They were ordered immediately to give up all private property on board the vessel 'not being in the Nature of merchandize' (TNA, Treasury Solicitor [hereafter TS] 8/6, Charles Bishop to the Commissioners of the Treasury, 14 December 1807).

[2] Note, verso: 'Letter from the Deputy-/Marshall of the Admiralty | on the subject of the | Iceland Ships detained | at the Port of Leith. | *Received* 23 Nov*embe*r 1807.'

[3] See Document 41.

[4] See Document 40.

[5] The inventory of Petræus's private property is in TNA, TS 8/6 f. 394. It includes iron and wooden bedsteads, a 'Sopha', a close stool, four washing tubs, crockery and casks of salt beef, salt fish and butter.

[6] On 8 November 1807, Vashon wrote to the Lords of the Treasury enclosing a petition from Petræus 'praying that he may be permitted to remove from the said Vessel a part of his private Property consisting of Household Furniture &c' and asking they comply to the request, his family being 'in much Distress and want'. The above-mentioned inventory was attached and on 14 December J. Nicholl, the King's Advocate-General, delivered his 'Opinion' that the articles 'should be immediately delivered up to the Petitioner'. All documents regarding this case are to be found in TNA, TS 8/6, ff. 392–396; and see further TNA, Treasury [hereafter T] 4/13, George Harrison, referred to King's Proctor 16 November 1807. Thus Vashon's request met with success.

[7] A *Droit of the Admiralty* means certain rights or perquisites of the proceeds arising from the seizure of enemy ships, wrecks etc., formerly belonging to the Court of Admiralty, but now paid into the Exchequer.

Ships, in the Exception made in the Capitulation of Copenhagen, of the private property of Individuals.

A Mr. Petrius, I understand is owner of the two Vessels and Cargoes, from Iceland.[1] He had resided many years in that Island; and acquired what he thought, a competency to make him return to his native country, Denmark: For this purpose, He laid out his little All! on the produce & commodities of the Island; and embark'd with his wife, six small children, his servants, and household goods, not knowing of hostilities, between the two Nations. This excellent man, is at Lieth in the greatest distress, reduced from affluence to abject poverty; and he himself confined to bed with sickness.[2]

Mr J F Denovan, at the Council Chamber Lieth, is interpreter, for the German, Danish & Islandic Languages. He tells me, that Mr. Petrius speaks both German & French[3] that he is a well informed man; polite & easy in his manners; That his house was open to strangers, who visited Iceland; and that he accompanied them, on excursions, or tours in that country. He perhaps, was not there in your time; but probably, in that of our Friend M^r [John Thomas] Stanley 1789.[4]

I have been with the Printer of the Edin*burgh* Evening Courant, & you will receive that Paper by Tomorrows Post: stating the hard case of this Gentleman, and his forlorn situation.[5] I hope our Magistrates, and the Public at large will afford them some temporary assistance.

When it is determined, who of these Icelanders are to go up. I shall most cheerfully advance the Money to carry them to London. The Admiral and several others, have offered to procure them a free passage in a Lieth smack.[6]

In the meantime, I dare say that all proceedings in the Court of Admiralty, will be stayed, untill some of the concerned make their appearance in Lond*o*n.

I beg you will lay your commands on me how farther to proceed. I have the Hon*our* to be with The greatest respect.

My dear Sir. most faithfully yours
[*Signed*] Willm Wright

SL, Banks Papers, I 1:5. Autograph letter.[7]

---

[1] In the *Edinburgh Evening Courant* of 28 November 1807 (p. 3) Petræus is described as the owner of three ships, one brought to Leith, one to Cork and the third to Yarmouth. In actual fact he owned the *Seyen* which was taken to Cork and probably owned part of the cargoes in the other two.

[2] In this paragraph the words are underlined in the original letter by the writer and also by Banks, as some lines are in ink and others in pencil.

[3] Though it seems unlikely on the face of it that Petræus was proficient in French, he was an experienced seaman (who may have been among those sailing from Iceland to Bordeaux) and Denovan certainly had command of several European languages (see Appendix 13).

[4] The Stanley Expedition visited Iceland in 1789 (see Introduction, p. 24 above). Petræus would not have met them as he did not begin trading in Keflavík until 1790 (Andrésson, *Verzlunarsaga Íslands*, II, pp. 489–90).

[5] Petræus's 'distressing case' is described in the *Edinburgh Evening Courant* as mentioned in n. 1 above.

[6] They sailed in the smack *Sprightly*.

[7] Note in Banks's autograph, on the bottom of the first page of the letter: '[Received] Nov*embe*r 30 [1807]'.

44. Dr William Wright to Banks, 27 November [1807]

[Edinburgh]

Dear Sir

In addition to the within, I have to say that I gave Mr Denovan[1] a copy of your letter,[2] and he was to read it to the owner & Masters of the Iceland ships that they might know the interrest [sic] you take in their case. The Public too will know & learn from the Courant[3] of your noble and generous conduct towards the Unfortunate Stranger.

Wm Wright.

[P.S.]   Since writing the above yours of the 24th current[4] is come to hand (one o'clock) I go to Lieth [sic] to deliver the order to the Collector of the Customs.[5]

SL, Banks Papers, I 1:6. Autograph letter.[6]

45. A Project for the Conduct of *Sir Joseph Banks* in Respect to Iceland [late November 1807[7]]

as we hear Every day more & more of the high degree of Irritation under which the Government of Denmark Proceeds towards this Country, the expediency, of attempting any intercourse with Magnus Stephensen, who is now Resident in Copenhagen,[8] becomes doubtfull, Least by the Letter being opend, as all Letters are said to be at the Danish Post office, some information of the intentions of this Country towards Iceland should be disclosd or at Leest be suspected.

may it not be better then that the following Steps be taken.

that a message be sent to Leith intimating to the masters of the Iceland Ships detaind there, that they will do wisely if they depute some of their Body whom they can most Rely upon & in Preference the Two merchant friends of Magnus Stephensen owners of the *Tvende Söstre*,[9] with another /who knows/ in case neither of them Speaks Either the

---

[1] Denovan, see preceding document.

[2] See Document 41.

[3] The newspaper is the *Edinburgh Evening Courant*: see Document 43.

[4] This letter does not appear to be extant

[5] In spite of diligent work on the part of research officers in the Scottish National Archives and the National Archives, London, there do not seem to be extant any papers from the Collector of Customs in Leith.

[6] Note in Banks's autograph, left-hand bottom of page: '+ Enclosing the Pasports'; and in Banks's autograph, verso: 'R*eceived* Nov*embe*r 30 1807'.

[7] This project was obviously conceived soon after Banks received Stephensen's letter, dated 17 October 1807 (Document 40), which would probably have reached Banks in late October, and after Denmark's declaration of war on Great Britain on 4 November 1807, but before he wrote to Lord Hawkesbury. See Hawkesbury's reply in the following letter.

[8] See Document 40. Banks was determined to come to the rescue of the Icelanders and now entered upon his most active period respecting Icelandic affairs, writing letters, memoranda and preparing projects dealing with the release of the Iceland ships, the plight of the Iceland trade and, most interestingly, a British annexation of the island. He became more convinced than in 1801 (see Document 38) that the only cure for the ills of the Icelanders was a British annexation.

[9] Bjarni Sívertsen and Westy Petræus.

English the French or the German Language to Lay their Case before Government in the hopes of Obtaining a Release of their Private Property as was done at Copenhagen[1] at a Period Later than the Capture of Some at Least of the Iceland Vessels.

that these persons when they arrive /in London/ Should be advised to apply to *Sir Joseph Banks* as a Person well inclind towards the Country of Iceland in Consequence of the Civilities He has Receivd from the inhabitants and one whom Government is in the habit of Permitting to lay Cases before them on such Subjects as they think him well informed in.

'as merchants of Iceland, owners of a Ship & Persons who have deservd well of their native Country by having with no Small sacrifice savd the Lives of Thousands of their Countrymen' (so M*agnus* Stephensen represents them),[2] these persons must have Some weight at home & be able to give a good account of the dispositions & wishes of their Countrymen respecting this Government.

of them Enquiry /*might*/ may be made whether the General Opinion of the Islanders remains now as it Certainly was when Sir J B. was on the Island much inclind to become dependent on the Government of England, in Preference to denmark Considering England as a Countrey whose trade is so widely Extended that She has a better Prospect of Finding a market for the Produce of Iceland than denmark can Possibly have.

They /*might*/ may then be told that their Tallow & hides of their Sheep /& Cattle/ & Salt Provisions whenever they leern the way of Curing them properly, will always find a market Either in England or her Colonies, that their Brimstone & Probably other Volcanic Productions, will be brought by England into the market cheaper than the Same Articles from Italy can be, & it may be held out to them that their Stockfish which is Certainly the Cheepest animal food for man which Europe Produces, may become some part of the subsistence of the negroes in our Colonies, in which case Iceland would Quickly recover her Ancient Population & the Place She formerly held among the more respected Countries[3] in Europe.

if in Conversation with these people a disposition satisfactory to the Kings Servants is found *The Tvende Söstre* /*and*/ or all the Iceland Vessels might be Liberated as an act of Conciliation on the Part of this Country.[4]

by these Vessels *Sir Joseph Banks* would write Letters to the Family of Stephensen, /the children of the Person who was amptman[5] when Sir J B was in the Island & Stiftsamptman or Governor soon after & the Brethren of Magnus Stephensen/[6] now in Iceland Pointing out to them the opportunity which now offers of Placing their Island under the dominion of England, the great Likelyhood that after a voluntary Surrender in the Part of the people,

---

[1] Banks is referring to the articles of the Capitulation of Copenhagen, signed on 7 September 1807.

[2] At best paraphrased by Banks. If compared with the aforementioned letter of Stephensen, it is certainly much exaggerated.

[3] It looks as though Banks began to write *Nations* and then reconsidered; Iceland was not a nation to him.

[4] This was eventually achieved in June 1808. But this mention of conciliation crops up in the following letter to Hawkesbury.

[5] Ólafur Stephensen, then a district governor.

[6] This is added in the margin to give a precise description of 'the family of Stephensen' and it is of importance in discovering who Banks wished the recipient of Document 61 to be. Magnús Stephensen had two brothers, Stefán Stephensen (1767–1820), district governor of the Western Amt, and Björn Stephensen (1769–1835), a court official. Magnús had mentioned in his letter to Banks that he was 'the eldest son' (Document 40).

England would feel little inclination to Part with /her/ at a Peace, a People so Eminently well affected to the English nation /and also/ the various benefits they could derive from a Connexion so much more natural to them than their Present one with Denmark, of the Good disposition of England towards them /& their inclination to take Iceland under her Protection/[1] the Liberator of their vessels /*would*/ must be Considerd as a Sufficient Pledge.

if the Return to Iceland is Still Practicable as *Sir* J B believes it to be, a small arm'd vessel may be sent in Company with the Liberated Icelanders, to see them safe home and on board of her a Person Capable of negotiating may be sent, who in Case of any Conduct in the Part of the Icelanders that gives him Cause of Jealousy or suspecting a delay in Calling together the Thing or Folkmote, which is still held in Iceland,[2] may Return with the Vessel, Report what he has Seen, & how he thinks measures may be best taken in Spring for fulfilling the wishes of his Majesties Government whatever they may then be.

Iceland in its Present State is not worth conquering, the inhabitants however cannot continue to exist without the Protection of England if they Continue subjects of Denmark their Ships will be seizd by our Cruisers & they will find it wholly impracticable to Furnish themselves with the Flour Clothing brandy, wood &c which are absolutely necessary for their Comfortable existence.

There appears then to [be] no alternative for these people than that of voluntarily offering themselves as subjects to England. This measure will no doubt conciliate the English Government & incline them to take measures, as they may appear to be expedient for Preventing the Danes Should they make any attempt to Resume the sovereignty they have Lost.

under the Protection of the English Nation /the/ Iceland/ers/ may & will become a Prosperous People. England Can Consume or dispose of all the spare Produce of that little Island & in Return give her a vast increase of Comforts & advantages such indeed as no other Nation has at present the Power to bestow. Besides Iceland and the Ferroe Isles appear to be naturaly an absolute Continuation of the British Isles and of Course will after a Little usage be considerd as an integral Part of them.

new Sources of adventure & of Profitable Trade will soon present themselves. The Ice which annually loosens itself from the Greenland Coast & come upon or near that of Iceland is no doubt Loaded with seals as well as with white beers. Of these seals the inhabitants of newfoundland have in one year killd more than 100,000 the Skins & oil of which probably produced more in the European Market than all the skins tallow & Flesh of all the oxen & Sheep killd in Iceland would & in the same time.

SL, Banks Papers, I 1:47. Autograph draft.

---

[1] This is added in the margin.

[2] In this Banks was wrong. The Althing (*Alþingi*), the Icelandic assembly founded in 930, was still being convened when he visited the island in 1772, though in a much reduced state having lost all powers of legislation. It was subsequently abolished in the year 1800.

46. Robert Banks Jenkinson, Lord Hawkesbury, to Banks, 29 November 1807

London

My dear Sir

I return you Mr. Stephenson's Letter.[1] I have communicated it to the King's Servants; and they are of opinion that it might be very adviseable that you should have some further communication with him, with a view of ascertaining whether, through him or through any other Channel, the Island of Iceland could be secured to His Majesty, at least during the continuance of the present war.[2] In that case the Fisheries and Trade of Iceland would be protected; and I should hope that we might be able to obtain the services of some of their mariners. There will be no objection, as a Measure of conciliation, in releasing the few Icelanders' Ships which are at Leith.[3]

Believe me to be
My Dear Sir
very sincerely yours
Hawkesbury[4]

NHM, DTC 17, f. 72. Contemporary copy.
Hermannsson, 'Banks and Iceland', p. 36.

47. Dr William Wright to Banks, 29 November 1807

Edinburgh

My dear Sir

I have the honour of introducing Messrs Westy Petreeus and Biarne Severtson.[5] The owners of the *two Sisters* lately captured and brought in here.[6] The first has his Wife and Six Children & two Serva*n*ts at Lieth. Both Gentlemen speak the Danish and German Languages, and are otherwise well informed – Mr Severtson remembers Mr Stanley[7] very well and his relations shewed him and his attendants very mark'd Attention. I believe you

---

[1] Document 40.

[2] From this it is clear that Hawkesbury was already acquainted with Banks's plan as outlined in the 'Project for the Conduct of Sir J B in Respect to Iceland' (Document 45), for instance the suggestion that the Icelanders' ships be released as a conciliatory measure.

[3] The ships were eventually released in the spring and summer of 1808.

[4] This letter is of immense importance as it is the only written evidence of a secretary of state contemplating the possibility of a British annexation of Iceland. As will be seen, this encouraged Banks to secure the best possible and most accurate information on Iceland and plan the country's annexation in detail (Document 63).

[5] Westy Petræus and Bjarni Sívertsen.

[6] See Document 40 and notes.

[7] John Stanley visited Iceland in 1789. He visited Hafnarfjörður but there is no record of Sívertsen in the journals. As Sívertsen was born in 1763, he would have been 26 at the time of Stanley's visit. As he lived in the vicinity of Hafnarfjörður it is not unlikely that they met. The three diarists of the Stanley expedition tended to record their relations with 'the high society' of Iceland, which would not have included a merchant.

are acquainted That Mr Stephenson[1] is brother in law to Mr Petreeus – Independant of the kindness and Hospitallity these good people have shewen to all Strangers. Their misfortunes and hard Case call aloud on the Generosity of the British Nation for relief. Their Household furniture is given up and as the captured Vessels are a droit of the Admiralty;[2] there is every hope that the ships or a compensation to the sufferers will be given them.[3]

I sent you by yesterdays Post the Edinburgh Evening Courant[4] giving some Account of Mr Petreeus but he has resided Twenty One[5] years in Iceland.[6]

May your noble and generous efforts in behalf of these Virtuous and innocent people be crowned with success.

Adieu my dear Sir
I have the honour to be
with the greatest respect
your faithfull & obliged Servant
Willm Wright

[P.S.] A Subscription is opened both at Edinburgh & relief for the immediate benefit of those Gentlemen & Families [*sic*]

BL, Add MS 33981, ff. 265–266. Autograph letter. [7]

---

[1] Banks would have understood this as referring to Magnús Stephensen (see Document 40). In fact Petræus's second wife was Katrín Margrét Hölter, whose sister Marta María was married to Magnús Stephensen's brother, Stefán Stephensen (the district governor of Western Iceland). They were the daughters of the merchant Didrik Hölter. Thus, the families were connected.

[2] A *droit of Admiralty*: see Document 43, p. 231, n. 7 above.

[3] Petræus was successful in regaining his property (see Document 43, p. 231, n. 5).

[4] Accompanying this letter is Wright's handwritten excerpt from the *Edinburgh Evening Courant*, f. 267, stating:

'The following singularly distressing Case of an individual now in Leith deserves the attention of a Benevolent Public.

The Unfortunate Sufferer alluded to has for 17 years resided in the frozen Regions of Iceland, where by industry & hard labour he earned what he deemed a competence for the future Support of himself & family, converting his all into the produce of the country & unconscious of the Danger which the unlooked for Rupture had occasioned to embark the whole in 3 vessels & dispatched them for his much longed for home Copenhagen. The one in which he sailed *with* his wife 6 Children & 2 Servants was captured & brought into Leith another has been carried into Cork & the third into Yarmouth. This unfortunate Person remains in Leith confined to Bed by Ill Health neither he nor any of his family can utter a word of the Language of this country & all they have to Support them is the Bounty of Government.

Several Benevolent Persons have joined in an application to government for aid in this distressing Case SIR JOSEPH BANKS we understand in grateful remembrance of the civilities shewn him in Iceland takes a leading Part in the Business.'

On comparison with the original in the *Edinburgh Evening Courant*, 28 November 1807, Wright has his own method of using capitals and punctuation, though he is faithful to the text.

[5] Underlined in the document.

[6] According to Petræus himself, he had spent 16 years in Iceland, which as he arrived in 1790 is much closer to the truth.

[7] This letter was sent to Soho Square inscribed with 'Favour of Messrs Petreeus & Sivertson and Bredal', but there is no Banks endorsement.

48. James Frederick Denovan to Banks, 1 December 1807

Town House, Leith

Much honored Sir,

The Bearers hereof are Messrs Westy Petræus, Biörne Sivertsen, and Hans George Bredal[1] of Iceland, who, in consequence of the very lively interest you have been pleased to take in their Case, leave this Port today, in the Smack *Sprightly*, James Taylor Master, for London, and, I trust will make a quick passage. – They are in great spirits, and very anxious to meet so respected a character who has deigned to regard their misfortunes with the tenderness of a Father.

Mr. Petræus speaks German & Danish, and will act as Interpreter for the other two. – Petræus is a most amiable Man; and I trust that while his misfortunes have thus claimed your notice his many Virtues will both engage your esteem & sanction your benevolence. Indeed, altho' I have traversed the greater part of the Continent of Europe, I must acknowledge that these poor Icelanders are the most worthy, inoffensive people I ever yet saw. – I visited Mr. Petræus in his greatest distress: – when he, his Wife, and their infant Children were at the same moment, stretched on a Bed of Sickness – when the Sailors belonging to the Ship by which he came passenger, & his own domestic Servants, were marched to Prison, and an officious intruder informing him all his property was condemned, & that he & his Family would be sent to Prison so soon as they were in a state of convalescence – I saw him go thro' this very trying scene with something more than I can describe – I never saw so much fortitude & resignation depicted in the face of Man.

Feeling as I ought to do for such a worthy good Family I opened a Subscription among my private friends, for their relief, as well as that of the other Icelanders at Leith, & I flatter myself they shall feel no want while they remain here. It is only a very few days since poor Petræus was able to leave his Bed, otherways would probably, have been in London ere now, as he obtained a Passport Sometime ago, along with the other Icelanders, thro' the Friendship of Rear Admiral Vashon, to whom I had strongly recommended, or rather pressed their Case. The Admiral was good enough also to write to the Lords of the Treasury, craving that the Private Effects of the Icelanders, (not being the Cargoes) might be given up to them, which I am happy to inform you has been granted – the Order reached this [place] on the 28th current.[2]

It is very much to be regreted that at this Port of the Metropolis of Scotland We have no Alien Office, nor any official channel thro' which unfortunate Strangers can apply for relief, or even make known what they are. – I need not remark to a Gentleman of your very great experience of the World 'the Insolence of Office' when opposed to matters not exactly under their cognizance. – Altho' it does not belong to me, to interest myself with Aliens, yet from my Knowledge of the continental Languages & the public situation I hold, I am generally obligd to do so & very frequently to assist them in their distress – I

---

[1] The merchants Banks had asked to be sent to London to place their case before the government.

[2] This is rather puzzling. Vashon wrote to William Huskisson, Secretary of the Treasury, on 8 November 1807. On 16 November 1807 George Harrison of the Treasury sent the case to the King's Procurator General, Charles Bishop, for his opinion. But it was not until 14 December that the decision was taken by the Lords of the Treasury to release the private effects of Petræus (TNA, TS 8/6, Bishop to the Lords of the Treasury). Thus it seems that the Treasury officials pre-empted the reply expected from Bishop.

find a pleasure in doing this, but would find much more pleasure had I sanction from Government for that purpose, or any rate to walk by, indeed I should be proud if Government thought proper to give me instructions on that head, without any Kind of gratification whatever. I have the honor to be with the greatest respect & Esteem

Much honored Sir
your most ob<i>edient</i> & faithful humble
Servant
J F Denovan
Assist<i>ant</i> Town Clerk

SL, Banks Papers, I: 1:7. Autograph letter.

49.   Gunnlaugur Halldórsson to Banks, 2 December 1807

Leith

Reight honourable
Baron Joseph Banks
London

I have Several times been exposed to the uncheurfull Humour of fate but newer so cruel as et precent whilst on my first emigration from mi mother Country Island where I[1] Left 3 Months ago my Wife /and 3/ litle Children[2] it has brought me as a Passenger from the Ship *De Tvende Sostre*[3] along with an English man of War[4] hither to a Country I doe not understand the Language, deprived of money and credit and exposed to the utmost whants, instead of my destination was to trawell to Danemark and there to get support and a Compagnion to begin a Small trade in Island for own account.[5]

Now I have been detain here allreadi 10 weeks and being obliged to change the few danish money I braugth along with me from Island into English Money with about 50 pc^t loss for to produce the most nesessary as living & House here wich cost here a great deal of money, now for about 14 days ago when I had almost consumed the Money, was I along with some of the others Passengers from *De Tvende Söstre* granted a Passport from the English Government wich permit me to proced on my Voyage to Danemark[6] at the same time there was a Hamborger Ship lying in the Road here taking in a Cargoe for Sweden extremely glad with my Pasport I went on board immediately agreed with the Master for to bring me to Gothenborg and then from thence to get forwards to Danmark but a few days ago just when /the/ Ship was most readi to sail she was so much dammaged

---

[1] Halldórsson always writes *j* instead of *I*, as was common at the time, and used umlauts. The spelling in this letter leaves much to be desired, but he is clearly writing it himself though, as he does not understand the language, he is copying the text.

[2] Halldórsson had numerous children, both in and out of wedlock.

[3] This ship belonged to Bjarni Sivertsen.

[4] The *Peacock*: see Document 40.

[5] According to the 'Free Trade' regulations, promulgated by the King of Denmark, an Icelandic merchant had to be in company with a Danish merchant. This obligation was abolished in 1816.

[6] This was the usual treatment meted out to prisoners-of-war after the decision to release them had been taken.

by striking on the ground that the Cargoe must be taking out again and the Ship remain here all the Winter.

This new and unexpected fatal event has now extinguished the Spark of any hopes and brought me into the most dark desperation, not finding myself able withot assistance to live here all the Winter still less to obtain the purpose of my Voyage from Island, this being the last Ship wich was going away from this Place and I most likely condemd to Wait for an Opportunity to next Spring.

The immortal Praise wich you dear Sire! has left behind you in my Mother Country by your charitableness and liberality when you trawelled there, as also the recent Proof of your generous mind towards us unfortunate Islanders, Wich I have been so happy to learn by your letter to Dr Wright of wich I hav seen a Copy by one of mi Country men. This encourages one to Venture to inform you of my Present unfortunate Situation: least in the hopes that if you will interfear with the Causes of my Countrymen by Government You will likewise be Pleased graciously to remember me. My indigence with the tough[1] to be the trouble of other mean Withdraws irresistible desires to trawell along with my Country men to London.

I humbly beg your Lordship will forgive the Boldness of a Stranger and further to permit me with the first opportunity to be informd if any Assistance from Government may be exspected[2] for yor most humble Servent.

G: Haldorsen

SL, Banks Papers, I 1:8. Autograph letter.[3]

50. [Questions put to the Icelandic merchants when they arrived at Soho Square, about 9 December 1807][4]

1. When did you Leave the Island
2. Tell me the Circumstances of your Adventure
3. had any Ships arrivd at Iceland when you Left it & what did they bring
4. were any Ships expected when you Saild & what were they expected to bring
5. is the Island allowd to trade to any Port of the world except denmark
6. a List of these articles which are necessary for the Comforts of the Island that are annually imported.
7. what are the Largest boats or vessels usd by the Icelanders
8. how far do they Go to See in them

---

[1] Doubtless 'thought'. Halldórsson often inverts 'th'.

[2] Banks as usual did what he could. Halldórsson and Petræus were paid the arrears of subsistence owed to them and the Admiralty also ordered that they would be continued to be paid subsistence while in the United Kingdom. Moreover, enquiries were ordered to be made for a vessel to convey them and other Danish non-combatants to a port in Denmark (TNA, ADM 99/183, Admiralty to Lieut. Flinn, 28 December 1807).

[3] Note in Banks's autograph, bottom left of first page: '[Received] December 21 [1807].'

[4] By asking these questions Banks gained information on the current state of Iceland, which he subsequently made use of in his memorials. It was a habit of Banks's to make up such questionnaires regarding his projects. There is a note scribbled by Banks preserved in the Sutro Library which lists the following: 'Tonnage | Nature of the Cargo | Number of men | How many of them | are Icelanders', probably an aide-memoire to other topics to be discussed (SL I 1:37).

9. do they take Whales or any Fishes but Cod
10. is their Fish always Curd by frost in the winter which we Call Stock Fish or do they sometimes Cure in other ways
(How much Fish was Cured Last year beyond the Supply of the Country)[1]
11. are the Icelanders Good Seemen are there any in Either the mercantile or the military marine of Denmark
12. in what manner is the Revenue necessary for the internal Government Recvd
13. What is its amount.
14. Who is Stiftsampman[2] & what is his salary
    Do. amptmen[3]
    Do. Judges[4] & how many
    Do. Bishops of Holum & Scalholt[5]
    What other officers are Paid
    how are the Clergy Paid
15. What is the Present State of the Island in Respect to the Annual supplies necessary for the Comfort of the inhabitants
16. are the harbors in Iceland Shut in the winter & when do they open in the Spring
17. what number of Danish or norwegian Sailors in Each Ship
18. What number of Danes or norwegians domicialiated in Iceland
19. are there many Icelanders as seemen in the mercantile or military marine of Denmark
20. has Graaf Trampe[6] any Guard of Soldiers about him
21. are there any soldiers in Iceland
22. do any persons who hunt wild beests use Fire arms

SL, Banks Papers, I 1:38. Banks's autograph.

[New page][7]
When did you leave Iceland
how many Ships Saild from Iceland this year What were their Cargoes
What were their Masters names & these of their owners at Leith
May be given a Licence
What ships had arrivd at I or were expected when you Saild with an account of them & of their Cargoes
I meen by this to Obtain as good an account as you can give
of the Trade of Iceland in the Present Year

[1] This line is written in the margin, an addition.
[2] The *Stiftamtmaður* or governor of the island at the time was the Danish aristocrat Count Trampe (see Appendix 13).
[3] The two district governors were Stefán Þórarinsson and Stefán Stephensen.
[4] Magnús Stephensen was the Chief Justice and the two assessors, Benedikt Gröndal (1762–1825) and Ísleifur Einarsson (1765–1836).
[5] Banks was not aware that both bishoprics, which had existed in 1772, had been abolished in 1801 and Iceland had become one bishopric, with the bishop, Geir Vídalín, residing in Reykjavík.
[6] At this point he has received the answer (to question 14) that the governor was Count Trampe.
[7] As will be seen many of the questions are the same, but this one is more detailed and 'rougher'. In the next document it is clear that he interviewed the merchants one by one.

The names of the Ships at Leith
   their crews⎫
   their masters⎬ Destin*ation*
   their Cargoes⎭
What is the Present state of I*cel*and Compard with that of 1773[1] in Point of
   Cattle
   Sheep
   Fish
   Sulphur
any other mercantile produce
What is the disposition of the People towards England
I fear the Island cannot be worth our taking[2]
at what times is the Island inaccessible to Ships
at what times is the Thing[3] held
do they take into Consideration matters of State
Who is amptman
Who Stiftsamptman
Who is B*isho*p of Holum
Who   „   of Scalholt
What danes are inhabitants of the Island
have the danes any Icelanders in their warlike navy or in the Mercantile Navy
is the trade of Iceland now Subject to a monopoly or free
are the people at liberty to go to any country but Denmark or Norway
are there any Ships that winter in Iceland
What are the Largest boats usd there
how far do they go out to Sea to Fish
do any people use hunting for wild beests & do these use Fire arms
how are the white Foxes Taken
how are the White beers killd when they Land from the Ice[4]
how much Stockfish was made Last year
is it not made intirely in the winter by the Frost without Salt
what number of Danes or other Sailors & what of Icelanders in Each Ship
has Graaf Trampe any Guard of Soldiers or arms at his house
do the Cows Eat the Fish heeds & are they dressd or preservd for them.[5]
have the people any arms or are there any Public arms

SL, Banks Papers, I 1:39. Banks's autograph.

[1] Banks often makes this mistake. He of course went to Iceland in 1772.
[2] Note the feeling of hopelessness.
[3] Thing = *Alþingi*, the Icelandic assembly founded in 930 and abolished in 1800. Banks was unaware of the assembly's fate.
[4] He is referring to polar bears that drifted over on icebergs to Iceland from Greenland (which still happens).
[5] Banks was genuinely interested in agriculture and especially in fattening up cows.

51. [The answers given by the merchants to Banks, c. 9 December 1807.]

[Westy Petræus]
1   5 Sept Hafnefiord
2   Left his wife & 5 Children at Leith Left on*e* child in Iceland
he meant to Settle his Family in Copenhagen & trade to Iceland
he has 3 Settlements of mercantile in Iceland Reckawick, Keplevick, Westmanoe[1] where he has agents
Ships Cannot go in winter
3   18 Ships had arrivd & dischargd their Cargoes in [Iceland]
4 belonging to Iceland
The Supply is compleet for the winter in Spring new supply is expected
Mr Bredal went from Norway [as] Super Cargo[2]
<u>Leith 3 [ships]</u>
Two Sisters
Conference Raad Prætorius
Roedefiord wharf
<u>London [ships]</u>
Bosand
New Trial[3]
<u>Stornoway [ship]</u>
5 Brothers[4]
/Icelander Siversen Reckavig Hafnefiord/[5]
<u>Plymouth [ships]</u>
Johanne Charlotte Sivertsen & Petræus
The expected number of Ships was 70 the medium 50

[*Exports*]

Stockfish	Train oil[6]	Wool
Tallow	Sheep Skins	Fox Skins
Swan Skins	Eiderdown	Salt fish
Salt meet	Stockings & Gloves	Cloth jackets

[*Imports*]

Rice & Flower[7]	Grits	Biscuit
Iron	Hemp & Lines hooks	Salt
Tobacco	Coals	Tar
Wood of all kinds	Spirits	Hats
Sugar & Coffee	Boots & Shoes	some wine

[1] Reykjavík, Keflavík and the Westman Islands (Vestmannaeyjar).
[2] Bredal was the supercargo on the *Conferentz Raad Prætorius*.
[3] This is the ship *Den Nye Pröve,* which means *New Trial* in English.
[4] The ship *De Fem Brödre.*
[5] This is probably deleted here as he interviews Sívertsen later, in same document. For more details of Sívertsen and a portrait, see Plate 11.
[6] Train oil was the oil of fish, including shark, and whale.
[7] Doubtless *flour.*

7. 14 men Few Whales no fishery of them
Mr Seversen has 5 or 6 fish yatchs, which stay out all night 24 Tons
12 The King Pay 10,000 Dollars
The Landvoght[1] Collects & the Sysselmen[2] under him
The Payments are chiefly in Kind[3]
14 Graaf Trampe 1500 rd[4]
Stephenson 1200
Thorensen[5] 1200
Holum
Scalholt

[Sivertsen]
Bishop Geir Wetalin[6]
Reikevic
60 houses
2 Icelanders have been lately brought in more he thinks will be
he believes that 20 Saild in the middle of october some ofthem he thinks maybe Capturd
40 Ships he beleives were sent this year from Iceland
14 are in Britain
Norway has scarce Corn for itself it imports from the Baltic none goes from there to Iceland except in very Good years. Barley & oats are the cheef grains
There are few if any Icelanders in these Ships
There are not 100 Iceland Seemen in the Danish Service
In the *2 Sisters*[7] 11 Sailors 22 Passengers
When I was in Iceland the trade was in the hands of the King
it is now open to the Danes & carried on by danish Families domiciliated in Iceland
24 Danish Families are establishd there
They had begun to burn Kelp
Thorlever Gudmundsen[8]
Olaver Magnussen[9]
Peter Svane[10]

---

[1] The *landfögeti*: see Glossary. Among his duties was the collection of taxes.

[2] The *sýslumaður* (plural: *sýslumenn*) was the county magistrate: see Glossary.

[3] Very little coinage was in circulation in Iceland, the Icelanders making frequent complaints, but by using barter the merchants kept the upper hand in trade.

[4] *Rigsdaler* or rixdollar, the Danish currency (see 'Currency, Weights and Measures').

[5] Stefán Þórarinsson (1754–1823), the district governor of Northern and Eastern Iceland.

[6] Geir Vídalín: see Appendix 13.

[7] *De Tvende Söstre*, Sivertsen's ship.

[8] He is listing some of the passengers. Þorleifur Guðmundsson was from Borgarfjörður. He was on his way to Copenhagen for hospital treatment. See further Stephensen, *Ferðadagbækur*, p. 22n.

[9] Ólafur Magnússon is unidentified.

[10] Peder Ludvig Svane, a merchant, partner to Petræus in the Westman Islands 1793–1801. He is registered in the 1801 census as a merchant in Reykjavík (aged 28 and unmarried). He was Petræus's partner/assistant and appears to have been considered among the passengers.

Peter Adriensen[1]
1 in 20 of the Sailors are Icelanders
½ of the Property may be Danish Property /*the Rest*/
but this is in Past belonging to agents resident in Iceland

SL, Banks Papers, I 1:40. Autograph note.[2]

52. [Note written by Banks, 9 December 1807][3]

De*cembe*r 9
1 Westy Petreeus Brother in Law to Stephensen[4]
2 Biarne Severtson
3 Hans Georgé Bredal a Norwegian
No 1 has wife & 6 children with him has been ill is now only Recovering[5]
2 Remembers Stanley[6]

SL, Banks Papers, I 1:79. Autograph note.[7]

53. Bjarni Sívertsen, Westy Petræus and Hans Georg Bredal to Banks, 15 December 1807

London

Sir
Your benevolent participation of our misfortunes and that of our unhappy country, your condescending attention to our humble persons and affairs, and the benign reception with which you have personally honoured us[8] have left upon our hearts an impression which time can not efface, and called forth sentiments of gratitude which we would labour in vain to express. Permit us, /Sir,/ to thank you from the inmost of our hearts, that in all we can do, for the favour you have shewn us; and permit us, thus emboldened

---

[1] Unidentified, but presumably a passenger on *De Tvende Söstre*.

[2] These sets of answers and the answers given by Sívertsen on 3 January 1808 (Document 67) are all written on the same sheet of paper, folded to give four sides.

[3] These notes were doubtless taken in conversation with Petræus and Sivertsen (see previous document). This note gives the information of when the interview actually took place.

[4] Stefán Stephensen, district governor of the Western Amt of Iceland. Stefán and Petræus were married to sisters (see Document 47, p. 237, n. 1).

[5] The number of Petræus's children in Leith were five (see further Documents 51 and 56) and then there was an infant left in Iceland, the total number of children being six. However, there are some discrepancies in the various accounts (see Documents 47 and 53 and the *Edinburgh Evening Courant*, 28 November 1807). Mrs Petræus was expected to give birth in March or April 1808.

[6] Sir John Thomas Stanley who visited Iceland in 1789.

[7] Note in Banks's autograph, verso: 'a subscription has been opend at Leith' [see Documents 47 and 48].

[8] They had visited Banks at Soho Square on about 9 December 1807. They had probably been told to state their cases properly in writing. In the Sutro Banks Collection there are rough notes made by Banks based both upon this letter and his interview with the three merchants (I, 1:44).

by your goodness, to lay before you a full statement of our circumstances, soliciting your propitious intercession in behalf of ourselves, and that suffering country to which we belong. –

We are all citizens and merchants of Iceland, trading to Denmark and Norway according to the regulations enacted by the Danish government, exporting the produce of the country, comprehending principally fish, tallow, feather, wool & woollen goods, and importing timber, deals, implements of fishery and other necessary articles, without a supply of which the inhabitants of our poor and barren island can not subsist.

I Bjarne Sivertsen, a native of the country, have carried on a commercial concern at Havnefiord, under the district of Reikevik,[1] since the year 1791. In the month of September last, having shiped in different bottoms, namely: in the *Tvende Söstre* and the *Johanna Charlotta*, both belonging to myself, and in the *Seien*, belonging to Mr. Petræus, goods to the amount of more than £3000, I left my country and family, consisting of wife and four children, besides four adopted orphans, intending after having disposed of my property to return to my home in the Spring.

I Westi Petræus have been settled since the year 1790 as a citizen and merchant of Reikevik, where I have constantly lived these 17 years. Being obliged for the sake of arranging my affairs to make a voyage to Denmark, and wishing to place the eldest of my children where they could obtain a better education, than our ill-favoured region affords, hoping also to derive some relief for my sickly wife from the temporary stay in a milder clime, I determined to take the whole of my numerous family of six children with me. I accordingly departed from Iceland in September last, without however abandoning my concern at Reikavik, which is still going on, and to which it was my intention to return as soon as my affairs would permit. In the three vessels mentioned above I have also property to a large amount, upwards of £5000.

I Hans Georg Bredal, a native of Finmarken in Norway, born a few miles inland from Nord Cap, went to Iceland last spring with a view to obtain a settlement there, in which I succeeded and became a citizen of Wetaaer.[2] I was now returning to fetch my family over from Norway, and brought with me in the *Conferentzraad Prætorius* a cargo of the produce of the country, worth £3000.[3]

Such, Sir, were our circumstances and our views on leaving our homes; but Providence had decreed otherwise. As we had received no intelligence from Copenhagen later than the month of May, we were totally ignorant of war having broken out between England and Denmark until our Ships our Ships [sic] were detained and captured at sea, a few leagues off the cape of Lindesness, and brought into Lieth [sic].[4] The same has also been

---

[1] In 1788 six privileged or market towns had been established in Iceland, each with its trading district.

[2] This must be Vatneyri, a trading station in the Western Fjords. In his journal of 1807 Stephensen mentions they believed the Norwegian ship had come from Bíldudalur, which is in the vicinity.

[3] Bredal is not mentioned in Icelandic sources. This ship came directly to Iceland from Norway – probably as a speculator as it was not issued with a *söpas* (sea licence) from the Danish government in 1807.

[4] In his journal entry for 19 September 1807, Stephensen describes it thus: 'Everybody was in the best of health and spirits and Mr. Petræus gave a party ... to celebrate a successful voyage to Cape Lindesnes [off the coast of Norway] ... No! the ship which had been sighted was ... a large English warship which in the evening took us prisoner, announcing that a state of war existed between England and Denmark' (Stephensen, *Ferðadagbækur*, pp. 21–2).

the fate of the other ships in which our property was embarked, which have likewise been captured and brought into different ports of this country. If we are to suffer the loss of these vessels and their cargoes, our condition will be miserable indeed; as will all be reduced from a state of decent competency to that of absolute poverty, and for ever be incapable of retrieving our ruined affairs.

But Sir! what will weigh peculiarly with your generous mind, it is not only us and other individuals that will be ruined, our poor country will be overwhelmed with the same ruin as ourselves. Situated under the most inclement sky, unproductive of corn, and we might almost say of any vegetable food, destitute of wood & metals and even of the means of pursuing the principal lifelihood of its inhabitants, the fishery, nothing better than the terrors of famine can await it, if its scanty commerce be interrupted, and it shall for any time be prevented from disposing of its few commodities, and thereby procuring a supply of those articles without which it can not subsist.

Under these circumstances we /take the liberty/ humbly /to/ solicit your kind intercession in our favour; and firmly confident that the cause of suffering humanity can not be pleaded in vain by you, we fondly flatter ourselves with a hope, that, through your representations the British government may be induced generously to withhold, or in some measure to withhold the severities of war from a country, and a tribe of human creatures already pressed so hard by the severities of Nature.

We also humbly pray, that we may be permitted to draw your attention to the peculiar embarrasments to which in our present situation we are individually exposed. Having come, by so unexpected an event into this foreign country the language of which we do not even know,[1] where we had never any correspondence or connexion, & where we are unable to communicate with our correspondents in Denmark or to draw upon them, we have till this time had no other resource of supporting ourselves, than by disposing, to the greatest disadvantage, of what Danish coin & paper-currency we had about us, and must soon feel ourselves embarrassed in the extreme. It would therefore be to us a most desirable and highly wanted relief in our misfortunes, and we should deem it a most valuable fruit of your benevolent intercession, if some support could be allowed us for the time that we shall be under the necessity to pass in this country.[2]

---

[1] Not one of the Icelanders had any knowledge of the English language at the time and the Danish consul in Leith, George Home, spoke 'very little except English'. He, however, obtained 'unattractive' lodgings for them ashore, where they had trouble sleeping because of the 'English tipplers'. On his arrival in Copenhagen in October 1807, Magnús Stephensen hired a clergyman to teach him English (see Stephensen, *Ferðadagbækur*, pp. 27–8).

[2] Banks did indeed offer them pecuniary support. An autograph note in the Banks Papers in the Sutro Library lists the names of four Icelanders who wrote to him in December 1807 (G. Haldorsen, W. Petreus, Hans Thielsen and Biarne Sívertsen) and records that Banks lent Petræus £30 (probably because of his large family) while Bjarni Sívertsen received £20 (SL, Banks Papers, I 1:97).

Also among the Banks Papers in the Sutro Library there is a printed flyer regarding a meeting of 'merchants trading to Denmark and Norway' at the Banks Coffee-House in Cornhill on 4 December 1807. Georg Wolff (see Appendix 13) chaired the meeting which agreed to open a public subscription for the relief of the 'several hundred Masters and Mates of detained and condemned Danish and Norwegian Merchant Vessels', now prisoners-of-war in England. Jens Wolff was the treasurer. The fact that this is among Banks's papers indicates his interest in the matter (SL, Banks Papers, I 1:82).

Trusting, Sir, to the generosity of your heart we hope & pray that we may not have abused your goodness, or tired your condescending interest in our concerns by this long exposition of our circumstances, and desires; and we beg leave to finish by repeating that the most grateful acknowledgment of your kind and honouring attention shall never be extinguished in our hearts.

We are Sir, with the deepest sense of gratitude
Your most humble and obedient Servants
[*Signed*] B: Sivertsen W: Petræus H:G: Bredal

SL, Banks Papers, I 1:9. Letter, written by an Englishman, but with their autograph signatures.

54. Hans Thielsen to Banks, 17 December 1807

St. Catherine Square, N°. 8 Towerhill [London]

Sir!

An Iceland Merchant has informed me of Your generous Determination to employ all Your Influence in favor of poor unfortunate Iceland. – That wretched Country will indeed be exposed to Starvation & Destruction if it be not in the ensuing Year supplied with provisions and the Inhabitants with the Materials indispensibly necessary for their Occupations.

It has come to my Knowledge that the Two Iceland Merchants Petræus and B: Sivertsen who are at present in this City have been particularly recommended to You by Mr. Councellor of Justice Steffensen.[1]

I have myself also a share in 2 Establishments in Iceland namely in Reikevig and Havnefiord & I am convinced that upon my giving you a faithful Statement of my Case You will pardon the Liberty I take in calling your attention thereto.

My Father died about Two Years ago leaving my Mother with Seven Children, most of them not brought up, myself being the Eldest.[2] – He possessed a share in the abovementioned Establishments at Reikevig & Havnefiord which he bequeathed to me with the Task of providing for the family left behind him. By the Loss thereof I certainly am deprived of the whole of my present Property, this however is not so painful as to see such a large Family without the prospect of a future Maintenance. This dismal prospect for my Family induces me to repeat to You Sir my Solicitation that should Your Endeavours for poor unfortunate Iceland meet with the desired effect of which there can be no doubt I may expect from Your benevolent Heart that You will also have me in Remembrance as well as the Two Merchants recommended to You by Councellor Steffensen.

By the above Representation of my Case I flatter myself that I have [put][3] to You that I am not altogether unworthy of Your Interposition.

---

[1] Magnús Stephensen. Banks's fame as a benefactor was spreading.

[2] Hans Thielsen was the son of a merchant of the same name who had conducted a trade from Flensburg to Hafnarfjörður since 1789, in company with other Flensburg merchants. Later he also began trading in Reykjavík (Andrésson, *Verzlunarsaga Íslands*, I, p. 387, II, p. 476).

[3] The letter is torn here, with a piece missing.

My Ship the *Rennthier* was Captured near Elsinore on the 19th of August last having onboard a Return Cargo from Iceland. After lying with her in the Roads of Copenhagen until the 3d. of October she was conveyed to Yarmouth under Royal English Convoy and arrived the 28th of the same Month.[1]

With the utmost Veneration & Esteem
[*Signed*] Hans Thielsen

SL, Banks Papers, I 1:10. Autograph letter.[2]

55. Note from John Christopher Preidel to Banks and note by Banks, [19 December 1807[3]]

J. C. Preidel, Merch*ant* Waits on Right Honourable Sir Joseph Banks, N. 23 Great Winchester Street with Capt*ain* Thielsen.

[*Verso*][4]
Thielsen has $\frac{1}{15}$ of his Ship
Most of them have Shares
he is Burgher of Reikevic 3 years[5] never wintered there
Flensburgh the Fish belongs to different owners
Cannot clear for any place but Denmark have Licences[6]
These are paid for [*indecipherable*]
a Shippund is 320 lb, avoird*upois*: Matt fish (?) £6.
Lispund 16 lbs
Wool 18–20 LS [£s?] per Shippund. Eiderdown 20/– a lb.

SL, Banks Papers, I 1:16. Autograph note, recto from Preidel, verso by Banks.

---

[1] The ship had set sail from Reykjavík on 1 August 1807, bound for Copenhagen. Hans Thielsen junior was the 26 year-old master of the ship, which hailed from Flensburg in Jutland, and was owned by 15 merchants of that town. The seven mariners were all Danes and six had been set ashore at Copenhagen, where the ship was first taken before going to Yarmouth. The passengers were also sent ashore at Copenhagen. According to the sworn affidavit, Thielsen owned $\frac{1}{15}$ of the cargo which consisted of stockfish and klipfish, salted salmon, lambskins, wool, train-oil, down feathers, bed feathers, tallow and woollen gloves. His total loss was about £600. The papers of the *Rennthier* are in TNA, HCA 32/1645, no. 5076.

[2] Note in Banks's autograph, bottom of the page: '[Received] De*cembe*r 17.'

[3] This note is undated and difficult to decipher. However, in Thielsen's letter to Banks of 22 February 1808 (Document 73), he mentioned that he waited on Banks on the 19 December and Preidel in Document 70 mentioned that he waited on Banks about three weeks earlier. Thus Thielsen was almost immediately granted an interview with Banks.

[4] Notes jotted down in Banks's autograph, doubtless during the interview, and difficult to decipher.

[5] Hans Thielsen the younger became a burgher of Reykjavík on 2 June 1802, thus five years.

[6] These would be licences issued by the Danish Exchequer (Rentekammer).

56. Westy Petræus to Banks, 22 December 1807

London

Sir

As you have kindly promised to procure the liberation of the Icelandic passengers in the *Tvende Söstre* and the *Conferentsraad Prætorius*, and particularly of Thorlev Gudmundsen[1] also a passenger in the *Tvende Söstre*; it is my duty to inform you without delay, that I have just learnt by a letter from my wife, dated Lieth the 16th instant, that the /said/ passengers have already been liberated, and obtained passports. I therefore humbly and sincerely thank you for the favour you have shewn me and my countrymen in this respect, and beg to give you no farther trouble concerning them.

Your goodness, Sir, and favour towards us have been so great, that I venture, while I am here trespassing on your time, to communicate a few more particulars, than I have hitherto done, of my own peculiarly embarrassed situation, and humbly to implore your advice respecting it.

That I was proceeding with my family to Denmark, when I was detained by His Majesty's brig the *Peacock*, and sent to Lieth in the beginning of October last, I have taken the liberty to state before. The family with which I have thus unexpectedly been brought into a foreign country consists of no less than twelve persons including myself & my wife; namely: 5 children the eldest 8 years of age (a boy of 18 months was left in Iceland till we should return) two young girls of 15 & 17 years, the children of poor families in Iceland whom I have undertaken to provide for, an old servant maid, a commercial assistant by name Peder Svane,[2] and the above mentioned Thorlev Gudmundsen, an Icelandic peasant who is a cripple, and whom I brought with me for the purpose, if possible, to see him cured at Copenhagen. All these persons I have to provide for, myself in this expensive metropolis, the rest at Leith. But my means of so doing must fail to a very allarming degree. For October and part of November I received from the British Government through the agent of the Transport Office at Lieth[3] one shilling a day per head for myself & my dependants, which though insufficient to support us all, was still a great assistance, and added to the provisions I had left from the ship, and the little money I happened to bring with me (which I was obliged to exchange to great disadvantage) has enabled me to go through my difficulties until the present moment. But when the Danes were declared prisoners of war, that support was discontinued, and I had no other means of living, than to drain the scanty resources that remained. And now when these resources are almost totally exhausted, when all my property, which in my own country would make me a man of independent circumstances, is detained in various ports of these kingdoms, and I am unable to avail myself of the assistance I might otherwise expect from my friends at home, I can not without horror think of my situation in this foreign and expensive country, with so numerous a train of

---

[1] Þorleifur Guðmundsson was the sick passenger on board *De Tvende Söstre*, mentioned in the answers given to Banks by the merchants (see Document 51). Stephensen wrote in his journal of 1807 when discussing the surrender of Copenhagen to the British forces: 'English bombs razed to the ground with fire ... and what was most distressing of all the good Fredriks Hospital. And thus that refuge was closed to Thorleifur from Fellsöxl' (*Ferðadagbækur*, p. 26).

[2] Peder Ludvig Svane: see Document 51, p. 244, n. 10.

[3] The agent of the Transport Office in Leith was a Lieutenant Flinn.

dependants, without money or property or credit or connexions.¹ Indeed my embarrasments can better be imagined than described. – It is true, when my allowance from government was discontinued, I at the same time obtained passports for leaving the country, and by so doing I might have escaped these embarrasments. But the season was then past that a passage could be obtained for the Baltic, and I was in consequence under the necessity of remaining till the Spring. This stay was rendered still more necessary by another circumstance, and must even be delayed by it for a few months longer, a circumstance at any time a source of bliss, but now an addition to my sorrows & distress, namely the pregnancy of my wife who is expected to lie in the month of March or April.

Sir, having engrossed your precious moments by this tale of my woes, give me leave humbly to ask you with the greatest confidence in your undissembled kindness, and with the greatest deference to your opinion, whether I could with any propriety and hope make an application to the British government for some pecuniary assistance during the time I shall be obliged to remain here under the present circumstances; and if you are of opinion that such an application could with propriety be made, I also venture to solicit your propitious support to its success.

Trusting that your generous heart in my misfortunes and distresses will find an excuse for the liberty I have herewith taken, to which nothing else in the world could induce me, I remain

Sir, with the greatest acknowledgement of
your unexampled kindness
your most humble and
obedient Servant
W: Petræus

SL, Banks Papers, I 1:11. Letter, written by an Englishman, signed by Petræus.²

57. Banks to Alexander McLeay [Transport Office], 24 December 1807

Soho Square

Sir:
In compliance with your request I transmit to you the annexd Extracts of Letters receivd by me from Burghers or Natives of Iceland relative to the present situation of such of them as have been liberated as non combatants under the late humane order of your Board, but are unable to procure the means of proceeding to Denmark at present & in the mean time are in the utmost necessity for that subsistance which the generosity of this Country never fails to afford even to our Enemies when in distress.

---

[1] On 13 October 1807 George Home, consul in Leith, wrote to the Danish authorities, informing them that Madam Petræus, a female friend, her five children and two female servants had been sent ashore. He had applied to the Provision Agent on their behalf, but his reply was that he could not give them any allowance; 'as W. Petræus is unable to support them from the circumstance of his having no correspondent in this country'. Home asked the British government be approached so they would receive the same allowance as other detained passengers (RA, DfdUA, Ges. Ark. London III 1806–1807, no. 424, Home to J. Rist).

[2] Note in Banks's autograph, bottom of last page: 'Sent him £30'; and in Banks's autograph, left bottom of first page: 'December 21'.

Extract of a Letter from G. Haldorsen Native of Iceland dated Leith 2ᵈ. December 1807, receivd December 21.
[*See Document 49 (p. 239 above)*: extract begins 'Now I have been detain' and ends at 'to next Spring'.]

Extract of a Letter from Westy Petræus – Burgher of Iceland dated London 22 December 1807.
[*See Document 56 (p. 250 above)*: extract begins 'That I was proceeding' and ends at 'months of March or April'.]

Under the Circumstances Sir I trust your Board will not think my application an improper one in requesting that the Case in which the Native Icelanders & Burghers of Iceland now find themselves may be considerd by them and a proper subsistence issued to these unfortunate Men till they shall be otherwise disposd of. The Case of Mr Westy Petræus seems to me of an unusually interesting nature, as such I think it possible that the allowance to him on account of his Family & the situation of his Wife may be in some degree increasd, if this can be done in the ordinary course I shall be thankfull, if not & if you will do me the favor to let me know, I will make a special application to Government in his favor & I have great hopes that I shall not be disappointed.

SL, Banks Papers, I 1:13. Contemporary copy, amanuensis (William Cartlich).[1]

58. Westy Petræus to Banks, 24 December 1807

London

Sir

Fearing to become too troublesome by the frequency of my visits, and labouring under the disadvantage of expressing to you my sentiments only through the medium of a third person, I beg you will permit me to convey in this manner my thankful acknowledgment of the receipt of your letter of yesterday, and the enclosed proof of your unbounded generousity.[2] I accept with heartfelt thanks this new affecting mark of your benevolence, which causes a very considerable and very acceptable relief to my embarrasments. Only give me leave, Sir, humbly to say, that it was very far from my ideas or my wishes to become in this manner a burden to your generousity, of which I have, independently of this, experienced such proof, as must for ever claim all the gratitude of which my heart is capable.

I remain Sir, deeply impressed with the obligations I owe you
Your most humble &
obedient Servant
[*Signed*] W: Petræus

SL, Banks Papers, I 1:12. Letter, written by an Englishman, and signed by Petræus.

[1] The original has not survived.
[2] Petræus is thanking Banks for the £30 Banks either lent or gave him (see previous letter).

59. Alexander McLeay to Banks, 28 December 1807

Transport Office

Sir,
I have had the Honour to receive your Letter dated the 24th of this Month,[1] and having laid the same before the Commissioners for the Transport Service &c., I am directed to acquaint you, that Lieut*enant* Flinn, the Agent for Transports at Leith, has been ordered to pay to the Persons mentioned by you the Arrears of their Subsistence, from the Time they were last paid, and to continue to subsist them, until an Opportunity shall offer for them to proceed to a Port in Denmark, which the Board will do their Endeavour to procure for them, and all the other Danish Noncombattants who have recived Permission to return Home by the way of Leith.

Should Mr. Petræus be in London,[2] I am to request, that you will be so obliging as to desire him to call at this Office.

I have the Honour to be,
Sir,
Your most obedient
humble Servant,
[*Signed*] Alex. McLeay Secre*tary*

SL, Banks Papers, I 1:14. Autograph letter.

60. Banks to Georg Wolff, 28 December [1807][3]

Sir
having made application to the Transport board and laid M. Petreus's case before them requesting that his Subsistence may continue to be paid to him & if possible it may be made in consequence of the Special Circumstances of his Case more suitable to his situation in Life & the necessities of M^r. Petræus they have granted the first part of my Request & have desird that Mr *Petræus* may attend their Board in order – That the Commissioners may make themselves masters of the Case I beg therefore that you will be so good as to Communicate this to Mr. Petræus & give him instructions that he may properly present himself to the board

SL, Banks Papers, I 1:14v. Autograph letter. Draft.

---

[1] Document 57.

[2] Petræus was still in London. He was writing from London on 5 February 1808 (Document 72) and according to Document 58 he had not been travelling to and from London.

[3] This draft is written on the bottom half of the second page of the letter from Alexander McLeay (Document 59).

61. Banks to 'Mr Stephensen of Reikiavick [?]',[1] [late December 1807][2]

Sir

As the Son of a Father, who, when I visited your Island received me with kindness, entertained me with hospitality, & admitted us[3] into a Participation of his Friendship; and as a most respectable Burger of an Island, where the National character for integrity of Conduct, purity of mind, & attention to religious observances is not exceeded in any part[4] of the whole circumference of the globe which I have seen; I cannot hesitate, at a moment interesting as the present is, to renew with you the correspondence I maintained with your Father during his Lifetime.[5]

---

[1] To one of the sons of the leading family in Iceland, a brother of Magnús Stephensen. In his letter of 17 October (Document 40) Magnús mentions that he is the 'eldest son'. There were two younger ones, Stefán Stephensen, district governor of the Western amt, and Björn, an official of the High Court of Justice. Banks knew that Magnús himself was in Copenhagen, contact with him was thus too risky, but as he wrote in the 'Project' document (Document 45) he would try to contact 'the brethren of Magnús Stephensen, now in Iceland'. Banks did not know their Christian names, thus he simply addressed the letter to 'Mr. Stephensen', who was to receive this letter from the hands of the commanding officer of a fleet of ships sent to Iceland to offer the inhabitants the option of 'voluntarily' becoming British subjects. This letter was never sent, quite simply because this plan was never put into effect, and exists only as a draft. However, the draft letter figured prominently in the arguments of both Halldór Hermannsson, who erroneously dated it to 1801 ('Banks and Iceland', p. 32), and Helgi P. Briem, who interpreted it as proof of the firm intention of the British government to annex Iceland (*Sjálfstæði Íslands*, pp. 498–505). Its importance, on the other hand, lies in the fact that it shows how detailed Banks's plans for the annexation of Iceland had become by the end of December – he was actually drafting the letter for the commander of the invading fleet. The government, however, eventually took the decision not to annex the island (Agnarsdóttir, 'Great Britain and Iceland 1800–1820', pp. 43–4).

[2] On the DTC copy the following is written: 'To follow Sir J.B.'s Remarks on Iceland, 1801' (Document 38). Hermannsson did not query this (p. 30), neither did Dawson nor Edward Smith and it is filed as such in the 12th volume of the DTC, the letter accordingly dated as written in 1801. This is, however, a clerical error, as Briem pointed out in his doctoral thesis *Sjálfstæði Íslands*. The contents of the letter give abundant clues to the fact that it could only have been written after the outbreak of hostilities in 1807 and the capture of the merchant ships. The letter for instance asks that the governor Count Trampe be sent on board the British vessel now in Reykjavík harbour. Trampe first became governor of Iceland in 1806, in 1801 he was a law student. The letter could, moreover, not have been written until after the arrival of the Iceland merchants in London in early December 1807 because only then did Banks learn the name of the current governor. And it is unlikely that it was written after 3 January 1808, because then Banks notes down the information from Sivertsen that 'Graaf Trampe is in Denmark' (Document 67). However, it is just conceivable that it was written later, as there was always the possibility that by the time the letter reached Iceland – and it was obviously expected to reach Iceland in 1808 (viz. 'the very ships & Persons who traded with you *last year* from Copenhagen will this year return to you from London') that Trampe would have been expected to be back at his post. (But he did not return until June 1809.) The conclusion here, however, is that the letter in question is most likely to have been written in the second half of December 1807 after Banks had questioned the merchants, in connection with Hawkesbury's letter of November 29 (Document 46) and is probably 'part of the detail' mentioned by Banks to Hawkesbury in his memorandum of 30 December 1807 (Document 63).

[3] In the draft it says 'me'.

[4] Banks crossed out 'country' in the draft.

[5] Banks spoke of Ólafur Stephensen as dead, while he did not die until 1812. This misunderstanding was doubtless based on a remark made by Magnús in his letter of 17 October 1807 (Document 40) where he wrote: 'I hope, Dear Sir, you will favour with a kind thought the Manes of your former friend and my father, Olav Stephensen'. Magnús Stephensen (or whoever helped him with the letter as Stephensen knew little English at the time) used the word *manes* incorrectly ('shade of departed person') and Banks, not surprisingly, took this as

Denmark has now, by the inveterate animosity of her public declarations, by the rigor of her domestic restrictions on the trade & by connecting herself with the inveterate enemies of Britain, dissolved for ever those ancient bonds of Friendship & alliance,[1] which for so many centuries held together the People of these Two countries in an amicable unity of Interest. Britain may now look at the map of the Danish dominions with a view of detaching from the possession of her Enemy & seizing for her own use whatever she may deem likely to advance her future views, or to diminish the resources of an enemy with whom her contest is not likely soon to terminate.

No one who looks upon the Map of Europe can doubt that Iceland is by nature a Part of the Group of Islands called by the Ancients 'Britannia', & consequently that it ought to be a part of the British Empire, which consists of every thing in Europe accessible only by Seas; as such possessions can never be wrested from her by the powers of the combin'd world. England has occasion also for an increase of Population, to enable her to occupy with advantage the Trade of the World, which, spite of all opposition, she must in the end possess; but that Iceland will, if allowed to participate in British Freedom, quickly resume its ancient stock of people, & when a Trading Country, become populous in proportion to its Prosperity, are matters that admit of no doubt. – These considerations will easily, Sir, convince you that Iceland ought for its own advantage to become subject to the dominion of Britain, from whose mild government Icelanders[2] may expect that liberal protection which has been so long witheld from them by Denmark, without a fear of being disappointed; that it must become so without the least delay is self evident, as Britain can command the conquest of the Island by sending a single Regiment of Soldiers.

Unwilling, however, to subject a Country, which she wishes to maintain in every degree of prosperity, to the horrors & the desolation of war, she hesitates in taking any step likely to hazard a consequence so destructive to a people she wishes to preserve, secure, & ever hereafter to regard with Sentiments of Brotherly Friendship & affection.

I am allowed to tell you, Sir, that it is in your Power & in that of your Countrymen to avert the imminent danger that awaits you even upon your threshold. Britain has consented to delay the Blow for a short time, and to offer you by an authorized agent[3] the alternative of voluntarily becoming her Subjects; but your Resolves must be instantaneous; for even the least delay cannot be brooked in matters of so great national importance.

When I was in your Country, the wish, I may say, of every man with whom I conversed, was to be placed under the Government of England. This must now happen, either by conquest or by Revolution; it is therefore sincerely to be hoped that the same disposition may continue to prevail; an instant acknowledgment of homage & Fealty to the Crown of the United Kingdom, ratified by sending on board the British vessels now lying in your

---

an indication, that his old friend was dead. On 28 May 1809 Banks wrote a letter to Ólafur Stephensen (Document 118), which began: 'It has given me sincere pleasure to hear from various persons lately, that you are still alive and comfortable'. His source was doubtless Jörgen Jörgensen (see Appendix 13), who visited Banks in April 1809, soon after his arrival in England from the first trading voyage to Iceland.

[1] '& Alliance' added in draft.
[2] 'whose mild government Icelanders' added in margin in draft.
[3] 'an authorized agent' added in draft.

Harbour, your Stiftsamptman,[1] Baron Trampe, & such others, if such shall be found, as are pertinaciously adverse to the measure, will do away the necessity of Conquest, preserve you from the excesses of an invading army, which no degree of discipline can wholly prevent; your women will remain inviolate, your Children in Security, your Lives in Safety, your Property unaltered, your Laws unchanged, & your Religion untouched, & besides, such is the Generosity[2] of the British nation to those who refuse to be her enemies, the property of Natives & Burghers of Iceland, seized in the Seas by the King's Cruisers, will, as far as can legally be done, be restored to their owners, & the very ships & Persons who traded with you last year[3] from Copenhagen will this year return to you from London.

As a Friend of your deceased Father, as a hearty well wisher to a virtuous and an interesting people, & to the son of the man I so sincerely esteemed, I feel the deepest interest in the result of your deliberation. May the all-Gracious Heaven direct your Councils, preserve you from your Enemies, & deliver you & yours undiminished & unimpaired into the Friendly hands to which by this Letter I recommend you.

NHM, DTC 12, ff. 167–170. Contemporary copy.
Adelaide, Geographical Society of South Australia, Icelandic Manuscripts Collection, MS 6c, pp. 1–4. Draft in Banks's autograph.
Hermannsson, 'Banks and Iceland', pp. 30–32.

62. Project [for the annexation of Iceland, late December 1807][4]

Give Parole to all Iceland People in the Iceland Vessels whether Native or domiciliated but to no Person who cannot Prove that he has Spent at Least one winter in Iceland allow them a week to be Releasd if the cargo is restord

Send all the Danish Seemen & Masters to the Same Custody as other danish Prisoners now here

Restore the Property belonging to Icelanders whether Native or domiciliated with a Promise of a Licence to trade to Iceland in the Spring under Conditions that will then be declard to them

In case of Danish Property of Persons not domiciliated in Iceland the Liberated merchants & masters to be allowd to take Posession of it on being bound in Proper Penalties to be accountable for the amount to H.M. account of Droits of Admiralty, this to Enable the Icelanders to Provide a sufficient supply of necessaries for the next seeson for the Island

The first Condition to be the navigating their Ships with English or Icelandic Sailors & no others

---

[1] The governor of Iceland, Count Trampe (see Appendix 13).
[2] 'Generosity' written in place of 'magnanimity', which is crossed out.
[3] That is to say 1807. Obviously the proposed invasion was to take place in 1808.
[4] This is yet another rough draft of a project to annex Iceland. Banks is further along in his planning compared to his first draft of the Project (Document 45). It must have been written after he had met the merchants from Leith, as he knows about the sick man Þorleifur Guðmundsson and has better information on the captured ships. It is closely connected with the previous letter to Mr Stephensen.

To Sail in a Fleet under the Convoy of Two Frigates Each Ship to Carry soldiers di*tt*o on board the Frigates if Thought necesary to Compell the Surrender of the Island & the removal of the Stiftsamptman

To Replace the Danish with an english Stiftsamptman & Leave all other things in their Places a Proper Person should be sent out who, I think, should be a Civilian & not a Soldier. A Continuance of Laws, Customs & Courts of Justice to be Stipulated only. If any danish Laws have been fixed upon the Island, to allow the Islanders to Repeel /*it*/ them

The Island to be Considerd as a Conquerd Island & no Law made by the Thing[1] to have power[2] till approvd by H M. in Council

The Country to submit to a Militia Enrollment of Every man from 18 to 45 –

Johanne Charlotte Plymouth	*Sejen* Cork
Bedre Tider London	*de Fem Bredere*[3] Stornoway
Swanen ditto	
Renthier London I beleive	[*Marginal notes*] 14 vessels[4]
Regina here	200 Seemen
Bildahl	10 of them Icelanders[5]
Christine Maria here	Brimstone
Den Neje Prove[6]	
Bosand	
De tvende Sostre Leith	
Conference Ra[ad] Prætorious di*tt*o	

If Force is Necessary to Subdue the Island all Danish Property of Persons not domiciliated to be Prize to the Captors.

Send a vessel to reconnoitre the Island. Fogs are the only Obstacle

an Icelander or two to go with her.

Thorleiver Goodmensen[7] the Sick man

SL, Banks Papers, I 1:46. Autograph draft.[8]

63. Memorandum from Banks to Lord Hawkesbury, 30 December 1807

My Lord

In obedience to your Lordship's commands, I have collected together such information relative to the present state of the Island of Iceland as I have been able to obtain, either

---

[1] *Alþingi* (see Glossary).
[2] This word is all but illegible.
[3] *De Fem Brödre*. This is a list of the captured ships in 1807.
[4] These notes on the right-hand margin are in pencil. Actually this is a list of only 13 vessels, the *Charlotte Amalia* (see Document 99), is missing, but Banks was unaware of this ship at this point.
[5] The fact was that there were few Icelanders manning the Danish merchantships.
[6] *Den Nye Pröve*.
[7] Þorleifur Guðmundsson: see Document 56.
[8] Called 'very rough draft' in the filing cards of the Sutro Library.

from the recollection of what I saw when I visited the country in the year 1773,[1] or from the information of the Natives & Burghers of that Island now in this Country.[2]

It is scarcely possible, even for an unthinking observer who casts his eye over the chart of Europe with a view of examining what relation the respective kingdoms into which it is divided bear to each other, not to be struck with the evident propriety of Iceland being considered as a part of the Group of Islands called by the Ancients Britannica, Great Britain, Iceland, Orkney, Shetland, Ferroe, & Iceland form together an archipelago, naturally offering to the mind a Section eminently fitted for the establishment of a Naval Empire, incompleat, if the whole are not united under one monarch; but capable, when joined together of furnishing inhabitants sufficient by the manifold advantages deriveable from maritime strength, to keep the whole world in awe & if necessary in subjection.

The Island of Iceland itself is large enough, was it situated in a Temperate Climate to maintain a large body of population: it is said to contain 86,000 Square geometrical miles;[3] & tho' a large portion of it consists /of/ inaccessible mountains, covered with perpetual Snow; & of Volcanoes, & of the effects of their irruptions. Still, as there are many rivers whose Banks are widely covered with verdure & valleys furnished with excellent soil, the quantity of food it would be capable of producing under a temperate atmosphere is very considerable.

The Climate is the most ungenial of any inhabited by civilized man. Trees will not grow upon it, nor is any kind of Corn sown there;[4] Cabbages, Turnips, & some other excellent vegetables are cultivated; but very sparingly. Potatoes also frequently succeed;[5] but the summer season is never safe from Frosty nights, which this tender vegetable cannot endure: in some years these Frosts occur with so much severity as to destroy the grass in the meadows, & damage that in the Pastures; such seasons have sometimes occasioned Famine.

---

[1] As he often did, Banks wrote 1773, instead of 1772, throughout the document.

[2] This is the culmination of Banks's grand project of annexing Iceland to the British Crown, which he sincerely believed was the only hope for the Icelanders' salvation. Since receiving the letter from Magnús Stephensen in October 1807, Banks had been working towards this end. His informants on the current situation in Iceland were of course the Iceland merchants. In the Banks papers in the Sutro Library there is to be found a list, written by Banks, entitled 'Contents', undated but doubtless from the end of 1807 (SL, Banks Papers, I 1:81), which may have been a list of documents enclosed with this lengthy memorandum to Hawkesbury. At the very least these were the documents used by Banks as sources for this memorandum. This list shows how well-organized Banks was regarding the planning of the proposed annexation. The papers listed were: 1. Printed Declaration of Denmark with a Subscription among the merchants for captured Danish Seamen & Norwegians but no Icelanders mentioned; 2. Magnus Stephensen's Letter recommending 4 Vessels for Liberation [Document 40]; 3. Ld. Hawkesbury's Correspondence [Document 46 – this is significant as it suggests that there were more letters]; 4. Correspondence with Dr. Wright & Letters from Edinburgh [Documents 41, 42, 43 and 44]; 5. Information from the Marshall of the Admiralty [31 December 1807, Bishop to Fawkener with list of detained Iceland vessels, TNA, PRO 30/42/14/8]; 6. My Remarks on the matter [Document 38]; 7. Information from Westy Petreus and Hans Thielsen [Documents 53, 54, 56 and 58]; and finally, 8. Letter from Icelanders & concerning them [Document 49]. In the Sutro Banks Papers there are a small number of rough notes for this memorandum (I 1:44).

[3] Iceland has a total area of 103,000 km², which is closer to 40,000 square miles.

[4] The production of grain ceased in the 16th century due to the deteriorating climate.

[5] Potatoes were first introduced into Iceland in 1758, but took time to gain acceptance. In his 1801 memorandum Banks wrote that potatoes 'were not known' (see p. 221 above). This is an example of the new information he had gained from the merchants.

Nevertheless, it appears that in early ages the seasons were not to ungenial: that wood once grew in the Island of considerable size[1] is clear both from trunks of trees being found in Bogs, & from the sortebrand /(surturbrand?)/,[2] a kind of half form'd coal consisting of the Trunks & branches of Timber Trees, which on the North side of the Island is abundant: the marks of the Plough & of ancient tillage are said also [to] be visible in a vast number of places, especially on the North Side /of the island/ where the best soil is found.

The inhabitants attribute this increase of summer Frosts to the enlargement of the body of Ice which for many years has adhered to the East Coast[3] of Greenland, where a danish Colony was formerly settled, but which has now been wholly inaccessible to shipping for 2 centuries at the least.[4] If this opinion is just[5] the ancient state of the climate may in due time be restored, & will be so whenever this newformed Barrier of Ice shall melt & disappear, for we cannot easily believe that any such novel accumulation of what was not originally destin'd to occupy the place it now fills up, can be permitted by Nature[6] to be perpetual.

At present, the vegetable food of the Islanders consists of the seeds of three kinds of wild plants, all of them known in Britain, tho the seeds of them are in no place thought worth the collecting: of Flour, made from the Leaves of a kind of Moss, used in England as a nutritive food for persons in a decline; a sort of sea weed, which, like the laws of this Country is a miserable diet in its crude state; & the Leaves of a few wild vegetables, for gardens are extremely uncommon. Their actual support is chiefly derived from Fish & milk, which they prepare in many ways, & the Flesh of their sheep & Neat cattle;[7] hence it is that the scurvy is a prevalent & often a fatal disease among them, degenerating sometimes into a horrible kind of Leprosy.[8]

The Population of Iceland is said to have gradually diminish'd for a long space of time: in 1773 it was estimated at 60,000.[9] The letter from an Icelander living at Copenhagen which I had the honor of submiting to your Lordship's perusal, considers them as at present as consisting of no more than 47,000 Souls.[10]

---

[1] When the first settlers reached Iceland c. 870 the island was wooded from the mountains to the coast according to *Íslendingabók* (Book of Icelanders) written by Ari Þorgilsson, or Ari the Learned, in 1122–33.

[2] Banks was presented with specimens of *surtarbrandur* (a variety of lignite or brown coal) by Claus Heide, (Document 2). The word *surturbrand?* is an addition in another hand, above the line.

[3] In the Kew draft Banks first wrote 'west side' of Greenland, which is in fact correct as the two Norse colonies in Greenland, called *Vestribyggð* (Western Settlement) and *Eystribyggð* (Eastern Settlement) were both on the west coast. The King of Denmark sent several expeditions to the east coast of Greenland in search of the Eastern Settlement and it was not until the end of the 18th century that men first realized that it was in fact situated on the west coast, albeit to the east of the Western Settlement.

[4] The last extant documentary source of the Norse in Greenland records a wedding in the mid-15th century. The fate of the Norse settlement in Greenland has been the subject of intense debate, but there is no consensus of opinion. For one of the more recent theories, see Seaver, 'Norse Greenland on the Eve of Renaissance Exploration in the North Atlantic'.

[5] Banks wrote 'true' in the draft.

[6] Banks first wrote 'Providence' in the draft.

[7] *Neat cattle* is a collective term for all types of cattle.

[8] Leprosy, or *elephantiasis* as it was identified by visitors, was widespread in Iceland from the 16th century until the 20th century. Both von Troil (*Letters on Iceland*, letter XXIV) and Henry Holland ('On the Diseases of the Icelanders', the first appendix in Mackenzie's *Travels in Iceland*) wrote on the subject.

[9] According to the first census of 1703, the total number of inhabitants was 50,358, by 1801 they were down to 47,240.

[10] The letter from Magnús Stephensen of 17 October 1807 (Document 40).

They have abundance of Horses, which are very necessary to carry their merchandize from the inland to the harbors where it is to be ship'd, as they have few, if any, navigable Rivers. They have also Neat Cattle & Sheep in tolerable plenty. The Skins, the wool, & the Tallow of which with salted mutton, make a valuable part of their exports.

Their Stock of Cattle was, not many years ago, much lessened by a Calamity unheard of before among them,& unknown in all other parts of the world – a Volcano at once burst forth from a part of the Country that was either inaccessible or had not been visited by the people.[1] The magnitude of the Streams of Lava that march'd forward into the inhabited Country, filling up wide & deep valleys & damming up rivers as it flowed, was upon a Scale of which the eruptions of Vesuvius or of Etna are mere miniature representations; but this flood of Lava was a small part of the mischief occasioned by the eruption. An immense quantity of the volcanic glass, called vitrum obsidianum,[2] appears to have been formed in the focus of the Volcanic heat: & this being blown up into the air by the projectile shower of the crater, became divided into Filaments almost as fine as the web of a Spider: these, floating upon the wind, were carried over an extensive district, & falling upon the grass gave the appearance to the Fields that we call Gossamer in Autumn.

The Sheep or Cattle that ate of the grass covered with these threads certainly died in a few days; nor was it possible to save any of them; for before the danger was suspected the animals had every where eaten more than enough to destroy them; nor, had the danger been apprehended, would it have been possible to preserve many of them, as the district subjected to this calamity was several days journey across.

The Trade of Iceland was formerly open to Europe: in the 16th Century the English used it,[3] & much Stock fish was consumed, especially by our seafaring people: long before that time, the Icelanders had been active & adventurous Sailors; they clearly discovered America long before Columbus sail'd in quest of it,[4] & have recorded the discovery very clearly in their annals.[5]

In the beginning of the 17th Century,[6] the Danes, taking pattern no doubt from the Colonial Policy established by the Europeans who at that time possessed territory in America & Asia, prohibited all nations but their own from every kind of intercourse with Iceland; & from that time it appears that the hardy & enterprising nature of the Icelanders has gradually degenerated to their present torpid character.[7] Unable to build Ships for the want of Timber, their boats being built of Danish[8] wood, they fell intirely upon the mercy of their mother country; &, soon finding that the price offered for their produce by their danish visitors, was in all cases proportion'd to the stock of Commodities prepared by them for their annual market, so that they (however large their stock of fish, of wool,

---

[1] The great eruption of Laki was in 1783–4: see e.g. Document 28.

[2] *Vitrum obsidianum*: obsidian (in Icelandic *hrafntinna*) is a black, wholly glassy acid, volcanic rock (rhyolite).

[3] Actually the English trade in Iceland probably began in the late 14th century (see Introduction, pp. 11–12 above).

[4] Traditionally it is believed that Leifur Eiríksson (Leif the Lucky) discovered America in the year 1000.

[5] Mainly in *The Saga of Erik the Red* and *The Saga of the Greenlanders*. There were no annals written in this period.

[6] The monopoly of trade was introduced in 1602.

[7] This is the general consensus and several contemporary Icelanders were of the same opinion.

[8] Norwegian. Denmark is not known for its commercial forests, but then this was the Kingdom of Denmark-Norway.

of Tallow, Feathers, &c might be) they were never able to obtain more than a limited quantity of the necessaries they requir'd, they became quite indifferent to all kinds of exertion, each providing as much of marketable commodities as would at a medium price provide him with his annual supply of meal, Biscuits, Hats, &c. and no more.

In this state they were found to be in 1773,[1] religious, moral, honest, & virtuous in the highest degree, but without animation or sagacity: dancing & all other social amusements were unknown in the Island, in fact no Icelander was seen to laugh; not that he had any particular inclination for gravity, but because nothing in the detail of his mode of living seemed ever to excite him to gaiety, much less to merriment or laughter.[2]

At this time the Trade of the Island had been granted by the Crown of Denmark to a Company of Merchants at Copenhagen, who paid to the King about £8000 Sterling a year for the privilege,[3] & provided also for the Salaries of the /First/ Functionaries of the Island: the oppression of this Company, in fixing their own prices upon all they bought & upon all they had to sell, was beyond what had ever before been felt, & had reduc'd the people to /that/ inactive & torpid State in which we found them not long after the Trade was opened to the Subjects of Denmark as it formerly had been:[4] at present the people are beginning to become more industrious, & no doubt in another generation, if liberty is restored to them, will recover the active & intelligent character of their early ancestors.

The Trade is chiefly carried on by Danes, who, on the trade being thrown open, vested capital in that line of Business, proceeded to Iceland, & having been admitted burghers of the Island; erected proper warehouses. These persons by themselves or their agents collected together materials for loading their ships, which annually visit the Island from Copenhagen in the Summer. Of these there are said to be 24 families; but several Icelanders[5] have already been able to enter into this business, & have Shares in Ships and in cargoes, a sure proof that their industry is increasing, & that in time with due encouragement they will become animated, active & zealous subjects, useful to themselves, to their King, & to their Country.

It is said that 40 vessels[6] were this year loaded in Iceland: one of these was bound to Bordeaux[7] and one to Bilbao,[8] being Danish property: this is less than the customary

---

[1] 1772.

[2] Pietism had had a great deal of influence in Iceland in the 18th century, from the 1740s onward.

[3] The Royal Chartered General Trading Company of Copenhagen (see Document 38 p. 219, n. 2 above) was awarded the monopoly trade from the beginning of January 1764 until the end of 1783. The fee for the privilege was 7,000 rixdollars (15 August 1763, 'Octroy paa Islands Beseiling i tyve Aar for det almindelige Handels-Compagnie', *Lovsamling for Island,* III, p. 468). Banks's figure is grossly exaggerated (see p. xxi above).

[4] 'Free trade' in Iceland was introduced in 1 January 1788.

[5] Bjarni Sívertsen is the best example. Other acknowledged pioneers were Guðmundur Scheving (1773–1837) in Flatey and Ólafur Thorlacius (1762–1815) in Bíldudalur.

[6] According to the *söpas* registers in the *Rentekammer* records in the Rigsarkivet in Copenhagen 42 passports were issued and 41 ships appear to have sailed.

[7] This was the *Johanne Charlotte,* owned by Wulff, Petræus and Bjarni Sívertsen. She left Hafnarfjörður bound for Bordeaux on 19 August 1807. The vessel was captured by the *Amethyst* on 14 September 1807 and taken to Plymouth. All papers in TNA, HCA 32/1976, part 1, no. 3348.

[8] This was the *Seyen,* owned by Petræus, which sailed from Reykjavík bound for Bilbao on 24 August 1807, captured by HMS *Virginia* and sent into Cork. All papers regarding this matter are to be found in TNA, HCA 32/1176, no. 5398.

number of Ships,[1] & the trade is certainly on the increase: 13 or 14 of these Ships are now in English Ports,[2] more may be, but the Marshal of the Admiralty[3] has not yet supplied the necessary information.[4] The Articles exported from Iceland are Stock fish, Salt fish, Salted Salmon, Train Oil, Wool, Tallow, Sheep Skins, Salted mutton, Fox skins, Swan Skins, Feathers, Eider Down, Stockings & Gloves, Coarse cloth & Kelp, a new article. Brimstone may also be procured of the best quality & in any quantity. The Icelanders take in return Rice, Flour, Groats, Biscuit, Salt, Sugar, Coffee, Spices, Spirits, Some wine, Tobacco, Wood, Hemp, Lines & hooks for fishing, Tar, Coals, Iron, hats, Boots, Shoes.

They are at present governed by a Stiftsamptman, who is a Dane,[5] & two Amptmen[6] who are native Icelanders, the Stiftsamptman or Governor has no court or State about him or guards of any kind. We are told that there is not or has been in the memory of any man a soldier of any kind on the Island,[7] & that the Natives are so ill furnish'd with Fire arms,[8] that when assaulted by the white Bears, who land from the Greenland Ice, when it drifts on Shore in the summer season on the Northern Coast, they use pikes, instead of Guns, to destroy them.

By the Constitution of their Country they assemble once a year in the month of July, at a place called Thingvalle[9] where their Stiftsamptman, Amptmen, Lagmen,[10] Systlemen,[11] Landvogts[12] & indeed all public functionaries attend, & all such persons who for that purpose of the decision of Causes, Civil or Criminal, are called upon to give their attendance: here is held what they call an /all thing/ Althing[13] or general Court of Justice, & this is, no doubt, constitutionaly the Parliament or Convention[14] of the nation, in which the opinions of the people may be legally collected & their resolutions declared. The remains of a similar custom in the country is preserved in the hustings at Guildhall, which is evidently the Court held in a house or Housething of our ancestors, the all thing[15] is still held abroad in Iceland in Thingvalle, where a small building, only here sufficient to hold the Judges & Lawyers, is prepared, & the multitude of suitors must remain either in tents or in the open air.

[1] Around 1800 the norm was 50–60 ships a year.

[2] These are the figures noted by Banks in his memorandum (Document 62). As Banks rightly wrote, not all the information had been gathered at this point of time. The total would appear to have been at least 18 vessels (Agnarsdóttir, Great Britain and Iceland, pp. 47–9).

[3] In the Kew draft Banks has added here 'does not distinguish Iceland ships from Danish' and then crossed it out. The Marshal of the High Court of Admiralty was John Crickett who served in this post 1788-1811.

[4] But Charles Bishop did later. The list, dated 31 December 1807, is in TNA, PRO 30/42/14/8.

[5] Count Trampe.

[6] The district governors at the time were Stefán Stephensen in the West and Stefán Þórarinsson in the North and East, while the governor Count Trampe had jurisdiction in the South.

[7] This remains true.

[8] This is correct. In 1809 at the beginning of the Revolution (see Introduction, p. 32 above) Hooker reported 'the arms of the inhabitants were secured, which did not amount in the whole to above twenty wretched muskets, most of them were quite in a useless state, and a few rusty cutlasses' (Hooker, *A Tour in Iceland*, I, p. 56).

[9] Þingvellir, the site of the ancient parliament. What is surprising is that Banks did not know that the *Alþingi* had been abolished in 1800, but then none of the questions he put to the Icelanders (see Document 50) pertain to it.

[10] There were two *lögmenn* (lawmen) in Iceland: see Glossary.

[11] *Systlemen* (phonetically written, *sýslumenn*): county magistrates: see Glossary.

[12] *landfógeti*, the royal treasurer. There was only one: see Glossary.

[13] Corrected in a different hand, 'all thing' in the Kew draft.

[14] Banks first wrote 'convocation' in the Kew draft.

[15] Underlined in the original.

Far from deriving any revenue from Iceland, the Danish government is said to furnish to the Island an annual supply of at least the value of £10,000 a year, for the maintenance of the different Functionaries of Government & the Church
the Salary of these

Stiftsamptman is	1,500 dollars[1]
Amptman Thoransen[2]	1,200
Amptman Stephensen[3]	1,000
a Bishop[4]	[blank]
The Resident[5] Clergy of 189 Parishes	from 4 to 500 Rixdollars each say 76,050[6]

Howsoever desirable the acquisition of an Island circumstanced as Iceland is may be to the Imperial Crown of the United Kingdom (& that it is highly desireable no one who has the glory of the British Crown at heart can doubt)[7] it is not certainly worth the expence of an expedition, however small it may be; nor would it be conformable to the humane & generous dispositions of Englishmen to subject such a Country to the horrors of Conquest, which, as it is unfortunately ever accompanied with plunder, would naturaly in a Country where every one lives from day to day, & where no accumulation of Property can be met with /be/ succeeded by Famine.

The very humane & generous disposition manifested in every conversation I have had the honor to hold with your Lordship on this subject, leads to a very different result: to offer to a people, virtuous, religious, honorable, honest, & poor, the option of placing themselves under the dominion of this Country by a Resolution rather than subjecting them to the miseries of unnecessary warfare, cannot but be considered as a Testimony of the magnanimity as well as of the humanity of the British nation, which no virtuous man of any other nation can hesitate in approving.

As a project for offering this option to a Country, for which I confess I feel a degree of interest which may possibly mislead me, the following is submitted to your Lordship's consideration & correction.

That the detained Ships and their Cargoes the property of native Icelanders or of Burghers of the Island of Iceland establish'd in that Country, be considered as Pledges & the Icelanders and Burghers of Iceland, belonging to them now in this Country, as hostages for the conduct of their Countrymen in Iceland.

---

[1] This is correct. See 6 June 1806, 'Kongelig Resolution ang. Bestyrelsen i Island', *Lovsamling for Island*, VII, pp. 29–30.

[2] Stefan (Torarinsson) Thorarensen. According to Hooker's 'Account of Salaries and Pensions paid yearly in Iceland by the Landfogued Frydensberg' in his *A Tour in Iceland*, I, pp. xxxv–xliv, who had his information from the treasurer of the island, Rasmus Frydensberg (1778–1840), in 1809 Thorarensen had a salary of about 700 rixdollars, but he also received revenue in kind (Hooker, *A Tour in Iceland*, I, xxxvi).

[3] Stefán Stephensen. This is correct.

[4] Geir Vidalin was the first bishop of all Iceland. Hooker notes that the bishop had 1,248 rixdollars per annum with an additional 600 rixdollars (*A Tour in Iceland*, I, pp. xxxv–xliv). Icelandic sources confirm that the bishop's salary was 1,200 rixdollars per annum, plus 48 rixdollars for housing (NAI, Skjalasafn landfógeta, VI, 71, p. 57). From 18 May 1805 he had been awarded an extra 600 rixdollars (*Lovsamling for Island*, VI, p. 747), thus confirming the reliability of Hooker's useful list.

[5] 'President' in the DTC copy, which is obviously a slip of the pen; 'Resident' in Kew draft.

[6] According to Hooker's useful list, the total yearly expenditure in salaries and pensions, paid by the treasurer, was 18,713 rixdollars and 63 skillings. This does not include the clergy.

[7] These parentheses are not in the Kew draft.

That in the case of the Icelanders declaring without delay & under proper authority that they will transfer their allegiance from the Crown of Denmark to that of the united kingdom & of their sending their danish Stiftsamptman to England & receiving an English Stiftsamptman in his stead that the said ships & Cargoes shall be restored to their respective owners, & allowed to sail to Iceland, & return with Cargoes in the autumn to England.[1]

In the meantime, the owners may be put into a conditional possession of the Cargoes, & allowed under proper control to sell them, & to purchase with the Proceeds Cargoes fitted for the Iceland market for the next season, & to have such Cargoes stowed on board; these Cargoes, however, not to be considered as the Property of those to whom the Crown has /derived/ decreed[2] this temporary interest in them, unless on the contingency of the Island of Iceland submitting itself in the first instance to the crown of the united kingdom.

If this idea meets with your Lordships approbation, it is probable that an offer will be made to the Icelanders of retaining their religion, their Laws, & their customs, intire; only that in future all Legislation shall, as in our Colonial System, be submitted to the will of the King & Council in England, and all the Prerogatives of the Crown of Denmark over Iceland be transferred to that of the united kingdom.

Under such a regulation, the whole change that would take place would be the displacing of the Danish Stiftsamptman, & placing an English one in his stead; & I cannot help auguring that a conduct so generous & indeed so magnanimous on the part of the English, would give an instantaneous energy to the character of the Icelander. The people were in 1773 universally desirous of being placed under the dominion of England, the applications made to me personaly by natives of the best quality were continual:[3] their project was, that England should purchase from Denmark the dominion of Iceland, which, as it then produced a Revenue of £8000 a year[4] only to its mother country, they concluded the wealth of England could easily purchase, & the poverty of Denmark would willingly sell.[5]

Access to the Southern & South Western part of Iceland in the winter is principally prevented by the continual fogs which prevail there: as, however, fogs seldom, if ever, are met with when winds high enough to make navigation hazardous are blowing, this will not probably be an obstacle so insurmountable as to prevent some small vessels from attempting an intercourse even at this season: on board such vessels, some of the best of the Native Icelanders now prisoners here might be sent, in case, on Conversation, they prove inclined to approve with the warmth it deserves the system proposed above: in Iceland, where there is little difference in rank, every man thinks as much at least as is done in England of the general policy of the Country, & talks with his neighbour about

---

[1] Here he refers to the letter to 'Mr. Stephensen' (Document 61).

[2] Only 'derived' in the Kew draft.

[3] We only have Banks's word for this.

[4] According to Professor Gísli Gunnarsson, the net 'gain' to the Danish government during the monopoly period 1743–88 was about 5,200 rixdollars a year. See his book *Upp er boðið Ísaland*. Banks believed it to be nearer 4,000 rixdollars, as he noted in 1801 (see p. 219 above).

[5] In Kew draft: 'dispose of'.

it, in Iceland Education is more general than in other Countries;[1] & the lower ranks are clearly much better informed than in other parts, in fact the lowest ranks of European Society scarcely exist in Iceland: the farmer there who does his work chiefly by his Family is in a state similar to that of an English yeoman; & the mob of Labourers, mechanics, & their Servants & dependants, are not in existence.[2]

Further details on this subject are evidently unnecessary. In case your Lordship should think the project Utopian & visionary, which it may possibly be. In the other case, if it should be determined to act upon it, I shall be most ready to communicate to your Lordship every part of the detail,[3] which, in thinking a considerable time on the subject, has occurred to me, & in executing any commands with which your Lordship may at any time think fit to honor me on this or any other subject.[4]

NHM, DTC 17, ff. 78–89. Contemporary copy.
Kew, Banks Correspondence, II, JBK/1/7 ff. 329–332.[5] Draft.
Hermannsson 'Banks and Iceland', pp. 36–40 (extracts only).

64. Bjarni Sívertsen to Banks, 31 December 1807

Wellclose Square

Sir

Though I must fear to tire out your patience with verbal and epistolary applications, with such repeated complaints, and prayers, yet the unparalelled kindness you have shewn towards me and my companions in misfortune encourages, and the distressed circumstances under which I groan impel me once more to address you on a subject to which I have also before, in common with my fellow sufferers, in our letter of the 15th instant,[6] ventured to draw your attention, namely: The embarrasments of my present situation, the wants and miseries with which I am immediately threatened, against which I have no other refuge than your kind and compassionate assistance.

It was my misfortune to have in my ship *the Tvende Söstre* no goods of any value that were not set down in the bills of loading, and consequently must follow the cargo; so that I could not obtain permission to take any thing out of the ship, the sale of which might

---

[1] Other British travellers remarked on this. Stanley wrote, optimistically, that the Icelanders could mix 'without any fear of Inferiority, amongst the informed & well bred of any society in London, Paris or any other Capital' (*The Journals of the Stanley Expedition*, I, p. 208, n. 201).

[2] It is certainly true that there was little difference in rank compared to English society, but there was of course an élite – the officials appointed by the King. These officials were, however, also farmers. See Introduction, p. 14 above.

[3] Banks's letter to 'Mr Stephensen' (Document 61) is a good example of how detailed his plan had become.

[4] Hermannsson interpreted this memorandum as demonstrating that 'the government apparently was on the point of carrying out the plan of 1801 ... Evidently there was, however, some hitch or hesitation which delayed the matter so that nothing was done' ('Banks and Iceland', p. 40). Although this document quite clearly contains a project that Banks is proposing to the government, this does not in itself prove that 'the government was on the point of [acting]', as Hermannsson wrote. If anything, Banks is somewhat doubtful, using the words 'Utopian and visionary' at the end of the document. In fact the British government never attempted to carry out the annexation of Iceland.

[5] Here the fair copy in the Natural History Museum has been transcribed, with discrepancies noted in footnotes.

[6] Document 53.

have enabled me to bear the expences of my stay and travels in this country. I had some money in Danish bank-notes, which I was necessitated to exchange at three forths of their value; that is all I have had to subsist on till the present moment, and with which I have also been obliged to assist the Captain and crew of my ship as much as I could. This money is now all expended, and I am even involved in some debts here for the discharge of which, as for my future sustenance I have no prospect or resource. If the communication with Denmark had not been stopped, I should not have been under the necessity of applying in this manner for relief; but notwithstanding all my attempts, I have not been able to obtain the smallest assistance from thence, not even to receive an answer to my repeated letters. My only hope, Sir, and my only resource therefore is your kindness, and the humane interest which you have taken in our misfortune, to which I humbly and with confidence appeal, praying that you would add to all your past favours also that of representing to your government my distressed situation & my calling want of assistance, and procuring me from the same, if possible, such support for the time that I shall be obliged to stay here, as may enable me to live, and to discharge those debts into which necessity has plunged me. Or if this, possibly, can not, even by your mediation, be effected, I would feel no less indebted to your kindness if, through your recommendation, I could obtain the loan of a sum of money, adequate to my necessities which I would thankfully repay as soon as my circumstances shall admit.[1]

Sir, I have addressed you with that frankness, with which it is natural to a native of my rough country; If I apprehended to offend you by this, I should do an injustice to the tried generosity of your heart, in a firm reliance of which I pray and hope that you will kindly pardon my importunity, to which nothing but the most calling necessity could have induced me.

I remain Sir, with the deepest sense of gratitude
Your most humble & obedient
Servant
[*Signed*] B: Sivertsen

SL, Banks Papers, I 1:15. Autograph letter, written by an Englishman, but signed by Sivertsen.

65.   Banks to Bjarni Sívertsen, 1 January 1808

Soho Square

Sir

I shall be glad to see you here if you will Call upon me on Sunday morning[2] at any hour that best suits you I request you to bring with you Some one who can interpret for you[3] & I will Enquire into the Particulars of your Case & do my Endeavour to provide a Remedy for the inconveniences you Suffer.

SL, Banks Papers, I 1:15. Autograph draft.

[1] Banks did indeed lend money to Sívertsen (see Document 53).
[2] Written by Banks on a Friday; the appointment was for Sunday, 3 January 1808.
[3] Sívertsen's English would gradually improve during his stay (see Document 106 which he is obviously writing himself without help).

66. Banks to Robert Stewart, Viscount Castlereagh, 2 January 1808

My dear Lord

I did not till night obtain the information your Lordship desired me to request from the Marshal of the Admiralty; from this it appears that 11 Iceland Ships have been detained for adjudication & that 2 of them have been already condemned to His Majesty,[1] & that process is going on against the remainder.[2]

I trouble your Lordship with this information in order to suggest for consideration, whether it may be proper to suspend further proceedings against the condemned Ships, Lest Sales of their Cargoes should take place; & whether, in case it is very likely that his Majesty's Ministers will recommend the restoration of the whole,[3] it may not be expedient to suspend proceedings against all of them.

NHM, DTC 17, f. 90. Contemporary copy.
Hermannsson, 'Banks and Iceland', p. 41.

67. Note [Banks in conversation with Bjarni Sívertsen, 3 January 1808]

Jan*uary* 3 Sievertsen
Graaf Trampe[4] is in Denmark
Einarsen assessor at the Thing[5] Frydensburgh Collector of Taxes[6] are commissioners for Stiftsamptman
Stephen Stephensen is amptman[7] of the Western ampt
Magnus in Copenhagen is Justice of the high Court
Biorne in Iceland is secretary[8]

SL, Banks Papers, I 1:40.[9]

---

[1] These were the *Johanne Charlotte* and *Svanen* (TNA, PRO 30/42/14/8, Charles Bishop to Fawkener, Doctors Commons, 31 December 1807).

[2] These ships were the *Regina, Christina Maria, Bassant* [*Bosand*], *Haabet, Sejen, Two Sisters* [*De Tvende Söstre*], *Conferentz Raad Prætorius, Rödefiord, Johanne Charlotte, Bildahl* and *Svanen*. (For source, see preceding note.)

[3] This was a crucial factor in Banks's plans for the annexation of Iceland (see Document 63).

[4] All this information is in addition to that given in the memorandum to Lord Hawkesbury (Document 63). The Governor of Iceland, Count Trampe, was in Denmark. Banks probably noted this because part of the plan in the memorandum and in the letter to Mr Stephensen (Document 61) was to arrest the governor.

[5] Ísleifur Einarsson was assessor not at the Thing but at the High Court of Justice (*Landsyfirréttur*), which had replaced the *Alþingi* in 1800.

[6] Rasmus Frydensberg was the *landfógeti* (King's treasurer), who indeed collected the taxes among other duties.

[7] Here Banks is noting down information on the three Stephensen brothers, which one of them would be most suited to carry out Banks's wishes as outlined in his letter to 'Mr Stephensen' (Document 61).

[8] The older brother, Stefán, was in western Iceland, but Björn Stephensen, who was secretary to the High Court of Justice and lived at Lágafell in the vicinity of Reykjavík, was in Iceland and would thus have been a likely candidate.

[9] More or less the same information is repeated in Sutro, Banks Papers, I 1:41, in a note written by Banks in pencil.

68. John Christopher Preidel to Banks, 4 January 1808

No 23 Great Winchester Street

Sir!

Inclosed I take the Liberty of handing You Copy of a Memorial¹ which I have presented this day to the Right Hon*oura*ble the Lords Commissioners of His Majesty's Treasury in favor of three Danish Masters.²

I humbly request that you will be pleased to interest Yourself in their Behalf so that their Request may be granted. & remain with profound Respect,

Sir! Your most obedient and
devoted humble Servant
[*Signed*] J. C. Preidel

SL, Banks Papers, I 1:17. Autograph letter.

69. Petition, John Christopher Preidel to the Lords of the Treasury, 4 January 1808

Copy³

To the Right Honorable the Lords Commissioners of His Majesty's Treasury.

The humble Memorial of John Christopher Preidel of London Merchant in Behalf and as Agent to the hereundermentioned three danish Masters.

Sheweth:

That <u>Phillip Christian Hansen</u>⁴ Master of the Danish Ship *Fortuna*,⁵ detained at Plymouth by His Majestys Frigate *Lavinia*, Lord William Stuart Commander, is besides his private adventure on board the said Ship, proprietor of one eighth part of the Ship and Cargo of Salt and Eight Bales of almonds on board the same as declared by him on Oath in the annexed affidavit marked **A**.

That <u>Hans Thielsen</u> Master of the danish Iceland ship *Rennthier* captured off Copenhagen by His Majesty's Ship the *Cruizer* Pringle Stoddard Esq*uire* Commander and brought to Yarmouth, is also, besides his Private adventure, owner of one Fifteenth part of the said Ship and Cargo of Fish, as declared by him upon oath in the annexed affidavit marked **B**.⁶

That <u>Jacob Jacobsen</u> Master of the Danish Iceland Ship *the Probitas*⁷ detained by His Majesty's ship *Shannon* P. B. V. Broke Esq*uire* Commander & brought to Yarmouth is

---

¹ This is the following document, which explains Preidel's role in the matter.

² The Treasury sent the memorial on to Charles Bishop, His Majesty's Procurator General, or King's Proctor, for his opinion on 'what is fit to be done therein' (TNA, TS 8/29, Harrison to Bishop, 7 January 1808).

³ The memorial was referred to the King's Proctor on 7 January 1808 by George Harrison (TNA, T 4/13. Précis: Praying restoration of the private adventures of Danish ships *Fortuna*, *Rennthier* and *Probitas*).

⁴ All the underlining is done with a pencil, probably by Banks.

⁵ The *Fortuna* was not an Iceland ship, as is confirmed in the following letter.

⁶ The papers of the *Rennthier* are in TNA, HCA 32/1645, 26 April 1808, no. 5076. The private adventure of the master and mate was the following: 10 ship pounds of stockfish, 5½ of klipfish, 2 casks of salted salmon, 66 lambskins, 50 pairs woollen gloves etc.

⁷ The *Probitas* was not an Iceland ship, as is confirmed in the following letter.

likewise and besides his private adventure on board of the said Ship owner of one twelfth part of said Ship her Freight Primage & Hatmoney[1] as declared by him in the annexed affidavit marked **C**.

That His Majesty has been most graciously pleased to direct that the private adventures of the masters of the several Danish ships condemned to the Crown should be restored to them & the masters adventures have accordingly been ordered to be restored at the Time when the Ships above named were condemned & your memorialist is by this gracious act of the Crown induced to hope that His Majesty may be further prevailed upon to grant the Masters above-named the amount of their Trifling shares as part owner of the Ships and Cargoes respectively, which amount, tho' inconsiderable in itself, would go a great way in alleviating the Distresses under which they now Labour.

Your Memorialist therefore most humbly prays your Lordships will be pleased to take into Consideration the Situation of those unfortunate Men & to recommend it to His Majesty to direct that out of the Nett Proceeds of the said three Ships & Cargoes the Shares abovementioned may be paid to your Memorialist for the use of the said Danish Captains.

And your memorialist as in duty
bound will ever pray
[*Signed*] J. C. Preidel.

SL, Banks Papers, I 1:18. Contemporary autograph copy.

70. John Christopher Preidel to Banks, 5 January 1808

[No. 23] Winchester Street

Sir

In Reply to Your kind & condescending Letter to me of Yesterday I have the Honor to acquaint You that out of the three Ships mentioned in my Memorial to the Right Hon*oura*ble the Lords of His Majesty's Treasury. *The Rennthier* Hans Thielsen[2] Master, (with whom about three weeks ago I had the Honor to wait on You) is the only Ship from Iceland amongst them.[3]

The Probitas, Jacob Jacobsen Master (erroneously /stated/ in the Admiralty Court to be from Iceland) was bound from Bergen in Norway with dried & Salted Fish to Barcelona.

The Fortuna, P. C. Hansen Master, was bound from Alicant to Copenhagen laden with Salt & 8 Bales Almonds.

But all these three Ships were captured & brought in previous to any Declaration of War & issuing of Letters of Mark on the part of this Government.

They have been since proceeded against in the High Court of Admiralty and condemned as Danish Property to the Crown, as having been captured previous to the

---

[1] Primage was a customary allowance formerly made by a shipper to the master and crew of a vessel for the loading and care of the cargo – also called hat-money.
[2] All underlining in the document is made in pencil.
[3] See Document 54.

Declaration of War, whereas if they had captured after the Declaration of War & the issuing of the Letters of Mark, they would have been condemned to the Captors.

At the time those Ships & Cargoes were condemned the Masters private Adventure were by Motion of Council decreed to be restored, but not their Shares each had in the Ships, Cargoes & freights, as it is not in the Power of the Court to restore that without a special order from Government.

It was my Wish & Desire to have presented my Memorial to His Majesty in Council, but was in that Respect overruled by my Proctors who advised me to present & address it to the Lords of Treasury, as the Disposal of the Net proceeds arrising [*sic*] from the Sale of Danish Ships & property condemned to the Crown, came under Their Direction. Being apprehensive that my Memorial will be returned to His Majesty's Advocate & Procurator general in the Admiralty Court, who as usualy [*sic*] will reply that my Request can not be complied with; I took the Liberty to send you Sir the Copy of it, that by an early Application on Your part to the Treasury in favor of those unfortunate Men, the Consequences before stated may be prevented & their Request granted.[1]

My Interference in their Behalf does not proceed from any selfish Views but from the purest Motives of Humanity, under which Assurance I beg leave to subscribe myself most respectfully,

Sir!
Your most obedient & most
devoted humble Servant
[*Signed*] J. C. Preidel

SL, Banks Papers, I 1:19. Autograph letter.

71. Adser Christian Knudsen to Banks, 16 January 1808

Stornoway

Sir

I had the misfortune on my passage from Iceland to Copenhagen, to have been drove to this place by a continuation of Gales of Wind during my Voyage and in consequence of the War betwixt this Country and Denmark, my vessel is detained by the Customhouse officers here[2] –

By a letter I received from my Brother in Law, M[r]. Petreeus,[3] and by a Paragraph which has appeared in the Edinburg Newspapers,[4] it appears that your goodness has laid you to

---

[1] Eventually this petition proved successful. On 10 September 1808 it was documented that the *Rennthier* had already been 'restored as an Iceland Vessel, that the other two Vessels not having been usually engaged in trading to the Country nor having been taken on a Voyage to or from a British Port and there being no Circumstances peculiarly distinguishing these Cases from various other Cases the Masters do and appear entitled to the special Indulgence from the Court of Admiralty' (TNA, TS 8/29, J. Nicholl, Opinion, f. 12 followed by other documents relating to the case, including a report from Charles Bishop, dated 25 October 1808, ff. 16–18).

[2] His vessel was *De Fem Brödre*. All papers regarding this ship are in TNA, HCA 32/1032, no. 2223.

[3] Westy Petræus. Knudsen was married to Jóhanna Margrét, yet another daughter of the Iceland merchant, Diðrik Hölter of Skagaströnd (see Document 47).

[4] An article appeared in the *Edinburgh Evening Courant* on Saturday 28 November 1807 (see Document 47).

take M^r. Petreeus under your protection. – By a letter I received from M^r. Petreeus from London, dated the 23rd Ult*im*o, he writes me to know if I wish him to procure a Pass for me, but does not say whether the Pass is for Leith or Copenhagen; any Pass that M^r. Petreeus can procure for me must I presume be through your Goodness. – And from his last letter I am uncertain whether M^r. Petreeus is at Leith or /*Copenhagen*/ London;[1] and from that circumstance, I take the liberty of /*pay*/ writing you –

I this day applied to the Collector & Comptroller of the Customhouse here, for a letter to Certify that my situation is really what I have taken the liberty of stating to you, so as that letter might accompany me to Leith, and prevent my being interrupted on my way, but the Collector and Comptroller of the Customs here did not think themselves authorised to give me such a letter, and I conceive it dangerous Sir for me to leave this place without some sort of Pass

I now take the liberty (through what your goodness has prompted you to do for Mr. Petreeus) to beg of you to procure such a pass as is necessary for me to have to Leith; and be so good as send it to me enclosed as soon as possible, As I am most anxious to go to Leith.[2] –

I trust that my distress of mind will be easily conceived, and that it will, at least in some measure apologise for my putting you to this trouble.

I have the honour to be
Sir
Your most ob*edient* Serva*n*t
[*Signed*] Adser Christian Knudsen

Canberra, National Library of Australia, Banks Papers, MS 9/106. Autograph letter.
NHM, DTC 17, ff. 196–197. Contemporary copy.
Hermannsson, 'Banks and Iceland', pp. 41–2.

72. Westy Petræus, Bjarni Sívertsen and Hans Georg Bredal to Banks, 5 February 1808

London

To the Right Honourable Sir Joseph Banks, B^rt K. B.
Sir
When we took the liberty some time ago to wait upon you for the purpose of expressing our thanks for new benefits derived from your kindness, we heard with the deepest

---

[1] Petræus was in London at least until early February, lamenting the fact that he could not get back to his family (see Document 72).

[2] On 8 February (doubtless through Banks's intervention) the King's Proctor (Charles Bishop) had written a report which, with a letter from the Lords Commissioners of the Treasury, had been sent to the Collector & Comptroller of Stornoway, with directions to proceed accordingly (TNA, ADM 98/114, Commissioners of Treasury Office to Pole, 31 March 1808). Knudsen appears in the records again in March. He had been sent on parole to Peebles. In March Jens Wolff requested he be released (TNA, ADM 98/114, Commissioners of Transport Office to Pole, 31 March 1808); MacDougall, *Prisoners of War in Scotland*, p. 146). It is clear that this letter reached Scotland (The National Archives of Scotland [hereafter NAS], Customs and Exchequer [hereafter CE] 1/41, Minutes, 23 May 1808).

affliction that you were indisposed,[1] and could not receive strangers. Being however on a second call consoled with the information, that you were in a good way of recovery, and that you could receive a letter, we humbly beg leave in this manner to convey to you our thanks, and implore your propitious advice concerning the present aspect of our affairs.

We have been informed from the Transport Office, that the English Government has been pleased to grant us a continuation of that allowance of one shilling a day per head which we enjoyed during the first part of our stay in this country, and that the agent of the Transport Office at Lieth had been ordered to procure us, as soon as possible, a passage to Denmark.[2] Convinced, Sir, that it is solely to your kind interference in our behalf that we are indebted for this alleviation of our distress, we thank you from our inmost hearts for this new favour bestowed upon us, & beg you be assured of our deepest and everlasting gratitude. – The latter part of the above information, pleasing as it is, and great as is the relief to us of obtaining a free conveyance to the place of our destination, puts us however, at the present moment, to a dilemma, respecting which we venture to solicit your kind advice. In consequence of that information it seems to us that we ought immediately to repair to Scotland, in order to be ready to avail ourselves of the opportunity that may, very soon perhaps, be provided for us. But on the other hand it also seems to us, that we can not, during the present undecided state of our affairs, leave this place, where, in case those hopes which your goodness, Sir, and your influence bid us entertain, should in the end be realized, our presence may perhaps be indispensably required. In this respect Sir, we therefore humbly solicit to be guided by your opinion to that course which will be the best for us to follow.

A circumstance of which we have lately been informed at the counting house of M^r. Wolff, the Danish Consul[3] has not a little tended on one hand to enliven our hopes, but on the other to increase the uncertainty and embarrasment in which we find ourselves concerning the last mentioned point. We have namely learnt on very good authority, that an order has been given to discontinue the condemnation of icelandic vessels, and not to proceed to the sale of such as have already been condemned.[4] – What may be the cause or the intention of this order we have not been able to learn; but our sanguine hearts have not omitted to suggest to us, that this measure can only have originated in your benevolent representations, and may, most likely, prove the forerunner of a decision favourable to our interests; and we think in this instance to have another cause of blessing divine Providence for throwing us under the protection of a man who to the will unites also the powers of rescuing from misery. Sir, if it be not too much presumption and importunity, could we venture to ask an additional mark of your condescending benevolence in obtaining from yourself a confirmation of our flattering hopes in this respect.

---

[1] Banks was afflicted with gout. He first suffered from this disease in November 1787. There is a summary of Banks's gout history in Carter, *Sir Joseph Banks*, pp. 525–36, with a diagram showing his periods of incapacitation by year.

[2] See Document 59.

[3] Jens Wolff became assistant consul to his father Georg in 1793, designated to be his successor (Kjölsen and Sjöqvist, *Den danske udenrigstjeneste 1770–1970*, I, p. 290).

[4] So far only the *Rennthier* appears to have been condemned.

On the subject of these our prospects, Sir, may we still be permitted to submit with the greatest humility, that if the British government could be induced generously to exempt us and our poor country from the calamities of the war, by restoring to us our property, and permitting us to continue without interruption that insignificant trade which is our only support, the sooner this can be decided the greater will be the benefit thereby bestowed both on us and our country.[1] Our cargoes at present detained in the ports of these kingdoms consist principally of fish and train-oil, which are subject to great damage and loss by remaining so long in the ships. – Our trade in Iceland is so circumstanced, that if we do not arrive there early in the Spring, when our purchases are to be made, & our ships to be employed in the fishery, nothing can be done for that year. The persons also in whose hands we have left the conduct of our affairs at home, cut off as they are at present from all communication with us and the rest of the world, are utterly incapable of undertaking any thing, if we do not return in proper season. – Hence, Sir, it is evident how great the loss must be to us, and to the country in general from every delay in the decision of our fate. Indeed every day that we remain in the present inactivity is a heavy drawback on the boon that will be conferred on us, if we are to see the accomplishment of the hopes with which your high & active protection has filled our hearts; and if ruin should prove to be our doom (which may God forbid) every day it is deferred will be an additional ruin to us. Forgive, Sir, the liberty we have taken in thus urging to you our impatience and its causes; it does not proceed from a want of confidence in your benevolent intentions towards us; what you have done already is more than sufficient to convince us, that you will leave nothing undone that can forward the success of our cause. But apprehensive that you might not see this part of it in its true light, we have made bold to state the circumstances, humbly entreating you to represent in our behalf of what urgent importance it is to us, and even to our country, that our fate be soon decided.

As to me, Westi Petræus, the present afflicting state of my family at Lieth [sic] renders my stay in London painful in the extreme, both to them & me, and makes me ardently wish to join them, if possibly I can without neglecting our common interests. Almost the whole number of my family have long been very ill with the measles, which have carried off my eldest son & one of my foster-daughters,[2] and left my next eldest son in a very dubious state, in which he still continued when my wife sent her last letter. She being of a weak constitution, and very sickly during her pregnancy, is quite worn out by such a complication of misfortunes and griefs, and every letter I receive from her conveys her anxious impatience for my return. If therefore it should take some time yet before any thing can be finally determined concerning our affairs, and if I can, in case things should end in a favourable manner for us, be absent from here without detriment to my affairs, it is my earnest wish to return to my family with the least possible delay. But in this Sir, I can do nothing without your approbation, and I shall feel unspeakably indebted to your kindness for being guided by your opinion in

---

[1] Banks in fact was instrumental in drafting the orders permitting the Iceland–Copenhagen trade to continue. In the Sutro MS there is a note drafted in his hand and that of one of his amanuenses dealing with this (Sutro I: 1: 43). See Document 86.

[2] His 8-year-old son (see Document 56). Measles was common in Scotland at that time and there was a high mortality rate among infants and children.

this painful alternative.¹ If I should go away M^r [Bjarni] Sivertsen, my friend and companion, will still remain here, he might perhaps act in my behalf, and he is ready to do it, in case it should be necessary. –

Humbly entreating your forgiveness for troubling you so often and so much with our concerns, we remain,

Sir, with the highest respect & gratitude
Your most obedient
and devoted Servants
[*Signed*] W. Petræus, B. Sívertsen
H: G: Bredal

BL, Add MS 33981, ff. 277–278. Letter, written by an Englishman, signed by them.
DTC 17, ff. 124–128. Contemporary copy.
Hermannsson, 'Banks and Iceland', pp. 42–5.

73. Hans Thielsen to Banks, 22 February 1808

No. 8 St. Catherines Square

Sir

I beg leave to bring to your recollection, that I had the honor of waiting on you personally on the 19th of december last past.² Since which time political circumstances have made me a Prisoner of war, and accordingly was forced to proceed to Reading in the County of Berkshire on Parole: fortunately, my detention there was of no long duration; and was set at liberty on the 6th of this month in consequence of the Capitulation concluded at Copenhagen, provided with a passport of the Honorable Transport board, allowing me to proceed to Denmark in a flag of truce prepared for that purpose.³

I have no money nor credit. Some of the mercantile houses here to whom I preferred my hopes, upon their good & noble disposition in regard to my Ship & cargo & used them for credit to pay my necessary maintenance have compleatly refused me.

---

¹ A note in Banks's hand in the Sutro papers mentions that Westy Petræus was lent or given £30 while Sivertsen received £20. No sum is mentioned next to the names of G. Haldorsen [Gunnlaugur Halldórsson] and Hans Thielsen (SL, Banks Papers I, 1: 97). Eventually Petræus was paid 'Arrears of Subsistence at the Rate of One Shilling per Diem from the 12th of November last' (TNA, ADM 98/308, McLeay to Petræus in Leith, 10 September 1808).

² With his agent J. C. Preidel (see Document 55).

³ Hans Thielsen, master of the *Rennthier*, had been captured on 19 August 1807, the ship being declared a merchant vessel. He was received on parole in Reading 25 January 1808 from London and discharged 7 February 1808 (TNA, ADM 103/579, General Entry Book of Danish Prisoners of War on Parole at Reading, f. 14). On 2 February 1808 the Admiralty ordered the Transport Board to release Hans Thielsen and his son, Hans Thielsen junior (TNA, ADM 99/184). The following day the Transport Office sent on the order to furnish them with passports to proceed to Denmark whenever they should be ready to depart (TNA, ADM 99/184, Transport Office to Mr Lewis). Thielsen returned to Copenhagen only to come back to England in September 1808 to pick up his released ship (Stephensen, *Ferðadagbækur*, pp. 103–5).

Now, Sir, my prayer is, if you can not assist me with means to satisfy my necessary subsistence, being with out money & credit.[1]

Sir
Your Most Humble & Obedient Servant
Hans Thielsen

NHM, DTC 17, f. 130. Contemporary copy.

74. Banks to Lord Hawkesbury,[2] 24 February 1808

Soho Square

My Lord

As the Question asked by the King's Proctor,[3] 'what is now to be done in the Case of the detained Iceland Ships,' which your Lordship has done me the Honour to order to be referred to me,[4] can only be decided[5] by His Majesty's confidential Servants, I am under the Necessity of troubling your Lordship with this Letter.

In consequence of Information obtained under Your Lordship's Directions, from Masters of Iceland Ships detained at Leith, which I had the Honour to state to Your Lordship at large, about Two Months ago,[6] I gave it as my poor Opinion, that it would be a humane, as well as a wise Policy, to separate entirely, the Case of the detained Danish & Norwegian Vessels, from those of Iceland, and to restore to the Icelanders the whole of their Ships & Cargoes, and this for the following Reasons.

I conclude it will be judged necessary to annex the Island of Iceland to the British Crown, without the least Delay; in this Case, it will be returning[7] to our own Subjects the Means of Prosperity, of which we certainly cannot wish to deprive them.

A Measure generous, and I may say magnanimous, as this will be, cannot fail of making a deep Impression on the Minds of the People, inclining them to surrender their Country on the first Summons, & to become attached as well as dutiful Subjects.

It will restore to the poorest People in the European World, a Part of their little Capital in Trade, the whole of which they must otherwise lose, as all the Ships and Cargoes of the Island are either in Denmark, or in England.

[1] Banks sent this letter on to Alexander McLeay; see McLeay's reply (Document 75).

[2] Hermannsson believed this letter was written to Lord Castlereagh, the Secretary for War ('Banks and Iceland', pp. 45–6). That is not surprising as he is using the DTC copy where the transcriber has written 'Sir Joseph Banks to Viscount Castlereagh' at the top of the page. Dawson made the same error (*The Banks Letters*, p. 792), but he only had the Royal Society draft (where a clerk has pencilled in 'Ld. Castlereagh (?)' and it is filed as to Lord Castlereagh) and the DTC copy, but not the original. However, the original, beautifully written by one of Banks's amanuenses and signed by Banks, is clearly addressed to 'the Right Honourable the Lord Hawkesbury'. Hawkesbury was then the Home Secretary, though it is to be found in the War Office files of the TNA. The fact that it is there is not surprising and easily explained. Hawkesbury sent it on to his colleague Castlereagh because of its contents. Banks sent his lengthy memorandum on the wisdom of annexing Iceland to Hawkesbury at the end of December (see Document 63) and the subject matter is much the same. Banks's contact within the government at this point was his close friend Lord Hawkesbury. Henceforth Edward Cooke, Castlereagh's Under-Secretary, takes on the case.

[3] In 1808 the King's Proctor was Charles Bishop.

[4] This letter is not extant.

[5] The draft of the letter in the Royal Society Archives, is almost identical to the final version. However, here Banks first wrote 'answerd' and changed it to 'decided'.

[6] Here Banks is referring to his memorandum to Hawkesbury (Document 63).

[7] In the draft Banks wrote 'restoring'.

It will save them, I believe I may say from Famine; as they have no Kind of Corn or Grain in their Country, except what they import.

It will teach our People at once, in what Manner the Commerce of Iceland is carried on; & how the Ships are advantageously employed in fishing, during their Stay; which for the first Year at least they could not otherwise learn.

In Consequence of my having urged these & similar Reasons to Your Lordship, You were so good as to order some Time ago, that the Iceland Ships should not for the present be proceeded against, in Order that You might have Time to consider what it would be proper to do with them; – The Time is now come, My Lord, when it is necessary to decide, whether they shall be condemned or restored.

Their Cargoes have begun to suffer; & the Time of Preparation is come: The Ships, if restored, must sail in the first Week in April: In the mean Time their Cargoes must be sold, new ones bought, & their Damages repaired.

In case of your Lordship's Decision in their Favour, a Frigate, and Two small Vessels may accompany their Ships: If a person properly empowered to negociate goes with them, it is probable the Surrender of the Island will at once be accomplished; especially if the Stiftsamptman, Count Trampe, who was at Copenhagen in the Autumn, has not got back to the Island: Should this fail, one of the light Vessels will in a Fortnight return, with Intelligence of the Force necessary to compel a Surrender, which in an Island where there are neither Soldiers nor Arms, cannot, We may believe, be either large or expensive.

I have the Honour to be with
due Respect & sincere Regard
Your Lordship's most faithful
and most humble Servant
[*Signed*] Jos: Banks

TNA, WO 1/1117, ff. 65–69.[1] Amanuensis with Banks's signature.
NHM, DTC 17, ff. 131–133. Contemporary copy.
The Royal Society, MM/6/29. Draft (major discrepancies are footnoted).
Hermannsson, 'Banks and Iceland', pp. 45–6.

75. Alexander McLeay to Banks, 25 February 1808

Transport Office

Dear Sir

I am glad that you referred the enclosed Letter[2] to me before you complied with the Writer's impudent request. – When he wrote to you pleading so much misery he knew perfectly that a Vessel lay at Deptford ready to receive him and about 20 more of his Countrymen for the purpose of conveying them to Denmark, and that they might have embarked last week if they chose; but it appears that M^r. Thielsen and several others preferred remaining on shore upon their Allowance as Prisoners of War to being on board the Cartel where they might have lived without any Expense whatever. – A peremptory Order, however, has been given to all of them to embark without further delay, and as the

[1] Note, verso: '[Received] 24 February 1808 | S. J. Banks'.
[2] The letter from Thielsen, Document 73.

Cartel will probably go down the River tomorrow, I do not think that Mr Thielsen will trouble you again.¹ –

I have the Honour to be, with
great respect
Dear Sir
Your much obliged
and faithful Servant
[*Signed*] Alex. McLeay

TNA, WO 1/1117, ff. 73–74. Autograph letter.²

76. Banks to [Edward Cooke], 26 February 1808

Soho Square

My dear Sir
by the Enclosd Letter³ it would seem as if my poor Icelanders are to be Packd off to Copenhagen before Their Case has been decided if they have any chance of Restoration you will no doubt order the Cartel to be delayd⁴
  oh the Gout the Gout⁵

always yours
[*Signed*] Jos: Banks

TNA, WO 1/1117, f. 71. Autograph letter.

77. Edward Cooke to Banks, 26 February 1808

Downing Street

Dear Sir
Government decides to restore the Iceland Ships⁶ & I have written to stop the Cartel.⁷

Most truly
& obediently
[*Signed*] E. Cooke

Canberra, National Library of Australia, Banks Papers, MS 9/46. Autograph letter.
NHM, DTC 17, f. 134. Contemporary copy.
Hermannsson, 'Banks and Iceland', p. 46.

¹ On receiving this letter Banks appears to have found the Transport Office's treatment of 'his' Icelanders too harsh and sent Edward Cooke the following letter.
² Note, verso: '*Sir* J Banks as to | the Icelanders'.
³ The letter in question is Alexander McLeay's to Banks, Document 75.
⁴ Banks got his wish; see following letter from Cooke.
⁵ See Document 72, p. 272, n. 1.
⁶ Though a few Iceland ships would be released in the next few months the majority were not liberated until June.
⁷ Cooke, as he says here, wrote immediately to the Admiralty: 'it being intended to liberate the Iceland vessels which have been seized desired that the Cartel containing their Crews should not proceed'. The following Monday, 29 February 1808, the Admiralty replied to Cooke 'that although Orders were sent out immediately on receipt of his letter they arrived too late, the Cartel having previously sailed' (TNA, ADM 99/185, 26 and 29 February 1808).

78. Bjarni Sívertsen to Banks, 26 February 1808

No. 2 Wellclose Square

Sir

As you have before kindly supplied me with the means of conquering my present embarrassments, and even offered me your further assistance in case it should be necessary, I again take the liberty to have recourse to your goodness,[1] and solicit of you the loan of £30,[2] which I hope, shall enable me to discharge the debts, I have again contracted, and to live until something be finally decided respecting our affairs. This money together with what I have received of you before, I shall thankfully repay; either before I leave this country, in case my property be liberated, or otherwise after my return to Iceland as soon as an opportunity can possibly be found. Yet the kindness you have shewn me, & the benevolent interest you have taken in my misfortunes, must for ever leave a debt upon my shoulders, which no prosperity could ever enable me to discharge.

I am truly ashamed to trouble you with applications of this nature, but the difficulties of my present situation, in which I am entirely cut off from every source of supply, have left me no other choice than an appeal to that goodness which you have yourself offered me, and of which I have seen so many proofs. Hoping therefore that you will forgive my importunity, I remain with the deepest gratitude

Sir, your most humble and obedient
Servant
[*Signed*] B: Sivertsen

BL, Add MS 33981, f. 279. Letter, written by an Englishman, signed by Sivertsen.
NHM, DTC 17, ff. 135–36. Contemporary copy.
Hermannsson, 'Banks and Iceland', p. 47.

79. Bjarni Sívertsen to Banks, 29 February 1808

London

Sir

I have the honour thankfully to acknowledge the receipt of your letter of Saturday, in which you have had the kindness to send me a draught of thirty Pounds,[3] and informed me that your Government had decided to restore the Iceland ships. Respecting the latter I have not yet heard any thing officially;[4] but hope very soon, according to your kind communication. For both, Sir, I am more infinitely bound to you, than I can express by words. Nay, not only I, but all my compatriots too are infinitely bound to you for this

---

[1] Sívertsen had visited Banks on 3 January (see Document 67) when he probably received the previous loan.
[2] Banks did, see following letter.
[3] As requested by Sívertsen in his previous letter to Banks.
[4] Not surprisingly, as the Admiralty was dealing with the matter. The Lords of the Treasury had agreed, because of an application from Corbett, Borthwick & Co. in Leith, that 'a quantity of Rye, Biscuit or Bread, of Hulled Barley, had been allowed to be shipped in the Tvende Söstre for Iceland, under a Licence from Lord Hawkesbury'. This was an exception as these articles were 'not exportable' (NAS, CE 1/41, Minutes, 23 May 1808). Sívertsen would be granted a licence on 23 July 1808 (TNA, Privy Council [hereafter PC] 2/177).

favour, which, without your protection & interest, we could never have hoped to attain; and I beg leave to give you the most heartfelt thanks in their as well as in my own behalf. I remain with the deepest gratitude

Sir,
your most humble & obedient Servant
[*Signed*] B: Sivertsen

BL, Add MS 33981, f. 281. Letter, written by an Englishman, signed by Sivertsen.
NHM, DTC 17, f. 137. Contemporary copy.
Hermannsson, 'Banks and Iceland', pp. 47–8.

80. Banks to [Edward Cooke],[1] 3 March 1808

Soho Square

My dear Sir,
in all the Conversations I have had with the Icelanders detained here, I have understood from them, that no Vessel ever sails from Denmark for Iceland till the first week in april, or wishing to arrive at the Island till the Last week of that month or the first in may, as this information coincided with that I obtain when I visited the Island I was satisfied with it.

yesterday I happend to stumble upon the 15th Ch. 2nd Ch. 16, an act for Regulating the Herring & other Fisheries &c, in which among other things it is Enacted, that no vessel shall sail from England for Iceland sooner than the 10th of March in any year, hence it is Clear that Vessels had then saild for Iceland at an Earlier period Indeed I believe we then had a winter Fishery there

Whether the Island is now as accessible in the Early Spring as it used to be I do not Know, the Ice is said to have increasd much on the North side of the Island of Late years, but not on the South, the South is said however to be subject in the Spring to dense Fogs which the Natives are afraid of Encountering in their Craft

I Thought it necessary to give you this[2] information for Lord Castlereagh as It is new to me[3] and as I have before given to Lord Hawkesbury all the information I could Collect on the Subject[4]

I am my dear Sir
Your Most Faithfull
Humble Servant
[*Signed*] Jos: Banks

[1] This must be to Edward Cooke, Castlereagh's Under-Secretary.
[2] In the Royal Society draft Banks has added 'statement for the information of'.
[3] Hermannsson believed that this letter was addressed to Bjarni Sívertsen ('Banks and Iceland', p. 48). This is not surprising as the DTC copy, which Hermannsson used, is so annotated by the transcriber. However, the original in the War Office papers in the TNA is not addressed, but simply says 'My dear Sir'. The contents of this letter make it clear that the recipient could be no other than Edward Cooke, Under-Secretary to the Secretary of State for War (Castlereagh), and a frequent correspondent of Banks.
[4] See Banks's memorandum to Hawkesbury, Document 63.

TNA, WO 1/1117, ff. 79–80. Autograph letter.[1]
NHM, DTC 17, ff. 238–239.
Royal Society, MM/6/30.[2] Draft in Banks's autograph.
Hermannsson, 'Banks and Iceland', p. 48.

81. Boulton & Baker to Banks, 7 April 1808

Wellclose Square

Sir

We have for some time been exerting our utmost Endeavours to procure the Liberation of three Iceland Ships, whose Captains, knowing our extensive Connections in Denmark & Norway, have applied to us for advice & assistance in their present distressed Situations: –

From Messrs Petræus & Sivertsen, we have learn'd with what Benevolence & Kindness You, Sir, have espoused their Cause, and that, thro' your Application & Representation, Strong Hopes of Success are held out to them, and we trust the other unfortunate Men will equally benefit thereby. We have now ascertain'd that the whole rests with the Kings Proctor,[3] who says he cannot go into these Cases, untill he knows the Grounds on which your application to the Government was made, in fact, untill we hand him a Copy of the Memorial or Statement you had the Goodness to present on their Behalf. We have to entreat, therefore, you would, do them & us the favour to cause a Copy or the Particulars thereof to be sent either to the Kings Proctor, or to us, whereby we doubt not but your kind Intentions will be carried into effect. The Ships we allude to are

The Nye Prove – Nielsen Master[4]
„ *Bosand* – Aas[5]
„ *Husevig* – Petersen late Bergman[6]

---

[1] Note, verso: '3 March | S. J. Banks | Icelanders'.

[2] On the Royal Society draft (and copied in the DTC) Banks has added:
memoranda
See Collins' Salt & Fishery, 1682*
1405 The Danes forbid our Fishing in Iceland.
1595 Elizabeth ask'd & obtained leave to renew it.
1615. our Iceland Fishery employed 120 ships & Barks Westmony Westman, &c.
*Banks's copy of Collins is in the British Library. Finding no satisfactory account of Iceland, John Collins had talked to the dogger men of the Royal Fishery Company as well as consulting other sources in order to write his own account. Pages 75–90 are devoted to Iceland, containing a general description of Iceland, the Icelanders for instance being described as 'a lusty, comely, affable People ... addicted to Learning' (p. 79). He mentions the weather, imports, exports, as well as the fisheries and the fact that there is no militia in the island. All this would have been helpful to Banks regarding his annexation memoranda. The Royal Society draft differs quite a lot in the use of words but the meaning is the same.

[3] The Procurator-General was his full title. At this time the post was held by Charles Bishop.

[4] The *Nye Pröve*, Hans Nielsen master, was captured on 26 October 1807 by the *Spencer* and brought first to Yarmouth and then to London. The owners were the Iceland merchants Ørum & Wulff.

[5] The *Bosand*, Hans Andersen Aas master, was also captured by the *Spencer* and eventually brought to London. The owner was the Iceland merchant Höwisch.

[6] The *Husevig*, Lorentz Andersen Bergmann master, was captured on 30 October 1807 by the *Banterer* and taken to Yarmouth. The owners were Ørum & Wulff.

We trust our Motives will prove our best apoligies for the Liberty we now take, and Remain with great Respect

Sir,
Your very obed*ient* humble Serva*nt*s
[*Signed*] Boulton & Baker

BL, Add MS 33981, f. 283. Autograph letter.
NHM, DTC 17, ff. 163–164. Contemporary copy.
Hermannsson, 'Banks and Iceland', p. 49.

82. Banks to [Edward Cooke], 7 April 1808

Soho Square

My dear Sir
it was on the 25th of February[1] that you were so kind as to inform me, that Government had decided to Liberate the Iceland Ships & I had then some Reason to Flatter myself that H. M. Ministers had Come to an Earlier decision on the matter than they would otherwise have done, in Consequence of an application from me, Stating in answer to a Reference I had the honor to Receive, that the Cargoes were then incurring daily damage by being detain'd on board the Ships

I was Greivd to Learn yesterday,[2] that their Lordships Generous & Liberal intentions Towards a harmless & virtuous people, at this Moment in serious hazard of Missing their annual Supply of the Necessaries of Life, has not yet advancd one single step towards being Realisd. The King's Proctor as I understand declares that he is unable to Proceed for want of the necessary orders.[3]

as the Season is now advancd, as warm weather has actualy taken Place, which will Quickly Corrupt & Spoil the Stockfish if not unloaded & Placd in airy warehouses & as the actual time of Sailing for Iceland with the Customary necessaries, the Condition on which I Conclude the Liberation is made to depend, is very near at hand;[4] I take the Liberty to Trouble you with this to Request that you will bring this matter to the Recollection of the Good Lord Castlereagh & Solicit his Lordship to give orders[5] that Proper authority may be Issued to Enable the Kings Proctor to Proceed with dispatch, for a Day is now of serious importance and may in Some Cases produce a deterioration of 5 or 6 per Cent in the value of a Cargo, which if it once begins to Ferment will be Quickly & wholly destroyd, Perhaps the Kings Procter has Ears not well in Tune to harmonise with the Tone of Liberation, a hint from his Superiors may I beleive be of use to him[6]

I am my dear Sir
always yours
[*Signed*] Jos: Banks

[1] Actually on 26 February 1808 (Document 77).
[2] He must be referring to the previous letter from Boulton & Baker, dated 7 April, the same day in fact.
[3] 'being issued' is added in DTC.
[4] 'a great deal in comparison' added in DTC.
[5] See letter from Cooke to Banks, dated 13 April 1808 (Document 84).
[6] Note, verso on War Office letter: '7 April S J Banks's | Iceland ships'.

[P.S.] I hope in a Few days to be able to Call upon you I have a multitude of Points on which I Stand in need of your Council.

TNA, WO 1/1117, ff. 87–89. Autograph letter.[1]
Canberra, National Library of Australia, Banks Papers, MS 3/46. Draft in Banks's autograph.[2]
NHM, DTC 17, ff. 161–162. Contemporary copy.[3]
Hermannsson, 'Banks and Iceland', pp. 49–50.

83. Banks to [Boulton & Baker], 7 April 1808

[Soho Square]

Sirs

The King's Proctor[4] must not be [in] a State of Sound Mind if he thinks it Possible that I Should Lay open to any man the Representations I made to H M ministers on the Subject of the Iceland Ships unless Directed so to do Besides it is impossible for me to know whether the Grounds on which the Cabinet came to the Generous Resolution they adopted 6 weeks ago were the arguments usd by me or not.[5]

I have this day written to the Quarter I think most likely to take up the matter,[6] & have most seriously recommended an immediate /Exception to the/ Issue of orders which no doubt have been delayd by the Continual occupation which Parliament has afforded to these [who] guide the State I have represented the Iceland Cargoes as suffering a daily deterioration[7] by the warm weather The near approach of the Season for Sailing to Iceland, & other matters likely to Produce an immediate decision, & have hinted the Propriety of an order by the Kings Proctor to Proceed with more speed than is usual in his office. I cannot but hope that the Effect of this Letter will be felt in a Few days if another week Elapses before the Kings Procter will stir, do me favor to let me know.

BL, Add MS 33981, f. 284. Autograph letter, draft.
NHM, DTC 17, f. 165. Contemporary copy.
BL, Add MS 56301 (3), f. 67. Warren Dawson transcript.
Hermannsson, 'Banks and Iceland', p. 50.

[1] Note, verso: 'April Sir J Banks's | Iceland Important'.
[2] This draft has the same meaning but in places different wording.
[3] The syntax has been improved but this is a copy of the Canberra draft.
[4] Charles Bishop.
[5] See the letter from Edward Cooke to Banks, Document 77.
[6] The previous letter, to Edward Cooke, who would refer the matter to Lord Castlereagh.
[7] Dawson writes 'degeneration' instead of daily 'deterioration'.

84. Edward Cooke to Banks, 13 April [1808]

Downing Street

Dear Sir,
I inclose to you the Result of my Inquiry respecting the Iceland Vessels upon which I shall be obliged for your sentiments.[1]

Believe me dear Sir
your most faithful
& humble Servant
[*Signed*] E Cooke

SL, Banks Papers, I 1:78. Autograph letter.

85. Bjarni Sívertsen to Banks, 18 April 1808

Wellclose Square

Sir
Great as the kindness is that I have experienced from you, and the pains you have so benevolently taken to do me good (but not only me, my fellow-citizens my countrymen in general) yet it is my misfortune to be still in such a situation, as to have no other resource left me, than another appeal to your kind protection, and one more call upon your philanthropic intercession. The order of the British Government for the liberation of Icelandic Ships meets with a very slow and very difficult execution, because no ships are considered as Icelandic but such as belong to persons really settled in, or natives of the country.[2] Accordingly no more ships have till this time been released, than the *Two Sisters*[3] belonging to me, and the *Seien*[4] belonging to Mr. Petræus, and very few more will share the same good fortune if the present plan is persisted in, which will almost totally defeat the benevolent intentions of your government towards my country, as it can not possibly by two or three ships be provided with necessaries.[5] But, Sir, what particularly makes me now encroach upon your time, is another difficulty that has lately started, which threatens

---

[1] This letter does not appear to be extant. However, Cooke wrote to the Lords of the Treasury, 'respecting the Restoration of Iceland Vessels; and stating that Lord Castlereagh is of opinion that the measure of Restitution should have Effect with respect to such Vessels only or Cargoes as are the Property of the Inhabitants & are bona fide residing in Iceland'. He was asked to act accordingly (TNA, T 15/1, Harrison to Charles Bishop, the King's Proctor, 13 April 1808). Thus the ships belonging to the inhabitants of Iceland were to be restored, thanks to Banks's intervention.

[2] Most of the Iceland merchants actually resided in Denmark during the winter, visiting Iceland during the summer trading season and employing factors during their absence. They were, however, formally 'burghers' of their trading districts. The ships were eventually released on the orders of Lord Castlereagh (Secretary of War), dated 23 June 1808. Earl Bathurst, the president of the Board of Trade, had urged him to do so.

[3] The *Tvende Söstre* (Two Sisters) was captured on 19 September 1807 off the Naze of Norway and taken to Leith.

[4] The *Seyen* was captured on 3 September 1807 on her way to Bilbao.

[5] In the years 1788–1807, 56 merchant ships sailed on average to Iceland with an aggregate tonnage of 2,300 commercial tons (*kommercelæster*, abbreviated kmcl: see 'Currency, Weights and Measures'). See Rubin, *Frederik VI's Tid*, p. 118. After 1802 the number had steadily declined; in 1807 42 ships applied for licences to sail to Iceland.

the destruction of every hope I have entertained for myself and my country. We can not obtain the liberation of our people. An application to that purpose has been made to the Admiralty through the Transport Office, but the answer has been an absolute refusal to liberate a single man.[1] If the British Government can not be induced to relax from this severity, it will certainly be impossible for us to get away with our ships, and the great favour bestowed upon us and our country, which you, Sir, have taken so much trouble to procure us, will be almost entirely destitute of any benefit to either. Respecting this distressing circumstance I most humbly beg leave to ask your advice what step I can properly take, and to implore your kind intercession in our favour, if you can possibly grant it in this case.[2] – If through your kind recommendation this difficulty should be happily removed, the question is still undecided, where we can go to with out cargoes; for to dispose of them here is utterly impossible, as the greater part of our goods will not in this country fetch half of their original price, and what little can be sold to least disadvantage, will hardly be sufficient to cover our expenses on the ships, and discharge the debts we have contracted. – If therefore Iceland shall derive any relief from our liberation, it will be highly desirable, and indeed absolutely requisite for us to obtain licences to proceed to some Danish or other port, where a market could be found, and again to return from thence with a cargo of provisions to Iceland. How far this could possibly be attained I also venture, with the greatest submission, to solicit your opinion.[3]

Humbly requesting your pardon, and hoping your kind forgiveness for thus troubling you again and again with our unfortunate affairs, I remain, Sir, with the deepest sense of gratitude

Your most humble and obedient
Servant
[*Signed*] B: Sivertsen

SL, Banks Papers, I 1:20. Letter, written by an Englishman, and signed by Sivertsen.

[1] As usual Banks came to the rescue. On 25 May 1808 Alexander McLeay, Secretary of the Transport Office (see Appendix 13), wrote to Sívertsen that whenever the Commissioners for His Majesty's Transport Service would be 'officially informed of the Danish Ships belonging to Burghers and Inhabitants of Iceland being ordered to be restored their Crews will be ordered to be immediately released'. On 2 June 1808 McLeay again wrote to Sívertsen, that 'Orders will be given to the Agents for Prisoners of war at Peebles, Ashburton, Greenlaw, Plymouth and Greenlaw, to release such of the Crews of these Ships in Question as are at these places ...' But Sívertsen's troubles were not yet over. Some of his crew had made their escape, as on 6 August 1808 McLeay informed him that enquiries had found 'that [as] no Persons belonging to your Ship the *Tvende Söstre* have entered into the British Service, no Danes can be released in the room of the Three referred to by you'. All documents, TNA, ADM 98/308.

[2] For a change Banks was not enthusiastic. His attitude here was that of an Englishman faced with enemy Danes – a totally different breed from his favoured Icelanders. Sir Joseph felt that only a sufficient number of Danish seamen should be released as was necessary to sail the ships from England to Denmark, where extra men could be taken on for the voyage to Iceland. So it was not until November that Hans Thielsen was permitted to go to Chatham to select a crew for his ship (TNA, ADM 1/4979, Preidel to the Admiralty, 28 November 1808). The result was that no ships made the voyage to Iceland in 1808, a dozen in 1809, but Thielsen did not return to Iceland until 1810, as it was not until the Order in Council of 7 February 1810 that those still imprisoned were released. See further Agnarsdóttir, 'Great Britain and Iceland', pp. 67–70.

[3] Here Sívertsen is asking that the Iceland trade be incorporated into the British licensed trade. Banks was instrumental in the wording of the licences (see following note). He eventually persuaded the government to release the Icelandic merchant ships that were technically enemy vessels. They were issued with British licences and permitted to carry on their traditional trade with Denmark in relative safety.

86. Banks: Note on Licences, [undated, April 1808?]

after the words to proceed from Leith without unloading[1]
– with liberty to proceed in Ballast from the port of her discharge to the port of her loading
– from Denmark to proceed direct to Iceland with liberty to touch at Leith to compleate her cargo
– To return to Leith or London with a cargo of all kind of Iceland produce
– As the Captains might be altered it would be very convenient if the names were left out.
– The ships can not return to Leith from Iceland before the autumn which will make the whole voyage last about 9 months. –

[*Added in Banks's autograph*]
a Licence for a Ship to be purchased in Norway or Sweden to proceed in ballast to some Port on Norway North of the Naze to take on board Timber /*in Norway to proceed from thence to England for inspection & from England*/ and to proceed from thence to Iceland /*in some port North of the Naze & to proceed from thence to Iceland*/.
They declare that they will not make use of Ships of more than 300 Tons Term[2] & from thence to proceed without unlading having Completed her Cargo on application for that purpose.

SL, Banks Papers, I 1:43. Draft by Banks and an amanuensis.

87. John Christopher Preidel to Banks, 20 April 1808

N. 23 Great Winchester Street

Sir!
The *Rentheir*, Hans Theilsen Master bound from Iceland to Coppenhagen was captured off that place & sent to Yarmouth in October last,[3] & on the 15. December following was condemned in the High Court of Admiralty as Prize to His Majesty which Condemnation through your kind Interference in Behalf of the unfortunate Icelanders was suspended till Government should have decided whether that Ship & Cargo should be restored to the Owners or the Condemnation be put in Execution.

Having been informed that Government had given Orders to give up & liberate the Iceland Ships & Cargoes, I applied to the King's Proctor for the Liberation of the above Ship & her Cargo, but was told that the *Rentheir* did not come under the Orders which My Lord Castlereagh had given. That the *Rentheir* could not be considered as

---

[1] Banks was instrumental in drafting the licences permitting the Iceland–Copenhagen trade to continue, as this note shows.
[2] 'Term' here means limit.
[3] There are several spelling mistakes in this letter. On the *Rennthier* (Preidel always misspells the name) see Document 54, and p. 268, n. 6 and p. 270, n. 1.

Iceland, but as Danish Property – because the Master of her at the time of his Examination had deposed that besides himself & Mr. Boy P. Holdt of Flensburg, several others in Danemark were part Owners.[1] To convince myself of the facts [I] requested to see the Depositions but was refused. But I am given to understand that the Kings Proctor has applied to my Lord Castlereagh for further Explanations of his Orders.

meanwhile as my friends in Yarmouth inform me that the Act of Condemnation has been sent down to the Danish Commissioners Agent, who in a few days would proceed to the Unlivery of the Cargo & most likely soon after to the Sale of the Cargo & the Ship. To the Unlivery of & sale of the Cargo I do not object on the part of the Owners, as it will save it from spoiling & diminishing in its intrinsic Value, but should be sorry if the Ship was sold.

I therefore implore your kind Interference in this Matter, & as the *Rentheir* stands on the List of Iceland Ships & has been represented to You Sir, as such by the Deputies from that Island, & I have reasons to belive there is amongst the Ships papers (to which I am refused to have Access) a Burger Breif shewing that Capt*ain* Theilsen[2] & Mr. Boy P. Holdt[3] are Burgers of Iceland; that thro' your Interest with my Lord Castlereagh, orders may be given to the King's Proctor to give up to the Proprietors said Ship and Cargo.

The Master and Mates private Adventure has been restored and at the time he claimed the same upon Oath, he likewise declared being one fifteenth part owner in the Ship & Cargo. Should it so happend that on Account of the Objections made in the Admiralty the Condemnation should not be rescinded then I hope Sir that thro your Influence at least poor Capt*ain* Theilsen's $\frac{1}{15}$. Share may be paid him out of the Net proceeds Ship & Cargo may sell for.

I am most respectfully.
Sir!
Your most obedient & humble
Servant
[*Signed*] J. C. Preidel

P.S. I will do myself the Honor of waiting on Sir Joseph Banks tomorrow.[4]

SL, Banks Papers, I 1:21. Autograph letter.

---

[1] Boy Petersen Holst and 16 others from Flensborg in Schleswig owned joint trading establishments in Reykjavík and Hafnarfjörður.

[2] Hans Thielsen had become a burgher of Reykjavík on 25 May 1789, his son of the same name on 13 January 1802 (Björnsson, *Kaupstaður í hálfa öld,* pp. 53, 58).

[3] There is no record of Holst obtaining a burghership, at least not in Reykjavík.

[4] An undated note among the Banks papers in the Sutro Library (I 1:16) touches on this case, Document 55.

88. Boulton & Baker to Banks, 29 April 1808

Wellclose Square

Sir

We felt highly gratified by your early & very kind Reply to our Letter of the 6th Inst:[1] respecting the unfortunate Iceland Captains who had applied to us for Advice & Assistance in their present lamentable Situation.[2]

We had hoped that the Applications & Recommendations you were so humanely making to the higher Powers, might have induced the King's Advocate[3] & Proctor to have looked with a favorable Eye on the Statement & Affidavits which we had caused to be laid before them. –

But we are sorry to say that we are now inform'd by our Proctor, that the Opinion given is <u>unfavourable</u>,[4] and that we have little hope of Success, in consequence of the Owners being now, to the best of our Knowledge, in <u>Copenhagen</u>, whither they had proceeded in the usual course of their Trade, to dispose of their Goods, & lay in such Supplies as they proposed to take back with them. –

When, therefore the Captains were asked, where the Owners were they replied, <u>in Copenhagen</u>, nor were then aware that any further Explanation as to their /usual/ <u>Residence</u> was needful: – and now altho' very clear & full Affidavits have been deliver'd in, They are thrown on their backs, by referring to their <u>Depositions</u> when <u>examin'd</u>, and told the Proofs are insufficient. – Messrs <u>Örum & Wolff</u>,[5] appear to have left Iceland with three Ships; the <u>Eskefiord</u> on board of which was <u>Wolff</u>,[6] supposed to have got into Norway or Denmark; the <u>Husevig</u>, with <u>Örum</u>[7] on board taken by Ad*mir*al Gambier's Fleet, himself & the Captain put on shore at Copenhagen; and the /~~Bosand~~/ <u>Nye Prove</u> taken & brought into Yarmouth; whither also, Ad*mir*al Gambier sent the <u>Husevig</u>, and where the <u>latter</u> still remains; but the /~~Bosand~~/ <u>Nye Prove</u> was brought round to London. – We humbly conceive that these facts, proved by the Captain of the <u>Bosand</u> and one or two more Persons now here, establish the Fact of the Property being <u>Iceland</u>.[8] –

The /~~Nye Prove~~/ <u>Bosand</u>, also brought into Yarmouth & since removed to London belong'd to a Mr. Howitz[9] who had been establishd in Iceland above Twenty years, and was /also/ in the Habit of coming annually to Copenhagen, & returning in the Spring, until the last two Winters, when from violent Rheumatic Complaints which threatened his Life had he return'd to so cold a Climate, he had been obliged to remain at

---

[1] Presumably Document 81.
[2] i.e. Hans Nielsen of the *Nye Pröve*, Hans Andersen Aas of the *Bosand* and Petersen of the *Husevig*.
[3] The King's Advocate was Sir Christopher Robinson (1766–1833).
[4] All underlining is in the letter itself.
[5] Ørum & Wulff had at this point trading establishments in Húsavík in the north (since 1804) and Eskifjörður in eastern Iceland (since 1798). From 1799 to 1805 they also conducted the trade of Seyðisförður (eastern Iceland).
[6] Jens Andreas Wulff was a merchant in Eskifjörður from 1798.
[7] Niels Ørum.
[8] This is certainly true. All the merchants mentioned had trading establishments of long standing in Iceland.
[9] Johan Gottfred Höwisch had been trading in Iceland from c. 1780 and now owned the trading establishment at Hofsós in Northern Iceland. He was among the wealthier merchants trading in Iceland.

Copenhagen: The Captain has known him <u>seventeen Years</u>, and swears to these Facts. – Mess'rs <u>Örum & Wolff</u> /& Howitz also/ have not so much as a <u>House</u> in Copenhag'n whereas in Iceland they have very extensive landed Property, and Establishments; – raising, preparing, & shipping their Cargoes entirely from their own Possessions and with their own People. – If nothing in the <u>Ship's Papers</u> contradicts the Affidavits which these People have leisurely & solemnly sworn to, 'tis hard, very hard, that we /*should*/ are now to be told we must bring <u>farther Proof</u>, when no Communication can be had with the Parties! –

Could those in the Administration of Government be induced to look at these Affidavits (now before the King's Advocate). We firmly believe the Result would be favourable to us! but this we fear is impracticable. We entreat your Pardon for intruding so largely on your time & attention, and remain

Sir Your obliged & obedient humble Servants
[*Signed*] Boulton & Baker

SL, Banks Papers, I 1:22. Autograph letter.

89.   D. B. Baker to Banks, Wednesday Morn'g [29 April 1808]

Wellclose Square

Sir:

Having thro'out being very desirous of the Honor of an Interview with your good self on the Subject of the Iceland Property, I the more lament being now prevented from embracing so favorable an Opportunity: – But I am not able to leave my House. –

The Bearer Mr. Boyesen, one of our Counting House, a Norwegian by Birth, but well vers'd in the English Language, will be able to give all the Information we possess as well /as/ or indeed <u>better</u>[1] than, myself; – as he always attended the Captains at the Commons[2] as Interpreter, assisted in framing & translating their Affidavits, and had many long Conversations & Consultations with them on the Subject both <u>here</u>, & at our <u>Proctors</u>. –

I trust you will find him capable of affording all the Information required, – and remain Sir

Your obliged & obedient Servant,
[*Signed*] (for B & B[3]) D. B. Baker

<u>He is rather deaf.</u>

SL, Banks Papers, I 1:23. Autograph letter.

---

[1] All underlining is in the letter itself.
[2] Doctors' Commons was a building in Paternoster Row in London, housing the Admiralty court as well as ecclesaiastical courts.
[3] Boulton & Baker.

90. [Note written down by Banks in conversation with Boyesen,[1] 29 April 1808]

Wellclose Square

<u>Condemned</u>: *Bosand Nye Prove Husevic*[2]
have Copenhagen Burghers Letters Licence of Gen*eral* Trade to Trade to Iceland a B*urgher* of Copenhagen must have 6000 Dollars & Iceland Burghers Letters[3]
Persons who have no establishment or even house in Denmark but who have Establishments in Iceland.[4]

SL, Banks Papers, I 1:23v. Autograph note.

91. Boulton & Baker to Banks, 16 May 1808

Wellclose Square

Sir

As we took the Liberty on the 29th Ult*imo*[5] to lay before you all the Particulars that had come to our knowledge respecting the three Iceland Ships whose Capt'ns had applied to us for our Advice & Assistance, we should not have presumed to have intruded on you again on this Subject had not our Proctors[6] this Instant sent us notice that, 'on the List of Motions for Condemnation at the Adm*iralty* Court <u>tomorrow</u>, they observe the <u>*Nye Prove*</u>, <u>*Bosand*</u>, & <u>*Husevig*</u>,[7] and require our Instructions.' –

We have desired Counsel may be instructed to <u>oppose these Motions</u>, and at least obtain some Respite, in hopes the Interest you have so Humanely manifested towards these Men, may work out something for their Benefit. –

We should feel greatly obliged by your favouring us with a few Lines, & giving us your Opinion & Advice as to any Hope of Success you may entertain, or any Step you could recommend to us: –

If it is not the Intention of Government to relax at all from the general Rule, in favour of the Iceland Ships,[8] it is useless for us to throw away Money in <u>Litigation</u>, which we have no means of recovering: – But we would not abandon these poor fellows, while any

---

[1] See previous letter. The representative of Boulton & Baker.

[2] See following letter.

[3] The Danish merchants and captains sailing to Iceland should all have had Copenhagen burgher letters *and* Iceland burgher letters. An Iceland burgher was expected to own property worth 3,000 rixdollars ('Anordning ang. Kjöbstæderne paa Island', *Lovsamling for Island*, V, p. 345). See Introduction, p. 15 above.

[4] That would apply to the native Iceland merchants. They, however, had to be in company with a Danish merchant.

[5] See Document 88.

[6] Those lawyers who conduct the cases in court.

[7] The *Bosand* (Aas), the *Husevig* (Bergman) and the *Nye pröve* (Nielsen) were indeed all condemned to the Crown, both ship and cargo, on 17 April 1808 (RA, KK 1797–1816, Handels- og Konsulatsfagets sekretariat. Fortegnelser over de til England opbragte danske skibe, 1807, 1810–11 og udat).

[8] What they are referring to is the rule that the only ships liberated must belong to *native* Icelanders. The three ships mentioned belonged to Danes owning trading establishments in Iceland.

Hope of Success remained. – Once more entreating your Pardon for being thus importunate & troublesome.

> We remain, with great Respect,
> Sir
> Your very Obedient Servants
> [Signed] Boulton & Baker

SL, Banks Papers, I 1:24. Autograph letter.

92. Banks to Boulton & Baker, 16 May [1808]

[Soho Square]

Sir

On the Receipt of your Last Letter I made an immediate application in Recommendation of an Extention of the British humanity in Favor of my Iceland Friends but I have not yet owing to stress of Business receivd any answer. on the Receipt of yours of this day, I /*wrote instantly*/ have written a Letter Requesting that if any further indulgence is intended the Proceedings against the three Ships you mention may be /*stopd*/ delayd[1] till I am allowd to urge further arguments which Letter will Reach its destination[2] this Evening I Sincerely hope it will produce its Effect

SL, Banks Papers, I 1:24. Autograph letter, draft.[3]

93. Boulton & Baker to Banks, 20 May 1808

Wellclose Square

Sir

We feel greatly obliged by your immediate Attention & Reply to our Letter of the 16th,[4] – and have now merely to add, that the 3 Ships & Cargoes, respecting which we have thus often troubled you were condemn'd in the Admiralty Court on the following day:[5] – the perishable or rather <u>perishing</u> state of the Cargoes being advanc'd as an Argument for their immediate Disposal.

Still the <u>Nett Proceeds</u> thereof, as also the <u>Ships</u> and Crews, might be restored to the original Proprietors, should Government feel so inclined. – If you should think a Petition to His Majesty in Council, or the Lords of the Treasury, would be

---

[1] See previous letter.
[2] Banks wrote to a member of the Cabinet, probably Castlereagh as usual. Banks, however, was too late. The following day the ships and cargoes were condemned in the Admiralty Court (see following letter).
[3] This draft reply is written on the next empty page of the previous letter from Boulton & Baker.
[4] Document 91.
[5] The *Nye Pröve*, *Bosand* and *Husevig*.

prudent or adviseable, we will, on being favour'd with your Reply, cause one to be made out.¹

We remain with great Respect,
Sir
Your very obedient Servants
[*Signed*] Boulton & Baker

SL, Banks Papers, I 1:25. Autograph letter.

94. Banks to Boulton & Baker, 20 May 1808

[Soho Square]

Gentlemen
as the Letter I wrote of which I advisd you in my Last² has not producd any effect I am obligd to Conclude from that Circumstance that my intercession in favor of an extention of the mercy of the Cabinet towards the Iceland Ships is not acceded to. I cannot therefore advice you to take any further steps for as I took pains to have the Question fully & Fairly brought into the view of H.M. ministers I have little hope that the Resolutions they have passed on the Subject will be changd

SL, Banks Papers, I 1:25. Autograph letter, draft.³

95. Jörgen Jörgensen to Banks, 27 May 1808

London

Sir!
Happening the day before yesterday to hear you being very anxious if possible to relieve some of the distress of the poor Ice and Greenlanders,⁴ I⁵ have inclosed sent you the following, entirely for your own inspection.⁶

I do not perceive that the inhabitants in those parts can receive any relief, but in the manner inserted, I have a litle knowledge of their trade and way of living; what I have inserted may not be very correct and practicable, and could therefore wish,

---

¹ Banks advised them not to; see following letter of the same date.
² Document 92.
³ This is a draft on verso of previous letter.
⁴ Bjarni Sívertsen, who was still in London, was doubtless Jörgensen's informant. Jörgensen used to go to the Royal Exchange, the meeting place of Danes and Icelanders, where he met Sívertsen (see BL, Department of Manuscripts, Egerton [hereafter Eg.] MS 2067, ff. 221–222, and Eg. MS 2068, f. 3). Even if Banks did not discuss his plans for annexation openly with the Iceland merchants – though there is little reason to believe he did not, considering the fact that he needed their co-operation – the questions asked (Document 50) were of such a nature that the Icelanders could not have failed to reach that conclusion.
⁵ Jörgensen, in this letter, always writes I as i and uses umlauts. For the sake of clarity this has been amended.
⁶ See the following document.

that anything of the kind, may not be known as coming from me.¹ I am with the greatest respect

Your humble ob*edient* Servant
[*Signed*] Jörgen Jörgensen

SL, Banks Papers, I 1:48. Autograph letter.

96. Enclosure to Banks: Jörgen Jörgensen's proposal to help the Icelanders and Greenlanders.

The best and most effectual way² to procure relief to the Ice and Greenlanders would be for the English to take posession of those places,³ which formerly did belong to the King of Dannemark, but this would be attended with no benefit to England, as the expense would be very great, for if the Ice and Greenlander and inhabitants of David's⁴ Straits were supplyd from this country, they would have no gold, silver nor any kind of merchandize to return. The fish procured would not be of any Service to the country; the coarse stockings, mitts and other woollen goods would here be of no kind of Service, the salted mutton which is cured in Iceland would likewise here be useless. The oil procured in Greenland and in the Straits would here sell for little or nothing, as the price of black oil is but very low.

The only means of subsistence for those places then is their woollen goods, salt mutton, fish and oil, with which they procures from Dannemark in return, <u>Corn, Iron ware, coarse linen</u>, and other of the most common necessaries of life.

Now as the present state of affairs will not allow those inhabitants to trade with Dannemark in the usual way, they will in a short time be reduced to the greatest misery and want. The Ice and Greenlanders then as subjects of the danish government have no right to look to England for redress, but on the other hand as those poor people suffers as much by the war as if they actually were blocaded, it will appear that the english ought to take possession of Iceland and the colonies, which would remove their misery, but this is [*sic*] before observed will be attended with inconvenience and expense, without any benefit to England. Thence it appears that those colonies is in a more lamentable state than any other country, which is actually threatened with invasion, & or in a state of blockade, for in those cases the inhabitants, may get relief by throwing themselves under the protection of their very enemies. But here the case is different, they are full as badly off, as if they were really invaded or blocaded without having a power of throwing themselves in the arms of the ennemy, for what would be the consequence, if they were to send to England and say: come take possession of our places and we will obey you. In the first place english Government might perhaps not accept the invitation, and next, after a general peace was once established, the inhabitants would be threated as traitors and punished accordingly.

Now as it appears that England will not suffer them to trade to Dannemark, nor take posession of their colonies, they ought in a manner to procure them relief, which can only

---

¹ Probably because of his status as a Danish prisoner-of-war and thus what he was up to was treasonable.
² As was the Danish custom, Jörgensen always puts an umlaut over the *y*.
³ Exactly as Banks had planned to do (see Documents 45, 62 and 63).
⁴ Davis Strait is the correct term.

be effected by permitting the Ice and Greenlanders to send their small vessels to Dannemark for such absolute necessaries, as might be wanted. The situation of Iceland will allow the people to send their vessels to the northern parts of Norway, without fear of being captured; but there they will get no corn, no linen, no fabricated iron work, as the Norwegians are in want themselves.

Permitting the Icelanders and colonies /to send vessels to Dannemark/ will not be attended with any inconvenience to England, but some good may result from it. It may be done on certain conditions.

1.) That no Ships or Vessels, except as those immediately belonging to the colonies shall bring necessaries from Dannemark.

2) That such Ships or vessels do not bring any goods or merchandize from Dannemark, but such as actually is necessary to the support of the inhabitants, and such articles as shall be permitted by the government of Great Britain to be sent to these.

3) That no Iceland or colonial vessel do send any where for goods, except in such places, as shall be named by the british government.

4) That no Ships from the Colonies either to Dannemark or from Dannemark, attempts to go in to any Norwegian Port, of any other harbor, except in case of emminent distress, nor to any other port, except that which is specified in her papers.

5) That if any Ship or vessel from the colonies steers any other course, but that directly for her port of destination, she shall be liable to capture, and taken in to a british port, except such steerage is occasioned by bad weather, contrary winds or distress.

6) That if any ship or vessel should attempt to go any where but to her destined port, or if she should carry any other goods, but what's permitted, or in any other manner deviate from the common rules, she shall then be liable to capture, and deemed a lawful prize to the Ship by whom captured.

It will not be any injury to England, that Dannemark parts with corn, which the danes wants so much themselves, and in other respects much good may result from Ships going forwards and backwards, and by those means in these very Ships much british goods may find its way in the northern countries. For as those Ships are too far distant from France, french influence will not do any material hurt to the trade, and the vessels and Ships from the colonies may in a manner be deemed neutrals.

I have observed before, that it can not be expected, that Iceland and the colonies will unsolicited throw themselves into the arms of Great Britain, so it will be necessary for a couple of english fregats [frigates] to shew themselves on the coasts to enforce the inhabitants to do a thing beneficial to themselves, without exposing themselves to the displeasure of their own government.[1]

SL, Banks Papers, I 1:49. Autograph.

---

[1] This letter is significant as it demonstrates a direct link between Banks and Jörgensen, the Revolutionary Chief of 1809 (see Introduction, p. 32 above), on the subject of a British annexation of Iceland. Like Banks, Jörgensen saw the only solution for the ills of the Icelanders in a British annexation, agreeing that the benefits to England would not be great. It must be assumed that Jörgensen was fully aware of Banks's interest in annexing Iceland, doubtless from his informant, the merchant Bjarni Sívertsen. Jörgensen's ideas are quite simply the same as Banks in his 'Project' (Document 45), note especially the last paragraph here with the mention of frigates and 'voluntary' surrender.

97. Banks to [Earl Bathurst], 27 May 1808

[Soho Square]

In Obedience to your Lordships Commands,[1] I have made Enquiries into the Situation of the Iceland Vessels detained here, of which the following is the Result.

Sixteen ships that had cleared out from Iceland have been brought in; Fifteen of these were taken before the Commencement of Hostilities. One has been since captured by a Cruizer.[2]

Of the Fifteen Ships in which the Captors have no vested Interest,
Four have been Liberated:
Three have been condemned:
Eight, being under similar Circumstances with the Last, must be condemned in a short Time, unless His Majesty's Ministers are pleased to interfere in their Favour.

The Four Ships that have been liberated, are the Property of Native Icelanders, who in some, if not in all Cases, were found on Board them, on their Passage to Copenhagen, or other Places, for the Purpose of selling their Cargoes, and purchasing proper Returns, to be carried back to Iceland in the Spring.

The Owners of the Ships that have been condemned are Danes, whose Houses of Residence & Mercantile Establishments are in Iceland; but who are in the Habit of going very Frequently to Denmark with their Ships in Autumn, to dispose of their Cargoes, and purchase others with which they return to Iceland.

In Two of the Cases of Ships condemned, the Affidavits sworn to in the Court declare, that the owners are Burghers of Iceland – have considerable landed Property there – have Houses there, where they reside with their Families; and when absent, leave their Business in the Hands of Clerks; and have no Houses in Copenhagen, but live in Lodgings there, hireing [*sic*] a Warehouse for their Merchandize.

By the Regulations of the Danish Government, no one is allowed to trade with Iceland, unless he is a free Burgher of the Island, and possesses Property there to the value of 6000 *Rixdollars*. Nor can any One trade in Copenhagen, unless he is a free Burgher of that City; On this Point the King's Advocate seems to have laid much Stress, in the Condemnations that have taken Place.

Should it please His Majesty's Ministers to extend their humane Dispositions towards the helpless Inhabitants of Iceland, by liberating the Ships belonging to this last Description of Owners, there is every Reason to believe, that the whole of them are entirely desirous of investing all the Proceeds of their Cargoes, in Necessaries for Iceland, & proceeding there as soon as possible, and of returning from thence to England, provided they can be protected from British and from Swedish Capture.

It is clear however, that the whole of the Proceeds of these Fifteen Ships, can by no means raise a Sum sufficient to provide, even a moderate Supply, of the Necessaries the

---

[1] This letter is to be found in the papers of the War Office, where Earl Bathurst sent it on Monday 30 May 1808 in the evening, calling it the answer to the letter he wrote Banks on the subject of the Icelanders. He added: 'I am very much for the remaining Vessels being released – Their value is inconsiderable: and their release may be advantageous. When they are released, they will not be able to depart without a licence; and the conditions on which we shall give them licences will be such as to occasion them going <u>direct</u> to Iceland' (TNA, War Office [hereafter WO] 1/883, f. 347).

[2] The *Charlotte Amalia*: see Document 99.

Icelanders are accustomed to receive from Denmark: The usual annual Importation of Farinaceous Food alone, amounts to about 15,000 Quarters of Rye, which they receive chiefly in the Form of Rye Meal, but also in that of Biscuit Grotts [groats], Rice &c. The liberated Icelanders are of Opinion, that all the Cargoes of the detained Ships, will not be sufficient to Purchase one Half of this necessary Supply, when the Expences of Law, and the Fees of Detention are deducted; But even this Half they say will be a Relief of the utmost Importance to the Icelanders, who if they are left wholly unsupplied during the ensuing Winter, must experience the deepest Distress, and indeed many of the horrors of Famine.

Nothing can be less suited to a favourable Disposal of their Cargoes, than the present State of the British Market; Stock Fish, the chief Article, has not for some Centuries been at all consumed here; and salted Mutton (which they call Beef) though esteemed an Article of Luxury in Denmark, is utterly unknown in England: This may, it is hoped, be sold to some Advantage for Exportation, as the Swedes are likely to be acquainted with the Use of it; But as the Season for consuming Stock Fish, the Winter, is passed; and as this Article is believed to have sustained much Damage by the late Wet Weather, the Advantage to be derived from the Sale of it, must be very small.[1]

[*The next section is on a loose page within the Sutro draft.*]
There were in all 15 Iceland ships brought in before the declaration of war & one after [*Marginal note*: The Crews should be Liberated They are Danes]
      Two Sisters Sivertsen
      Johanne Charlotte
      Sejen, Westy Petræus.
<u>Liberated</u>
      Bedre Tüder
may be liberated by the Rules of the Court but Capt. Knutsen is on his Parole and has not appeard.
[*Marginal note*: alleviation of expences]
      3 Condemned
      Nye Prove
      Bosand
      Husewig
      11 in all condemned.
[*Marginal note*: Cargoes deteriorate]
      3 to 4 £ Sterling a barrell.
      30,000 Barrells.
      15,000 Quarters.
      £ 12 ——lb. 320 wool
      10 ——

Sejen's cargo was sold in Feb last but no information Relative to Price can be obtaind. [*Loose page in Sutro draft ends here.*]

Her cargo is wool, tallow, <u>Salted Beef</u>, Brimstone &c.

---

[1] Here the copy in TNA: WO 1/883 ends (with no endorsement), ff. 351–55. The draft in the Sutro Library continues.

[*Marginal note*: *Nye Prove*, Hans Nielsen, Master of Eskefiorde in Iceland]

Niels Ohram[1] of Eskefiorde & Jens Wulff of Husewig in Iceland are the true owners of the ship and cargo.

Niels Ohram has been merchant & inhabitant of Iceland more than 16 years & Jens Wulff more than 7 years.

They?[2] in purchasing an Estate in Iceland to the extent of 12 English miles & have their Dwelling houses & Establishments there. The Cargo is the Produce of this Estate.

Niels Ohram with his son & only child saild from Iceland on board the Ship *Husewig* belonging to Ohram & Wulf in Oct last in order to sell the Cargoes of the *Nye Prove*, the *Husewig* & the *Eskefiord* belonging to Ohram & Wulf & to purchase third ship to extend their business. He was Captured on board the *Husewig* off Copenhagen & sent in Shore there & the *Husewig* sent to Yarmouth. His wife had been left by them at Copenhagen in Autumn 1806, & it was his intention to Return again to Iceland in the Spring with his wife & the ship he intended to purchase.

That Jens Wulff was married in autumn 1806 at Copenhagen & in Spring 1807 Returnd to Iceland to prepare for the Reception of his wife and was still there when the *Nye Prove* saild from thence but intended to Sail in the *Eskefiorde* & has we beleive arrivd at Copenhagen. That his intention in visiting Copenhagen was to sell his Cargoes & bring back his wife & he has no doubt that both men Ohram & Wulff will Return to Iceland as soon as they are able, they having no house of Trade or Establishment at Copenhagen but Living in Lodgings there.

Their business in Iceland is conducted by Clerks, they are Burghers of Iceland & also of Copenhagen.

That had he well understood the Preparatory Questions of where Mess. Ohram & Wulff resided, he should not have answered as he did, that they resided at Copenhagen & their house & Establishment is in Iceland & their residence at Copenhagen only temporary.

The Invoice amount to	*Rixdollars*	10339.58
Capt*ain* Nielsen's		5904.50

The whole at Prime Cost.

[*Marginal note*: Bosand Hans Andersen Aas Master of Hofsos, Iceland]

A cargo of Tallow, Salted mutton (Calld Beef), hides, &c. That I. G. Howiesh[3] of Hofsos is the true owner of the Ship & her apparell & of the whole Cargo on board.

That J. G. Howiesh is owner of land & houses in Iceland & has his own dwelling house & Establishment there. That he is the owner of three Ships & usualy Sails with one of them every year to Copenhagen & having sold the Cargoes of his Ships & Provided them with Cargoes, returns with them to Iceland & that he has no house of Trade at Copenhagen or other Establishment save a Lodging for himself and a Cellar for his Goods.

It is uncertain whether he is a native of Iceland but he has been Establishd there 17 years & is beleived to have possessed his estates there more than 20.

---

[1] Ørum is the correct spelling.
[2] Difficult to decipher.
[3] Höwisch is the correct spelling.

The Cargo is the Produce of these Estates. /*Howiesh is a Burgher of Copenhagen & also*/ Howeish is a Burgher & inhabitant of Iceland & also a Burgher of Copenhagen.

By the Danish Regulations no one can Trade to Iceland unless he is a Burgher of Iceland & Possessd of Property in Iceland to the /*amount*/ value of 6000 Dollars,[1] but no one can trade in Copenhagen or trade as a Danish subject with any other Place unless he is a Burgher of Copenhagen.

That J. G. Howiesh went from Iceland to Copenhagen in the Autumn 1805 with the intention of Returning, but being afflicted with the Rheumatism he continued there for advice and did not Return to Iceland either in the Spring 1806 or in that of 1807. His health is not yet re-established but the Depon*en*t verily beleives he intends to Return as soon as he is well. His business in Iceland is conducted by Clerks.

He in his preparation Examination understood the Question put to him he should have answerd it otherwise as Howiesk's real residence & Establishment is in Iceland & not at Copenhagen where he is only a temporary inhabitant.

TNA, WO 1/883, ff. 347–349 [incomplete]. Contemporary copy.
SL, Banks Papers, I 1:26.[2] Draft in Banks's autograph.

98. Hans Georg Bredal to Banks, 31 May 1808

No. 47 Wellclose Square, London

Sir,
I humbly presume to lay before You a Copy of an Affidavit which I have sworn to, setting forth the Interest I have in the *Conferenz Rath Prætorius*.[3] This Affidavit has some time since been laid before the Kings Advocate[4] under the full persuasion that I should be considered entitled to the benefit of the order of the British Government for the Release of Iceland Ships & Cargoes. I am now however given to understand that His Majesty's Advocate has intimated an unfavourable opinion, and declared that further application can be of no avail. Under these discouraging circumstances, I take the liberty to solicit Your kind opinion upon my case, whether I may not entertain a hope of still being included in the benevolent intentions of the British Government towards the Citizens of Iceland, and what steps I /*must*/ should take, if hope must not be entirely given up in

---

[1] The following is crossed out (after 6000 dollars): 'but a Burgher of Iceland Can only trade with Copenhagen unless he is also a Burgher of Copenhagen'.

[2] The draft is more or less identical, with differences in punctuation and capital letters.

[3] The *Conferentz Raad Prætorius* was owned by Norwegian merchants, perhaps speculators, which may be the reason for Bredal's difficulties. This ship had sailed directly to Iceland from Norway (it had no sea-licence from the Danish government) and was captured by the *Peacock* and brought to Leith prior to the commencement of hostilities in 1807. Norway at this point was in a state of blockade.

An affidavit was sworn by the master 'Fiorsoaag' (Forsvaag) before John Crawford JP at Leith on 15 February 1808, the cargo consisting of oil, wool and feathers. All were 'in a perishable state' (TNA, HCA 32/1197, with the papers of *De Tvende Söstre*, no. 5778).

[4] The King's Advocate was Sir John Nicholl (1759–1838) until 1 March 1809, when he was succeeded by Sir Christopher Robinson (1766–1833). On 9 August 1808 the release of the prospective mate was refused (TNA, ADM 98/308, Transport Office to S. Forsvaag, Master of the Danish Vessel *Conferentz Raad Prætorius*).

order to obtain the liberation of my Property.¹ The benevolent Protection, Sir, that You have granted the citizens of Iceland, whom the fortune of war has brought into this Country, has emboldened me to make this application to You, and I humbly ask Your forgiveness for the liberty I have herewith taken. I remain with the deepest gratitude & esteem

Sir:
Your most obed<i>ien</i>t humble Servant
[<i>Signed</i>] H. G. Bredal

SL, Banks Papers, I 1:27. Letter, written by an Englishman, signed by Bredal.

99. Kjartan Ísfjörð to Banks, 23 August 1808[2]

Copenhagen

In the year 1802 I set up a Trade in the town of Eskefjord[3] in my native Country Iceland – the endeavours of many Years and uninterrupted Assiduity from the earliest times of my Youth procured me a small fortune, which however was not sufficient to this long wished for establishment; a noble and goodhearted man, believing me honest and industrious, intrusted me with a Share of his fortune in order to enlarge my Trade; from 1802 untill 1806 bad fisheries and Dearths afflicted Iceland, whereof the particular consequence for me was such, that I scarcely was indemnified for the Trade of the 4 first Years in Iceland.[4] In the Year 1807 I made every exertion in order to carry Provisions to Iceland and expected to get a gainful Return-Cargo; I possessed but one Vessel, whereon I bestowed a melioration or reparation before she went from hence in the Spring 1807, amounting to above 3000 R<i>ix</i>doll<i>ar</i>s. – The Vessel set <i>off</i> to her destination, was sent from Iceland with a full Cargo in order to sail to Copenhagen: Carsten C. Juul, the Master of the Vessel named *Charlotte Amalia*,[5] ignorant of the War broken out between Denmark and England, was taken under the continuation of his voyage by an English Brig of War the *Pelican*, Wahr[6] Commander, near Skagen and from thence carried to London.[7]

From the Merchant Knudsen[8] of Iceland lately arrived to this Place from London, that the Baron Sir Joseph Banks, so universally revered for his generosity and humanity hath from compassion with my deplorable Country Iceland prevailed on the British

---

[1] On 1 October 1808 the Privy Council granted this ship, being 'one of the vessels which has been liberated by order of Government upon condition that she shall return with a Cargo of articles suited to the wants of the Inhabitants of Iceland', a licence (TNA, PC 2/178).

[2] This letter is a translation from the Danish, as it says at the top.

[3] Eskifjörður.

[4] On Kjartan Ísfjörð see Andrésson, *Verzlunarsaga Íslands*, II, pp. 594–6.

[5] In the left margin in pencil there is written '5th Novr 1807', which is the date of capture. He was extremely unfortunate to be captured only a day after the outbreak of hostilities on 4 November 1807.

[6] William Ward was his name.

[7] He was first taken to Yarmouth and from there to London. All the ship's papers are in TNA, HCA 32/985, no. 993.

[8] Adser Christian Knudsen (see Appendix 13), then in London, who had also written to Banks (see Document 71).

Government to set at Liberty all Icelandish Ships and Properties; nevertheless I have been extremely grieved by learning from the said Knudsen, that my very said Ship, and no other Icelander; shall have been sold and condemned at London. How alarming to me this intelligence must be, is easily conceived, as not only my whole fortune lies in this Ship and Cargo, so that by the annihilation of the same I am quite ruined and put altogether out of the way of maintaining myself and my family, but also, what is the most painful for a honest man, I am rendered unable to pay off a Capital lent me by a nobleminded Man before the War for the benefit and continuation of my Trade set up in Iceland. On account of these sad circumstances I shall foster the hopes, that the wellknown Philantropist the Baron Sir Joseph Banks, who at his stay at /*Örebak*/ /Havnefiord/[1] in Iceland has given my family there so many causes yet on this day to remember his presence with reverence and the most agreeable feelings,[2] and who, through his powerful influence, hath under the present circumstances procured so much good for the benefit of Iceland will not, if possible, permit that I as an /*single Man*/ Individual and a born Icelander shall suffer alone, nor that the Icelanders, benefited by my Trade, shall now be exposed to Hunger and Death. On account hereof I have, with the Knowledge of my most gracious King, empowered Mr. Frederik Kolvig to set off for England, and in consequence thereof I do hereby give and grant unto said Mr. Kolvig full power to claim the said Ship *Charlotte Amalia* with her Cargo, or the amount thereof, and to receive the released Ship or Vessel *Regina*[3] together with her Cargo or the amount thereof if sold, which said Ship and Cargo are belonging to me agreeable to the Purchase Agreement of the 22^d of August this Year, of which Agreement or Indenture Mr. Kolvig hath a Notarial Copy.

As the said Mr. Kolvig hath undertaken this voyage from friendship towards me, and from true feelings for my Interest, and as I know, that he will behave in this respect with particular care and probity, I do hereby grant him unlimited authority respecting all what may concern this business as well in England as in Iceland, be it in claiming the Ship's and Cargo's by petitioning the Government, the Council boards and others or otherwise, in receiving, and giving proper acquittances for, the Amounts thereof if sold, in getting true accounts and viewing the same, in buying Goods in England for the Produce already had or to be had for the Provision of Iceland, in hiring Ship's Officers and Crew on the Spot; and forwarding and favouring the benefits of the expedition in Iceland, the whole without any limitation, wherever and whatever it may be, as all what he may do or cause to be done in conformity with the laws and Grants is to be looked upon, as if I myself was personally present and did it or caused it to be done.

And as this Letter of Attorney binds me as long time as it is lawfully in Mr. Kolvig's possession, the same is also binding for me respecting every other person without exception, with whom my said Attorney may interfere in this behalf and as soon as occcasion offers of advicing [*sic*] me in a lawful manner of whatever has been executed respecting this present business, he is bound so to do as also his duties towards me are then to be decided according to the later instructions which I might impart him.

(Signed K. Isfiord) | (L.S.)

---

[1] Hafnarfjörður instead of Eyrarbakki, where Banks stayed during his 1772 visit.

[2] Ísfjörð was the son and stepson of county magistrates in Iceland and a wealthy merchant.

[3] The *Regina* was actually solely owned by Just Ludvigsen of Copenhagen: see the *Regina*'s papers in TNA, HCA 32/1162, no. 5087.

I the underwritten Gottsche Hans Olsen, Notary Public Royal in and for this Royal Residence City of Copenhagen, do hereby attest, that Mr. Kiartan Isfiord, Burgher and Icelandish Merchant, hath himself personally signed and sealed the aforegoing Letter of Attorney.

Witness my Hand and Seal of Office. Actum, presente teste, Joh. Christ. Riisbrich Havniæ d. 23tio. Augusti 1808[1]

(L.S.) G. H. Olsen Not*ary* Publ*ic*

Pro vera versione ex Originali, Havniæ d. 24to Augusti 1808

Quod attestor[2]

G H Olsen Not*ary* Publ*ic*[3]

Wisconsin, Banks Papers, ff. 25–26. Autograph letter.

100. William Wellesley-Pole to William Fawkener,[4] 7 January 1809

Admiralty Office

Sir

Having laid before my Lords Commissioners of the Admiralty[5] your Letter of this Day's Date inclosing a Copy of a Memorial from Messrs. Boulton and Baker[6] praying that they may be allowed to select from among the Danish Prisoners of War two Masters, Nine Mates, and thirty five Seamen for the purpose of navigating Seven Iceland Ships liberated by Order of the British Government, and recommending the Prayer of the said Memorial to the favourable and early Consideration of their Lordships; I am commanded to acquaint you, for the Information of the Lords of the Council, that in consequence of that Recommendation, Orders have been given for liberating the Danish Prisoners required for the abovementioned purpose.[7]

I am &c

(signed)[8] W. W. Pole

NHM, BC, ff. 20–21. Contemporary copy.

---

[1] Translation: Done in the prescence of a witness, Joh. Christ Riisbrich, Copenhagen, 23 August 1808.

[2] Translation: I bear witness that this is a true translation of the original, Copenhagen 24 August 1808.

[3] A seal is affixed to the document, which is written on official paper.

[4] This was a copy sent to Banks. 'Copy' is written in top left-hand corner.

[5] The Lords Commissioners of the Admiralty were seven in number at this time and were appointed by the Crown. In January 1809 they were: Lord Mulgrave, Sir Richard Bickerton, William Johnstone Hope, Robert Ward, Viscount Palmerston, James Buller and William Domett. The Hon. William Wellesley-Pole was the First Secretary of the Admiralty, while John Barrow was the Second.

[6] This memorial does not appear to be extant.

[7] See decision in the Minutes of the Admiralty (TNA, ADM 2/155, Lords of the Admiralty to the Transport Board, 7 January 1809).

[8] Thus in copy, i.e. not an autograph signature.

101. Markús Magnússon[1] to Jörgen Jörgensen, [18 January 1809]

[Garðar, Álftanes peninsula]

Sir!

I[2] dare not on account of the threatnd hostilities yesterday[3] attempt to receive your friendly invitation to dine on board to day.[4] I wish every thing could be setled for the best,[5] to which I wish to assist as much as possible, and in like manner it is the wish of the inhabitants to serve you, and deal with you in a friendly manner, without being inclined to shew the least sign of Hostility, All comes from the danish Authorities being overcautious,[6] and their wrong explanations of the law, which I and more sensible people seems not to be only quite superfluous under the present circumstances,[7] but entirely unbecoming, nay even dangerous. All the /*Inhabitants*/ Instruction I can give you, and all the support I am able to give you, and consistent with my situation, you shall have, but you know political affairs do not come within my Sphere.[8]

I subscribe myself
in the most friendly manner your
M. Magnusen

Captain Jörgensen on board the *Clarence* } Provst in Kiöse and Guldbri*n*ge Syssel[9] – Stiftprovist[10] in all Iceland – Priest to Gouly[11] and Bessastede.

---

[1] Warren Dawson wrongly attributed this letter to one Arni Magnusson (c. 1752–1810), unknown in Icelandic sources. The Revd Markús Magnússon was the author.

[2] He always writes i instead of I, which has been amended here for clarification.

[3] At the end of December 1808, Phelps, Troward & Bracebridge, of Cuper's Bridge, Lambeth, a firm of soap manufacturers, sent the ship *Clarence*, armed with a letter-of-marque, on a trading venture to Iceland in search of tallow. Jörgensen was on board as interpreter. They arrived in Iceland on 12 January, where the British were met with hostility, the Danish authorities being intent on upholding the ban against foreigners trading in Iceland. After fruitless negotiations Savignac, the mercantile firm's supercargo, decided to put pressure on the officials. He ordered Captain Jackson of the *Clarence* to seize the *Justitia*, a Norwegian brig, which had arrived in October 1808 from Trondheim loaded with necessities at the King of Denmark's expense. This took place on 17 January 1809 and this is what Magnússon is referring to. On their arrival in Iceland Jörgensen and Savignac had stayed with Magnússon at his home in Garðar, near Hafnarfjörður. This would be the pre-cursor to the Icelandic Revolution, see Introduction, p. 32 above

[4] On board the *Clarence*.

[5] Magnússon got his wish. An agreement was signed the next day, 19 January 1809, between the Danish officials and the British merchants, permitting the latter to trade in Iceland. The original of the document is to be found in the National Archives of Iceland [hereafter NAI], Bæjarfógetinn í Reykjavík [hereafter Bf. Rvk.] A/2-1, Bréfdagabók 1806–13, 20 January 1809, no. 2305.

[6] They stuck rigidly to the law prohibiting foreigners from trading in Iceland.

[7] The *Justitia* was the only merchant ship to reach Iceland in 1808, so the need was great.

[8] He was a clergyman.

[9] Provost to the counties Gullbringusýsla and Kjósarsýsla (in south-west Iceland).

[10] *Stiftsprófastur* was the substitute for the bishop. He served in this capacity only in 1796–7 in the place of the bishop in Skálholt, the bishopric of southern Iceland.

[11] Gouly makes no sense, a misreading (Jörgensen is translating a letter doubtless written in Danish and in Gothic script). This must be Garðar as Magnússon served both the Garðar and Bessastaðir churches.

NB. This letter was sent to me supposing me to be interested in the Expedition, which was not the case.[1]

NHM, BC, ff. 40–41.[2] This letter is in Jörgensen's autograph and obviously translated by him.

102. Boulton & Baker to Banks, 8 February 1809

Wellclose Square

We had fully hoped & assured ourselves, that all would now have gone on smoothly with the Iceland Ships,[3] and that we should have had the Satisfaction of seeing them all depart on their Voyages without having any occasion further to trouble Sir Joseph Banks, or further to intrude upon his Time, Patience, & Humanity. –

But Sir, We are now thrown into a considerable Degree of Difficulty & Concern, and must once more entreat your Assistance, thro' which we trust we may be relievd. In all the Licenses hitherto granted to these Ships, Permission is granted for them to proceed from Denmark to Iceland with a Cargoe of 'Provisions and other Necessaries'. – A few days since we applied in the usual Terms for the *Bosand*; but on receiving the License, it specified 'Provisions Only'.[4] On consulting with our Friend Mr. [Jens Andreas] Wulff (who has frequently waited on you respecting his ships *Husevig* & *Nye Prove*). He pointed out that this was a most material & fatal Omission, as there were such a Number of Articles of which the Inhabitants stood nearly as much in need as Provisions. Hoping it might be merely an Oversight in the Clerk who filld up the Blanks. We sent up Mr. Alsing,[5] the Person who has had the Honor of waiting on you occasionally from this House, to the Privy Council Office, to endeavour to get it rectified. – But Mr. Fawkener[6] assured him the Lords felt no Disposition to make the wishd for Addition and advised us to hand in a Petition stating what those Necessaries were, that we were desirous to carry to Iceland. In order to give us an Idea thereof, Mr. Wulff last night handed us the List which he had himself made out prior to his quitting his Home, of the Articles wanting at those Places under his own immediate Superintendance, which List, I may venture to assert, does not enumerate less than Three Hundred Articles![7] Being, in fact, just that kind of Assortment that a Person would send out to a Settlement newly form'd on some hitherto uninhabited Island,

---

[1] This is in Jörgensen's handwriting.
[2] Note, f. 41 verso: 'Translation of Mr Magnussens letter'.
[3] All underlining is by the letter-writer.
[4] This is correct. The licence had been granted on 16 January 1809 (TNA, PC 2/179) 'upon condition that the cargo carried in the said vessel from Denmark to Iceland provisions only, and that if any part of the return cargo consists of naval stores and be destined for a port south of Hull, the vessel shall, unless under protection of convoy, stop at Dundee or Leith and there obtain a fresh clearance for her port of destination, and must sail with convoy for one voyage only'.
[5] Alsing is obviously a clerk of Boulton & Baker's, but apart from that nothing is known of him.
[6] William Augustus Fawkener (1750?–1810) was the Secretary of the Privy Council for Trade and Plantations during the years 1786–1811.
[7] This list does not seem to be extant.

where they had not begun to raise or manufacture anything for their own Use, and without which they would be nearly as much distressd, as if in Want of <u>Corn</u> only. Now Sir, You will have the Goodness to recollect that this Vessel <u>Bosand</u>[1] is the one more immediately under our own Care & Direction, from the Circumstance of there being here no Owner or Captain to assist or advise us on the Subject; & of course we feel doubly anxious that all should go as well with her, as with those whose Owners or Captains are here on the Spot to watch their own Interests: and should we fail therein, Mr Howisch the Owner, now in Copenhagen, will naturally conclude that such failure arose from our own Innatention & Lukewarmness where we had no one at our Elbow <u>to spur us on</u>. The Reasons of our not applying for the Licence at the same time that, thro your kind Interference we got those for Mr. Wulff's two Ships, was, that we could not then fix on a proper Person to commmand her: – as soon as that was done & his Release obtain'd, we applied, and this has been the Result!

But we yet hope, Sir, that thro' your kind Attention we shall not be denied the same Indulgence for this Ship, as has been granted for all the others: – we shall immediately set about a Petition, & state the principal Articles which Mr. Wulff conceives his Neighbour Howisch would be desirous to send out in the <u>Bosand</u> and wait on you therewith early on <u>Friday Morni*n*g</u>! These Necessaries may we think be reduced and in the following Heads: Ironmongery, Cutlery, Hardware, Woollens, Cottons, Linens, Glassware, Haberdashery, Hats & Shoes, Groceries, Drugs, Books & Stationary, Cordage & Lines for the Fishery, a little Hemp, Tar, Iron, Steel & Lead, Powder & Shot, Spirits, Wines, Tobacco, Timber, Deals, Staves & other Materials for Coopers & Carpenters Work. –

Much lamenting that we should be placed in the disagreable Necessity of again troubling you on these Matters, – but trusting that you will not desert us in our Difficulty.[2] We remain with great Respect,

Sir
Y*ou*r very obed*ien*t Ser*van*ts
[*Signed*] Boulton & Baker

[P.S.] The Nett Proceeds of this Ship's Cargoe will but just fit her out for Sea.

NHM, BC, ff. 22–23.[3] Autograph letter.[4]

---

[1] Belonging to Höwisch, merchant at Hofsós, and liberated 23 June 1808, but was unable to make the voyage that year.

[2] Banks probably came to their aid. At any rate the *Bosand* arrived in Iceland in the late summer, and came to Yarmouth in November with a cargo imported from Iceland (TNA, BT 6/197, no. 11,968). A licence was granted on 21 December 1808 permitting the vessel to proceed to Norway or Sweden without the Baltic to load grain there for Iceland (TNA, BT 6/197, no. 13,008).

[3] Note in Banks's autograph, left-hand margin of first page: '[Received] Feb*ruary* 8 [1809]'.

[4] Note in Banks's autograph in pencil at top of first page: '*Eskefiord* in Norway 150 Tons | *Bosand* | *Regina* Wolf & Dorville'.

103. Petition from Boulton and Baker to the King in Council, 20 February 1809

Wellclose Square

To The King's Most Excellent Majesty in Council
The humble Petition of Boulton & Baker of London Merchants
Sheweth

That Jens Andreas Wulff[1] of Iceland Merchant, at present in London, has clearly represented to your Petitioners that he has constantly employed Three Ships for supplying the different places in Iceland where he has Establishments, and even these three Ships have not been sufficient for that purpose, as he has not been able to send either of them, during the last two years for a Cargo of Timber, Deals etc., of which the Island is therefore now nearly as much in want as of Provisions – And that the third of his Ships which has constantly traded to Iceland, called the '*Eskefiord*' being now in Norway, he is very anxious to obtain Your Majestys most gracious Permission to load in the said Vessel an assorted Cargo of Timber, Balks,[2] Deals and such other Articles of Wood, which are the produce of Norway, and absolutely wanted in Iceland, and at present not to be had in Denmark; also a small quantity of Iron, Ironware, Tar and Salt which at the same time might be procured in Norway.

Your Petitioners therefore humbly pray that Your Majesty will be graciously pleased to grant a Licence for permitting the said Ship called the '*Eskefiord*' to proceed from a Port in Norway with an assorted Cargo of Timber, Deals, Balks and other Articles of Wood also of Iron, Ironware, Tar & Salt for Iceland, and to return with a Cargo of such Goods as are permitted by Your Majestys Order in Council of the 11 Nove*mbe*r 1807 to be imported from Iceland to any Port of Great Britain, and that the Master may be permitted to receive his Freight and depart with his Vessel and Crew to any Port not blockaded.[3]

And your Petitioners shall ever

pray

(signed)[4] Boulton & Baker.

NHM, BC, ff. 24–25. Contemporary copy.

104. Boulton & Baker to Banks, Wednesdy Eve*nin*g 22 February 1809

Wellclose Square

Sir

When we again wrote you on the 8th Inst*ant*[5] respecting the Difficulty we were in about our Ship *Bosand*,[6] we surely thought that we might promise not to be any more

---

[1] Jens Andreas Wulff: see Appendix 13.

[2] A balk is a roughly squared timber beam.

[3] This petition was taken into consideration by the Privy Council on 29 March 1809, which 'hereby ordered in council, that a licence be granted' (TNA, PC 2/180). Among the stipulations was a total ban on bringing oil or fish from Norway.

[4] Thus in copy, i.e. not autograph signature.

[5] See Document 102.

[6] The *Bosand*, released on 23 June 1808, was issued a licence on 16 January 1809 (TNA, PC 2/179). The petition of Boulton & Baker, agents for Johan Gottfred Höwisch, owner of the ship, requesting 'a licence for the

troublesome to you on the Score of our Iceland Friends: – But, Sir, We now with much Concern, see that Obstacles have also fallen in the way where, from what you were kind enough to say to Mr. Alsing when he waited on you from us, we as little expected them: – that is, in the Attainment of a License for Mr. Wulff's third Ship *Eskefiord*,[1] now laying in Norway, permitting her to go from <u>thence</u> to Iceland with a Cargoe of the Produce of that Country (as specified in the Petition, of which we make free to hand a Copy) and of which Articles the Island is by all Accounts in great want, and which can be had from Norway only, or at least the greatest part of them. The Petition was delivered at the Office on Monday last, and on our sending up this morning to know the Result we learned that it had been <u>unfavorable</u>.[2] We hope, Sir, that in taking, this our Prayer into your Consideration, you will feel disposed once more to give your assent & Furtherance thereto, as we trust it can in no shape interfere with the commercial Interests of this Country, or be considered as 'transporting naval or warlike Stores from one Enemy's port to another', – as the Articles wanted are merely for building & repairing Houses & Warehouses, making <u>Casks</u> for the Exportation of their Produce (without which, of course it cannot be brought hither) and such like Household or mercantile Purposes. – You Sir, are doubtless aware that <u>Ironworks</u> & <u>Saltworks</u> are both carried on in Norway, and that <u>Tar</u> likewise is there to be procur'd at a low Price, – so that these Articles, as well as those consisting of Deals, Timber & other woodenware, were usually supplied from thence; altho' it has occurr'd to us, that some of the Lords in Council, might object to those Articles, from its not occurring to them that they were actually the <u>produce</u> of Norway. – And should these still form the principal Ground of Objection, Mr. Wulff must be content to confine himself to Articles of <u>Wood</u> only. – Suffer us Sir to indulge the Hope, that this <u>our last</u> Prayer be not wholly refused to us, but that thro' your kind Interference, We may, if not indulged to the full extent, at least have the Pleasure to be instrumental in these unfortunate People's getting from Norway by this means, what they are greatly distressed for, and can obtain from no other quarter, & thro' no other Channel. – We are almost afraid to attempt any more Apologies! – so merely declare ourselves,

with unfeigned Respect,
Sir
Your very humble & obedient Servants
[*Signed*] Boulton & Baker

[P.S.] Should you advise a Petition <u>differently worded</u> we shall be happy to receive your Instructions.

---

vessel to proceed in ballast or with a cargo of such goods as are permitted by virtue of His Majesty's Order in Council of 11 November 1807 to be exported from London to a Danish port to proceed from such port with a cargo of provisions and other necessaries for Iceland, and to return with such goods as are permitted by virtue of H.M.'s said Order to be imported from Iceland to Great Britain, and that the master's may be permitted'.

[1] It may have been unfavourable then, but was later granted. The *Eskefiord*, belonging to Jens Andreas Wulff, was granted a licence to import a cargo (but neither fish nor oil) from Norway and then to proceed to Denmark to pick up a cargo for Iceland. The ship must sail in convoy and the cargo loaded in Denmark must only consist of provisions, clothing and 'necessaries for the fishery' (TNA, PC 2/180, 29 March 1809).

[2] This has not been found.

[*Note by Banks on letter*] their Lordships objections to opening a Trade between Norway & Iceland are in my opinion insurmountable besides Deals are not a matter of Prime Necessity

NHM, BC, ff. 26–27.

105. Boulton & Baker to Banks, 25 February 1809

Wellclose Square

Sir

We duly receiv'd your favour of the 23rd & explain'd the same to Mr. Wulff:[1] – He must of course think no more of this Part of his Plans for the Relief of his Countrymen; – at the same time we trust your Candour & Liberality will pardon our merely stating to you the Impression your Letter made on our Minds, on first perusing it. –

In Iceland, we understand no Timber grows, – not enough for Fuel: – The Inhabitants must require a constant, tho' small supply of various Articles in that way, for building & repairing Dwellings & Warehouses, Boats, and some Implements of Machinery, altho' their <u>Agriculture</u>[2] or <u>Manufactures</u> are but trifling – also Casks & other Packages for the Exportation of their Produce. – We really should not think it too much to say that a small Supply now & then must therefore be of <u>the first</u> <u>Necessity</u>. This Supply <u>we</u> can /<u>not</u>/ spare them, nor can they without <u>our</u> Permission (unless indeed by mere chance, and at a great Risque) get it elsewhere. – Is it therefore a greater Indulgence to permit <u>one small Vessel</u>, now shut up in a port in Norway,[3] to take thither such an Assortment as they are most in want of, than to permit several Ships to go from hence to Denmark & there lay in Cargoes of Provisions & other Necessaries for their use, – or is the request more unreasonable? or does the one come more under the Idea of <u>opening a Trade</u> with Norway,[4] than the other does with Denmark? –

These Queries immediately presented themselves, and altho' we did not at first intend to make so free as to trouble you with them, or even <u>now</u> expect you to pay them any attention, yet we thought you would pardon our so doing, and that it would in some measure shew that we did not merely acquiesce in the Wishes of Mr. Wulff, without well weighing the nature of the Application, and forming an Opinion (tho' to our Concern, it appears <u>erroneously</u>) that we were not acting unreasonably or inconsistently.

We do certainly much regret having asked anything your good self and the other Members of the Hon*oura*ble Board should deem unfit to be granted, and feel highly gratified & honour'd by the Attention we have thro' out experienced. – Permit us once more to subscribe ourselves with great Respect, Sir

Y*ou*r very obed*ien*t Serv*an*ts

[*Signed*] Boulton & Baker

NHM, BC, ff. 28–29. Autograph letter.[5]

[1] Jens Andreas Wulff.

[2] All underlining in this letter is by the letter-writer himself.

[3] The *Eskefiord*, owned by Ørum and Wulff, did eventually sail to Iceland in 1809.

[4] Timber had long been exported into Britain from Norway and by 1640 England had become Norway's major customer. From the 17th century Norwegian merchants had settled in London to conduct the trade. Wolffs & Dorville was the most prominent firm of timber importers from 1767 until it went bankrupt in 1812.

[5] Note in Banks's autograph, left margin of the first page: '[Received] Feb*ruary* 25 [1809]'.

106. Bjarni Sívertsen to Banks, 28 February 1809

Leith

Right honourable Sir Joseph Banks
Sir

Although I am wery Sensible of to have Several times been accessioned [occasioned][1] to make you so much Trouble concerning myself and my Native Contry Island, and have enjoyed so many Favours by Your Kindness, my continual Disasters and your said Kindness, make me so confident to Relate to you my accidents since I left London, and Solicit some Reliew [relief] by your good intervention, from your Government;

We[2] left this place 26 of September last, and 80 mile from Island[3] got a Hurrican ther Keept us 3 dais and 3 Nights, carried away many of our Sails and makte the Wessel wery leck [leak], so we was obliged to turn back i gain, and after much Sufferings was happy to come to this place i gain the 4 of Nov*ember* last, we was obliged to take out our Cargo for to get the Wessel Repaired, ther now is ready so we hope to be clear to begen a New Woyage the 20 of next month, my only Trouble concerning this Matter is to know wher I shall get money to pay for the Wessels and Riggings amending, and other Expences, ther Will amount about £300. We had when we went away so many money in London as was due to You for to repay the money You so Kindly advanced us in London last Winter,[4] I am now so free humbly to desire that you Wowld have indulgence with us till we come back again and you would allow us to use this Money as far as it can reach.

My other Wessel *Johanne Charlotte* left Plymouth about the end of Aug*ust* last year, in order to goe to Copenhagen and take Cargo of Nesessaries for Island, accordant to her leicence,[5] and was forced to seek harbour for bad wether in Aalborg in jutland, ther the master is been ordered, without my Concent, of my Agent in Copenhagen to take in his Wessel som Ray [rye][6] and Barley for Island, and was farther ordered to proced to Copenhagen for lay /to/ ther the Winter over, and take in more and other Nesessaries, as Ropemakers work and Tar etc. ther was not to be got in jutland; on her Woyage from Aalborg to Copenhagen the Wessel is been taken, as I suppose of Swedish Privaters, and brought to Carlscrona[7] and ther condemned; thus I have lost my good Wessel and Island a Cargo of Nesessaries by my Agents misconduct in this Matter, the Cargo of this Wessel when it was taken to Engeland had cost in Island £1600 and I was obliged to pay £150 besides the nette provenn of this Cargo for to get /the/ Wessel Clear from Plymouth with Ballast; I feel this last loss so much more as I am not able to get Nesessaries for my own hushold from this Contry, and for want of Wessel, from Danemark.

---

[1] Sívertsen is progressing with learning English. He is writing this letter himself, mixing the English he had heard and Danish spelling. He uses e.g. 'w' instead of 'v' or 'u'.

[2] Sívertsen and Petræus were travelling on the *De Tvende Söstre*, captured in 1807 and released in 1808. It arrived in Iceland at the end of April 1809, carrying 400 barrels of rye (NAI, Stiftamstjournal II, Nr. 7. 6.2.1809–22.6.1809, Koefoed to *Stiftet* (the administration), 24 April 1809, no. 46.). It would later sail to Liverpool in October (TNA, BT 6/197, 6 October 1809).

[3] He doubtless means Iceland, *Ísland* being the Icelandic word for Iceland.

[4] See Document 53, p. 247, n. 2.

[5] The *Johanne Charlotte* was one of the first Iceland ships to receive a licence (TNA, PC 2/177, 23 July 1808). Sívertsen was permitted to sail to Copenhagen in ballast to pick up 'cargo of provisions' for the Icelanders.

[6] Rye was the main corn import into Iceland.

[7] Carlscrona in Blekinge on Sweden's Baltic coast, seat of the Royal Swedish Navy.

That is therfore my only Comfort, humbly to Solicit you for to Procure me a <u>new Leicence for a Wessel of 100 a 120 Tons Burthen ther may be Bowght for my Account in Danemark, for to go with proper Cargo to Island in next Sommer and come back i gain her</u>,[1] if your Government will be so indulgent to grant this my Request, I will sende the Leicence with one of the Wessels ther goe from London to Danemark[2] –

Sir [space] I may at last beg your Pardon, not only for the Trouble I make you and farther more for the bad Writing of this letter thes is the first I have taken me the fredom to write to you with own hand, and after own thowght.

I remain honoured Sir
your most humble and most
obedient Servant
[*Signed*] B: Sivertsen

NHM, BC, ff. 30–31. Autograph letter.[3]

107. Boulton & Baker to Banks, 27 March 1809

Wellclose Square

Sir,

In a few days M^r J: A: Wulff[4] will depart from hence in the *Husevig* with sundry Goods to the full amount of the Net Proceeds of the Two Cargos restored to him by the British Government.[5] He is therefore very anxious to know, before his Departure, whether a Licence will be granted him for his Iceland ship the *Eskefiord* in Norway on any Terms; if Their Lordships should not have favourably reconsidered our former Petition dated 8th Ins*tan*t which was for the entire Voyage of the Ship from Norway to this Country thence to Danmark & Iceland and back to England, and there should unfortunately be no prospect of our benefiting by Your Personal Interference; would you then recommend us to repeat our Petition, praying in the First Instance to grant a Licence for permitting the *Eskefiord* to come direct from Norway to Great Britain with a Cargo of Timber and Deals, and after the arrival and unloading of the Cargo to petition for a further Licence.[6]

---

[1] Underlined in pencil.

[2] Sívertsen got his wish. On 30 March 1809 the Lords of the Privy Council granted a licence permitting him to buy a vessel in Denmark and proceed with a cargo for Iceland, stopping at Dundee or Leith for a fresh clearance (TNA, PC 2/180). The vessel was the *Orion*.

[3] Note in Banks's autograph, verso: 'Sivertsen | [Received] March 3 – [18]09'.

[4] The other ship restored to Jens Andreas Wulff was *Den Nye Pröve*.

[5] The *Husevig* had been granted a licence on 12 December 1808 to sail to Denmark, from thence to Iceland and to return to Great Britain (TNA, PC 2/179), but on her return was forced by gales into Dundee (see Documents 132 and 133).

[6] See above, p. 305, n. 1.

Mr. Sivertsen is also on the point of proceeding on his Voyage and we much lament our not being able to give him a favorable account of our endeavors in his behalf.[1] –

We have the honor to subscribe ourselves most respectfully
Sir,
Your very humble Servants
[*Signed*] Boulton & Baker

NHM, BC, ff. 32–33. Autograph letter.

108. Boulton & Baker to Banks, 30 March 1809

Wellclose Square

Sir,
We beg leave to inform you, that on Tuesday last, The Lords of Trade[2] were pleased to grant us a Licence for the Iceland Ship '*Eskefiord*' in Norway to come to England with a Cargo of Timber and Deals, then to proceed to a danish Port, there to load Provisions for Iceland and from thence to return with Iceland=Produce to this Country.[3] Their Lordships at the same time granted a Licence to the *Regina*.[4] But They would farther consider our Petition for B: Sivertsen.[5]

The *Husevig* will depart from hence in the course of the next week, and when she is gone we shall have no other Opportunity of communicating with M^r. B: Sivertsens Agents in Denmark; we have therefore humbly repeated our former Prayer of the 8th Ins*tan*t and take the Liberty to hand you enclosed a Copy thereof, wishing most sincerely that your Health may afford us your powerful Interference in favor of the Petition, especially as M^r. Sivertsen has been the most unfortunate of all the Icelanders who have benefited so largely by the Benevolence of the British Government; he poor Man, lately returned to Leith with an other of his Vessels in a damaged State[6] and is now enabled, after heavy Repairs, to proceed direct to Iceland but has only a small Quantity of Rye on board; and of course is a most fit Object of Commiseration.

We are most respectfully
Sir,
Your very humble Servants
[*Signed*] Boulton & Baker

NHM, BC, ff. 34–35. Autograph letter.

[1] Probably the answer to his request for a new ship had not yet arrived (viz. TNA, PC 2/180, 30 March 1809).

[2] The Lords of Trade are the members of the Committee of the Privy Council for Trade and Plantations, established in 1786 (until 1870). In 1786 the Board of Trade had taken the form of a committee of the Privy Council, the members being either holders of certain offices *ex officio* or specifically appointed by the Crown, headed by the President, at this time Earl Bathurst (1807–12) with George Rose as his Vice-President (1807–12). The members varied in number. Sir Joseph Banks had been appointed to the committee on 29 March 1797. Their role was to recommend policies relating to trade and the colonies but eventually they became a predominantly administrative body concerned with commercial matters.

[3] See TNA, PC 2/180, 29 March 1809.

[4] In the *Regina* licence it was stipulated that the cargo from Denmark to Iceland should 'consist wholly of provisions, clothing & necessaries for the fishery' (TNA, PC 2/180, 29 March 1809).

[5] See following document.

[6] This was *De Tvende Söstre*.

109. Petition from Boulton & Baker to His Majesty in Council, 30 March 1809

Wellclose Square

To the Kings Most Excellent Majesty in Council
The humble Petition of Boulton & Baker of London Merchants

Sheweth

That on the 8th of this Month[1] Your Petitioners humbly pray'd Your Majesty would be graciously pleased to grant a Licence for permitting B: Sivertsen of Iceland to purchase a Vessel in Denmark, in lieu of his former Vessel called the '*Johanne Charlotte*' for which he had obtained Your Majesty's Licence and proceeded from Plymouth to Copenhagen in order to load Provisions and other Necessaries for Iceland, when by stress of Weather the said Vessel was obliged to put into Aalburg in Jutland where the Master, by directions from the Agents in Copenhagen (who it appears did not rightly understand the Tenor of the Licence) – but wholly unknown to B: Sivertsen himself, took in some Corn and was proceeding to Copenhagen there to compleat his Cargo with such Articles as could not be procured at Aalburg, when he was captured and brought to Carlscrona in Sweden where Ship and Goods have been condemned;[2] by which the said B: Sivertsen not only suffers the Loss of his Property, but the Inhabitants of that part of Iceland where he is resident,[3] himself and Family and his numerous Tenants, will sustain a more serious Disappointment by not receiving the only Supply of Provisions which was in his Power to procure for their Relief.[4]

That Your Petitioners now venture to repeat the above Prayer humbly representing, that in the course of next week the last Iceland Ship will depart from hence to a danish Port, and that should it please Your Majesty, through Commiseration for the unfortunate Icelanders, to grant the same; Your Petitioners will not have an other Opportunity of forwarding the Licence to Denmark after the Departure of the said Iceland Ship –

Your Petitioners therefore humbly pray that Your Majesty will be graciously pleased to grant B: Sivertsen of Iceland a Licence for permitting him to purchase another Vessel in Denmark, in lieu of the *Johanne Charlotte* and to proceed from a Danish Port with Provisions and other Necessaries to Iceland and to return with a Cargo of Iceland Produce to any Port in Great Britain and that the Master may be permitted to receive his Freight and depart with his Vessel and Crew to any port not blockaded.[5] –

And Your Petitioners
shall ever pray –
[*Signed*] Boulton & Baker

NHM, BC, ff. 36–37. Autograph letter.

[1] Not extant.
[2] See Document 106.
[3] Hafnarfjörður.
[4] The licence was granted by the Privy Council that same day (TNA, PC 2/180, 30 March 1809).
[5] Sívertsen returned to Hafnarfjörður on *De Tvende Söstre* in the spring. No record of another new ship, belonging to him, arrived that year in Iceland, nor in 1810. The ship he eventually sent in 1811 was the *Anna Dorothea*.

110. Jörgen Jörgensen to Banks, 14 April 1809

[London]

Sir!

I hereby take the liberty of sending you a correct account of the Brig *Clarences*[1] proceedings in Iceland,[2] with a sample of Tallow, down and Wool. You will find Sir in the packet inclosed in latin and danish a description of the foursquare peice of coin.[3] I have sent you two Charts, the only which were on the Island for your Inspection, but can not spare them long.

If I thought it could interest you, I would in the course of a fortnight give you an account of the Iceland commerce, /of/ from the year 1764 to 1784, with all imports and exports annually. The name of harbors for trade, which are by no means known here.[4]

I shall do myself the honor of attending you to morrow morning Sir, to explane anything you should wish to know

I am Sir with the greatest
Respect
your most humble ob*edient* serv*ant*
[*Signed*] Jorgen Jorgensen

NHM, BC, ff. 38–39. Autograph letter.
BL, Add MS 56301 (5) f. 53. Dawson contemporary transcript.

111. Jörgen Jörgensen, [Memorandum on Iceland], [late April – early May 1809?][5]

[1] First of all is a proclamation in which the King gives up the trade of Iceland, but at the same time reserving to himself to impose such restrictions as he may think proper. Therefore between page 1 and page 91 is nothing but royal decrees for that purpose.

---

[1] This was the ship sent by Phelps, Troward & Bracebridge to Iceland in December 1808. Jörgensen was the interpreter and returned to England on the ship to report what had happened in Iceland. Umlauts have been removed.

[2] This account does not appear to be extant, but Banks certainly made use of it in his letter to Earl Bathurst (Document 113).

[3] This description is not extant, but foursquare coins were minted in Denmark during the period 1560–1670. This would have been a rarity and none are preserved in the Coin Collection of the Central Bank of Iceland. Information from the curator, Anton Holt.

[4] See the following document, which lists all the harbours.

[5] This memorandum is in English, a summary by Jörgensen. It is lengthy so that the page numbers of the actual document have been inserted in bold. If Jörgensen kept his promise from the previous letter he would have delivered this memorandum to Banks in late April or early May.

Jörgensen is obviously translating from a book or manuscript. After a fruitless search through all likely printed books the conclusion is that this is translated from a book in manuscript form, written by Finnur Magnússon (1781–1847) listing many aspects concerning Iceland. This must have been a massive volume as the price was high. Among Magnússon's private papers in the Rigsarkivet in Copenhagen is his cash-book where he notes that on 18 March 1809, just before Jörgensen returned to London on the *Clarence*, Magnússon sold him a book for 50 rixdollars. Jörgensen probably commissioned Magnússon to gather all kinds of information for him during his January to March stay in Iceland. It is clear in all his later dealings in Iceland that Jörgensen was extremely well informed on the state of the country (see e.g. his proclamations). At that time Magnússon was a young man

Page 91
Iceland is divided in*to* three Amts;[1] <u>one</u> 1. The Southern quarter, 2 the Western quarter, 3 the Northern and Eastern Quarter. In each of those Amts are two towns: In the Southern amt Reykevig and Westmanöe; In the Western amt Grönnefiord and Isefiord. In the Northernamt Öefiord and Eskefiord.[2]

1. The District of Reykavig.

This District is divided in three Sysseler: 1 Guldbringe, 2 Kiöse and Borgefiords, the two last stretching as far as Hvitaaen, where the Westerland commences.[3] [*Margin*: Guldbringe Syssel.] The principal support of this District is the Fisheries, which is carried on not only by the resident Inhabi [2] tants, but by people from other /parts/ going t*h*ither in the beginning and the latter end of the year. The Island Vidöe[4] gives a great deel [*sic*] of Eiderdown. Sheep are not here in great quantities. Page 92. Near Krisevig[5] is good Sulphurmines[6] which were formerly drove to a good advantage. This Syssel had with Kiösesyssel in the year 1769 in all 3470 Inhabitants. The number of Farms 129. Boats here and in Kiösesyssel 616. Sheep 3225. Horncatle [horned cattle] 850. Horses here and in Kiösesyssel 1411.[7]

[*Margin*: Kiösesyssel.] Has good and plenty grass a pretty good quantity of Sheep. Fishery in certain places during the whole of the year. Number of Farms 89. Sheep 12750. Horncatle 1280.

[*Margin*: Borgefiord Syssel.] Good grass and a number of Sheep and Catle, produces a deel of woollen manufactur'd goods. Eiderdown in some places. Here [3] is found a number of Rivers and Lakes, which gives abundance of Salmons. The number of Farms 220. Inhabitants 1696. Boats 93. Sheep 20878. Horncatle 1186.

[*Margin*: Page 93.] Horses 1350. The harbors in Reykavig district are. 1) Reykavig, a very good Summerharbor, and may do in the Winter. 2) Havnefiord, good in all Seasons. 3) Kieblevig, a pretty good harbor in the Summer, but can not by any means receive Ships

---

living in Reykjavík, working both as a lawyer for the High Court of Justice (the cases were few) and as a clerk for Frydensberg the treasurer. In 1812 Magnússon moved to Copenhagen where his career blossomed and he became a professor and the Royal Archivist. See Briem, *Sjálfstæði Íslands,* pp. 369–71; Rafnsson, 'Oldsagskommissionens præsteindberetninger fra Island. Nogle forudsætninger og konsekvenser.', p. 229; Helgason, 'Finnur Magnússon', pp. 171–96. I am indebted to Professor Sveinbjörn Rafnsson for putting me on the right path.

[1] *Amt*: district (see Glossary).
[2] Modern place-names: Reykjavík, Vestmannaeyjar, Grundarfjörður, Ísafjörður, Eyjarfjörður (Akureyri), Eskifjörður.
[3] Gullbringusýsla, Kjósarsýsla, Borgarfjarðarsýsla, Hvítá (a glacier river).
[4] Viðey.
[5] Krísuvík.
[6] Original footnote: 'See Ole Henchels description and account of the Sulphurmines in the Appendix to Olavius Voyages.' The reference is to O. Henchel, 'Underretning om de Islandske Svovel=Miiner samt Svovel=raffineringen sammesteds, 30 Januar 1776' published in Olaus Olavius, *Oeconomisk Reyse igiennem … Island,* pp. 668–734. Banks owned a copy.
[7] Original footnote: 'See the account of Mr Amtmand Stephensen in the Icelandish lit. Society's 6 Vol. page 96, the quantity of Sheep and the manner of feeding them, before the Sickness came among them, and from that you may judge, what it may be again, now the Sickness is extinguished. See the Tables.' The reference is correct. The article is the following: Ólafur Stephensen, 'Um not af nautpeningi'[On the use of cattle], *Rit þess (konunglega) íslenzka Lærdómslistafélagsins,* 6, 1785, pp. 20–96, with the table mentioned, listing the number of boats, sheep, goats, cows, cattles and horses in 1770. In this memorandum this table is used for the number of boats and livestock in the year 1770.

in the winter. 4) Bosand a pretty good Summer harbor. 5) Grindevig, which cannot receive any but small Vessels.[1]

2. Westmanöe district

Divided in four Sysseler, 1) Arnæs Rangevalle and Westmanöe Syssel of the Southern quarter and western quarter. 2) Skaptafells Syssel of the Easternquarter.[2]

[*Margin*: ArnæsSyssel.] Lowland and fruitful, much Hay. The Inhabitants principal Support is their [*Margin*: Page 94] Sheep and Catle. Great deel of fresh Water Fishery, especially in Thingvallvatn, which is the largest Sea[3] in Iceland, and contains Abundance of Salmon. – The number [4] of Inhabitants 4828. Farms 385. Boats 68. Sheep 57043. Horncatle 4686. Horses 4623.

[*Margin*: RangevalleSyssel.] Best in Iceland. Here is Salmon. The Fisheries are carried on. Ruins of former commercial Houses. Number of Inhabitants here and in Westmanöe 4449, Farms 268. Boats 30, Sheep 91000, Horncatle 5152 and Horses 7202.

[*Margin*: Westmanöerne.] That is the Westman islands consists of 14 in Number, lies abrest that part of Rangevalle Syssel called Land-Eyar.[4] Only four produces Grass, and only the largest, or Heima Island[5] is inhabited. A great Number of Birds and Eggs. The Number of Farms 28, Boats 16, Sheep 900, Horncatle 44, and Horses 56.

[*Margin*: Wester Skaptafels Syssel.] Produces a deel of what is called Sandoats or Wildcorn. In some places good Sealing.

[*Margin*: Page 95.] Number of Inhabitants 2707, Farms 148. Boats 10. Sheep 15928. Horncatle 1835. Horses 2337. [5] [*Margin*: Harbors.] 1)Westmanöe is a good harbor in the Summer, but difficult for large Vessels in the Winter. 2. Örebaks[6] harbor, is very difficult. Ships must be moored by Warps[7] on Shore, besides their proper Anchors down. It is necessary to have buoys on the Cables otherwise they are liable to rub, for the ground is foul, and no Ship must pretend to go in above 13 feet deep. No Ship can remain there after Michaelmasday. The commerce of this harbor is very considerable respecting mutton and tallow.

3. Eskefiords District

Divided in three Sysseler, which all belongs to the Easternquarter: 1) Öster Skaptafels, and 2) both Mule Sysselerne.[8]

[*Margin*: Osterskaptefels Syssel.] Some Eiderdown, and Seal. Farms 52. Boats 11. Sheep 14005. Horncatle 725 and [*Margin*: Page 96.] Horses 1004.

[*Margin*: Mulesysselerne,] or as they are calld at times Östfiordene.[9] Here is a number of Sea Calves, and at times Whales seeking the coast. A deel of Thimber [6] thrifts[10] to here. The adjacent Eilands affords a deel of Eiderdown and feathers. The Islands are in

---

[1] Hafnarfjörður, Keflavík, Básendar, Grindavík.
[2] Árnessýsla, Rangavallarsýsla, Vestmannaeyjasýsla, Skaftafellssýsla.
[3] He means Lake, but in Danish *sø* means lake.
[4] Landeyjar.
[5] Heimaey, where a huge eruption took place in 1973 and over 5,000 people were moved safely to the mainland in one night.
[6] Eyrarbakki.
[7] 'Warp': rope used in towing.
[8] Austur-Skaftafellssýsla, Múlasýslur. Norður-Múlasýsla and Suður-Múlasýsla.
[9] Austfirðir (the Eastern fjords). The north and south Múlasýslur cover the east coast of Iceland.
[10] He must mean driftwood.

considerable numbers. Inhabitants 3470. Farms 363. Sheep 44967. Horncatle 1276. Horses 1824.

[*Margin*: Harbors] are 1) Berrefiord, a pretty good harbor at [*Margin*: Page 97] all Seasons. 2) Rödefiord a midling good Summer Harbor, exposed to the Winds at NNW. 3) Vapnefiord a midling good harbor at all times, only exposed to the SW.[1]

4. Öefiords District

This District contains the whole of the Northernquarter, and divided. 1) Tingöe or Norder Syssel. 2) Vadle or Öefiords Syssel, 3) Hegranes or Skagefiords Syssel. 4) Hunevands Syssel.[2]

[*Margin*: Tingöe Syssel.] Some Burning mountains. A deel of Seal. A deel of fresh Water fishery. Here is a great number of Islands, which gives a good deel of

[*Margin*: Page 98.] Grass and feathers. A deel of thrift Thimber. Some forest. Considerable Sulphurmines. Inhabitants in the whole northern quarter [7] 11985. The farms 314. Boats 78. Sheep 19960. Horncatle 841. Goats 712. Horse 937.

[*Margin*: VadleSyssel.] A Number of Seal. Here is to be seen the Ruins of former commercial houses, and a fort. Farms 360. Boats 56. Sheep 19255. Horncatle 1533. Horses 2286.

[*Margin*: Skagefiords Syssel.] Contains a Number of Islands, that produces plenty of feathers. Number of Farms 374.

[*Margin*: Page 99.] Boats 55. Sheep 50000. Horncatle 1455. Horses 1634.

[*Margin*: Hunevands Syssel.] The farms 344. Boats 8. Sheep 41680. Horncatle 1692. Horses 2011. The four harbors in the district are all Slaughterhavens. 1) Huusevig[3] a pretty good [*Margin*: Page 100.] Summerhaven, exposed to Winds from W to N, on the Eastern Side is a Sandbank, which make it somewhat difficult to go in. 2) Öefiord, is a good and large harbor during all the year. 3) Hofsos a pretty good Summerharbor, exposed to SW Winds, and is not entirely secure from heavy swells, and the Greenlandice. 4) Skagestrand is neither quite secure from the Ice and heavy swells [8] with the Winds from SW to NW.[4] From all those four last ports are exported a considerable deel of Sharkoil.

5 Isefiords District.

Divided in two Sysseler: Strande and Isefiord Syssel, with a Deel[5] of Bardestrandsyssel, the two last are called Vestfiordene.[6]

[*Margin*: StrandeSyssel.] Here is Seal. Here is found ruins of former commercial Houses. The adjacent [*Margin*: Page 101.] Islands gives a deal of Eyderdown. Inhabitants 954. Farms 120. Boats 40. Sheep 8000. Horncatle 565. Horses 577.

[*Margin*: Isefiords Syssel.] The Seas about here abounds with Seacalves, and some whale. Here is found [*Margin*: Page 102] a great deel of thrift thimber. Ædöe Island[7] gives a quantity of feathers. Formerly a Saltwork on Reykanæs.[8] Inhabitants 3338. Farms 251. Boats 110. Sheep 21740. Horncatle 695. Horses 406.

---

[1] Berufjörður, Reyðarfjörður, Vopnafjörður.
[2] Norður-Þingeyjarsýsla, (Vaðla) or Eyjafjarðarsýsla, (Hegranes) or Skagafjarðarsýsla, Húnavatnssýsla.
[3] Húsavík.
[4] Eyjafjörður, Hofsós, Skagaströnd.
[5] *deel* means *part* in Danish.
[6] Strandasýsla, Ísafjarðarsýsla, Barðastrandarsýsla, Vestfirðir (the Western fjords).
[7] Æðey.
[8] Reykjanes.

[*Margin*: Harbors.] 1) Reikefiord, a midling good Summerharbor, but not secure from driftice. The **[9]** bottom is very uneven from 10 to 19 a 25 a 35 fathoms. 2) Isefiord, a good harbor during all the year. 3) Dyrefiord a good Summerharbor, not bad in Winter. 4) Bildal a good Summerharbor. 5) Patrixfiord, a good harbor both in Summer and Winter.[1]

[*Margin*: Page 103]   6 Grönnefiords District.

This district contain the rest of Bardestrands Syssel, with four other Sysseler. 1) Dale 2) Snæfelnæs. 3) Hnappedals 4) Myhre Syssel.[2]

[*Margin*: Bardestrand Syssel.] This part contains remarkable good Eiderdown. Inhabitants 2288. Farms 177, Boats 108, Sheep 16000, Horncatle 1115. Horses 571.

[*Margin*: Dale Syssel.] It seems here as if part of the land had formerly been cultivated. Inhabitants 1704. Farms 185. Boats 56. Sheep 16645. [*Margin*: Page 104.] Horncatle 1505. Horses 919.

[*Margin*: Sneefelsness] Syssel. The Islands on the northern Sides contains great quantities of Feathers and Eyderdown. Farms 213, Boats 318, Sheep 16288. **[10]** Horncatle 1686. Horses 1071. Inhabitants here and in Hnappedal 3414.

[*Margin*: HnappedalsSyssel.] The number of Farms 67, Boats 6, Sheep 7000, Horncatle 457 Horses 660.

[*Margin*: Page 105]

[*Margin*: MyreSyssel.] Gives a deel of Salmon. Haalsöe, or Haalsisland[3] gives a great quantity of Eyderdown, and on the Islands in the Sea Langevatn[4] are great numbers of Swanns. Inhabitants 1918, Farms 165. Boats 52. Sheep 14660, Horncatle 1517. Horses 1810.

[*Margin*: Harbors] 1) Flatöe a pretty good harbor. 2) Stikkelsholm a very good Summerharbor. 3) Grönnefiord may receive in the Winter Ships of a midling draft of Water. 4) Olufsvig, requires very good Anchors and Cables 5) Stappen is rather a difficult place.[5] 6) A very good harbor, but no Ship deeper than eight feet can go in.

NB. It must be remarked where farms are mentioned it will only say, a certain piece of Land belonging to one owner, where Sheep are feeding, for here is no cultivation.

[11] From page 109 to page 113, contains a list of such goods as was then required in Iceland, but the times are so greatly alterd, that it can not be of any use at present, but will refer to the Requistion List in our possession for what goods may be of Service.

Page. 113. Gives an account of what merchandize is exported from Iceland and the amount. Gives an account that the Fish exported amounted to the Sum of 1900000 Rixdollars from the year 1774 to 1784, when at the same time the inland produce only amounted to 700000 Rixdollars, owing to the Sickness among the Sheep. From page 113 to 121 Gives an account of the manner of curing fish, and where to be exported, and a remark that feathers and down are now exported in greater quantities than before, and further remarks respecting woollen goods and the quantities exported.

[12] Page 123 Gives an account of what quantity of goods of each Sort a Ship /according to tonnage/ is able to contain, and the masters are obliged to slow his ship accordingly.

---

[1] Reykjarfjörður, Ísafjörður, Dýrafjörður, Bíldudalur, Patreksfjörður.
[2] Grundarfjörður, Dalasýsla, Snæfellsnessýsla, Hnappadalssýsla, Mýrasýsla.
[3] He must mean Hvalseyjar.
[4] Langavatn is a lake in Mýrasýsla.
[5] Flatey, Stykkishólmur, Grundarfjörður, Ólafsvík, Stapi.

Page 124. Shews what marks must be put on Packages, destined to each and from each port in Iceland.

Page 132. Gives an account since the year 1777, the danish way of housekeeping has gained great ground on the Island, and others owned far more Wine, Coffee, /*Thea*/ Tea, Sugar, Cloth, Linen and other sorts of goods to be imported than formerly.

Page 137. Gives an account of the exports from the Island consisting of. Different Sorts of Tallow, Butter, Hides, Sheep and Goat Skins. Sealskins, Foxskins white & grey, Swannskins, [*Margin*: NB. At present no Sulphur exported.] Swannsfeather, other feathers, Eiderdown, Refined Sulphur, Wool, Woollen manufactured goods, Oil of different Sorts, Beef and mutton, Different Sorts of Fish Salted and dried.

[13] Page 138. Gives an account of that at present Iceland merchants, finds it most convenient to send out Ships of their own, and not to freight them. Page 139. Gives that the whole amount of the Cargoes imported from the year 1764 to 1784 was 2,560000 Rixdollars, and on the contrary the returngoods or exports amounted to 4,665000 Rixdollars, so that when you lay to the first freight and Insurances 840710 Rixdollars, and 10 pr Cent to other expences, there will still remain a profit of 10,00000 Rixdollars. Page 140, remarks, that /the/ result will be, that if the commerce had been carried on by private merchants in the last 21 years, would have given a much greater profit /*if the trade had been carried on private Merchants*/ for the Expences, and great number of Kings servants has made it very expensive to the crown, and consequently less profits. From page 142 to page 156 is given /*an account of the goods imported gives*/ an account of the yearly quantity af goods [14] imported in Iceland. From page 158 to page 166, is the fixt prices on imports and exports, during the King had the trade. From page 168 to 176, will be seen what was exported from Iceland during one year. From page 178 to 184 gives an account of the goods or exports from the year 1764 to 1774 and from 1774 to 1784. From page 185 to 194 contains an account of freight and Insurances. Page 196 and 197 gives an Account of the whole amount of imports and exports during the 21 years from 1764 to 1784. Page 198 Gives an account of the worth of the houses and Inventories in all the harbors in the year 1784.

NB. One Remark Page 118, i have omitted and may be of service, it says. Of Tallow is more exported, than in last ten years from 1764 to 1773. The reason is the Sickness among the Sheep raged greatly, but from the year 1772, when all sick Sheep were killed, and as the Sheep which were sent on the Island begun to be more numerous, this branch of commerce [15] begun to flourish more.

Now if we turn over to the foregoing page 13, We will find, that notwithstanding numerous Servants, and bad years the clear gain to danish /Goverment/, was 10,00000 Rixdollars in 21 years. Hence we might reasonably suppose Iceland to be a rich and opulent Island, but the case is widely different, for if the Icelanders had been permitted, to export and import their own goods they would then been great gainers. But it will be found, that the trade was carried on in such a manner that the natives were by a certain Taxt[1] obliged to sell goods and receive it at such prizes [prices], so the country gained nothing for the money /*just*/ paid the Civil officers and Clergy, /*was taken from it*/ and if any remained it was sent to Copenhagen. From 1784. The King did not trade any

---

[1] *Taxt* means table of tariffs.

more, but that did not alter the case, for the poor inhabitants were then under the lash of six or seven [16] danish Merchants, which was still more insufferable than before. However the Exports and Imports are much greater at present than before, for by the different Merchants [and] Factors being on the Island, there is kind of emulation between them, and they exert themselves a deel more than the former kings Factors. That is then to say the Icelanders, slaves and works at present a great deel more, but is still as poor as ever.

All the beforegiving Accounts and Calculation, of Men, Sheep, Imports, Exports, has entirely to do with the years from 1764 to 1784. At the present the Exports are exactly of the Same Kinds as before, but the Imports as seen in the Requisition List.

At present the Icelanders are very awkward situated, they are exposed to the insults of every lawless privateer,[1] wether danish french or english. The Island as it now stands can not be of sufficient importance for british goverment to take possession [17] of it, the Icelanders themselves are sensible of it, however it may in course of time, be a good place for trade, if the inhabitants can be any ways encouraged. Therefore the only way to serve them would be to permit a small quantity of nine, or twelwe pounders to go out, about 20 in Number, and 150 Stand of Arms[2] and Cutlasses so the inhabitants in Reykavig might be enabled to keep off a good smart privateer, and they would without doubt soon declare themselves independant. It would /without/ in some cases be an act of rebellion, but not in this case, for if a goverment is willing and able to defend one of its territories, and then still the inhabitants should prove obstinate and ungovernable it would be rebellion and no honest man ought to engage in the business, but where a miserable people are daily put in disorder of fear for want of protection, they are certainly justified in declaring themselves independant, nay even to thanks from that very [18] government from which they are parting, for ridding it of a great deel of uneccessary trouble. The whole establishment of Iceland will not then receive at present, allowing every person to remain in full pay, more money than the profits of the goods imported and exported will bring in the Country, and by imposing a Tax upon all imports and exports there will still be a balance in favor of the Island, which it needs to have, to give the Icelandish banknotes value. The tax of 5 p^r Cent out and in /upon import and exports/ would be quite sufficient. On the contrary should british goverment take possession of the Island, it would be found that an Establishment would be far to havy,[3] and do more harm than good, for the country could never pay. By the Island declaring itself neutral other advantages might still accrue, for different powers may suffer Ships to go there to unload car [19] goes there, to be reshipped to other countries. It is true some powers such as france and Dannemark might object to all cargoes coming from them, but others might again suffer it. A british frigate from the Greenland Station looking into the ports now and then or making its appearance on the coast will be of infinite Service.[4]

Wisconsin, Banks Papers, ff. 214–224. Autograph document.

[1] For example, the visit of the British privateer the *Salamine* in 1808.
[2] *Stand of arms*: complete set of weapons for one man.
[3] Jörgensen probably means 'too heavy'.
[4] Compare the last paragraphs with what Jörgensen wrote to Banks in 1808 (see Document 96).

112. [Notes on the voyage of the ship *Clarence* to Iceland written down on 15 April [?] 1809][1]

and he [?] Guaranteed by Government that no Persons on the Island Should be molested for having Purchasd Goods brought from England[2]
The Farmers have 2 years Produce on their hands & in their Posession the Factors having Refusd to make their usual advances
1000 Tons of Tallow, 500 Tons of Stock fish Price 24 dollars (48s[hillings]) pr. Pound, 320 being 3 Cw^r, & nearly British weight this may be Shipd on a British Ship for Messina or Trieste.[3]
There is Plenty of Salt on the Island the people have not usd their Last supply for want of Sale
Tallow 8 d [pence] a Pound wool 6 d*itto* [pence]
Eider Down 10 shillings by Barter 100 [?] Tons 160 lb Each of Feathers at 1/lb [a shilling a pound]
piratical Seizure has been the Cause of much Evil he sent as here uncivil & indeed Ferocious persons[4] who Compelld the Stiftsamptman[5] to Rise in the middle of the night & to Give an account of all his Presents indeed to Exhibit all that was valuable
Einardsen the amptmand is invested with the Powers of Stifsamptmand[6]
Frydensberg is Landfoged Kings Receiver[7]
Koefoed Sysselman are dane[8]
10lb. Tallow, 1 lb Coffee 12 lb Tallow 1 lb tobacco
Gilpin, Hompeschs Vessel[9]
Mr Mitchel a Scotsman Resident there[10]
26 Jan a Comet seen near Jupiter above the Planet
Latter End of 1808 Sept. a Norwegian vessel a brig arrivd from Drontheim[11] laden with Provision & Fish Tackle on account of the King

---

[1] These notes were doubtless penned by Banks during his conversation with Jörgensen on 15 April and were to appear again in the letter to Bathurst written the following day. Although difficult to decipher, they are interesting as they contain more detailed information, especially on prices, than Banks uses in the following letter. The two should be read in conjunction and the notes to this document only deal with material omitted by Banks in his letter to Earl Bathurst.

[2] This refers to a promise the acting governor, Ísleifur Einarsson, was forced to make by Jörgensen, backed by the guns of the *Clarence*.

[3] The Danish factors tried to prevent the Icelanders from trading with the British. Icelanders had begun trading with the Mediterranean in the late 18th century. The numbers are difficult to decipher.

[4] Sailors belonging to the crew of the *Salamine* privateer, commanded by Thomas Gilpin.

[5] Ólafur Stephensen, former governor of Iceland. This episode is well documented.

[6] Ísleifur Einarsson (1765–1836), actually assessor to the High Court of Justice, acting governor in Count Trampe's absence.

[7] Rasmus Frydensberg, the Royal Treasurer in Iceland 1804–13, when he returned to Denmark. See Document 63.

[8] Both Frydensberg and Hans Wölner Koefoed, the *sysselman* (county magistrate), were Danes.

[9] The *Salamine*.

[10] Andrew Mitchell was a Scottish engineer who came to Denmark in 1788 on behalf of the Danish admiralty because of his expertise regarding steam engines. In 1803 he became a burgher of Reykjavík. He had apparently studied medicine in his youth and was granted the right to practise medicine in Iceland, but only in the absence of a doctor. In Icelandic sources he is titled either merchant or factor. He left Reykjavík in 1812.

[11] The *Justitia* of Trondheim.

She chargd for her Rye 21 Rixdollars [*Margin*: 42 shillings Per–] per Barrell of 160 lb the Icelanders would not buy it.

The *Clarence* from Liverpool arrivd 12 Jan being a Letter of Marque She took the norwegian[1] [ship] but Releasd her on a Convention[2] made with the Island for the supercargo[3] to leave [?] & cease to Trade now & hereafter he is Left on the Island & is Guaranteed by government there

another Norwegian is expected[4]

Einarsen the Stifsamptman

5 day after Leaving England we made Iceland on the 12 [January] anchored in Hafnefiord

They Producd American papers but were not allowd to trade.[5]

They then produced a British Licence but this was disallowd from the fear of the Goods bought of us being Confiscated

on this they produced their Letter of Marque & Captured the Norwegian brig the next day they agreed to Release her on Condition of being Permitted to Land & dispose of their Cargo[6]

Hompesch was not in Iceland[7] Gilpin the Capt*ain* Treated the People well paid for what he had from individuals & Took nothing but the Kings Chest[8]

All the clergy Bishops & all are paid out of this Chest also the functionaries of Government the money is Raisd by a Small Tax which is not Sufficient

10 to 12.000 R*ixdollars* are sent annualy from Denmark.

Danish Traders
        Thomsen[9] Factor to Norberg Jutland
        Strube[10]
        ~~Petreus~~[11]
        Knutsen[12] factor to Petreus
          not [?] 10 Danes more in the Island.

---

[1] The above-mentioned vessel.

[2] The agreement was signed on 19 January 1809.

[3] James Savignac, the supercargo on the *Clarence*, who stayed on in Iceland while Jörgensen returned to England.

[4] This was the *Providentia*.

[5] It was usual during the Napoleonic Wars for ships to carry false American papers, the Americans being at the time a neutral nation.

[6] All the information on the *Clarence* is essentially correct, well documented in Icelandic sources.

[7] Baron Charles von Hompesch was a Hungarian noble and served in the Prussian and Austrian armies. At this time he was lieutenant-general in the British army, where he commanded a corps of Hussars. He had hired the privateer *Salamine* and obtained a letter of marque, but did not himself go to Iceland. He had, however, conducted the raid on the Faroe Islands in 1808. On returning to Lerwick with their booty he left the ship and thus did not join the next expedition to Iceland which was captained by Thomas Gilpin.

[8] The *Jarðabókarsjóður*: see Ingimundardóttir, *Skjalasafn landfógeta 1695–1904*.

[9] Jesse Thomsen, merchant in Reykjavík.

[10] Christian Conrad Strube, merchant in Reykjavík.

[11] Petræus was one of the captured merchants, who had not yet arrived back in Iceland. Crossed out by Banks probably because he was now considered an 'Icelander'.

[12] Lars Knudsen, merchant in Reykjavík.

no danish vessel had arrivd in 1808 but the norwegian. She had /not/ Sold little of her Cargo having set an unreasonable Price.

all the merchants in the island had Enterd into a Combination to Raise the prices of all Danish & Colonial articles

NHM, BC, ff. 64–65. Banks's autograph.

113. Banks to Earl Bathurst, 16 April 1809

Soho Square

My dear Lord[1]

I enclose the information[2] I have been able to obtain relative to the present situation of Iceland & the British Voyage to the Island, undertaken under your Lordships licence, I shall have I hope more to communicate on the subject in due time.

The *Clarence* saild from Liverpool Dec. 29 & made Iceland the 5th day after losing sight of the British Isles.[3]

The Danes had not visited the Island since Autumn 1807 except the Norwegian Brig[4] mentioned in the letters which had not much corn on board, & no other articles of necessity, she belongd to the Danish government and had orders to sell her Cargo at so excessive a rate that she had not as is allegd sold 8 barrels of Rye, when the *Clarence* arrivd, the *Clarence* carried Potatoes & Barly Meal, the Potatoes were ready sale, the Barly Meal was not at first relished but before the Ship saild the Icelanders had found the way to make very good Bread of Barly Meal & Potatoes. The small quantity of Colonial

---

[1] The recipient of this letter has erroneously been thought to be William Wellesley-Pole. In *The Banks Letters*, Warren Dawson stated the letter was written to William Wellesley-Pole, secretary of the Admiralty (brother of the Duke of Wellington and the Marquess Wellesley) as is pencilled on the letter itself in the BC Collection in the Botany Library of the Natural History Museum. This is incorrect, as the letter is addressed to 'My dear Lord' and Pole was not ennobled until 1821, becoming Baron Maryborough and later 3rd Earl of Mornington in 1842. In 1809 he was usually addressed as 'Dear Sir' (e.g. TNA, ADM 1/692, Nagle to Wellesley-Pole, 5 May 1809). One must assume Banks knew the correct way of addressing people. There is no doubt that Banks was writing to Earl Bathurst, President of the Board of Trade. The letter advocates the annexation of Iceland to 'his Majesty's ministers' and in the Portland ministry the lords most likely to be recipients were Liverpool (Home Office) Castlereagh (War) and Bathurst (Board of Trade). Liverpool is immediately ruled out as he is mentioned in the letter. Bathurst, rather than Castlereagh, is doubtless the recipient for the following reasons: first, the letter deals with trade matters (see especially the postscript). Second, the letter speaks of 'your Lordship's licence'; licences were usually issued in the name of the Home Secretary, in this case Liverpool, who cannot be Banks's correspondent. However, these licences were trade licences, so it would not be unnatural to speak of them as licences of the President of the Board of Trade. Third, in the Historical Manuscripts Commission, *Report on the Manuscripts of Earl Bathurst*, there are two documents relating to Icelandic affairs: James Savignac's letter to Phelps & Co., dated 19 March 1809, which gives information on the *Clarence* voyage, and an extract of Andrew Mitchell's journal, which describes the Gilpin raid (pp. 54–6), doubtless the enclosed information Banks speaks of in the first sentence.

[2] This would be information from Jörgensen: see letter of 14 April 1809 (Document 110) and the above-mentioned documents in the Bathurst papers.

[3] The *Clarence* was a 310 ton brig sent by Phelps, Troward & Bracebridge. It was furnished with a letter of marque (TNA, HCA 25/192, no. 167). The ship had arrived in the harbour of Hafnarfjörður on 12 January 1809.

[4] The *Justitia* from Trondheim had arrived in October 1808.

produce consumd by the Islanders sold at high rates as did Spirits & coarse British Manufacture.¹

There are in Iceland about 10 Danes, two in office <u>Freydenberg</u> is Landfoged or Kings Receiver² <u>Koefoed</u> is Sytelman [*sic*]³ of Hafnefiord an Office somewhat resembling our High constable or rather what our High /Constable/ was anciently when the Office was more respected. The rest of the Danes are Factors for Danish merchants, who have Factories or Stores on the Island both the Danes in Office signd the Convention.⁴

There is one Scotsman <u>Mitchel</u>/⁵ by name who has long been resident there & is well acquainted with the language & customs of the country.

Baron Hompesch did not go ashore in person on the Island,⁶ he sent his Captain Gilpin & two Irish sailors & it seems their force was sufficient to conquer the Island, for these Men made the Governor rise in the middle of the night, open his Chests & his Cupboards, exhibit all his Money Plate & Goods, the most valuable of which they carried off to the Ship, on due consideration however of the enormity of this proceeding the Goods or at least such of them as the plundering party had not converted to their own use, were returnd.⁷

The King's Chest⁸ as it is calld which is supplied by a small Tax levied on the Island & a remittance of about 12,000 Danish Rixdollars annually to make up deficiencies about £1,200, was the chief prize of Hompech, he made however many Icelanders exchange Bank Notes found in it for Silver by force and threats.

The Kings Chest is the source from whence all the Clergy in the Island & all Public functionaries are paid, the whole of these are therefore in a state of lamentable poverty & most willing to be relievd.

I have no doubt on this occasion to repeat the opinion I gave in the Autumn 1807 to my friend Lord Liverpool,⁹ that it is expedient Iceland should be annexd to the Crown of England & never hereafter separated. The owner of the *Clarence*¹⁰ offers to make the conquest /with a privateer/,¹¹ it may certainly be effected by a Gun Brig & protected by a Frigate being orderd to look in once or twice in a Summer from the Greenland Station,

---

¹ The cargo list of the *Clarence* consisted of 6,444 lb of coffee, almost 3,000 lb of sugar, 5,975 pots of rum, 3,598 lb of tobacco, 1,743 hats, 1,800 barrels of Liverpool salt and 278 barrels of barley (Þorkelsson, *Saga Jörundar hundadagakóngs*, p. 205).

² Rasmus Frydensberg.

³ Koefoed was a *sýslumaður* or county magistrate, residing in Hafnarfjörður.

⁴ This refers to the trading agreement of 19 January 1809 permitting the British merchants to trade despite the official prohibition. A British show of force obliged the two officials, Koefoed and Ísleifur Einarsson, to sign the agreement.

⁵ On Andrew Mitchell see the previous document, p. 318, n. 10.

⁶ Hompesch (see above, p. 319, n. 7) did not go to Iceland, as indeed Banks noted in the previous document, though the wording might suggest he did. Captain Thomas Gilpin was the protagonist in Iceland.

⁷ On Gilpin's Raid see Agnarsdóttir, 'Gilpinsránið 1808', pp. 60–77.

⁸ The *Jarðabókarsjóður*, also called in English contemporary accounts 'the King's Money Box' (e.g. Document 112).

⁹ Document 63.

¹⁰ Banks is referring to Samuel Phelps, even though the ship was chartered and belonged to a Mr Hurry of Liverpool.

¹¹ These important words are penned in Banks's autograph.

in truth was any attempt to be made by the Danes to interrupt our occupation of it, the News would be brought to England in a Week & a Force to recapture it sent out in another week.

The Natives are desirous of becoming subjects to a State which is able & willing to protect them in the enjoyment of national liberty, they will make no resistance & ten Danes surely can make none, were the Icelanders allowd to bring their Goods to a fair Market they would soon increase, & as there are not in this world a more hardy set of Men, would in time afford a resource of no inconsiderable importance for Manning the British Navy.[1]

I need not enlarge on this subject; to his Majestys Ministers & to their wisdom I commit it to be dealt with as they see fitting, if my opinion meets with approbation I shall be diligent in collecting information of all kinds.

I have the honour to be
Your Lordship's most faithful
& obedient H*umb*le Servant
Jos: Banks

[P.S.] The Norwegians are I understand well supplied with Rye which they got from Archangel if they can get round the North Cape they can carry it by a Navigation within Islands to the South of Norway & have gotten some of it into Denmark.

NHM, BC, ff. 42–43. Contemporary copy, amanuensis (William Cartlich) with an addition in Banks's autograph. The original must have been sent to Bathurst.

114. Samuel Phelps to Banks, 21 April 1809

Cuper Bridge [Lambeth]

Mr. Phelps has the honor of inclosing S*ir* Joseph Banks the petition which /they/ he wishes with his approval to present to the Treasury. Should it be possible to obtain a Convoy or protection for their trade with Iceland[2] it would he hopes be of national as well as individual benefit to him & his partners,[3] & Mr. Phelps would in that case certainly embark for Iceland.

NHM, BC, f. 44. Autograph letter.[4]

---

[1] See Lord Hawkesbury's letter to Banks of 29 November 1807 (Document 46) where he wrote: 'the Island of Iceland could be secured to His Majesty, at least during the continuance of the present war. In that case ... I should hope that we might be able to obtain the services of some of their mariners.'

[2] See below, p. 323, n. 4.

[3] Phelps's partners were Abraham Bracebridge and Richard Troward.

[4] Note in Banks's autograph, verso: '[Received] April 21 [18]09'.

ICELAND CORRESPONDENCE AND DOCUMENTS

115. Samuel Phelps to Banks, 22 April 1809

Cupers Bridge [Lambeth]

Mr Phelps takes the liberty of informing Sr Joseph Banks that he intends to wait at the Treasury (Mr. Lacks office)[1] from one to two o'Clock this day in hopes of receiving a favourable answer to the petition for a licence which he had the honor of sending Sr Joseph Banks yesterday, or for the purpose of making any alteration /in it which/ Sr Joseph Banks will please to point out as necessary.[2]

[P.S.] It would be a great protection to the trade of Iceland if /*government*/ their Lordships would allow Mr Phelps in case of need to land a few Guns, Arms & ammunition on the Island, & should Government send out a Vessel for the protection of the trade Mr. Phelps would be very thankful if he could get a passage out in that vessel, which would place him in safety in case a vessel of war might reach Iceland from Norway or Denmark before Messrs Phelps & Co.'s Vessels arrive there – unless such vessel could convoy Messrs Phelps & Co's vessels out.[3] Mr. Phelps would beg leave to recommend some sufficiently powerful vessel of war to be dispatched from Liverpool or Greenwich as it is the nearest passage, if not from the Leith station;[4] but the best security /for the trade/ would be to convoy their merchant vessels out.

NHM, BC, f. 45. Autograph letter.[5]

116. Sir George Steuart Mackenzie to Banks, 20 May 1809

Coul, Dingwall

Sir,

I take the liberty of addressing you on a Subject which I am sure will be very interesting to you. I have accidentally met with a young Icelander while visiting a part of my estate on the

---

[1] There is no record in TNA, T 4/13, a register of petitions for this period, of a petition from Phelps on this subject and indeed this petition has not been found in the TNA. That is hardly surprising as the Treasury did not issue licences. (The only petition from Phelps from this period found in the Treasury papers is a memorial requesting to be allowed a drawback on 3 chests of tea to be exported to Iceland, which the Treasury refused to grant. The Treasury dealt with this request on 26 May (TNA, T 11/48, no. 5,836) when Phelps was making preparation for his trading venture to Iceland and this reply is typical of the problems merchants faced during the wars.) In his book *Observations on the Importance of Extending the British Fisheries*, Phelps does not mention the Treasury but wrote that he had the backing of Banks, the Board of Trade and the Admiralty which is amply documented (pp. 53–4).

[2] Banks would have advised him to approach the Privy Council for Trade and Plantations, which did indeed issue him with licences for two ships. Phelps was granted the licence to trade with Iceland on 2 May 1809, to which was added lists of the permitted articles for export and import (TNA, PC 2/181). He was further granted a licence for the *Flora* (TNA, PC 2/181, 18 May and 1 June 1809).

[3] Phelps travelled out to Iceland on his merchant-ship, the *Margaret and Ann*.

[4] In this they were successful, with Banks's help (see Document 119). The Admiralty agreed to send a sloop-of-war to be stationed off Iceland until the end of the trading season. The purpose of the sloop being sent into Icelandic waters was 'to protect the vessels laden with supplies for the inhabitants of Iceland, as these supplies are sent in vessels licensed by the British government' (TNA, ADM 2/1368, Wellesley-Pole to Rear-Admiral Sir Edmund Nagle at Leith, 9 May 1809). And on 22 May 1809 the *Rover* sloop-of-war was despatched from Leith for Iceland (TNA, ADM 53/1103, 'A Journal of the Procedure of H.M.'s sloop *Rover*'). The Admiralty also offered a convoy (see Document 119).

[5] Note in Banks's autograph, on left margin of last page: 'Mr Phelps | [Received] April 22'.

west Coast of Ross. This young mans Story is sufficiently interesting to have determined me to protect him & to give to him every advantage in prosecuting his Studies. – About 6 years ago, a professor[1] was sent to Iceland to instruct some young native of ability in Medical Science & in surgery, & in order to introduce the vaccine inoculation. My new aquaintance from his ability, was selected for the purpose, & he lived 5 years with the Professor – He was now determined to send him to Copenhagen, & his all was embarked in a ship bound for that City. – As he could not depart by that ship, he waited & took the opportunity of going in a Merchant Ship, which was captured, & driven by stress of weather into the Island of Lewis about a year ago.[2] William Loptson,[3] (for so my young friend is called) was much astonished at the kindness of the people in Lewis & he desired to remain there, which some people who took an interest in him accomplished for him. – He has acquired the English & Gaelic languages. – I have had much conversation with him respecting Iceland & its present state appears to be such as to make it very desirable for our gracious King to take it under His protection. I made several enquiries as to the articles of produce & found that that there were many remaining quite unnoticed, which would be sources of great wealth. – I have written to Lord Castlereagh on the Subject, & have proposed that a Sloop of War shall be sent to take possession.[4] I have offered to go & to make every enquiry, thro' Loptson, so as to make up a report with which I may return before Winter. – I have also offered to go back next Spring & to spend a Winter there; & have hinted that I should be thankful to be employed as Governor.[5] I would bring Loptson back, in order that he might attend the medical classes &c. at Edinburgh & be prepared to accompany me in my second voyage, when I shall take measures, if Government are favorable to the Scheme, for some scientific observations – I hope that you will have the goodness to second my application to Lord Castlereagh, by waiting on His Lordship as soon as you can. When I get a footing in Iceland, I will engage to explore it perfectly, & for the sake of Science. I hope you will be kind enough to endeavour to influence Lord Castlereagh in the cause. – Loptson assures me that the Danish Government has long neglected the Island, & kept the inhabitants in great distress. – And that they will most gladly place themselves under the care of our Sovereign to whom he has sworn allegiance – I would dwell longer on the subject & enter more into the detail of such sources of wealth

---

[1] This is a reference to Thomas Klog, state physician in Iceland 1804–13. He was to instruct students in medicine.

[2] The ship was *De Fem Brödre* (*The Five Brothers*), which had been driven by gales into Stornoway. See further on this ship in Document 71.

[3] His real name was Ólafur Loftsson (b.1783?). In 1810 he was Mackenzie's guide and interpreter in Iceland, during which period Loftsson's true character was revealed. He 'practised upon us numerous deceptions ... we ... had much reason to be dissatisfied with his conduct as a guide & assistant to us in our travels – His manner & behaviour ... were highly ridiculous ...' etc. (Holland, *Iceland Journal*, pp. 85–6, 263–4; and see further Mackenzie, *Travels in Iceland*, pp. viii–ix).

[4] This letter is not in the TNA, though it may be extant as Castlereagh papers are to be found in Durham and Northern Ireland. What is remarkable is the fact that both Banks and Mackenzie, independently, were reaching the same conclusion about the benefits an annexation of Iceland would have for Britain. Mackenzie was encouraged by his fortuitous meeting with Loftsson while Banks, as has been seen, had been of this opinion since 1801. Banks would have been loath to aid him. He did not have a high opinion of the Scot (see Document 177) and had decided to send his protégé William Jackson Hooker to Iceland with Phelps & Co. to make a collection of plants. Mackenzie, however, set off for Iceland on a successful expedition the following year, in 1810.

[5] Like the Hon. John Cochrane (see Document 29), Mackenzie, a father of 10 children (though only 4 or 5 born at this time), was looking for a position in society.

as I know are to be found in Iceland[1] – If I am not mistaken you gave to the British Museum a number of Icelandic Mss.[2] Loptson will soon be fit to translate them, & I should be very glad to assist him. – I shall not trouble you more at present – but I earnestly request that you will write to me soon. – There is yet ample time for a voyage if I set out by the end of June. – I have the honor to be with much regard

yours sincerely
[*Signed*] G. S. Mackenzie Bt.

NHM, BC, ff. 46–47. Autograph letter.

117. Banks to Sir George Steuart Mackenzie, [20? May 1809]

Government gave up the Capturd Iceland Ships & their Cargoes to the owners on Condition of their Proceeding to the Island with supplies & returning with their Cargoes to England.

This does not Carry the appearance of any disposition on the Part of Government to take Posession of Iceland nor do I think it Likely they Should Resolve to do So I have had many conferences with H.M Ministers about it & as I Saw all the masters of the Capturd Iceland Vessels that were resturd I conclude I have given them the best information on the State of the Island that can be Obtaind besides which an English Ship which Saild in the winter has returnd leaving her Cargo there[3] & is going back with two[4] more to bring back the Proceeds.

NHM, BC, f. 47v. Draft reply in Banks's autograph on letter from Mackenzie dated 20 May 1809.

118. Banks to Ólafur Stephensen, 28 May 1809

London

Dear Sir & my very old Friend!

It has given me sincere pleasure to hear from various persons lately, that you are still alive & comfortable.[5] The friendly reception I met with from you & your family 37 years ago I never have nor ever shall cease to remember with pleasure & gratitude.[6]

I have tried to return the acts of friendship I received from you and abundance of other good Icelanders when I was among you, to those of your countrymen, who were brought here as prisoners of war & tho' I have not succeeded in all matters as I could have wished,

---

[1] Such as sulphur, the fisheries etc.
[2] This was correct, see Introduction, p. 21 above.
[3] The *Clarence*.
[4] The *Margaret and Ann* and the *Flora*.
[5] Banks had misunderstood Magnús Stephensen's letter of 17 October 1807 (Document 40) where he mentioned 'the Manes' of his father. *Manes* means shade of departed person. The reason is probably Magnús's imperfect knowledge of English. Jörgen Jörgensen, recently returned from Iceland, had evidently told Banks that the elder Stephensen was alive and well. The British visitors were to enjoy an incredibly sumptuous meal at Stephensen's invitation the following summer (Hooker, *A Tour in Iceland*, I, pp. 58–78; and Document 121, p. 329, n. 5).
[6] See Introduction, pp. 29–30 above.

I have, I hope, done them some acceptable service. I shall never cease to consider those Icelanders, who may come to England, as friends who have a right to ask for, and to receive every act of friendship & assistance I have the power to confer.

Allow me to recommend to your particular protection the bearer of this M^r. Hooker,[1] a young English Gentleman of property, who visits your country from the same motives of curiosity as induced me to visit you in 1773:[2] Do me the favor to recommend him to all my friends in Iceland, and request them to promote his scientific researches, which are chiefly botanical, and provide for him the assistance he may have occasion for in travelling over your Island.

I beg also to recommend to your friendly offices the proprietor of the ship – M^r. Phelps[3] – he, also, is a scientific, as well as a commercial man, and will, no doubt, visit the curiosities of your Island and bring me valuable accounts of many matters which the shortness of my stay on the Island prevented me from investigating.

Sir John Stanley who visited your Island since I was there,[4] and who received those hospitable civilities which Icelanders always bestow on strangers whose conduct they approve, is well and happy, and desires to be remembered in the kindest manner to you, and to his other friends. I send you by M^r. Hooker some prints of Geyser engraved from the drawings made by Sir John's Artist.[5]

I beg my dear Sir and very old friend that you will believe me
Your assured Good Friend & faithful Humble Servant
(signed) – 'Joseph Banks'

RA, Rtk 373.133. Contemporary copy.
Þorkelsson, *Saga Jörundar hundadagakóngs*, pp. 147–8.
Hermannsson, 'Banks and Iceland', p. 59.

119. John Barrow to Banks, 2 June 1809

Admiralty

Dear Sir Joseph

Sir E Nagle[6] has been directed (in consequence of a letter from you I believe) to give protection to the trade of Iceland & he has constantly a Sloop of War stationed near that

---

[1] William Jackson Hooker, later Director of Kew Gardens, was at this time Banks's protégé and had been sent to Iceland by Banks to gather plants (see Appendix 13). He mentions this letter in *A Tour in Iceland*, I, p. 61.

[2] Banks's expedition to Iceland took place of course in 1772. Banks made this mistake frequently. Yet the figure of 37 years he mentions at the beginning of the letter is correct.

[3] Samuel Phelps (see Appendix 13). Note that Banks is here recommending the man who would be held responsible for the Icelandic Revolution.

[4] Stanley visited Iceland in 1789. He had succeeded to his father's title in 1807, becoming Sir John Thomas Stanley, 7th baronet of Alderley Park in Cheshire.

[5] Nicholas Pocock (1741?–1821), a well-known marine painter. John West wrote in his introduction to the third volume of the *Journals of the Stanley Expedition* that both Stanley and Baine made sketches while in Iceland. Later Nicholas Pocock and Edmund Dayes (1763–1804) made paintings and watercolours from Baine's sketchbook, many of which have been preserved.

[6] Rear Admiral Sir Edmund Nagle (1757–1830) began his naval career in 1770, was made rear admiral in 1805, eventually becoming an admiral in 1819 and an aide-de-camp to the Prince Regent. At this time he was the commander-in-chief at Leith.

Island;¹ but no application has been made for Convoy from the Owner.² Will you therefore have the goodness to state what Ships are ready to proceed & to what place you would wish to have them protected and we will direct the Admiral at the Nore to furnish the necessary Convoy.³ As they are going to an Enemy's Island it is presumed they are furnished with proper licences.⁴

I am Dear Sir Joseph
very truly yours
[*Signed*] John Barrow

NHM, BC, ff. 48–49. Autograph letter.

120. Charles Dietrich Eberhard Konig to Banks, 24 June 1809

British Museum

Sir

I have had an opportunity of showing your volume of Icelandic Poetry⁵ to Mr Herbert⁶ (brother to Lord Carnaervon)⁷ who, I suppose, is the only person in this country

---

¹ On 1 May 1809 the Lords of the Admiralty had decided to send a sloop to protect Iceland waters. The vessel would be relieved from time to time so that a sloop-of-war would be constantly stationed off Iceland until the end of the season. The sloop would protect 'the vessels laden with supplies for the inhabitants of Iceland, as these supplies are sent in vessels licensed by the British government it is supposed they may be molested by the Danish privateers' (TNA, ADM 2/1368, file on Secret Letters, Wellesley-Pole to Nagle, 9 May 1809; see also: TNA, ADM 2/1368, special minutes, 1 May 1809, and TNA, ADM 1/692, Nagle to Wellesley-Pole, 5 May 1809).

² Samuel Phelps.

³ The following day Barrow ordered the *Flora*, one of Phelps's ships to sail in convoy (TNA, ADM 2/1103, John Barrow to Vice Admiral Wells, Sheerness, 3 June 1809; and Wellesley-Pole to Captain Jones, Senior Officer, Sheerness, 5 June 1809). No letters regarding Phelps's other ship, the *Margaret and Ann*, have been found.

⁴ The *Margaret and Ann* received a licence on 2 May 1809 and the *Flora* on 18 May 1809 (both in TNA, PC 2/181).

⁵ This letter is bound in volume ADD MS 6121 'Icelandic Poetry presented by Sir Joseph Banks'. On the first page it says 'Presented by Sir Joseph Banks Feb. 8 1812', doubtless the day Banks presented it to the British Library. In pencil above, in Banks's autograph, is written 'cottles islandic poetry 1797'. This is a reference to a paraphrastic translation by A. S. Cottle of Magdalen College, Cambridge, of *Icelandic Poetry or The Edda of Saemund*, published in 1797 with an 'epistle' from Robert Southey (1774–1843), poet laureate 1813–43, and (along with Wordsworth and Coleridge) one of the 'Lake Poets'. I am indebted to Professor Andrew Wawn of the University of Leeds for great help with this literary letter.

The 'List of contents' in Latin (f. 3r–v.) is by G. Vidalinus (Geir Vídalín, at the time bishop of Iceland, see Appendix 13), and also the same list in Danish translation (f. 4–v.). The volume contains 120 poems, beautifully written over 87 folios. Letter, poems and lists are all bound together. The first page is entitled 'Fra Starkade og Vicare kge', (from the *Gautreks saga*; English translation by Pálsson and Edwards in *Seven Viking Romances*, pp. 138–70). There are sometimes marginal notes and comments written on the pages between the poems.

⁶ The Revd and Hon. William Herbert, DD (1778–1847), 3rd son of the 1st Earl of Carnarvon, was a classical scholar and a prolific translator not only from Icelandic but from a variety of European languages. He also wrote poetry based on Norse subjects. He was an MP 1806–12, but left politics to enter the Anglican ministry, eventually (like his brother) becoming Dean of Manchester. He was also a noted botanist.

⁷ Henry Herbert (1741–1811), 1st Earl of Carnarvon and Dean of Manchester.

conversant with the subject & /who/ has lately published a work on it.[1] This gentleman was very much pleased with the sight of the book, it being so beautifully written & perfectly legible; which is more than can be said of most other Icelandic MSS; /which/ particularly on account of the many ab*b*reviations that occur in them. N[os]. 16, 17, & 18,[2] Merlin's prophecy, are quite new to Mr. Herbert and he supposes they have never been published; the same he thinks of N°. 15:[3] the prose, which is written in a running band[4] between the stanzas, is an exposition or explanation of the verse. N° 5[5] is a celebrated poem which he has never before seen; but he believes it is printed. The last half of N° 1 is printed, with some variations in the prose parts, in the the 'Saga af Gautreki oc Hrolfi', Upsal 1664, from whence Mr. H. translated it: part of it is also to be found in a work of Bartholinus[6]

N. 6[7] is probably not printed; it differs from Harald Harfagens Saga and the verses quoted are not the same; but one of the principal passages is printed in Heimskringla Har. Harf. Saga, from whence Mr. H. has translated it. He does not know whether N° 12[8] be printed or not – N[os]. 13 & 14[9] were written by Egill Skalagrimson, whose Saga has been printed by the Icelandic Society at Copenhagen;[10] he has not seen the volume, but he doubts not that both these poems are printed therein. – N. 14 is also printed in Worm's Literatura Runica[11] & from thence reprinted by Bishop Percy.[12]

---

[1] William Herbert published *Select Icelandic Poetry translated from the originals with notes in 2 parts* in London 1804–6. His Icelandic translations were acclaimed by both contemporary reviewers (including Lord Byron) and modern scholars 'as marking a new departure towards the professionalization of Old Norse studies in Britain' (Ross, *The Norse Muse in Britain 1750–1820*, p. 183).

[2] No. 16 is *Merlinus spaa*, no. 17 is *merlinus spaa* [sic] and no. 18 is *Fragmenta af Merlinsspá*. Apparently three versions of the same work. *Merlínusspá*, a poem by the monk Gunnlaugur Leifsson (d. 1218 or 1219), a scholar, writer and poet of the Benedictine Monastery at Þingeyrar. The *Merlínusspá* is a Norse translation of *Prophetiae Merlini* by Geoffrey of Monmouth.

[3] No. 15 is *Biargbúa þáttr*. One of the poems in the *The Edda of Saemund*, the *þáttr* is a prose narrative that includes a 12-stanza poem. The poem itself is not an 'Eddic' piece though it contains mythological references. There is an English translation entitled 'The Tale of the Mountain-Dweller' (*Bergbúa þáttur*) by Marvin Taylor in *The Complete Sagas of Icelanders*, ed. Viðar Hreinsson, II, pp. 444–8.

[4] Running band means that prose and poetry alternate.

[5] No. 5 is Lióþ Örvar Odds (*Ljóð Örvar-Odds* or *Örvar-Odds saga*), the poem or saga of Örvar-Oddur. There is an English translation by Pálsson and Edwards, *Seven Viking Romances*, pp. 25–137.

[6] Thomas Bartholin (1659–90), antiquarian to the Danish King. The work referred to is the *Antiquitatem Danicarum de Causis Contemptæ a Danis adhuc Gentilibus Mortis. Libri Tres* from 1689.

[7] No. 6 is Af Haralldi kgi enom Hárfag[w] (Of King Harald Fairhair).

[8] No. 12 is Kvæþi Gísla Súrsson[r] (a poem on Gísli Súrsson, the hero of *Gísla saga Súrssonar*). 'Gisli Surssons's Saga' has recently been translated by Martin Regal in *The Complete Sagas of Icelanders*, ed. Viðar Hreinsson, II, 1–48.

[9] No. 13 is Sona Torrek Egils Skallagr (*Sonatorrek Egils Skallagrímssonar* 'The Irreparable Loss of Sons'), and No. 14 is Haufutlausn Eg SkgrS. (*Höfuðlausn Egils Skallagrímssonar* 'The Head Ransome of Egill the Scald'). These are the two major poems of Egill Skallagrímson, the hero of *Egils Saga* (English translation by Bernard Scudder in *The Complete Sagas of Icelanders*, ed. Viðar Hreinsson, I, pp. 33–177), the former composed after the death of his two sons, and the latter written in praise of King Erik Blood-Axe.

[10] This is *Egils saga, sive Egilli Skallagrimii vita* ..., edited by Guðmundur Magnússon and Grímur Thorkelín in Copenhagen 1809 (Icelandic and Latin texts).

[11] Always referred to as *Literatura Runica*. Ole Worm, *Seu Danica Literatura Antiqvissima, Vulgo Gothica dicta, luci reddita, opera O. Wormiii; acc. de prisca Danorum poesi dissertation*, Amsterdam, 1636.

[12] Bishop Thomas Percy (1729–1811) was a pioneering translator of Norse poetry, e.g. *Five Pieces of Runic Poetry*, London, 1763. See Ross, *The Old Norse Poetic Translations of Thomas Percy*; Wawn, *The Vikings and the Victorians*, passim; also Groom, *The Making of Percy's Reliques* and Kristmannsson, *Literary Diplomacy. The Role of Translation in the Construction of National Literatures in Britain and Germany 1750–1830*, esp. I, pp. 122–205.

This is all Mr. Herbert has communicated to me respecting this beautiful MSS. I hope it may be satisfactory to you I remain

with great respect
Sir
Your most obed*ien*t & obliged Servant
[*Signed*] Chas Konig

BL, Add MS 6121, ff. 1–2. Autograph letter.

121. Ólafur Stephensen to Banks, 10 August 1809

Widoe[1]

Right honorable Patron![2]

I had the satisfaction and honor to reseive from your noble hand, a Letter dated the 28 May[3] this year, and some bewtifull Pictures of Geiser,[4] which oblige me wery much, and bring me in New Remembrance of your noble friendship, for this, and farther more for the distinguished goodness you have s*h*ewn against so many of my Country Men who were brought as prisoner /of war/ to Great Britain, I have the honor to give you my most humble thanks, and the said my Countrymen, implore the supreme being to preserve you and bestow on you its Blessings. –

That I have done my duty in promoting Mr. Hookers journey through several Districts in this Island, I am sure he him selv will tell you,[5] I only wishet to get another opportunity to make you a much more welluable Servise, likewise I wished to have done the great Merchant Mr. Phelps some servise. –

---

[1] Viðey, an island off Reykjavík, the residence of Ólafur Stephensen. The island can be seen in the distance in Plate 14.

[2] There are some spelling mistakes in this letter. It was probably translated by Magnús Stephensen, his son, who had been studying English during his enforced stay in Copenhagen 1807–8, when no ships sailed to Iceland.

[3] This letter is a reply to Document 118.

[4] Hooker wrote that he presented Stephensen with the above-mentioned letter of introduction, and a present of prints of the Geysers taken from drawings made by Stanley in 1789, and books from Banks, 'whose very name made him almost shed tears' (Hooker, *A Tour in Iceland*, I, pp. 61–2).

[5] The day after the seizure of power in the Icelandic Revolution, namely 27 June 1809, Phelps, Jörgensen and Hooker were invited to visit Ólafur Stephensen, who met them in his full-dress uniform as Governor of Iceland, though without the sword. Hooker described it thus: 'His coat was of scarlet cloth, turned up with green, and ornamented with gold lace: his pantaloons of blue cloth, with gold trimmings; and he had half-boots with gold bindings and tassels, and a three-cornered hat, likewise ornamented with gold tassels, and trimmings of the same, and with a long white feather.' There Hooker and his companions 'suffered from the hospitality' of Stephensen. Before taking a walk around the island they were offered rum and Norway biscuit. On their return they sat down to a dinner consisting of a soup of sago, claret and raisins, followed by salmon with melted butter, then each guest was presented a dozen hard-boiled eggs of the Cree (the arctic tern) with a sauce of cream and sugar, followed by well-roasted mutton with 'a mess of sorrel' and finally two huge waffles. Each guest had a bottle of claret by his place, to be followed by coffee and rum punch. Hooker continued: 'If at any time we flagged in drinking, "Baron Banks" was always the signal for emptying our glasses, in order that we might have them filled with bumpers, to drink to his health; a task that no Englishman ought to hesitate about complying with most gladly, though assuredly, if any exception might be made to such a rule, it would be an instance like the present.' Three cups of tea each brought 'this extraordinary feast' to an end (Hooker, *A Tour in Iceland*, I, pp. 58–72, covers the visit in detail).

I dare humbly beg you, by opportunity, to bring my most obliging compliments for Mr. Jon Stanley,[1] whom I have formerly had the honor to see in my own house. –

Dear Patron! I wish you for the Future would bestow on me the same your estimated Friendship as before, and give me the honor to be

Most honored Sir
your most humble Friend and Servant
O Stephensen[2]

Wisconsin, Banks MS, ff. 30–31. Contemporary translation.
The original in Danish in the autograph of Ólafur Stephensen is ff. 61–62.

122. Magnús Stephensen to Banks, 25 August 1809

Reikevig in Iceland

Right honorable Benefactor of Iceland!
It is to you dear Sir! that I in my own, and my whole Family's name, here in Iceland, which you have honored with many years unalterable Friendship,[3] ney in the whole mankind's name I hereby give you my most humble and sincere thanks for the many unforgetly [unforgettable][4] Benevolences and succours you have reputed us, under the present War so much unfortunate, and under the severity of the Nature, in the corner of the World Groaning and poor Nation, by acquiring a Release for the Icelandish vessels and Liberty for the same to provided to Iceland with necessaries for to relief the inhabitants from famine and further more for the carefulness you Sir. have shewn by procuring Iceland assistance from your own Country. –

Dear Sir! It was my the first son of your old friend and former Governor of Iceland Olav Stephensen who, when I in the year 1807, was been taken as Prisoner of War, wentured in October the same year to implore the noble lover of mankind Sir Joseph Banks for intercession for my Suffering Country[5] and every son had /the/ Satisfaction to hear that this my Desire easily got admittance to your noble heart. This was a doing of you, which in the cronicles of every Country is pressing a New stamp to the Charming, history of your noble and immortal exploits, which in every Country in /the/ whole World will make the name of Sir Joseph Banks most estimable and illustrious and for ever beloved, from the soud [south] sea to Icelands Nord Caps; Sir. what you in this period have done for Iceland does as much honor to your noble Heart as it is a certain prov [proof] of the British Government's Magnanimity. – I had the honor last year in Copenhagen personally, to represent for my Gracious King,[6] your elevated Merits who with astonishing admirability and honour mentioned the name of Sir Joseph Banks! and

---

[1] Sir John Thomas Stanley. *Jón* is the Icelandic spelling for John.
[2] Note in Banks's autograph, bottom left of first page: ' [Received] Sept*ember* 26.'
[3] As is evident from the surviving letters from Ólafur Stephensen to Banks (Documents 7, 16, 21 and 25).
[4] Stephensen's English is not perfect and he often uses 'w' for 'v'.
[5] This is Document 40.
[6] Frederik VI was born in 1768 and reigned 1808–39, but due to the mental illness of his father King Christian VII, he had acted as regent while crown prince (*kronsprinsregenten*) since 1784. Stephensen lived in Copenhagen from October 1807 until September 1808, during which period he met the new King.

was so intirely glad to hear that his suffering Subjects in Iceland had got a relief from so noble a hand.¹ – Besides this I have in my own officially published writings maked known, for the Danish Norwegian and Icelandish people, your noble exploits as a clear mark of Gratitude, of the said Writings, I have delivered some few /to Mr. Hooker/ on purpose if they could be honored with a place in your so much estimated Library, amongst this I suppose a Book called Island i det attende Aarhundrede, (or Iceland in the 18d Seiecle [century]²) who is printed in the Danish, and Translated in the German Language³ could possibly deserve a Translation in the English,⁴ as I hope it would be interesting for some learned men in your Country to get a more Clear and Correct Statistic and History of the in some degree notable Island of Iceland, than is published in Adams (Rector university of Edingborg),⁵ his Book⁶ called Summary of History and Geography ancient and modern, some other few i have writen in the Icelandish Language I did not consider of so much welew [value] that I could venture to send them to so elevated a mans Library as yours.⁷

Dear Sir! You have done all what a noble and humane man could do, to relief for my unhappy Country and for to make it exempted from Hostilities, nevertheless is the Island been treated with many injuries particularly from two English Vessels who visited Iceland last and this year: in July 1808 arrived the Privater *Salamine* Gilpin Master to Reikevig, where he with armed force took of way 34,994 Rd. belonging to Holum Bishopric and the only Latin School in Iceland,⁸ the here by annexed Extracts from the public Cashes accounts⁹ I hope will fully ascertain that the money really are the property of sush

---

¹ At the end of the war Banks was rewarded for his efforts by the King. He received a personal letter of thanks from Frederik VI (see Document 295) and three cases of books, including the *Flora Danica*.

² Stephensen had some knowledge of French.

³ This German translation was never published. However, another of Stephensen's works, on the eruption in Laki in 1783, originally published in Danish, was published in German in Altona in 1786 in *Philosophisches Schilderung der gegenwärtigen Verfassung von Island: nebst Stephensens zuverlässiger Beschreibung des Erdbrandes im Jahre 1783...*, in an edition by C. U. D. Eggers (pp. 307–86).

⁴ *Island i det Attende Aarhundrede, historisk-politisk skildret* (Iceland in the Eighteenth Century from a Historical and Political Perspective in Danish), Copenhagen, 1808. Stephensen did, eventually after the war, try to get this book published in Scotland (see letters to Constable, Archibald & Co., booksellers in Edinburgh, National Library of Iceland, Lbs. 2395, 4to II, 17 September 1814; and in the National Library of Scotland [hereafter NLS], MS 789, the Constable Letter Book). However, all these books were lost when the *Margaret and Ann* went down on the way to England (see Document 123).

⁵ Alexander Adam (1741–1809), chiefly known as a writer on Roman antiquities. In 1768 he became rector of the High School in Edinburgh, not of the university as Stephensen believed.

⁶ *A Summary of Geography and History, both Ancient and Modern ...* was published in Edinburgh in 1794. The second edition from 1797 is consulted here, where it says on Iceland: 'The chief town is Scalholt. The inhabitants are supposed to be about 80,000, most of them Christians, but some Heathens. The most remarkable thing in this island is Mount HECLA, which, although covered with snow, is always throwing up flames of sulphur, and torrents of boiling water, which renders it unsafe to approach it' (pp. 583–4). There are quite a number of mistakes in this short text, namely the population (then c. 45,000) and the false 'fact' that there were still heathens in the island.

⁷ In fact Magnús Stephensen was an amazingly prolific writer and publisher; for a list of most of his published works, see Sigurðsson, *Hugmyndaheimur Magnúsar Stephensens*, pp. 186–96. Many of them are available in the British Library.

⁸ The School at Bessastaðir (Bessastaðaskóli) was founded in 1805 and remained there until it was transferred to Reykjavík in 1846. See further Hugason, *Bessastadaskolan*.

⁹ These are not included.

Institutions, although I can not at present get a time to deliver this extracts in the English language I hope a Translation will be easely had in London.[1] There is no doubt you have in the most noble Design encouraged the Merchants house Phelps & Co. to begin a Trade with Iceland[2] and bring a necessaries to support for the country and that your Government has in the same noble wew [view] Granted a Licence to sush Traders, but without Authority (as I suppose)[3] this said Mr. Phelps and one him, as interpreter, following Danish Mand[4] by name Jorgen Jorgensen have acted in the most hostile manner against my poor Country, the said Jörgensen has under the present War acted by chance /acted/ med [for][5] and against England and now he appeared as a Shameful Traitor of his Country, has practised the most violent Despotism against a Defenceless, peasable and innocent Nation, Seised the most beloved and elevated Governor of the Island Chamberlain Count Trampe,[6] and brought him as Prisoner on board and kept him in strong Custody (: without the least Fault on his side :) for 9 wiks [weeks] – The Betrayer Jörgensen soon after maked him selve Ruler of the country, Detested and hated of every good citizen loked upon with adversion as usurper of the Island, who has attemted with armed hand and thretening of dead and confinement without Lawsuit judgement to every one who would not obey him in breaking their oath of allegiance to their Rightful King and native Country, Robbed all what he could as well belonging to private as public institutions, money and goods, from cashes and Warehouses, not only in Reikivig, but whole over the Island as far his power could reach and all this without to give the least account ther of – Yet this unfortunate Country was remaining sighing under the same constrained thraldom and hostilities, which [if] not providence has sendt the honourable Captain Jones Commander of His britannic Majesty's Sloop of War *Talbot* to whom I attempted (although with much fear for the above-mentioned threatenings of immediately death) /etc./ to solicit an interwew with him which he willingly Granted[7] when I according to my Duty told him the matters concerning the before mentioned shocking affairs, soliciting he, as the only commanding Royal officer in a Icelandish harbor,

---

[1] Gilpin's raid was a major incident in 1808 and Count Trampe, with the help of Bjarni Sivertsen, eventually managed to recover most of the money taken (Agnarsdóttir: 'Great Britain and Iceland', chap. IV; and for a fuller account 'Gilpinsránið 1808').

[2] Banks was indeed instrumental in this (see e.g. Documents 118 and 119). In his *Observations on the Importance of Extending the British Fisheries*, Phelps wrote about his trading venture: 'Sir Joseph said we could not undertake a more humane, or better thing' (pp. 51–2).

[3] It was generally believed at the time that Phelps's trading expedition had the support of the British government. Rumours that the English had invaded and occupied Iceland reached Copenhagen with the result that the Danish government banned all exports to Iceland ('Reskript til Rentekammeret, ang. Forbud mod al Tilförsel til Island', *Lovsamling for Island*, VII, p. 265). That the British might seize Iceland during the war had been a possibility considered by the Danish authorities as Magnús Stephensen's defence (*Varnarrit*) makes clear (letter to Kaas, minister of justice, 19 September 1815, printed in *Ísafold*, 1882, nos. 2 and 4, p. 13). Phelps and Jörgensen appear to have convinced the Icelanders that they had the backing of the English government. According to the Icelandic chronicler Espólín, Phelps said he had been granted the exclusive right to the Iceland trade for the next five years and had pretended to have a letter from the King of England saying he could treat the Icelanders as he pleased (*Íslands Árbækur*, XII, p. 27).

[4] Danish spelling here, *mand* for 'man'.

[5] *Med* in Danish means 'for'.

[6] This is generous considering their mutual animosity.

[7] This interview took place on 21 August 1809 (TNA, ADM 1/1995, M. Stephensen to Captain Jones, 21 August 1809).

would use his power for to bring the affairs in a Right Order which he to my highest Satisfaction did by altering all the matters and bring¹ every thing in the best order, as far his Power could reach, and for this his noble doings this Gentleman's name will be remembered with due Respect in the Icelandish annales as a estimated Deliverer² of the Country, a convention is been agreed to, /*bet*/ betwen him on the one Side and my and my Brodder³ [brother] on the other, that the country should be Governed in the same manner as before the revolution, this convention⁴ with some other papers he is now dispatching by one of his first officers to the British Government,⁵ this said papers I am sure of will contain a true Narrative of what is passed in Iceland this sommer, which I hope will come to your hands and you will be astonished when you read it, what I only whish is that you might get a proper Knowledgement of the particulars respecting the said affairs in Iceland before you had fixed any confydence to Phelps or Jorgensens recountings, who supposingly will be wery uncorrect and containing several excuses of no truth.⁶ –

Chamberlain Count Trampe will Supposingly bring you a more detail narration of the mentioned affairs in Iceland to which you can fix a full confidence,⁷ and I hope you will according to your known Noblesse and patriotisme for Iceland do every thing in your power to assist him in optaining Reparation for the suffered Damage of private usurpation, a Neutrality for the Island, with Liberty to make its Trade uninterrupted, and, protected of your Government, and at last to get further Suppley of necessaries from Great Britain whereof we have hitherto got but a indifferent part in the Cargoes sendt hither from your Country who for a great deal were consisting of wares wery unfit for the use of the country⁸ –

Eternally honored and perceived in the remaining and future Generations shall the name of Sir Joseph Banks be and remembred with Gratitude in Iceland and Europ's annales; and to me last Breath shall I wish me the honor to be reckoned amongst this immortal Iceland Benefactors

most humble and Obed*ient* Serv*ant*
M Stephensen
Royal Danish Councellor [sic] of State and Chief justice in Iceland

Wisconsin, Banks MS, ff. 32–34. Contemporary translation, possibly by Stephensen himself. The original in Danish in the autograph of Magnús Stephensen is ff. 35–38.

---

¹ Corrected from 'brought'.
² 'Releaser' has been written above 'Deliverer'.
³ Stefán Stephensen, district governor of the Western Amt.
⁴ Dated 22 August 1809, Appendix 9.
⁵ Lieutenant Stewart of the *Talbot* was about to leave, taking the documents and the offensive Iceland flag to Leith. He was to travel on the *Margaret and Ann*, which was ready to sail for England. But in fact the ship went down, though all were saved by Jörgensen, captaining the *Orion* (Trampe's ship, now Phelps's prize). New copies of the documents were made but the Iceland colours had disappeared for ever.
⁶ Banks was indeed influenced by Stephensen's account, taking a very negative view of the whole proceedings.
⁷ This is somewhat surprising considering their animosity, but they were united in their loathing of Jörgensen and the Revolution.
⁸ The ultimate result was the Order in Council of 7 February 1810 (Appendix 10).

123. Banks to William Jackson Hooker, 1 October 1809

Revesby Abbey,[1] Boston

My Dear Sir

I condole with you Sincerely on the Failure of an Expedition,[2] from whence you must derive however in the opinion of these who knew you no inconsiderable reception in theer Good opinion of your Talents & your Resources. The decision with which you enterd into the undertaking, the Promptitude with [which] you Carried your Preparations into Excution, & alacrity with which you Encounterd the hazard of a voyage, are I beg you to be assurd, deeply impressd upon my mind & have decided me to attempt to gain your friendship which in Future I shall omit no opportunity to Cultivate; I Condole with you for the Losses you have sustaind, but I condole much more seriously with your Friends who must lose the Participation you had prepard for them in Scenes of natural wonder & Delight which you However have seen & impressd on your memory in characters as durable as the memory in which they are impressd.

I Feel however sincere Joy in Congratulat*ing* you on your Escape from the most tremendous Calamity to which a Sailor is subject.[3] Fortunate you have been & Fortunate I trust you will be in Future if worth deserves to be Fortunate.

Recommend me to my friend Dawson Turner[4] when you see him he will greive at the Loss of Specimens & drawings, all all my Little ones Lost together. Some one must go next year to Iceland & I am of opinion that many will go & that the Specimens at leest will not be Lost to Science I had once a Proposal for a Cargo of Scotsmen with Sir Geo*rge* Mackenzie of Cawl at their Lead[5] but I advisd them to wait for another seeson next seeson I expect these Gentlemen will Start[6] not displeesd I conclude in the inside of their minds at your having Lost the Start you had so Gallantly taken of them.

Adieu my dear Sir
beleive me I pray you Faithfully &
sincerely your H*um*^ble Servant
[*Signed*] Jos: Banks

Kew, Hooker's Correspondence, Director's Correspondence, I, f. 30. Autograph letter.

---

[1] The Banks family seat where he spent September and October of each year.

[2] *The Margaret and Ann* was burned at sea and with it all Hooker's botanical specimens, books etc.

[3] That is to say drowning, when the *Margaret and Ann* went down on the way back to England.

[4] Dawson Turner (1775–1858), FRS, banker, botanist and antiquary. He lived in Yarmouth near Hooker and later became his father-in-law. After Banks's death he was for a time a candidate for writing the biography of Banks. Many volumes of bound manuscripts were sent to him, which remained with him for 12 years. It was during this time that his two daughters, Mary and Hannah, transcribed the letters which we now know as the Dawson Turner Collection (DTC), no less than 23 bound volumes. Dawson Turner, however, never wrote a biography of Banks but he did write four volumes, dedicated to Banks, on *Fuci, or Coloured Figures and Descriptions of the Plants Referred to by Botanists to the Genus Fucus* during the years 1808–19 (Carter, *Banks. The Sources,* pp. 19, 164). See also the Textual Introduction above, p. 35.

[5] See Document 116.

[6] Sir George Steuart Mackenzie of Coul mounted an expedition to Iceland the following year.

124. Memorandum from Count Trampe to Earl Bathurst,[1] 6 November 1809

London

To His Excellency the Right Honourable Henry Earl Bathurst one of his Majesty's principal Secretaries of State

My Lord[2]

His British Majesty has, during the present war with the King my master, on different occasions proved it to be His gracious will and intention, that the island of Iceland, placed under the ice of the Pole, and in so many instances labouring, under the unkindness of nature, the government of which the King my master has trusted in my hands as Stiftbefalingsmand,[3] should, notwithstanding the present state of war, be regarded and treated with forbearance.

Not only have hostilities been withheld from her shores, and the internal tranquillity suffered to remain undisturbed by the British marine, but by the release of the little number of ships, detained in British ports, which the rising commerce of the country had put in motion, a humane interest was shewn in the fate of the inhabitants, for which they

---

[1] The original to Bathurst has not been found. In the 'List of Papers sent to the Board of Trade relating to the Transactions which took place in Iceland at different periods from the 12th of January to the Month of October 1809' in TNA, FO 40/1, one of the listed documents is 'Count Trampe to Earl Bathurst – dated November 6th 1809 – With an Enclosure' and this is doubtless the original, the whereabouts of which is unknown. The Wisconsin contemporary copy has been transcribed here. It was the one in Banks's own possession and obviously written by a Dane (the 900 umlauts over each *y* and the 1,400 accents over each *u* have been removed for the convenience of the reader). It is almost identical to Trampe's draft (written by two copyists) preserved in the National Archives of Iceland, though it lacks the date. On the draft is written 'To his Excellency the Right Honourable Henry Earl Bathurst one of His Majesty's principal Secretaries of State'. The draft was sent by Trampe to Iceland from London on 4 April 1810, being registered in the right file on 6 June 1810 (NAI, Stiftamtsjournal II, Nr. 7 6.2.1809–22.6.1809, no. 2). A fair copy would not have been necessary – the draft is legible and straightforward, but Trampe obviously wanted copies of all important papers during his governorship (he was sending quite a few relating to his actions in London) to be filed in the Icelandic archives.

There are several copies of this memorandum extant. In the Manuscript Department of the National Library of Iceland there are two whole sets (with all enclosures), JS 111 fol. (copied by Þorsteinn Helgason in Reykholt) and Lbs. 168 fol. (the copyist is unknown), both in Danish. In both cases the copies are dated and addressed to Earl Bathurst. There is a copy in the British Library in Eg. MS 2067, which is addressed and dedicated, very fulsomely, to the Marquis Wellesley: 'these my humble efforts to give the world a correct account of a late Revolution on the island of Iceland, to Your Excellency'. Wellesley took over as Foreign Secretary on 6 December 1809, so this was obviously copied after that and possibly as an attempt by Jörgensen to ingratiate himself with the new foreign secretary. The fact that it is bound with other works of Jörgensen in the Egerton manuscripts indicates that it was never sent, unless of course there is another copy somewhere in Wellesley's papers but he probably inherited the original addressed to Bathurst. In those days secretaries of state frequently regarded documents as their private possessions and took them home with them upon leaving office.

Trampe also swore an affidavit in the High Court of Admiralty on 5 May 1810, where he gave a short version of the events in Iceland during the summer of 1809 and emphasized the fate of the ships and cargoes seized by Phelps and Liston. In this document Jörgensen is referred to as 'the Linguist', rather than as the interpreter (TNA, HCA 32/1083).

[2] This is clearly stated on the draft. What follows is the Wisconsin copy, which was the one in Banks's possession. It is lengthy and paragraphs have been added wherever there are gaps in the written narrative, for easier reading.

[3] Governor. Trampe is referring to himself.

Figure 6. Count Trampe (1779–1832). Count Trampe was the governor of Iceland during the Icelandic Revolution of 1809. Banks met him when he came to London in the autumn and helped him gain redress for the Icelanders. Image PO–010379. Courtesy of Norges Teknisk-Naturvitenskapelige Universitet, Trondheim.

will ever with gratitude remember the exalted philanthropist, Sir Joseph Bancks [*sic*] who on this occasion undertook to advocate their cause.

But at the same time that His British Majesty's forbearing views were so unequivocally shewn, I have deeply to lament, that private ships of war,[1] belonging to His British Majesty's subjects, presumed to act upon a very contrary plan against this defenceless country.

I will not here detain Your Lordship[2] with the acts of Cap*tai*n Gilpin, the commander of a private armed vessel, that came to Iceland and carried away a sum of money of more than thirty thousand Rixdollars, belonging for the greater part to schools and hospitals and Poor Chests in the country. This fact I beg leave to reserve myself the liberty to represent in another place;[3] But my object in these leaves is to lay before Your Lordship the account of proceedings, on the part of British Subjects – equally as shameless as unheard in history and fraught with consequences more dreadful, than those to be apprehended from open hostility.

---

[1] Both the *Salamine* and the *Margaret and Ann*.

[2] 'Excellency' is written in the draft.

[3] Trampe and Sivertsen swore before the Admiralty Court that the money taken by Gilpin in Iceland was private property and eventually the High Court of Admiralty ruled in Iceland's favour and most of the money was returned in 1812.

In the middle of January last, the *Clarence*, an armed merchantship, entered a port on the south country of Iceland[1] under the colour of the United American states,[2] reporting, to bring a cargo of provisions and other necessaries and requesting permission to trade with the inhabitants.

The existing laws of the country, strongly prohibiting all trade with foreign nations, it was the duty of the officers in whose hands I had at that time, during my absence on a voyage to Copenhagen, left the conduct of the public affairs, to refuse this application, and it was accordingly refused.[3]

Notwithstanding this refusal was accompanied by the most convincing reasons, it was yet received with marks of dissatisfaction, and M^r. Jackson, the master of the armed ship, now changed his tone, declared himself a British subject and alledged to be furnished with authority to commit hostilities, of which he gave an immediate earnest by taking possession of a Brig, just then lying in the harbour, that was arrived from Norway with necessary supplies for the island.[4]

Occasioned by this, and in order to avoid farther acts of violence with which they were threatened, the said officers found themselves under the necessity of granting the permission,[5] formerly refused,[6] and a mercantile concern, thus violently forced upon them, was established in the town of Reykevig under the management of Mr Savignac, who came over in the *Clarence* as supercargo.

This little town was now overstocked with luxuries[7] of all descriptions that could not but find a slow and tardy sale at a season of the Year when the commerce could only be carried on with the inhabitants of the town and its immediate neighbourhood, for it is only in the month of June that a degree of communication, intercourse and barter between the more distant towns and parts of the country begins to take place in Iceland.[8] Of real necessities, on the contrary, such as the country wanted, and for which there was at that time in particular a demand, only a very inconsiderable quantity was imported.[9]

Partly owing to these causes, partly also to extravagant sacrifices and expences, to rash and imprudent speculations and mismanagement in general, it did not last long, – as it was reported – before this new establishment turned out to be a losing concern.

The managers of the same had beforehand been warned of the prospects that were now realizing themselves, but still the civil officers of the country, who in truth could not, across impervious deserts, bring forth buyers of goods, but little wanted by an Icelander in his house, must find themselves accused of having by clandestine insinuations, keept back the inhabitants from any intercourse with the establishment; still the rest of the citizens and merchants of the town, who after the pending differences were settled, met the strangers with all the undissembled hospitality which the inhabitants of Iceland have

---

[1] Hafnarfjörður.

[2] During war it was common for ships to carry flags of neutral nations.

[3] By Koefoed, the county magistrate (*sýslumaður*) of the district.

[4] This was the *Justitia*, sent at the expense of the King of Denmark.

[5] The Agreement of 19 January 1809.

[6] As proscribed by law.

[7] According to the cargo-list of the *Clarence*, these 'luxuries' included coffee, sugar, rum, tobacco, hats and 1,800 barrels of Liverpool salt.

[8] The fact that the *handel* took place in the summer was the real reason why the trade proved so disappointing.

[9] This is borne out by the fact that Savignac sent a letter to his employers listing the goods 'in current consumption here' (Bathurst MS, Savignac to Phelps & Co, 19 March 1809, p. 86). The list itself is not extant.

made their law, must find themselves accused of having by every sort of intrigue endeavoured to bring about the miscarriage of the enterprise. The whole of this will be more fully elucidated and confirmed by a report presented to me by 5 of the established citizens and merchants of Reykevig under the date of the 1st of September this Year, of which I beg leave to produce a translation in the voucher marked **A**.[1]

It is here evidently proved that the above mentioned accusations against the civil officers and the citizens of the country were untrue and unfounded, and I dare maintain, that they were invented and set afloat by the manager of the concern [Savignac] only with a view to provide a cloak for themselves in the account they were going to render to their principals for the losses occasioned by their own misconduct.

In order not to interrupt the thread of the history in the following events, I cannot omit to state in this place, that, besides M^r. Savignac the supercargo, a Dane of the name of Jörgen Jörgensen also came to Iceland in the *Clarence*, acting in the capacity of an interpreter. This person, who was born at Copenhagen, where I have known his father and a brother of his to be established citizens, who after being in his younger Years brought up to the sea in the British service, has during the present war commanded a danish privateer which was taken by the English to whom he was made a prisoner of war.[2] This person, I say, here exhibited himself in the most ambiguous light. By taking an active share in the proceedings of Cap*tai*n Jackson he soon proved himself to be disengaged from every tie that binds a man to his native country; but at other times again he discovered himself, as corresponding but in a small degree to the confidence which appeared to be placed in him. Upon the whole he was seen to be a person without principle and of manners the most low.

From Iceland, where his follies had rendered /him/ most conspicuous, he at length returned to England in the *Clarence*, in order, as the event has proved to prepare himself for acting a still more treacherous part.

Having in the Year 1807 undertaken a tour to Copenhagen, as I have taken the liberty to mention before, I set out in consequence of the most gracious orders of the King my master, on my return to Iceland in the month of May this year. On my arrival there on the 6th of June following I was informed of the convention which Cap*tai*n Jackson had forcibly compelled the officers, authorized to act in my absence, to conclude with him.

I felt mortified at a convention of this kind, concluded with an armament unauthorized to enforce it; but acknowledging the sacredness of contracts, I had no idea of curtailing in any respects the rights thereby granted to British subjects, notwithstanding M^r. Savignac, the supercargo, had himself acted contrary to the convention and thus sufficient cause had long existed to dissolve it.

I only wished that information might be conveyed to His British Majesty of the acts of violence committed by private individuals among his subjects against an island, against which His Majesty himself forbore to act as enemy, which having, with the mother country, through a long succession of years enjoyed uninterrupted peace and having

---

[1] The merchants are Gísli Símonsen, Jesse Thomsen, Ph.[Filippus] Gunnarsen [Gunnarsson], C[hristian] C[onrad] Strube and L[ars] Knudsen.

[2] After the outbreak of war between England and Denmark in 1807 Jörgensen served as commander of the largest Danish privateer, the *Admiral Juul*. In March 1808 the ship fell in with a British man-of-war, the *Sappho*, off Flamborough Head, and after a short engagement the Dane was obliged to surrender. He became a prisoner-of-war, but was released on parole.

internal tranquility firmly established and maintained by the reverence of the laws without wanting the aid of military force, was found in an unarmed and defenceless state.[1]

This object I also found an opportunity to attain a few days after my arrival in the country, when His British Majesty's sloop of war the *Rover* entered the port of Reykevig, where I delivered into the hands of Captain Nott,[2] the commander of that ship, a memoir, which I hope this gentleman, in consequence of my earnest request, has not omitted to forward and cause it to be laid before His British Majesty.[3]

I must here beg leave to observe, that from the existing warlike relations I did not view with indifference the arrival of an armed force belonging to His British Majesty, with the objects of which in those parts I was unacquainted, the peaceable proceedings of which no convention secured, and my duty imposed it upon me to take every possible measure of precaution; but having been assured, that Captain Nott was far from intending any hostility against the country, I could not but wish, under the existing circumstances, that a compact, entered into with a man acting under public authority, might establish a firm and understood relation between the inhabitants of the Island and those British subjects who were settled there already or who might come there hereafter for the purposes of trade, hoping at the same time, that such a convention would meet from British subjects with that respect which was due to stipulations into which the contracting parties had not entered without a knowledge to the principles of their respective governments, which consequently British subjects could not venture to violate without running the risk of acting contrary to rules, already perhaps sanctioned by His British Majesty, after the convention having been presented to him.

Captain Nott appeared to be equally inclined with myself in this respect, and after I had received his proposals in a letter dated June the 16th last, as will be more fully seen by the voucher marked **B**.[4] a convention was after some negociation concluded, of which I have the honor to produce a verified copy in the voucher marked **C**.[5]

The perusal of the same must convince every person, that I gained nothing on the behalf of the island of Iceland except the attainment of the wish which, on my part, had given rise to the convention. To British subjects, on the contrary, conditions and conveniences were granted, that could not fail to render their trade both secure and advantageous.

---

[1] Trampe had tried to get Iceland fortified in some manner, writing to the Danish authorities on 22 October 1808. After all there was a fort with a garrison of 30 in Torshavn in the Faroe Islands. But his entreaties fell on deaf ears (*Lovsamling for Island*, VII, pp. 243–4).

[2] Francis John Nott was made a commander of the *Curacoa* sloop on the Jamaica station at the end of 1801, subsequently appointed to the *Childers* and *Rover* brigs. He continued in the latter until promoted to post-rank 21 October 1810. On 19 June 1822 he obtained the out-pension (a pension paid to non-residents) at Greenwich Hospital.

[3] It was certainly sent to the Admiralty (TNA, ADM 1/692, Trampe to Nott, 12 June 1809). It is lengthy and in Danish Gothic handwriting. No translation is attached. However, Nott clearly somehow made himself understood to Koefoed, so he probably managed to get the letter translated. Bjarni Sívertsen may have been the translator. He had recently returned from England and lived in Hafnarfjörður, but, being a merchant, would not have been enthusiastic about the British mercantile competition.

[4] This letter, from Captain Nott to Governor Trampe, is dated 16 June 1809, written on board His Britannic Majesty's Ship *Rover* in Reykjavik Bay (NAI, Jörundarskjöl).

[5] The Nott-Trampe convention with the addition (which is not in all copies) 'the above convention has been shewn to me and I am acquainted with the content. J. Savignac'. The originals with Savignac's signature are in NAI, Jörundarskjöl and TNA, ADM 1/1995 (see Appendix 9).

As soon as the convention was concluded I immediately ordered the printing press, which was distant some miles in the country, to be put in motion with printing copies of it for circulation and publication in every part of the island; but in order that its publicity might in the meantime be forwarded as much as possible, I did not omit at the same time to cause notifications of it to be issued from my office to both the Amtmænd in the country, M^r. Thorarensen and M^r. Stephensen,[1] and I also directed both the civil officers, living nearest to me, M^r. Frydensberg, Landfoged,[2] and M^r. Koefoed, Sysselmand,[3] who had both been employed in the negociations that took place before the convention was concluded, who had been present at the signing of it, and were most thoroughly acquainted with its nature, to cause its contents to be made known in the neighbourhood. To M^r. Savignac I ordered the original convention to be produced for his perusal at my office; This he has acknowledged with his own signature, a copy of which will be seen in the voucher marked **C**. to which I have just taken the liberty to refer.

Everything was now peaceable and quiet; M^r. Savignac having signified to me that he wanted a dwelling house and a warehouse, I had agreed to let him have the use of two such houses that were in my disposal, and I promised the country, I promised to myself undisturbed tranquillity and mutual approximation.

The arrival at Reykevig, a few days after the departure of Capt*ai*n Nott, of the *Margarete and Anna*,[4] an armed merchantman from London, having on board M^r. Philp[5] merchant and owner of the cargo left by *Clarence*, with the above mentioned Jörgen Jörgensen and some other passengers, occasioned in the first instance, no change in this state of affairs. It was reported indeed that M^r. Philps had expressed much displeasure at the state of his commercial establishment, at the great sacrifices that had been made, and in short at the losses with which he had been obliged to pay for his first attempt in the Iceland trade; but who would have imagined, that commercial reverses were to be retrieved by political designs, that should go to subvert a convention concluded for mutual benefit, and afterwards introduce the most dreadful state of things, totally destroying that peace which the inhabitants, happy in their poverty, enjoyed in this island so unkindly treated by nature, while all the rest of Europe was desolated by the war? – still, such the event actually turned out.

On the fourth day after M^r. Philps's arrival, namely on the sunday the 25th of June, scenes began to be acted which perhaps should not be credited if they could not be established by the most incontestable proofs.

On that day, after divine service, while I remained in my house, all of a sudden I saw it surrounded with 10 or 12 men armed with firelocks and swords, and at the same instant an armed man, whom I afterwards found to be Capt*ai*n Liston, the master of the *Margarete and Anna*, opened the door of my room and declared me to be a prisoner of war.

He was followed by M^r. Phelps, and as I have since been informed, also by M^r. Savignac, armed with pistols, to whom his post had been assigned in the hall before the room. Capt*ai*n Liston's declaration, that I was a prisoner of war, was at first replied by the

---

[1] In the draft there is an explanatory note: '(the sub-governors of the Western and Northern districts)'.
[2] In the draft there is an explanatory note: '(cashier)'.
[3] In the draft there is an explanatory note: '(judge and master of police for a particular district)'.
[4] Actually the *Margaret and Ann*.
[5] Phelps is sometimes called Philps in this document and always so in the draft.

question: what was his right to declare me so? I put them in mind of the convention that had been concluded with Cap*tai*n Nott a few days ago; I required to know what were their objects, and what they wanted of me; but all these questions and remonstrances were avoided partly by M^r. Phelps signifying that there was no time for conversation, partly by Cap*tai*n Liston's admonitions that, in case I did not immediately go voluntarily on board his ship, he would feel himself under the necessity to compel me in a more uncivil manner. I will not here dwell upon the circumstance, of M^r. Savignac and likewise M^r. Jörgensen being allowed during this transaction to rush into my room and attack me with the grossest abuse, which I reserve myself the liberty to notice more at large hereafter in my statement of the many personal insults and abuses to which I have been obliged to submit.

After having in vain represented to M^r. Phelps my sensation at being thus precipitately dragged away from my public functions and my very important private affairs[1] – after having in vain insisted upon being treated with that respect which I felt to be due to the rank and dignity of my office, I found myself, unarmed as I was, alone with M^r. Koefoed (who happened to be with me at the time) surrounded by a band of armed men, and having no assistance to expect from the unarmed and defenceless inhabitants.[2]

I found myself, I say, in the mortifying – nay more than mortifying, necessity of surrendering to the dishonorable perpetrators of this shameless outrage. Having locked the door of my office, where all my papers were, and set my seal to it, I was conducted under an armed escort onboard the *Margarete & Anna*. Immediately after, on the same day, the *Orion*, a bark-Ship which I purchased in the spring of the year of M^r. Knudsen,[3] an Iceland merchant, and which was furnished with His British Majesty's special Licence[4] (the same vessel in which I had come over from Christiansand[5] with a cargo of provisions and other necessaries, part of which had not yet been unloaded, and among the rest a quantity of goods amounting to 6000 rixdollars danish Currence, which were intended for gratis distribution among the poor and distressed Icelanders) was taken possession of and the English colours hoisted on her by armed men from the *Margarete & Anna* after a gun had been fired from the latter, with English colours flying.

Of the events, which after this took place on shore, I can only speak from written documents which are in my possession and from accounts afterwards communicated to me by others, as I was myself on board the Letter of marque subject to the strictest inspections and confinement and could but seldom find an opportunity to procure secretly some superficial information of what was going on.

As soon as the plot against my person had been carried into effect, it was given out by Jörgensen, and circulated in the town, that Cap*tai*n Nott had committed an astonishing blunder by concluding the convention of the 16th of June; that he was unauthorized to take any such step, and had not known what he was about; that he had acted in a very

---

[1] It is important to note that Trampe had a personal stake in all matters concerning trade.

[2] According to Hooker, the residents of Reykjavík, who were 'idling about' had 'looked on with the most perfect indifference' (*A Tour in Iceland*, I, p. 54). Icelandic sources, however, claim that Trampe himself, knowing he could not expect any help from the inhabitants, had asked the people gathered not to endanger themselves by trying to rescue him (Espólín, *Íslands Árbækur*, XII, p. 28).

[3] Adser Christian Knudsen: see Appendix 13.

[4] The *Orion*'s licence.

[5] 'Copenhagen' is written in the draft. Christiansand is a port in Norway.

stupid manner; that after his departure from Scotland orders had been sent for him to Leith by the British government, which, failing to find him there, had been returned to London and given in charge to M^r. Phelps (who from his influence with the Brittish ministry and his connexions with different of its members was described as extremely powerfull) with injunction, in case Capt*ai*n Nott were no longer found in Iceland, to open the sealed orders and execute them; that in consequence of the same, the island of Iceland was immediately to be taken in possession in His Brittish Majesty's name untill two of His Majesty's frigates which were shortly expected should arrive; that a bank of more than 100000 rixdollars danish currency was to be erected; that the country should flourish in plenty and prosperity, with other falsities and boasts like these.

The multitude is generally easy to be dazzled; but the reflecting man entertained the most well-founded doubts of the authority and the views, of the execution and the issue of what had been begun. The next day a proclamation, signed Jörgen Jörgensen, was seen publicly stuck up in Reykevig, of which I have the honor to present a translation in the voucher marked **D**.[1] Without any preamble, without discovering a glimpse of cause or object, it is here decreed in the 1st. article that all danish authority is dissolved in Iceland; in the 4th. article that every kind of arms and ammunition shall be delivered up: in the 6th. article, that the keys of private warehouses and shops, that money accounts and paper belonging to or concerning the interests of the king my master or danish merchants, shall likewise be delivered up; in the 2d., 3d. and 5th. articles all born Danes are ordered to remain within their own houses, and all others are prohibited to communicate with them; in the 9th. article bloodshed is threatened against all who should act contrary to these decrees, who were to be brought before a military court and shot within two hours; whereas in the 8th., 10th. & 11th. articles undisturbed tranquillity is promised to all native Icelanders, and a felicity hitherto unknown.

The execution of the 4th. article was begun without delay by Jörgensen, who at the head of Capt*ai*n Liston and a number of armed men from the *Margarete & Anna* marched from house to house, carrying off all sorts of arms with which the inhabitants had provided themselves for the chase.[2]

The day was not elapsed before this proclamation was followed by another, likewise signed Jörgen Jörgensen, of which I have the honor to present a copy in the voucher marked **E**.[3] It begins in the same manner as the first and is made up with a mixture of folly and nonsense, of madness and cruelty; it clearly pointed out the path to impending miseries, it discovered at the same time the ultimate object of the whole plan and immersed every one in grief and sorrow. In the 1st. article it is again decreed that Iceland shall be independent of Denmark; in the 5th. that a new and republican constitution shall be introduced like that which existed before the country was united to Norway in the 13th century, and in the mean time, untill this new constitution could be founded by the chosen representatives of the people, the existing authorities, are in the 6th. article set out

---

[1] Jörgensen's first proclamation of 26 June 1809. This proclamation, unlike the other two, was not printed (there was no need to as it pertained only to the actual execution of the usurpation) but there is a copy in Hooker, *A Tour in Iceland*, II, pp. 67–70.

[2] The 4th article read: 'All sorts of arms, without exception, such as muskets, pistols, cutlasses, daggers, or ammunition, shall instantly be delivered up.'

[3] Jörgensen's second proclamation of 26 June 1809, which was later printed; a few copies survive.

of force. The 8th. & 9th. articles are too ridiculous to be noticed,[1] were I not on another occasion called upon to make some mention of them.

In the 10th. article it is declared that a regular defence shall be established; in the 12th. that all debts to danish merchants in the country or abroad are made null and secret payment of the same is prohibited under pain of paying a fine to the same amount to the new government as it styles itself.

In the 13th. article the prices of provisions are declared to be too high and that they shall be reduced; In the 14th the taxes levied upon the inhabitants are diminished by one half, in the 18th all communications with danish ships is forbidden; For the rest it is seen by the 2d., 3rd & 19th article, that none but natives could continue to fill public employment and in the 6th article such of the natives, as did not wish to remain in their employments, are left to their own fate.

When a faithless subject, forgetting the blessings for which he stands indebted to his government and his country, has the impudence, as here, to intrude himself, and pretend to speak and act in the name of fifty thousand individuals out of whom I can venture to assure by every thing that is sacred, that not a single person had called upon him for it, but that every honest man was shocked and astonished, the truth of my former assertion no doubt will be evident and undeniable, that this kind of proceeding, this manner of revolutionizing and severing a people from their lawful prince is without example in all the revolutions which history records; And when the proclamations of the 26th of June threaten military justice and the punishment of death to all who should act contrary to these decrees dictated by stupidity and madness; When the republican form of government is recommended notwithstanding the numerous instances with which history shews what has become the fate of nations under it in every period of the world; When a poor island like Iceland, by nature denied internal strength and the means of supporting an independence, is proclaimed a state; When all debts due to danish subjects are declared null, and in the 19th article[2] of the latter proclamation it is supposed possible that the natives might avail themselves of the existing state of things to the injury of the Danes settled in the country, whereby to every principle of public good, it is endeavoured to sow the seeds of discord between citizens and citizens; when in the 13th article[3] of the said proclamation, the rights of private individuals are openly trampled under feet: only a glance at this catalogue of confused ideas is wanted, in order to foresee what would prove the end of the happiness promised to the Icelanders.

My house in Reykevig was now occupied by Jörgensen; the seal on the door of my office was broken and all my papers without exception laid hold of; not even my most private letters were spared; In this house, where he established his government-office (Regjerings Contoir) as he called it, he was daily seen in the company of M^r. Phelps attending to employments that formed a singular contrast to those seaman's occupations in which and for which he had been brought up, and spreading about him a magisterial dignity for which he did not appear at all to have been born.

---

[1] 'Iceland has its own flag. Iceland shall be at peace with all nations, and peace is to be established with Great Britain, which will protect it' (Hooker, *A Tour in Iceland*, II, p. 72).

[2] 'No Icelander must, on account of the late liberty being granted, presume to offend or assault a person for being a Dane, nor for having held a situation under the king, nor for having been in the employment of, or connected with, a Danish mercantile house, provided they do not interfere with the political affairs of the island.'

[3] 'All kinds of grain shall by no means be sold at exorbitant prices.' Here Trampe is writing as a merchant.

In order to obtain celebrity with an unreflecting multitude by an appearance of mildness, or perhaps rather because he wanted, for the more convenient execution of the schemes, conceived a free disposal of a public building called the House of Correction (Tugthuus) where dangerous criminals were confined and kept [sic] to hard labour for a shorter or longer period, four such malefactors, who were at the time in the house, were let loose upon the public. Of the report presented to me of this transaction by the two overseers of this institution, I beg leave to produce a translation in the voucher marked **F**.[1]

An exception however was afterwards made on the earnest representations of some civil officers, of one of these malefactors who is marked in the report with No. 2, who was confined under the suspicion of having committed a murder and was known to be a dangerous person.

As an instance (besides so many others) of the usurper Jörgensen's incapacity of conducting magisterial functions I can not here omit to add in the voucher marked **G** a translation of the discharge which was given to one of the above mentioned malefactors on his release, dated June 28th this Year.[2] His name was Sigfus Jonsen; from the voucher marked **F** it will be seen, that he had been sentenced to be publicly whipped and to work in irons for his life; in this discharge he is granted his liberty, but with an admonition, that if he should again commit anything contrary to the laws, his punishment should be doubled. In what manner this double punishment was to have been executed, I confess I do not comprehend.

It was again pretended to be mildness and solicitude for the Icelanders, but more truly perhaps it was because no inconsiderable sums of money of which otherwise the house of Phelps did not seem to have superfluity, were required for the execution of such extensive plans coupled with an unbounded prodigality, that goods with which the King my master had provided the country in the time of scarcity, and which in consequence of the proclamation were now considered as confiscated, were reduced to prises far below their current value.[3]

To acquire celebrity and followers by such means did not cost the usurper any great pains, a few strockes with the pen were sufficient, as will be seen by the translated documents which I have the honor to present in the vouchers marked **H, I, K**.[4] These are three orders from Jörgensen, in which he most seriously enjoins the respective merchants, to whom the sale of the goods had been given in commission, to sell certain sorts of corn at the reduced prices he had fixed. The voucher marked **H**. I venture in particular to recommend to attention in so far as it discovers the avidity with which the authors were

---

[1] This is 'Fortegnelse over Tugthuuslemmernes Antal ved det islandske Tugthuus d: 26de Junii 1809' signed by Sívertsen, the Oeconomus, noting four prisoners, three (Sigfus Johnsen, Haukon Haukonsson [Hákon Hákonarson] and Thuridur Nicolasdatter [Þuríður Nikulasdóttir]) being released on 27 June on Jörgensen's orders while the fourth, Jón Jónsson from Arakot, was never freed. This was signed on 3 September by Sivertsen and Ole Björnsen *Tugtmester* (prison warden).

[2] Passport for Sigfus Johnsen, dated 28 June and signed by Jörgensen allowing him to 'pass free and unhindered' wherever he wished to go.

[3] Trampe had brought a lot of rye with him to Iceland, one of his first actions once back in Iceland being to increase the price of a barrel of rye to 16 rixdollars (Espólín, *Íslands Árbækur*, XII, p. 26).

[4] All letters signed by Jörgensen and dated 27 June. Voucher **H** is Jörgensen's letter to the merchant Gísli Simonsen, ordering him to sell only rye, a barrel to cost 10 rixdollars; Voucher **I** to Bjarni Sívertsen deals with the *Justitia*; and Voucher **K** to the same deals with the cost of a barrel of peas and barley (Þorkelsson, *Saga Jörundar hundadagakóngs*, pp. 155–6).

looking out for the most current and for M^r. Phelp's interest most profitable export articles of Iceland, wool and tallow, and thus contributes to elucidate the commercial views that were at the bottom of these political revolutions.

The history of the succeeding days is particularly a chain of usurpations and violence. To confiscate to the treasury of the state (Statens Kasse) as it was called all that could be confiscated in order to furnish resources for an all devouring prodigality was the occupation and the object of every day. Shops and warehouses in Reykevig belonging to Danes not living in Iceland or even to danish merchants who were only absent at the time, were from the first day put under guard and to the distant towns and settlements emissaries were sent to execute the same errand. I should never have done were I to relate the proceedings step by step in full detail; I must therefore confine myself, to the complaints of the injured parties on the losses they have suffered in so far as such complaints have been presented to me, of which therefore I beg leave here to produce translations, namely:

In the voucher marked **L**.[1] Mr F. Biornsen of Reykevig his report of the manner in which he was stripped of a sum of 2600 Rixdollars danish currency, belonging to M^r. Adzer Knudsen, merchant, who was in Denmarck at the time but otherwise is settled in Iceland.

In the voucher marked **M**. M^r. Christian Conrad Strubes account of the seizure of stock on hand consisting of tallow, train-oil, fish and woollen goods belonging to a trading company in Flensburg whose factor M^r. Strube is.[2]

In the voucher marked **N**.[3] M^r. Jess Thomsens similar account of the seizure of goods belonging to the Nordburgish trading establishment, as it is called, one of the most considerable in Iceland, the proprietor of which is M^r. Jess Thomsen, merchant, who lives at Nordburg.

These three reports are all attested by the Notari Publici[4] of the place as to the signatures of the persons reporting and their competency so to do. It will not be in vain to throw a look at the two last mentioned papers in particular, the authors of which are so trust-worthy men, that I may with the greatest confidence pledge myself for every word they have said, in order to see in what manner Jörgensen dealt with the plundered goods.

For the farther elucidation of this matter I also beg leave to add in the voucher marked **O**.[5] the translation of a note in Jörgensens own handwriting, in which M^r. Phelps' supercargo, M^r. Savignac, is authorized to take out of the above mentioned Jess Thomsens goods whatever he thought proper.

Though the factors, who were accountable to their principals, foresaw the necessity of having a specification drawn up of the different sorts of goods and every thing in particular, which in time might serve for their indemnification, yet no such thing was ever executed. No, relying as much on the present possession of power, as on the impossibility of his being called to account, the usurper thought it sufficient either verbally or with a little note, like that referred to above in the voucher marked **O**, or like another

[1] Finnbogi Björnsson, factor to Adser Christian Knudsen. This is a lengthy letter and dated 1 September 1809.
[2] Dated 2 September 1809, this is another lengthy letter.
[3] Dated 2 September 1809, yet another lengthy letter.
[4] Namely Rasmus Frydensberg.
[5] Simply two lines to this effect, the goods being those that had been confiscated, dated 11 July 1809 and signed by Jörgensen.

of the same description of which I beg leave to present a translation in the voucher marked **P.**,[1] to despatch M.ʳ Phelps's clerks,[2] ready for all sorts of expeditions, to plunder whatever could prove a profitable cargo to the *Margarete & Anna*, leaving it to the persons concerned to see how far either nothing at all or certificates like that of which I take the liberty to produce a translation in the voucher marked **Q.**[3] could afterwards be sufficient to settle claims and responsibility between them. It was impossible therefore at the time when the above reports were drawn up to make out any account of the goods plundered and their value, which is the real loss the owners have suffered through Jörgensen and his coadjutors, not to mention the injury they have sustained by the want of goods which by traffick in different ways might have been turned to good account, by the total stagnation of their business during the best season of the Year, by the loss of their account books which were taken from them, by the discontinuance of the payment of debts by debtors, who perhaps in the mean time may have become insolvent.

After having stopped to relate the acts of plundering and rapine that were committed, of which more will hereafter occur in their order, I will now follow the progress and the course of the revolution itself.

The dread with which the proclamations of the 26th of June had struck the minds could not in a moment be removed, and means were also taken to keep it up partly by daily scenes of violence, partly by an armed force from the crew of the *Margarete & Anna* who perpetually patroled [sic] in the streets and particularly round the houses where Phelps and Jörgensen were hatching their pernicious schemes. The new state was by most people considered as a bubble; It was the public officers of the country in particular who, in consequence of the proclamations, were to pay the first homage, but it was not done; some of them laid down their functions,[4] others declared that they would, for the good of the country, continue in the execution of them, by virtue of the same authority under which they had hitherto acted.[5] To what translation in the voucher marked **R.**,[6] in which letter he, Mʳ. Jörgensen, also appeals to His Brittish Majesty's cruisers and ships of war against every civil officer who did not declare his willingness to conform to the contents of the proclamation.

But alas! while every honest man remained unshaken mindful of the numberless benefactions which through a long succession of years have bound the sons of Iceland to the Royal Family of Denmarck with the most sacred ties of gratitude, others were found who suffered themselves to be deluded. The sums of money which the usurper had already amassed by his plunder, enabled him to be lavish beyond measure; the pompous promises which were every day retailed on paper or held forth in the harangues of Jörgensen did not fail of some effect; idle persons and men of ruined fortunes did not neglect to offer

---

[1] 1 July 1809. This is the letter in Þorkelsson, *Saga Jörundar hundadagakóngs*, pp. 156–7, though he dated it provisionally as written on 27 June.

[2] Actually an Icelander called Óli Sandholt, who was employed by the British merchants.

[3] 26 July 1809, Jörgensen on the confiscation of Jesse Thomsen's goods.

[4] Only four county magistrates decided to resign their offices as well as the district governor of Northern and Eastern Iceland. Most other officials sent in letters of allegiance. They had little choice as the payment of their salaries depended on this and it was the common belief that Phelps had the backing of the British government so there was no point in resistance.

[5] For example, Magnus Stephensen.

[6] 27 June 1809 to Bishop Vídalín (Þorkelsson, *Saga Jörundar hundadagakóngs*, pp. 161–2).

themselves and in a few days he was seen surrounded by a contemptible band well fitted for spies and slanderers, whose influence was soon discovered in the persecution and detriment of innocent citizens. Thus he was seen to procure for himself out of the refuse of society, a guard, as he called it, consisting of 6 vagabonds, malefactors, liberated by himself, and other persons of the same stamp, amongst whose number was also one whom I had formerly employed in my office, but been obliged to dismiss on account of dishonesty, and who now requited my former kindness by incensing Jörgensen against every person who was suspected of feeling for the treatment I received. A pretended rumour that M[r]. Einarsen, one of the judges in the upper Court of justice (Overret)[1] of the country, was engaged in a conspiracy that had for its object to surprise the usurper,[2] afforded M[r]. Jörgensen the first opportunity of employing the militia he had brought together by arresting M[r]. Frydensberg, the Landfoged who is at the same time Byefoged (master of police) of Reykevig,[3] and after him likewise M[r]. Einarsen the latter of whom he keept confined for 10 days.

Both these public officers have themselves in their reports presented to me, described so circumstantially the ill-treatment they underwent, that [I] have nothing to add, but only beg leave to present a translation in extenso of the said reports, that of M[r]. Frydensberg in the voucher marked **S**,[4] that of M[r]. Einarsen in the voucher marked **T**.[5] accompanied by 5 special vouchers to which they refer. It is here evidently seen, what kind of judicature was about to be established in the country; the usurpers will was to go for trial and sentence at once. Having thus removed an obstacle he dreaded, but at the same time found how little a system coupled with violence and rapine was calculated to gain the hearts (for if we except a few clergymen of the neighbourhood, who, on a day when they were assembled in a yearly synod at Reykevig, were surprised into a declaration in his favor, all the rest remained unshaken) Jörgensen now thought it time to add the finishing stroke to his work by adding new usurpations and new enormities to the former.

Thus issued on the 11th of July a third proclamation publicly stuck up in the town of Reykevig, of which I have the honor to present a translation in the voucher marked **U**.[6] After the usurper has here explained how much he has been solicited by the people to take upon himself the conduct of the public affairs, and how the inhabitants had by hundreds offered themselves for the defence of the country (assertions which thousands of people if necessary, will prove to be grossly untrue) he has the effrontery to decree: in the 1st article We Jörgen Jörgensen have taken upon ourselves the government of the country untill a regular constitution can be established, with power to make war and conclude peace with foreign potentates, in the 2d. article it is made known that the soldiery, which it is to be observed consisted of the 6 vagabonds mentioned before, had chosen him to be their leader and to conduct the whole military department; in the 3d.

---

[1] *Landsyfirréttur*, the High Court of Justice.

[2] It is not at all clear what actually took place. Einarsson was, however, arrested and imprisoned in the Government House for ten days. Furthermore he was removed from his position.

[3] According to Espólín, Frydensberg, because he was a Dane, was arrested while taking his child for a horseback ride and imprisoned in the house of correction, but his wife's entreaties secured his release (*Íslands Árbækur*, XII, p. 29).

[4] 2 September 1809, Frydensberg's report.

[5] 2 September 1809, Ísleifur Einarsson's report on his arrest and the five enclosures pertaining to this.

[6] Jörgensen's third proclamation of 11 July 1809. See e.g. Hooker, *A Tour in Iceland*, II, pp. 70–74.

art: a new flag is instituted for Iceland, the honour of which he, the man without honour, promises to defend with his life and blood, in the 4th. art. the ancient seal of the country is abolished, instead of which his own private seal is to be used, untill the representatives of the people should have fixed upon a new one; in the 5th. art. the time granted to the civil officers for declaring their obedience or resignation is prolonged to 10 days for the nearest and 4 weeks for the most distant parts of the country, after the expiration of which time all who had not given in their declarations were to be suspended from their employments, notwithstanding 7 or 8 weeks were required before the message could be conveyed to most of the distant parts; in the mean time it is in the 6th. art. announced to all who should resign, that they were to be exiled to Westmannoe (an Island where their stay would bring upon themselves and their families the greatest sufferings) untill an opportunity could be found to convey them to Copenhagen. in the 7th. art. it is promised to that part of the clergy who had been willing to declare themselves, that their circumstances should [be] bettered. According to the 8th. art. it is still intended to place the country in a state of defence; in the 9th. art. this insolent man was preparing to send an ambassador to His British Majesty to conclude peace; the 10th art. contains something relative to the duties and rights of British subjects that were living in Iceland; in the 11th. art. it is declared that only Icelanders should be qualified to fill public employments, whence all the public officers in the country who were Danes by birth are of necessity included in the cruel decree of the 6th. article; by the 12th. it is seen that the usurper intended to continue his office as he calls it, untill a regular constitution could be established; the 13th. article again declares what had already been executed, seen and felt, the confiscation of danish property; and in the 14th. the Amtmænd[1] (Subgovernors) whether they would continue in their offices or not, are enjoined to execute this violent measure; in the 15th. art. it is pretended, though with very little probability, that some civil officers, in order to secure themselves against the displeasure of the King their master, had expressed a wish that they might be compelled to exercise their public functions; the 16th. Art. has for its object to uphold the dignity of the usurper, by forbidding all irreverence towards his person. His stay of two months at Reykevig during the winter of 1809, when he came over in the *Clarence* had exhibited him as a person of very low manners and breeding. Impressions made on the minds of men by time and habit can not easily be changed at once into the contrary; hence it may have been very necessary, but was dreadfully cruel at the same time, that the punishment of death was decreed to all who should not comply with the altered demands of the times.

With the exception of this case and others concerning the happiness of the country, it is assured in the 17th. and last Art. that laws and regulations shall remain as before untill the new constitution be established; Two exceptions however are immediately after made, by one of which it is permitted to all persons to travel in the country without any passport; by the other, it is prohibited to execute any verdict or sentence procured at the courts of justice, untill it is signed by the usurper.

The former of these exceptions is very favorable for malefactors and suspicious persons; the latter is only a means of rendering it more difficult to every Man to obtain his right.

Thus a new order of things presenting to view all the miseries that can spring from boundless despotism, was forced upon an innocent people loyal and faithfull to their

---

[1] The plural of *amtmand* is *amtmænd*.

King. The Danes that had been in public employments, who were now deprived of their places and labouring under a suspicion, otherwise honourable to themselves, of detesting the introduced changes, and meditating schemes for the fall of the usurper and who were on that account exposed every moment to the same persecutions and ill-treatments of which so many instances had been seen, resolved to depart from a country where with their best wishes, for want of means and assistance, they found no possibility of being useful.

Many natives in public functions followed their example in resigning, whose offices were filled with the most unqualified persons, by notorious drunkards and flatterers, who were indebted only to their officiousness as spies and calumniators for the favor and protection of the new ruler.

In the mean time Jörgensen and his associates daily continued their pitiful and foolish exhibitions; sometimes they caused the above mentioned Iceland colours to be hoisted in Reykevig, on the top of a warehouse occupied by M^r. Phelps under a salute of eleven guns from the *Margarete & Anna* with English pendant and colours flying; at another time the erection of two grand offices was proclaimed, of one of which Jörgensen placed himself at the head giving it the name of Regierings Contoir (Government office) whence were issued orders and resolutions like that to Bishop Vidalin, of which I have the honor to produce a translation in the voucher marked **V**.[1] the conduct of the other office he gave in charge to a person who in his letters scrupulously observed to give the usurper the tittle of Excellency, which was also – so he caused it to be announced by the report, the only one by which he could be addressed in letters and applications if they would expect to be attended to, and which a placard stuck up in Reykevig, of, which I beg leave to add a translation in the voucher marked **W**.[2] shews that his own creatures publicly attributed to him.

Respecting the barracks mentioned in the voucher last referred to I must observe that they were the house of correction mentioned above, which had now been converted into quarters for Jörgensens band of hirelings and on that account been denominated barracks; the former servants of the house had been expelled and the authors of this placard now figured as its masters in their stead. Farther a battery was seen to be constructed near Reykevig, called Fort Phelps; this work was carried on with much activity by the assistance of men from the *Margarete & Anna* and under the personal inspection and direction of M^r. Phelps. When finished, it was mounted with 6 guns that had been sent over to Iceland 140 Years ago and had for a number of years been lying covered in the ground,[3] to the use of which the whole of the little quantity of gunpowder and lead, so necessary for the destruction of foxes, was appropriated, which had been imported in the last merchantships. Plunder however still remained the grand object of the usurper and always continued to mark his career.

A vessel arrived protected by His British Majestys licence, which brought over a cargo of the most necessary articles for the supply of the country, and likewise a sum of money of 10000 rixdollars Danish currency to pay the salaries of the public officers and other

---

[1] 1 August 1809, Jörgensen to Geir Vídalín, about marriage licences (Þorkelsson, *Saga Jörundar hundadagakóngs*, p. 184).
[2] 19 August 1809. Proclamation signed by Peter Malmquist and J. Klog (in both Danish and Icelandic).
[3] At Bessastaðir, the former residence of the governor of Iceland.

necessary expences, for which money were wanting since the chest called Jordebogs Cassen had been carried away by Cap*tai*n Gilpin;[1] this vessel could not fail to tempt the rapacity of the usurper, and of course it became his prey. In what manner this was done is fully and circumstancially explained in a declaration of the master Cap*tai*n Holm made before a regular Court of justice, of which I have the honor to produce a translation in the voucher marked **X**;[2] receipts for the seized goods and monies are also existent, given partly by M^r. W. Petræus, who was engaged as M^r. Phelps' agent, partly by Jörgensen as will be seen from a notarial deed of which I beg leave to add a translation in the voucher marked **Y**.[3]

The above mentioned declaration, voucher marked **X**, shews that Jörgensen compelled the master to deliver up all papers and written documents that were found in the ship, which extended not only to what is understood by ships papers, but also to every private letter which was found containing the most indifferent things; and I am informed on the best authority that M^r. Jörgensen assisted by M^r. Phelps and M^r. Savignac devoted two days to open these letters and examine them most closely, before they were delivered to the owners.

All public sums of money that could by any means be rendered subject to confiscation were constantly keept in view. On the 11th of June M^r. Frydensberg, the Landfoged (cashier) who had the general chest of the country in his hands, received an order to deliver up the same, of which order a verified copy will be seen in the voucher marked **Z**.[4] This order not being voluntarily obeyed by M^r. Frydensberg, what little cash remained in his hands amounting to 2700 rixdollars was taken from him by force of arms on the 25 of the same month. M^r. Koefoed, Sysselmand, who had likewise some money in his hands partly belonging to the government, partly to public institutions, was repeatedly ordered to deliver up the same, as will be seen from two letters of which I beg leave to add translations in the vouchers marked **AA** & **B.B.**[5] Whatever money was acquired in this manner fell into the hands of the usurper. The system of plundering was most indefatigably followed up, whenever booty was to be expected; Jörgensen was seen accompanied by his armed attendants to scour about the country even up to its northern parts and on these excursions enforce by fire and sword whatever he wanted; Emissaries were sent and orders dispatched to other places, where requisitions could be made with advantage. In proof of this I beg leave to state the four following historical facts.

1st. The memorial, with the two special vouchers belonging to it, which I have the honour to submit in the voucher marked **CC**.[6] relates in what manner M^r. Savignac (a man whose deceit and malice rendered him every day more detested) accompanied by another person in M^r. Phelp's employ came armed to a settlement called Öreback belonging to M^r. Lambertsen, merchant, bringing with him goods chiefly consisting of

---

[1] See e.g. pp. 331, 336.
[2] The cargo of the *Tykkebay* listed by Petræus (Þorkelsson, *Saga Jörundar hundadagakóngs*, p. 177).
[3] 1 September 1809, very detailed report of the *Tykkebay* by Frydensberg.
[4] 11 July 1809, Jörgensen to Frydensberg (Þorkelsson, *Saga Jörundar hundadagakóngs*, p. 172).
[5] To Koefoed, 7 August 1809.
[6] 2 September 1809, memorial from N[iels] Lambertsen to Svend Sivertsen, factor at Eyrarbakki, with further enclosures, including accounts, in lengthy detail, including a trading permit for Sveinn Sigurðsson (Svend Sivertsen) signed by him and Samuel Phelps on 22 July 1809. Jörgensen discusses this in detail (see BL, Eg. MS 2070, Jörgensen to Hooker, 23 November 1810).

tobacco, coffee and other articles of luxury in a greater quantity, than could [be] disposed of at that place, which they forced M^r. Lambertsens factor to receive, though far from ordering any such goods for his house, he had even applied to M^r. Jörgensen to be exempted from receiving any; nevertheless he must now submit to the necessity not only of receiving the said goods in such quantities and at such prises [*sic*] as the special voucher marked No. 1[1] will shew but also of giving up a quantity of goods of Iceland produce which M^r. Jörgensen, having previously ordered a return to be made of the same, had directed him to sell to M^r. Phelps.

2^d. Mr Westy Petræus, merchant, the above mentioned agent of M^r. Phelps, had a trading establishment in Westmannoe, the management of which he had committed to the charge of a factor who lived at the place. Against the factor, whose name is Poulsen,[2] an accusation had been fabricated either by the inhabitants misled thereto by a person who had formerly conducted the establishment, but whom M^r. Petræus had been obliged to dismiss on account of misbehaviour, or by this disaffected person himself affecting to speak in the name of the inhabitants, in which accusation M^r. Paulsen [*sic*] was charged with oppression and extortion. It was known that there were a good deal of tallow and feathers on hand in Westmanöe, and the articles were weighty proofs of the truth of the accusation, weighty arguments for the exemplary punishment of the accused. Without the last inquiry into the grounds of the accusation, without listening to any defence, Jörgensen immediately ordered in a letter to M^r. Thorlewsen,[3] Sysselmand, dated the 24th of Julii, of which I have the honor to present a translation in the voucher marked **DD**,[4] that all stores of merchandize in the island, to whomsoever they might belong, should be confiscated to the public treasury of Iceland, that M^r. Poulsen, the accused factor, should without delay be conveyed to Reykevig, and finally that out of the stores found in the island all the tallow that was at hand should be sent and feathers for the rest.

On this act of violence though not the only one of its kind, I must beg leave to observe, in the first place, that if even the pretended accusation had been founded, innocent persons could not be made responsible for the wrongs complained of, whence the confiscation of property that was ordered was an act of cruel injustice that could only originate in the most detestable tyranny; in the second place, that the seizure of M^r. Poulsens person and his transportation as a criminal to Reykevig (where he was kept confined for several days till at length, without any examination of his case, without any compensation whatever being made to him, he was signified that he might return to his house) was a flagrant violation of the very laws which it had lately been proclaimed should be kept inviolable; for the laws of the country neither sanction the seizure of any citizens person without examination, nor his being forcibly dragged away from forum domicilii or forum delicti.

3^d. In the latter days of July a vessel called *Seyen*, Petersen master, came to Reykevig laden with necessaries for the account of M^r. Clausen a merchant who is settled in Iceland. This vessel had in the Year 1807 on her passage from Iceland been brought into a British port and detained, but afterwards been released and obtained his British Majesty's licence. As

---

[1] List of goods from Phelps to Lambertsen, signed by Petræus (at the time agent to Phelps).
[2] Grímur Pálsson, factor to Petræus, in the Westman Islands.
[3] Jón Þorleifsson (1769?–1815).
[4] 24 July 1809, Jörgensen to the county magistrate (*sysselmand*) in the Westman Islands.

soon as she approached the harbour, armed men were sent on board of her from the *Margarete & Anne* who forced the Captain to surrender all the papers of the ship and likewise all the letters he had brought, with which the former process of opening and examining was again repeated.

Much deliberation, it is said, now arose on the best way in which the cargo of this vessel could be made a prize to M[r]. Phelps by virtue of the letters of marque with which the *Margarete & Anna* was provided, notwithstanding both the ship and the cargo had been examined at sea by several of his British Majesty's ships of war and suffered to pass unmolested. During this deliberation which was much delayed by the associate's M[r]. Jörgensens absence on one of his expeditions to the interior of the country, the ship, which was bound for another port of Iceland, was detained at Reykevig, where M[r]. Clausen, the owner of the cargo, that he might be induced to accede to some conditions that were required of him, chiefly concerning the renunciation of his right to freight the vessel for her returning voyage, was told every day of the many and various grounds on which M[r]. Phelps had it in his power to make the ship and the cargo his prize, a threat the reality of which M[r]. Clausen was so far from apprehending that in order to see an end of this unwarrantable delay of the prosecution of his voyage, he even challenged the final detention of the ship and cargo, only requiring that the condemnation should take place, before a regular court and at a proper place but not in Iceland which M[r]. Phelps is said to have contended was the right place.

Jörgensen however at length came to the assistance of his embarrassed friend with the expedient he has employed so often; it only cost him a few lines to Cap*tai*n Petersen, the master of the vessel, and the cargo was confiscated to the state.

This event the more fortunate termination of which must probably be ascribed to the arrival of His British Majesty's sloop of war the *Talbot*[1] and the interest which the upright and honourable Cap*tai*n Jones took in the fate of the country, I have here related as it was notoriously known in Reykevig in conformity with the statements of M[r]. Clausen, a trustworthy man. The distance of his place of abode[2] from Reykevig made it impossible for him to hear of the fall of the usurpers in time to send before my departure any account of the transaction furnished with proper certificates, but it is highly probable that sometime he may call the persons concerned to private account for the manner in which he was treated and that the written documents necessary to elucidate the case will then be produced, whereas I must here content myself with the production of the only one which I was enabled to procure, of which I request permission to submit a translation in the voucher marked **E.E.**;[3] it is a letter from Jörgensen in which he directs the factor, M[r]. Biering, to receive the whole of the cargo which Cap*tai*n Petersen had been ordered to unload without delay.

4th. a similar occurrence will be found related in a letter from M[r]. Lambertsen, merchant of Öreback,[4] of which I have the honor to produce a translation in the voucher marked

---

[1] The *Talbot* sloop-of-war. See all particulars in TNA, HCA 30/52, ff. 524–26. Captain Alexander Montgomery Jones (1778–1862) was the commander, the master was Arthur Richard and Thomas Stewart was the 1st lieutenant.
[2] Ólafsvík in western Iceland.
[3] 9 August 1809, Jörgensen to Factor Bjering, relating to the *Seyen*.
[4] Eyrarbakki.

F:F: with the special vouchers belonging to it.[1] This gentleman came to the country confident in the security afforded him by His British Majesty's licence with which his ship was furnished. Nevertheless he was by Jörgensen ordered from his home to Reykevig and there declared a prisoner and the cargo he had brought to the country was confiscated. For the whole of this I venture to solicit attention to M^r. Lambertsens own relation in the above mentioned letter.

So manyfold and so insufferable were the outrages to which an innocent country and an oppressed people were obliged to submit. Hopes therefore were raised in every bosom on learning the arrival of His British Majesty's Sloop of war the *Talbot*. Convinced that no naval officer could or would approve of proceedings like those which the usurpers had adopted, the inhabitants seized the first opportunity to explain the situation of the country to the Commander of the said sloop, the honorable Capt*ai*n Jones, applying to him for that protection and deliverance which, in their defenceless state, they could not obtain for themselves. Capt*ai*n Jones did not disappoint their hopes; he interfered in a manner that bore witness to his warm feeling, for the rights of nations and the principles of public justice, and maked it my duty here publicly to express my high and undissembled esteem for him and the officers under his command, who animated with the same sentiments zealously co operated with him in the cause. As I have no doubt but Capt*ai*n Jones has already presented to his government an account of the convention concluded between himself on one part and the two principal civil officers of the country M^r. Stephensen counsellor of state and justitiarius in the upper court of justice (Lands Overret) and the Amtmand (Subgovernor) M^r. Stephensen on the other part,[2] as also of the circumstances connected with the same, I consider that I may here omit making any farther mention of it, which would only be unnecessary repetition.

As my own fate and treatment, from the first moment the attack upon my person was made, is so closely connected with the whole of these events, and so much the occasion of my having the honor of personally to present this mcmoir, I dare request the permission to devote a few lines to the consideration of it.

The indecent manner in which M^r. Phelps and Capt*ai*n Liston treated me during the declaration[3] of my destined captivity and dragged me onboard without even allowing me a few minutes of time for the previous arrangement of my domestic affairs, I have before taken the liberty to describe.

I have also taken the liberty just to mention, that M^r. Savignac and M^r. Jörgensen were allowed on the same occasion to assault me with the coarsest abuse.

Though there existed no necessity nor any object, as far as I can conceive, of M^r. Savignac being present in the room during the transaction of the moment, yet he came rushing in exclaiming, that I was a man without honor and faith, that I had broken the convention and that he had proofs of it in his hands, then, after I had called upon him to produce those proofs, immediately running out again struck dumb perhaps by self-convicted falsehood and base ingratitude. In the same manner M^r. Jörgensen without any occasion whatever that I can conceive, came rushing in, declaring himself to be a British subject and telling me that he knew with certainty I had lately been in Scotland as a spy,

---

[1] 2 September 1809, Niels Lambertsen to Magnús Stephensen in Reikevig. The vouchers are from Jörgensen, dated 9 August, and V. Gunnarsen of Eyrarbakki, dated 2 September 1809.

[2] See Appendix 9.

[3] Over this word Banks has written 'deliverance?'.

then immediately hurried out again of the room, only allowing me time to shew him the way with a look of contempt.

How entirely Mr Phelps approved of these contemptible indignities his own succeeding conduct in my personal treatment abundantly proves. Bent down under the weight of so much grief and affliction united, it now became my lot, so he had planned and directed, to be kept confined in a narrow and dirty cabin, and sometimes when Captain Liston took it in his head, even shut up in a smaller room or rather closet where I was deprived of the light of the day; constantly I was obliged to put up with the society of drunken and noisy mates and with them for my companions I was reduced to subsist on fare which even the men complained of as being more than commonly indifferent; in short I was deprived, for the space of 9 weeks of every convenience and comfort of life to which I had been used, subjected to all the sufferings which the oppressor had it in his power to inflict. His contempt of decorum and humanity even went so far as to refuse a request that was made on my behalf by one of my friends, Bishop Vidalin, that I might be allowed now and then[1] to take exercise on a small uninhabited island near which the ship was lying; I would even have submitted to be under an armed escort of the ship's crew, if it had been thought necessary, whom I offered to pay for the trouble. Yet, this request M^r. Phelps refused through Jörgensen, of whose letter to that purpose I have the honor to add a translation in the voucher marked **GG**;[2] it is remarkable in particular for the assurance it contains that M^r. Phelps could not justify his conduct to his own government were he to adopt any other measures than those which had been taken.

My possessions were arbitrarily disposed of. Some of them they delivered up to me, but the rest they applied as they thought fit. A barkship, called the *Orion*, which with her cargo I had bought for the relief of the distressed Icelanders was seized as a prize, with such part of the cargo as was left in her; the rest of the cargo that had been unshipped was confiscated. My houses were occupied, my goods and property were plundered, all my public papers and my private letters were ransacked and huddled about in the most robberlike manner.

It is impossible for me at this moment to state with any degree of accuracy the amount of my loss, because the shortness of the time before my departure did not permit me to take any account of what was left remaining and saved when Captain Jones's interference put a stop to these proceedings, but I know that I must have suffered to a very great amount; and, what is the worst of all my papers having partly been destroyed or thrown away, partly mutilated and disordered, I shall never perhaps be able to retrieve the loss and inconvenience which will thence arise. After I had for nine weeks endured such usage, it was at length thought proper time to send me to England for His British Majesty's nearer orders which I had long since required they should do. The departure took place on the 25th. of August last, but unfortunately on the 3d. day of our voyage, the 27th. of the same month, the ship took fire by which she was totally consumed after we had saved ourselves on board the *Orion* which was sailing in our company as the prize of the *Margarete & Anna* [see Plate 15].

---

[1] In the draft the following has been crossed out: '(even, if it were thought necessary, under an armed escort from the crew of the ship, whom I offered to pay for their trouble)'.
[2] 1 July 1809, Jörgensen to Bishop Geir Vídalín.

We now returned to the roads of Reykevig where we arrived on the 29th. following. Here Mr. Phelps granted me the liberty to go on shore on my parole and to remain in the house of a friend untill we should again be ready to sail, which it was expected would take place in two or three days. I was impatiently waiting for the expiration of this time, when on the 31st. at noon, to my great astonishment, I received through Mr. Stephensen, the counsellor of state, the compliments of Mr. Phelps, requesting me to accept my perfect liberty, which request was shortly after verbally repeated to me by Mr. Phelps with an addition, that he wished I would accept of the offer and return to my office. This request was made to me in the presence of the Honble. Captain Jones and Mr. Stephensen counsellor of State.

The testimony of the former I may confidently refer to and of that of the latter I have the honor to produce a verified copy in the voucher marked **HH**.[1] I required a couple of hours for consideration and having in this interval procured the very kind promise of Captain Jones to give me and the persons I intended to take with me a passage in the sloop of war under his command I returned Mr. Phelps an answer accordingly.

I entertain the most confident assurance, that His British Majesty will not refuse me and the Island, the civil government of which has been committed to my charge, and its wronged and injured inhabitants such satisfaction and redress for the offences, outrages and piracies committed by His Majesty's subjects, as justice and equity in conformity with the law of nations dictate and which in consequence of these I consider myself entitled to claim.

In order to make representations on this hoped for satisfaction and redress and likewise in order to make representations on the protection of the island in future from such abuses on the part of british subjects, more afflicting and dreadful than any that could be apprehended after a conquest by force of arms under a regular military force, I have charged myself personally with this mission in the cause of the country.

I am in possession of the original papers of which I have taken the liberty to present verified copies, the said originals of which can at any time, if desired, be produced, and if further proofs should be required of the truth of the facts here related, I shall no less be enabled to furnish such proofs in the evidence of trustworthy men. I therefore now proceed to the deduction[2] of the following propositions, which I beg leave to submit to Your Excellencys Consideration.

1. That Mr. Phelps, the owner of the *Margarete & Anna*, Letter of mark and Mr. Liston the master of that Ship, as British subjects who in consequence of the 5th. Article of the convention concluded with Captain Nott, on the 16th of June were bound during their stay in Iceland to submit to the authority of the laws of the island, but during such stay rose as rebels against the said authority, may (:as it was only the want of an adequate force that made it impossible for the civil authority of the country to secure their persons on the spot:) be delivered up to me, in order to be sent over to the island, to be placed before a court of justice and suffer punishment according to the laws of the country; or, in case Your Excellency should find reason to cause the said persons to be tried before a British court of justice for the crimes of which it is presumed they have been guilty against the highness of his British Majesty, then such British court may be ordered at the same time

---

[1] In English, 2 September 1809, Magnús Stephensen.
[2] In the draft 'representation' has been substituted for 'deduction'.

to hear and to pronounce upon representation and claims made in the name of Iceland on the subject of the above mentioned violation of the law of nations, by breaking the convention and other rebellious conduct.

2$\underline{^d}$. That M^r. Jörgen Jörgensen, the Dane born at Copenhagen, formerly captain of a danish privateer, who, being a prisoner of war, allowed by the British government to be at large on his parole of honor, broke his parole in order to accompany M^r. Phelps on his voyage to Iceland, may as a malefactor and a felon guilty of high treason against his own country and his lawful sovereign the King my master, be delivered up to my authority, in order that I may cause him to be sent over to his own country.

3$\underline{^d}$. That M^r. Phelps may be adjudged to refund and pay immediately the sum of $19^{x}220$ [*sic*][1] Rixdollars 86 skillings danish currency which in a specification or note drawn up by himself, of which I have the honor to add a verified copy in the voucher marked **JJ**.[2] he states to have forcibly seized in ready money, with an interest of 4 per Cent from the day of seizure untill the restitution is made.

$\underline{4}$. That these individuals who have already made complaints on the injuries they have suffered, as also every other citizen of Iceland, in general who may conceive himself to be injured and wronged whether in his person or his property by M^r. Phelps or by any other person employed by him in these transactions, may have their right referred to them and be allowed a proper time to come forward with their claim and proofs in order to recover damages and restitution, and that, as it is presumed that the damages and restitutions which will thus be claimed, will amount to more than the value of M^r. Phelps' property at present remaining in Iceland, he may be caused to give proper security in his possessions in great Britain for such eventual payment as he may be adjudged to make – and lastly

5th. That through Your Excellency's kind representations are as two persons may be authorized by His British Majesty and furnished with full powers to negociate with me a definitive convention founded on the principles of the convention of the 16th of June, by which the neutrality of the country may be established and the mutual relations placed on a firm footing.

My authority for concluding on the part of the Island such a definitive convention I have the honor to produce in the unlimited power with which the King my master has furnished me in and for the civil government of the island under the date of the 10th of August 1808, a translation of which I have added in the voucher marked **K.K.**[3] –

I have the honour to remain
Frederik Count Trampe[4]

Wisconsin, Banks Papers, ff. 192–213. Copy in Banks's possession.
BL, Department of Manuscripts, Eg. MS 2067, ff. 191–213. Copy in Jörgensen's autograph.
NAI, Jörundarskjöl. Draft. Other copies of this memorandum are in the National Library of Iceland, Lbs 168 fol. and JS 111 fol., both in Danish.

[1] This is the way it is written, the superscript x is odd, probably some kind of abbreviation for *rixdollar*. It should simply read 19,220.

[2] 2 September 1809, Magnús Stephensen: 'This is the original list, delivered to me by Mr. Phelps, without any other account for the taken ready money in this island, without restoring of any things and without any account for all the confiscated privat [*sic*] property.'

[3] 10 August 1808, Frederik VI to Trampe, *Lovsamling for Island*, VII, pp. 217–18.

[4] This is added in the Jörgensen copy.

125. Banks's Notes on Trampe's Memorandum and draft of letter to Lord [Wellesley] asking for some sort of protection for the Icelanders [November 1809][1]

June 11 the Landfogd or Treasurer of the Country Receivd an order from Jergensen to deliver up the Public Chest on his Refusal 2600 Dollars in his hands were taken from him.

June 25 the Systelman or Sherif Receivd a Similar order to deliver all /*Public*/ money in his hands belonging to Government or to any public institution, similar measures were suffered [?] in many parts of the Country & the Factors of Danes were Compelld to Receive Goods from the *Margaret & Anne* which they had not orderd & to give in Return Iceland Produce at Prices fixd by the new Government

a Factor Calld Poulsen was accused of Extortion on this accusation & without formal Trial all his Goods were orderd to be Confiscated to the Public & the Sytelman of the district where he Lived was orderd to send the Tallow & Feathers to Reikevic where the English Ship Lay

another Ship with a British Licence arrivd this also was Seizd & Confiscated to the new Government

soon after Capt Jones of the *Talbot*[2] arrivd & put an end to these Revolutionary proceedings he brought hence the Governor Count Trampe who Solicits justice at the hands of His majesties ministers the Count is invested by his Government with the most Extensive Powers I have ever seen He deems himself authorised by these to sign such Conventions as his Majesties Ministers may be Pleasd to grant to Iceland in order to Regulate in Future the Conduct of Such British traders as may be permitted to visit the Island

Your Lordship will I hope give [?] this Trouble the very kind Treatment the Innocent & interesting[3] Icelanders met with from H M. ministers Last year gives me hopes that their Case will be this year also attended to & redress granted them their disposition cannot be more advantageously pointed out than by saying that during the Whole of the terrible Scenes I have stated Above all of them except a few and a very few miscreents remain Quiet in their homes Jergensens bodyguard is said not to have exceeded 7 or 8 men

The Count is as far as I have seen him a man of Quiet disposition, well informd moderate & Reesonable & Certainly has Talents, if I can be employd by your Lordship in any way towards forming an arrangement with him for the Consideration & approval of your Lordship & your Colleagues I Shall most willingly undertake the Task

The *Margaret & Anne* Letter of Mart was burnd Soon after she Left the Island the Crew were savd. Mr Phelps the chief owner & Mr Jergensen /his interpreter/ the Revolutionary Cheef are now in England. The Latter is a Prisoner of war & at this time on board a Prison Ship where he has been put for breaking his Parole Iceland was when Count Trampe Left it freed by the Sailing of the Letter of mart from all Revolutionary Proceedings & is beleived to be at Present in the State it was in before She arrivd The Count meens to Return as soon as /*he can*/ the Season will allow if he can prevail upon the English Government to Put Iceland into such a state as may in Future Protect the/*se*/ good inhabitants from the oppression they have experienced from Englishmen who

---

[1] These were obviously noted down during his perusal of Trampe's memoir. The same mistakes appear.
[2] For Captain Alexander Jones, see Document 124, pp. 352–5. The *Talbot* sloop-of-war was relieving the *Rover*.
[3] Banks uses 'interesting' again in his letter to Lord Liverpool (Document 134).

/ought not to have acted as they have done/ were Licensed to Treet them as Friends but in no Shape authorisd to interfere with their constituted authorities

Wisconsin, Banks Papers, f. 156. Autograph draft.

126. Lorentz Holtermann to Banks, 6 November 1809

Bergen in Norway

Hon*ora*ble Sir

Your Honours magnanimity and generous sentiments exhibited towards Counsellor [Magnús] Stephenson[1] and the whole of the inhabitants of Iceland have rendered You dear to them and Your name ever memorable both in Iceland and in this country.

At the request of Counsellor Stephenson, who represented the then supposed miserable state of Iceland, I and several others of this place sent a Vessel from hence in March last with a Cargo of necessary stores to Iceland by which Mr. Stephenson went passenger;[2] the Vessel is now safely returned after having made her escape from thence the 20th of July.[3]

According to what Counsellor Stephenson hath told the Master of my Ship,[4] Iceland is now well supplied with provisions, but in the greatest want of Wood and Iron, not having so much of either as was necessary for their implements of fishing and Agriculture, and that he requested me to do all in my power to send a supply of Articles so much wanted.

In confidence to Your Honours generosity and known humanity I am emboldened to come to Your Honour with my humble and most respectful petition.

If Sir you would add another instance of Your goodness and protection to the people of Iceland in their great necessity by furnishing me with a Passport or Licence in consequence of which I could be enabled to freight a Danish Vessel to go from hence in March or April ensuing with a Cargo of Wood and Iron to Iceland and return here again with the products of that country, unmolested by His Britannick Majesty's Ships of war or Subjects.

If Your Honour should condescend in granting my humble request, the Licence may be transmitted to Messrs Armstrong Thompson & Co. Newcastle,[5] who will forward it to me.

With the profoundest respect
[*Signed*] Lorentz Holtermann

Wisconsin, Banks Papers, ff. 39–40. Autograph letter.

[1] Magnús Stephensen had spent the winter of 1808–9 in Norway arranging for supplies to be sent from there to Iceland. See Stephensen, *Ferðadagbækur*, pp. 113–44.

[2] *The Providentia*. This vessel had arrived in Iceland on 4 April 1809, the second vessel that year, the *Clarence* being the first.

[3] He is referring to an escape from the Icelandic Revolution.

[4] Captain Lund. Magnús Stephensen prepared a licence for him (dated 12 June 1809) to return to Bergen in Norway. The original is in Rigsarkivet (the Danish National Archives) and written in tolerable English. Perhaps he enjoyed some help from either Savignac or Mitchell, the two native English speakers then in Reykjavík (printed in Þorkelsson, *Saga Jörundar hundadagakóngs*, pp. 148–50). This licence is interesting as it makes quite clear the role played by Banks, dubbed here 'the generous Lover of mankind', in helping the Icelanders in 1807 and 1808.

[5] Armstrong, Thompson & Co. merchants, Cowgate, Newcastle.

Plate 1. Sir Joseph Banks, by Thomas Phillips, 1809. This portrait shows Banks as President of the Royal Society in 1809, the year of the Icelandic Revolution. By Courtesy of the Trustees of the Royal Society.

Plate 2. Iceland: View of the Danish merchant houses in Hafnarfjörður, by John Cleveley Jr. This was where Banks and his party stayed during their visit. The boat on the right is probably the one used in the chase to find a pilot, described by Banks in his journal. BL, Add MS 15511, f. 13. Courtesy of the Trustees of the British Library.

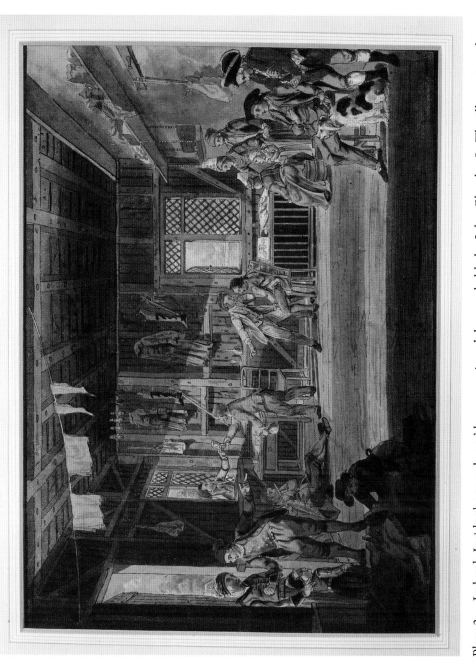

Plate 3. Iceland: Inside the merchants' house, unsigned, but probably by John Cleveley Jr. The illustration shows women and children in Icelandic dress, specimens on the table, a flute-player and Banks being handed a glass of wine. By permission of the Linnean Society, London.

Plate 4. Iceland: Banks and his travelling caravan, by John Cleveley Jr. The snow-covered Mount Hekla is in the background, with lava and a volcano. BL, Add MS 15511, f. 48. Courtesy of the Trustees of the British Library.

Plate 5. Iceland: The traditional farm of Þorsteinn Jónsson, Banks's guide, at Hvaleyri, by John Cleveley Jr. The image has been cropped to show the details in the foreground, including an Icelandic pony with a woman's saddle and a woman in full riding habit to the left, and to the right the artist's impresssion of people processing fish. BL, Add MS 15511, f. 19. Courtesy of the Trustees of the British Library.

Plate 6. Iceland: Woman in her bridal gown, with decorations in silver, by John Cleveley Jr. This is believed to be Sigríður Magnúsdóttir, the wife of Ólafur Stephensen, district governor in 1772. BL, Add MS 15511, f. 17. Courtesy of the Trustees of the British Library.

Plate 7. Iceland: Eruption of the Great Geysir, by John Cleveley Jr. The expedition made exact measurements of this 'volcano of water'. BL, Add MS 15511, f. 37. Courtesy of the Trustees of the British Library.

Plate 8. The frontispiece of 'A Journal of a Voyage to the Hebrides Iceland and the Orkneys by James Roberts. 1772'. A rather fanciful image with Hekla apparently looming in the background. By W. Brand, Boston, Lincolnshire, October 1796. A 1594. Courtesy of the State Library of New South Wales.

Plate 9. Jörgen Jörgensen (1780–1841), the 'Revolutionary Chief' or 'Protector of Iceland' in 1808, by Christoffer Wilhelm Eckersberg. Courtesy of the Museum of National History, Frederiksborg Castle.

Plate 10. Magnús Stephensen (1762–1833), Chief Justice of Iceland, by Andreas Flint. It was Stephensen who wrote to Banks in October 1807 asking him to come to the aid of the Icelanders, which led to Banks taking on the role of protecting Iceland. This is a copper engraving from c. 1800. Courtesy of the National Museum of Iceland, number 4978.

Plate 11. Bjarni Sívertsen (1762–1833), one of the leading Icelandic merchants, ascribed to Rafn Þorgrímsson Svarfdalín. He was among the first of the stranded Icelanders to visit Banks in London. Courtesy of the National Museum of Iceland, number 886.

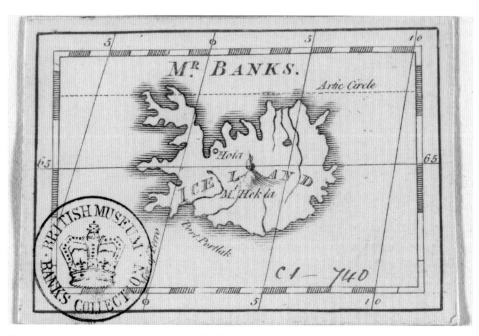

Plate 12. Banks's visiting card. Banks had this visiting card made on his return from Iceland. Note the pride of place given to Mount Hekla. British Museum, C.1–740, Sarah Banks's collection of visiting cards. © The Trustees of the British Museum.

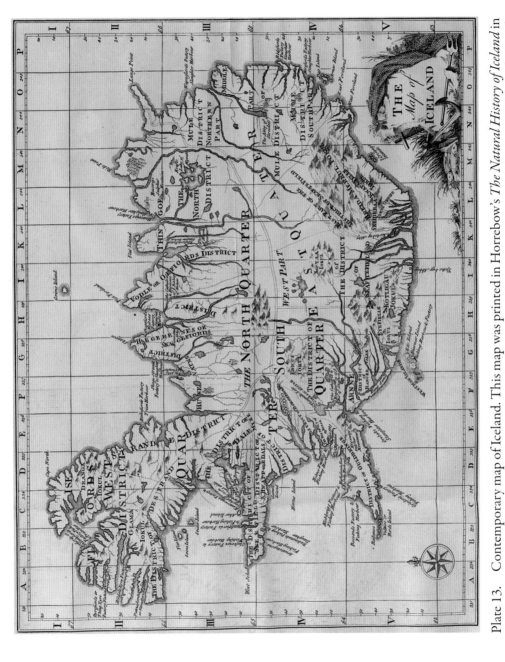

Plate 13. Contemporary map of Iceland. This map was printed in Horrebow's *The Natural History of Iceland* in 1758, close to Banks's visit to Iceland. *District* here means *sysla* and *quarter* is used for *amt*. Hekla is depicted. Courtesy of the National Library of Iceland.

Plate 14. A view of the town and harbour of Reykjavik c. 1820, by the Danish governor of Iceland, Count Moltke. The cathedral is on the left. The farmers bringing their produce into town, camping in tents, can be seen in the centre. Viðey is the island to the right and the Danish flags are a reminder that at that time Iceland was a dependency of Denmark. Courtesy of the National Library of Iceland.

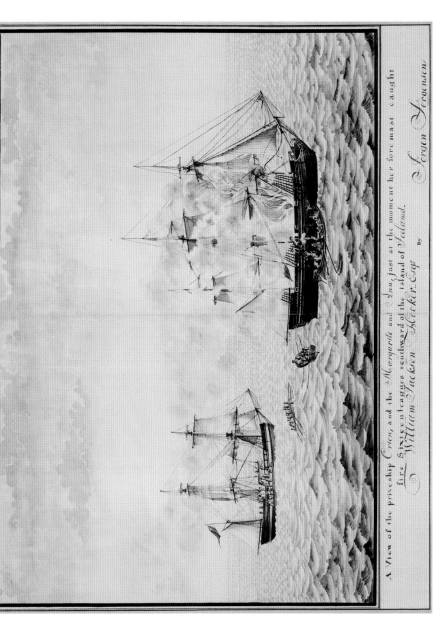

Plate 15. 'A View of the prize ship Orion and the Margarite and Ann just at the moment her foremast caught fire Sixteen leagues southward of the island of Iceland'; by Jörgen Jörgensen. Drawing with grey wash. © National Maritime Museum, Greenwich, London, object PAH8461, image B7482. The ship's name was actually *Margaret and Ann* and this event took place on 27 August 1809, just after leaving Iceland. This image was apparently in the possession of William Jackson Hooker, whose life was saved by Jörgensen, on the *Orion*, when the *Margaret and Ann* went down. Hocker, unlike Banks, always remained loyal to Jörgensen, as their correspondence in the British Library attests.

Plate 16. 'The Fly Catching Macaroni', satirical print of Banks from 1772 balancing on two globes, one lettered 'Antartick Circle' and the other 'Artick Circle', denoting his travels on the *Endeavour* voyage with Cook and his expedition to Iceland, respectively. 1868.0808.4476. © The Trustees of the British Museum.

127. Boulton & Baker to Earl Bathurst, 23 November 1809

Wellclose Square

My Lord,

Apprehending it may not be known to your Lordship, We humbly beg leave to state, that on the 8th: Inst*ant* we petitioned for 40 Licences to import Corn to Norway from Denmark,[1] which was refused on account of the many Speculators who, on the same day, petitioned for a great number of Similar Licences; we however were permitted by the honorable Board of Trade to state the Fact that 50 danish Vessels with Corn for Norway had been captured the latter end of last month & brought to Gothenburg, which we proved by an Original Letter from thence; yet we were not granted the Licences we had prayed for. We therefore beg leave humbly to represent to your Lordship, that many years previous to the War between this Country and Denmark, we have, as Merchants, transacted the commercial concerns of above Two Hundred Correspondents in Norway, who have consigned to us for Sale their Cargoes of Timber and Deals.

From the 14th Sept*embe*r to the 18th of last month we have prayed for, and obtained, 130 Corn=Licences,[2] several of which have not reached their destination, owing to the difficult mode of Conveyance from Sweden to Norway; and many will expire before they can be used.

Since the Re-commencement of Commerce between this Country and Norway, we have insured at Lloyds, on Norway Ships and Cargoes of Timber and Deals from Norway to Great Britain, and on Return=Cargoes of Salt, Colonial Produce, and British Manufactory, from this Country to Norway, according to the Policies of Insurance in our hands, a Capital amounting to £132,370: Sterling, and we are happy to know that mostly all the Vessels and Cargoes have arrived Safe; This we presume is a Strong proof that the Norwegians are very desirous to trade with this Country; but the Inhabitants of Norway are still in great distress for Corn, and by every Mail from Gothenburg we are requested to procure our Correspondents more Corn=Licences, as they are bound to import a Cargo of Corn from Denmark to Norway for every Cargo of Timber and Deals they are allowed to export. Now, as it can not be the Wish or Intention of the British Government to cause such distress to a Nation that is so very anxious to continue its Trade with this Country, we humbly hope, that your Lordship will be pleased to cause the 40 Licences for Corn,[3] which we have prayed for, to be granted to us; otherwise, an equal number of

---

[1] TNA, Board of Trade [hereafter BT] 5/19, 30 November 1809. The decision taken by the Lords of the Privy Council was to write to Bathurst, to whom the parties had addressed a similar application a short time before, 'on the Ground of a large Fleet going with grain to Norway, having been taken or dispersed, thought it right to refuse the same, as more Licences had been granted for the Importation of Grain into Norway than could possibly be wanted for the purpose of relieving the Distress of the people of that country, and because it appeared that only two of the Vessels taken, were stated to have had licences from His Majesty'. Their former application was on 8 November 1809 and was suspended the following day (TNA, BT 6/197, no. 11,735).

[2] As an example of this Boulton & Baker had petitioned successfully for licences for 40 vessels to proceed in ballast from Norway to Denmark, there to load cargoes of corn for Leith etc. This was granted on 10 October 1809 (TNA, BT 6/197, no. 10,323). In general, timber imports were permitted.

[3] There is no record of this petition. On this day (23 November) they were granted two licences for a total of 14 ships to import timber and deals from Norway to any port in the United Kingdom and to return to Norway with cargoes of salt, provisions etc. (TNA, BT 6/197, nos. 12,227–8).

Licences to import from Norway to this Country, which we have also petitioned for, will prove useless.

> We have the honor to Subscribe ourselves
> My Lord,
> Your Lordships
> most obed*ient* hu*m*ble Serva*n*ts
> Boulton & Baker

Wisconsin, Banks Papers, ff. 41–42. Contemporary copy.[1]

128. Letter from Jens Andreas Wulff, [undated, probably November 1809][2]

Mr. Wulff is particularly anxious to obtain Permission for the Iceland Ships to touch at a Port in Norway prior to their entering any Port in Denmark for the following Reasons:
<u>First</u>, to procure empty Casks, or materials for making the same, for packing both the outward & homeward Cargoes.
<u>Secondly</u>, There is seldom an Instance of a Vessel, during the Winter Season, being able to make a Port in Denmark without putting into Norway, it being generally utterly impossible to avoid it, owing to the Winds, Weather, and strong Current that prevail at that Season on the Coast: – They also obtain thereby Information as to the State of Ice on this Danish Coast. –
<u>Thirdly</u>, That, alltho' the King of Denmark is highly gratified by the Protection & Assistance afforded by the British Goverment to the Iceland Trade, the General Order 'that no Ship be admitted to Entry in a Danish Port, coming direct from England /with Goods/', which was issued in consequence of the present unfortunate Misunderstanding is still in force, but would prove no Obstacle to the Iceland Ships, if they obtain Permission to touch in Norway, and there pay the Import Duties and take fresh Clearances. –
<u>Fourthly</u>, altho' <u>absolute Necessity</u> (from Winds Currents &c) might be an undeniable Excuse for a Ships' putting into Norway, still might she meet much trouble of detention if discoverd by the British Cruisers in going from thence to Denmark, without the Licence contain a special Permission for that Purpose.

Wisconsin, Banks Papers, ff. 43–44. Autograph letter.[3]

129. Jörgen Jörgensen to Banks, 24 November 1809

> Chatham, *Bahama* prisonship

Sir!
When I had the honor of receiving your last letter,[4] I was then by order of the Transport board confined in Tothillfield prison in Westminster, and should have wrote to you Sir if

---

[1] As it says clearly on the letter.

[2] In the Wisconsin Papers this letter follows the previous letter. As both deal with Norway, it seems logical to place it here.

[3] Note in Banks's autograph, verso: 'My Lords will not Grant Licences from Denmark | they will for Norway Sweden & Iceland to Return to G.B'.

[4] Probably inquiring about his part in the Icelandic Revolution. There are no letters extant from Banks to Jörgensen.

you had been in town.¹ Since that I have been removed to this prisonship in Chatham, and now I expect you are in town, I have taken the liberty to send you this letter.

When I last wrote to you Sir,² I merely intented to have given you an account of the last two voyages to Iceland with all the circumstances relating to those voyages, for your own private amusement and inspection, and not in the least to persuade you to interfere in the business,³ for I would never think to give you the least trouble that way, for I have allways carefully avoided to solicit your interest in a public way, as I have obtained so many different favors in a private, that I should have thought myself rather impudent if I had.

Notwithstanding I immediately on my arrival in town wrote a letter to the Admiralty acquainting them with my arrival⁴ and wrote down my address, and next day went in person myself and gave my address, I was sixteen days after arrested for having broke my parole and secreted myself somewhere in the City.⁵

Being on board here I have no means to justify myself, nor contradict any thing that can be said again*st* me, and my enemies may say or do whatever they please, and as I flatter myself with having been once in possession of your good opinion,⁶ I intent at all hazards to retain it if possible, and therefore hope Sir Joseph Banks will suspend passing his own private opinion about the Iceland business ontil [*sic*] I shall have an opportunity to lay the whole case open to you Sir; for believe me nothing can be more distressing to a man than to be lowered in the opinion of a*n*other person, in whose esteem a man wishes to be – and I have hitherto reckoned it no small credit to myself to have been taken notice of by you Sir, and can not well support the Idea, of not being so longer; for I have no right to lose a friend without having committed some act that deserves it.⁷

One thing is it to have a man face to face, and assert a fact, and another thing is to secretly whisper about things, which never existed, especially when a man is put in a situation, where he can not know the extent of the insinuations again*st* him. It's singular people the least courageous in the field are the most alert at any secret work. – When I

---

¹ During September and October 1809 Banks was at his residences first at Overton, then at Revesby Abbey. On returning to Liverpool on 20 September, Jörgensen had made his way to London: 'On my arrival in London I immediately drove to Sir Joseph Banks, and acquainted him with my arrival, and dined with Lady and Miss Banks the same day; he was of opinion that Captain Jones had done wrong in destroying the fort, and enjoyed much what I told him concerning I /*my*/ our proceedings' (BL, Eg. MS 2070, Jörgensen to Hooker, 3 October 1809). There has been some debate whether Jörgensen was telling the truth, but the fact that he met Banks is quite clear from this letter.

² Document 110 deals with the voyage of the *Clarence*. However, Jörgensen repeatedly wrote accounts of the English trading ventures and the Revolution preserved in the BL, Eg. MSS 2066, 2067 and 2068.

³ i.e. the Icelandic Revolution.

⁴ This is correct. As mentioned above, the *Orion* arrived in Liverpool on 20 September 1809. Jörgensen's undated letter to Wellesley-Pole at the Admiralty stated that 'he was ready to attend any summons at any time' (TNA, ADM 1/4773; and see further TNA, ADM 2/657, Barrow to Bagot, 28 September 1809).

⁵ He was arrested, not surprisingly, at the instigation of Count Trampe (TNA, ADM 2/893, Barrow to Trampe and to Wolffs & Dorville, 17 October 1809).

⁶ Phelps wrote in his book *Observations* when planning the first expedition to Iceland on the *Clarence*: 'This person, named Jorgensen, represented himself to be very well known to Sir Joseph Banks, and I applied personally to Sir Joseph ... [who] thought Jorgensen was a very proper person to conduct the enterprise, (and I have no doubt he thought so.)' (pp. 51–2).

⁷ Jörgensen understandably became very upset by Banks's not unreasonable attitude. He wrote to Hooker: 'Nothing so sincerely vexes me, as the opinion of Sir Joseph Banks ... If I had no true regard for Sir Joseph Banks, and had received no favours from him, then should I not care so much about his sentiments respecting me' (BL, Eg. MS 2070, 22 December 1809).

saw you Sir the first day in London after my arrival last, I might have told you many facts, but did not think it honorable to prejudice any gentleman in my favor, when the persons in question were absent.[1]

On board of this prisonship *Bahama* which is a old Spanish sixty four, we are in all above eight hundred, and no distinction between officers and men. We are confined from four o'clock in the afternoon to eight o'clock next morning between two decks. You may have an idea of our Situation if you have ever seen in your voyages, a Ships hatches all closely secured in a very heavy gale, and no person able to go on deck, with exception of there being so many people on board that it makes it extremely disagreable to be on board here, but it would not be very delicate to mention what I mean. God knows as for breaking my parole,[2] I did not even rightly understand it was one when I signed it, for Mr. Adams at the Transport board knew I did not read it, and as for secreting myself I solemnly declare I never did, and attended strictly every summons from the office. First of all I remained eight or nine months in London without signing any thing, and in that time formed such connections as made it my inclination as well as interest to remain in this country; Next I heard some disagreable news, which was no less than I could not return to my own country for I would be surely hanged for having struck my flag to a british privateer of much inferior force without firing one Shot,[3] this was afterwards confirmed by count Trampe; for which you may refer to Mr. Hooker.[4] This was not true, but still I never intented to risk my life by standing a continental trial. All this in addition to my former principles, would not allow me to return ever to Denmark. At fourteen I entered the british Service, and in the year 1806, when I first had the honor of being acquainted with you Sir,[5] I left it, and surely between the age of fourteen and twenty six, a mans principles are in general fixed for life. No wonder that I, who had far distant from my native country, friends and relatives, wandered under the british flag from one end of the world to the other, and traversed the globe in all directions,[6] that I solitary and unprotected as I stood, and still found myself humanely, generously threated [*sic*], that I should deem England a second country more dear to me than my own. But more, when I returned to England a prisonner of war I found my wants amply relieved by Englishmen, when my own countrymen denied me assistance. In leaving me in town so long without signing a parole, every one might see I did not wish to escape, for I might have done that as easy as going to Iceland. All other persons on parole, and who have broke it are released, only myself. The particular reason therefore government detain me on must be on account of the Iceland business, and there I should not feel vexed if I had a fair chance to answer every accusation again*st* me. I should not feel vexed if the complaint came from the poor, the fatherless, the widows or the clergy in the Island, but not one person can come forth in honor and in truth to assert that he is the loser by my proceedings, not like Baron

---

[1] Meaning Count Trampe who had travelled to England on the *Talbot* sloop-of-war, arriving at Leith, and did not arrive in London until the beginning of October.

[2] Which he undeniably did by leaving England to go to Iceland.

[3] Jörgensen had commanded the Danish privateer *Admiral Juul*, which had surrendered to the *Sappho* after a short engagement in March 1808. As the commanding officer he was paroled.

[4] In his book Hooker gave a sympathetic account of Jörgensen, who after all had saved his life. See his later letter to Banks: Document 167.

[5] See Document 39.

[6] Jörgensen had certainly travelled widely. See Appendix 13.

Hompesch,[1] taking away the support of those classes of people, but even having found means to pay all difficiencies to all. I hope some day or other to be able to give you a detailed account of the whole,[2] and should have done it already, if I had not received your last letter.

What make it still more dangerous for me to be on board here, is, that some officious person had been on board the Ship before I came down, and prejudiced the mind of all the danish prisoners, and made it exceeding dangerous for me to be here, and since that some letter has been sent from another prison Ship to the people on board here, and exaggerated every circumstance relating the Iceland expedition. So my ennemies in London have not neglected any thing that could rob me of Life and liberty.[3]

I have entirely wrote this letter in order to prevent you Sir from judging any thing ill of me, before you have heard all, for I am extremely anxious not to lose your esteem. It's not because you are in a exalted situation, it's not because your interest is great that I write, for should one of those unforeseen and wonderful revolutions take place, in which the greatest and wisest of men are reduced and put on a level with the lowest,[4] I should be equal anxious to preserve your good opinion, I should then if you was poor and low be able to shew my gratitude in a more proper manner, than I could at any other time, I should willingly labor from Morning to Night, and from Night to Morning to supply your wants. I hope Sir you will not think it flattery, when I write in this Strain; If you do you are not sufficiently acquainted with my Character. In deed it will please me much better, that you never say any thing in my behalf, than if you do, for it is not your interest I desire, but your good opinion, and if I have that I have no doubt, but what government will not be unjust when a proper explanation has taken place. I only wish you will recollect I am confined and cannot take any steps to clear up any point, and therefore it would be ungenerous to believe any thing as a fact, ontil I can do that. Lastly I would /not/ wish the wisdom and penetration of Sir Joseph Banks should have regarded any person, that not constantly deserve it. In deed if I had not invariable when in the british service tried to follow up a course of undeviating rectitude, I should then not deserve any lenity. Did ever I when after having hard earned my money, spend it in riot, drinking or debauchery? or did I not freely part with to relieve the distresses of my fellow creatures, and where my own power was not sufficient to do all the good I wished, I have been bold enough to address some exalted Character in order to do what I could not perform myself. Just and good, I hope you will take my letter in it's right light. It is entirely private, not meaning any complaint in my present situation, for allthough I have not the opinion of Candid,[5] or Voltaires, all for the best. It is my duty notwithstanding to bear misfortunes with some degree of fortitude. Next I hope; if You Sir should deem it, me writing in such strong terms either as flattery, or [*page torn*][6] selfinterested views, that you will not even think me worthy of answering my letter, if on the contrary you should be good enough to relieve

---

[1] The owner of the *Salamine* privateer.
[2] This he eventually did (see Document 263).
[3] His enemies were principally Count Trampe and Georg Wolff, the former Danish consul.
[4] This Jörgensen had done in his proclamation of 11 July 1809 where he stated that the 'poor and common' had the same rights as 'the rich and powerful'.
[5] *Candide*, Voltaire's most popular philosophical novel, published in 1759.
[6] Here the letter is torn, with some words missing.

my mind, and as I am now a prisoner it would be double goodness, I should think myself extremely happy.

Any letter to me, if it was sent to Mr. M'clea[1] at the Transport office enclosed will no doubt be directed to me unopened, otherwise when sent immediately down here; it will be opened by Cap*tain* Hutchinson the Agent for prisoners of war in Chatham. I am Sir!
with the greatest respect
Your most ob*edien*t & humb*le* Servant
[*Signed*] J Jörgensen

Wisconsin, Banks Papers, ff. 171–174. Autograph letter.[2]

130. Bjarni Sívertsen to Banks, 25 November 1809

47 Southampton Row, Blomsberry [*sic*], London

Dear Sir

By this I have the honor to return your Letters and a Translation of the same,[3] and humbly beg you would excuse that the Letters have remaind so long time in my hands, and I have taken me the Liberty to deliver the other Documents to Count Trampe as he told me that you had been so good to allow him to use the said Documents for reclaiming the money Robbed from Iceland in 1808,[4] nevertheless this Documents shall be returned to you, or if you would be contented with a Translation allone therof when you please to command it. –

Concerning the Translation I hope to have keept the meaning or analogy, and beg pardon for the Style and writing. –
I have the honor to remain most honoured Dear Sir!
Your most humble and Obedient Serv*an*t
[*Signed*] B Sivertsen

Wisconsin, Banks Papers, ff. 59–60. Autograph letter.

131. Samuel Phelps to Banks, 29 November 1809

Cupers Bridge [Lambeth]

Sir,

I have the honor of inclosing you Mr. B. Sivertsen's draft on Phelps & Co. at 6 mo^s 6 s^h [5] for £44 – which we have accepted having this day received advice from Mr. Severtsen.[6]

---

[1] Alexander McLeay, Secretary of the Transport Board: see Appendix 13.

[2] Note in Banks's autograph, on bottom left margin of the fifth page of letter: '[Received] Nov*ember* 27'.

[3] The letters in question are documents concerning the Gilpin Raid. It is also possible that some concerned the Icelandic Revolution. Trampe, the Danish governor of Iceland, in his lengthy memorandum to the Foreign Office included a large number of documents, mostly written in Danish, which would have required translation (see Document 124).

[4] The High Court of Admiralty ruled in Iceland's favour and the money seized by Gilpin was reclaimed in 1812. Bjarni Sívertsen emerged as the hero of the case. In recognition of his efforts the King of Denmark made him a knight of the Dannebrog, the first Icelander to be awarded this honour.

[5] This is difficult to decipher, possibly a bill due for payment in six months at six shillings' interest.

[6] This draft is not extant.

It would give me great pleasure at any time to give you every information of the proceedings of the Iceland transactions but in the present state of the business I conceive it would be very indilicate [*sic*] on my part to trouble you upon the subject, until facts appear before the public – for so many false assertions (and even false affidavits) having been brought forward that I have not the confidence to believe any assertion of mine would have credit, until the proceedings are brought to trial, when facts will be proved.[1]

I had the honor of receiving Your letter of the 12 October from Revesby Abbey,[2] and am sorry to say we were not insured, as you concluded, to the full extent of our loss by many thousand pounds; – & the underwriters even refuse to pay that portion which was insured on the Ship & Cargo in consequence of a representation made to them by Count Trampe and his adherents, & (as they declare) on no other account.[3] – We must therefore be subject to great delay & a lawsuit. – Independant of this is the immense /property/ we have in Iceland perhaps never to be recovered. – We can prove by our Books & receipts that we have embarkd & have actually paid full £45,000 – in this concern, and as yet we have not recovered a single shilling.

I have sent a statement of facts to Mr. Hamilton[4] of the Foreign Office[5] and I trust it will have the effect of proving that I have been very much injured, and that under all circumstances I was fully justified in the conduct I adopted (which is y^e [the] full opinion of the Lawyer, I have consulted) & that even if some errors may

---

[1] Phelps later regretted not having been quicker to approach the government, thus permitting Trampe and the others to make their accusations 'as numerous and voluminous, as they were false and infamous' (Phelps, *Observations*, p. 68). Not until 1817 was Phelps to produce his own detailed version of the events, defending his actions, in this book.

[2] Not extant.

[3] The *Orion* was liberated in the High Court of Admiralty on 1 December and restored to Trampe. All papers concerning the *Orion* trial are to be found in: TNA, HCA 32/1614, no. 4657 and TNA, HCA 8/14, restitution of ship and cargo, 20 January 1810. Much more important, however, was Phelps's case against the underwriters for the recovery of the insurance of the *Margaret and Ann* (Campbell, *Reports of Cases Determined at Nisi Prius in the Courts of King's Bench ... 1809–1811*, II, pp. 350–51, case called Phelps v. Auldjo; Phelps, *Observations*, pp. 62–6). For further legal sources on this case, see Marsden, *A Digest of Cases relating to Shipping, Admiralty and Insurance Law*, p. 1216. According to this letter the unrelenting Trampe was here again to blame. The verdict went against Phelps and the insurance was forfeited to the tune of about £19,000 (TNA, T 1/1121, Phelps to the Treasury, 3 March 1810; Phelps, *Observations*, pp. 65–6). In his book the sum mentioned is £39,500, but that is probably the estimate of the full cost. In this letter Phelps, however, reckoned that the full loss of his enterprise was in the region of £45,000. Phelps renewed his lawsuit with the underwriters of the *Margaret and Ann* in the winter of 1810–11 (BL, Eg. MS 2070, Jörgensen to Hooker, 7 October, 8 and 11 November and 13 December 1810). Mackenzie, who published his book on his travels in Iceland in 1811, had to omit the appendix (no. VIII) which dealt with the Icelandic Revolution because it was felt that his account might influence the court (Mackenzie, *Travels in Iceland*, 1811, p. 483). However, Mackenzie cunningly hid his account in a lengthy note (pp. 80–82). Phelps lost again (Campbell, *Reports of Cases*, pp. 350–51).

[4] William R. Hamilton (1777–1859), Under-Secretary for Foreign Affairs 1809–22.

[5] This is doubtless 'Statement of Mr. Phelps – dated November 23rd 1809 – With an Enclosure', one of the missing papers 'relating to the Transactions which took place in Iceland at different periods from the 12th of January to the Month of October 1809' sent to the Board of Trade and afterwards sent on to the Admiralty. There is a list of them in TNA, FO 40/1. Despite searches by more than one scholar these important papers have not yet been located. A further five letters written by Phelps in November and December are among these missing papers.

be found, allowances may be made for a man in an agitated situation – /*for*/ it is not reasonable /to suppose/ I should have made the desperate attempt but from desperate necessity.[1]

I have the honor to be
Sir with greatest respect
[*Signed*] Saml Phelps

Wisconsin, Banks Papers, ff. 63–64. Autograph letter.[2]

132.   Boulton & Baker to Banks, 5 December 1809

Wellclose Square

Sir
Our friend Mr. J. A. Wulff[3] is again arriv'd in this Country in the hope of superintending the disposal of the Return Cargoes of the two Vessels which, thro' your humane and unceasing Interference were permitted to take out Necessaries to Iceland.

It gives us much Concern, that on his first entering our Door, we should have to make known to him, the Misfortune his Ship /*Freden*/ *Husevig* had met with:[4] – By which his Plan for supplying that part of Iceland[5] has been wholly frustrated, the Ship & Stores greatly injur'd, and the principal part of the Cargoe compleatly spoild, consisting too (most unfortunately) of those very Articles our Underwriters will make no Compensation for, unless in case of total Loss! He had however the Comfort to hear, in the course of a few days, that the *Freden* was safely arrived in the Thames, after having landed her Cargoe at her destined Port.[6] – Much Distress we fear will ensue from the failure of the *Husevig*'s Expedition, and a heavy pecuniary Loss to himself. Other Afflictions too appear to have fallen rather heavy on that illfated Island, which we understand have already been laid before you, and which we are assured your Philanthropy will endeavour to remedy or alleviate. – Mr. Wulff was desirous to have waited /on/ you immediately on his reaching London, but learn'd that your Health was too indifferent to admit such an Intrusion:[7] – We hope & trust this will not long continue to be an Obstacle to his Wishes: – In the mean time, he has requested us to lay before you the accompanying Statement[8] of what has taken place regarding the Ships the British Government so graciously permitted to be employed for the Service of Iceland, and which, tho' rather in the Danish Idiom, we trust you will readily apprehend. Nothing certainly on his Part appears to have been wanting to carry into effect the Task he engaged in; altho' in one Point, his Purpose has by adverse Winds & Weather been so completely defeated: But this no human Foresight can guard

---

[1] Phelps made similar statements at the time to Captain Jones.
[2] Note in Banks's autograph, bottom of first page: '[Received] Nov*ember* 30.'
[3] Jens Andreas Wulff. See Appendix 13.
[4] See p. 308, n. 5 above.
[5] The north and east of Iceland (Húsavík, Seyðisfjörður, Eskifjörður).
[6] Probably Eskifjörður.
[7] Banks was probably laid low by gout.
[8] See following letter.

against. Entreating your Pardon for thus taking up your time we subscribe ourselves, with great Respect.

Sir
Your very obedient Servants
[Signed] Boulton & Baker

Wisconsin, Banks Papers, ff. 45–46. Autograph letter.

133. Jens Andreas Wulff to Banks, 5 December 1809

[London]

Sir

Being among the number of those who last year, in consequence of the humane views of the British Government towards Iceland, and your benevolent representations,[1] obtained the liberation of my ships and property that had been detained in this country, I feel it to /be/ a sacred duty to you, Sir, whose philanthropic care of Iceland was so beneficent, and of whose kindness and assistance I received so many uncommon proofs, most humbly to state to you what use I have made, as to the provisioning of Iceland, of the great favours that were conferred on me.

As soon as I arrived in Denmark I immediately began fitting out both my ships, *Freden* and *Husevig* with all the expedition which the repairs of which they both were wanting rendered possible. *Freden* first set sail for our settlement on the East Country[2] having received a cargo of double the usual quantity of rye, flour, pease, bread, groats etc. by which that part of the country has been well supplied with provisions and other necessaries at least till the middle of next summer; all the provisions will be sold there this winter at my own purchase prices only with the addition of freight insurance and other expences, though a great part must be given on credit to the poorest of the inhabitants by which considerable losses are unavoidable. The Ship is now safely arrived here with a returning cargo, which however does not amount to one half of the value of the cargo she /brought over/ carried out.[3]

The other ship was destined for our settlement on the north of the island.[4] But unfortunately she met in their northern latitude such severe frost accompanied with dreadful storms, that she could not attain her destination, which has also been the case with other ships. Having long struggled against continual gales of North-Easterly winds, being completely covered with ice which almost totally destroyed her tackle and stores, and getting at the same time very leaky, she was at length, nearly in a sinking state, compelled to seek what port she could reach, and arrived, in conformity with the tenor of the licence, at Dundee in Scotland, where she now lies. On taking out the cargo more than two thirds of it was found to be completely spoiled which will cause a clear loss to me of 20,000 rixdollars, for though I had effected an assurance here at

---

[1] As usual everyone is aware of Banks's efforts on behalf of the Icelanders.
[2] Eskifjörður.
[3] The *Freden* had arrived from Iceland with wool, skins etc.
[4] This is the *Husevig* which arrived in Eskifjörður on 15 September 1809 bringing among other goods coffee and bread to the delight of the inhabitants (Thorlacius, *Endurminningar*, pp. 74–6).

a premium of 12 *Per Cent* yet I understand that no compensation is made for spoiled corn. I regret to say that, that owing to the failure of this voyage, that part of the country for which the *Husevig* was destined is threatened with great distress during the winter.

I wished very much to have had the honour, Sir, of personally presenting these Statements to you; but having been deeply concerned to hear that you are indisposed, I beg leave to solicit your permission to pay my humble and grateful respects to you at any future time you may be kind enough to appoint as most convenient. I beg the honour to subscribe myself, most respectfully

Sir
your obedient and humble
Servant
[*Signed*] J. A. Wulff

Wisconsin, Banks Papers, ff. 47–48. Autograph letter.

134. Banks to Lord Liverpool, 11 December 1809

Soho Square

My dear Lord[1]

Allow me under favor of your Lordships kind permission to offer to you a very brief account of the conduct of the Crew of the British Letter of Mart *Margaret & Elizabeth*[2] in Iceland, which has been of such a nature as to determine the Governor of that Island Count Trampe to come in person to England to seek redress from the justice of His Majesty's Ministers; a full detail of this business has been communicated by the Count to the Secretary of State for Foreign affairs.[3]

Four days after the arrival of the Letter of Mart, an Armd force was sent ashore to the House of Count Trampe the Governor, his Person was seizd, & he was carried Prisoner on board the Vessel, where he was confind nine weeks.

The next day[4] Jergen Jergenson the Interpreter of the Ship a Dane, issued a Proclamation in his own name, declaring all Danish authority to be at an end, commanding all Arms of the Natives to be deliverd up, & also the Keys of all Warehouses Containing Public Property or that of Danish Merchants, & that all Danes should remain in their Houses & no Persons communicate with them.

Another Proclamation signd by Jergenson soon after appeard, declaring Iceland independent of Denmark & that a new Republican constitution, similar to the ancient one of Iceland should be introduced, in the interim, that the present Authorities are suspended, all Debts due from Icelanders to Danes annulld, the secret payment of them

---

[1] Banks is sending a précis of Trampe's memorandum (Document 124) to his friend Lord Liverpool to keep him up to date with what had been happening in Iceland.

[2] Actually *Margaret and Ann*.

[3] Henry, Earl Bathurst, was Secretary of State for Foreign Affairs from 11 October to 6 December 1809. Trampe's memorandum was dated 6 November 1809 (Document 124). The Marquess Wellesley had now been appointed Bathurst's successor.

[4] 26 June 1809.

made a crime, the price of Provisions to be lowered, & the Taxes Reducd to half of their present amount

Jergensen now took possession of the Government House, & removd the Seal the Governor had placd on the Apartment where his Public & Private Papers were kept, & he establishd in it what he calld a Government Office, where the *Margaret & Elizabeth*'s People attended

Jergensen next seizd the House of Correction & let loose the Malefactors confind in it, his next measure was to place a guard of the few[1] Islanders all of bad character who acted as his Soldiers, upon the Warehouses of Danes

Confiscations were now begun, Mr. Biarnyen[2] was compelld to deliver up 2600 Dollars the Property of Adzer Knutson[3] then in Denmark, Mr Struber[4] was made to deliver up Tallow Oil Fish &c belonging to a Trading Company of which he was Factor;[5] others were treated in the same manner, & Mr Phelps was authorizd by an order from Jergensen, to take what he thought proper of those confiscated Goods

Jergensen soon after appointed a Guard to his own Person & arrested two of the Public Functionaries on pretence of a Plot against himself[6]

July 11 Jergenson Issued a Proclamation as follows, We Jergen Jergenson have taken upon ourselves the Government of the Country till a regular Constitution shall be establishd, with power to make War & to conclude Peace &c a new Flag is appointed for Iceland which Icelanders are to defend with their blood, the ancient Seal of the Island is annulld, our own private Seal to be usd in its stead till another can be procurd, the time for the Civil Officers of the Island to acknowledge the new Constitution is prolongd & in case of their contumacy they are suspended, he then declares his intention to send an Ambassador to England to make Peace.[7]

All this time & to the end of this silly business, it does not appear that any Icelanders, except a very few of bad character gave the least countenance to these Revolutionary proceedings, these quiet people remaind silent & suffering spectators of the strange scene which was acted.

The Crew of the Letter of Mart now erected a Battery & mounted on it 6 old Guns, which had been brought to Iceland 140 years before & had been buried ever since, all the Gunpowder in the Island usd for killing Foxes was put into requisition to supply this Battery

Soon after a Vessel arrivd having a British Licence, she brought Necessaries for Iceland & 10,000 Rixdollars[8] to pay the Salaries of the Public Officers whose pay usd to be provided out of the Public Chest, which Capt*ain* Gilpin took away, this Vessel was seizd & Receipt given for the Property by Petræus Mr. Phelps Agent.[9]

---

[1] Their number was limited to eight due to the lack of weapons.
[2] Hans Peter Bjering.
[3] Adser Christian Knudsen: see Appendix 13.
[4] Christian Conrad Strube.
[5] The Flensburg Company.
[6] Ísleifur Einarsson, the assessor of the High Court, and Frydensberg, the royal treasurer.
[7] Indeed Article 9 in the proclamation of 11 July 1809 stated: 'That a person shall be invested with full power to conclude a peace with his Majesty, the King of Great Britain.'
[8] The ship was the *Tykkebay*.
[9] Up to this point there is a draft in Banks's autograph in the Wisconsin Papers.

June 11[1] the Landfogd or Treasurer of the Country[2] receivd an order from Jergensen to deliver up the Public Chest, on his refusal 2600 Dollars in his hands were taken from him.

June 25 the Systelman or Sherif[3] receivd a similar order to deliver up all Money in his hands belonging to Government or to any public institution, similar measures were enforcd in many parts of the Country & the Factors of Danes were compelld to recive Goods from the *Margaret & Anne* which they had not orderd & to give in return Iceland produce at prices fixd by the new Government.

A Factor called Poulsen[4] was accusd of Extortion, on this accusation & without form of trial, all his Goods were orderd to be confiscated to the Public & the Sytelman [*sic*] of the District where he lived was orderd to send the Tallow & Feathers to Reckavic where the English Ship lay.

Another Ship with a British Licence[5] arrived, this also was seizd & confiscated to the new Government.

Soon after Cap*tain* Jones of the *Talbot* arrivd & put an end to these Revolutionary proceedings, he brought home the Governor Count Trampe/at the Governor's request not as a Prisoner[6] he now/[7] /who/ solicits justice at the hands of His Majesty's Ministers, the Count is invested by his Government with the most extensive powers I have ever seen,[8] he deems himself authorizd by these to sign such Convention as His Majestys Ministers may be pleasd to grant to Iceland, in order to regulate in future the conduct of such British Traders as may be permitted to visit the Island & that of the Natives.

Your Lordship will I hope forgive this trouble, the very kind treatment the innocent & interesting Icelanders met with from His Majestys Ministers last year, gives me hopes that their case will be this year also attended to & redress granted them, their dispositions cannot be more advantageously pointed out than by saying that during the whole of the horrible scenes I have stated above, all of them except a few & a very few Miscreants, remaind quiet in their houses, Jergenson's body guard is said not to have exceeded 7 or 8 men.

The Count is a as far as I have seen him a man of quiet disposition well informed moderate & reasonable & certainly has talents, if I can be employd by your Lordship in any way towards forming an arrangement with thim for the consideration & approval of your Lordship & your Colleagues I shall most willingly undertake the task.[9]

---

[1] This should be 11 July.

[2] Rasmus Frydensberg.

[3] Hans Koefoed, the *sysselmand* (county magistrate), who had accompanied Trampe to London as a witness. The order to him was dated 11 July, not 25 June.

[4] Grímur Pálsson, factor to Petræus in the Westman Islands (Vestmannaeyjar).

[5] The *Seyen*.

[6] This is incorrect, as Banks realized later (see below, Document 140, p. 380). Trampe was a prisoner, as the log of the *Talbot* makes clear. Phelps had offered him liberty but Trampe refused, wishing to go to England to lay the case of the Icelandic Revolution before the government.

[7] This insertion is in Banks's autograph.

[8] Because of the difficulties in communications during wartime, the Danish king, commending Trampe to God, actually invested him with a share in his own absolute royal power declaring with the royal *We* Trampe's right to take decisions and adopt such measures as he considered best (see p. 356, n. 3).

[9] The result was the Order in Council, Appendix 10.

The *Margaret & Anne* Letter of Mart was burnd soon after she left the Island the Crew were savd[1] Mr Phelps the chief owner & Mr Jörgensen his Interpreter, the Revolutionary Chief, are now in England, the latter is a Prisoner of War & at this time on board a Prison Ship,[2] where he has been put for breaking his Parole. Iceland was when Count Trampe left it freed by the sailing of the Letter of Mart from all Revolutionary Proceedings & is believd to be at present in the state it was in before she arrivd, the Count means to return as soon as the Season will allow,[3] if he can prevail upon the English[4] government to put Iceland into such a state as may in future protect the good Inhabitants from the oppression they have experiencd from Englishmen, who were Licencd to treat them as Friends, but in no shape authorizd to interfere with the constitutional authorities.

[*Unsigned*]

TNA, FO 40/1, ff. 5–8. Contemporary copy, amanuensis (William Cartlich).[5]
Draft of first part in Wisconsin, Banks Papers, f. 49.[6]

135. Banks to [Culling Charles Smith], 13 December 1809

Soho Square

My Dear Sir[7]

I send with this for your Private use a Copy of the Paper I mentiond to you yesterday the /which/ original I have sent to Lord Liverpool,[8] who will be the better prepard to discuss the matter in Cabinet it Cannot be calld a Précis of the business it is [in] Fact no more that a very breif Statement of /some of/ the Atrocities (Such they appear to me) Committed by British Subjects & their agent Jergensen in a Countrey under the Protection of Great Britain, as is Testified by the Granting of the very liberal Licences which issued Last year from the Board of Trade.

The *Orion* Knudsen[9] Touchd at Leith on May 20 1809 as my memorandum tells me, what Could a Spy have to do there, who was to Proceed to Iceland before he went home

---

[1] Thanks to Jörgen Jörgensen, who saved them all.
[2] The *Bahama* prison ship at Chatham.
[3] Count Trampe was never to return to Iceland.
[4] 'English' in FO, 'British' in Wisconsin draft.
[5] This is a copy, identical to one sent to Banks's closest friend in government, Lord Liverpool, sent to the Foreign Office to the Marquess Wellesley. Banks had obviously read it through before sending, as he has made a correction to it. See further following letter.
[6] Note, verso: 'Copy | Sir Joseph Banks to | Lord Liverpool Dec*ember* 11 1809 | Proceedings of Phelps | and his Interpreter Jergen Jergenson in | Iceland, in June & July | 1809. Count Trampe the Governor now in | England – prepared | to enter into terms | with the British Govt. | Sir Joseph anxious to give his assistance.'
[7] Culling Charles Smith was under-secretary at the Foreign Office from 13 December 1809 until 27 February 1812 when he was succeeded by Edward Cooke (TNA, Foreign Office [hereafter FO] 366/671 f. 402, and TNA, FO 366/380). Banks is obviously making it his business to make sure that the new Foreign Secretary, the Marquess Wellesley, is well acquainted with the affair. On Smith see Appendix 13.
[8] See previous document.
[9] The *Orion* belonged originally to Adser Christian Knudsen, who sold it to Count Trampe earlier in 1809. The ship had received a licence on 27 June 1808 (TNA, PC 2/177). It eventually arrived in Iceland on 6 June 1809 with Count Trampe on board.

to Denmark to tell his discoveries, I shall make out[1] whether the Count was on Shore & how Long the ship remaind there.[2]

I have no doubt in my mind that the Count is an honest man I have grounded this opinion Cheifly on his having no interest as Far as I can see in being a Rogue his Powers, of which you have a Copy admit that Iceland is severd from Denmark & /that/ no Communication is Likely to take Place during the War. This admitted which Cannot Easily be denied, What can the Count do injurious to this Country with any hope of Success & how dares he to trust himself among us if he is a Rogue.

His situation Just now either is or may be interpreted into that of an alien Enemy resident among us, as such I do not see how he can appear in an English Court of Justice, this inability may I conceive be taken off & I Sincerely wish it was, as I am sure that Every man in England ought to be allowd Fair Play & a Fair hearing, as a humble advocate for the Icelanders, who in Fact have been Receivd under the Protection of this Countrey & who must literely be Starved if they are put out of the Kings Peace I venture to Suggest this.

Believe me my Dear Sir
always Faithfully yours
[*Signed*] Jos: Banks.

P.S. as the whole of my information is derivd from Count Trampe being in Truth extracted from the Representations[3] he has Lodgd in your office you must give it no other credit than you give to the Count I may be wrong but I give the Count Full Credit his Station & the Powers with which his Sovereign has intrusted him in my Judgment demand it.[4]

Possibly the marquis [Wellesley] may do me the honor of seeing me on the Subject of the Spanish Sheep[5] I shall most Thankfully wait upon him whenever he will honor me with an appointment.[6]

TNA, FO 40/1, ff. 3–4. Autograph letter.

136. Banks to Boulton & Baker, 26 December [1809]

Soho Square

Gentlemen
I shall take the Earliest opportunity of the meeting of the Board of Trade after the very short Recess of Christmas to lay before them your Letter but it is fit that I should apprise

---

[1] The page here is bound so tightly that this is a guess, i.e. 'make out'.
[2] Banks is here trying to refute Jörgensen's allegation that Trampe was a spy. There is no reason to think he was.
[3] See Document 124. Banks owned the copy transcribed here.
[4] See previous letter, p. 370.
[5] Banks is referring to the importation of Spanish merino sheep shipped from Cadiz (Carter, *H. M's Spanish Flock.* chapters XI and XII).
[6] Note, verso: 'Sir Joseph Banks | [Received] Dec*ember* 13 1809 | Inclosing copy of a letter to Lord Liverpool on the late disturbances in Iceland'.

you of my opinion on the Subject in the hope that a further adjustment may take place between you & me before the application is made as I think it has very Little chance of Success in its present Shape

I cannot Conceive the necessity of taking Crews larger than those which navigate the Ships from Iceland to Denmark for the Short trip between England & Denmark what I must recommend is that a Less number than /usual/ the original one be askd for as I am by no means clear that Even that will be granted the present application looks too much like a Pretence to Liberate more danish seamen than are necessary for the occasion & I hear it will be considerd as such which will sink the character of the Icelanders in the opinion of the public a matter which would give me Serious concern as I have hitherto in all my representations held it up as virtuous & merits much above the Generality of mankind[1]

I should also advise that no person who has in any way Commanded or born command in a Ship of war of any description be asked for by name Such a request would Savor of an intention to furnish our Enemies with persons who have been Employd in active Service against us the description of people which we more particularly wish to Retain till a Cartel can be adjusted between the two Countrys

P.S please to remember that if the whole of what may be wanted are not applied for at first it will be very improbable I should be able to obtain a second indulgence of a similar nature

NHM, BC, f. 66. Autograph letter. Draft.

137. Banks to Boulton & Baker, 27 December [1809]

Gentlemen
I shall lay your Letters before the next meeting of the Lords Commiss*ioners* for Trade[2] but I am Still of opinion that it is not necessary for these Icelanders to Provide themselves in England with a Crew sufficient to navigate their ships to Iceland unless they were bound to that Island direct as they have been indulged with the great favor of being allowd to Touch in Denmark they should I think in all Cases be Contented with men Sufficient to navigate their vessel to that Countrey, where they may procure as many additional ones as shall be found necessary to carry the Ships to Iceland & to Carry on the Fishery there.

NHM, BC, f. 67. Draft in Banks's autograph.

---

[1] This is certainly true; see esp. Document 63, also Documents 41, 61.
[2] Banks was a member of the Privy Council for Trade. In December 1809 Boulton & Baker had sent in several petitions for licences regarding the Iceland ships *Husevig, Bosand, Svanen* and the *Freden*. On 21 December all these vessels were permitted 'to proceed to Norway and any port of Sweden without the Baltic with goods permitted by Law. There to load grain, biscuit and flour, thence to Iceland and to import from Iceland to Great Britain Iceland produce' (TNA, BT 6/197, nos. 13,005–8).

138. Corbett, Borthwick & Co. to Banks, [undated, 1810][1]

[Leith]

Sir

We were much gratified to learn from your obliging letter of 26th,[2] that the information we were enabled to send you about Greenland & Faroe had been useful and are very much pleased to observe the prospect you now have, of making some general arrangement for the Trade of the Danish Settlements at these places, & Iceland, which will allow the inhabitants, the enjoyment of the indulgence you have procured for them, without subjecting you to so much personal trouble in future.

We have been well aware of the jealousy existing against the introduction of Klip or Salted Fish from Iceland,[3] & for which there is some ground, if it was permitted to be warehoused for general exportation, now that we have some direct access to Spain, the principal market for this commodity, but in either allowing it to remain in the vessel, & to be carried out in the same Ship to Denmark, or rather (as it may not be always convenient for the Iceland ship to proceed direct to Denmark) in giving the option to do so, or to warehouse for Exportation to Denmark alone, this interference with the English Fish Curer is entirely done away, because Denmark has never been, and can far less at present, be a market for Shetland or Newfoundland Fish, & from Denmark they cannot be sent to Spain or France. This method of curing Fish is by no means new in Iceland, but has always been practiced there. – By the Fishery alone, are a very great proportion of the Inhabitants enabled to pay for the Provisions brought them from Denmark, nor can they cure a much greater proportion as Stock Fish, because from the great Ship room it takes up, Stock Fish cannot be brought here as a principal part of the cargo, except at too great an expence.[4]

As to the Fish Oil you are quite correct, that it is almost entirely made from small Fish, & that in Iceland they have no Whale Fishery, we have not been able to get any correct information as to the quantity of this Oil usually exported from Faroe, but believe it cannot annually exceed 1 to 200 Barrels (of 30 Gallons each).[5] From Greenland we

---

[1] They are obviously asking for new favours following the Order in Council. The letter is dated from Banks's note in the margin, denoting that he received this letter from Leith on 3 January 1810.

[2] Not extant, but Banks obviously replied on 3 January (see above). The letter came from Leith so it was most probably written at the very end of 1809.

[3] Underlined in pencil. The licensing system was far from popular in the British ports, where the feeling was that the neutrals had been allowed too large a share of British trade and shipping (Heckscher, *The Continental System*, p. 209). The British merchants and shipowners engaged in the Greenland and Davis Straits' fisheries opposed the newly licensed trade, not surprisingly as their chief products were the same: oil and blubber. Those based in London believed the importation of oil in Danish ships, even though they were not permitted to be imported for home consumption, would lead to 'the entire destruction of these fisheries' (TNA, BT 1/46, William Mellish et al. to the Lords of the Treasury, 11 Aug. 1809). Their colleagues in Hull agreed, fearing 'utter ruin' if Danish ships were licensed to bring oil into England (TNA, BT 1/46, G. J. Egginton et al. to the Lords of the Treasury, 3 Aug. 1809).

[4] Thus klipfish – dried salted cod – was the major export, from the point of view of the English at least, rather than stockfish.

[5] In the margin here (on the second page of the letter) Banks has written in pencil: 'Say 4500 gills/20 tons'.

believe, the importation consists partly of <u>Whale Blubber</u>, but principally <u>Seal Blubber</u>, & by limiting the Licences to that settlement to two or three vessels, & allowing them to proceed to Denmark without discharging, the interference with our Whale Fishery must be extremely immaterial, or rather, not worth notice.

As the shipping places in Iceland are distant, & the number of Merchants considerable, we do not see how the shipment of Salted Fish & Oil, could be restricted to a limited quantity, without fixing a quantity for each harbour, & to prevent the first shippers from using the whole quantity licenced, /& *in even this way there*/ would be /*an impossibility*/ very difficult.

As the particular application for each cargo has been attended with much trouble, and from the detention of the Ships for two or three months in this country, had caused much loss to the Iceland Merchants without any benefit to this country, we are most happy to find that you now propose to have certain enumerated articles which can be shipped and landed here, without application, nor do we find any remarks necessary upon the List of goods except, that by the <u>Salt Laws</u> <u>Liverpool Salt cannot be brought here</u> for Exportation, & our common <u>Pan Salt is not strong enough to be used for curing Fish</u>, or Mutton in Iceland, & therefore it would be necessary to get a Treasury order, to warehouse here for Exportation to Iceland, Faroe & Greenland <u>2 to 300 tuns</u> Liverpool Salt annually, this is an object of the first importance, because Liverpool is so much out of the Iceland course,[1] and without this permission no supply can be got here; & also to allow the exportation of a small quantity of <u>Gun Powder & Bird Shot to each of these</u> <u>settlements</u>, for the killing of the <u>Swans & wild Fowl</u>, which are a considerable object in their trade, & to Greenland a small quantity <u>of Lead</u> is also necessary for the Bear Hunting. – As Hulled Barley is a considerable article in Iceland, & is by our Customhouse Laws, not classed as Corn, but subject to a prohibitory duty, we would recommend your inserting it, in the list of goods from Denmark. The Board of Trade have of late refused Licences to Iceland & stated that Iceland ships could clear from Leith & London under the order in council March last[2] without Licence, & we have last season sent out a vessel in this way; we notice you now mean to revert to the use of Licences, and as the Permission to include Denmark in the voyage is restored, it will probably be necessary, but we foresee some difficulty in fixing who shall be intittled to receive these Licences if it is meant strictly to confine the trade to the Inhabitants of Iceland formerly in it, or to limit the number of Ships employed, but on this point Mr. Clausen[3] can probably assist you & will we trust do it impartially. With regard to Faroe & Greenland the trade is confined to <u>one company for whom we act</u>,[4] & if two Licences are granted to us for <u>Faroe & three for Greenland</u> annually, we believe it will be sufficiently satisfactory to them, the following are the ships now here from Iceland & Faroe for which we see you are in hopes to procure

---

[1] This is not strictly true as most of the British merchants trading to Iceland sailed from Liverpool (e.g. Phelps & Co, Horne & Stackhouse and Titherington & Allanson).

[2] Actually on 7 February 1810 (Appendix 10).

[3] Holger Peter Clausen, a leading Iceland merchant, who became their spokesman. See Appendix 13.

[4] In 1776 the King of Denmark declared the Greenland trade from the 60th to 73rd parallel to be the exclusive privilege of one commercial institution only, the Royal Greenland Trade Department. In the Faroes, the Royal Faroese Monopoly was in operation until 1856.

Licence[1] to proceed to Denmark with the Salted Fish & Fish Oil now in them & with other goods permitted by law.

Bildal	Captn. Kielsen [Kieldsen]	116 Tons	
Unge Gause[2]	Capt. Berulf [Bæruldsen]	87 „	
Swan	„ Michaelsen [Michelsen]	142 „	from Iceland
Seyen	„ Pedersen [Petersen]	140 „	
Husevig	„ Jessen	131 „	
Jubelfast	„ Bonitz		from Faroe

As some of these ships have already been detained nearly 3 months we trust the Licence for them to proceed, will now be soon procured.

We have the Honor to be your m*os*t obed*ien*t *hum*ble Ser*van*ts
[*Signed*] Corbett, Borthwick & Co.

SL, Banks Papers I 1:58. Autograph letter.[3]

139. Letter to [Culling Charles Smith], Foreign Office, 12 January 1810

In mercy to the helpless inhabitants of Iceland, who have been twice plunderd[4] by persons acting under pretended British authority, it may be deemd expedient to permit Mr. P.[5] to do away by compromise[6] the foul charges exhibited against him of having extorted money

---

[1] The *Swan* and *Husevig* received licences (TNA, BT 6/198, nos. 13,293 and 13,886; TNA, PC 2/185, February 15 1810), the *Bildahl* did not receive a licence until May, when the ship left Liverpool on the 10th (Holland, *Iceland Journal*, p. 140), *De Jonge Goose* certainly made it to Iceland, as she returned to Leith from Iceland in December 1810, and the *Seyen* left Copenhagen on 8 March 1810 for England where an application for a licence from Corbett, Borthwick & Co. elicited the answer 'No licence necessary if the ship is an Iceland ship' (TNA, BT 6/200, no. 22,940). After the Order in Council of 7 February 1810, licences were obviously believed to be unnecessary, but as it says in this letter 'we notice you now mean to revert to the use of licences'. The *Jubelfest* must have been granted a licence to Copenhagen as she was back in Leith in the first week of August 1810, loaded with grain, and was granted a licence to return to the Faroes (TNA, BT 6/201, no. 24,062).

[2] *De Jonge Goose* (*unge* (Danish) and *jonge* (Dutch) both mean young).

[3] Note in Banks's autograph, left-hand margin of last page: 'Corbett & Borthwick | [Received] Jan*uary* 3 [18]10'.

[4] Once by Gilpin in 1808 and by the British mercantile expedition of 1809.

[5] Samuel Phelps.

[6] Wellesley, egged on by Banks, decided that the conduct of Phelps and his adherents in Iceland should be examined by the law officers. In early February 1810 the papers concerning the Icelandic Revolution were sent first to the King's advocate, Sir Christopher Robinson, who was asked to give his opinion. Robinson's opinion was that there were grounds for prosecuting Phelps, who was throughout presented as the man who bore the principal responsibility for the misconduct, but not Jörgensen, who was 'only a subordinate agent' in the affair. Wellesley decided that the King's proctor should institute a judicial inquiry in the High Court of Admiralty. During the following months the various law officers studied this case of 'very special circumstances' (TNA, ADM 1/3899).

The case was far from straightforward. The crux of the matter was deciding the formal status between Iceland and Britain at the time when the crimes were committed, 'as being in Amity or at War'. The case was complicated and legal proceedings would be difficult. The Admiralty Court, after having considered all the evidence and legal opinions, came to the final conclusion in May, that 'the offences stated to have been committed in Iceland

from Public Functionaries and seizd upon the goods of private individuals; but it is impossible to suppose that the lenity of Government will be further extended on such an occasion. The alledgd crimes of seizing & imprisoning the Person of the Governor, of issuing Proclamations subversive of the just rights of the Danish Crown, of hoisting a pretended National Flag,[1] and of saluting it when hoisted with the artillery of the Letter of Mart, with other charges of a similar tendency, must be investigated & if they amount as is allegd to the *Crimen Læxi Majestatis*,[2] must be publicly punished,[3] Crownd heads owe this duty to each other & Britain has never yet been found deficient in the fulfillment of her Public duties.

TNA, FO 40/1, f. 17.[4] Banks amanuensis.

140. Memorandum from Banks to [Marquess Wellesley], 12 January 1810

A brief recital of some of the enormities lately committed in Iceland, by the Crew of the Letter of Mart *Margaret & Anne*; chiefly abstracted from the representations of Count Trampe, Governor of that Island, presented by him to His Majestys Secretary of State for Foreign affairs.[5]

1809
Jan. 12th Mr. Phelps' Brig *Clarence,* a Letter of Mart, Jackson Commander, Savignac supra cargo, & Jergensen a Dane, a Prisoner of War on his Parole, Interpreter, arrivd at Hafnefiord[6] in Iceland.
14th. Savignac demanded Licence to Trade, & threatend to use force, if a written permission was not sent to him without delay.
15th. The Deputy Governor[7] of the Island & other Functionaries being assembled for this purpose, Jackson Savignac and Jergensen, attended them, & exhibited the British

---

cannot be made the subject of criminal prosecution in this country' (TNA, FO 83/2293, Robinson and the other law officers to the Admiralty, 8 May 1810).

Trampe and Phelps agreed to settle their dispute by a compromise agreement. Phelps's paramount consideration was to continue his trade with Iceland. The government would not insist on prosecuting, which in any case was fraught with legal difficulties, if Phelps showed a willingness, pecuniary or otherwise, to make amends for what had taken place in Iceland. This Phelps did, sending another ship, the *Elbe*, to Iceland. On the whole the wronged Icelanders were satisfied. Thus Phelps and Liston were never brought to trial, as originally demanded by Count Trampe. See Agnarsdóttir, 'Great Britain and Iceland', chap. VI, 3.

[1] Three white stockfish on a blue background. The flag was lost when the *Margaret and Ann* went down.
[2] Treason.
[3] This never happened. Jörgensen was imprisoned on board the *Bahama* prison-ship, ostensibly for breaking his parole, but was never, either in Iceland or Denmark, brought to account for his actions in Iceland.
[4] Note, verso: 'In Sir Joseph Banks': | of Jan*uary* 12. 1810.'
[5] Trampe's memorial was addressed to Earl Bathurst (Document 124), but by this time Wellesley had taken over the Foreign Office. As Banks derived this document ('chiefly abstracted') from Trampe's memorial, footnotes are only added when necessary for clarification. It is clear that Banks had other sources, as he dated events and added some details not to be found in the Trampe memorandum.
[6] For instance, Trampe does not mention the town, Hafnarfjörður.
[7] Ísleifur Einarsson.

Licence of the Ship. The Governor declard that he had no power to allow Strangers to Trade, but that he had no means of preventing them, Savignac answerd that if he was interrupted in his lawfull intercourse, (legalisd as he contended by his Licence), he should consider such interruption as an insult to the British Government & resent it accordingly.

17th A Danish Vessel[1] laying in the Harbor was seizd by the Crew of the *Clarence*, & English Colors hoisted over the Danish flag.

18th Savignac sent a peremptory message to the Government, requiring a declaration of free trade from them, on the receipt of which he would liberate the Danish Prize.

19th An agreement for free Trade was signd by the Government of Iceland & /by/ Jackson, the Prize was releasd.

March 24th The *Clarence* saild for England, having landed & warehoused her Cargo which was left under the care of Savignac.

June 21st Mr Phelps himself, in his Letter of Mart, the *Margaret & Anne*, Liston Commander, and Jergensen again Interpreter, arrivd, they found that the Danish Governor of the Island, Count Trampe had arrivd on the 6th of June, also, that Cap*tain* Nott Commander of His Majestys Ship *Rover* had visited the Island, that the Governor had annulled the Agreement enterd into by his Deputy with Jackson Commander of the *Clarence*, on account his not having a Commission, but had enterd into another equally favorable to the English, with Cap*tain* Nott, who being a Kings Officer, the Governor considerd himself at liberty to treat with.

25th Mr. Phelps, with Liston his Captain, Savignac & others of his Crew, came armd to the Government house & seizd the person of Count Trampe, scarce allowing him time to lock & seal up one Room, in which his Papers Public & Private were deposited, they then carried him as a Prisoner on board the Letter of Mart.

On the same day the Crew seizd the Iceland Vessel *Orion*, having a British Licence, this Vessell has lately been restored by a Decree of the Admiralty Court.

The next day a Proclamation appeard fix'd up in the Street of Reikawic, signd by Jergensen to the following purport

All Danish authority in Iceland is dissolvd,

All Arms & Ammunition are to be deliverd up,

The Keys of Warehouses that contain Money or Goods the property of the State or of Danish Merchants are to be deliverd up.

All Danes are to remain quiet in their Houses.

Disobedience to these Orders are on Conviction before a Military Tribunal, to be punishd by the Offender being shot within two hours after sentence.

Jorgensen & Liston with a body of armd sailors began immediately to collect the Fowling pieces of the Inhabitants & their Ammunition the only Arms they were possessed of, in the Evening a new Proclamation was issued by Jergensen as follows

Iceland shall be independent of Denmark.

A Republican Constitution similar to the ancient one of Iceland shall be introducd.

The existing Authorities are annulld, till new ones can be elected by the People.

A System of defence for the Island shall be establishd.

All debts due to Danes are void, the payment of them privately shall be severely punishd,

---

[1] The *Justitia*.

The price of provisions shall be lowerd,
Only half of the present Taxes shall be collected.

Jergensen soon after, removd to the Government house & occupied it, having broken the Seal placd by the Governor on his Papers & seizd the whole of them, he immediately establishd in it what he calld a Government office, in which Mr. Phelps & himself continually were in attendance.

Jorgensen next seizd upon the House of Correction & liberated all the Malefactors confind in it, except one accusd of Murder, he next seizd the State Treasury & put the Cash & Goods of Individuals into requisition. F. Biornsen[1] was compelld by him to give up 2,600 Rixdollars the property of Adzer Knutsen[2] resident in Denmark, Conrad Strober[3] was forcd to deliver Tallow Oil Fish & Wool, the property of a Danish Company to whom he was Factor & many others were treated in the same manner.

Jergensen gave orders to Savignac to take of those Goods, what he thought fit for the Cargo of the Ship, but no regular Inventories of them were made.

The People of Iceland remaind all this time unmovd. The Public Officers refusd to ackowledge Jergensen, & two of them were imprisond. None declard in his favor,[4] but a few Clergymen Assembled in their annual Synod at Reikawic, the only Icelanders he could obtain as his assistants, were a few persons of ill character & desperate fortunes with the Malefactors whom he had liberated from confinement.

On the 11th of July Jergensen issued a Proclamation, stating how much he had been solicited to take upon himself the conduct of Public affairs he proceeds

We, Jergen Jergensen have taken upon ourselves the Government of the Country, until a regular Constitution can be establishd, with power to make War & Peace with foreign Potentates; we have establishd a new Flag for the Country which we will defend with our blood. The Seal of the Island shall be broken, & our Seal usd in its stead, till another shall have been provided by the Representatives of the People. The Public Officers must declare their obedience to us or resign their places, those near us within ten days, & the most remote within four weeks, all who resign must be confind in Westmensoe [the Westman Islands] till they can be sent to Denmark. The Clergy who have declard in our favor shall be rewarded, & we will send Ambassadors to England to treat for a Peace &c &c

Many of the Public Officers now resignd their Functions, while Phelps and Jergensen amusd themselves by hoisting the new Flag upon Phelps's Warehouse & saluting it with 11 Guns from the Ship. The Crew soon after began to erect a Battery on which four Guns brought to Iceland 140 years ago & since buried were mounted, this was called Fort Phelps, an Order was then issued from the Government Office, commanding all persons to address Jergensen by the title of Excellence.

A Vessel now arrivd from Denmark having a British Licence, she had a Cargo of Necessaries & 10,000 Rixdollars for the Liquidation of the Salaries of the Clergy & Public Officers who had not been paid since Gilpin, Baron Hompesch's Captain plunderd the Island. This Vessel was seizd and confiscated to the new State, Receipts however were given for the Cargo by Petreus, Phelps's Agent & Jergensen, More Public Money was now

---

[1] Finnbogi Björnsson.

[2] Adser Christian Knudsen.

[3] Christian Conrad Strube.

[4] Most in fact agreed to continue in their offices; for instance Magnús Stephensen, the Chief Justice, himself (see 14 July 1809 in Þorkelsson, *Saga Jörundar hundadagakóngs*, p. 173).

discoverd & seizd, the Treasurer of the Island was compelld to give up 2700 Rixdollars in his hands & the Systleman was repeatedly orderd to pay in the balance in his possession.[1]

Jorgensen now began a progress through the Island & actually visited several remote parts attended by his armd Banditti, which are said not to have exceeded 8[2] in number, he plunderd however & confiscated property wherever he could find a pretence, in the mean time the persons in Phelps's employ nearer home compelld the Merchants to take Colonial produce with which they had before supplied themselves at a price fixd by the Seller & to pay for it in Iceland produce at the rates fixd by Jorgensen.

This system of revolutionary violence of which abundant instances are omitted for brevity sake, continued without abatement or interruption till the middle of August when the *Talbot*, commanded by the Hon*our*able Capt*ain* Jones arrivd, to him the Icelanders applied to be releasd from the tyranny of Jorgensen & his English employers. At this time Count Trampe the Danish Governor had been confind nine Weeks on board the Letter of Mart, treated with indignity & refusd permission to take air & exercise on an uninhabited Island altho he offerd to pay the Cost of Centinels to guard him.

Before Capt*ain* Jones had time to decide upon the merits of the case laid before him, the Letter of Mart saild with all her Crew except Savignac, having the Governor Count Trampe, against whom Phelps declard to Capt*ain* Jones that he had private accusations, as a Prisoner on board. On the third day after sailing she took fire & was burnt, fortunately the Prize [*Orion*] since liberated by the Admiralty Court was in company, by which the Crew was savd with their cloths only, every other part of their property being lost.

She returnd to Iceland, where Mr. Phelps immediately restord Count Trampe to his liberty;[3] before this time, Capt*ain* Jones had Demolishd Fort Phelps,[4] & restord the Government of the Country to the state it was in before Jorgensen's Revolution, the whole of his proceedings may be seen in his dispatches to the Admiralty,[5] he found in the Count a man of fine manners & more than usual talent & offerd to him & his necessary Attendants a passage to England in his Sloop, which the Count accepted & arrivd in October, he immediately laid before the Secretary of State for Foreign affairs a detaild account[6] of the conduct of Mr Phelps & the Crew of his Letter of Mart, from which the above statement is in great measure extracted, he seeks redress for the injurd Inhabitants from the justice of the British Government, but he more than all implores from the benevolence of the King, some arrangement in favor of the remote Colonies of Denmark, which may in future secure them from a recurrence of the evils, which his Majesties lawless subjects have twice inflicted on an innocent & unprotected People.

On a due consideration of the above stated conduct, it is not unjust to conjecture, that the motive of the Owner of the Letter of Mart & of his Agents & Officers, which inducd them to labor as they did in the attempt to involve the innocent Icelanders in the horrors

---

[1] Frydensberg and Koefoed.
[2] Trampe wrote 6.
[3] Phelps offered him liberty, but Trampe refused this offer, and chose to sail on the *Talbot*, as a prisoner-of-war.
[4] With the help of the crew and the citizens of Reykjavík.
[5] In TNA, ADM 1/1995.
[6] Document 124.

of a Revolution was the hope of being able to purchase from a newly establishd Government, the confiscated property of the resident Danes, at the low price at which illgotten property is generally disposd of; till a more laudable motive is assignd for conduct apparently unjustifiable, it is not easy to maintain a different opinion.[1]

TNA, FO 40/1, ff. 13–16. Contemporary memorandum (amanuensis William Cartlich).
NHM, DTC 18, ff. 20–27, tentatively dated, wrongly, 'probably about May 1 1810'. Contemporary copy, entitled 'Remarks by Sir Joseph Banks in his own writing.' The whereabouts of the original are not known.
Draft in the Royal Geographical Society of South Australia, Banks, Sir Joseph. Notes on Iceland. MS 6.
Hermannsson, 'Banks and Iceland', pp. 54–8 (not annotated).

141. Sir George Steuart Mackenzie to Banks, 23 January 1810

Edinburgh

Dear Sir

The Bearer of this letter is Mr Isfiord,[2] a native at Iceland, & a Merchant. A vessel of his the *Charlotte Amelia*, was taken, as you will find set forth in the enclosed Power of Attorney, granted by Mr Isfiord to Mr Kolvig. The latter, after getting this power, went to Sweden, where he was detained by Sickness for some time. He then sailed for England in a Swedish Vessel which was taken by the French & carried into Dunkirk. On Kolvig being recognized as a Dane, he was set at liberty. He went to Holland with the intention of getting to England, but on arriving there he found the blockade so strict, & the prohibition so severe that he could not get away, & was under the necessity of returning to Copenhagen. Mean time the *Charlotte Amelia* was condemned & sold. Mr Isfiord has lost all he gave to Kolvig, & is the only Icelander who has not obtained redress.[3] Your goodness which has been so largely spread among the Icelanders, will excuse me for taking the liberty of introducing Mr. Isfiord to you, but will then get you to give him such advice & assistance as may lead to the Recovery of the proceeds of the Sale of his Ship & cargo, or at any rate some Indemnification. His father saw you in Iceland & his family speak of you with respect. Mr. Isfiord has a snuffbox which you will probably recognise as one you gave to his Father[4] – I need say no more to induce you to interest yourself in this case. – It is probable that I shall go to Iceland after all.[5] –

I should be much obliged by you in favouring me, how the London Royal Society manages their elections. Owing to Partywork (I blush to mention it) it is almost impossible for a candidate to gain admission – Yesterday there were 11 Candidates all of whom, except perhaps one, were unexceptional: but only one was admitted – I wish to

---

[1] Note, verso: 'From Sir Joseph Banks on the violence in Iceland | [Received] Jan*uary* 12 1810.'
[2] Kjartan Ísfjörð: see Appendix 13.
[3] This is correct. This ship was captured the day after the formal declaration of war between Denmark and Great Britain, i.e. 5 November.
[4] This is highly unlikely; Þorlákur Ísfjörð (c. 1748–81), who became a county magistrate in south-east Iceland, was attending the University in Copenhagen 1771– 6.
[5] Mackenzie left for Iceland on 18 April 1810, returning to Stromness at the beginning of September.

have this vile partywork done away, otherwise our Society¹ will be ruined; & for that & I wish much to know how you proceed at Elections.

I am always with great respect
Dear Sir
Your Obliged & Obe*diant* [Servant]
[*Signed*] G. S. Mackenzie

Wisconsin, Banks Papers, ff. 54–55. Autograph letter.

142. Banks to [Culling Charles Smith?], 25 Jan*uary* [1810]

Office for Trade

My Dear Sir,

a Small puzzle has arisen in their Lordships Councils here respecting the proper disposal of the Iceland papers² which cannot be Cleard up till the advice of Sir Wm Scott³ has been Taken I am therefore to Request that you will Delay the Transmission of The Papers till you hear from me again which I trust will be very Soon.

I have the honor to be
Your Most Faithfull
H*um*ble Serv*an*t
[*Signed*] Jos: Banks

TNA, FO 40/1, f. 18. Autograph letter.⁴

143. Note from Banks to Foreign Office, 1 February 1810⁵

[What proceedings may be had in this Case /either/ for punishing the Offender and for obtaining Redress to the Inhabitants of Iceland, for the Injuries they have received,]⁶ by

---

¹ The Royal Society of Edinburgh.

² The Iceland papers referred to probably include the list of papers to be found on the first folio page in the TNA, FO 40/1 dossier, which are now missing. They should be in either the Foreign Office, Admiralty or Board of Trade but have not been found despite detailed searches.

³ During his 10-year tenure as judge of the High Court of Admiralty, Sir William Scott (1745–1836) drafted numerous opinions for the Foreign Office regarding neutral rights, trade with enemy colonies etc. After he was appointed judge of the High Court of Admiralty in October 1798 (also being an MP) the government frequently sought his opinion (Bourguignon, *Sir William Scott*, p. 49).

⁴ Note, verso: Sir Joseph Banks
Dated } January 25 1810
Received }
Desiring the Iceland papers may not be sent to the Board of Trade untill further communication.

⁵ The note itself is undated, but this date is verso. However, this is more or less verbatim from a letter Smith wrote to Sir Christopher Robinson, the King's advocate, on the following day, 2 February 1810 (TNA, FO 95/371). This indicates that Banks is behind the scenes guiding the aftermath to the Revolution.

⁶ These brackets are in pencil, for insertion into the above-mentioned letter, Banks was urging Wellesley not to let the conduct of Phelps go unpunished.

the seizure of their property on Shore, /or/ & by the Capture /& Plunder/¹ of the licenced Ships, /the [blank] and Sejen/²

[A copy of the Order in Council follows this note, f. 21.]

TNA, FO 40/1, ff. 19–20. Banks amanuensis.

144. Draft of the Order in Council of 7 February 1810,³ [December 1809–February 1810]

By the King

It having been humbly represented to us, that the Islands of Feroe & Iceland and also certain Settlements on the Coast of Greenland, parts of the Realm of Denmark, have,

---

¹ '& plunder' is inserted in Banks's autograph
² Note, verso: 'Iceland – from Sir J. Banks | [Received] Feb*ruary* 1 1810'.
³ As has been seen, Banks had become absolutely horrified by the Icelandic Revolution, condemning it in no uncertain terms. This 'attempt to involve the innocent Icelanders in the horrors of a revolution' was a far cry from the sort of proper peaceful annexation, desired and supported by the natives, Banks had had in mind (see Document 63). Trampe had asked for a convention to regulate Anglo-Icelandic relations (see Document 124). Moreover, Magnús Stephensen had also approached Banks directly, asking for 'a neutrality For the Island, with Liberty to make its Trade uninterrupted, and, protected of [sic] your Government' (Document 122). It was thus clearly in the interests of all parties: the Icelanders, the Danish authorities in the island and the British merchants, that such a convention be drawn up. Banks was of the opinion that a repetition of 'the horrors of a revolution' or Gilpin's raid must be avoided at all costs and Anglo-Icelandic relations placed on a firm official footing, as Trampe and Stephensen had requested. This draft shows that Banks was the principal author of the subsequent Order in Council, government having taken up his offer of assistance in drawing up a convention, which he made on 11 December (Document 134). This draft was doubtless ready at the end of December, when *The Times* described it fully in a leading article on 23 December 1809:

> 'It gives us pleasure to be able to state, that the subject has been considered with that attention to the interests of humanity, the exercise of which will prove extremely beneficial to the poor Inhabitants of those frozen regions, and reflect no small portion of credit on the benevolent character of this country. If we mistake not, the Gazette of tonight will contain a proclamation, which extends our friendly protection to the natives of Iceland, and to the inhabitants and settlers in the islands of Faroe and certain possessions on the coast of Greenland ... The relations existing between them and this country are to be re-established, on the footing upon which they stood previously to the commencement of hostilities with Denmark. When any of the inhabitants of these places arrive in this country, they are on no account to be treated as enemies; but are to be considered with the same regard which is shown to the subjects of these states with which we are in amity.'

This draft was subsequently polished by Council officials. On 7 February 1810 the Privy Council finally issued the Order, which stated the British government's official policy towards Iceland and the other Danish dependencies (Appendix10). Thus Iceland's position in relation to Britain was in effect changed. Iceland, as opposed to Denmark, was now in a state of neutrality and amity with England. The Order acknowledged that the sovereignty of Iceland was still vested in the Crown of Denmark but that Iceland was under the protection of Britain. While the Order was in force Iceland could not legally be annexed by England. Moreover, as the dependencies were placed under the protection of the British Crown, it was clear they would be restored to Denmark after the war and would thus not play any role in future peace negotiations.

The Order in Council had much in common with Nott's convention of 16 June (Appendix 9) but went much further. Nott, for example, had not been able to place Iceland in a neutral state. It also clearly demonstrated that

since the commencement of our War with that Nation been deprivd of all intercourse with the Mother Countrey & that the Inhabitants thereof are, in consequence of the want of their accustomd Supplies from thence reducd to extreme misery, being deprived of many of the Necessaries & of most of the Conveniencies of Life.

We being movd by compassion for the unmerited sufferings of these defenceless people, but wholly averse to the seizure of the detachd Dominions even of our Enemies, merely because their Sovereigns are unable to protect them, have thought fit to declare our Royal Will & Pleasure.

That the relations of amity between our loving Subjects & these unoffending People which have not been interrupted by any act of hostility on their part, be considerd as unchangd from the state in which they were, when our War with their Sovereign commenced, That the Islands & the Settlement above mentioned be for the present receivd under our Royal Protection /& *all the Inhabitants free from the disabilities consequent upon a State of War with us*/.

That the Inhabitants of all these places be considerd when resident in our Dominions as Stranger Friends under the safeguard of our Royal Peace & intitled to the protection of the Laws of our Realm & in no case treated as Alien Enemies.

That the ships of our United Kingdom navigated according to Law, be permitted to repair to these Places & to Trade with the Inhabitants, & that Ships belonging to the Inhabitants of these places be permitted to Trade with the Port of London in England the Port of Leith in Scotland & the Port of Dublin in Ireland, & also with such other Ports either within /&/ /or/ without our Realms, as we by our Royal Licence shall from time to time be pleased to permit, under such restrictions & regulations respecting the whole of this intercourse as we with the advice of our Privy Council shall think fit in future to prescribe.[1]

That Ships belonging to any of the aforesaid places shall when they proceed to Sea be navigated under the Danish Flag with addition of /the British/[2] /a Red/ Lion on a White Carton in the uppermost corner of the Colors nearest the truck of the Flag Staff, in order

---

Iceland (and the other two dependencies, the Faroes and Greenland) were of some importance to England as regards the possibilities of trade. Britain needed new markets and trade with Iceland was thus officially encouraged.

The King's advocate, Sir Christopher Robinson, was at once ordered to prepare instructions for the commanders of HM's warships and privateers to protect the Danish dependencies from the attack and hostility of British forces and subjects (TNA, PC 2/185, Council Office to King's advocate, 7 February 1810). Ten copies of *The London Gazette* were dispatched to Iceland, where they were sent to the county magistrates around the country (NAI, Jörundarskjöl, Trampe to Frydensberg and Einarsson, 5 April 1810). It was translated and displayed in public in Reykjavík (NAI, Bf. Rvk. A/2-1, Bréfadagbók 1806–13, 9 June 1810, no. 2580) to the 'infinite satisfaction & delight' of the people (Holland, *Iceland Journal*, p. 99).

The Order in Council gained the official approval of the Danish authorities, being a successful answer to Trampe's fifth demand (see letters between Trampe, King Frederik VI and the Danish Foreign Office in RA, DfdUA, Island og Færöer). It was a benevolent measure from the British government and undeniably of great benefit to the Icelanders. It promised them freedom from hostilities, thereby making trade between Iceland and Denmark a viable option again and guaranteeing supplies.

[1] On left margin in Banks's hand is written: 'Should not the Temporary Legislation of an intercourse which depends wholley on the King's will & Pleasure be in the King Solely'.

[2] Added in Banks's autograph (see Figure 8).

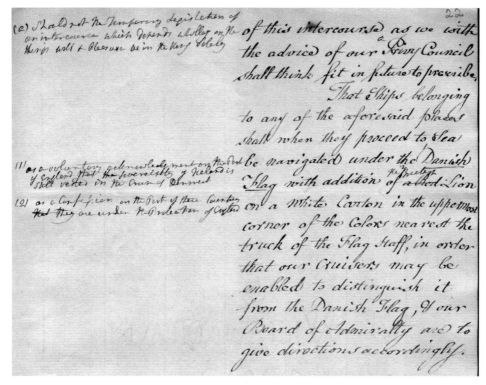

Figure 7. Last Page of the Draft of the 1810 Order in Council (Document 144). Banks was instrumental in drafting the Order in Council of 7 February 1810. In the left margin in his well-known hand Banks has noted: 'as a voluntary acknowledgment on the Part of England that the Sovereignty of Iceland is still vested in the Crown of Denmark' and secondly 'as a Confession on the Part of these Countries /that/ they are under the Protection of England'. Courtesy of the University of Wisconsin Digital Collections.

that our Cruisers may be enabled to distinguish it from the Danish Flag, & our Board of Admiralty are to give directions accordingly [see Figure 7].[1]

Wisconsin, Banks Papers, ff. 21–22. Draft. Amanuensis (William Cartlich).[2]

145. Kjartan Ísfjörð to Banks, 10 February 1810

88 Queen Street Cheapside, London

Right Honorable Sir

Convinced that it is thro' your powerful & humane intercession that the Island of Iceland my native Country, has been saved from famine & misery, and the Merchants from utter ruin, – I (as the only Merchant of that Island who has not got his property restored which

---

[1] In the left margin in Banks's hand: 'as a voluntary acknowledgment on the Part of England that the Sovereignty of Iceland is still vested in the Crown of Denmark' and 'as a Confession on the Part of these Countries /that/ they are under the Protection of England.' This is more bluntly put than the government officials could accept, but shows how far Banks was willing to go.

[2] Note in Banks's autograph, verso: 'Declaration in Favor of Ferroe Iceland &c | Rough Scetch'.

had been detained in this Country) take the Liberty hereby to sollicit your kind support which is become of the utmost consequence to me. –

The history of my Late misfortune is as follows.

I am a native of Iceland & a Citizen or burgher of the town of Eskefiord in the said Island, where I have been established as Merchant since the Year 1802.[1]

In the Year 1807 the only Vessel, I was possessed of was, with its Cargo, which also belonged to me, taken by an English cruizer[2] near the Skag[3] on a Voyage from Iceland to Copenhagen. – It was a Brig called *Charlotte Amalia* – C. C. Juul, Master.

When the account was received at Copenhagen that the English Government in consequence of your humane Intercession had restored all the Iceland Ships & Property, I applied to the Danish Government & received permission to send an Agent Mr. Frederik Kolvig, to England, in order to claim my property, which I considered as also restored. – This Agent Left Copenhagen in the Month of September 1808 for Gothenburg, but was detained at the Latter place by illness about a fortnight. – After this he sailed in a Convoy for Yarmouth, but the Convoy being dispersed in a heavy gale of Wind, the Ship on board of which my Agent had taken his passage, was unfortunately taken by a French privateer and carried to Dunkirk[4] – all his papers were taken from him & sent to Paris & he himself was dragged into a prison of Arras. – After a Lapse of eight Weeks he was set at Liberty & the papers restored to him on account of his being a Danish subject. – He now proceeded to Holland in hopes of getting a passage for England & continued some time at Amsterdam for that purpose, but was not fortunate enough to meet with one in consequence of the strict blockade of the Dutch ports. – He therefore returned to Copenhagen, where he arrived in May 1809 – without having effected any thing. –

These reverses occasioned a very considerable expence to me & have been felt severely. –

Count Trampe, the Governor of Iceland, will no doubt bear Witness to the truth of what I have taken the Liberty to state.[5]

I am now come to this Country in hopes, thro' your benevolent Intercession in my behalf, if you should deem me worthy of them, to get my Little property restored to me, having learnt that it has been condemned in the Court of Admiralty & sold – I chiefly rest my expectations on the supposition that as all other property of the inhabitants of Iceland has been restored to the owners, the generousity of the british Nation will not allow mine to form a solitary exception.[6] But if I should not prove successful it will have been the occasion of the utter ruin of myself & family.

I have taken the Liberty to inclose a Letter from Sir George Mackenzie Baronet of Edinburgh, to your direction, which contains further particulars of my unfortunate Situation.[7]

[1] This is correct (Andrésson, *Verzlunarsaga Íslands*, II, p. 608).

[2] The sloop *Pelican* was commanded by William Ward. The brig was captured about 20 miles from Arendal in Norway 'by reason as he was informed of a difference having taken place between Great Britain & Denmark'. It was carried first to Yarmouth and then to the River Thames (TNA, HCA 32/985, Deposition in the High Court of Admiralty of Carsten Christensen Juul on 22 December 1807, no. 993).

[3] Skag, i.e. Skagen, the northernmost tip of Jutland.

[4] This ship is unknown.

[5] Count Trampe was still in London pursuing the aftermath of the Icelandic Revolution.

[6] Which he was; see Bjarni Sívertsen's account in 'Ágrip af ævisögu Bjarna riddara Sivertsen', *Sunnanpósturinn*.

[7] Probably Document 141.

I beg Leave most respectfully to apologize for the Liberty of this Letter & will do myself the honor to wait on you whenever it will be convenient for you to receive me. I remain with great Respect

Right Honourable Sir
Your obedient humble
Servant
[*Signed*] K. Isfiord

Wisconsin, Banks Papers, ff. 50–51. Autograph letter.

146. Rooke & Horneman to Banks, 10 February 1810

88 Queen Street, Cheapside

Right Honorable Sir,

Mr Kiartan Isfiord being particularly recommended to us, we beg leave to join our respectful sollicitations in his behalf. We believe that nearly his all was embarked in his ship, the *Charlotte Amalia*, C. C. Juul master,[1] and that great distress awaits him, unless he should be fortunate enough to get his property restored in this Country.

We feel great interest in his success the more so as our H. F. Horneman[2] is a native of Denmark and desirous to alleviate the misfortunes of his Countrymen, as far as it lies in his power. He will be happy to attend with Mr Isfiord, if you should grant this Gentleman an interview being more conversant in the English language. We remain with the greatest respect

Right Honorable Sir,
Your obed*ient* and humble Servants
[*Signed*] Rooke & Horneman

Wisconsin, Banks Papers, ff. 52–53. Autograph letter.

147. 'Statement from Mr: [Henrik Christian] Paus, after an extract of his letter to H[olger] P[eter] Clausen', [Spring 1810?][3]

The ship which I had in partnership with Mr. Busch[4] was brought in to England in Oct*o*ber 1807 but graziously liberated the following Oct*o*ber 1808 and (furnish'd with a

---

[1] The case papers: TNA, HCA 32/985, no. 993. The main problem for this ship was the fact that she was the only one captured *after* the outbreak of hostilities. The master was Carsten Christensen Juul. See further in the following document.

[2] Hans Frederik Horneman was a Danish merchant trading in London. The Danish Board of Trade (Kommercekollegiet) appointed him to look after the interests of the Danish prisoners-of-war during the war. Charles Rooke was an Englishman who became Horneman's partner in November 1809.

[3] This is as written on the document: 'Statement from Mr: Paus, after an extract of his letter to H: P: Clausen'. The letter is undated. From the capture of the ship in February 1810 until the granting of the purchase of a new ship in July, and taking into account the approximate time for these matters to take in the licensing system, dates this letter to the spring of 1810.

[4] Jens Lassen Busch had at least five ships sailing to Iceland. Paus is not mentioned in the Danish *söpas* records, but the two were in partnership (Andrésson, *Verzlunarsaga Íslands*, II, p. 531). The ship in question is doubtless the hooker *Rödefiord*, as mentioned in Banks's note on the letter.

Licence) permitted to go direct to Iceland and from thence to return with a cargo of Iceland produce.¹ – A heavy storm in which the ship was very much injured forced the master to put in to Norway in November in the same year, from whence he, after having got the ship repaired, proceeded the next spring 1809 to Iceland where he arrived in the month of June. – After a months stay the famous revolution² broke out in Iceland, and informed that Mr. Jorgensen had put himself in possession of my warehouses the master resolved to fly³ with the empty ship. – As care was taken to rumour that the whole transaction was founded upon His Britannic Majesty's Authority⁴ he durst not go to Leith but made his escape to Bergen in Norway where he arrived 1809 in September.⁵ – After so many fruitless expences; great loss, and 2½ years voyage, without being one step nearer his design the master resolved in my absence to accept a freight offered and took in a cargo of stockfish, herrings & oil for Copenhagen, but now again he was so unhappy to be captured by an English Frigate,⁶ brought to Leith and there condemned. – I have therefore no ship now, and my desire is, that the English Government graziously will permit me to purchase or hire a ship by which means I could continue my trade upon Iceland.⁷ –

I have a particular confidence to the generosity of Sir Joseph Banks, if you will state my case to him.⁸ –

Wisconsin Papers, ff. 128–130. Contemporary copy.

148. Banks on Jörgen Jörgensen, 2 April 1810

[Report⁹]

Jorgen Jorgensen a Danish Prisoner of War, capturd by a British Cruiser early in 1808 was landed in the River /Thames/¹⁰ & took his Parole at London under the Orders of the Transport Board, but was soon after missing having absconded & he could not be found for the purpose of being sent to the Quarters assignd for such Prisoners.

Soon after this he enterd into Mr. Phelps' service as an Interpreter in the Armd Brig *Clarence* & with her he went to Iceland where he arrivd in January 1809 & soon after

---

¹ The *Rödefiord* was issued a licence on 29 September 1808 (TNA, PC 2/178).

² The Icelandic Revolution. This is correct, the vessel eventually arrived in Iceland from Bergen in 1809 with rye and other necessaries, only to be seized by the 'revolutionaries', i.e. Jörgensen and Phelps (Þorkelsson, *Saga Jörundar hundadagakóngs*, p. 113).

³ He means 'flee'.

⁴ This was the common belief in Iceland at the time (see e.g. Espólín, *Íslands Árbækur*, XII, p. 27).

⁵ This ship was the first to bring news of the Icelandic Revolution to Europe.

⁶ The *Rödefiord* was captured by the *Egeria*, Commander Lewis Hole, on 5 February 1810 (*The London Gazette*, 14 August 1810, p. 1229).

⁷ On 9 July 1810 a petition was granted, at the request of Corbett, Borthwick & Co. for a vessel to carry rye and necessaries from Leith to Iceland. This was because of the *Rödefiord* having been driven away by Jorgemann [*sic*] and 'in consequence thereof captured'. This ship was allowed to sail 'in the room of the Rodefiord' (TNA, BT 6/200, no. 22,703).

⁸ A note in Banks's autograph, verso: 'Mr Paus has traded too much with Norway | What was the Ship's name | Rödefiord?'

⁹ This report was doubtless written in support of Count Trampe's efforts to get his hands on Jörgensen. Banks was greatly influenced by Trampe and this is in effect a precis of Trampe's memorandum to Bathurst (Document 124). The facts are on the whole correct.

¹⁰ All the corrections are made by Banks himself.

returnd to England; he again shipd himself on board the *Margaret* [*and Ann*]¹ Letter of Marque in the same capacity, in this Ship Mr. Phelps himself proceeded to Iceland.

On the arrival of the *Margaret* in Iceland, the most violent measures were pursued, with a promptitude which implied a previous consideration and arrangement,² to subvert the public authorities of the Island & to /create/ effect a Revolution there, in this /the name of/ business Jorgenson was /made use of/ deeply implicated and he actually assumed the Government of the Country, established a military force, & directed confiscations to be made of Property to a considerable amount, principally as is presumed, with a view of enriching the concern in which he was employd, as confiscated property to a considerable amount was carried on board Mr. Phelps's Ship the *Margaret*.

The first act of violence committed by the Crew of the *Margaret* was to seize the person of Count Trampe the Governor of the Island & confine him as a Prisoner of War on board the *Margaret*.

The next day a Proclamation was issued signed by Jorgenson in which among other things it was declared – that Danish authority in the Island is at an end, that Public property & that of Danes must be delivered up, & that all who oppose the execution of these Orders shall be subject to military execution, & this Proclamation was without delay carried into execution by Jorgensen who on this occasion took the command of the Crew of the *Margaret*, paraded the Streets of Reykavig & entered private houses at his pleasure.

This Proclamation was followed without the least delay by another likewise signed Jorgen Jorgensen, in which it is declared that Iceland shall be independent of Denmark, that a Republican constitution similar to the ancient one shall be introduced there, that all Debts due from Icelanders to Danes shall be void, the Taxes reduced to one half, & none but Icelanders shall fill Public Offices.

Jorgensen then took possession of the Government house & broke the Seal placed by the Governor on his Public & private papers, & daily transacted publicly the business of Confiscation &c there, in company with Mr. Phelps & the Officers of the *Margaret*, his next measure was to seize upon the House of Correction & establish in it Barracks for armed Men levied by him & considered as his Guards, which however are said never to have exceeded eight in number.

On the 11th of July he issued a Proclamation which was fixed up in the town of Rykevig signed by his name in which were the following articles

We Jorgen Jorgensen have taken upon ourselves the Government of the Country until a regular Constitution can be established, with power to make War & Peace with foreign potentates, a new Flag for the Island is appointed, the ancient Seal of the Island is no more to be used, ours is to be substituted in its stead, all existing authorities must declare their obedience to us within a limited time,³ or resign their Offices, we are preparing to send an Embassy to the British Government.⁴

Signed, Jorgen Jorgensen.

---

¹ Throughout this document the privateer is simply called *Margaret*.

² This comment is interesting as it implies that Banks now believed that the Revolution had been planned beforehand, though there is no documentary evidence whatsoever for this.

³ The draft in the Wisconsin Papers ends here.

⁴ Article 9 of Jörgensen's Proclamation of 11 July 1809 read: 'That a person shall be invested with full power to conclude a peace with his Majesty, the King of Great Britain.'

In this manner did Jorgensen proceed till the arrival of the *Talbot* Capt*ain* Jones who instantly on hearing the complaints of the Icelanders, ordered the constituted authorities of the Danish Government to resume their stations, soon after which the *Margaret* sailed & carried off with her the Revolutionary Jorgensen.

On his arrival in England Jorgensen was found by the Transport Board, he was immediately seized by their order & sent on board a prison ship at Chatham where he now is confind.[1]

This is the man whom Count Trampe in his capacity of Governor of Iceland requests to have given up to him in order that he may be sent to Denmark there to suffer the punishment due for the crimes he has committed in the Dominions of his Sovereign, against the Laws of his Country; the Count however in whose character forgiveness of injuries receivd & tenderness even towards the least deserving of all characters that of a Revolutionist, are predominant features, begs leave to /add/ /declare/ that it will give him the most heartfelt satisfaction if in rendering up to justice the offender Jorgensen, a request for sparing his life should be added by the benevolent inclinations of the British Government. The rendering up a fugitive subject has always been deemd a matter of serious consideration & has never it is presumd been done, unless by immediate application from one crowned head to another through the means of representative Agency; this however is a case of a very different nature, such a one as is not likely to have any Precedent & which need not be allowed to form one.

Jorgensen is not domiciliated in this Country, he is a Prisoner of War actually at this time /*imprisond*/ /in confinement/, he may therefore be given up in the way of Exchange; but should he as may be the case object to return to his own country, he will then become an object within the purview of the Alien Laws, under their provisions he may be compelld to leave the Country, & surely if any character of mankind is deservedly obnoxious to a regular Government it is the Man who has instigated & in person conducted a Revolution to the utmost extent to which the people he had to deal with could be carried by every possible instigation on his part to abandon their lawfull Sovereign & establish a new Government according to their own fancies and whims.[2]

TNA, FO 40/1, ff. 28–31. Banks amanuensis (William Cartlich).
Wisconsin, Banks Papers, ff. 65–66. Banks autograph draft, dated 30 March.

---

[1] The prison hulk was the *Bahama*. It might be worth adding that five days later Jörgensen wrote to the Lords Commissioners of the Treasury worrying about the 6th article in the Phelps/Jones/Stephensen convention of 22 August 1809 (Appendix 9). Even though it stipulated that all persons who had 'taken part with Mr. Jorgen Jorgensen, shall no longer be in employment, but their persons and property ... shall be respected and protected' (Hooker, *A Tour in Iceland,* II, p. 101), he expressed his worries that this agreement would not be 'sufficiently binding', saying that Trampe had already sent a list from London to Copenhagen of more than 20 civil officers 'who will be exposed to the resentment of Danish government'. They were not 'real enemies to this Kingdom'. He asked their Lordships to write to Count Trampe ensuring that this article would remain in full force 'which will greatly reduce my anxiety' (TNA, WO 1/844, 7 April 1810). Notice was taken of Jörgensen's request and nobody was punished.

[2] Verso: 'Sir Joseph Banks | communicated April 2 1810 | Iceland | That Jergen Jorgenson | should be given up | to Justice, having | been principally | concerned in the | violences which | were committed on the | inhabitants of | Iceland in the | year 1809.'

149. Kjartan Ísfjörð to Banks, 18 April 1810

London

Sir,

Before I leave this place I feel it to be my first duty to thank you most noble Sir for all your kindness to me during my stay in London, but not having the english Language so much in my power as to be able to fulfil this my duty verbally I beg leave to do it by way of writing. That I & every Icelander with the warmest gratitude acknowledge how much you have done since the commencement of the war between Denmark & England for the good & welfare of our defenceless & indigent Island. I beg you most noble Sir to be assured of, & conscious as you must be yourself of your philantropic deeds it is in these you may seek the first & noblest reward for your exertions in our behalf.

On my arrival hither I took the liberty to acquaint you most noble Sir, with the case of my Ship the *Charlotte Amalia* that has been condemned & though I have spent much time & money for the sake of this Ship, yet I have not been able to bring it farther than thereto: that Mr Wilberforce[1] has taken my petition into his hands & promised in course of time to advocate my cause. – I chose this channel according to Count Trampes[2] advice in order that I might not trouble you most noble Sir too much & the Count also told me that I might make sure of that you would back my petition when the cause came on & which I venture hereby most humbly to intreat for, flattering myself that as my welfare a good deal depends upon which turn this matter takes that you will not deny to add this favor to the many others I in common with every Icelander have shared. –

Lastly I beg leave to observe that the house of Messrs. Rooke & Horneman are my Agents here, & that these Gentlemen in course of time will make free to wait on you respecting this affair of mine.

I have the honor with great respect to remain
Most noble Sir
Your most obedient
humble Serv*an*t
Kiartan Isfiord

Wisconsin, Banks Papers, ff. 67–68. Autograph letter.

150. Banks to Count Frederik Trampe, 23 April 1810[3]

Soho Square

My Dear Count

I have seen M[r]. C Smith[4] & have had a conversation with him on the Subject of your request on the part of your Gov*ernm*ent, to have the person of Jersen[5] Jergensen [*sic*] delivered up

---

[1] William Wilberforce (1759–1833), philanthropist, abolitionist and MP at this time (1784–1812). He helped other Icelanders besides Ísfjörð.

[2] Trampe may have felt that Banks had his hands full clearing up the aftermath of the Icelandic Revolution.

[3] This letter is important, as it is the answer of the British government, through the offices of Banks, to Trampe's demands for the extradition of Jörgensen to Denmark. As the two nations were at war Banks proved here to be a useful channel.

[4] Culling Charles Smith, Under-Secretary of the Foreign Office: see Appendix 13.

[5] In the RA copy Trampe has corrected this to Jergen.

to You as a State prisoner, to be carried to Denmark, there to be made answerable for Crimes committed against the Laws of his Country in the Island of Iceland.[1]

The answer of The Marquis Wellesley which Mr Smith was so good as to communicate is as follows, in times of Peace and Amity between The Sovereigns of different Nations, the act of delivering up a subject free from suspicion of having infringed the Laws of the Country in which he is Resident, to answer for Crimes alledged to have been committed by Him in the Dominions of another Sovereign, has always been considered as one of the most delicate and difficult points of Diplomacy, by every nation of Europe, in fact it has seldom been carried into effect; in a case however like the present, when the Relations of Amity between the two Countries are unfortunately suspended, The Marquis [Wellesley] apprehends, that no such request has ever yet been made, and is unable to discover any sound principle under which it can with propriety be acceded to.

On a full review of the unfortunate business of Iceland, which induced the Count to appear personally in England, The Marquis after having taken into consideration the whole of the question, the evil conduct of certain Individuals there, as well as the beneficial arrangement in favor of the Island granted by The Government of England, is of opinion, that the protection of England to the Inhabitants of Iceland against the Miseries of War,[2] should be considered as an expiation and complete satisfaction for all that happened, before the existing Relations between England & Iceland could have been well understood in either Country, and that in order to render the future tranquillity which England means to bestow on the distant dependencies of the Danish Crown perfect and uninterrupted, a compleat amnesty should take place for all that has passed, and He trusts to the gratitude of the Parties concerned that no one in Iceland or elsewhere shall be questioned criminally, for things done during the anarchy, which unfortunately prevailed there in the course of the last Year.

I have the honor to be
with due Consideration
& with Sincere Esteem & Regard
My dear Count
Your Most Faithfull
& most Humble Servant[3]

I entirely approve
this answer
W.[4]

The original is not extant. However, there are several copies.
TNA, FO 40/1, ff. 35–38 transcription – as the letter was sent with the approval of the Foreign Office.
Wisconsin, Banks Papers, ff. 27–28, fair copy, written by the Banks amanuensis William Cartlich, which Wellesley approved and Banks added the closing words. Both have been added here.
Wisconsin, Banks Papers, f. 29. Draft in Banks's autograph.
RA, Rtk 2214.55, a copy in Trampe's autograph sent to the Danish authorities (the wording is the same though there are differences in spelling, capital letters and punctuation).

[1] This is a reference to Count Trampe's memorial to the Foreign Office (Document 124).
[2] This is a reference to the Order in Council of February 7 1810 (see Appendix 10).
[3] This is written by Banks on the draft in the Wisconsin, Banks MS.
[4] The Marquess Wellesley. This note is written in his hand on the draft in the Wisconsin, Banks Papers.
Note, verso on the FO 40/1 copy: '(Copy) | Draft | Sir Joseph Banks Communicated April 23 1810 | This intended Letter to Count Trampe, upon The Iceland Subject. – | Approved by Lord Wellesley'.

151. Culling Charles Smith to [Marquess Wellesley?], 23 April 1810

It having been represented that Count Trampe, on his return to Iceland,[1] might be inclined to punish those persons who had been aiding what is called the British Cause. I have stated to Sir Joseph Banks that it will be advisable to impress upon the mind of Count Trampe the necessity of a general amnesty.[2] And that the Consul, going out to Iceland,[3] must be instructed to watch the conduct of the Governor that measures may be taken to secure to /the/ whole of the Icelanders the benefits of British protection.

I also stated to him that it would be better to withhold further proceedings against Jorgen Jerghenson /*no regular demand having*/ for no regular demand has been made for his production – & he is, besides, an Enemy's subject[4]

[*Signed*] CCS

TNA, FO 40/1, ff. 33–34. Autograph letter.[5]

152. Banks to [Culling Charles Smith], 24 April 1810

Soho Square

My Dear Sir,
as the Letter you desird me to write to Count Trampe explaining to him the Refusal to deliver up Jergensen & the Marquis's desire then an amnesty should take place appears to me of rather a delicate nature I enclose a Draught requesting you to Peruse and Correct it to the Satisfaction of the Marquis after which I will dispose of it in such manner as shall be deemd Proper

The name of the Person referrd to in my note as very desirous of being nominated to the Consulship[6] is Brown his Christian name I have not yet Recoverd he was acting Lieutenant[7] in the *Talbot* Sloop Commanded by the Honourable Captain Jones when the Revolutionary proceedings of Phelps & Jergensen were put an End to & the Legitimate Government Restord

---

[1] In actual fact Count Trampe never returned to Iceland. He was appointed governor of Trondheim in Norway in 1810.

[2] The reason for this talk of a general amnesty was a letter Jörgensen sent to the Lords Commissioners of the Treasury. He was worried about the fate of the Icelanders who had supported the Revolution, having heard that Trampe had assembled a list of those who would 'be exposed to the resentment of Danish Government'. This petition was sent on to Lord Liverpool at the War Office, who despatched it in turn to the Foreign Office, where it received this prompt attention (TNA, WO 1/844, 7 April 1810).

[3] John Parke: see Appendix 13.

[4] Jörgensen remained a prisoner, eventually being released in mid-September 1810.

[5] Note, verso: 'Memorandum to | Sir Joseph Banks | on Iceland | April 23 1810 | the War Department on this subject.'

[6] It was Samuel Phelps who asked for a resident consul in Iceland. As he told the Lords of the Treasury, he had already invested a large amount of money, £40,000, in his trading venture and would not take further risks until a consul were appointed 'to interfere officially in the protection of the British trade' (TNA, T 1/1121, Phelps & Co. to the Treasury, 3 March 1810). On 14 April 1810 Earl Bathurst decided to appoint a consul for Iceland (TNA, FO 40/1, Bathurst to Foreign Office). See further Agnarsdóttir, 'Great Britain and Iceland', chap. VIII, section 1.

[7] According to *Steel's Original and Correct List of the Royal Navy* of 1809, this would be Thomas Browning. Lieut Stewart had been the 1st lieutenant, but he had been sent to England on the *Margaret and Ann*.

my Principal is not averse to allowing your Tyrolean Deputy[1] to take away a Limited Quantity of Bark The Quantity intended to be reported to the board, for their Consideration, his Lordship however has prudently suggested that the greater the Quantity of Bark[2] is that he is allowd to carry off, the Less ought to be the sum paid him, if such is to be the case, for his Armies here

always yours
[Signed] JB

office for Trade
½ past one O'clock
allow me to Recommend the enclosed letter to yours & mr. Broughton's[3] Good offices.

TNA, FO 40/1, ff. 38a–38b. Autograph letter.[4]

153. Banks to [Edward Cooke], 18 May 1810

Soho Square

My dear Sir

I have now Read the Clause of the Prize act referred to in the opinion which you were so good as to Shew to me yesterday, the Tener of it is merely to Place under the Jurisdiction of the admiralty Court, all Persons accusd of a Breach of his Majesties instructions, or a violation of the Law of nations. [*Margin*: Iceland]

I have also seen Mr. Barrow[5] one of the Secretaries of the Admiralty who intirely Confirms your opinion, That it is the business of the Board of admiralty to give orders for Commencing & Carrying on the Proceedings in the King's name as soon as the measure has recivd the sanction of the approbation of H.M. Sec*retary* of State.[6]

I apprehend therefore that all you have now to do to clear your Desk from Every Remains of Icelandich [*sic*] affairs, is to signify to the Board of admiralty the marquis's intire Concurrence in the opinion of the Law officers of the Crown[7] & his wish that the Proceedings recommended by them may be Instituted without further delay[8]

I have the honor to [be]
you*r* most Faithfull
& most H*um*ble Serva*n*t
[*Signed*] Jo*s*. Banks

[1] Unidentified. The name is on the endorsement but is difficult to decipher, perhaps Miller. See n. 4 below.

[2] Bark, especially of the oak, produces tannic acid used for the tanning of skins and production of leather and was of especial importance in wartime.

[3] Charles Rivington Broughton (d. 1832) was Senior Clerk of the Foreign Office, appointed in 1803.

[4] Note, verso: 'Sir Joseph Banks | *Delivered/Replied* [?] /*Received* [?] April 24. 1810 | Enclosing his | intended letter to | Co*un*t Trampe & | on permission to | Major Muller [?] to | take some Bark | to the Tyrol. | The Enclosure approvd'.

[5] John Barrow (1764–1848), Second Secretary of the Admiralty: see Appendix 13.

[6] The Marquis Wellesley, Secretary of State for Foreign Affairs.

[7] The Law Officers of the Crown were: Sir Christopher Robinson (the King's Advocate and a Judge of the Admiralty, Sir Vicary Gibbs (Attorney-General), Sir Thomas Plumer (Solicitor-General), William Battine (Admiralty Advocate General) and H. B. Swabey (Proctor of the College of Advocates).

[8] See Document 139.

TNA, FO 83/2293, f. 6. Autograph letter.[1]

154. Sir George Steuart Mackenzie to Banks, 20 May 1810

Reikiavick

My Dear Sir,
Before I set out on my first trip thro the Guldbringe Syssel,[2] I sit down with much pleasure to inform you that I have been twice at Vidöe with the old Governor Stephenson[3] who is in very good health for a man of his age – He speaks of you with rapture & remembers you with affection – He is a delightful old Man, & has been uncommonly polite to me. – Indeed, I am so taken with him that I really love him – I have now got better acquainted with his son[4] than I could have expected; as he did me the honor to request my assistance in adjusting some difference among the merchants, & in settling some matters respecting some claims against Phelps & Co – <u>All which are in the point of being amicably</u>[5] & fairly settled – I have copies of all my letters[6] – The conversations were taken down & I am perfectly prepared to show that my Conduct as a mediator has been satisfactory to the Dep*u*ty Governor,[7] altho' I assure you it was much ag*ain*st my inclination to interfere even in that Capacity – But I hope I have prevented what would have been very disagreable. I was the first to present the King's Proclamation[8] to the Good old Gentleman at Vidoe, & recollecting what happened last year,[9] I was very anxious to prevent any disturbance which thank God I have accomplished – The particulars of this I have not time at present to detail – I set off tomorrow & shall stop at Havnefiord[10] as the Bishop has requested my presence at the Examination of the School at Besestadir.[11] The Ship *Elbe*[12] will probably sail before I return, & I cannot neglect the opportunity of writing a few lines to you, altho' I have yet some to write to my family – My friend & I have

---

[1] A fascimile of this letter is printed in *Law Officers' Opinions*, vol. 42, pp. 277–8. Note on the left-hand corner of first page in Banks's autograph: *Received* [?] May 18 1810.'

[2] Gullbringusýsla.

[3] Viðey, an island off Reykjavík, the home of Ólafur Stephensen. Mackenzie's party was met by the former governor dressed in 'the uniform of a Danish colonel of the guards'. In this Mackenzie was mistaken: it was in fact Stephensen's uniform as Governor of Iceland (see p. 329, n. 5). (Hooker's account is interesting for the sake of comparison.) Mackenzie was not impressed with the house but accepted an invitation to partake in a sumptuous banquet. His 'sufferings' were not over until he had drunk a tumbler of 'smoking punch' (*Travels in Iceland*, pp. 84–5, 88–9).

[4] Magnús Stephensen, the Chief Justice of Iceland: see Appendix 13.

[5] This is underlined in pencil and was repeated in a letter to Hooker (Document 164).

[6] There are also copies in the National Archives of Iceland, Jörundarskjöl. Of especial interest are the 'Notes of Conversation between Mr. Fell and Mr. Petræus in the presence of the Governor and Sir George Mackenzie', dated 16 May 1810.

[7] Magnús Stephensen was still governor, according to the Agreement of 22 August 1809 (see Appendix 9). There is a useful account of subsequent troubles in Holland, *Iceland Journal*, pp. 101–2.

[8] The Order in Council of 7 February 1810 (Appendix 10).

[9] The Icelandic Revolution.

[10] Hafnarfjörður.

[11] Bessastaðir, which he attended (see Mackenzie's book *Travels in Iceland*, pp. 102–4).

[12] This ship was sent by Phelps & Co. with Liston, the captain of the *Margaret and Ann* in 1809, as captain. The Mackenzie expedition eventually sailed from Iceland on the *Flora*, yet another ship sent by Phelps & Co., at the end of August.

examined the rocks in this neighbourhood & of Vidoe & we have seen some very curious Geological facts in /part of the/ Country hitherto considered quite uninteresting –

We have been detained here by the difficulty of procuring horses, a vast number having perished of disease thro' the winter. I heard a report that an English Consul was coming out[1] – One is much wanted[2] –

I saw some plants which had been collected for M[r]. Hooker in very indifferent order – We propose to collect as many as we can for him, tho' I fear we shall do but little, as we shall be so fully occupied with Mineralogy.[3] –

If Count Trampe is still in London, or if you have any means of communicating with him, present him, if you please, with my best Compl*iment*s[4] & tell him that I have been obliged to occupy his house as I could find no other – I am doing it no harm, but probably much good by keeping fire in it – Believe me to be

My Dear Sir
Yours faithfully
[*Signed*] G. S. Mackenzie

BL, Add MS 33982, f. 1. Autograph letter.[5]
NHM, DTC, 18, ff. 31–33. Contemporary copy.
Hermannsson, 'Banks and Iceland', pp. 93–4.

155. Hans Frederik Horneman to Banks, 30 May 1810

Queen St*reet*, Cheapside

Right Honorable Sir,
The enclosed letter from Mr. Clausen[6] arrived this morning.[7] He informs me in a letter to my house that he has sent two volumes of Flora Danica[8] by the *Bildahl*, now detained at Yarmouth,[9] which appears to be the only book which he has been able to procure at present – The moment I receive them they shall be forwarded. –

Yesterday a petition was delivered in for the *Keblevik* & the *Skalholt*, belonging to Mr. Jacobæus[10] & now at Leith[11] – The former is one of the original Iceland vessels, the other

---

[1] John Parke was appointed consul in 1810, arriving in Iceland in 1811.

[2] Phelps had already written a letter to the Lords of the Treasury, stressing the insecurity of the British trade if no consul were appointed (TNA, T 1/1124, 23 March 1810).

[3] Banks sent this information on to Hooker (see Document 164).

[4] Mackenzie dedicated his book to the former governor, Count Trampe, thanking him for 'his kind exertions to render the journey through Iceland agreeable and successful'. They must have met in London.

[5] Note in Banks's autograph, on bottom left-hand corner of the first page: 'July 23 [1810]'. Note in Phelps's autograph, on back of letter: 'forwarded by Phelps & C[o]. – Cupers | Bridge Lambeth 20 July 1810.'

[6] Holger P. Clausen, Iceland merchant: see Appendix 13.

[7] The enclosure is not extant.

[8] The *Flora Danica* was a huge work in folio-size, describing and depicting all plants in the Danish kingdom. The first volume appeared in 1761, the last, number 54, in 1883. It contains 3,250 copper engravings as well as hand-coloured tables. Banks was sent the published volumes of this work by the King after the war as a token of gratitude (see Document 294).

[9] The *Bildahl* was later issued with a licence, leaving Liverpool on 10 May, arriving in Iceland on 6 June with Bjarni Sívertsen on board (Mackenzie, *Travels in Iceland*, pp. 129–30; Holland, *Iceland Journal*, pp. 140–41).

[10] Christian Adolf Jacobæus, Danish merchant in Keflavik.

[11] These two ships did not apply for sea-licences from the Danish authorities in 1810.

has been lately brought into the trade – Mr. Thomsen is also arrived at Leith with the ship *Norberg*[1] – This Gent*leman* has no more than this ship at present, & a considerable quantity of goods in Iceland – If the two former are granted he could be very thankful for a similar favour.

Mr. Clausen is arrived in Norway & intends to proceed to Iceland direct – He is very anxious to know the quantity of Cod Liver Oil Shark liver oil, as well as the quantity of Stockfish & Klipfish which may be allowed this year to be exported from Iceland. I named to him that I understood 300 Tuns of Oil & perhaps 300 Tuns fish would be granted.[2]

With the greatest respect I have the honor to be,
Right Honourable Sir
Your obliged humble Serv*an*t
[*Signed*] Hans Fred^k. Horneman

SL, Banks Papers, I 1:50. Autograph letter.[3]

156. Banks: 'Observations on Jergensen's Narrative',[4] 9 June 1810

1. The Licences were presented to Iceland traders in June–aug*ust* 1808. No ship proceeded with licence to Iceland that year but usd their Licences in trading with Norway. No Licencd ship had arrivd in Iceland in Nov*ember* 1808.

2. Having Embarkd a Cargo of the value of £1000 he was caled upon for Payment when on the Point of sailing the first time in Mr Phelps's service. Mr Phelps advancd the money for him.

3. Mr Phelps's first Cargo Consisted of Barley meel Potatoes Salt a Small proportion of Rum, Tobacco Sugar & Coffee. He avoided British manufacture.

4. Savignac[5] Enterd into a Convention with the Government of Iceland. He complains that after this the people purchasd only such part of his Cargo as was wanted for their Private use tho he sold Barley meel for ten dollars a Barrell while Rye meel was sold for 22. The Danish Factors terrified the Icelanders from Purchasing & he Returned without an <u>ounce</u> of Produce of any kind

[*Margin*: He had promised his Employers 150 Tons of Tallow offerd to him on the Royal Exchange.]

---

[1] Jesse Thomsen had arrived in London in order to try to prosecute Jörgensen. This would become a lengthy complicated case, not being resolved until 26 June 1829 (see Þorkelsson, *Saga Jörundar hundadagakóngs*, pp. 157–61).

[2] This was correct, see instructions to the Consul, Document 202 below.

[3] In Banks's autograph, verso: 'Mr Horneman [Received] May 30 [18]11 [Answered] June 1 | Keblewic Scalholt.'

[4] These observations are made on Jörgensen's 'Historical Account of a Revolution on the Island of Iceland in the Year 1809', BL, Eg. MS 2067, 368 handwritten pages, which Jörgensen sent to Banks as a vindication of his actions in Iceland. The numbers have been pencilled in, quite possibly by Banks himself. The marginal notes are snippets of information which come later in the narrative and Banks places these notes in the margin in the logical chronological place. Banks mostly paraphrases Jörgensen in these observations. Occasionally he expresses his personal view, those cases are footnoted. These notes are often difficult to decipher. Some full stops and capital letters have been added to aid comprehension.

[5] James Savignac (fl. 1809) was employed by Phelps & Co. as a supercargo, both on the *Clarence* and on the *Margaret and Ann*. Little further is known about him than appears in these documents except that he was of Huguenot descent.

5 When the *Clarence* took in Balast she was made to pay 235 dollars for the Stones.
6 Jergensen who had taken an interest in the first voyage Relinquishd all Share in the Profits of the second.
[*Margin*: With a view to the Revolution he attempted which was Plannd before the Second vessel Sailed he avows his his [*sic*] intention of acting in a new Capacity. I conclude that of Sovereign page 27.[1]]

The moment Count Trampe arrivd he Posted up a Paper prohibiting under Pain of Death all intercourse with the british. Cap*tain* Nott however had Compelld the Count to Enter into a new Convention allowing the British free trade Subject to the Laws of the Island before the arrival of the *Margaret & Anne* but this was a private agreement Expressing only Cap*tain* Nott & his officers.
7 Mr Phelps waited some days Expecting this agreement to be sent to him or Posted up but neither was done & the Counts first Proclamation Remaind unrepeeld.
[*Margin*: He [Phelps] waited from Wednesday to Sunday.]
8 The *Orion* in which Count Trampe Saild violated her Licence by Touching in Norway. She went from thence with a Ship of Truce to Leith & London 3 British prisoners.[2] The Count went by the name of Adzer Knutsen[3] to whom the Licence had been granted.
9 As he was sent out for the Purpose of putting the Island in a State of Defence as appeard by his Papers that were Seizd and as he had a Captains Commission he Should have been hangd for Entering an Enemies Port under a False Name.
10 The Greenland Fishery Company provided a Cargo for the *Orion* in Norway the Count appeerd as the owner tho the Ship as well as the Cargo was theirs.
11 Count Trampe is the Son in Law of Colbiornsen[4] the head of the French party at Copenhagen. He himself is a French partisan. In 1806 he prevented two French Frigates in an Iceland port[5] from falling into the hands of English Frigates.[6] At the time Denmark was neutral.
[*Margin*: These /French/ Frigates were in an Iceland port.]
12. For these Reesons Mr Phelps directed his Captain to seize the Count when unguarded by any kind of militia at his house.

---

[1] Jörgensen wrote that when he went the second time he went 'to act in a different capacity, to what I had done before'. That is to say not as the interpreter. Banks deduced from this that the Revolution had been planned beforehand and Jörgensen's aim had been to make himself sovereign over the island. The evidence, however, suggests that the Revolution was spontaneous.

[2] This is a bit cryptic: Jörgensen wrote (p. 34): 'it could not be supposed the *Orion* would pass uninterruptedly on her voyage, and therefore three British prisoners of war, were sent on board, under a flag of truce, and ordered to proceed to Leith in Scotland, thereby thinking to elude the vigilance of English cruizers'.

[3] Adser Christian Knudsen: see Appendix 13.

[4] Trampe's father-in-law was actually Edvard Røring Colbiørnsen, (1751–1792), brother of Christian Colbiørnsen (1749–1814), the lawyer and royal official to whom Jörgensen is referring.

[5] Jörgensen wrote 'Borgarfjörður'.

[6] They came to Hafnarfjörður. According to Jörgensen, they had burnt 17 British whalers on the Greenland coast. The correct version appears to be that in July 1806 two French frigates, *La Revanche* and *La Syrène*, had come to Patreksfjörður in the Western Fjords to get supplies and tend their sick, having succeeded in destroying 6 whalers (not 17). Later two British frigates, the *Phoebe* and *Thames,* arrived at Hafnarfjörður in an unsuccessful search for the French ships (RA, DfdUA, Island og Færöer, Rentekammer to the DfdUA, 18 April 1807).

13 He had a Right to do this under the authority of the Letter of marque Even [if] the Capt*ain* [Nott] not had made a Convention.
[*Margin*: Government and the few officers here though*t* otherwise.]
14 Letters of marque authorise the holders to act against Dominions Territories vessels[1] & Subjects of an Enemy formerly Prisoners were deemd the property of the Captors & Ransomd. Kings have been thus Capturd & their Ransoms given to a Privateers Crew. In our days this warfare is not proper but it is Lawfull a Letter of marque having not only Equal to but Superior to that of a Kings Ship.
15 Wood*e*s Rogers[2] laid Contributions on Conquerd [Terri*tory*?][3] in the South Sees to Enrich himself & his Crew he was nevertheless made Governor of Bahama Island.
16 The last year of the past kings Reign Two Privateers were fitted out to attack Rio Plate. They enterd into an alliance with a Portugues*e* Frigate to attack it which no man of war could do.
17 Ten years ago a Privateer was fitted out from London & Commanded by an american. She seizd a Spanish Governor going from Lisbon to *south* America in a Portuguese vessell & made him Pay a Ransom therefore persons Capturd by Private Ships of war are the Property of the Captors and not of Government. A pamphlet was publishd by Stephensen Chief judge to Rouse the natives against the English another by Amptman Thor*a*rensen in North Iceland prohibiting the people from assisting any Ship of the Robbers (Englishmen).This neerly amounted to a declara*tion* of war.
18 Can the Commander of a Ship of war Enter into a Convention with a nation at war? If So he cannot bind by it a Fellow Subject he could not Subject them to the Laws of the Country they were allowd to trade with. This Convention then was not binding nor had Count Trampe attended to it.
[*Margin*: How can one person reside in & trade with a Foreign Country without being subject to its Laws unless by previous Stipulation.]
19 Mr Phelps could not Sell his Cargo. Count Trampe must of Course prevent him[4] £50000 must be lost or the Count seizd.[5]
20 Count Trampe has Come to the Island to Place it in a Posture of Defence & Levy a militia. The Licence of the *Orion* had been forfeited by touching in Norway & again by a Contract Enterd into by the Count with Capt*ain* Hibolt[6] in Norway to furnish the Danish Gun boats in Norway with woolen cloth & Salt mutton before winter set in for which purpose the *Orion* has to have proceeded to Norway.
21 The Counts Letters requested arms ammunition & warlike Stores.
22 The natives present when Count Trampe was Seizd made no shew even of Resistance tho the Captain & 12 Seamen were all who were Employd.

---

[1] Jörgensen always writes v*a*ssals.
[2] Woodes Rogers (c. 1679–1732), was an English sea captain and privateer, subsequently the first Royal Governor of the Bahamas. He was the captain of the vessel that rescued Alexander Selkirk, believed to be the prototype for Defoe's Robinson Crusoe.
[3] Difficult to decipher. Jörgensen writes *people* but that does not seem to be the term used here.
[4] Jörgensen is probably referring to the fact that Count Trampe had arrived in Iceland with a cargo to sell to the inhabitants and would not have appreciated the competition from the British merchants.
[5] Difficult to decipher. Jörgensen says Phelps was ruined to the amount of £50,000 (pp. 55–6).
[6] Captain Hibolt is unidentified.

23 Mr Vancouver[1] advisd that the Island should be plunderd. Mr Phelps Refusd to proceed further than the imprisonment of the Count.
[*Margin*: He Receivd on board Abundance of Tallow etc Confiscated by Jergensen for some part of which he gave Receipts but not for all.]
24 Jergensen by the Will and Sanction of the Icelanders assumed the cheif Command of the Island & Issued proclamtions.
25 The overall Expence of the Government of Iceland does not exceed 27,000 dollars.
[*Margin*: See Eliz*a*beth ch*a*pter [*blank*][2] an act for the increase of mariners & for the maintenance of navigation & more Especialy for recovering the Trade of Iceland that formerly Employd more than Two hundred Ships a year. p. 81 he says that he Recievd from the Public Chest 2700 dollars that the arrears due for the Last year were 27,000 that he recd the whole being procurd money on his own Credit].
26 Only 2 danes beside Count Trampe held public offices.[3]
27 The natives attackd & ill usd some danish Factors on the Day the Count was seized.[4]
28 A consequence of the Earnest Solicitation of the inhabitants, I consented to Retain the cheif Command till the first of July 1810 the time appointed for the Representatives to assemble.
29 No one was hurt none arrested Except Einarsen who told the chief judge that he had 50 men & wishd him to Procure 50 more to seize the Ship & Crew.[5]
30 The new arrangements here approvd, the Bishop & all the Clergy at a Synodal meeting signd a paper expressing their Readiness to support the Revolution[6] & to Exhort other classes to do the same. The cheif judge & all officers whatever Signd & Seald their Consent to Continue in office. All Classes appeard Placid. Volunteers offerd their Services few were accepted for want of arms. 150 Set down their names to Serve.[7] I proceeded to the north & sent with 5 armd natives I needed no more being Receivd with cordiality Every where. When he returnd he had only one armed man with him. A proof that his revolution was approvd?[8]
[*Margin*: Between 30 to 40 musquets[9] were seizd on the first day p. 65]

---

[1] Charles Vancouver (1756–1811), older brother of George Vancouver (1757–98) the famous explorer who, among other exploits, sailed with Cook on his second and third voyages. By 1785 Charles had moved to the United States, settling in Kentucky. He returned to Britain in 1793 where he wrote two volumes (on Cambridgeshire and Essex) for the Board of Agriculture series of General Views of agriculture of counties. Between writing the volumes on Devon (1808) and Hampshire (1810) he appears to have been working for Phelps in some capacity, clerk perhaps, and was accompanied on the *Margaret and Ann* by his wife, Louisa Josephina van Coeverden (Vancouver is derived from this name), a cousin. He was disliked by Jörgensen.

[2] Jörgensen wrote: 'We shall find in the reign of Queen Elizabeth an act passed in favor of Iceland trade.' There was a great deal of correspondence between Queen Elizabeth and the Danish kings regarding English fishing and trading in Icelandic waters; see especially the second chapter of Þorláksson, *Sjórán og siglingar*.

[3] This would have been Frydensberg the treasurer and Koefoed the country magistrate.

[4] There is no corroborating evidence for this statement.

[5] Einarsson was imprisoned and so was Frydensberg for a short time. The facts surrounding Einarsson's 'conspiracy' are unclear but Jörgensen wrote: 'I had reason to suppose the information without foundation, and proceeding from malice' (p. 96). It was no secret that Stephensen and Einarsson were far from being friends.

[6] Printed in English in Þorkelsson, *Saga Jörundar hundadagakóngs*, p. 166.

[7] This is not corroborated by other evidence.

[8] This is Banks's question. Jörgensen certainly thought so. He wrote: 'During all this, there appeared not the smallest dissatisfaction among the people in general' (p. 102).

[9] Twenty according to Hooker (*A Tour in Iceland*, I, p. 56).

31. No dissatisfaction appeard anywhere. Judge Stephensen told Mr Hooker he had no Complaint to make against Jergensen

[*Margin*: He would have been imprisond as Einarsen had been had he complaind.][1]

32 Count Trampe would have been allowd Some Liberty had not Stephensen told Mr Phelps that it would Create General dissatisfaction.

33. Mr Vancouver approvd of all I had done & Calld me a Second Bonaparte

34. On a Ship appearing in the offing the natives mannd Mr Phelps Battery while the English went on board their Ship

35 Capt*ain* Jones had been Condemnd for a breach of discipline & Receivd a Royal pardon.[2]

36 When he came ashore he said the union was a destructive measure that his admiral Nagle[3] was a Damned hag & that Englishmen made themselves hated wherever they came.

37 Capt*ain* Jones sent for Count Trampe from the Letter of Marque. Mr Vancouver allowd him to so Mr Phelps told him he had done wrong in permitting a secret intercourse between the Danish Governor & Capt*ain* Jones.

38 Capt*ain* Jones Issued orders to the Letter of Marque that he considerd the conduct of the Crew as a violition of the Law of Nations that Jergensen was an improper person to be at the head of Government that Count Trampe ought to have been sent home when seizd with an account of all proceedings, & he orderd one of the vessels to proceed with him & Jergensen in 24 hours.[4]

39 Capt*ain* Jones proposd that Vancouver should be placd in Jergensen's Place. Mr Phelps declard that no citizen of a State that had Rebelld against its Sovereign Should hold power where he was. Capt*ain* Jones is nephew to the american General Montgomery[5] & respects his memory.

40 Capt*ain* Jones made a Convention to Restore Danish Government & Place Stephensen at the head of it & destroy the Iceland Flag – he had not the sanction of the inhabitants. A remonstrance was made against Stephensens appointment but it was not attended to.

41 Pride, Private pique & disaffected Principles were Capt*ain* Jones's motives.

42 Capt*ain* Jones by declaring against M{t} Phelps destroyd his trade.

43 The Captain orderd the Danish Flag to be hoisted & Recivd 3 Cheers under it while his own was flying on board.

44 A letter of thanks to Capt*ain* Jones was signd by many Danes. The natives refusd to sign it & it was destroyd.[6]

---

[1] A comment by Banks.

[2] The Honourable Alexander Montgomery Jones (1778–1862) was the captain of the *Talbot*. This took place in 1804 on the *Naiad* frigate. Jones was court-martialled for striking a superior officer (who had behaved 'in an ungentlemanly manner' and was dismissed from the Navy) and sentenced to be shot. However, he received the King's free pardon and was restored to his rank. In 1806 he was promoted to commander and ended life as a vice-admiral (John Marshall, *Royal Naval Biography*, vol. X, p. 390).

[3] Admiral Edmund Nagle of the Leith station (see Document 119). It is unlikely that Jones would have said this in Jörgensen's hearing.

[4] This did not take place.

[5] Richard Montgomery (1738–75) was an Irish-born soldier who became a major general in the Continental Army during the War of American Independence. He led the invasion of Canada in 1775. His sister was Sarah Montgomery (d. 1812) who married Charles Wilkinson Jones, 4th Viscount Ranelagh (d. 1797). Captain Jones was their ninth son.

[6] There is no corroborative evidence for this statement.

45 Jergensen might have retird into the Country & defended himself. He had no fear of leving. He objected to being made accountable for the Public money he had spent. He Receivd

from the Public Chest   2,700 [rixdollars]
from the *Orion* –   10,000

46 The whole Expenditure of the year with the new military Establishment amounted to 4 times that sum being

12,700 x 4 –   <u>50,800</u>
The 3 [?]rts 12,700 x 3   38,100

He procurd on his own Credit no one then has a right to enquire into the Expenditure.[1]
[*Margin*: Of eight soldiers 5 of them went out with *Jorgensen* & one only released.]
47 Cap*tain* Jones wishd to insert in his Convention an act of amnesty for Jergensen & this Jergensen Objected to. There is however such a Stipulation in favour of all who took part with Jorgensen.[2]
48 Cap*tain* Jones who had orderd Liston to see[3] denied it in Court. Liston had no order to Shew Phelps. [Phelps] lost his Suit. Lieut*enant* Steuart[4] brought Jones's order on board Listons Ship. He was in court & not examind.
49 The *Orion* was releasd through the ignorance of Mr Phelps's Lawyers.
[*Margin*: Mr Troward a Partner in the adventure is neither ignorant nor silly.]
50   <u>Jergensens Life</u>
At an early age he was apprenticd to an English Collier.
At 25 he Quitted the British sevice.
He then became known to *Sir* J[oseph] B[anks].
He arrivd at Copenhagen in Aug*ust* 1806.
When Copenhagen was attackd he saw Gen*eral* Wellesley[5] & was accusd of having given him improper information.
He had the Command of a vessel of 28 guns & 83 men a Brig. She was the Property of a private adventurer but was mannd & armd by the King. She had on board Double Commission one as the *Admiral Juel*[6] the other as the *Christine Henriette*.
[*Margin*: He told he had had the Command of a Squadron of 6 Frigates to Cruize on our East India Trade.]
51 He took 3 Ships a british a Prussian & a Swede that were Condemnd. He was taken by the *Sappho* Cap*tain* Langford near Flambro heed[7] having fough*t* 40 minutes in which She Dischargd 17 Broadsides & had not powder enough for one more.
52 He was accusd of having given up his Ship.
53 His vessel was Condemnd under the name of *Christina Henriette* the name of her crew knew her by that name [*sic*]. He has stated the Reason why she had double papers to the admiralty.

---

[1] These amounts are on p. 164. But the 3 × 12,700 is not in Jörgensen's account.

[2] Article 6 read: 'All officers or other persons either armed or unarmed, who, during the late events, have taken part with Mr. Jorgen Jorgensen, shall no longer be in employment, but their persons and property in every respect (whatsoever or of whatever nation they may be) shall be respected and protected the same as other persons and natives.' See Appendix 9.

[3] 'Sea' is meant.

[4] The 1st lieutenant on the *Talbot*.

[5] Arthur Wellesley (1769–1852), later 1st Duke of Wellington.

[6] More correctly *Admiral Juul*.

[7] Flamborough Head off the coast of Yorkshire. This took place on 2 March 1808.

54 He always refusd to Receive Prize money. He has 2000 in his name on the books of Mr Hogan at the Cape[1] being Share of a Prize which he has not taken.

55 The Ship was sent into Yarmouth. Gen*eral* Wellesley got Jergensen Leave to Come to London.

56 He soon after Enterd into a mercantile Speculation to Send goods to the Continent. He had got a Prize of £20,000 in the English Lottery. During his Stay in London he saw S*ir* J[oseph] B[anks] several times.

[*Margin*: Borrowd of his [Banks's] Small Sums at 4 different times from april 19 1808 to april 12 1809 in all £107.9. tho he had first won £20,000. He Borrowd of Phelps £1000.]

57 His Passport allowd him only 4 days in London. He stayed here 8 months before he was calld upon to sign parole & preceed to Reading.

58 Mr Adams[2] laid before him a Paper & told him he need not Read it. He signed it and afterwards found it was his Parole of honor. Mr Adams told him he would in the Evening Send him a Letter containing his Pasport.

In the Evening a Letter arrivd directed to him as Captain of a French Privateer. He took advantage of this did not open the Letter & Remaind in London. He acknowledges, that he was aware of the mistake.

He went to Iceland without permission as his intention were to make a Revolution there & he would not have it Said that a Danish Prisoner was releasd with such an intention.

I let Sir Jos*eph* know that I was Conceald from the Transport board when I arrivd from my first Voyage & I took care he should not know I had Embarkd on the second till I was at sea.

He says p. 247 no one will in Future Trade successfully with Iceland.

Did not Mr Phelps load a large Ship had he not Cargoes for five or six more.[3]

[*Margin*: of goods Confiscated under Jorgensen's administration.]

Phelps neglected to give in a State*ment* of the Iceland business in detail. This would have justified him & me he rather chose to try a Compromise.

[*Margin*: he gave in to the Foreign office an immence volume of Papers.][4]

Count Trampe has no Land in Denmark his Family have been Lately imported from Germany. He is in the 3^d Rank in Denmark. His key may be purchasd for 5 or 600 Dollars.[5] The Government of Iceland is the Lowest in Rank of all.[6] He is Captain in the Danish Militia & his whole income 1500 Rixdollars a year say £100 Sterling.[7]

---

[1] Hogan was 'a great merchant at the Cape of good Hope' (Jörgensen's account, pp. 219–20). Jörgensen had been a sailor on his privateer the *Harriet*, and insisted that he had not accepted any prize money.

[2] G. Adams, Extra Clerk at the Department for Prisoners-of-War at the Transport Board.

[3] Though this is a question from Banks, the information is in the account.

[4] It is not clear whether this refers to Jörgensen or Phelps. Probably the former, as Jörgensen copied a great many of these papers, preserved in BL, Eg. MS 2068. There is a box of unsorted papers relating to the Icelandic Revolution in TNA, FO 95/648.

[5] Jörgensen is writing about Trampe calling himself a Chamberlain to the King of Denmark, pointing out that it is only a title, which anyone can buy.

[6] Jörgensen wrote (pp. 260–61 in his account): 'Being governor (Stiftamtmand) of Iceland is nothing of any consequence at all. It is reckoned the lowest of all districts in Denmark or Norway, and hitherto the governors have not had a single man to command, in a military capacity on the island.'

[7] More likely £300 (see 'Currency, Weights and Measures').

The Iceland transactions originated in a mercantile speculation and all that took Place afterwards from its disappointment. My first voyage was undertaken with a view of gain my Second to Relieve a wretched people from hunger & to Establish Freedom in the Island.

I have sacrificed my whole worldly welfare because I hated to serve against Great Britain.

Wisconsin, Banks Papers, ff. 145–150.

157. Banks: Note on Jörgensen's 'Historical Account of a Revolution on the Island of Iceland in the Year 1809'

That I did not apply to the British Government for Permission to leave England for Iceland proceeded from, that, the first time I went out, I thought it [entirely] unnecessary & the second I did by no means wish for Such Favor as my intentions were to Compell Danish Government to Retract their oppressive measures towards the inhabitants of the Island & I did not wish it be said that the British Government Releasd a Prisoner of War for the purpose of stirring up insurrection against a Power at war with England page 237.[1]

Wisconsin, Banks Papers, f. 151.

158. Banks: Memorandum on Jörgensen's 'Historical Account of a Revolution on the Island of Iceland in the Year 1809'.[2]

Jergensen appears to be grossly ignorant of the Law of nations & the Customs of war among the Civilisd nations of Europe
he argues at Length the Legality of the Capture of an unarmed Governor upon the Principle Prisoners being the property of the Captor & Liable to pay a Ransom
This barbarous Custom /of our ancestors/ has been long Abolishd all prisoners of war now belong to the nation & those not taken under arms are liberated
Government here /& the Law officers/ were of opinion that M^r. Phelps' conduct in this Particular was unjustified.
he argues in favor of the capture of the *Orion*
The Court of admiralty restored the Ship
he Censures the decision of the Court respecting the Insurance
We shall not Easily believe that Mr /*Phelps*/ Trowards attorney was sufferd to Lose the Cause

---

[1] BL, Eg. MS 2067, pp. 237–8.
[2] The account Banks had been reading was BL, Eg. MS 2067, 368ff., probably written in late 1809 or early 1810. Following it is a copy of a letter to the Marquess Wellesley, the Foreign Secretary, where Jörgensen wrote: 'Right Illustrious and Most Noble! | I have without having the honour to be personally known to Your Excellency, or without Your Excellency's permission, presumed to dedicate these my humble efforts to give the world a correct account of a late Revolution on the island of Iceland, to your Excellency, and which can have no other merit than the truth contained therein.'

he argues in favor of his own Conduct in /use/ organizing as far as he was able a Revolution in Iceland

he will Certainly be hangd for this as he ought to be if he is Caught within the Dominions of his Lawfull sovereign

he faild from his want of Talent or rather he would not had he possessed any have made the attempt to nothing but horrors & public executions could be the End of a Revolution in which an Island of 40,000 inhabitants had dared to declare themselves independent of a great nation to whom they owed allegiance

I never saw Jergensen till he came to me on his Return from N*ew South W*a*les* with two Otaheite boys which the master of the Ship in which he Saild had Abandond in the [*indecipherable*]

he had a double Commission one for the *Christine Henriette* the other for the *Admiral Juul*.[1] 17 broadsides Exhausted his Powder

he told me he had a squadron of Strong Frigates intrusted to him by his Government to Cruize in the South Seas[2]

he won a 20,000 Prize in the English Lottery & did not Pay me the £100 he owes me

he says he never Rec*eiv*ed his undertaking of Parole when he signd it & that Mr Adams did not tell him what it was?

his countrymen say that he gave up his Ship

he went from England with the intention of Revolutionising Iceland Phelps must have known this p. 238[3]

Wisconsin, Banks Papers, f. 152. Autograph memorandum.

159.   Banks to William Jackson Hooker, 15 June [1810]

Spring Grove

My Dear Sir

I have Read Jorgensens Narrative[4] & I cannot say it has raisd him in my opinion. The Revolution he attempted was Evidently & indeed Confessedly Plannd by him before he Left England[5] in the *Margaret & anne*. Of Course we cannot Doubt that Mr P*helps* was a Party. The Presence of Count Trampe was not Expected & in Fact destroyd the

---

[1] In actual fact the *Christine Henriette* was a merchant brig donated by private citizens and converted into the warship *Admiral Juul*.

[2] This is not mentioned by Jörgensen in Eg MS 2067 in the British Library.

[3] In the Sutro Banks Papers there is an identical scribbled undated note: 'he went from England with the intention of Revolutionising Iceland Phelps must have known this'. This is Banks's assumption.

[4] Three slightly different versions are preserved in the British Library, Eg. MSS 2066, 2067 and 2068. Eg. MS 2066 is a partly fictionalized account, but it does have notes pencilled in by Hooker. These three manuscripts containing Jörgensen's narratives of the Icelandic Revolution remained in the Hooker family until sold to the British Library in 1868 by Sir Joseph Dalton Hooker, director of Kew Gardens (information from J. Conway of the BL, Department of Manuscripts, letter to Agnarsdóttir, 14 June 1995). Eg. MS 2070 contains the letters Jörgensen sent to Hooker. Full stops and capital letters have been added for the convenience of the reader.

[5] Why Banks came to this conclusion probably rests on the conversation he had with Phelps and Jörgensen on 15 April 1809 before they left for Iceland. In his letter to Bathurst (Document 113) he certainly mentions the fact that they are prepared 'to make the conquest'.

Enterprise for had he been absent it would have been Easy for Phelps to have persuaded us that the Revolution was a measure of the Islanders own Contrivance.[1]

His arguments are labord but they stand on most unsound Foundations. He tries to prove that the *orion*[2] was a Lawfull Prize, Sir Wm Scot[3] declard otherwise. He says that Mr. Phelps's Cause of insurence was Lost because a witness present in Court was not Examind. I can Trust Mr Troward[4] for Leaving no Legal measure untried that could have Carried any Point of his. He argues on the Propriety of a Revolution & Forgets that a Danish Subject who had no Iceland greivance to complain of was certainly a most improper Leader. His allegiance was broken where he had no concern in the Case.

He says he obtaind a 20,000 Prize and was at the time borrowing £20 at a time from me, on Pretence of not having the means of support here, is Either a Lie or a great deception practicd upon me. He says nothing of breaking bulk[5] of the Prizes he seizd, which is expressly forbid to all Captors. They are to bring home their Prizes for adjudication. To break bulk before Judgment is Passd, /or the Prize is Delivered to the Court/ is a piratical act & worst of all he defends the abominable Practice of individuals Taking Prisoners & Compelling them to Ransom themselves. By his Reasoning Count Trampe was Phelps's Prisoner & could not be Releasd by Government without his Consent.[6]

We no Longer Capture unarmd Persons. Even these taken on board Ships who are not of the Crew or are Civil officers as Pursers, Surgeons &c. are dischargd as being noncombattants. His work has however been usefull to me as it exhibits no doubt all the charges he could muster against Count Trampe.

I am glad I have now read both sides[7] & am of course able to make up my mind upon a more Certain Foundation than I could do before. My mind is, that Jergensen is a bad man Phelps as bad, & that Count Trampe is a good man as Good I mean as Danes are when they are good which is by no means so good as a Good Englishman.

I am my dear Sir
your very Faithfull
& very H*um*ble Servant
[*Signed*] Jos: Banks

Kew, Hooker's Correspondence, Director's Correspondence, I, f. 38. Autograph letter.[8]
Hermannsson, 'Banks and Iceland', pp. 61–2.

---

[1] It is ironic that this is Banks's original plan. He felt it necessary to involve the Icelanders, that the annexation be voluntary and not accomplished with force (see Document 45 above).

[2] The *Orion* was restored to Count Trampe.

[3] Sir William Scott, the King's Advocate (see Document 142, p. 382, n. 3).

[4] Richard Troward was one of Phelps's partners. He was a lawyer and defended Warren Hastings (1732–1818) in his famous impeachment trial (notes on BL, Eg. MS 2066, f. 347, probably pencilled in by Hooker).

[5] 'Break bulk' means to begin unloading the cargo.

[6] Captain Jones of the *Talbot* sloop-of-war believed this too.

[7] Banks had now read all three accounts of the Revolution by the major protagonists. Trampe's narrative by the wronged governor (Document 124), a letter from Magnús Stephensen who had put across the view of the Icelandic elite (Document 122) and finally he had now read Jörgensen's of a very different tenor.

[8] Addressed to Hooker at Halesworth Suffolk.

160. Banks to William Jackson Hooker, 16 June 1810

Soho Square

My Dear Sir

You will have Receivd by yesterday's Yarmouth Coach Jergensens papers & have Learnd by my Letter that my opinion of Jergensen is not improved by the perusal of them

Respecting the paper in Question I certainly was not aware that I had laid it with those I put into your hands. I in no Shape regret your having seen it but as it never was intended for Publication & is for many reasons not a proper paper to see the Light I must beg to have it Returnd at your Leisure[1]

The papers Containing memorandums of some matters I saw in Iceland[2] you are welcome to use as remembrances but as I have declind to publish them myself you will Easily understand that I had no view when I put them into your hands of their being publishd in my name or with any kind of Reference to me as far as they may be usefull to you in your own work they are intirely at your Service. I meant them as some recompence for the Loss of your own papers[3] & in this Point of view I am in hopes they may be of Some use

adieu my dear Sir
in haste but Faithfully yours
[*Signed*] Jos: Banks

Kew, Hooker's Correspondence, Director's Correspondence, I, f. 31. Autograph letter. Hermannsson, 'Banks and Iceland', p. 90.

---

[1] Possibly one of his projects for annexing Iceland, given the clue in the following document 'situated as that country [Iceland] is now with regard to us, you are right in witholding it'. After the Order in Council of 7 February 1810 it would have been diplomatically improper for England to annex Iceland, which was now in a state of neutrality and amity with England. Hooker published the Order in Council in his own book, *A Tour of Iceland* (II, pp. 59–61). He discusses the aftermath to the Revolution, remarking: 'Hence then it appears that a mercantile speculation the most unfortunate, and a revolution the most singular in its nature, have been the means of placing the island in a greater state of security than formerly. Thanks to 'the humane intentions' of the British government the Icelanders' situation had been much improved. Hooker used the opportunity to echo Banks's sentiments concerning a British annexation. Like Banks before him, Hooker wrote of 'every native Icelander's' willingness to come under English rule, seeing this 'as the greatest blessing that can befal*l* his country'. Hooker castigated the Danish King as 'an inefficient but presumptuous tyrant' and argued that the condition of the Icelanders could only deteriorate. England's 'taking possession of Iceland' and incorporating the island into the British Empire would be an action of mutual benefit to the two countries. The Iceland of the future, 'guarded by the protection of our fleets, and fostered by the liberal policy of our commercial laws', would be a prosperous country and England would be repaid by an extension of her fisheries (II, pp. 61–3).

[2] He must be referring to his journals. Hooker did make use of them in his forthcoming Iceland book (see following document).

[3] Hooker lost his papers when the *Margaret and Ann* went down in August 1809.

161. William Jackson Hooker to Banks, 20 June [1810]

Halesworth

My dear Sir

Owing to a little error in the direction of your letter it has been round to Yarmouth & I have, but this moment received it. I lose no time in returning your Iceland paper,[1] & altho' I regret it is not to be published, yet situated as that country is now with regard to us, you are right in witholding it.

Some of the Iceland poetry will be an interesting specimen of the abilities of the natives at making Odes & as such I hope I may be allowed to bring them into my little work, either in a note or in the Appendix.[2] It will be necessary for me to say something about Hecla,[3] altho' the state of the weather would not allow me to climb it, & in your memoranda are some interesting remarks which you made on your way thither. – These would be extremely useful to me; but I do not see that I can possibly make use of them without saying to whom I am indebted for them, even altho' I should intermix with it some matter from Von Troil.

I have received your Letter & the Narrative of the Iceland revolution.[4] Had I waited another[5] day before I wrote to you I should not have troubled you with my saying so much about Jorgensen in my last.

There was certainly much ill conduct in the late affair & it has always struck me that the worst part of it was the Treatment of Count Trampe.[6] – But whether the blame was most to be attributed to the Merchant or the Usurper I don't know.[7]

Jorgensen must wait with patience till an exchange of Prisoners & ought to feel happy that he is so well off.

I remain, Dear Sir,
with great sincerity
Your faithful & very obliged humble Serva*nt*
[*Signed*] W. J. Hooker

Kew, Banks Correspondence, II, JBK/1/7 f. 334. Autograph letter.
NHM, DTC, 18, ff. 42–43. Contemporary copy.
Hermannsson, 'Banks and Iceland', pp. 90–91.

---

[1] Possibly a copy of his account to the Foreign Office (Document 156).

[2] Hooker did indeed include the Odes, both those presented to Banks as well as Captain Jones, in the second volume of his book, Appendix D.

[3] Hooker also included an extensive chapter on firstly Hekla and then on other volcanoes. He referred not only to Uno von Troil's *Letters on Iceland* but to many other scholars (*A Tour in Iceland*, II, Appendix C, pp. 105–269).

[4] Hooker also included a 63-page account of the Revolution (Appendix A).

[5] In the DTC copy 'and then day' is written instead of 'another', obviously a mistake.

[6] Trampe was kept in a dirty narrow cabin on the *Margaret and Ann* for nine weeks (see his memorandum to Bathurst, Document 124, p. 354).

[7] Hooker is here referring to Samuel Phelps (the merchant) and Jörgen Jörgensen (the usurper). The Law Lords had ruled that the reponsibility for the decision to arrest the governor and effect a revolution must rest with Phelps, Jörgensen being only 'a subordinate agent' (TNA, FO 83/2293, Robinson to Wellesley, 10 February 1810).

162. Banks to William Jackson Hooker, 29 June 1810

Soho Square

Dear Sir

I See no Objection to your using my name in the intended work when it Cannot be Avoided in notes I have the Leest Objection to See it Somewhat more am I disinclind to an appendix. For the Text I hope it will not be Found necessary to Place me there[1]

Whether the merchant or the usurper[2] best deserve to be hangd is a matter that may be doubted, but that both of them deserve it Richly for hazarding the destinies of an innocent nation in order to facilitate their Trade is what I do not doubt & a matter of which I think you will in due time be Convincd if you are not already

as you do not Correct the Error of my Last direction I fear I may again Err

in haste but faithfuly yours
[*Signed*] Jos: Banks

Kew, Hooker's Correspondence, Director's Correspondence, I, f. 32. Autograph letter.
NHM, DTC 18, f. 44. Contemporary copy.
Hermannsson, 'Banks and Iceland', pp. 62–3 (excerpt).

163. William Jackson Hooker to Banks, 22 July 1810

Halesworth

My dear Sir

As I know you feel deeply interested in every thing that relates to Iceland, some information which I have just received from Mr. Phelps on that subject may not be uninteresting to you.

Mr. Phelps says that Mr. Savigniac,[3] Petræus,[4] & the Tatsroed Stephensen,[5] were about to ship off all his property in Iceland, to America & had it not been for the arrival of the *Elbe*[6] this spring the whole would have been gone – Stephensen & Savigniac then locked up the warehouse doors & declared the property theirs – Liston[7] by the advice of Sir George

---

[1] Banks is referring to Hooker's book *A Tour in Iceland in the Summer of 1809*, in which Hooker made use of Banks's notes from his voyage to Iceland in 1772, as is clear in previous letters. Banks appears to be very sensitive at his name appearing, perhaps because of his conscience. After all he played not an insignificant part in facilitating the 'revolutionaries' expedition to Iceland.

There is a draft of the first paragraph of his reply scribbled on the 20 June 1810 letter from Hooker (Document 161), dated 24 June: 'I See no Objection to your using any part of the information as Coming from me unles it is of a nature that Renders the disclosure of the Source from whence it has been Obtaind is necessary it will be the more agreable to me the more it is done in notes where my name must come forward | The appendix is next then the Text I conclude my name need not appear'

[2] Banks is referring to Phelps (the merchant) and Jörgensen (the usurper).

[3] Savignac first arrived in Iceland in January 1809 on board the *Clarence* and was an active participant in the Icelandic Revolution.

[4] Westy Petræus: see Appendix 13.

[5] Magnús Stephensen had been honoured by the king with the title *etatsraad* (see Glossary) in August 1808. The British titled him Tatsroed.

[6] A ship sent by Phelps, carrying the Mackenzie Expedition.

[7] John Liston was the captain of the *Elbe*.

Mackenzie & Mr. Fell[1] landed his armed Crew & broke open the doors – Savigniac headed a great number of Danes & /Stephensen/ said they should be cut in pieces – but they did not shed blood. Stephensen added that Phelps should not trade nor ship an ounce of Goods from the Island – Mr. Fell (a gentleman who accompanied Sir George) traded & shipped goods during this bustle – at that time an order by the vessel from Liverpool commanded the dismissal of Stephenson,[2] who was appointed temporary Governor, & three other persons where [*sic*] charged with the management of the government – these are steady friends to Sir George & Mr Fell, & on the *Elbe*'s leaving Iceland every thing was in a quiet state.[3]

All these matters shew the necessity there is for the presence of a consul,[4] by whose interference things may be likely to continue in a peacable condition. I had forgot to say that Mr Phelps seems to have had this information from Sir George Mackenzie.[5]

A Vessel is now about to return to Iceland[6] & by her I intend sending out some books as presents to the Bishop, Tatsroed &c. which I have had some time by me for that purpose. You will I dare say recollect my telling you that Tatsroed Stephensen gave into my charge two books for you, which treated on Icelandic History.[7] Altho' I had the misfortune to lose them with the rest of my things, you will still perhaps write a few lines to the Tatsroed mentioning that you heard from me of the intended present – by so doing you would highly gratify him. I can enclose the letter in my parcel, which will go in about a week.

Your wishes with regard my making use of your Mss shall be faithfully attended to.

I have the honor to remain
Dear Sir
Your very sincere & obedient Serva*n*t
[*Signed*] W. J. Hooker

Halesworth Suffolk not near <u>Yarmouth</u>[8]

Kew, Banks Correspondence, II, JBK/1/7 f. 335. Autograph letter.[9]
NHM, DTC 18, ff. 48–49. Contemporary copy.
Hermannsson, 'Banks and Iceland', p. 66.

---

[1] Michael Fell, an employee of Phelps & Co. and not a member of Mackenzie's expedition, though they both travelled on the *Elbe*.

[2] This was a letter from the vindictive Trampe stripping the Stephensen brothers of their powers, conferred on them in the Agreement of 22 August 1809. Instead, a triumvirate of Stefán Þórarinsson, the district governor of Northern and Eastern Iceland, Ísleifur Einarsson, the first assessor of the High Court of Justice, and Frydensberg, the treasurer (*landfógeti*), were appointed to govern the country until the arrival of a new governor, Johan C. T. v. Castenschiold, appointed later that year.

[3] See Document 154. Holland offers a lively account in his *Iceland Journal* remarking 'The whole of the transactions displayed in a striking light the feebleness & inefficacy of the government of the island.' Mackenzie proved instrumental to restoring 'tranquillity' to Reykjavík (pp. 101–2, 105). Papers regarding this case are in NAI, Jörundarskjöl.

[4] John Parke: see Appendix 13.

[5] Not surprisingly, as Mackenzie took on the role of mediator between the warring factions.

[6] Probably the *Rennthier*.

[7] One of them was Stephensen's *Eptirmæli Átjándu Aldar eptir Krists híngad-burd, frá Ey=konunni Islandi*, mentioned under its Danish title in Document 122, p. 331. This is Stephensen's history of the 18th century in Iceland. However, as it was published in two editions that same year with different page-numbering this may be the two books mentioned. There are copies in the British Library (both Icelandic and Danish versions) but neither are stamped as coming from Banks's library.

[8] This note correcting Banks's address on a previous letter is omitted in the DTC copy.

[9] Note in Banks's autograph, in left-hand corner of first page: '[Received] July 23 [1810].'

164. Banks to William Jackson Hooker, 26 July [1810]

Soho Square

My dear Sir

I thank you for the notices you Give me of Transactions in Iceland which are very interesting, if I understand the Question which as the Statement Comes from P–[1] I am not sure I do. I conclude that Savignac & Co. meant to have appropriated the Goods Left in their charge to their own use But it does not appear from yours what vessel Liston had the Command of I conclude it was one Sent out by Phelps[2]

The Tatsroed you mention I take to be the deposd Son of my old Friend[3] I do not Feel much inclination to write to him,[4] but if you think I ought I will & can send you my Letter by return of Post when yours is Receivd

Sir Geo*rge* [Steuart Mackenzie] writes me a Short Letter[5] in which he says that he has undertaken the office of mediator between the Icelanders & the Phelpians & that the Claims of the former for Plundering during the Phelpian Revolution are on the Point of being Amicably & Finaly settled unless Phelps says something to give on that head it may be better not to mention it to him

Sir Geo is Living at the Government house & seems to occupy himself much in Iceland affairs I hope he will do well

Faithfully yours
[*Signed*] Jos: Banks

[P.S.] Sir Geo says
I saw some plants Collected for Mr Hooker in very indifferent order we prepare to collect as many as we can for him tho' I fear we shall do but little being fully occupied by mineralogy

Kew, Hooker's Correspondence, Director's Correspondence, I, f. 42. Autograph letter.
Hermannsson, 'Banks and Iceland', p. 67.

165. Banks to Grímur Jónsson Thorkelín, 26 July 1810[6]

Soho Square

My Dear Sir

allow me to Recall myself to your Recollection by a Request which I am sure you will be inclind to comply with as I have had so many opportunities of knowing the Liberality & Kindness of your disposition

---

[1] Samuel Phelps: see Appendix 13.
[2] This was the *Elbe*.
[3] Magnús Stephensen, Chief Justice of Iceland, son of Ólafur Stephensen: see Appendix 13.
[4] See Hooker's request in previous letter.
[5] Document 154.
[6] Banks addressed him as 'Mr Thorkelin Professor of History and Antiquities Copenhagen'.

it is now Some years ago Since you observd in the Privy archives of your Country of which you are Keeper many records relative to the Earl of Bothwell[1] & others of the Scots nobility in the Reigns of James Mary & the Regent[2]

as matters of history these are interesting to us & particularly to my friend M^r. Chalmers[3] to whose pen history is alreedy so deeply indebted

if my dear Sir you can indulge me with Copies of as many of these papers as you think interesting to the illustration of the history of the interesting times when they were written & with some account of the nature of the Remainder, in order that we may judge whether any of them will also be usefull; you will confer a Great Obligation on me, the Expence of Transcribing I shall hold myself accountable for in whatever monney you will permit me to Remit it to you

I am my dear Sir
With much regard & Esteem
Your most Faithfull
Humble Servant
[*Signed*] Jos: Banks

Edinburgh, University Library, La.III.379.f.25. Autograph letter.

166. Banks to [Culling Charles Smith], 26 July [18]10

Soho Square

My dear Sir

I conclude that you have receivd intelligence from Iceland sufficient to shew the great want there has been of the Establishment of a British Authority, in Some Shape there, & I also conclude that the Consul has before this time arrivd in the Island & occupied his Post[4]

Sir Geo[rge Steuart] Mackenzie a young man of more than ordinary Talents who visits the Island with Scientific views, has, as he tells me, at the desire of the Icelanders, undertaken the office of a mediator for settling the claims of those who were Plunderd

---

[1] The Earl of Bothwell (James Hepburn, the 4th Earl) was the third husband of Mary Queen of Scots. After Mary surrendered to the Scottish nobles at Carberry Hill, Bothwell escaped to Denmark where he was imprisoned in appalling conditions for 11 years, dying insane in 1578.

[2] James I (VI of Scotland), Mary Queen of Scots and 'the regent' probably refers to James, Earl of Arran, who ruled as Lord Governor, the Scottish equivalent of regent, Mary of Guise having failed to become regent to her daughter. Arran was head of the powerful house of Hamilton and heir to the throne should the few-weeks-old Mary not survive.

[3] George Chalmers (1742–1825), FRS, clerk to the Board of Trade, a man of letters and a frequent correspondent of Banks's from 1792 to 1818. They often discussed historical subjects: for instance Banks asked him for information on the internment of Charles I, the condition of Henry VIII's coffin (when the vault of St George's Chapel was opened in 1696) and the mistresses of Charles II. In 1818 Chalmers sent Banks a 'volcanic present' (Dawson, *The Banks Letters*, pp. 210–11). Chalmers was the author of various works including the three-volume *Caledonia: or an account, historical and topographic, of North Britain ... with a Dictionary* published in 1807–24 and *The Life of Mary Queen of Scots* published in 1818.

[4] John Parke finally reached his destination in July 1811.

during the Phelpian Revolution 'all of which are on the Point of being amicably & Finaly settled' by him as he tells me[1]

I learn from another Quarter[2] that a battle had nearly taken Place between Phelps's Resident agents aided by the Danes & free Burghers & the Crew of his Ship which lately arrivd but that no blood was Spilt

No Steftsamptman or Governor appears yet to have arrivd from denmark[3] but the Persons Provisionaly intrusted with the Government[4] were Delighted with the Favorable provisions allowd to the Island by England & traded freely with our People with the Exception of Phelps' Resident agents

I have a Letter from Count Trampe at Copenhagen on the Subject of an Exchange of Prisoners,[5] a business which the higher powers will know to Conduct, so I say nothing on the Subject, if you have any Iceland news which you think I ought to know I am sure you will be so good as to order it to be Communicated to me

I am my dear sir
most humbly
& most Faithfully yours
[*Signed*] Jos: Banks.[6]

TNA, FO 40/1, ff. 53–54. Autograph letter.

167. William Jackson Hooker to Banks, 27 July 1810

Halesworth

My dear Sir

I was so anxious to give you the earliest possible information respecting Icelandic affairs, that I wrote my last letter[7] to you in too great a hurry & I fear did not explain myself as I ought to have done.

All that I know of the transactions I received the very day I wrote to you in a letter from Mr Phelps, who seems to be in excellent spirits & speaks in the highest possible terms of Sir George Mackenzie, without whose assistance & interference it appears most plainly that the 'Phelpian'[8] affairs would be in a more unfortunate state than ever. The

---

[1] See Document 154.

[2] Namely William Jackson Hooker (see Document 163).

[3] Castenschiold arrived in the summer of 1810 but only appointed as district governor in the south of Iceland, but this news would not have reached Banks. He did not formally become governor until 1813. See Appendix 13.

[4] See Document 113. On 5 July 1811 Parke presented his credentials to Castenschiold.

[5] During his stay in London Trampe worked diligently at gaining the release of Danish prisoners in England.

[6] Précis on back of document: 'Sir Joseph Banks | D[elivered]/R[eplied] July 26 1810 | The want of a | British Agent in | Iceland – threatened | conflict between | Phelps' Resident | agent, aided by the Danes | & free Burghers | and the crew of Phelps' | ship which had lately | arrived at Iceland. | Sir G. Mackenzie | had undertaken to | settle the former | differences'.

[7] See Document 164.

[8] Underlined by Hooker.

Tatsroed[1] is, as you suspect, the son of your old & truly faithful friend [Ólafur] Stephensen – a Man of greater talents or a more artful politician I never met with. Yet the advancement of his own family & the aggrandizing of riches are I fear his greatest objects. I am well aware, & so I should think was Count Trampe (tho' I am sure he had not the information from me) that there was not a single man in the whole country who so truly rejoiced at the deposition of the lawful Governor than the Tatsroed, hoping by that means to have possession of the Government himself.[2] His great learning, however, & his knowledge of the laws of the Country induced the Count, Cap*tai*n Jones & Mr Phelps to nominate him to be head of the affairs untill the will of our Government was known, or untill the release of Count Trampe. How well he has conducted himself during that period, notwithstanding he was bound to protect[3] the English goods there, may be seen by the information Sir George sends over, namely that he & Savigniac were shipping off all the English property to America.[4] Such a man you cannot wish to correspond with & I think there will be no necessity for you to write to him about the books, which I had told him shared the fate of my own collections. From the very great civility I received from the Tatsroed & his family I think myself bound to make them some acknowledgement in return; which is the only reason of my correspondence. I think I may very fairly ask him to write to me how affairs have been going on since I was there – We shall then be able to say 'audivimus alteram partem'.[5] With regard to Savigniac I have always suspected his honesty & Count Trampe readily acknowledged that he thought him at the bottom of all[6] the mischief that took place.[7]

I have given you I believe one wrong piece of information when I told you that Mr Fell went out with Sir George Mackenzie; he went on account of Mr Phelps, to superintend the affairs. I shall not, nor do I think there is any necessity for me to tell Phelps of anything which I hear from you or from any other channel, unless I think it would be serviceable to him, consistently with what was right.

I must, Sir Joseph, once more crave your indulgence, whilst I bring forward the name of Jorgensen.[8] I am fully sensible that you have done for him more than could reasonably

---

[1] He is referring to Magnús Stephensen (see Glossary).

[2] It was generally assumed that Stephensen wanted to follow in his father's footsteps.

[3] Underlined by Hooker.

[4] Another American ship had arrived in Iceland, the *Vigilant* from Boston, belonging to Edward Cruft. The ship had been sent at the urging of the Danish consul in North America, Pedersen. The trading venture had been successful and the following year Cruft had sent another ship, which was captured by the British on its return journey from Iceland ('Kongelig Resolution ang. Tilladelse for Kjöbmand Cruft af Boston, at handle paa Island', dated 22 March 1815, *Lovsamling for Island*, VII, pp. 539–41). It might be mentioned that in August 1809 an American ship *The Neptune and Providence* had arrived from Providence, Rhode Island. Magnús Stephensen had made arrangements for it to sail during his stay in Norway during the winter of 1809 and had prudently asked Count Trampe to agree to this venture, which he had (see Agnarsdóttir, 'Great Britain and Iceland', pp. 117–18).

[5] Quotation from St Augustine, *De duabus animabus contra Manicheas*, chap. 14, meaning 'hear the other side', more familiarly *Audi partem alterum*. That is to say: we shall be able to say we have listened to both sides [in this matter].

[6] Underlined by Hooker.

[7] Hermannsson was certainly of this opinion and in no doubt that Savignac was 'the real villain in the play' ('Banks and Iceland', p. 69). Savignac is a shadowy figure and little is known about him (see Document 156).

[8] Hooker kept up a life-long friendship with Jörgensen. His letters to Hooker are preserved in BL, Eg. MS 2070.

be expected & more too than he perhaps deserves. I think too from some hints that have escaped you, that you pity him in his present situation & that you would not object to hear of his immediate release; but at the same time that it would be neither consistent nor proper for you[1] to do any more for him. Under this idea I have written to a friend in Town, to ask him whether he thought his speedy release might be procured without making a bustle, at the same time consistently with justice.[2] To this I have yet received no answer, nor I [*sic*] shall I proceed further till I hear from you whether or not you object to my doing so, for I should not only be vexed but ashamed to interest myself in an affair, which has occupied a great deal of your time & attention, without your knowledge & approbation. I feel my situation with regard to Jorgensen a very peculiar one – His exertions on board the *Orion* most undoubtedly saved my life & those of the rest of the passengers of *Mary & Ann* [see Plate 15][3] besides that his pleasant manners & goodness of heart have excited in me a friendship for him which I should be glad to make use of in his behalf, but which I will never do if contrary to your wishes, for that would be contrary to reason & justice.

I remain
Dear Sir,
most faithfully
your sincere & obliged
[*Signed*] W. J. Hooker

P.S. I think it better to send you Phelps's letter.[4] My plants I expect to receive daily.

Kew, Banks Correspondence, II, JBK/1/7 f. 336. Autograph letter.[5]
NHM, DTC 18, ff. 50–52.
Hermannsson, 'Banks and Iceland', pp. 63, 67–8.

168. Banks to [Culling Charles Smith?], 28 July 1810

Spring Grove

My Dear Sir

I Enclose the Two Letters[6] I have Receivd the one from Iceland the other from a gentleman who visited that Island Last year in Phelps's Ship, as Sir Geo*rge* Mackenzie is a young man of Some Talent & a good disposition and as the affairs of the Country Respecting England appear to have been placd in his hands by the People, I judge it by no means necessary that any Extraordinary means should be Resorted to hasten the Consuls[7]

---

[1] Underlined by Banks.

[2] Sarah Bakewell, Jörgensen's biographer, suggested this was 'a release engineered by Macleay [*sic*] and Hooker with the tacit approval of Banks' (*The English Dane*, p. 127).

[3] A slip of the pen – the *Margaret and Ann* is intended.

[4] This letter has not yet been found among the Banks Papers.

[5] Note in Banks's autograph, margin, bottom of first page of the letter: '[Received] July 29 [Answered] 30.'

[6] These two letters must be one from Hooker, probably the previous letter, and another from Mackenzie, though copies are not to be found in TNA, FO 40/1, where they should have been preserved.

[7] John Parke: see Appendix 13.

departure, I Conclude however that the Ship of Buske¹ now at Leith, I believe, to which we granted a Licence a fortnight ago & which is bound to Iceland direct will sail very soon indeed if she has not Saild already

Believe me my dear Sir
your Faithfull Serv*an*t.
[*Signed*] Jos: Banks

TNA, FO 40/1, ff. 59–60. Autograph letter.²

169. From Jens Andreas Wulff, 30 July 1810

On board of the *Freden*

Sir,
Your generous Exertions to promote the general Wellfare of Iceland, and the personal Kindness you have shewn to me individually, I shall ever entertain a grateful Remembrance of.

I take the liberty to acquaint you with the Departure of the Ship *Freden*³ from Denmark & Leith for Iceland with a Cargo of Corn and other Necessaries in lieu of the cargo which was unfortunately lost last year; I could not accomplish this Shipment sooner, from the length of time necessarily occupied in repairing the Vessel, and from the great and various difficulties which at this period interupt all commercial Affairs; tho' the King of Denmark has been pleased to grant every facility consistent with present Circumstances.

There are several Persons totally unconnected with Iceland, who under assumed names are pretending to dispatch Vessels thither they obtain British Licences for the Importation of Corn, and then petition the Privy Council for Licences allowing them to proceed to Iceland.

The liberal Treatment which Iceland and its Merchants have been favored with under the present Circumstances, demands that your kind Protection should not be abused;⁴ I

---

¹ A ship owned by Jens Lassen Busch, probably *De Jonge Goose* (or possibly *Helleflynderen*) which received a licence on 13 July 1810 (TNA, PC 2/188, f. 141). There is a blank for the ship's name in the document.

² Note, verso:
Sir Joseph Banks
Dated/July 28th } 1810
Received/August 4th
Answered/ – 7th –
Enclosing two Letters relative | to Iceland by which it appears | not to be absolutely necess | -ary for a ship to be imme | -diately prepared to convey | Mʳ. Parke to that Island.

³ All the underlinings in this letter are in pencil, presumably made by Banks.

⁴ Abuses attending the use of licences were frequent. As Earl Bathurst pointed out in the House of Lords in 1810, it was impossible to provide against many of the irregularities (*Cobbett's Parliamentary Debates, House of Lords*, London, 1810, XVII, p. 168). The Iceland merchants were no exception to this rule, and chanced a profitable side-trip across the Sound to Norway, then sailing quickly back to Copenhagen to pick up a fresh cargo under shelter of their licences (numerous sources attest to this including Holland, *Iceland Journal*, p. 305; Mackenzie, *Travels in Iceland*, p. 73; and, BL, Eg. MS 2070, Jörgensen to Hooker, 23 November 1810). Often, however, the Iceland ships received permission to sail to Norway to load timber. One of the newly appointed British consul's first actions on arrival in Iceland in 1811 was to send a report describing licence abuse to the Foreign Office (see Document 203). A similar report to Document 170 was sent to Wellesley (TNA, FO 40/2). Thus both Banks and the Foreign Secretary were kept up to date on what was happening in Iceland.

therefore thought it my duty to state this Circumstance for your Consideration, recollecting that in order to prevent such Abuse you had during my stay in London required a List of all the Ships which were real bona fide Iceland Traders.

I am most respectfully
Sir
Your very humble Servant
[*Signed*] J: A: Wulff

SL, Banks MS I, 2:10. Contemporary English translation of the Danish original which is SL, Banks Papers, I 2:9.
Erhvervsarkivet, Århus, Denmark, Købmandsfirmaet Ørum & Wulff, København, 1805–12, kopibog, ff. 187–190. Copy.

170. Report from Corbett, Borthwick & Co.

The *Rennthier*, Hans Thielsen Master from Iceland, 3 September 1810.

Leith

In bulk, about 12 Tons Stockfish
In bulk & 1 Matt[1] about 8 tons 11 Cw^{t2} Salted fish
98 Bags, about 14.$^{T(ons)}$ 5 ^{C3}. 2 lb Sheeps Wool, some of the Bags opened
14 Barrels & Casks 1.^{Ton}10^C. 1 lb Tallow
15½ Casks Fish Oil
13536. pair worsted Mittens
2540. pair ditto Stockings
249. woolin Night Shirts
9554. Swan quills
72. Yards Woolen Stuff
2. lb worsted yarn   } in 53 Bags 1 barrel 1 Sack
91. Fox skins   1 Box & 1 Seal Skin & in Bulk
850. Lamb Skins in the wool
17. Swan Skins
13. Bags about 490 t Eyder Down
3. Bags about 111 lb Bedfeathers
180. lb Cod Sounds[4]
5. Chests Wearing Apparel

---

[1] *Matt*, an alternative spelling of 'mat' which in this context refers to a matt of salted fish, meaning a bag or sack made of matting filled with the fish.
[2] Hundredweight.
[3] Hundredweight.
[4] A fish sound was the swim bladder of fishes, especially those that are dried and used as food, or for the preparation of isinglass. Dried swim bladders of fish form a kind of collagen used mainly for the clarification of wine and beer. Isinglass was originally made exclusively from sturgeon, until the 1795 invention by William Murdoch of a cheap substitute using cod. This was extensively used in Britain in place of Russian isinglass. Thus the Icelandic stockfish would have been of use in this process.

1. Chest of Drawers & a large Trunk belonging to Count Trampe supposed to contain his & his Lady's[1] Cloaths.[2] Silver plate &c to be forwarded to him in Norway.[3]
40. Empty Sacks

Ship Stores

20 lb Sugar	20 lb. Tobacco
4 „ Tea	160. lb Salted fish
10 „ Coffee	50 lb Candles
12 „ Gall^s Corn Brandy	1½ Anker[4] Fish Oil
10 „ Bottles Wine	1½ fathoms fire wood
	6 Deals

(Signed) Hans Thielsen
Corbett, Borthwick & Co.

NHM, BC, f. 58. Contemporary copy.

171.　Rooke & Horneman to Banks, 6 September 1810

Queen Street, Cheapside

Right Honorable Sir

Your Letter to Sir George Mackenzie[5] was forwarded from Leith to Iceland without any Loss of time there being an immediate opportunity for that purpose, otherwise we should with pleasure have delivered it to Mr. Parke,[6] who did us the honor to call upon us.

We /have/ received papers from Denmark which prove the perfect protections Vessels engaged in the Iceland trade enjoy, & our H*ans* F*rederik* H*orneman* took the Liberty of calling at your House in town with an intention of Laying them before you. –

Mr. Sivertsens Ship *The Two Sisters* which was detained at Hull, arrived on the 18th Aug*us*t safe at Copenhagen.[7] –

---

[1] Trampe was thrice married. Trampe had first arrived in Iceland in 1804 and his wife may have accompanied him then though there are no records of her being in Iceland. See Appendix 13.

[2] Presumably a variant of 'clothes'.

[3] Trampe was appointed the governor of Trondheim on 23 July 1810.

[4] (Cask containing) quantity of about 8 gallons.

[5] Mackenzie's papers have not been preserved, but as Mackenzie left Iceland on 19 August, Banks's letter would not have reached him – in Iceland at least.

[6] John Parke, the British consul to Iceland.

[7] At the beginning of March 1810 *De Tvende Söstre* was granted a licence to export a cargo of salt & tobacco etc. from Liverpool to Norway or Sweden and return with a cargo of permitted goods to any port in the United Kingdom. But it was stipulated the ship must apply for a further licence to Iceland on her return (TNA, BT 6/198, no. 15,667). A licence was then granted for the ship to sail from Hull to Copenhagen and sail in ballast to a port in the Baltic or Norway to take in cargo for Iceland and return from there to the United Kingdom (TNA, BT 6/200, no. 23,664). No Iceland licence was issued in Copenhagen so it is clear that Sívertsen's ship never sailed to Iceland in 1810.

A. W. Aaröe[1] is arrived with a Vessel at Leith /from Denmark/ of which an Icelander of the name of Gudmund Petersen[2] is owner & a Burgher and inhabitant of the Island of Iceland. – The Ships name is *Maria Bonaventura*.[3] Mr. Petersen is at the same time a Magistrate (Sysselman) & owner of Kroseveig[4] and Wapnefiords Facture[5] at Iceland. – He intended to have gone to Iceland this Year if he could have procured a Licence for that purpose & if the Season which is so far advanced would have allowed it, but that being the case he intends only to sell a small part of his Cargo to defray his expences, & to proceed to Norway, from thence to Denmark, but next Spring in the Month of April he wishes to proceed with the same Vessel from Denmark to Leith & from thence to Iceland. – He wishes now if possible to procure the necessary Licence for that purpose & he has requested us to apply to you in his behalf as he deems it impossible to obtain it mearly applying at the Council Office without Your powerful Intercession in his behalf. – The poor Man has no less than Sixteen Children who with his Wife live in great distress in Iceland, as there has been no trade at that place for some Years.[6]

He arrived at Copenhagen short*ly* before the Bombardment of the town[7] & is said to have lost all his Property in the Conflagration.

We therefore particularly submit this mans situation to Your benevolent consideration & humbly sollicit in his behalf that the Licence he wants may be granted him, as unless he can resume the intercourse with his native Land, under the protection of this Government, he will be destitute at providing for his numerous family. –

Two Icelanders of the name of Thielsen & Thomsen[8] have also applied to us in order to assist them to recover their Losses sustained by the depredations committed by the privateer *Margrethe & Ann*, John Liston Commander belonging to Messrs. Phelps & Co. – We would deem ourselves much obliged to you if you have the goodness to inform us if there is any chance of recovering any of their Losses by Law, Iceland being then considered as an Enemies Country.[9]

---

[1] Unidentified.

[2] Guðmundur Pétursson (1748–1811), county magistrate in Norður-Múlasýsla 1786–1807. He had resigned his post in 1807, going to Copenhagen where he stayed from 1807 to 1810. The summer of 1810 he sailed to Leith where he spent the winter. The following summer he visited London when the Iceland ships were getting ready to sail. On returning to Leith, he apparently fell in a Leith street (sources say he was fond of his drink) and broke his leg, dying a couple of days later on 11 August 1811. He was buried in Leith following a respectable funeral paid for by some admiring Scots.

[3] The *Maria Bonaventura* went on to Copenhagen arriving back in Leith on 30 October 1810 (Edinburgh, National Archives of Scotland [hereafter NAS], E[xchequer] 504/22/48). On 13 August 1810 Pétursson had obtained a licence for the ship in Copenhagen (RA, Rtk 373.133, ff. 86–87) but there are no more records regarding this ship.

[4] Krossavík was where Pétursson lived.

[5] He was involved in trade matters but Ørum & Wulff were the merchants at Vopnafjörður at this time.

[6] According to Icelandic sources he had 11, but others may have died young. During his stay in Copenhagen he was joined by two of his sons (Stephensen, *Ferðadagbækur*, p. 95).

[7] He left Iceland in 1807 on the *Havkalven* from Vopnafjörður in north-eastern Iceland. Due to contrary winds the ship sailed to Norway and there he learned of the war. In the spring of 1808 the ship continued its journey safely to Copenhagen.

[8] The merchants Hans Thielsen (see Appendix 13) and Jesse Thomsen.

[9] Trampe and Phelps agreed to settle their dispute by a compromise. The law officers had made it clear that, even though the Danish owners could go to the Prize Court and obtain restitution, it would be a costly and difficult process. Phelps's agent, Michael Fell, reached an agreement with the Icelanders and some of the Danish

His Excellency Count Trampe we find is appointed Governor of the province of Dronton[1] in Norway.

> We have the honor to be
> Right Honorable Sir
> Your most obedient humble
> Servants
> [*Signed*] Rooke & Horneman

NHM, BC, ff. 54–55. Autograph letter.

172. John Christopher Preidel to Banks, 11 September 1810

No. 13 Coleman Street, City, London

The Right Honourable Sir Joseph Banks. K*nt* B. etc. etc.
Sir!
At the Request of Captain Hans Theilsen of the Iceland ship *Rennthier*,[2] who presents his profound Respects to you [I] beg leave to address these Lines to You.

In November 1808 You was so kind to procure him a Licence from His Majesty to proceed with his said Ship from Yarmouth in Ballast to a Danish Port, from whence to proceed with that Ship or any other not exceeding 120 Commercial Lasts to Iceland with a Cargo of Provisions & other necessaries of Life for the Support of the Inhabitants of that Island, & to return with a Cargo of permited Goods to a British Port.

Having obtained said Licence he went to Yarmouth to fit her out for the Voyage which from the bad Condition she was in on account of the long Detention there, took up a Long time, he arrived at last in March 1809 at Flensburg, when he came there it was found necessary that she should have a New Bottom & other considerable Repairs, which not only took up a considerable time, cost him £600 & precluded him from a Voyage that year to Iceland, as he was not able to procure another Ship. Last spring just as he was fitting her out for that Voyage the news arrived there that a Danish Captain a most infamous Rogue & Vagabond[3] under the Production & by Orders an Owner of a Privateer here in London had confiscated & robbed his Establishments at Iceland of Goods to the Amount of *Rixdollars* 7305.281.[4] – at last he set sail for Iceland with a Cargo of Rye & other Necessaries of Life & arrived safe there & conform to his Licence brought over a Cargo consisting of various Goods the Produce & Manufacture of the Island, & arrived safe therewith at Leith in Scotland the Beginning of this month. On

---

merchants to make good on Phelps's account property confiscated in 1809. As has been seen, Sir George Mackenzie had helped Fell as mediator (Document 154) and by July Banks had been able to inform the Foreign Office that all the claims were 'on the point of being amicably and finally settled' (Document 164). On the whole the wronged Icelanders were satisfied. Only one, Jesse Thomsen, was not, the case dragging on unsuccessfully until 1829. The papers on his case are in RA, Rtk 373.133; and see RA, Island og Færöer, DfdUA to the King of Denmark, 24 February 1829.

[1] Drontheim, the contemporary English spelling for Trondheim.

[2] Preidel always spells Thielsen 'Theilsen' and Rennthier 'Renntheir'.

[3] He is referring to Jörgen Jörgensen.

[4] Jörgensen confiscated all goods he considered to be Danish property, among them goods belonging to Thielsen. There is, however, no mention of Thielsen in Trampe's memorial listing all Jörgensen's misdeeds.

his Arrival there he made his report at the Custom house of his Cargo – the particulars of which is inclosed, being told that the Cargo must be unladen there, he came up to Town & requested me to present a Petition to the Right Hon*ourable* the Lords of His Majestys Privy Council praying that the unlivery of the Cargo may not /be/ insisted on & that a New Licence may be granted him to carry the same to a Danish Port &c. &c. as You will see Sir from the inclosed Draft of a Petition I have drawn up.[1] Which you will have the Goodness to peruse & if it is to your approbation will then present it and if you would at the same time have the Kindness to recommend it to their Lordships, for which Theilsen & myself would be infinitly thankfull; the Success would be certain. The Ground of his Prayer that the Goods may not be unladen at Leith are as follows.

1^{ly}. The Prices of those Goods are now so low in England that if they were sold here would be attended with a Loss of at least £800.

2^{ly}. that the Goods on being unladen at Leith would greatly suffer in their Quality, particularly the Fish & Wool, for should it rain at the time of unlading & relading & the wool get wet it would heat, and the same misfortune might befall the *Renntheir* as did last Year to a Vessel of M^r· Phelps the Owner of the Privateer who commited such Depredations in Iceland, which from the same Cause took fire at Sea & was totaly lost.[2] That besides the unlading and reloading at Leith would in Case of rainy Weather be attended with not only Injury to the Goods, but also with Loss of time, so that the Expedition next Spring to Iceland would be greatly retarded.[3]

You are therefore once more[4] requested to interest Yourself in Behalf of Captain Theilsen & to favor me with a few Lines in Reply to this.

I hope Sir this will find you in good Health & that you may have remained since I had the Honor of seeing you last free from that painfull malady the Gout. I had a most violent Fit of it a fortnight ago, and thanks to you, for without it I should not have any Knowledge of it, by taking the Eau Medicinal[5] got rid of it in less than a Week. What I have ordered for you from Paris is not yet arrived; but Thynin is amply provided & has raised the price to 10/6.

I am most respectfully
Sir!
Your most obed*ient* & humble Ser*van*t J. C. Preidel

NHM, BC, ff. 56–57. Autograph letter.

[1] The following document.

[2] The ship lost was the *Margaret and Ann*. Overheating of wool was blamed.

[3] Thielsen's petition to export a cargo of wool, stockfish etc. from Leith to Denmark and to return with a cargo of corn, building materials etc. to any port in Britain was granted at the beginning of October (TNA, BT 6/201, no. 26,088). His request regarding not having to unload the cargo was, however, refused (TNA, BT 6/201, no. 26,387). He left for Copenhagen in mid-November 1810 (NAS, E. 504/22/48).

[4] This was not the first time Banks had helped Thielsen (see Document 54).

[5] Extract of Colchin was used by Banks as medicine for gout with success. See Ring, *Treatise of the Gout ... and Observations on the Eau Medicinale,* published in 1811. For Banks's 'gout history', see Carter, *Sir Joseph Ban*ks, esp. pp. 524–35.

173. Petition of John Christopher Preidel on behalf of Hans Thielsen[1] to the Lords of the Privy Council, Sept*ember* [*blank*] 1810

No. 13 Coleman Street, City

To the Right Hon*oura*ble The Lords of His Majesty's most Honorable Privy Council The humble Petition[2] of John Christopher Preidel of London Merchant on Behalf of Hans Theilsen Burger of Iceland & Master of the Iceland Ship *Rennthier* Sheweth

That the above said Master by Virtue of His Majesty's Licence granted him the 24 November 1808. Copy of which is hereunto annexed sailed last Spring from Flensburg to Iceland with a Cargo of Provisions & other necessary Articles for the Support of the Inhabitants, & in Return brought back from thence a Variety of Articles the produce & manufacture of that Island, and arrived with his Ship the Beginning of this Month at Leith in Scotland, where he delivered his Licence to the Collecter [*sic*] of His Majesty's Customs and made a Report of his Cargo as per annexed Copy of all the Goods he has on Board which Report he Saith is faithfull & true, that the whole is of the Produce of the Island, safe & except some Wearing Apparel & Boxes &c. belonging to Count Trampe & that he will confirm the same upon Oath if required. Your Petitioner now humbly prays (that as the Goods he brought over, on account of the low prices in England would be attended with a Loss to him of at least £800 if he were obliged to sell them here, and if unladen at Leith, particularly the Fish & the wool would greatly suffer in their Quality, & more so if at the time of unlading and relading the Weather should be rainy The Wool on getting Wet, would heat, and the same Misfortune might befall the *Rennthier*, which befell last year to one of Mr. Philp's Ships[3] which from the Wool being heated took Fire & was totaly lost) that your Lordships will be most graciously pleased to grant him a New Licence permitting him to proceed with his Ship & the present Cargo on Board (without being obliged to unload the same at Leith) to a Danish Port – and after haveing [*sic*] arrived there and delivered his Cargo to be permitted to go in Ballast to Flensburg and there next Spring take on Board of his Ship the *Rennthier* of 44 Commercial Lasts Burthen or on Board of any other Danish Ships not exceeding 120 Commercial Lasts.[4] One half of her Cargo in Provisions & other necessary Articles of Life for the Inhabitants of Iceland, with Liberty to touch in Norway to compleat her lading in Wool Timber Iron Tar &c. to enable them to carry on their Fisheries & to repair their Houses, of which they are now in the greatest Want. As also to be permitted to return from Iceland to a British Port with Produce of that Island and in Case the same could not be sold here without a Loss to be permitted to proceed with the same to [a] Danish Port, and your Petitioner as in Duty bound will ever pray.

J. C. Preidel

NHM, BC, f. 59. Contemporary copy.

---

[1] Actually spelt Theilsen in document.
[2] The original is not extant.
[3] He is referring to Phelps's ship the *Margaret and Ann*.
[4] A commercial last (*kommercelæst*) equalled 2.6 tons.

174. Hans Thielsen to the Right Honorable the Lords Commissioners of the Board of Trade, 11 October 1810

8 Queen Street, Cheapside

Right Honorable Sirs

The Lords of His Majesty's Most Honorable Privy Council having on the 3ᵈ instant[1] granted me a Licence to export on board my Ship the *Rennthier* sundry Goods the produce and Manufacture of The Island of Iceland brought to Leith on the said Ship. I humbly submit to Your Lordships some points connected with the Order in Council of 7 Febr*uary* 1810 respecting the Trade of Iceland, the *Rennthier* being the first Vessel coming from Iceland with a Cargo for export since His Majesty's said Gracious Order in Council.[2]

With deep gratitude I as well as every Icelander must ever acknowledge the generousity of these Realms, which has opened to the distressed Inhabitants new Sources of hope for their regular Supplies with the necessaries of Life. The inhabitants of Iceland procure these by taking the Produce & Manufactures of the Island to the Merchants Warehouses where they generally by barter receive corn & other necessary Articles in return as very little money is in circulation. Therefore I may be allowed to say that the Subsistence of these poor & defenceless people which is so benevolently intended by the said Order in Council chiefly depends on their being able to sell in exchange the produce of the Island & its rude Manufactures. – For what else can or have they to offer to the Merchants in return for the necessaries which they must endeavour to obtain? The merchant accepts of these and if he did not he would in vain look for any other payment. To encourage the Iceland Merchant in a Continuation of this Barter of Trade, is, if I dare venture an opinion the real Intention of His Majesty's Government. But to bring the Manufactures & produce of the Island to Great Britain with a view to dispose of them there is utterly destructive to his interest, for the Duties, particularly on Train Oil, worsted Manufactures, Woollen Stuff, worsted Yarn & Stockfish the staple commodities of Iceland, are so high as to prevent a Merchant entirely from shipping at least the 4 first named Articles.

Klipfish or Dryed Saltfish & Sound Cod do not appear to be mentioned in the Act as Articles of import or export. Your Lordships will not consider this an exaggeration & my humble request is therefore that some drawback[3] or modification might be granted on the Cargo of the *Rennthier* & the Iceland Trade in general on exporting the goods to foreign Countries. – I also beg leave to request your Lordships would allow me to pay Duty on the Cargo according to Invoice which would prevent me from unloading my Ship the *Rennthier* now at Leith & save a great expence.

I remain most respectfully etc. etc!

[*Signed*] Hans Thielsen

NHM, BC, ff. 60–61. Autograph letter.[4]

---

[1] On 3 October 1810, see Document 172.
[2] Appendix 10.
[3] A *drawback* is 'amount of excise or import duty paid back or remitted on goods exported'.
[4] Note (difficult to read), verso: 'Copie of it | Board of Trade | 1 Dec? 1810.'

175. Culling Charles Smith to Banks, 15 October 1810

Foreign Office

Dear Sir

I am desired by Lord Wellesley to transmit to you for your consideration the inclosed Letter from Sir George Mackenzie,[1] and I am to add that The Marquis Wellesley will feel much obliged to you for any suggestions or opinion on the subject of its contents.

I am,
Dear Sir,
Your very obedient
humble servant
[*Signed*] Culling Charles Smith

NHM, BC, f. 62. Autograph letter.

176. Banks on Mackenzie to [Smith?],[2] c. 15 October 1810.

Sir G*eorge* M*ackenzie* is not as I judge a deep Politician if he Conceives the King's declaration in favor of Iceland[3] a matter of Policy intended to hold out the Prospect of Ideal benefits rather than to Confer real ones.

The Tenor of this /*proclamation*/ declaration is to Place Iceland in a State of amity with G*reat* Britain & to Confer upon the inhabitants /*all the*/ more Abilities of Trading /*with G B*/ that nations at amity with /*Great Britain*/ her enjoy.

The Framers of that Instrument never meant to /*give up*/ surrender the Protection Great Britain has by Law given to the Greenland & Newfoundland fisheries how Could this have been done if intended without the authority of Parliament

The difficulty M*r* Phelps meets with in regard to the Stockfish & oil that makes part of his Cargo are Serious.[4] He as a british Subject Should not have had the imprudence to Ship such articles. The Iceland Ships from their Situation had their Burthen Lightend as the Board of Trade /*always incline to*/ are in the Custom of Letting them proceed from hence to Denmark with such parts of their Cargo as they Cannot dispose of to advantage in G*reat* Britain

---

[1] The 'inclosed' letter is not extant. However, on 21 August 1810 Mackenzie sent a memorial to the Council for Trade and Plantations (TNA, BT 1/64) and it is not unlikely that it was a copy of this letter that Banks received. Mackenzie wrote that after having spent the summer in Iceland he was now familiar with the pitiful situation in the island. He professed to be 'well aware of the humane interest which Sir Joseph Banks has taken in the fate of the poor Icelanders'. He mentioned the 'heavy, in truth prohibitous, duties' being placed on Iceland produce and the difficulties facing the Iceland merchants. He added: 'While it is thought proper not to take possession of Iceland & consider it as a part of the British Empire', he pointed out that: 'Destitute of every comfort, & every necessity of life, the Icelanders are entirely dependent on the generosity of Great Britain.' The purpose of his memorial was to offer a plan to facilitate the trade from Copenhagen to Leith and on to Iceland. It is likely that this is the plan that Banks is being asked to comment on.

[2] Probably to Culling Charles Smith for the information of the Foreign Secretary regarding the Mackenzie letter mentioned in the two previous letters.

[3] The Order in Council of 7 February 1810, Appendix 10.

[4] Oil was in direct competition to the products of the British merchants and shipowners in the Greenland fisheries (see Document 138).

I do not myself see the necessity of any Regulations Explanatory of the Kings declaration. The Case of Iceland under the benificent provisions of that instrument is Precisely the Same as that of other nations at amity as far as Relates to /*Great*/ Britain's. Every Consul /*must*/ before he takes upon him /*to enter upon*/ his office must make himself acquainted with the nature of the intercourse allowd between England & the Countrey /*The Trade of which it is his duty to Regulate*/ he is sent to but the /*proclamation*/ Declaration states if my memory does not Fail me very fully the differences between the /*State of*/ our amity with Iceland & that with independent Countries.

/*if*/ British Ships have Some advantages in the Trade over those of the /*Danes*/ Burghers of Iceland but the Latter have the advantage of carrying Klipfish & fish-oil etc. to Denmark.

/*the only consequence I can discover is that*/ The advantage preponderates on our side the Icelanders will in Time be Fed & Clothd from Britain instead of Denmark. This I think will do them no harm

The General Question of taking Possession of Iceland was argued much at Length in my Papers adressd to Lord Liverpool in which the /*business*/ whole proceedings originated /*the Cabinet*/ H M Ministers decided against the measure I of course gave up all thoughts of it tho in my opinion it would have provd an advantageous one for this Countrey

What Sir George means by restoring the Trade between Iceland & denmark & regulating it by a Convoy I do not Comprehend. A Considerable Trade between Denmark & Iceland carried on through the medium of this Countrey is allowd by Licence which in my Judgment is the only practicable method

The Icelanders now Carry their Stockfish to the Danish market under Licence & so might the English if the Danes would allow them

NHM, BC, f. 63.[1] Autograph draft.

177. Banks to Culling Charles Smith, [15?] Oct [18]10

Secret

Revesby Abbey

My dear Sir

in Obedience to your wishes and the Marquis's[2] Commands I Enclose Some Observations on Sir *George* Mackenzie's Letter the whole drift of it is in my opinion explaind by his having as is the Case applied to me Two years ago to be recommended for the Post of Governor of Iceland[3] I answerd that Government did not intend to take Possession of it had it been otherwise Sir G would not in my judgment have made So good a Governor as any one of the Junior Clerks of your office

---

[1] Written on back of a letter from Smith dated 15 October 1810 (Document 175).
[2] The Marquis Wellesley, Secretary of State for Foreign Affairs 1809–12.
[3] 'That island' was what Banks originally intended to write. He struck out 'that' and superimposed 'ice' on 'Island'. He is referring to Document 116.

his Idea of Teaching the Icelanders Mathematics Rests wholey on a wish to have a Scots professor Employd in translating some book & well paid for his Labor[1] These two things will I Conclude be found master keys to open the views of Sir George's Letters Past present & to come

NHM, BC, f. 63. Draft.

178. Petition from Holger Peter Clausen to the Right Honourable the Lords Commissioners of the Board of Trade, [29 November 1810][2]

The humble petition of Holger Peter CLAUSEN merchant in Iceland, for himself & others interested in the Iceland trade, being duly authorized for that purpose.
SHEWETH

That by His British Majesty's order in Council of the 7th of February last, it is, in consideration of the defenceless state of the Islands of Iceland, Ferroe[3] & the settlements of Greenland, and their unavoidable destruction in case they should be cut off from the supply of the necessaries of life, most graciously permitted, that their produce & manufactures, conveyed in vessels belonging to the inhabitants of any of these islands or colonies may be imported into London or Leith without being subject to seizure or Confiscation.

Notwithstanding a stronger proof could not be given of his Great Britannic Majesty's most gracious view towards the said unfortunate islands & settlements, yet the undersigned citizen and merchant of Iceland must with others concerned lament, that this instance of His Majesty's forbearance has not been productive of those beneficial effects, which they are most humbly convinced were the object of this Royal Grace.

In consequence of the said Royal Order in Council and the particular licences being granted, several Ships and cargoes are arrived in the port of Leith from Iceland, and though the expressions in His Majesty's Order in Council, that 'all goods being the growth, produce or Manufacture of the said islands and settlements etc. etc. shall not be liable to seizure & Confiscation as prize' seem to justify the hope of an unshackled trade, yet several Iceland ships with cargoes have been detained at Leith a very long time, without being permitted to unload and the reason assigned, is: that a part of their cargoes consists of klip-fish which is not permitted to be warehoused in Great Britain for exportation. However as no exception has been made of Klip-fish in His Majesty's Order in Council of 7 Febr*uary* last and the vessels that arrived at Leith last year were permitted to proceed

---

[1] On 16 February 1810 Mackenzie had written to Hooker mentioning that a Mr Jardine, lecturer in mathematics, was to have gone with him to Iceland (Kew, Hooker's Correspondence, I, f. 364, renumbered 200). This is possibly James Jardine (1776–1858), born in Dumfries, mathematician and civil engineer.

[2] The dating of this letter is derived from the abstract dated London, 29 November 1810 with notes in Banks's autograph (SL, Banks Papers, I 1:53). There is also a reference in the Board of Trade files. Clausen had sent a petition 'praying relief regarding the trade between Great Britain and Iceland'. The verdict was: 'The Laws must apply to the Articles imported from Iceland, as they do on similar articles from countries in amity with His Majesty.' (TNA, BT 6/202, no. 27,570).

[3] There was a fort in the Faroe Islands but in 1808 a British cruiser, the *Clio*, had disarmed it.

with this article to ports not blockaded, the owners conceived themselves the more secure this year from any risk of confiscation, a loss which they would suffer most innocently from, as they can in no respect be considered to have acted intentionally contrary to His Majesty's Laws: –

Another cause of the long detention of the Iceland vessels at Leith is the very /heavy/ duty that has been laid upon several of the goods; particularly upon the train-oil; a duty which will draw upon the owners the most sensible and destructive loss, which is a peculiar hardship, as when the same article was brought to Leith last year, the owners were suffered to export it on the payment of 7 £ per Ton, and the present duty demanded is 28 £ per Ton, this was the real inducement for venturing to export Train-oil this year from Iceland, in which hopes the merchants were further confirmed by His Majesty's Order in Council, which has since appeared & seemed to promise a more advantageous trade. – According to the market price of Train oil, if this high duty is to be paid, there would not be left enough behind to pay the prime cost much less the freight & expences: – By this therefore, having no other resources; the merchants would be obliged to decline accepting of the inhabitants the articles of fish and Trainoil in payment for the necessaries which they bring to the island; and as the country is poor and most sparingly gifted by nature and the commerce & Trade is nothing but a barter of commodities, His Majestys most gracious intention towards that country must be rendered ineffective. In case the merchants should be placed in the necessity of refusing to receive the fish and Train oil the island produces; for, the inhabitants, destitute of any other means of paying the value of the commodities necessary for their maintenance, must then perish in want and misery. – We flatter ourselves most humbly with the hope, that His Majesty is still moved by the same gracious compassion for the unfortunate inhabitants of that country which induced the issuing of the Order in Council of the 7th of February last; should the hope of your petitioner be disappointed, the above description of their melancholy fate must prove as true as it is deplorable.

If the Iceland merchants should be obliged to unload their cargoes here, several of the articles they contain will be greatly exposed to damage, such as the fish & the wool, which will take heat if they should get wet; besides which, great loss of time will be occasioned by the unloading and reloading of the vessels. –

Urged by this distressing circumstances, your petitioner most humbly prays that it may be permitted to the Iceland Ships to import into London or Leith all the articles of commerce which that island produces; and that it may be permitted either to dispose of them in Great Britain; or to export them under His Majestys Royal Licence and your petitioner begs leave most humbly to represent that the quantity of those articles exported /from/ Iceland does not exceed, even in the best years, 300 Tons of Trainoil and about 300 Tons of Klipfish, which is considerable less this year, about one third of the above quantity. – Here your petitioner begs to observe that it is uniformly the wish & only object of the Iceland merchants not to import the produce of the Iceland fisheries into Great Britain for home consumption, but to pay a moderate transit duty, and to export them to ports in the Baltic, where no competition with the English fisheries can possibly take place, as the high prices of Train-oil & Fish in this country prevents any, even the smallest quantity, to be shipped for the Baltic with advantage; indeed the loss would be very considerable if ever attempted.

As the ships; by the process of unloading and reloading are detained here two or three months as in the instance of the ship *Bildal*,[1] Kieldsen master, from the 15th of September and the *Rennthier*; Thielsen master, from the 31st of August both in Leith in this moment, and thereby subjected to such expences; as the trade carried on in so distant a country can not possibly bear, and besides the ships arrive here at a season of the year, when so long a delay may render the continuation of the voyage to the Baltic impracticable, owing to the ice of the Kattegat.[2] Your petitioner most humbly prays that, on the arrival of the vessels here, they may be most graciously exempted from unloading their cargoes, on paying a duty, proportioned to the tonnage of the vessels, or according to Invoice, as your Lordships may think fit and equitable for this trade.

As Iceland can not subsist without a supply of corn, fishing lines, iron, coals, timber etc. your petitioner most humbly prays that it may be permitted to export from England to Iceland these and other necessaries, particularly as a small quantity of them will be sufficient to supply the wants of the country.

As the heavy duty imposed upon Train oil could not, from what is stated before, be supposed to take place before the shipment from Iceland were made, nor, that an exception is made of salted dry fish in the Iceland produce permitted to be imported into Great Britain, your petitioner most humbly prays that the cargoes on board of the ships which are arrived here and which are still expected this year from Iceland might have the same indulgence, with respect to duty, granted as those arrived last year, and that they may be permitted with Licences to proceed with their dry salt-fish[3] without unloading the same, to some port in the Baltic not blockaded, as otherwise a most lamentable loss will be sustained by the merchants concerned herein.

The duty on wool and particularly woollen manufactures, as mittens, Stockings and Frocks is also very high considering its inferior quality, being worth only about the eighth part of Spanish wool.[4]

Finally your petitioner also ventures most humbly to pray that your Lordships would be pleased to establish certain regulations by which the merchants of Iceland may be permitted in future to navigate and trade between Great Britain and that country, that they may be fully acquainted with the obligations under which this trade may be permitted them.[5]

And your petitioner will ever pray,[6]

Holg^r P. Clausen

SL, Banks Papers, I 1:52. Contemporary copy.

---

[1] Corbett, Borthwick & Co. had sent a letter to the Privy Council stating that the *Bildahl* had arrived at Leith from Iceland with a cargo of sundry goods from Iceland asking to enter some goods for home consumption, the remainder for exportation. The petition was sent to the Commissioners of the Customs in Scotland for their opinion (TNA, PC 2/189, 8 Oct. 1810, Fawkener to Morris West, Customs in Scotland). This was granted except for the fish and oil (TNA, BT 6/201, no. 26,118).

[2] The Kattegat is a body of water between Jutland and southern Sweden.

[3] Klipfish.

[4] Banks was extremely interested in Merino wool. His correspondence on the subject has been published by Carter: *The Sheep and Wool Correspondence of Sir Joseph Banks*; and see further: Carter, *H.M.'s Spanish Flock*.

[5] As usual Banks came to the rescue and 'Additional Instructions' were approved permitting the Icelanders to export 300 tons of klipfish and the same amount of train-oil from Iceland.

[6] Banks has made notes on the abstract of this petition, probably in conversation with Clausen, see the next note.

179. Banks's Note on Oil and Fish, [late November 1810]

The Train oil is From a Fish Calld in Danish Havkalv[1] in Icelandish Sea Calf[2] it is taken from the Liver of this Fish which is a kind of Shark from 10 to 14 Feet Long & is very large
    some of the Oil of Cod Livers
    no other oil is made in Iceland no vessels are Ever Sent probably few Seals are killd
    The first sort is usd for Lamps in Denmark because it does not Freeze
    300 Tons of these Oils is as much as is usualy made the Exportation of this would amply answer all purposes[3]
    The Klip Fish is Cod Salted & Dried it was originaly made for the Spanish market[4] the merchants who Reside in Iceland take Fresh Fish from the Fishermen in Payment, these they Cure with Salt, the Icelanders Cure the Stockfish.
    300 Tons of this is as much as could Satisfy the Island.[5]
    Rentheir[6]

SL, Banks Papers, I 1:53. Autograph note.[7]

180. Holger Peter Clausen Jr to Banks, 5 December 1810

39 Poultry

Sir!
After having had the honour of waiting on you yesterday I applied at the Council Office for an answer to my petition, which I take the liberty to send a copy of, viz:

'the Laws must apply to the Articles imported from Iceland as they do on similar Articles from countries in Amity with His Majesty.'

I was at the same time informed that the answer had been given on Thursday last, which was the day on which the petition was delivered in.

I trust you will excuse the liberty I have taken in communicating this, as I am convinced every circumstance attending this subject will be interesting to you.[8]

---

[1] The Greenland shark.
[2] The Greenland shark (*Somniosus microcephalus*) is called *hákarl* in Icelandic. *Sea-calf* is a direct translation from the Danish but not Icelandic (the direct translation would be *sjókálfur*, which does not exist in Icelandic).
[3] That was what Banks decided would be the quota in the instructions to the consul (see Document 202).
[4] Merchants began sailing to the Mediterranean with Icelandic salted cod (klipfish) in the 1760s.
[5] That was what Banks decided would be the quota in the instructions to the consul (see Document 202).
[6] He mentions the ship *Rennthier*, which belonged to Hans Thielsen.
[7] Note in Banks's autograph, verso: 'Abstract of Petition | Clausen'.
[8] However, Clausen had also approached Sir George Steuart Mackenzie who had sent in a letter of support to the Marquess Wellesley, which was forwarded to the Council Office. The Council Office refused both the landing of goods without payment of duties and granting an exception in the matter of unloading the cargo. The letter ends: 'I am to observe that the state of the Trade of Iceland will shortly be relieved by a general regulation' (TNA, PC 2/189, Fawkener to Hamilton, 18 December 1810).

With every sense of the obligation I feel to be under for your kindness, I submit with confidence my future proceedings to your guidance.

I remain with the profoundest respect, Sir, your most obedient and humble servant.

[*Signed*] Holg^r P. Clausen jun

SL, Banks Papers, I 1:51. Autograph letter.[1]

181. Boulton & Baker to Banks, 11 December 1810

Wellclose Square

Sir,

Enclosed we beg leave to hand you a Letter from Mr. Wulff, the Owner of the three Icelands Ship [*sic*] '*Freden*'[2] '*Huusevig*' and '*Eskefiord*' but which we thought not worth troubling you with while at your Country Residence.[3]

Of this Letter, being in the danish Language, we have made free to subjoin a Translation for your kind perusal. At the same time we are very sorry to inform you, that the Ship *Freden*, on her passage from Leith to Iceland owing to stormy weather & contrary Winds, has been forced into Bergen in Norway, very much damaged, and her Cargo entirely spoiled.

The Ship *Huusevig* is arrived at Leith[4] with a full Cargo from Iceland, but, unfortunately can not find a Market either there or here for the disposal of the same. We know an Iceland Merchant by the name of Clausen is now petitioning the Privy Council[5] on behalf of several Iceland Vessels now at Leith, the *Huusevig* included, to grant the Exportation of such part of their Cargoes as will not sell in this Country, without unloading the same and on payment of a mitigated Exportation Duty if it can not be granted free of all Duty. We are also now petitioning to obtain a Licence for the Ship '*Eskefiord*' to proceed from Boston or Newcastle to Norway Sweden and Denmark, from thence to Iceland & back to this Country;[6] In all these Points we humbly beseech your kind Interference and Influence on behalf of our friend Mr. Wulff, who has hitherto been very unsuccessful in all his Expeditions to & from Iceland.

We are most respectfully
Sir
Your very humble Servants
[*Signed*] Boulton & Baker.

SL, Banks Papers, I 1:54. Autograph letter.[7]

[1] Note in Banks's autograph, left-hand margin of last page: 'Mr Clausen | Answer from the board.'

[2] All underlinings in this letter is part of the original.

[3] Spring Grove, though Banks stayed at his residence in Soho Square for most of November and all December 1810 (see 'A polar diagram of the life of Sir Joseph Banks ...', back interior cover of Carter, *Sir Joseph Banks*).

[4] See NAS, E. 504/22/48; TNA, PC 2/189, Entry, 16 November 1810. They hoped that the cargo, consisting of 137 casks of salted mutton and 4 casks of salted lamb's tongues, could be entered for home consumption. This was permitted on the payment of the proper duties.

[5] See Document 178. On 12 January 1811 Clausen was granted a licence, permitting him to buy a ship in Sweden or Norway, where he would load a cargo of deals, lumber, tar and timber for Iceland and return with a cargo of Iceland manufacture (TNA, PC 2/190).

[6] It took quite some time until the *Eskefiord* was granted a licence. This petition was initially refused. A petition to simply sail (probably from Leith) with a cargo to Scandinavia was put off. Finally, in January 1811, a licence was granted to proceed with a cargo of coals etc. to Norway, Sweden and Denmark, there to load a cargo for Iceland (TNA, BT 6/202, nos 27,445, 28,250, 28,447).

[7] Note in Banks's autograph, verso: 'Boulton & Baker | Wulff | Dec*ember* 11 [1810].'

182. Jacob Nolsøe to Consul Wolff, London, 12 December 181[0]¹

Thorshaven in Faroe

Trusting, Sir, that You will pardon my boldness, I beg leave to address this Letter to You, and at the same time to request that You will be pleased to favour me with an answer as to the particulars set forth therein.

On the 29 June 1808, my Brother Paul Nolsoe² commanding the Ship *Rojndin Freya*³ which he had built himself, left the Feroe Islands, with a view to procure for the inhabitants Corn and other provisions from Denmark, or wherever he might meet with them. On his voyage he was captured by a British Cruizer,⁴ and carried into Gothenburg, and from thence himself and his 4 Seamen were conveyed to London, and it has been currently reported here that in the course of the same Year he got leave to export from London to Feroe a Cargo of provisions.⁵ But he has never reached these Islands, and we do therefore apprehend that he has been lost at Sea, however no enquiry having ever been made here by the people who are said to have furnished him with his Cargo, the whole of above report may probably be without foundation. I will not hope that the Government here in Feroe⁶ have endeavoured to present my Brother in a light which he does not deserve, tho' I am well aware that they have always been opposing my brother, and concentrated as much as in their power his endeavours at navigating with ships belonging to Feroe and manned with our own people which is the only way of securing sufficient supplies of provisions to the inhabitants, and in which laudable exertions for the good of his native land he has spent the greater part of his property. Strange that men should thus prefer their private advantage, to the Good of the Commonwealth. I do therefore earnestly entreat You, especially for the sake of his mourning wife and children to give me whatever information You can about my brother from the time of his arrival in London in the autumn 1808:

From Denmark we can under existing circumstances no longer expect sufficient supplies, and we look therefore to England for the necessary sustenance agreeable to the declaration of 7 February 1810.⁷ This stipulates that the supplies are to be shipped either in Feroe vessels, or in English Bottoms; but Feroe possesses no vessels fitted for this trade. The Ship by which I forward this letter, is the first English Trader to Feroe. By her we have certainly received a Deal of provisions and other goods from Leith in Scotland, but Mr Gibson the Supercargo rates his own wares so extravagantly high, and the Feroe produce so far under the current price in the Country, that the poorer of the inhabitants find it impossible to procure even the most needful articles for their sustenance. He tells us that

---

¹ Very probably an error as the Sutro Library catalogue says. From its contents and dates it is quite clear it should be 1810. It is, however, clearly dated 1811 (slip of the pen?).

² Poul Poulsen Nolsøe (1766–1808/9), or Nólsoyar-Páll, is the national hero of the Faroe Islands. He was 'the ablest farmer in Faroe as well as the best seaman', sailing to America, the West Indies, England, France and Portugal. He was also a talented poet and shipbuilder, and the undisputed leader of his countrymen (West, *Faroe*, pp. 50–51, 58).

³ Usually spelt *Royndin Fríða*.

⁴ HMS *Fury* seized the ship on 8 July, off the Skaw.

⁵ Nolsøe was sent to London, where he and his crew were released. The Privy Council gave him a new ship the *North Star*. He sailed in November 1808, never to arrive in the Faroe Islands and no trace was ever found of the ship (West, *Faroe*, pp. 66–7).

⁶ Major Löbner, the governor of the Faroe islands, and Poul Nolsøe were antagonists (West, *Faroe*, p. 67).

⁷ See Appendix 10.

the reason for his asking such high price for his goods is the very heavy duties levied in England on Feroe produce. But even in the event of the Duties being reduced, I fear that the inhabitants of Feroe would hardly find it possible to subsist unless they are somehow or other enabled to procure a vessel of their own, which would enable them to carry on themselves, a part at least, both of the Export and Import trade. With this view I have by desire of the inhabitants of the Island Nolsoe who are the only Seamen in the Feroe Islands, drawn up and forwarded an humble petition to His Britanic Majesty, praying that His Majesty would be pleased to provide us with a vessel from 30 to 50 Tons, and direct her to be loaded with a Cargo of Corn at Leith. That he would allow us 3 years for repaying the English Government the cost of Ship and Cargo; That he would reduce the Duties on Feroe produce; finally that he would grant us Licence for trading unmolested between Norway & Feroe, in order to procure the necessary Timber for building our Fishing boats, Norway being moreover the place where our produce would sell to the greatest advantage. We would wish the vessel to be delivered over to us at Leith, and hoping that this might be done by the end of May next year, we would meanwhile find means to collect a great part of the purchase money and pay it over on delivery of the Ship.[1]

To forward these our views with Your powerfull assistance, and to favour this important business with that attention which it so highly merits, is what I on behalf of my Countrymen, herewith humbly request, who remain &c &c

Signed J Nolsoe

SL, Banks Papers, I 1:90a. Contemporary translation and copy.

183. L. Lobnitz and H. Hammershaimb to Cap*tain* Bohnitze, 12 December 1810[2]

Thorshavn in Faroe

From Cap*tain* Alexander of the English Schooner *Adventure*,[3] by whom we send this, we learn, that it is probable you are still in Leith, & we therefor take the opportunity of writing you. –

---

[1] In July 1810 a schooner of 14 tons was built, which made two journeys to Leith in search of supplies. The first journey was a complete failure, since the high duty of the Faroese goods carried in the ship made them unsaleable, the same problem that the Icelanders had encountered. The second succeeded in bringing the dire situation in the Faroe Islands to the notice of a firm in Leith (Corbett, Borthwick & Co.) which had managed the Danish vice-consulate until its closure at the outbreak of war. This firm contacted the Admiralty and the gun brig *Forward* was sent to the Faroe Islands in July 1811. On its return the commander, Captain Bankes, made a detailed report on the state of provisions in the islands. As a result, two ships were allowed to carry corn from Denmark to the Faroe Islands, provided they called at Leith each year to have their licences renewed. For the rest of the war the islands were reasonably well supplied. Johan Henrik Schrøter, the clergyman in Suðuroy, established a trading connection with a Liverpool mercantile firm (West, *Faroe*, pp. 68–9). This was probably Horne & Stackhouse, which are known to have conducted a trade with the Faroe Islands in this period.

[2] L. Lobnitz is unidentified but was probably Emilius Marius Georgius Löbner (1766–1849) who was commandant of the fort in the Faroe Islands 1796–1825 and *amtmand* (district governor) there 1816–25. This is a translation and names are notoriously difficult to decipher (the copyist may have been mixing together Löbner and Capt. Bohnitz). Wenzel Hammershaimb (1744–1828), was the King's bailiff in the Faroe Islands 1765–1815. Capt. Jörgen Bohnitze or Bohnitz, was master of the *Jubelfest*, engaged in the Faroes trade.

[3] This ship was sent by Messrs Brown, Rogers & Brown and had permission to carry 30 barrels of American tar and 3 tons of hemp from Leith to the Faroe Islands on the British ship *Adventure*. The master was D. T. Alexander (TNA, PC 2/189, 13 October 1810).

We most earnestly entreat you, to use every endeavour to come here as early as possible in Spring with a cargo of Provisions – it is true that the Bearer of this, has been here with a cargo of different articles for sale, but also no provisions, except 3 to 400 Barrells of Potatoes, 10 Barrells of Meal & Pease, & some hundred Pounds of Bread, all of them exorbitantly dear, as he said, that on account of the high duty in Scotland, he could not take the Produce of this country in payment, unless at very low prices – Hemp is the only article, with which it can be said this vessell has supplied us – could you find any opportunity of letting us know the fate of your voyage, we are sure you will not omit writing us.[1] We live in the hope to see you here at latest by the middle of April, because our whole provisions, even with the greatest frugality, will by that time be consumed, & if God does not give us a better Fishery, than we have had since you left us, they will be exhausted before that time.

Signed L Lobnitz
H. Hammershaimb

SL, Banks Papers, I 1:88. Contemporary translation and copy.

184. Hans Frederik Horneman to Banks, 12 December 1810

88 Queen Street, Cheapside

Mr. Horneman presents his respectful compliments to S`r`. Joseph Banks, & begs leave to enclose a Copy of Duties paid by Capt*ain* Thielsen of the *Renthier*,[2] also a list of Iceland vessels now at Leith with a Report of their Cargoes.[3] Mr. Clausen is of opinion that <u>no other ship is expected this year</u> from Iceland but the *Seyen* belonging to Mr. Petræus.[4] – The whole number from Iceland this year would thus be six vessels.[5]

SL, Banks Papers, I 1:55.[6] Autograph letter.

185. Hans Frederik Horneman to Banks, 13 December 1810

Queen Street

Right Honorable Sir

I have had the honor to receive your Letter of the 12 inst. & hasten to reply to the contents.[7] – M`r`. Clausen informs me that the reason why so little Stockfish has been

---

[1] The *Jubelfest* made regular voyages to the Faroe Islands throughout the war. On 7 August 1810 it received a licence from Leith to the Faroe Islands and back (TNA, PC 2/188, called *Den nye Jubelfest*, J. Bohnitz master).

[2] Not extant.

[3] Not extant.

[4] The *Seyen* had arrived in Leith by 24 December 1810 (NAS, E. 504/22/48).

[5] They were: the *Rennthier, Bildahl, Svanen, De Jonge Goose, Husevig* and of course the *Seyen*. There were also 2 British ships, the *Elbe* and *Flora* from Phelps & Co., and the *Vigilant* from Boston.

[6] Note in Banks's autograph, verso: 'Mr. Horneman | [Received] Dec*ember* 12 – [18]10 | no other ships expected this year'

[7] Not extant.

brought from Iceland is, partly because the merchants will rather ship Klipfish on account of its very perishable state, partly because Stockfish takes nearly double the room at the same weight of Klipfish, which makes the freight so much higher on the former articles. An other reason is, that the inhabitants who are poor & who are generally much indebted to the merchants, can, from their poverty, not keep the fish which they ketch during Winter, till the Spring following, in order to cure it into Stockfish, they therefore deliver the newly caught Codfish to the merchants the moment after having caught them, who are obliged to cure it into Klipfish. Neither M^r. Clausen nor I have any exact report of the Cargo of the *Rentheir* Cap*tain* Thielsen but M^r. Clausen knows that he had about <u>10 Tons Klipfish & 8 or 10 Ton Liver Oil</u> on board. – He had about <u>22 Tons Stockfish</u> in his Vessel 11 out of which he has sold as a help to pay expences & has taken the remainder to Denmark.

The weights of the Barrels of Tallow differ considerably. – The Ship <u>*Jonge Gose*</u>[1] has 16960 lb. Danish about 8 Tons. –

The Ship *Bildal* has none. For the weight of the Tallow of the <u>*Husevig*</u> & <u>*Swan*</u> application has been made by this nights post. The Ship *Swanen* was detained in an English port in 1807[2] & went with Licence to Denmark in the spring of 1809,[3] & has been here Last year as well as this.[4] – M^r. Clausen knows not if the *Jonge Gose* has been here before,[5] but in the spring of this year she arrived with Licence from the Baltic & obtained permission to go to Iceland. He is not aware how it has been proved that the said Ship is Iceland property but he is sure it really is so.[6]

I have the honor to be
Right Honorable Sir
Your most obedient &
humble Servant
[*Signed*] Hans Fred^k Horneman

SL., Banks Papers, I 1:56. Autograph letter.[7]

---

[1] *The Jonge Goose.*

[2] The *Svanen* had been seized by the *Shannon* and brought to Yarmouth prior to the declaration of hostilities. It was later taken to London (TNA, HCA 24/165; TNA, ADM 1/5079, Wolffs & Dorville to the Lords of the Admiralty, 16 December 1808).

[3] A licence was granted on 10 December 1808 and permission given to sail from Denmark to Iceland and back to Britain (TNA, PC 2/179).

[4] The *Svanen*: a licence was issued to export a cargo of coals, leather etc. to Denmark, there to load a cargo of corn and other necessaries for Iceland and return with a cargo of permitted goods to Britain (TNA, BT 6/197, no. 12,691). Permission was also granted to sail from Britain to Norway to export salt and take on empty casks and eventually to sail from Britain to Iceland and return with Iceland produce, which 'under the particular circumstances of this case' would be warehoused for exportation only on payment of the proper duties (TNA, PC 2/189, 12 December 1810, TNA, BT 6/197, no. 13,007; and TNA, BT 6/198, no. 13,293).

[5] This ship was not among the captured vessels and did not apply for a Danish *söpas* (sea-licence) until early June 1810. However, there is no reason to doubt the veracity of this statement as ships' names are often omitted.

[6] *The Jonge Goose* had a *söpas* for 1810. This ship had been sailing from Copenhagen to Keflavík at least since 1807.

[7] Note in Banks's autograph, left-hand margin of second page: 'Mr Horneman | [Received] Dec*ember* 14 – [18]10 | Cargo of the Rentheir | information Concerning Stock | fish & Klip Fish'.

186. Corbett, Borthwick & Co. unto the Honble The Lords Commisioners of the Board of Trade, 21 December 1810

Leith

My Lords

Corbett Borthwick & Co. Merchants in Leith do most humbly represent, that previous to the year 1807 and the commencement of the War with Denmark Seven Vessels were employed from Copenhagen in carrying out supplies to the Danish Settlements in Greenland, & that during the last two years only one Vessel has been dispatched under Licence granted by your Lordships to his Excellency Count Trampe Governor of Iceland,[1] the Inhabitants of these settlements are therefore in the greatest want of Supplies. –

That upon the petitioners applying of late for Iceland Licences, your Lordships have been pleased to answer, that since the order in Council of March last,[2] no more licences would be granted, & they presume the same may be intended as to Greenland. By that order in Council Iceland may be supplied, as the Inhabitants thereof can receive unground Corn, which can be supplied to them from Foreign grain warehoused here for Exportation, but as the Inhabitants of the Greenland Settlements have not the implements for grinding Corn, & receive their supplies in Rye bread baked in a particular way, for preservation, which cannot be exported from this Country, the same opportunity of availing themselves of benefits of the order in Council does not exist.

Under these circumstances the Memorialists do most humbly petition, that your Lordships would grant Licence for two Vessels to proceed from Copenhagen to the Danish settlements in Greenland, with a Cargo of Provisions, calling at Leith for clearance, & with liberty there to fill up with other necessaries, & to return from thence to Leith, with a Cargo of Greenland produce – and the petitioners as in duty bound will ever pray.[3]

Corbett, Borthwick & C°

SL, Banks Papers, I 2:2. Contemporary copy.

187. Holger Peter Clausen to Banks, 24 December 1810

39, Poultry

Sir!
Inclosed I have the honour to send the information concerning the state of Greenland & Ferro, which I got from my agents in Leith Mess^r Corbett, Borthwick & Co. this day;

---

[1] This was the *Jubelfest*, which sailed to the Faroe Islands in August, returning to Leith in October (TNA, BT 6/201, nos 24062, 26695). Count Trampe was first refused his petition for licences for two vessels to sail to Greenland and the Faroe Islands with provisions and to import return cargoes (TNA, BT 6/199, no. 18,007).

[2] The Order in Council was actually dated 7 February 1810 (Appendix 10).

[3] The *Freden* was granted a licence to sail to Greenland in August 1810 with a cargo including gunpowder (TNA, BT 6/201, no. 24,013). Before the war a trade had been conducted by the Greenlanders and the captains on British whaling ships, especially those from Hull. A significant role was played by William Mellish, a London merchant, though this connection has not been researched. (See Saxtorph, 'Nordgrønlands Inspektorats britiske forbindelse under englænderkrigene 1807–1814'.) In the autumn of 1807 eight ships sent by the Royal Greenland

together with a petition to the Board of Trade concerning two wish'd for Licences, which I give me the liberty likewise to inclose for your perusal.¹

With the profoundest esteem I have the honour to remain, Sir,
your most obedient & humble servant
[*Signed*] Holg^r P. Clausen

SL, Banks Papers, I 2:1. Autograph letter.²

188. From Corbett, Borthwick & Co to Banks

[Leith] [25 December 1810]³

Sir

We have been informed by Mr. Claussen of your continued exertions in behalf of our Iceland friends, & we now use the freedom to address you, from the Interest we have for two years had in this trade & in consequence of your wish to have such information as we can give with regard to the Faroe & Greenland settlements, with a view to include them in the hoped for arrangement for the Iceland Traders.⁴ Mr. Claussen is so fully acquainted with all the particulars of the Iceland trade, & so able to explain and urge the necessity of some alteration in the late resolutions of the Board of Trade, as to the importation of Fish & Oil from these Danish settlements; if it is intended to allow them much advantage from the favour & commiseration promised them by His Majesties Government, principally thro' your benevolent exertions, that we do /not/ consider it necessary to enter into detail upon these points, more especially as Mr Claussen before leaving this, saw our Petitions to the Board of Trade, upon the arrival of the Iceland Ships now here, & addressed to us, & is fully acquainted with our opinion respecting them. In one of his late letters however he says, that the Board of Trade seemed to think, that it had been last year intimated to the Iceland Traders, that they could not be allowed to bring Fish & Oil, & as such an intimation would have materrially [*sic*] lessened their claims of indulgence for the present importations, had the shipments been made in opposition to it, & without any representation to the Board, we beg leave to call your attention particularly to the circumstance of our

---

Company had sailed for Denmark, three (*Hvidfisken*, *Jupiter* and *Sælhunden*) had evaded capture, the others were taken to British ports and condemned. In 1809 two ships had been sent with licences, the *Neptunus* which had a safe journey and the *Adventure*, which was wrecked off Kap Farvel. On the Greenland trade during the Napoleonic Wars see also Bobé, *Den grønlandske Handels og Kolonisations Historie*, esp. pp. 83–8; Tving, *Træk af Grønlandsfartens Historie*, esp. pp. 60–76; and Agnarsdóttir, 'Hjálendur Danakonungs', pp. 42–9.

¹ The former is probably the following letter; the second is not enclosed but, according to Board of Trade documents, Corbett, Borthwick & Co. petitioned for a licence for the *Anna Dorothea* to sail from Copenhagen to Iceland on c. 21 December 1810, which was refused the following day (TNA, BT 6/202, no. 28,024).

² Note in Banks's autograph, verso: 'Corbett & Borthwick'.

³ The letter is undated. The Sutro Library dates it to 1810. The major clue for dating is the phrase 'the *Seyen* arrived here yesterday'. She arrived on 24 December 1810 (NAS, E. 504/22/48).

⁴ Regarding the amount of klipfish and oil permitted to be exported.

having never heard of this, altho' we had the two Iceland Cargoes, under our charge at the time, & were particularly informed by his Excellency Count Tramp*e* about the Iceland order in Council,[1] & know that he was not aware, that any alteration was intended as to the articles of Fish & Oil, which we were last year in the course of his Excellencies discussion with you, allowed to warehouse for Exportation from the *Tykkebai* & the *Swan*[2] & that consequently the present importations of these articles were in ignorance of any difficulty with regard to them, & in the firm & natural belief that their admission into this Country would be the same as last year, it would therefore be very unlike the usual & acknowledged equity & liberality of this Government, to subject the Iceland Merchants to the loss of the Fish & Oil now in Leith Harbour, & which was brought there under the above circumstances. We beg leave to call your attention to another circumstance. That in the answers of the Board of Trade to our Petitions upon the subject, we have been informed that the duty must be paid upon the Fish & Oil; whereas Salted Fish is not enumerated in the Book of rates,[3] & must consequently be confiscated if not allowed to be warehoused for Exportation. We are confident that the Board of Trade when this is explained to them by you, will shew no wish to allow the innocent proprietors of these Fish, to be thus deprived of their property. We understand that the present difficulty arises principally from a fear that the importation of Fish & Oil from these Danish settlements, might injure the British Fisheries,[4] which is so much, and so properly the wish of our Government to encourage, but upon the exportation of the Shetland Salted Fish, there is a bounty of £3p. ton, which as they have the Salt duty free, & fully as cheap or cheaper, than the Icelanders seems upon an article only worth £20 p. ton to be a sufficient check to any very injurious interference with them, & this may if necessary be increased by a small duty on Iceland Fish, & further they have the whole sale of Great Britain, as it has not been asked to admit Iceland Fish for Home consumption. With regard to Fish Oil, we believe that there is not a single barrel of the British Oil exported, it being hardly sufficient for the consump*tion* of the country, you know that it is quite distinct from Whale Oil, (of which latter there is in common times a great exportation,) the Fish oil is principally used by Curriers, & Soap Manufacturers. In addition to the Iceland vessels, now here, as mentioned to you by Mr Claussen, the *Seyen* arrived here yesterday & has 100 Barrels of Fish Oil & 35 Tons Salted Fish – this cargo belongs to Mr Vesty Petreus [Westy Petræus] who with Mr Sivertson [Sívertsen], was the first Icelander, who in 1807, was by your benevolence supported under the misfortune of the Capture of his whole property, & which you got in part restored to him & he again looks to your kindness for the safety of a considerable property invested in the above articles.

---

[1] Appendix 10.

[2] The *Swan* had a *söpas* for 1810 and returned to Leith from Iceland in December (TNA, PC 2/189, 12 December 1810 and TNA, BT 6/202, no. 27,784).

[3] Perhaps *A New and Compleat Book of Rates, comprehending the rates of merchandize etc.* by Edward Burrow, Glasgow, 1774, a massive book of 666 pages without an index. Nothing else can be found under that title.

[4] See p. 374, n. 4 above.

The vessels used in the Faroe trade previous to the war were[1]

Jubelfasten { now in Leith on her return from Faroe under a Licence granted to Count Tramp last year[2]

Najaden[3] { now in Copenhagen, & has been there since the war broke out.

Anna Maria   Capt[n] Jurgen Findoe   { captured in 1807 at
name unknown   „ Niels Nissen   same time with the Iceland vessels and condemned

When the *Jubelfast* arrived at Farroe in the month of august, with cargo of grain, Captain Bohnitz found the Stock in the island almost wholly exhausted & the Inhabitants in great fear of famine, his cargo will it is hoped keep them from absolute want untill they receive next year further supplies by him or the other Faroe ship *Najaden* as no Licence is now granted for Iceland & Faroe, the grain must be imported here, <u>warehoused for Exportation, & then reshipped</u> which causes great delay in the voyage, & may make them late of arriving; It would therefore be of most material advantage to the Inhabitants could we procure Licences for the *Jubelfasten* & *Najaden* to carry a cargo of grain from Denmark to Faroe, calling here to clear as last year if you think such would be granted we would immediately make the necessary application.

The following Ships were employed in the Greenland Trade.

Brig *Hvidfischen*
„ *Selhun*   } are now in Copenhagen & have been there since the war.
„ *Jupiter*

Ship name unknown   Capt[n] Jessen
Brig *Fruhling*   } Captured when the war
„ *Hvide Biorne*   commenced & were condemned[4]
Geileat[5]   Unge Lars

Freden Captain Matthiesen[6] sailed from Leith for Greenland in August last, under a Licence granted to Count Tramp*e*.

---

[1] In the margin Banks has written 'Faroe'.

[2] The name of the ship varies, *Jubelfast* or *Jubelfasten*, though the correct spelling is *Jubelfest*. In August 1810 the vessel had been granted a licence to import necessaries, then at Leith, to the Faroe Islands and to return to Leith (TNA, PC 2/188, 7 August 1810; TNA, BT 6/202, no. 26,695). The ship eventually returned to Leith and in November the mitts, stockings and candles were permitted to be warehoused for exportation, but the ship also carried oil which could only be entered on payment of duty. There were further problems in November (TNA, BT 6/202, nos 26,645, 26,977 and 27,313). The matter was referred to the Commisioners of HM's Customs in Scotland (TNA, PC 2/189, Chetwynd to West, 21 November 1810). The reply, dated 15 December, was simply 'Nothing ordered' (TNA, BT 6/202, no. 27,313).

[3] This ship is always wrongly called *Hayaden* in this document, it should be *Najaden*, and the name has been corrected throughout to avoid confusion.

[4] The *Fruhling*, both ship and cargo, had been condemned to the Crown on 23 February 1808 (RA, Kommercekollegiet 1797–1816, Handels – og konsulatsfagets secretariat. Fortegnelser over de til England opbragte danske Skibe 1807 etc).

[5] *Galliot*.

[6] Lauritz Mathiesen. In August 1810 the *Freden* had been permitted to sail from Leith to Greenland with a cargo of tobacco, gunpowder, shot, lead and soap etc. Special permission was granted for the gunpowder (TNA, BT 6/201, no. 24,013).

From what we can learn of her passing Faroe, we hope she would get safely to the Danish settlements in Davids Straits, altho' she was too late of sailing, her cargo consisted principally of bread from Copenhagen, with some Tobacco, Gun Powder and Shot shipped by us here – a Captain of one of our Greenland Ships informed us, that he was ashore <u>last summer</u> at the Danish settlements of <u>Lievle</u>,[1] when the Inspector Mr. Mosfelt[2] said that they had about <u>2 months allowance of Bread at half allowances</u>, that the settlement to the <u>Southward, & at Frau Island</u>[3] were rather better off & that he would endeavor to send a small vessel there for some assistance if no Ship arrived from <u>Europe in July</u>, but that the distress would be most alarming, if they were not either allowed to leave the settlement, or get more regular supplies from Europe than during the last two years. The *Freden* was bound for this settlement of Lievle, & has we hope got there to relieve their distress. In our late applications about Iceland Licences, we have been answered by the Board of Trade, that in consequence of the order in Council March last[4] no further Licences would be granted, & we presume that the same is intended with respect to Faroe & Greenland, in so far as regards Iceland & Faroe, they may be in this way supplied with Grain warehoused for Exportation in this Country, but as the Shipments to Greenland, are never unground Corn, but Rye baked in a particular way, which cannot be supplied here, & as they cannot make use of unground Corn from want of Mills &c, we conceive it absolutely necessary that this should be allowed to be sent them from Copenhagen, in consequence of instructions from the Greenland Co*mp*any, we were about to apply for two Licences, & as we find from Mr. Claussen that you still feel that inclination to assist the unfortunate inhabitants of that unhospitable region, which procured the Licence for the *Freden*, we now use the freedom to inclose our Petition, with an earnest request, that for the above reason it may have your support at the Council office, we shall instruct our Agents in London to call & take out the Licences if granted, that you may have no unnecessary trouble.

We trust that whatever arrangement is now made for Iceland will, as in the last order in Council, extend equally to Farroe and Greenland, <u>when the article of Oil is as material a part of the Produce</u> as in the former, indeed unless <u>it</u> can be brought <u>from Greenland without loss</u> there is no other produce, to be got, for the supplies of Provisions sent them, & which of course cannot be taken there if no return can be made in payment.

As the misunderstanding of the Iceland order in Council has led this year to so much inconvenience, & a detention of nearly 3 months to some of the vessels, the arrangement of a fixed rule for that trade in future will be nearly as great a favour done its inhabitants, as the support and indulgence which they have already enjoyed thro' your exertions.

We have the honor to be
most respectfully
Your most obed*ient* Humble Servants
[*Signed*] Corbett, Borthwick & Co*mp*any

SL, Banks Papers, I 1:57. Autograph letter.

[1] Lievle or Lievlu was a name commonly used by the English for Godhavn on Disko Island. See Appendix 7 for the Danish settlements in Greenland.

[2] The Inspector was the highest-ranking Danish colonial officer in Greenland 1782–1924. There were two of them. Peter Hanning Motzfeldt (1774–1835) was inspector in North Greenland in the area between Holsteinsborg and Upernavik on the western coast 1801–14. Marcus Nissen Myhlenphort (1759–1821), a Norwegian, was the inspector for South Greenland 1802–21.

[3] Frau Island: present-day Upernavik.

[4] Actually the Order in Council of 7 February 1810, see Appendix 10.

189. Corbett, Borthwick & Co. to Banks, 1 January 1811

Leith

Sir,

We had the honor to write you yesterday & after the letter was dispatched, received from Cap*tain* Bohnitz of the *Jubelfast*, a letter addressed to him by two of the principal officers in Faroe, & forwarded by a small English Schooner[1] that had sailed from this [port] to Faroe, with a cargo of sundries about two months ago. We inclose a copy of the letter, & upon your sending it to the Board of Trade, we think their Lordships will readily grant licence for the *Jubelfast* to retain the small quantity of goods now on board p*er* annexed specification, to fill up with coals here for Denmark, and to return from thence with a cargo of provisions for Faroe, calling here for clearance, & to return with a cargo of Faroe Produce to Great Britain or this latter part to be filled up as may suit the new arrangements. – Unless this vessel is thus allowed to proceed immediately, it will be impossible that she can reach Faroe within the time mentioned in the letter at the latest period, to which by the greatest care, their present stock of Provisions can hold out. We understand M[r] Clausen has mentioned to you, the situation of an Iceland Merchant M[r] K Isifiord,[2] & as it is we believe among the many hard cases, of inhabitants of that island, in the war with Denmark, one of the most unfortunate, we doubt not it will meet with your usual sympathy, & that he will receive any assistance and support, which the nature of the case will admit of from your influence. In the year 1807 he was the only Iceland merchant, who entirely lost his property,[3] some of that had been sold as prize goods, but the proceeds repaid to the owners, his vessel the *Charlotte Amalia* with her cargo, even when sold in that way, which is seldom the most productive, left a free balance, after deducting all expences, of £2300 or about that sum. – Last year he had three Iceland Vessels with Licences,[4] which were all detained by the *Edgar* man of war, & altho' it was found in Court, that these vessels were proceeding in the Lawful fulfillment of the licences granted to M[r] Isifiord, & therefore restored; he was for each to pay very heavy expences, & his whole trade with Iceland has been stopped this year, which is of course a very great disappointment, & irreparable loss to him – his friends in London have explained Sir, we understand, to you the endeavours he is now making to recover the proceeds of the *Charlotte Amalia*; & all that we can add, is that we have the best means of knowing the hardships of the case, & that M[r] I*s*fiord is fully deserving of every assistance that you may be enabled to give him.

We have the honor to be
most respectfully
your obed*ient* hu*m*ble Serv*ants*
[*Signed*] Corbett, Borthwick & C[y]

[1] The English schooner is unidentified
[2] Kjartan Ísfjörð or Isfiord as he is called in English.
[3] See Document 99.
[4] The *Orion*, the *Regina* and the *Eliza*, were all issued with sea-licences from Denmark. On 27 April 1810 e.g. the *Orion* had been granted a licence to proceed from Liverpool with a cargo for Copenhagen, to a Norwegian port for empty casks and deals and to sail from Copenhagen to Iceland, touching at Leith, finally bringing a cargo of Iceland produce to Britain (TNA, BT 6/199, no. 18,691). They all seem to have been captured at sea.

[*Appended on separate page*]
Report of the *Jubelfast* J. Bohnitz M*aste*r from Faroe
57 Barrels Fish Oil   [*pencilled in by Banks*: '7s? £30 = 2340']
9½ Do. Do.
28 Anchors Do.   [*pencilled in by Banks*: 'about 10 Tons inc*lusive*']
9½ Do Do.
290 Sacks & 1 Cask Frocks
2471 Bundles Stockings
23 & ½ Barrels Feathers for Beds
25 Sacks Ditto
8 Boxes Candles, Frocks, & Stockings

SL, Banks Papers, I 1: 89.[1] Autograph letter.

190. Findlay, Bannatyne & Co. to Banks, 28 January 1811

New Broad Street, Monday Morning

Mess[rs] Findlay Bannatyne & Co present their respectful Comp*limen*ts to Sir Joseph Banks – at the request of their friends Mess[rs] Corbett Borthwick & Co of Leith, they take the liberty of soliciting Sir Joseph's good offices with regard to a petition for two Licences for the *Jubelfast*[2] & *Najaden*[3] to trade with the Ferro Islands, which was presented by F B & Co on the 23[d] Cur*rent*[4] and which has since remained, with several other suspended cases, before the Board of Trade.

Corbett Borthwick & Co have likewise requested F B & Co to make enquiries respecting two Greenland Licences, for which they say Sir Joseph was to present a petition.

SL, Banks Papers, I 1: 90.[5]

[1] Note in Banks's autograph, verso: 'Mess Corbet*t* & Borthwick | [Received] Jan*uary* 4 – [18]10'.

[2] Corbett, Borthwick & Co. petitioned for a licence for the Faroe vessel *Jubelfest* to proceed with her cargo of Faroe produce, without unloading, to Denmark and to return with a cargo of provisions for Faroe. This petition was received by the Board of Trade on 22 January 1811. It was granted to the Faroe Islands on the same conditions as the Iceland licences to Denmark-Leith-Iceland (TNA, BT 6/202, no. 28,710).

[3] On 30 January 1811, Corbett, Borthwick & Co. petitioned for a licence for the Iceland ship *Najaden* (this was a ruse, *Najaden* was not an Iceland ship) to proceed by way of Leith with a cargo of provisions from Copenhagen to the Faroe Islands (TNA, BT 6/202, no. 29,002). The following day this was granted on condition the ship came directly from Denmark to Leith before proceeding to the Faroe Islands. These were the same terms as Iceland licences.

[4] There is no mention of Findlay & Bannatyne presenting this in TNA, BT 6/202.

[5] Note in Banks's autograph, verso: 'Findlay & Bannatyne | [Received] Jan*uary* 28 – | [18]11 [Answered] Jan*uary* 29 – ditto'.

191. Corbett, Borthwick & Co. to Banks, 12 February 1811

Leith

Sir

We had the Honour to write you on the 8th Inst*ant*[1] & trust that you will excuse the continued trouble we thus impose upon you, in consideration of our motives, which are similar to those which have influenced you in supporting & aiding those Icelanders, who have been fortunate enough to come under your protection.

In one of our late letters we mentioned the case of Mr Karsten Iisfiord,[2] who was the only Iceland Merch*ant* that lost his ship & cargo; when captured in 1807, upon a voyage from Iceland to Denmark, under exactly similar circumstances w*h*ich those ships that were liberated in this country. – We also stated the misfortune that had befallen him, last summer, in the legal use of their Licences obtained by him thro Count Tramp*e*, for Iceland – the Three ships he sent out, the *Eliza, Orion* & *Regina*, were all captured by the *Edgar* man of War, & altho liberated by the Court of Doctors Commons, they are in consequence of this detention now lying in Yarmouth Roads, in place of having as the other vessels that sailed from Denmark, made a voyage to Iceland, & being now in this country on their return to Denmark with an Iceland Cargo.

This unfortunate detention is the most distressing to Mr Iisfiord, since he has not been able to send one Vessell to Iceland since the War, so that the whole goods collected then in his trade for these Years with the Boats, lay there useless to him, & by locking up his funds, prevent his being able to persecute any other business at home he now wishes thro' your influence, to obtain leave from the Board of Trade, to load the three Vessells above named upon their arrival in Iceland, & go direct to Denmark. There again to load, the goods permitted in the Licences lately granted & sail with them direct to Iceland without calling at Leith – by this means he would be enabled to go twice to Iceland this summer, & on the return in the second voyage would come to Leith in the same way as the other Iceland Vessells. – & therefore the only deviation from the present terms in the Iceland Licences, would be the insertion of liberty to go one Voyage direct to & from Denmark, in place of calling at Leith – & as this would only place Mr Iisfiord in the same situation, as if he had been last year /without/ interruption /been/ permitted to persecute the Voyage allowed him by his Licences, by which he would now have been ready to sail to Denmark with the very cargo he now wishes to carry direct, & as by a Court of Law he has been found to have made no improper use of his Licences, but to have suffered innocently, from one of their accidents which but too frequently occur in the present state of Trade, we hope you will be inclined to procure for him, this indemnication for his very severe Loss – Had the *Edgar* man of war sent these Ships directly to Leith, as the /Capt*ain*/ had only very doubtful grounds for supposing it possible to detain them, he might have satisfied himself as to the propriety of the Voyage, without injuring the owners of the ships, as this was the course decided in the Licence, but in place of this, the ships were first sent to Gothenburg, detained there a good many weeks, & thereafter sent to Yarmouth, thus has Mr Iisfiord lost the use of his Ships for six months, under a very heavy expence of Wages & Provisions, besides his Expenses in Doctors Commons – by permitting him one voyage direct to & from Iceland & Denmark, he would save 3 or 4 months, & therefore it would be

---

[1] Does not appear to be extant.
[2] The underlinings in this document are in pencil.

indemnifying him, for an interruption to that Protection of his Licences, which we are sure the Board of Trade wish always to be considered, as a perfect security to these who make a proper use of them.¹ – Mr Iisfiord is now personally in London & will probably if permitted by you, state fully the circumstances attending both his Losses, & his present Petition to the Board of Trade. As we have in our correspondence with him, found him at all times acting upon the most honorable principles, & struggling with the most persevering industry to overcome the misfortunes he has met with, of late years we trust you will excuse our being so much interested in his Welfare, as to trouble you with this Letter, in the hope that you may be able to render him a most essential service.

We have the Honour to be
Your mo*st* obe*dient* Serva*nt*
[*Signed*] Corbett Borthwick & *Compa*ny.

[P.S.] We have been to day called upon by the Customhouse to deposit a Letter obliging ourselves to pay Tonnage duty outwards, on the *Bildal*² if called for. May we request the favour of you to say if it is payable on Iceland ships going out with goods brought from Iceland without loading anything here, that we may if the intention of the Board of Trade imposes it.

Wisconsin, Banks Papers, ff. 75–76. Autograph letter.

192. Translation of Mr Wolff's answer to J. Nolsøe's letter of 19 December [1810]³

London 27 February 1811

M^r J Nolsoe
Thorshaven

I have in course received Your letter of 12 December last and hitherto delayed answering it, in hopes at the same time to be able to give You some satisfactory account as to the result of Your petition to Government. The day after I received it, it was handed over, together with a translation of Your letter, and my best recommendation to Mr Wilberforce,⁴ a Member of Parliament, and my good friend; who has assured me that he

---

¹ Kjartan Ísfjörð had indeed been unfortunate in his commercial affairs. At the end of February his petition for the three vessels mentioned to export cargoes of colonial produce etc. from Britain to Iceland and on to Denmark with Iceland produce and return directly to Iceland with provisions and other necessaries was refused (TNA, BT 6/202, no. 29,693). This is not surprising as the compulsory stops in British ports were the cornerstones of the licensing system. The British government could not have 'enemy' vessels sailing in the Atlantic with unsupervised cargoes. Ísfjörð applied that same day for the *Anna Dorothea* to proceed directly from Denmark to Iceland. This was also, understandably, refused (TNA, BT 6/202, no. 29,694). Petitions regarding other ships were also refused (TNA, BT 6/203, nos 29,941 and 29,942). However, his petition to export 10 tons of iron from Leith to Iceland on the ships *Orion* and *Regina* was granted (TNA, BT 6/203, no. 30,040). Ísfjörð was also permitted to export a cargo, including wines and spirits, from Leith to Copenhagen on the *Eliza* (TNA, BT 6/203, no. 31,840; TNA, BT 6/203, nos 31,922 and 32,064).
² The *Bildahl* had a Danish *söpas* for 1811.
³ This is a copy given to Banks.
⁴ William Wilberforce (1759–1833), MP, best known for leading the campaign against slavery. Due to his considerable wealth, Wilberforce had a steady throng of supplicants visiting him.

would with pleasure endeavour to promote Your views as far as circumstances might admit of. A few days since, I have again brought the business to his recollection, and thus every thing has been done to insure success. However as yet no answer has been received, and Government having such vast variety and number of applications to attend to, some time may probably yet elapse, before Your petition comes under consideration. I shall always keep it in view, and as soon as a decision is made communicate the same to You.

I am sorry to say, there is every probability that Your brother[1] has been lost on the voyage from here to Faroe. The Ship *Royndin Freija*,[2] having in 1808 been captured and carried into Gothenburgh,[3] it was there either thro' accident, or thro' the negligence of the Captors, totally lost; Capt*ain* Nolsoe and his Seamen where [sic] considered prisoners, and as such sent on board an English Man of war. After several fruitless attempts, Your Brother succeeded at length in bringing the English Admiral commanding in the Northsea acquainted with his reasons for leaving Feroe, and subsequent misfortunes. The Admiral treated him very kindly, promised to forward his petition to Government and ordered a Brig to carry him and his Seamen to London. Upon his arrival here, he called upon me, and I gave him with pleasure such advice and assistance as he stood in need of. Soon afterwards Government resolved to give him not only another and better ship in lieu of that lost at Gothenburg,[4] but likewise to load her with a full cargo of Corn for the use of the Inhabitants of Feroe. This Cargoe Your Brother engaged to sell on his arrival at Feroe for the current prices in the Country, and with the proceeds to purchase a return Cargo of Feroe produce to be sold in this Country for account of Government. During Captain Nolsoe's stay here, he endeavoured by every means in his power to promote the wellfare of his native land, he communicated to me the outrages committed by Baron Hompesch at Feroe,[5] and left with me before his departure a detailed written account of that transaction. The Ship having got all her cargo on board, and being furnished with every thing needfull for a voyage to Feroe, Capt*ain* Nolsoe and his Seamen were put in possession thereof, and he left us highly gratified with the result of his expedition, notwithstanding the commencement had been so very disastrous. It was late in autumn when he put to Sea, and having never been heard of since, we must apprehend that during the short days and boisterous weather so prevalent at that time of the year in the North sea, he has been totally lost,[6] thus ending his days in the service of his Country, at the very moment when his exertions were likely to be crowned with complete success.

SL, Banks Papers, I 1:91. Contemporary translation and copy.

---

[1] Poul Poulsen Nolsøe (see Document 182, p. 431, n. 2).

[2] This schooner, *Royndin Friða*, was the first seagoing vessel to be built in the Faroe Islands, from wreckage, and was launched in 1804 (West, *Faroe*, p. 55).

[3] Captured by HMS *Fury*.

[4] Admirals Keats and Bertie sent Nolsøe to London, where he and his crew were released and the Privy Council gave them a new ship, the *North Star*.

[5] Hompesch went in the privateer *Salamine* to the Faroe Islands, where he helped himself to public property. The *Salamine* subsequently sailed to Lerwick, where the booty was lodged in the Custom House, and the baron made his way to London where he applied to the High Court of Admiralty to have it condemned to him as lawful prize of war. However, in this he was unsuccessful. He later sent the *Salamine* to Iceland.

[6] Nolsøe had testified in court, regarding the raid on the Faroe Islands, on 17 November 1808, leaving shortly afterwards never to return, lost at sea.

193. Banks to William Jackson Hooker, 1 April 1811

Soho Square

My Dear Sir

I thank you for your Recollections which in General bear the Strongest traces of a Retentive memory in your State*ment* of the Causes Progress & Consequences of the Revolution however I Consider your memory as having not altogether done its duty to you.[1] I have read the book with much Pleasure and anticipate the Satisfaction I shall Receive from the Perusal of your account of Ceylon[2] if I Live to Read it if no Revolution happens there

under the Circumstances in which you are at present Placd I by no means advice [*sic*] you to hurry yourself, it is Quite impossible to you to Prepare yourself as you ought to be prepard for This years Fleets – in Truth I do not think a year from the present time too much to be Spent in acquiring the Knowledge of Exotic Botany which may be obtaind here & Cannot in Ceylon, To be well acquainted with Indian Plants before you go will Enable you to do in one year what without it Could not be Effected in Two or Three. To find on your Return that a Considerable number of Plants that have taken up much precious time in Delineating & describing as well Known & already properly Publishd, is the necessary Consequence of not being well prepard before hand my advice therefore is that you Lay aside all thoughts of proceeding this year that you hasten as much as Possible the Publication of your Jungermanniae[3] & Spend the Summer among the Gardens near London Examining Every East Indian Plant that Flowers & Perusing Every book you can obtain that describes the Produce of Ceylon & the neighboring Continent. a Summer & winter diligently Employd in This way will make you able to do more in one year in Ceylon than you would in Two or possibly Three if your preparation is dispensed with

I admire your Drawings of the Jungermannia genus & approve intirely of the Engravings I will send you back the Drawings as soon as I have heard from you whether you Chuse I Should Retain the Engravings which if Shewn in my Library will I think do you Credit & prepare for the Reception of the work. In any case I must advise you to Carry on the Publication here rather than in the Country. In May the Exotic Plants will begin to Flower & Surely if you intend to Study exotic Botany you Should not miss any opportunity of Seeing all the Season can give you

---

[1] Banks is thanking Hooker for his *Journal of a Tour in Iceland in the Summer of 1809* which was published in Yarmouth in 1811. The day before, Banks had written a letter to Dawson Turner in which he said: 'I have Receivd Mr Hooker's Recollections which Seem to have Stood him in good Stead except in his account of the Revolution where his memory I fear has faild him it is an Entertaining book & will do him Credit tho I fear he has interwoven Rather too much botany for a Common reader & rather too Little for a Botanist into the web of his Text' (31 March 1811, Chambers, *The Scientific Correspondence of Sir Joseph Banks*, VI, p. 53). Banks was not happy with the pro-Jörgensen slant in the narrative, but as Hooker had explained to him, Jörgensen had saved his life (see Document 167).

[2] In 1810–11 Hooker was preparing to accompany Sir Robert Brownrigg to Ceylon, but the fraught political situation in the island led to the abandonment of the proposed expedition. (Brownrigg became governor of Ceylon in 1813.)

[3] His *British Jungermanniae* was first published in London in 1816. This was Hooker's first scientific work. *Jungermannia* is a genus of mosses.

Let me hear from you as soon as Possible that I may not detain your Drawings too Long & believe me with much Esteem & Regard

your Faithfull H*um*ble Serva*nt*
[*Signed*] Jos: Banks

Kew, Hooker's Correspondence, Director's Correspondence, I, f. 33. Autograph letter. Hermannsson, 'Banks and Iceland', p. 63 (extract).

194. Jacob Nolsøe to Georg Wolff, London, 24 April 1811

Thorshaven

Your much esteemed favour of 27 Febr*uary*[1] was duly delivered unto me on the 20th inst*ant* I was extremely sorry to find that my fears as to the fate of my Brother were but too well founded.[2]

I am greatly obliged to You for the support and recommendation which You have given to my petition, and trust that it will be attended with success. The Ship which we have petitioned for and especially the Cargo which it would convey to us, would now indeed arrive very seasonably. Our stock of Corn remaining from last years supplies will be entirely consumed by the middle of May next, and in the event of no fresh arrival, in the meantime, we see nothing but famine before us.

I rely upon your kind assistance in aid of my distressed country & remain &c

signed   J Nolsoe

SL, Banks Papers, I 1:92. Contemporary translation and copy.

195. Hartvig Marcus Frisch, Friedrich Martin[i] and Hans Jensen to Corbett, Borthwick & Co.[3]

Copenhagen, 6 May 1811

Gent*lemen*

Your Letter of 6 April has been received, which disap*p*oints our hope of being enabled by a Licence, to send from this country to Greenland, the Provisions necessary for that colony – much as we lament this, it does not lessen our sense of Sir Joseph Banks's amiable, tho hitherto unsuccessful exertions in this matter, & wish that an opportunity might occur, for us to shew our gratitude to him.

There now seems to be no means of preventing the inhabitants of that deserted Country, from the miseries of famine, except they get assistance from England, unground Corn will not answer the purpose, because they are altogether without the necessary

---

[1] Document 192.

[2] His brother Nólsoyar-Páll had been lost at sea (see above, p. 431).

[3] Hartvig Martin Frisch was the director of the Royal Greenland Trading Company and lived in Copenhagen. Friedrich Martini was his co-director. Hans Jensen was a co-director of the Greenland administration. This letter is obviously a copy (and a translation) sent to Banks because they are relying on his assistance for a petition to the Lords of the Privy Council for Trade. See also RA, Rtk 373.121, f. 138. On all three, see Appendix 13.

implements for grinding it, & in a great measure are at a loss for fuel to bake it, it is therefore absolutely necessary that they should have their supplies in bread, & you state that the export of this would be attended with difficulties in England, & at any rate the purchase of it with you would be very expensive. – We therefore think that the supply of the Colony might be managed thus. –

The Ship *Freden*, belongs, as appears by the Papers she has with her to Mess[rs] Motzfeldt & Mylenphal,[1] both residing in Greenland, consequently this Vessell, by the English order in council 1810[2] is entitled to trade betwixt Leith & Greenland, she may be probably expected to arrive very soon at Leith[3] from there, & to enable her to return, might be allowed to lay up her Greenland Cargo with you – in the mean time a ship should be dispatched from this to Leith with a cargo of the most necessary Provisions, which would be transhipped in the *Freden*, & that vessell to complete her cargo in Leith with the other necessaries of Gun Powder, Lead, Tobacco, Coals, & other articles of Colonial Produce, that may be wanted in Greenland, & to proceed with as little delay as possible to that settlement in the ship which brings the Provisions from Denmark, could thus take in the Greenland Products landed from the *Freden* – we see however that the execution of this plan depends upon the following provisions

1st That you can procure permission to store the Cargo of the *Freden*, which will consist of Seal & whale Blubbers, Skins, Down & Feathers, to be afterwards shipped in a Danish Vessell

2[ly] That a Licence is granted, for a Vessell to bring from this, articles of Provision most wanted, namely, Bread, Meal, Butter, Pork, Stock fish, Salted Beef, Pease, Hulled Barley, Malt & Hops, also Wine for the churches, Mead, coarse Linnen, Tar, Fishery Lines, Fish Hooks, Twine, SettStones, & Medicines, to which must also be added Coopers materials, because without this, the Colonists can in no way take advantage of the Fish &c they kill, which is their only source of emolument. –

The only ship that would be employed with this Licence is the *Hvalfisk*,[4] of Copenhagen, burthen 82 Lasts commanded by Mr Jepsen Elberg,[5] & to be cleared out from Copenhagen.

We are in hopes, that Sir Joseph Banks, will by his benevolent exertions be enabled to prevail upon the Board of Trade, to adopt towards Greenland, the more liberal principles, which they have shewn towards Iceland & Færoe; when it is represented to them, that the Inhabitants of Greenland are the greater objects of pity, & more completely deprived of internal resources than any of them, that, if even the proposed plan is allowed, the assistance afforded to the Colonists, <u>by our ship</u> is insufficient, & that it is difficult to distribute it, over a Coast of nearly 200 Danish Miles – that by this means, the ship must indispensably, winter in Greenland & the crew consume a part of the supplies, intended for the

---

[1] The two Royal Inspectors in Greenland were the chief government officials in the island (see Document 188).

[2] Appendix 10.

[3] The *Freden* did eventually arrive.

[4] The correct name is *Hvalfisken*.

[5] Jörgen Hansen Lindberg according to official Danish records (RA, Rtk. 373.121, f. 139).

At the end of May, Corbett, Borthwick & Co. applied for a licence for the *Hvalfisken* to load provisions in Copenhagen for Greenland, touching at Leith to complete her cargo and returning from Greenland with a cargo to Leith. This was refused (TNA, BT 6/203, no. 32,783).

maintenance of the colonists, & their distress could not be effectually relieved unless two ships are allowed to go from this early in Spring, with the necessary supplies –

As the only payment which the colonists can offer, consists chiefly of Seal & whale Blubber; unless these articles can be landed at Leith, & reshipped without duty, to a harbour in Denmark, where provisions are so much cheaper than in England, the inhabitants will have nothing to pay with, & the Loss, if duty is paid in England, will be so heavy that the endeavour to support them must cease, the country must be abandoned, & its miserable inhabitants left to their dismal fate

signed H. Frisch, S.[1] Martin[i] and H. Jensen.

SL, Banks Papers, I 2:8. Contemporary translation and copy. Not autograph signatures.

196. Henrik Henkel to Hans Frederik Horneman, 13 May 1811

Wellclose Square

Mr Horneman
honored Sir
When I had the honor of waiting on the Noble minded Sir Joseph Banks, this friend to humanity enquired of me how the Iceland Merchants disposed of the Liver Oil exported from Iceland to Denmark, what price it was sold for at the Latter place, and how it was employed &c. – I am sorry to say that I am entirely unacquainted with the Language of this Country, therefore it is not in my power as much as I could wish to explain myself so fully as to render sufficient account concerning this Matter. This is the reason of my troubling you with these Lines in which I will endeavour to give my opinion thereof as I believe and am convinced of, it is consistent with truth & beg you will have the goodness to interpret for the said Benefactor of Iceland the true meaning of the Contents. – In order to make a true Calculation of the Profit this Article may produce for the Iceland Merchants, it will be needfull to commence from the purchase price in Iceland. As a standard for this there is an old Rule which has constantly been used /to/, which the Merchant as well as those with whom he traded were subject, namely a barter account, or the value thereof calculated in money viz. 2 Barrels of Rye for one Barrel of Liver Oil of 120 quarts, a Barrel of Corn 144 quarts danish measure:[2] this Rule has mostly been used where it could be practised, but on the south part of the Island where the fishery is best and the Goods in more abundance, more purchasers appeared and often too many Ships, which caused an exception in the common Rule, every one would willingly produce in the way of a barter as great a quantity of Iceland produce, as possible in order not to make a Voyage entirely in vain. – Then they offered or put up the Iceland Goods in a way that one Barrel of Liver Oil was paid with 3 to 5 Barrels Rye or flower, the consequence of these exorbitant prices was that most of the Merchants houses from Denmark or Norway; who had Establishments in the Island, drew themselves back with a considerable Loss, the remaining, real Iceland Burghers,[3] were by that means ruined so that they were not able

---

[1] His name was Friedrich Martini.
[2] A *quart Danish measure*, apparently 46.3 lb (21 kg).
[3] Those Iceland merchants actually living in Iceland and being burghers of one of the six towns.

to proceed in their Trade without a Loan from the Government. – Now I will take for granted that one Barrel Liver Oil is changed for two Barrels Rye, which this year purchased at moderate prices in the Low exchange of our money. – free on board at 12 rd. per Barrel   Danish Currency 24 rd.

    Freight 10rd per Barrel   20 –
    Insurance 10 pCt   4.38
    under measure 3 pCt   69
    unloading in Iceland Looses 3 pCt   69
    Warehouse Rent & other charges   2.
    One oak Barrel in staves   7.rd
    Hoops & cooperage   4.
the purchase price of one Barrel
  Train Oil in Iceland is thus   62.rd   80s
  Shipping charges in Iceland   48s
  Freight   12
  Insurance 10 pCt.   7.5
Warehouse Rent & other charges in Copenhagen   4.
Leakage in Iceland and on the Voyage about 10 pCt.   5.
as such with all expences a Barrel
    of Liver Oil will amount in Copenhagen   91.rd –37s

On the public Sales at the Exchange in Copenhagen this Spring a Barrel of Liver Oil, brought home in the Iceland Vessels was sold for   130 rd. –
which is the highest price there at any time has been paid for the article, and only arrises from the Low Exchange of danish Money –
    from the above Sum is to be deducted
Charges at Sale & brokerage 6 pCt. &
Commission 2 pCt to 8 pCt – 10rd.38s
added to before stated   91.37
                                                       101. 75.
                      remains surplus   29 rd. 21 s.

Out of this, is the Merchants Clerks & others in Iceland to be paid, and other Charges, which can not be foreseen, or determined, to be defrayed, which it is impossible to calculate.

The surplus which may remain is for the support and necessaries of Life for the Merchants and their Families – This sort of Liver Oil is only used to burn in Lamps for Lighting the Streets, but is also bought by poor people for their Lamps which is cheaper than Tallow candles, some is sold to the Inhabitants for the same use. –

Before the War broke out with Denmark and that Country had shipping in Greenland, some quantity could be exported to the Baltic but after that time the small quantity that arrives from Iceland has been consumed in Denmark. – In those years when there was free trade on Iceland and the fisheries went on well, there might be exported from the whole Island at the utmost 2000 Barrels yearly, of which the half <u>might be</u> calculated to be <u>havkalus Tran</u>,[1] the other half Cod, <u>Shark</u> and Seal Oil, but in unfortunate Years of

---

[1] *Havkal* is the Danish for the Greenland or gurry shark (*Somniosus microcephalus*). *Tran* means oil.

fishery Less than the half of this quantity. – This is all what I am able to state in regard to this subject. –

A Gentleman who came out from Sir Joseph Banks the day I had the honor to be there (Mr Parke)[1] asked me what time of the Year that fish was caught of which the Stockfish is produced, likewise the mode of curing Klipfish, to which question I can only say that the fish which is made into stockfish is a fish without Salt, caught in Winter, Spring and Autumn – it must be dryed at a time of the year when it may be free from Insects which otherwise would lay their Eggs in the fish while cured & thus the fish be spoiled by Worms. On the other hand the fish which is made into Klipfish must be salted when /is/ fresh, dryed in the Sun and constantly attended to, pressed and well presented. This fish is spoiled by Rain, and is caught in the Spring and the Summer otherwise it cannot be dryed but will be spoiled.

The fisheries cannot be carried on all over the Country at the same time of the Year, the various Situations of the country occasion a great deal of difference in that respect. – On the southern part of the Island where the Largest and greatest fish is caught, the fishing takes place early in the Winter; on the Eastern part viz by Oreback[2] the fishing commences Later, and the season called Værtime,[3] when the Inhabitants are absent from their homes in order to fish in this District ends 12th May.

On the Western part of the Country under Jökelen[4] which is an Ice Mountain constantly covered with Ice and Snow, the fisheries are likewise carried on during the Winter. Further to the North in the so-called Western inlets of the sea where I have my Establishments and which are called Patrixfiord, Bildal, Dyrefiord & Onundfiord[5] with a few more small inlets, whose Inhabitants trade on the before mentioned four places the Wærtime commences in the beginning of May and finishes in July when the Inhabitants begin their hay harvest.

The fish /that/ is caught is brought by the Merchants /who/ afterwards salt and make them into Klipfish as it can not be made into Dry or Stockfish without running the risk of having it spoiled by Worms. – Klipfish is therefore the principal Article of commerce that is exported from these places. – To the North of Dyrefiord & Onundfiord is Isfiord[6] a very Large Bay or inlet of the sea in which the fisheries are carried on both Winter and Summer untill the hay harvest begins. Here end the Western country and the fishing Districts. – The Klipfish which has in this Years been exported from Iceland to Denmark has all been consumed there. – This Article is in other respects not essentially necessary for Denmark, as that country is well provided with fish, as well salted as fresh water fish, and besides that there is produced Klipfish in some places in Jutland, independent of this Denmark can never be in distress for this Article, if even Iceland should be so unfortunate to Loose the advantage of curing & exporting it; because the fisheries of Norway are too considerable, as to suppose that a scarcity would be experienced. The same may be said about Train Oil, so that in regard to Denmark, that country will be supplied both from

---

[1] John Parke, the British consul in Iceland: see Appendix 13.

[2] Eyrarbakki in southern Iceland.

[3] *Vertíð* in Icelandic. The fishing season.

[4] Snæfellsjökull in western Iceland. *Jökull* means glacier in Icelandic.

[5] Patreksfjörður, Bíldudalur, Dýrafjörður and Önundarfjörður, all fjords with trading stations in the Western Fjords of Iceland.

[6] Ísafjörður in the Western Fjords.

Norway and Sweden with the Train Oil it may stand in need of. – I do not believe to be mistaken if I consider for certain that the British fisheries cannot by any means be injured by the Permission this Government has so nobly, under the present circumstances, granted to the distressed Inhabitants of the Island, namely to export the produce of their Country, in order by those means to obtain the first necessaries of Life.

The noble minded Men who have contributed to this Effect will besides being convinced in their own minds of having done a good deed constantly be thankfully remembered by people which has sense enough to appreciate a noble Action.

/signed/
H: Henkel

Wisconsin, Banks Papers, ff. 69–72. Contemporary translation and copy.

197. Corbett, Borthwick & Co. to Banks, 24 May 1811

Leith

Sir

We sometime ago,[1] presumed to address you, about the Importation of oil & Fish from Iceland & Faroe in 1811, from an anxiety to prevent the inhabitants of these Islands from sending goods here, which if the Board of Trade do not admit them, under the indulgence of last Year, may subject the owners to a very severe, & perhaps unexpected Loss; as many of them are entirely ignorant of the English Laws respecting these articles, & consider their Licences a Protection to all Iceland produce, whilst our commissioners of Customs interpret it, & we believe correctly, as only allowing them to import goods under the usual regulations of the country, which prohibits the importation of Salted Fish; we considered some explanation from the Board of Trade, upon this point as very desirable, that they might not in ignorance transgress upon the indulgence shown them by the English government. We have not since had the honour of hearing from you. – We consider, that if a /moderate/ duty upon these articles <u>for exportation</u>, is an Object with our Government, /she/ it would be much for the Interest of the Iceland Inhabitants, that it should be imposed provided their trade could by this means be put under precise regulation, so that their ships could come & go from Leith, without Special clearances, & delays, which are often more expensive, than a moderate duty.

in again troubling you with this Letter, we are somewhat afraid of encroaching upon your patience, & also upon your valuable time, & in a common case we would not have done it, but the poor Greenlanders have no other friend in this country, & seem more than ever in want of your assistance.

We inclose copy of a Letter from the Director of the Royal Greenland *Compan*y,[2] which ardently shews that their views are the support of the Colonists, & not profitable trade, & therefore we would hope that the Board of Trade will be inclind, to allow them to send out a ship from Copenhagen direct, calling at Leith to take in such goods as can be as well furnished from this country as from Denmark; as the duty upon oil their only

---

[1] See Document 188.
[2] Hartvig Marcus Frisch.

valuable return for Provisions, is about equal to the value in this country, we as individuals cannot undertake the supply of the settlement as proposed in Your last Letter, nor do we think, that without some instructions from Copenhagen their Bread could be got baked here, proper for their method of Housekeeping, as the season is now so far advanced, unless a Licence is immediately granted, it will not be possible to send a vessell from Copenhagen, & if no vessell has carried out Provisions from London this Year, it seems almost evident that the Inhabitants, or at least the Danish settlers, who cannot live as the natives do, must perish by famine, a circumstance which we are sure the Board of Trade, would not allow to happen, for any consideration, that could arise above the importation of a single Cargo of oil. We have by this Post transmitted to Findlay Bannatyne & Company a Petition to the Board of Trade, & they will intimate to you when it is presented in hopes that you may be able to attend at the board when it is considered, if any alteration in the Petition, or arrangement is wanted, we would be very much obliged by your sending a Note to Mr. Bannatyne to wait upon you, who will present it in the way you may think most proper, in the event of our being refused. –

We do not see that the plan proposed by the Greenland Company, could be accomplish'd, because before orders could go to Copenhagen the season will be so far advanced, that the time occupied in discharge & loading her would probably make them leave this too late for easily reaching Greenland before the Ice sets in – if no ship is allowed from Copenhagen, the *Freden* arrives soon,[1] we might endeavour to send her out this year again from Leith, by warehousing her Greenland cargo here, & being allowed to bake bread here, but it may be difficult to get the men to return immediately to Greenland, with the prospect of wintering there again, without seeing home; this is the only way in which we can propose to send supplies, if no Ship goes from Copenhagen or from London. –

The *Jubelfast* which was in Faroe last year,[2] is now in our Roads[3] with a cargo of Barley &c., & this with the cargo of the *Najaden*, which vessel saild when, we last wrote you, will secure the Inhabitants of the Faroe Islands from want of Provisions for some time to come.[4]

We are most respectfully
Your obedient Humble Servant
[*Signed*] Corbett, Borthwick & Co.

SL, Banks Papers, I 1:59. Autograph letter.[5]

[1] There appear to be two ships with the name *Freden*, one sailing to East Iceland for Ørum & Wulff, the captain being Peter Lorentz Petersen; this ship sailed to Iceland in 1810. The *Freden* sailing to Greenland, Captain Lauritz Mathiesen, began sailing in the Greenland trade in 1809. In mid-July 1810 the *Freden* was the only ship sent to Greenland. The voyage was extremely difficult and on 6 November 1810 the ship arrived in Egedesminde, where it wintered. The following summer it sailed south to Holsteinsborg. Leaving Greenland at the beginning of August 1811, loaded with oil, whalebone, skins and down, it was captured by an English ship in the North Sea and was taken to Leith. The ship was liberated in March 1812.

[2] In August the *Jubelfest* was granted a licence to proceed with the Faroe Islands cargo on board without unloading, to take on coals in England and sail to Copenhagen, then return to Leith with grain only and on to the Faroe Islands (TNA, BT 6/204, 34,892).

[3] Leith Roads.

[4] The *Najaden* and the *Jubelfest* both made it to the Faroe Islands in 1811 (TNA, BT 6/204, 33,272 and 34,892).

[5] Note in Banks's autograph, verso: 'Messrs Corbett & Borthwick | [Received] May 28 [18]11.'

198. Findlay, Bannatyne & Co. to Banks, 1 June 1811

New Broad Street, Saturday Morning

Messrs Findlay Bannatyne & Co present their respectful Compliments to Sir Joseph Banks, and beg to inform him, that the petition presented by them a few days /since/ at the Trade Office from Messrs Corbett Borthwick & Co of Leith, on behalf of the Greenland settlers has been refused by their Lordships.

F B & Co are desired by their friends at Leith, in the event of any difficulty preventing the attainment of their object, to have recourse to Sir Joseph Banks, and they trust he will excuse the liberty they take in venturing to solicit his early attention to this work of mercy.

SL, Banks Papers, I 2:11. Autograph letter.

199. Corbett, Borthwick & Co. to Banks, 10 June 1811

Leith

Sir

We have had the Honour to receive your Letter of 31 May,[1] & are glad to see that the English Government have resolved to supply the Danish settlements in Greenland on their own account, because the difficulty & delay attending the circuitous voyage /was such/ that in such a climate as Greenland, in addition to the expence, there was always a risk of their not being able to complete the Voyage after calling at Leith, & waiting in Denmark for the Licences; the *Freden* Cap*tain* Matthiesen[2] was in this situation last year as he did not leave this till the middle of August, & considerable fears were entertained of his being able to reach the coast, thro the Ice, as he has not yet reckoned, he has at all events been forced to winter in Greenland, which helps to consume the Provisions sent over. We are now most anxious to see him return, as we are not without fear of his safety in going out so late in the year, the moment he arrives we shall have the honour of sending you the information he gives us about the situation of the Colonists.[3]

We have at present to request your attention to the following particulars about the supply of Greenland, as we are confident that you are still willing to give your assistance to the unhappy inhabitants of this barren country.

The first ship that is despatched should be sent to the settlement of Disco,[4] to the Inspector there Mr Moesfelt,[5] who can thereafter distribute the supplies, through to all the inhabited parts of the most Northerly Inhabited Bay, and as this district is very soon shut up with Ice, & they can thereafter get no assistance from the other settlements, it is of the

---

[1] Not extant.

[2] Lauritz Mathiesen. The *Freden* was granted a licence in August 1810 for a cargo of tobacco, shot, lead, even gunpowder, to sail to Greenland and back to Leith. However, 'fish & oil to be excepted from the return cargo' (TNA, BT 6/201, no. 24,013). For the fate of this ship, see Document 188.

[3] Underlined in pencil, as are other underlinings.

[4] Probably Godhavn, the settlement on Disko Island in Disko Bay on the western coast of Greenland, southeast of Baffin Bay, founded in 1773. The region abounds with walruses, seals and whales.

[5] Peter Hanning Motzfeldt, the Royal Inspector of North Greenland.

utmost consequence that this ship should be dispatched with as little delay as possible; as Cap*tain* Matthesen if he got safely to Greenland, would only get last year to the more southern colonists, there is much reason to fear that the situation of the Colonists <u>at Disco is most deplorable</u>. If another Vessell is dispatched to <u>Mr Mylenphut</u>[1] <u>at Good Hope, near the Sugar Top</u>, he will undertake the distribution of the supplies along the coast in his neighbourhood, we understand this to have been the plan usually pursued by the Danish Greenland C*ompan*y, & as we formerly stated, it is absolutely necessary, that the supplies should be in Biscuit or other Baked bread, because they have not the means of either grinding, the Corn, or fuel for baking the Flour.

After the very great exertions made by you to assist & feed the inhabitants, of Denmarks ill fated colonies, it is truely painful to us, that in place of being able to communicate the gratifying accounts of the success of your unwearied attention in relieving altogether the fears of the innocent & helpless people, or of sending you, their grateful acknowledgements for the favour shewn them, without renewed claims for further trouble & exertions from you in their behalf, we are still forced to wait upon you, with fresh accounts of <u>misery & famine</u> – on the 2 May last, Cap*tain* Henrich[2] of the *Najaden* sailed from this [place] with a cargo of Barley for the Island of Faroe, Cap*tain* Henrich returned here Yesterday, & the Letter sent me by Mr. Munk[3] who received his cargo just says, '<u>we had only 10 days provisions in the Island</u>, when Cap*tain* Henrichsen arrived, you may judge of our feelings when we saw him' we have to day sent off a Petition to be presented to the Board of Trade praying for a Licence to send the *Najaden* as quickly as possible to Denmark with a <u>cargo of coals</u>, & to return from thence with a <u>cargo of Barley</u> for Faroe, & also to allow another Vessell to come over with a cargo of Grain, we use the freedom to inclose a copy of the Petition, that you may previously know the exact form of our request, we hope that it may be granted without your personal attendance, but in case it should not, we must entreat your assistance when by calling at the Council office or sending a note to Lord Bathurst our Petition will be given in to the Board <u>on the 13th Ins</u>t*ant*;[4] in it we have shortly stated the alarming situation of the Island, we now use the freedom to state a few more particulars to you, as we know the time of the Board is so precious that they cannot attend to long Petitions. –

Cap*tain* Henrich tells us, that he was witness of the <u>utmost hunger & distress</u> <u>amongst the inhabitants of Faroe</u>, their number is <u>about 6000</u> in all, <u>their Fishery had so entirely failed</u>, that he could not even procure fresh fish for his crew when there, <u>young children of 8 to 12 Years</u> old even, going about the shore <u>picking up such of the sea weed, as could be most easily eaten</u>, & previous to his arrival, the ratio of corn given out for each person was not more than would be given to a parcell of Fowls, & <u>that some of the poor people had died of absolute hunger</u>; that they had not only consumed their Seed Corn, but that <u>the Tallow</u> usually sent out of the Island, &c. <u>part of the oil</u> had been used by the poor

---

[1] Marcus Nissen Myhlenphort, the Royal Inspector of South Greenland.

[2] The 'sen' has been crossed out in 'Henrichsen'. According to official Danish sources his name was Boy Hendrichs.

[3] Mr Munk is unidentified.

[4] On 14 June 1811 the *Najaden*, on petition from Corbett, Borthwick & Co., was permitted to proceed with woollen frocks and worsted stockings from the Faroe Islands, adding coals in Leith and proceeding to Denmark to continue a journey Leith–Faroe Islands–Leith. This was granted 'under the special circumstances for the *Najaden*' (TNA, BT 6/204, no. 33,272; TNA, PC 2/191).

inhabitants from want of better food, & that the opinion on the Island was, that if the Autumn Fishery was not better, than that in Spring, 6000 Barells or about 3000 Quarters of Grain was the smallest quantity, with which the inhabitants could be saved from other famine. – The supplies already sent, are 500 Quarters which Cap*tain* Henrichs delivered this last month, the half of which was distributed from the ship side, & 700 Quarters by the *Jubelfast*, which Vessell sailed yesterday for Faroe.[1] – all the return sent by the *Najaden* for 500 Quarters of Barley, is some 20,000 pair of worsted Stockings & some woollen Frocks, the Poor inhabitants are not like them of Iceland who have something to pay for their supplies, but they have from the failure of their Fishery, absolutely nothing to send back this year, & consequently no British Merchant can send out anything there in the way of Trade, & therefore as no private individual can afford to support 6000 inhabitants it must be done either at the expence of the English or Danish Government, the latter we can assure you, cannot possibly derive any, the smallest advantage from it, & therefore we hope you will find little difficulty, in persuading the Board of Trade to allow the Danish Government to send from Denmark as much Grain, as is absolutely necessary to keep the inhabitants – this is one of the most distressing cases we have yet met with, in our correspondence with the distant Danish Colonies & therefore we entrust it without further apology to that benevolent care, which less urgent claims have met with from you.

We have the honour to be most respectfully,
Your obed*ient* hu*m*ble Serv*ant*s
[*Signed*] Corbett Borthwick & *Compan*y

SL, Banks Papers, I 1:98. Autograph letter.

200. Petition. [Corbett, Borthwick & Co.] to the Lords of the Board of Trade, 10 June 1811

Leith

Unto the Hon*oura*ble The Lords Commissioners of the Board of Trade
My Lords

The *Nayaden* Cap*tain* J: Henricksen arrived here yesterday from the Island of Faroe, where she delivered a Cargo of grain, carried out from Denmark under His Majestys Licence, when the *Nayaden* arrived, there were only 10 days provisions on the Island, upon the calculation of the much diminished rates, that had been given out for some months: the Inhabitants had consumed much of their seed corn, were eating the sea ware [*sic*] on the Shore, and some had even died of hunger, one half of the *Nayadens* Cargo was distributed in equal portions to the rich & the poor, immediately after arrival, & the remainder will be a scanty supply untill the *Jubelfast,* the other Danish Vessel, which has His Majesty's Licence for the supply of that Island, reaches Faroe, she sailed from Leith Roads yesterday with a Cargo of Barley &c.

These two Cargoes form but a very insufficient supply for the use of nearly Six thousand Inhabitants, & therefore we must humbly solicit in behalf the starving

---

[1] In May the *Jubelfest* was permitted to export from Leith to the Faroe Islands no less than 100 lb of gunpowder and a quantity of lead and bird shot (TNA, BT 6/203, no. 32,733).

Inhabitants of that Island, that the *Nayaden* of about 120 Tons Boye Henricksen Master may be allowed to retain on board, the few goods she has brought from Faroe, which consist of woolen Frocks and worsted Stockings, to fill up in this First with Coals for a Port in Denmark & to return from thence with a Cargo of Provisions, calling at Leith for Clearance, & if necessary to fill up with goods permitted by Law & return from Faroe with Faroe produce. The Spring Fishing having almost entirely failed this year, has added much to the presure felt in Faroe from the scarcity of Grain and the situation of the half starved Inhabitants as described by Cap*tain* Henricksen is so hugely deplorable that your Lordships protection in the granting of the Licence, is indispensably necessary to save the Inhabitants from immediate Famine.

As from the failure of the Fishery the inhabitants of Faroe that 3000 quarters grain is the smallest quantity with which their distress could be relieved & as the second Cargo of the *Nayaden* now petitioned for will only make about 1700 quarters sent over. We have further to request that your Lordships would in addition to this, issue a blank Licence by which the remaining quantity might be shipt from Denmark to Faroe calling at Leith to clear outwards.[1]

We have the honour to be [*sic*]

SL, Banks Papers, I 1:93. Autograph letter.[2] Unsigned, a copy for Banks.

201. Geir Vídalín to Banks, 9 July 1811

Reykjavík

Geir Vídalín, Bishop of the island of Iceland,[3] salutes the most illustrious and noble man, Sir Joseph Bangs.[4]

Were I to try, most noble and excellent Sir, to enumerate every good deed you have favoured my country with during these terrible wars, I fear I would try your extraordinary humanity and patience.

We will forever cherish in our grateful memory your exceptional benevolence, for only thus can we repay so many and great acts of kindness.

But, most noble and excellent man, I must beseech you to add yet one more to them. Allow me to entrust to your care my stepson, Halldór Thorgrímsen,[5] who, only provided

---

[1] The Lords certainly, as Banks noted, below, refused to issue a blank licence (TNA, BT 6/204, no. 33,272).
[2] Note in Banks's autograph in pencil (difficult to decipher), verso:
   The Lords refuse a blank Licence
   The Greenland Ship goes first to Disco She carries more meel than before or barley?
   The people there live on ?grain
   She sails with the Hudson Bay Ships
   3000 quarters of Corn is more than enough to keep 6000 Persons from Starving
[3] See Appendix 13.
[4] Slip of the pen for *Banks*.
[5] Halldór Guðmundsson Thorgrímsen (1789–1846). His stepfather, the writer of this letter, had taught him in his own school (clergy often had 'home schools' during this period), Halldór graduating in July 1811. Banks must have come to his aid as he went abroad to continue his studies in Copenhagen and was awarded a degree in Danish law in 1813. He was appointed the county magistrate of Kjósar and Gullbringasýsla in south-west Iceland in 1814.

with a meagre inheritance, intends to set off for Copenhagen to pursue his studies, during these dreadful times of wars. I can assure you that he will not prove unworthy of your patronage. –

May the best wishes of all the inhabitants of our island, as well as my own, be ever with you, most noble and excellent man. –

[*Signed*] Geir Vídalín

Latin original:
Virum perillustrem nobilissimum,
        Dominum Josephum Bangs
Salvere jubet Geirus Vidalinus Insulæ Islandiæ Episcopus.–

Si omnia Tua Vir nobilissime, optime erga patriam meam etiam inter hæc horrenda bella exhibita beneficia enumerare adgrederer, vereor sane ne singularem Tuam humanitatem nimis lædam.

Nostrum est singularem Tuam benignitatem grata diu colere memoria, nam hoc unum est, quo tot et tanta beneficia recompensare valemus.

Verum Vir nobilissime optime adde his adhuc unum; – Da mihi veniam privignum meum Haltorum Thorgrimsen, qvi nimis exiguo patrimonio suffultus, inter hæc horrenda bella Havniam studiorum excolendorum ergo profisci aggreditur, Tibi mandare persvasus illum tuo patrocinio se non indignum exhibiturum fore. –

Faustissima omnium Insulæ nostræ incolarum, meis additis, vota Te vir nobilissime optime semper proseqvantur.–

Reykiavicæ d.die IX^{no} Julii MDCCCXI

BL, Add MS 8100, ff. 125–126. Autograph letter.

202. Banks: Instructions to the British Consul in Iceland, [July?[1] 1811]

you are to make known to the whole of the mercantile Interest of Iceland, by a circular Letter Printed & Sent to Every merchant on the island, /*of which you are to Forward along to this office*/ That the Indulgence Extended by the English Government to the Iceland Trade must in Future be limited in Regard to the two articles of Klip Fish & Fish Liver Oil, both of which have a direct Interference with the Interests of the british Fisheries & that the amount of Klip Fish allowd to be exported /*to Denmark*/ from thence in the year 1811 must not exceed 150[2] Tons & that of Fish Liver Oil 300 Tons & that /*in Future*/ after the expiration of the year 1811 no more Klip Fish will be permitted to be Exported

you are in this Circular Letter to Require the Merchants to /*send*/ Furnish you with an account of the Quantity of Klip Fish & of fish Liver oil Each has in his warehouse

---

[1] This is not the first time Banks drafted documents relating to Iceland (see e.g. Document 144). Examples of Parke's 'Circulars to the Merchants of Iceland' for the years 1811–13 are to be found in Reykjavík, The National Library and University Library [hereafter Lbs.], Lbs. 355 fol. Parke wrote to the Foreign Secretary Marquis Wellesley enclosing a copy of the circular letter he had sent to the Iceland merchants. The circular differs in certain respects from the draft here, the major changes being footnoted. In August Parke received a letter from Marquis Wellesley, dated 12 July, containing his instructions (TNA, FO 40/2, Parke to Marquis Wellesley, 20 August 1811), hence the dating of this letter, early July.

[2] This was changed to 300 tons, which was a major benefit for the Iceland merchants.

intended for Exportation & if the Whole Quantity Exceeds the amount above Specified you are to apportion to Each merchant such Quantity of Each article in Proportion to the whole he is Possessd of, as will /*in the whole*/ make up the Quantity allowd,[1] & to furnish Each merchant with a Certificate of the Quantity he is Permitted to Put on board Each of his Ships, and at the same time to request him that all Klip Fish & Fish liver oil that arrives in the united Kingdom without your Certificate must be Landed there & Subjected to the high Duty[2]

you are also to Obtain from the Merchants a Particular account of all Linnens imported into the Island from Denmark by the Ships of this year Stating their Quantity & Quality of them. You Can by Enquiry make yourself a Judge of it, Particularly noting whether any of them are of Such a nature as might be supplied from our manufacturers in Ireland or in Scotland[3]

Wisconsin MS, Banks Papers, f. 66. Autograph letter. Draft.

203. John Parke to Banks, 9 July 1811

Reckaviig

Sir

As you did me the honor of communicating with me so freely respecting the affairs of Iceland, during my stay in London, I venture to address you now on the same subject, having rec*eiv*ed some information that I think you ought to be made acquainted with, & which I have communicated by this opportunity to the Marquis Wellesley.[4]

Jesse Thomsen[5] of Iceland petitioned in the Year 1808 for a Licence for his Brig the *Norberg* to proceed to Copenhagen, & thence to Leith, & Iceland, which he obtain'd, and with this protection on the Brig arriv'd in Denmark & has been employ'd since that time until the beginning of the present year as a privateer against British Ships, she has since return'd to Leith with the same protection,[6] & I am inform'd that Mr. Thomsen is now petitioning for the continuance of the same Licence or a new one. Mr. Thomsen is

---

[1] This is difficult to decipher but in Parke's circular letter the text reads: 'I am directed further to make known to you that I shall furnish you with a Certificate, setting forth the quantity you will be permitted to put on board each of your ships, & that all klip-fish & fish-liver oil, arriving in the United Kingdom, without my certificate of the Article being a part of the allowed quantity, must be landed & subject to the high duty' (TNA, FO 40/2, Parke to Wellesley, 20 August 1811).

[2] In the circular the following paragraph was added: 'You are however to understand that the klip-fish and fish-liver oil, so allowed to be exported to Great Britain with my certificate, is not to be unladen, but to proceed under licence to some port in Denmark, under such condition as may be therein expressed.' This was another major benefit for the Iceland merchants, relieving them of the time-consuming and expensive compulsory unlading in Leith.

[3] Note in Banks's autograph, verso: 'Draught of instructions to the Consul'.

[4] Parke's letter to Wellesley is not extant. The consul had only recently arrived in Iceland, presenting his credentials to the new district governor of southern Iceland, Castenschiold, on 5 July (NAI, Stiftamtsjournal V, Nr. 10, 11.6.1811–21.2.1820, Parke to Castenschiold, 3 July and 4 July 1811, nos 27 and 29).

[5] The underlinings are in pencil, presumably made by Banks.

[6] The *Norberg* came from Copenhagen to Leith and left for Iceland at the beginning of August 1811 (NAS, E. 504/22/52).

also the Owner of another Brig of which his <u>Son Nicholi Thomsen</u> is Master, formerly a constant Trader to Iceland, but now employ'd as a <u>privateer</u> against the British.

<u>Mr. Lambertsen of Orebac</u>[1] in Iceland obtain'd a Licence in 1809,[2] & instead of complying with its conditions, he sail'd with a full Cargo for Norway, <u>& has not yet visited this Country</u>. I am informed that he is now petitioning for a Licence.[3]

H. P. Clausen of Olafsviig,[4] who obtain'd a Licence on the 12th Jan*ua*ry last for his ship *Gratierne*,[5] has been in the habit of conveying Specie (Guineas) to Denmark & I am inform'd holds <u>a share in a privateer</u> which is now cruizing against the British Trade.

Orum & Woolff of Iceland, whose ship the *Huseviig*[6] is now licens'd, I am inform'd are also large <u>Shareholders in Privateers</u>, which cruize against British Shipping.

I am very sorry, I can assure you, to communicate such information to you, who have, on all occasions, us'd such exertions in their interest, that there are some very respectable Merchants is most certain, & I think I can venture to recommend to your notice Mr. Jacobæus,[7] who informs me that he is going to England for the purpose of procuring a Licence,[8] & of whom I have heard an excellent character. I may add that he has not yet experience'd the indulgence of the British Government, his application is only for a Vessel of about 100 Tons burthen.[9]

I think proper to inform you that one branch of the Commerce of these Merch'ts is to carry to Denmark the <u>Specie of this Country</u>,[10] for which purpose they came provided with the Bank Paper of Denmark, in June 1810 they gave <u>two</u> of the latter, & at this time <u>four</u>[11] are given for the Specie Dollar.

I have ventur'd, in consequence of the conversation I had the honor of having with you, to give Certificates with the Klip Fish, tho' I have not yet receiv'd the official instructions to that effect. I make the division according to the Tonnage of the respective Vessels, <u>giving a preference to the British</u>,[12] which I hope you will think correct.

---

[1] Eyrarbakki.

[2] This was the *Bedre Tider*.

[3] This appears to be wrong. There is no record of Lambertsen applying for a licence (*söpas*) in Denmark after 1809 until March 1813. He may, however, have been petitioning for a British licence.

[4] Ólafsvík.

[5] TNA, PC 2/190, 12 January 1811.

[6] *Söpas* 16 July 1811 (none for 1810). They received a British licence for the *Husevig* (TNA, BT 6/198, no. 13,456).

[7] Christian Jacobæus had commercial concerns in Reykjavík, Hafnarfjörður and Keflavík. Mackenzie was taken with him, describing him as 'a particularly intelligent man' who had long resided in Iceland (*Travels in Iceland*, p. 125).

[8] His vessel was the *Skalholt*. His first petition was refused because it was called a Danish vessel (TNA, BT 6/203, no. 32,860). In June he petitioned to sail two ships (the *Keblevig* and *Skalholt*) in ballast to Iceland, load them with Iceland produce and return to either London or Leith. This was referred to Customs in Scotland, the result being: 'if these vessels are in part owned by a Danish subject (not an Icelander) they cannot be allowed the benefit of H.M.'s Order-in-Council permitting a trade between this country and Iceland etc.' (TNA, BT 6/204, no. 33,178).

[9] The *Skalholt* was $53\frac{1}{2}$ *commercelæster*, just over 100 tons.

[10] Inflation was rife due to the war. There was a continual shortage of coin in Iceland, a perennial complaint against the Danish merchants.

[11] The 'two' and 'four' are underlined by Parke.

[12] These words would have been anathema to Banks, – though an imperialist, he had stated in his draft that the tonnage be distributed *fairly*.

I have the satisfaction of informing you that the Country appears to be in a very good state, from my own observations upon this place, as well as the information I have reciv'd from the more distant parts.

I will continue to take the liberty of communicating with you as I may collect information,[1] & I have the honor to be Sir

Your most Obedient & Humble Serv*ant*
[*Signed*] John Parke

SL, Banks Papers, I 1:60. Autograph letter.

204. John Parke to Banks, 13 July 1811[2]

Reckiviig

Sir

I had the honor of addressing you on the 9th Inst*ant*,[3] & I am persuaded that the information which my Letter contain'd wou'd give you displeasure, knowing how much you have exerted yourself in the interest of the Parties.

The observation I made respecting the Specie of the Country may appear trifling, but there is a much greater quantity in circulation than I cou'd possibly have suppos'd in so small a population, & the Merchants are draining it (sending five to Seven thousand Dollars in a Vessel) & issuing the Bank Paper of Denmark to an extent that, I fear will be very injurious to the Icelanders.

From the conversation that pass'd during the last interview I had the honor of having with you, I have concluded that 300 Tons of Klip Fish, & the same quantity of Oil, are permitted, tho' I have not yet receiv'd such instructions officially.

On my arrival here I found the Ship *Elbe*,[4] belonging to the House of Phelps & Co., & she had taken in 157 Casks of salted Cod Fish, containing together about 60 Tons, & I think it right to inform you, (as application may be made to the Board), that their Agent had purchas'd them before he had receiv'd any communications from the House of the difficulties they experience'd last Year in Landing the same description of Fish.

I have the satisfaction of informing you that Mr. [Ólafur] Stephensen[5] of Widey[6] is in good health, & has made many kind enquiries about you.

I have the honor to be
Sir
Your Ob*edient* & H*um*ble Serv.
[*Signed*] John Parke

---

[1] Parke's letter arrived in London at the beginning of August 1811. The Foreign Office immediately despatched the information regarding 'the improper use' several merchants in Iceland had made of their licences (TNA, BT 6/204, no. 34,910a). The matter was sent on to Customs in Scotland and to the Advocate-General.

[2] This document is marked 'Duplicate'

[3] See previous document.

[4] The *Elbe* made two voyages to Iceland in 1810.

[5] Banks's old friend and correspondent, the former governor of Iceland.

[6] Viðey.

SL, Banks Papers, I 1:65. Autograph letter.[1]
There is a duplicate (SL, Banks Papers, I 1:66), which was sent to Overton, Chesterfield, Derbyshire, another of Banks's residences and where he was staying at the time.

205. Jacob Nolsøe to Georg Wolff, 15 July 1811

Thorshaven

Mr Consul Wolff
London

Your honoured letter of 27 February was duly received, and on the 24 April, I had in reply the honour of thanking You for Your readiness in forwarding and recommending my petition. Not knowing whether my said letter is come safe to hand, I beg to assure You herewith anew of my unfeigned gratitude for Your exertions on behalf of my poor Countrymen. I had hoped to be favoured with an answer from Government to my petition with one of the vessels that have visited us this season, but have not as yet been so fortunate.

Perhaps it may not be out of the way to give You some account of our present situation. Two Ships with Corn from Denmark,[2] have been here this summer, they have both left us again, and being the only vessels intended for the trade to Faroe, they mean to make another voyage before winter.[3] Supposing they arrive safe, as we trust they will, still, unless it please God to give us abundance of Fish, we shall not have Corn enough for the support of the Inhabitants during the winter. But how easily may by some accident or other the arrival of the Ships be delayed or even altogether prevented, and then we are undone. The state we were in, before the arrival of these ships was wretched in the extreme. For whole months the Inhabitants fed upon Seaweeds and Shellfish which they gathered in the Sea shores, and thought themselves happy when to this they could now and then add a little Milk and Tallow, but this many were obliged go intirely without. A considerable number died in consequence of the miserable food, and 4 persons perished with hunger in the strictest sense of the words. All the inhabitants are greatly weakened by the poorness of the diet, and the more aged will hardly ever recover their health again. The following will give You an Idea of the scantiness of provisions. During the space of 6 months the inhabitants of Thorshaven received no more than 4 pints Corn each person P^r week, which is /*not quite*/ rather more than $\frac{1}{2}$ pint P^r day, and in the Country they had still less.

Signed J Nolsoe

SL, Banks Papers, I 1: 92. Contemporary translation and copy.

---

[1] Note in Banks's autograph, bottom left of first page: '[Received] 28 Au*gu*st [18]11'.

[2] The *Jubelfest* and *Najaden*.

[3] They doubtless received licences. At least records show that the *Jubelfest* was permitted to export 100 lb of gunpowder and a 'quantity' of lead and bird shot from Leith to the Faroe Islands in May 1811 (TNA, BT 6/203, no. 32,733).

206. Corbett, Borthwick & Co. to Banks, 19 July 1811

Leith

Sir

We have had the Honour to receive your Letter of 5th Inst*ant*[1] & aware of the very great pressure of important Public business, were unwilling to trouble you or the Board of Trade, with regard to the Faroe Licence, untill we knew that a final arrangement had been made with regard to Iceland. We this morning received the first of the Iceland Licences, & as in your last Letter you mention, that you had little doubt that a Licence would readily be granted upon similar terms for the Faroe vessell the *Jubelfast*,[2] now here, we have to day transmitted to the Board of Trade, thro' our correspondents in London Messr Findlay Bannatyne & C*ompan*y, a Petition for such a licence; from the letter we had lately the honor to transmit to you,[3] stating the short time, that the present supplies in the Island, would afford sufficient food for the inhabitants of Faroe, you know the urgent necessity, for the speedy dispatch of the *Jubelfast*, & we are satisfied that this is sufficient appology [*sic*] for our troubling you once more. The settlements of Faroe & Greenland, being originally included in the protection, you procured for the defenceless inhabitants of Iceland,[4] we have no doubt you have in this instance also urged their equal claims upon the compassion of the English Government, & that the new arrangement is also extended to them – but apprehension of any delay at the Council Office, we would sollicit in behalf of the suffering inhabitants, that your influence might procure the issue of the Licence now petitioned for – As this one vessel is not sufficient for the supply of about 6000 inhabitants in the Feroe Islands, we have in the same petition prayed for, a Licence to allow the other Faroe vesell the *Nayaden*[5] to proceed from Denmark with a cargo of Provisions also, this vessell was mentioned in our last Letter on this subject, as the only one beside the *Jubelfast* that escaped capture in 1807 – last spring the inhabitants of Faroe were so much in fear of Famine, that a small vessell was dispatched to Shetland praying for some assistance, when the *Jubelfast* arrived with the Licence you procured for Count Tramp*e*, with a much wanted supply; we trust that thro' your kind assistance the two Licences now petitioned for, will be granted to prevent them being this summer reduced to the same straits. –

You were so good as [to] present at the Council Office our Petition for two Greenland Licences,[6] which we trust will be granted, because the Danish settlers there, stand fully as much in need of assistance as either Iceland or Faroe, from the nature of the voyage it is of the utmost importance, that these Licences should reach Copenhagen so as to enable the Greenland ships to sail for this Port in April, & therefore we are most anxious about their issue, that we may forward them by the vessels now going, indeed last year, from

---

[1] Not extant.

[2] See previous document.

[3] The *Jubelfest* was granted a licence to import a cargo of grain from Copenhagen to the Faroe Islands after touching at Leith to fill up with tobacco, sugar, coffee and spirits and return with a cargo of Faroe produce to Leith 'under the regular Duties' (TNA, BT 6/204, no. 34,305).

[4] Order in Council of 7 February 1810 (Appendix 10).

[5] This will have been the *Najaden*, which was granted a licence (Leith–Danish port–Leith–Faroe Islands) on application of Corbett, Borthwick & Co. (TNA, PC 2/191, 14 June 1811). She was to take coals to Denmark.

[6] Corbett, Borthwick & Co. had applied for a licence for the *Hvalfisken* which was refused (TNA, BT 6/203, no. 32,783). The other may have been the *Najaden*.

Count Tramp*e*s late arrival in Denmark,¹ the *Freden* was so late of reaching Leith² that very few Capt*ain*s would have ventured to proceed to Greenland, we have desired Findlay Bannatyne & C*ompan*y to inquire for them at the Council Office & we now trust they will be issued, on a similar plan with those for Iceland.

We lately had an opportunity of writing Count Tramp*e*, & informing him of the very great trouble you had again experienced, & of the successful issue of your unwearied labours, in behalf of his defenceless & suffering countrymen.

We have the Honour to be
most respectfully
& your obed*ien*t humb*le* Serv*an*t
[*Signed*] Corbett, Borthwick & C*ompan*y

SL, Banks Papers, I 1:94. Autograph letter.³

207. Corbett, Borthwick & Co. to Banks, 2 August 1811

Leith

Sir

We had the Honour to write you on the [*blank*] Ul*timo*,⁴ & at same time transmitted to you a Letter from the Governor of Faroe, & to the Board of Trade a Petition for a Licence for the *Jubelfast* to proceed with the few articles now on board from Faroe, after filling up with Coals here, to Denmark there to load Corn for the use of the Inhabitants of Faroe; but am sorry that we have not yet obtained the resolution of the Board, (which we have little doubt will be to grant the Licence) as the Season is so far advanced, that the *Julbelfast* may not be able to complete the Voyage if she does not sail soon⁵ –

We now inclose the Letters received from Iceland, p*e*r the *Orion* which arrived here yesterday, belonging to M*ͬ* [Bjarni] Sivertsen, the first of the Icelanders that waited upon you in 1808. M*ͬ Sivertsen* is with the Vessell & we find from him; as also by a Letter from M*ͬ* Clausen who was in London last winter, that the Merchants there were much at a Loss, from not exactly knowing the intentions of the board of Trade, with respect to the Fish & Oil;⁶ M*ͬ* Parke had told them that nothing was yet fixed, but that he could give certificates for a small quantity of Fish. M*ͬ* Clausen adds that he M*ͬ* Park*e* was purchasing Fish himself;⁷ – as you stated in your Letter to us of 31 May, that M*ͬ* Park*e* had carried out

---

¹ Trampe arrived back in Denmark in the spring of 1810.

² On the difficulties surrounding the *Freden*'s voyage see Document 213.

³ Note in Banks's autograph, verso: 'Corbet*t* & Borthwick | [Received] Ju*ly* 22 – [18]11 | [Answered] Ju*ly* 23 – [1811]'.

⁴ The previous document.

⁵ This petition was granted, to proceed without unloading with the coals to Copenhagen and to return to Leith with grain only, where the ship would load other articles to the Faroe Islands. However, a licence for a return cargo was not granted 'unless it be for importation under the regular duties' (TNA, BT 6/204, no. 34,892).

⁶ See Document 178.

⁷ Parke (see Appendix 13) was the son of a Liverpool merchant, Thomas Parke. In this period consuls were almost invariably merchants who engaged in trade themselves. They were paid a token salary but part of their income was in the form of fees levied on the British trade in their port. Parke did indeed engage in the fishery, actually having a small fishing boat built in Iceland.

orders to Iceland to allow the Exportation of 300 Tons of Fish this Year, we were in hopes that our Iceland friends would have been at no loss upon this point after Mʳ P*arke*'s arrival[1] –

In Mʳ P*arke*'s certificate, that the Cargo of the *Orion* was Iceland Produce, we find that he refers particularly to the British Licence; as the Board of Trade have now resolved that no more Licences will be granted, & that the Trade shall be carried on under the order in council March[2] 1810; & as the Licences granted last year, only protect the vessels to Iceland, but leave them to bring here a Cargo of Iceland Produce without Licence to Leith or London, we hope Mʳ Parke is fully instructed of this being the intention of the Board of Trade, because we apprehend; that he may always hesitate about granting certificates to vessels without Licence, but entitled to trade under the order in Council –

We forwarded about 8 days ago a letter to Mʳ Parke, which appeared to be from the Secretary of States Office, & will have several opportunities of forwarding Letters to him during the next 14 days or three weeks, if the Board of Trade have any new communication to make to him –

We have the Honour to be
most respectfully
your obed*ient* hu*m*ble Serv*ants*
[*Signed*] Corbett, Borthwick & C*ompany*

Since writing the above we have learnt that a reply to our memorial to the Board of Trade is delayed by Lord Bathurst's absence from Town – May we use the freedom to beg that you will explain the circumstances of the case to Mʳ Rose[3] who cannot be intimately acquainted with it. –

[*Signed*] Corbett, Borthwick & C*ompany*

SL, Banks Papers, I 1:61. Autograph letter.[4]

208. John Parke to the Marquess Wellesley,[5] 20 August 1811

Reckevig

My Lord,
I have had the honor to receive Your Lordships Letter of the 12ᵗʰ Ult*imo* and in conforming with the Instructions therein given to me, I have addressed a Letter to all the Merchants in Iceland, of which the following is a Copy.[6] –

I am directed to make known to you that the Indulgence granted by His Britannick Majesty's Government to the Iceland Trade, must in future, be limited in regard to the

---

[1] Parke had not yet received the official instructions. See the following document.

[2] Actually February (see Appendix 10).

[3] George Rose (1744–1818) was vice-president of the Committee of the Privy Council for Trade and Plantations 1804–6 and had been reappointed 30 March 1807–29 September 1812.

[4] Note in Banks's autograph, on bottom left-hand corner of first page: '[Received] Aug*ust* 5 [1811]'.

[5] This is a copy, presumably sent by the Foreign Office to Banks for his information. The original is in TNA, FO 40/2.

[6] See Document 202.

Articles of Klip=Fish, and Fish Liver Oil, and that the whole amount of Klip Fish exported from Iceland in 1811, must not exceed three Hundred Tons, and that after the Expiration of the Year 1811 no more Klip Fish will be permitted to be exported from the Island. –

I have to request that you will furnish me, by the earliest opportunity, with an exact Account of the Quantity of Klip Fish, and Fish Liver Oil, which you may have in Your Warehouse, intended for Exportation, in order that I may appropriate to each Merchant, the Quantity he will be allowed to export;

I am directed further to make known to You, that I shall furnish you with a Certificate, setting forth the Quantity you will be permitted to put on board each of your Ships, and that all Klip Fish, and Fish Liver Oil, arriving in the United Kingdom, without my Certificate of the Article being a Part of the allowed Quantity, must be landed and subject to the High Duty.

You are, however, to understand that the Klip Fish, and Fish Liver Oil, so allowed to be exported to Great Britain with my Certificate, is not to be unladen but to proceed under Licence to some Port in Denmark, under such Conditions as may be therein expressed. –

I have further the request that you will furnish me with a detailed and correct account of all Linens, imported into Iceland from Denmark, by Your Ships and[1] the present Season, stating the Quantities and Qualities of each sort.

From the Information I have already collected, the Aggregate Quantities of Klip Fish and Fish Liver-Oil, now in this Country, and which has been shipped during the present Year, will not exceed in a great degree, the Amount permitted by His Majesty's Government to be exported; and I will take care to give your Lordship the required Information of the Quantity and the Quality of the Linens brought to this Country from Denmark, as soon as I shall receive it, in the meanwhile, I can so far inform Your Lordship, that, from the Enquiry I have been hitherto able to make, the Linens, most approved, are from the Manufactories of Scotland and Ireland. –

It may be proper to apprize Your Lordship, for the Information of the Lords of the Committee of Privy Council for Trade, that the Merchants, who can be considered strictly as Iceland Merchants, are W. Petreus, B. Sivertsen, C. A. Jacobæus, A. Thorlacius & N. Lambertsen[2] – The others, who have received Licences, are resident in Denmark, and it appears to me, that the object of the greater part of them, at present, is to withdraw their Property from this Country.

I have the honor to be &c
(signed) John Parke

SL, Banks Papers, I 1:62. Contemporary copy.[3]

---

[1] Should probably be 'in'.

[2] Westy Petræus (Dane), Bjarni Sívertsen (Icelander), Christian Adolf Jacobæus (Dane), Ólafur Þ. (not A) Thorlacius (Icelander) and Niels Lambertsen (Dane). The Danish merchants mentioned had long been settled in Iceland. The underlinings are in pencil.

[3] Note in Banks's autograph, verso: '7 October. A copy was sent the following day to the Council' [same information on FO original].

209. Corbett, Borthwick & Co. to Banks, 21 August 1811

Leith

Sir

We wrote you on the 19th Inst*ant*[1] with the information we had received from Greenland by one of the Whalers arrived here, about the Danish settlements in Greenland. May we now use the freedom for the first time to trouble you with the request of a little assistance on our own account.

In the month of June we shipped a valuable cargo of Colonial produce & British goods to Copenhagen & could probably continue to send some goods to Denmark, if we found it answer our expectations as to profit, but from the miserable state of the Exch*an*ge altho the goods have brought good prices there, there will be little or no advantage from them, if we are forced to get our payment by the purchase of Bills in England for the encouragement of the export of goods from Britain, in most cases a return Licence is granted for a ship carrying out Goods, even from Countries whence no Importation is allowed, without a previous Export from Britain, when the *Eliza* sailed with cargo from Leith,[2] which was worth about £7000, we applied for a return Licence which was refused, we did not then press it further, because by the then rate of Exch*an*ge we still hoped to be able to purchase Bills without Loss, but our last accounts which state the sale of the Goods, advise that the Exch*an*ge is within the last 3 Months, about 20 *per* Cent worse against remittances to Britain, & that our friends are much at a Loss how to convey our funds to us without Loss. Under these circumstances we have to day sent up a new Petition to the Board of Trade solliciting Licence to import from Denmark, & we shall be extremely obliged by your supporting our Petition, provided, you find it to be, as we conceive, an encouragement due to those who under the present difficulties endeavour to promote the Export of British goods – we have understood from your letters about our Iceland friends, that much of the difficulty about Danish Licences, arose from the inveteracy that appeared in the Public Edicts of that Government,[3] from our Intercourse with the subjects of that country, we have reason to believe, there is more form than reality in these Edicts, & that the Government are not unwilling to grant special exceptions to their strictest Edicts against Britain – when they can find a plausible pretext for it.

We have the Honour to be
Your most h*um*ble *Servants*
[*Signed*] Corbett Borthwick & C[y]

SL, Banks Papers, I 1:63. Autograph letter.[4]

---

[1] Not extant.
[2] The *Eliza* received a *söpas* on 27 July 1811 from Copenhagen to Eskifjörður. Kjartan Ísfjörð had in May petitioned successfully for a licence to export a cargo of wine, spirits etc. from Leith to Copenhagen (TNA, BT 6/203, nos 31, 840, 31,922 and 32,064). His petition to load grain in Denmark was not granted.
[3] After the bombardment of Copenhagen in 1807 Frederik VI had become extremely anti-British.
[4] Note in Banks's autograph, verso: 'Messrs Corbet*t* & Borthwick | [Received] Aug*u*st 26 [18]11.'

210. Everth & Hilton to Banks [August–September 1811?][1]

Thames Street
Saturday Morning.

Messrs Everth & Hilton present their respectful Compliments to Sir Joseph Banks & beg to inform him that their vessel the *Adventure* will clear for Iceland the middle of the ensuing Week, should he have any Letter or Package to forward to Mr. Parke, they shall be extremely happy to take charge of it for him.

They should also feel obliged by Sir Joseph informing them whether it is yet decided what quantity of Icelandic Oil may be imported[2] & if Duty free – & as they now look for the *Orion*'s[3] return if the Customhouse are in possession of the necessary information, to prevent any difficulties occurring should her cargo consist in Clip'd fish, which they were assurd might be imported duty free to the extent of 300 Tons. If it would not be intruding on his leisure Mr. Everth should be glad to wait on him on the subject.

SL, Banks Papers, I 1:67. Autograph letter.

211. John Everth to Banks, 3 September 1811

8 Thames Street

Mr Everth presents his Compliments to Sir Joseph Banks and encloses a Letter this Morning received from Mr. Parke, by the return of their vessel the *Orion*.[4]

Should Mr. Parke make any mention of having received his dispatches, he would esteem it a favor to be apprised, as among their return Cargo is about 40 Tons Klipfish, which is unaccompanied by any certificate, which they think must have arisen from the Instructions not having yet reached him.[5]

SL, Banks Papers, I 1:64.[6] Autograph letter.

---

[1] The Sutro Library has suggested this was written in September. The *Orion* arrived back in London at the beginning of September, thus this letter is written at the end of August or in the first two days of September.

[2] The decision was taken in July (see Document 208).

[3] There appear to be two ships of the same name sailing between Iceland and Britain. The master of this ship was Bennett Ireland and Consul Parke owned part of the cargo.

[4] See previous document.

[5] According to Document 208, the instructions did not reach Parke until late August. The instructions were despatched from the Foreign Office on 12 July 1811 and doubtless because of the difficulties in communication a duplicate was sent by Mr Lack on 23 September 1811 on Lord Bathurst's direction 'in a ship of Mr. Everth's'. Thus, Parke did not receive the instructions in time and Everth & Hilton were right in their assumption.

[6] Note in Banks's autograph, bottom left-hand corner: '[Received and answered] September 6–7 [1811]'.

212. Banks to William Jackson Hooker, 2 October [1811[1]]

Revesby Abbey near Boston

My dear Sir

Respecting Jorgensen all I can Say is, that I do not mean to give myself any Further Concern about him, his good or Evil Fortune will be Alike indifferent to me, it is not my wish to Excercise the office of an avenger & I thank my Good [sic] it is in no Shape my duty to interfere in his Case further than I have done[2]

many thanks for your hint on the Etymology of Word Sere[3] I have no doubt it is well founded & that your Countrymen 200 years ago had not wholly forgot the Saxon name of the month, which is Evidently the time meant by Tusser,[4] Johnson uses Tusser as an English author, but in many Cases miserably misunderstands his meaning in no Case more than in the Present one[5]

Sir Geo*rge* mackenzie has written to me to tell me that he had not when he wrote time to tell me any thing I wish he had waited till he had more time but he did not wait & I did not Learn any news whatever of Iceland or Elsewhere.

I place no dependence whatever on the accounts in the Paper relative to Park.[6] Mr Jackson who signd the Last Letter is the author of a book on Africa which has long ago beggard all my Stock of Credence.[7]

---

[1] Warren Dawson believed this to be written in 1813, but the contents of the letter indicate that it was written in 1811.

[2] Hooker's family, however, continued to take an interest in Jörgensen. Hooker's second son, Joseph Dalton Hooker (1817–1911), was a renowned botanist and explorer, a very close friend and collaborator of Darwin, and eventually succeeded his father as Director of the Royal Botanic Gardens at Kew. As a young man, in 1840, Joseph sailed on Sir John Franklin's expedition to Antarctica. He spent some months in Hobart where he was sought out by Jörgensen. He wrote to his father: 'I have seen him [Jörgensen] once or twice, but he is quite incorrigible; his drunken wife has died and left a more drunken widower; he was always in that state when I saw him, and used to cry about you.' Quoted by Dukewell, *The English Dane*, p. 233.

[3] Meaning *wood-sere*.

[4] Thomas Tusser (1524–80), agricultural writer and poet, best known for his work *A Hundreth Good Pointes of Husbandrie*, first published in 1557 and enlarged in 1573 with the title *Five Hundreth Pointes of Good Husbandrie*. In Hartley's 1931 edition of *Thomas Tusser. His Good Points of Husbandry*, woodsere is mentioned on pp. 65 and 77. See also Banks's own commentary and annotation of Tusser. This was produced in the years 1813 and 1814 by collating the 1585 edition with that of 1610: Kent History and Library Centre, Banks MS, U951/Z38. Banks's manuscript was, like so much else by him, never prepared for publication. See Chambers, *Scientific Correspondence of Sir Joseph Banks*, VI, pp. 53–4n.

[5] Wood-sere was used to denote the winter season (the time when there is no sap in the tree). Presumably the reference here is to Dr Samuel Johnson. See *Johnson's Dictionary*, where 'woodsare' is explained as 'the froth on herbs'.

[6] Mungo Park (1771–1806) was a Scottish explorer of the African continent. He received a surgical diploma from the University of Edinburgh and spent a year studying natural history under Professor John Walker. He was introduced to Banks, obtaining a post as assistant-surgeon on a ship sailing to Sumatra. Subsequently, in 1794, he offered his services to the African Association and was selected with Banks's support. He travelled to Gambia and after many tribulations reached the long-sought Niger at Segu, being the first European to do so. He returned to Scotland in 1797 to a rapturous welcome, having been presumed dead. He published his narrative *Travels in the Interior Districts of Africa performed under the direction and patronage of the African Association in the years 1795, 1796 and 1797*, in 1799. In 1805 he led a government expedition to Gambia, where he met his death.

[7] James Grey Jackson (1768–1840), Morocco merchant, author of *An Account of the Empire of Marocco and the District of Suse* (1809). Banks's own copy is in the British Library.

Sir Alexander Cannot have Saild yet.[1] I conclude he will not Sail till about Christmas or Later he will not Leave England I am Sure without Giving you due notice

beleive me my dear Sir
Very Faithfully yours
[*Signed*] Jos: Banks

Kew, Hooker's Correspondence, Director's Correspondence, I, f. 41. Autograph letter.

213. Corbett, Borthwick & Co. to Banks, 5 October 1811

Leith

Right Hon*oura*ble Sir J. Banks, K.B.
Sir
In the month of Aug*u*st we had the honour to communicate to you the contents of a letter then received from Greenland, which gave a very distressing account of that Colony, & stated the great joy felt, upon the arrival of the *Freden*, after all hopes of assistance from Europe was for that season given up. – The *Freden* has lately arrived in Leith,[2] & Cap*tain* Matthiesen fully confirms, the accounts we then received of the distress of the Colonies which he visited & from which, some of the other settlements, were forced to draw /supplies from/ such limited proportion of his Cargo, as could be spared to them by transportation partly in Sledges over the Ice, & partly by boats for some Hundred Miles; in some cases, the Natives had accompanied their demands of supplies from the Danes, with threats of violence in case of refusal, such was the miserable state of the Inhabitants, that most of the Europeans were anxious to escape from the settlements, then as thirty Passengers on board the *Freden*, & the Cap*tain* says that had he remained much longer, he believes his ship would have been filled with them. As the applications were continued from the clergymen, & all ranks of settlers – We trust the English Provision ship[3] is now with them, & that she carried out some supply of Powder & Shot, as we find, that what we sent for the *Freden* would be very soon consumed, & that from the want of it, the Native Greenlanders could not present their usual supplies for hunting, & became more burdensome upon the Colonists.

---

[1] 'Sir Alexander' is doubtless Sir Alexander Johnston (1775–1849), chief justice of Ceylon, who returned to Ceylon in 1811 after two years in England spent advising the Government on the administration of the island.
[2] There was great difficulty in obtaining permission for the *Freden* to continue to Copenhagen. In November a petition for permission for the ship with its Greenland cargo to go to Copenhagen was refused, 'the Oil being contrary to the Licence' (TNA, BT 6/205, no. 37,181). The ship then landed part of its cargo in Leith and wished to take the remainder to Copenhagen. This was refused by the Order in Council 'under which they have now wintered, the cargo must be landed' (TNA, BT 6/205, no. 37,416). Finally, on 30 December, the ship was permitted to proceed with her cargo of oil, blubber and other goods from Leith to Copenhagen, on condition that articles be exported for not less than £400 and to return with a cargo of grain, meal or flour only to Leith (TNA, BT 6/205, nos 37,729 and 38,008). The *Freden* did not set sail until the early months of 1812 after obtaining permission to take rye meal and hulled barley to export to Greenland (TNA, BT 6/205, no. 38,715).
[3] There were no English in Greenland so this was not a ship bringing provisions to them. In August–September 1811, however, two English merchantships came to Godhavn with supplies. A third ship came to Tassiussak in the district of Upernavik, which ended with the captain stealing blubber and skins and making off to sea. This was condemned and he had to pay restitution (see Gad, 'Tasiussaq-affæren 1811').

The *Freden* was upon her voyage hence detained by his Majesty's ship *Riffleman*,[1] & the Papers are now in Doctors Commons, in the Licence which she got here for carrying out Gun Powder & Shot &c from Leith, was inserted that she was to return from Greenland with a Cargo of Produce <u>except fish & oil</u>, this Licence was expired, & had it been retaind in /*Iceland*/ Greenland, no question could have arisen about the Cargo of the *Freden*, because by the Order in Council, she is allowed to import without Licence, Oil & Blubber subject /to/ the duties payable in Britain upon the articles of foreign fishing, if the /*Treasury*/ Board of Trade did not as has been done with Iceland & Faroe, (in consideration of its being a prohibitory duty, & in consequence of the views with which they opened a trade with Iceland, Greenland & Faroe) permit it to remain in the Vessell, & be sent forward to Denmark. –

As the whole of the assistance afforded to these settlements, has been the result of your humane exertions, you are fully acquainted with all the arrangements made while Count Tram*pe* was in London,[2] & since the Trade was opened, so that it is /*needless*/ unnecessary to mention that this Vessell is the same for which Licence was originally granted to Count Tram*pe* personally, when in London, & we presume that you would be sorry to see her condemned upon the voyage made under that Licence. The first Licence brought her to Leith, & to enable us to ship Gun Powder & Lead, we sent it to the Board of Trade & in place of it got the one /now/ with the Vessell, permitting to carry to Greenland, the Provisions there on board from Copenhagen with the addition of /Powder & Shot & other/ Articles from Leith; as the original Licence was lodged with the Board, you have no copy of it, we cannot be certain of its contents, but presume it was a copy of those granted at same time to Iceland, & Faroe, in which there was no restriction about Oil, & therefore we suppose that in the second Licence this Clause had been inserted by the clerks, as an invariable one in the common Licences, but without any Order from the Board, to alter the form of the original one, or to make this Licence different from those granted to the Icelanders.

The Trade of Greenland was formerly exclusively carried on by the Greenland Com*pany* in Copenhagen, & as the only means of supplying that settlement under the order in Council, they sent out the *Freden* belonging to M*essrs* Moienfeldt & Mylenphal,[3] in place of one of their own ships, & orderd these Gent*lemen*, the one of whom had been there 20 & the other 18 Years, that they should during the present state of affairs, use the Produce of the settlements for the supply of the Inhabitants in the best way they could, under the indulgence granted by England, selling it & repurchasing Provisions – upon the arrival of the *Freden* Mr. Motzfeldt received, the Licence upon which the Captors found their detention, but at [the] same time he received a printed copy of the order in Council in the one he found a restriction about oil, but in the other which was general, he found himself as an inhabitant of Greenland entittld to ship all kinds of Greenland Produce, to Leith or London, subject to the Revenue laws of this country; the vessel had arrived there after a most dangerous winter passage, & after retaining her there for some months distributing her supplies at the different settlements, he found himself forced, either to Collect stones to ballast the Vessell or to ship oil, (as besides this Greenland only

---

[1] Captured on 4 September 1811 and later liberated on 30 October 1811 (see Document 225).
[2] His activities included clearing up the Iceland Revolution business, trying to obtain the release of Danish prisoners-of-war and obtaining licences for Greenland and the Faroe Islands.
[3] The Greenland inspectors.

furnishes some few Skins) the shipment of that article was permitted by the order in Council, & from the motives under which the order was issued he could scarcely suppose, that a special Licence granted to the only vessell that had been able to take advantage of the humanity of the English government, could by any other means, than a mistake in writing it, take away one of the principal privileges granted in the general one – under the circumstances the oil was shipped, & we conceive that the condemnation of it would not only be very much in opposition to the particular views which induced you & the board of trade, to permit the communication with these defenceless & distressed colonies of Denmark, but contrary to the liberal interpretation which foreigners always expect to be given to English Laws – it was difficult for a foreigner in Greenland to judge exactly, betwixt the order in Council, & Licence, which he should most implicitly follow, & not surprising that under the particular circumstances of the case, he should put so much confidence in the general order in Council, as to trust this property under its protection.

As the trial of this case in Doctors Commons would not only be very tedious, but very expensive, we have stated it to you, in hopes that your usual interest in the distresses of these colonies will induce you to explain to the Kings advocate[1] the views of the Board of trade in the arrangements with Count Trampe, about the Greenland trade, which would probably satisfy him as to this Case – of the Passengers for *Freden* two Crew & their families are in Leith, & the remainder about 20 are on board the Flag Ship, in the roads, this unexpected detention is very distressing to them, & their situation will we hope also influence you in taking a little trouble in this matter – We could probably by application to the Admiral, get these from the Flag Ship, but we have no immediate means of sending them to Denmark & as they have no funds of their own it would show the support of the whole upon us, as during the detention as a prize, they cannot be received on board the *Freden* or allowed to use the Ships Provisions.

By the ships now arriving from Iceland, we are sorry to learn that last season has been very unfavourable for their fisheries, & that the Farmers have also suffered much from bad weather, amongst their sheep.

We have the Honour to be
Your mo*st* ob*edien*t H*u*m*b*le S*i*r
[*Signed*] Corbett, Borthwick & C*ompan*y

BL, Add MS 33982, ff. 20–23. Autograph letter.

214. Thomas Lack to Banks, 15 October 1811

Office for Trade, Whitehall

Mr. Lack[2] has been directed by L*or*d Bathurst[3] to forward the inclosed to Sir *Joseph* Banks with his Lordship's Compliments.[4]

SL, Banks Papers, I 1:68. Autograph letter.

---

[1] Sir Christopher Robinson was appointed the King's Advocate on 1 March 1809, a position which involved him in most prize cases.

[2] Thomas Lack was a clerk at the Board of Trade (*Holden's Annual London & Country Directory for 1811*, III, p. 34).

[3] Earl Bathurst was President of the Board of Trade: see Appendix 13.

[4] The enclosed was Parke's letter of 20 August 1811, according to the Sutro Library catalogue.

215. Corbett, Borthwick & Co. to Banks, 15 November 1811

Leith

Sir

We have had the honour to receive your Letters of the 11th Oct*ober* & 8th Inst*ant*:[1] & were much gratified, to find your kindness towards our northern friends unimpaired, by the constant trouble that their affairs & their wants impose upon you. We have now the pleasure to inform you that <u>the *Freden*</u>[2] & her Cargo have been <u>restored</u> in Doctors Commons,[3] to the Greenland Owners, as protected <u>by the order in Council Feb*ruary* 1810</u>[4] & altho strictly speaking the owners might be considered, as having themselves to blame, we are sure you will be well pleased with this decision, because it is not extraordinary that foreigners, not very well acquainted with our laws, should have made the mistake, of believing the general order in council of higher authority than the particular Licence for the ship; & finding that they must either allow her to go back empty after such a tedious & long voyage, or ship Oil, it was natural enough for them to trust a good deal to the known liberality of the English Government in interpreting such a law, which was only issued from pity to them in their forlorn situation. – <u>We have now petitioned the Board of Trade to allow this Vessell to proceed to Denmark, in the same way as the Iceland & Faroe ships,</u> & we have used the freedom to desire our Agents to intimate to you, when Our Petition will be under consideration of the Board, in the hope that you will be inclined to attend that day, & as the voyage of this vessel has already occupied nearly 18 months, & consequently incurred a heavy expence upon the owners, we trust the Board will not be inclined to make it almost a ruinous voyage by subjecting the owners to the duty on the oil, which is fully as much as its value; we have in our Petition, asked permission for this vessell to make another voyage with Provisions to Greenland, & to return with Greenland Produce, & have done so in ignorance of the exact views of government with regard to that settlement, & whether they mean to take the supply of its inhabitants entirely on themselves; & which from the difficulties attending the permission to import their only considerable and valuable article of Produce, <u>Whale Blubber & Oil</u>, we consider by far the simplest & best plan for the inhabitants; if this continues to be the view of the Board of Trade, we hope they will feel the less hesitation in granting the first part of our Petition, in allowing this single voyage under the order in Council to be completed to Copenhagen, as the inhabitants of Greenland have not previously had any advantage from it, nor will hereafter –

Within these few days we have received <u>a letter</u> from Greenland dated 14 Sept., by which we were made happy to learn <u>that both the ships from London had arrived safely; & discharged their Cargoes,</u> so that we are satisfied, that the serious apprehensions of famine are, now removed, tho the only material Articles omitted in their shipments, was by our letter, <u>Gun Powder & that, the want of which will be of serious inconvenience to the Native Greenlanders</u>, & almost preclude their obtaining for themselves their usual food. Our correspondent also mentions that there had been <u>no Butter</u> sent there from the return of these vessels to London, the Board of Trade will now be more able to judge, whether the supplies to Greenland can best be sent from England, or if they would be

---

[1] Neither is extant.
[2] All underlining in the letter is pencilled in, presumably by Banks.
[3] Doctors' Commons was where the Admiralty courts were housed.
[4] Appendix 10.

inclined to allow them to be sent from Denmark in the same way as to Iceland & Faroe; you will do us a particular favour, by writing us, if the information brought by these vessels induces the Board, to allow the plan adopted last Year, so that we may take such steps as may be necessary, in the event of the Colony being left to supply itself under the order in Council, to get their earlier dispatched than during the last two Years. – it would also be important for us to know, any particular plan the Board might adopt for the transmission of the supplies; if we even allowed Licences to bring the Provisions here from Denmark, we might then use English vessels from Leith to Greenland, if by them the return could be allowed to be made in oil or Blubber, & that to be warehoused for Exportation.

Our Letter complains that <u>an English ship belonging to Hull, had landed her crew in Greenland, & had taken away by force from the inhabitants 160 Barells of Oil, & some skins</u> – this infringement of the order in council which was meant to insure the defenceless inhabitants from all molestation seems to be most wanton & cruel, & very deserving of punishment, or at all events that the value of the Property should be restored, by the violators of the express orders of his Majesty's Government, as the circumstance has been represented from Greenland to Mr. Mellish,[1] as the person employed by the Board of Trade, to supply Greenland, he has probably intimated to the Board /of Trade/ perhaps it may be necessary to institute a process against the Cap*tain* or owners in the Admiralty Court, in which case it will be done either by us or Mess*r* Wolffs & Dorville[2] in behalf of the inhabitants if Mr. Mellish declines to interfere, the Cap*tain* of the ship, <u>must also we suspect have perjured himself</u> upon arrival at Hull, by declaring the whole of his oil to be of British fishery, to evade the foreign duty. –

We have also by this Post forwarded a Petition <u>for the Iceland ship *Swan*[3] to be permitted to proceed to Denmark from Leith</u>, & as one or two such Licences have been granted, we hope it is upon a general principle, & that the same will now be granted for the Iceland vessells that are still expected, without any particular dissension at the Board.

We are much obliged by your offer to assist the Greenland Emigrants but are glad we shall not have occasion to trouble you, as the Admiralty upon the case being stated to them, immediately ordered a Flag of Truce to carry them to Toningen,[4] so that they go home with a new proof of the humanity of the English government, they sailed about three weeks ago. Admiral Otway[5] was extremely attentive to their application & seemed anxious to give them as much convenience as possible on their voyage to Denmark, several of them had their wives & families.

We are with much respect
your mo*st* ob*edien*t &. H*u*mble S*ervant*s
[*Signed*] Corbett Borthwick & C*ompan*y

SL, Banks MS, I 1:99. Autograph letter.[6]

[1] William Mellish, a prominent London merchant.

[2] Wolffs & Dorville: see Document 105.

[3] The *Swan* was permitted to take her Iceland cargo to Denmark and return with grain to Leith to add permitted goods, proceed to Iceland and return to Leith with Iceland produce (TNA, BT 6/205, nos 37,182 and 37,480). They were asked about the quantity of fish oil and were obliged to export for the equivalent of £300.

[4] Tönning is on the North Sea coast of Schleswig-Holstein.

[5] Possibly William Albany Otway (1755–1815) who on his return from the Scheldt expedition in 1809 was appointed commander-in-chief in Scotland. He became a vice admiral in 1811.

[6] Note in Banks's autograph, right margin of last page: 'Corbet*t* & Borthwick | [Received] Nov*ember* 11'.

216. William Henry Majendie to Banks, 27 November 1811

Bevis Hill near Southampton

Sir,

I feel greatly flattered by the condescension with which you have noticed the anonymous letter which I took the liberty to address to you on the subject of Iceland,[1] and pledge my honor to hold in sacred secrecy every part of your candid and confidential communication. I was aware at the time I ventured to suggest the good policy and humanity of occupying that Island, that I was addressing one, whose thoughts had been benevolently employed, and whose influence had been steadily exerted in furtherance of this good cause; but I never contemplated the objection which you hint at as the most probable impediment opposed to the accomplishment of the scheme; and I may now express a natural though perhaps an ignorant surprise, that any consideration of the present defenceless state of Iceland should stand in the way of a project so promising in good consequences.[2] Adverting to the slender means of annoyance possessed by those who would most tenderly feel & most anxiously resent the measure recommended, and to the general poverty of so uninviting a country, I feel assured that every benefit which could result from its occupation might be secured, and much of the evil arising from the animosity of its late proprietors might be averted by placing the capital, and principal part of the Island in a state of such decent security as would furnish protection against the occasional visits of privateers. To that point the natives would resort to dispose of, & secure the produce of their country; & such slight works as would sufficiently protect this grand insular depôt from the insult & depredation of its enemies might be manned by a company or two of men raised on the Island, & maintained at an inconsiderable cost. I am at present without the materials requisite to form an accurate notion of the exact expenditure in which such an establishment would involve the Island; but it would not be burthensome enough to outweigh its prominent good consequences; and disused as the Icelanders are from all military service, it is surely not unreasonable to conjecture that a population of 50,000 people could furnish 150 men with all the essential qualities of soldiership.

Whilst I combat the solitary objection which seems to be stated to my project, and in conformity with my promise subscribe my real name, I hope it will not be deemed impertinent to add a few particulars which may explain the liberty I have taken in intruding my thoughts upon you. I am the eldest son of the Bishop of Bangor,[3] and a Captain in the Army of nearly 5 years standing; & having suffered in health by a residence in the East Indies, where I served on the personal staff of General Hewitt,[4] I am at this moment inclined to quit a profession into which I can no longer actively enter with any hope of being equal to its fatigues and hardship. As my prospects in life are handsome & perfectly independent the sacrifice of the profession in which I embarked is a matter of no moment; but I cannot in consequence reconcile myself to a life of mere vacancy and

---

[1] This letter does not appear to be extant.
[2] This letter is important as here Banks appears to have pinpointed the major obstacle to the annexation of Iceland, the question of its defence.
[3] Henry William Majendie (1754–1830), bishop of Bangor and Chester, of a family of Huguenot descent.
[4] Sir George Hewitt (1750–1840) 1st baronet, commander-in-chief, East Indies, 1806–11.

idleness: and as I have considered much the subject on which I have addressed you, I have not been without the hope that my humble services, offered without a single sordid motive (to which I thank God, circumstances make me superior) might be dedicated to some subordinate share in an undertaking which cannot fail to interest everyone with a grain of common feeling.

Hoping that you will consider the latter part of this letter as strictly confidential,

I have the honor to be Sir
Your most obedient
humble Servant
[*Signed*] W: H. Majendie
Cap*ta*in 17th Re*gime*nt

SL, Banks Papers, I 1: 69. Autograph letter.

217. Banks to Captain William Henry Majendie, 27? November 1811

all I can say Sir is that in Case the Island's taken possession of & I am Consulted relative to the Establishment your offer Shall not be forgotten[1]

SL, Banks Papers, I 1:69v. Autograph draft.[2]

218. Report of the *Gratierne*, 2 December 1811

Leith

Report, of the *Gratierne*, Thron Olsen, Master, from Iceland

Home Consumption

4 Bundles containing 8 salted sheep Skins
8 Casks & 2 Matts[3] containing 13 CW Tallow
1 Box 75 Foxskins & 7 Swan Skins
1 Barrel with 4600 Swanquills, 6 Foxskins & 40 Lambskins
1 Box 4800 Swanquills, 9 Foxskins & 1 Swan Skin
1 Box 140 Foxskins
24 Foxskins
8 dried Sheepskins
1 Box 1 Barrel & 1 Bag 1325 Lambskins
250, 0, 0 Stockfish

Exportation

125 Casks oil
577,0,0 Stockfish   20 fish each[4]

[1] No further correspondence between the two has been found and Iceland was of course never annexed.
[2] This is written by Banks at the bottom of the last page of Majendie's letter. This would be the draft of his reply.
[3] A *matt* is a bag or sack made of matting (in this case, of tallow).
[4] Pencilled in by Banks

32 Tons salted Fish   Klip fish   64,000 lb.[1]
1627 Worsted Frocks
7677 pairs worsted Stockings
60,160 pairs Mittens
170 Bags Wool
10 Bags & 3 casks Feathers
12 Bags Eiderdown   in all 600 lb[2]
1 Box with 50 Dogfishskins,[3] 42 Eaglequills, 30 pairs worsted Stockings
687 pairs worsted Stockings, 35 Frocks
4 Chests ⎫
1 Box   ⎬ Household Furniture
3 Casks   ⎭

SL, Banks Papers, I 1:71.[4]

219. From Corbett, Borthwick & Co to Banks, 7 December 1811

Leith

Sir

We were truly sorry to learn from our Agents in London, that our petitions with regard to the *Freden* Greenland vessell had been refused by the Board of Trade, both with regard to another voyage to Greenland; & for permission to sail to Denmark with part of her present Cargo in the event of a further voyage to Greenland, being refused, & that their Lordships seemed to insist upon the Payment of Duty on the Blubber & Oil.

    In our first letter to you, we stated the causes which probably influenced the settlers in Greenland, in making this shipment, namely from their want of sufficient experience in the use of Licences, to be able to appreciate the force of the Prohibition in the Licence, contrasted with the order in Council, which they saw gave them leave to import to Leith all kinds of Greenland Produce, whilst from their Ignorance of our Revenue Laws, they know not the particular situation of the Blubber & oil, when imported under the order in council, without Special indulgence from the Board of Trade, & also from their having no other Produce to Ship in payment for the Provisions they got, & therefore as they saw, that they could have no benefit from the Order in Council, without shipping the only valuable Article in the Settlement, it does not seem surprising, that they should have ventured to send it, trusting to the Order in Council which was avowedly issued in Commiseration of their sufferings, & which they would naturally expect, would not be interpreted very rigidly against them: Under these circumstances we trust that some indulgence will still be shewn to them in this case.

---

[1] Pencilled in by Banks.
[2] Ditto.
[3] A dogfish is a small species of shark.
[4] This list accompanied the letter from Clausen dated 20 December 1811 and the petition to the Privy Council of the same date. Note in Banks's autograph, verso: 'Clausens Cargo | Manifest.'

The Duty upon Blubber Foreign Fishing is £18.13.4 per Ton & the value of that on board the *Freden* if sold here is £10 or at the Utmost £12 per ton the duty on the oil is £28 & we scarcely suppose that the Quality of that in *Freden* would enable it to be sold, above £24 to £26 per Ton (these prices we calculate from the sale of a Cargo exactly similar sold here as a prize about 3 years ago) & therefore it would be a much smaller loss for the Proprietors to abandon it to the Customhouse as it now lays in the Ships for the duties, than to pay them, & land it here; we are convinced that this comparison of the value of this Article & the Duty payable upon it was, not in the View of the Board of Trade, when they felt inclined to insist upon the payment of duties, & we trust, judging from the usual liberality of their Views, that when informed of this the real situation of the Proprietors, the Board will not be inclined to impose upon them the total loss of the Blubber & oil which would be the result of its being subjected to the legal duties here, but which we believe they did by no means intend. –

As this is the first Shipment from Greenland under the order in Council & as from the resolution of the Board of Trade to supply that settlement with provisions from England it will be the only one whilst so many Iceland Ships have had the Benefit of the order in Council, & Carried the same Article of oil, We are inclined to think that you would regret that in the only shipment made from Greenland, the most valuable Part of the Cargo should be abandoned to the Crown as not Worth the duties because altho it may be said by a person fully acquainted with the Laws here, & the nature of the order in council & Licence that they suffer thro their own fault, yet to Foreigners it must rather appear a deviation from the Views held out as their object in Order in Council, that this property, when protected in a Court of Law, from the claims of a British cruizer, should be afterward taken from thence by the Revenue officers; under these circumstances we have ventured to give in another Petition to the Board of Trade, copy of which we use the freedom to inclose & we hope that if you are not able to attend, when it is under consideration, you may be induced to send a note to Lord Bathurst[1] recommending it to his consideration. As those in Denmark who have friends in Greenland are naturally in its present situation most anxious, about their means of getting supplies, we hope you will favour us with a few lines, to enable us to allay their fears by letting them know the intention of the English Government towards the settlers in that remote Corner

We have the Honor
To be Your obedient Humble Servants[2]

SL, Banks Collection, I 2:7. Autograph letter.[3]

---

[1] See Appendix 13.

[2] This letter is not signed.

[3] Note in Banks's autograph, on right-hand margin of last page: 'Corbett & Borthwick | Greenland Ship Freden'.

220. Petition. [Corbett, Borthwick & Co.] to the Lords Commissioners of the Board of Trade, [unknown date but 7–11 December 1811].[1]

Unto the Honourable The Lords Commissioners of the Board of Trade
My Lords
We learn with much regret, that your Lordships have refused both our Petitions on behalf of the Greenland ship *Freden* & that you seem inclined to subject the oil & Blubber to the payment of the duties Law.

We now beg leave with much submission, to state that the duty on Blubber of Foreign Fishing, is £18.13. 4 per ton, & that the value here is £10 & at the utmost £12 per Ton, that the duty upon the oil is £28 per ton, & the value here is £24 to £26 per ton, for such as is now on board the *Freden*; these valuations all taken from the sale, of a similar cargo here, about 3 years ago, & from the best information we can get of its present value, & therefore it will be much better for the Proprietors, to abandon the Blubber & oil to the Custom house, for the duty, as it now lays, than to pay the duty & land the Oil, & we further believe, that by the Law goods cannot be sold by the customhouse, at a smaller price than the duty, in which case the Blubber & oil must be destroyed; We do not suppose that your Lordships were aware, that the duty on the oil was more than its value, & in the belief, that you did not intend, to cause a total loss to the Proprietors; we now venture humbly to sollicit, under the consideration, that a licence may be granted, for the Trader to proceed with the Cargo now on board, & goods permitted by Law from Leith to Copenhagen.

It appears of this our earnest petition, in behalf of the Greenland proprietors, we beg leave further to state, that in that remote settlement, a very distinct judgement could not be expected, upon the point of the Clauses in the Licence, contrasted with the general order in Council, a Copy of which they received at some time, & from which they considered themselves allowed to ship Greenland Produce generally, that they were ignorant of the value here, & the duty, & had considered that the Iceland vessell had carried Fish Oil; – whilst the other settlements mentioned in the order in Council had many Ships; this is the first one from Greenland, & as your Lordships now intend to supply Greenland with provisions from England, it will be the only one.

Under these circumstances We humbly hope that your Lordships will yet be induced to grant the Licence petitioned for.

SL, Banks Papers, I 2:6. Contemporary copy.

---

[1] It obviously follows the letter from Corbett, Borthwick & Co. of 7 December 1811 (Document 219). The petitions were refused on 19 November (BT 6/205, no. 37,181) and on 29 November (BT 6/205, no. 37,416). This petition must have been written after then and before 11 December when 'nothing was ordered' but it was eventually granted on 30 December (BT 6/205, no. 37,729).

221. Banks to William Jackson Hooker, 18 December 1811

Soho Square

My dear Sir

I have seen Sir George & am Quite Satisfied that his Book will never be So Popular as yours[1] I am glad to see that he has Omitted to give us a map of the Island it is a thing at this moment very much wanted the map of 1771[2] is far from a Good one & may be Corrected by the French Observations of 1776[3] I Submit to you the Propriety of adding it to your book; the charge for Engraving I think would not exceed £50 That & Printing I think would be well Repaid by an additional Charge as it would Sell [to] many People who will not Trouble themselves About the book because there is no map nothing I think would lend so much to increase the interest of the Good People of England in the Island & of course tend So much to increase the Chance of our taking Possession of it[4] as the Publication of a map

adieu my Dear Sir
in much haste yours –
[*Signed*] Jos: Banks

Kew, Hooker's Correspondence, Director's Correspondence, I, f. 34. Autograph letter.
Hermannsson, 'Banks and Iceland', p. 91.

222. Holger Peter Clausen to Banks, 20 December 1811

St Paul's Coffeehouse

Sir!

According to your favourable permission I respectful*ly* inclose a copy of my petition to His Majesty's most honorable Privy Council, which, in consequence of the advice of Mr. Horneman is subscribed by myself.[5]

In consideration of the circumstances and principally on account of your influence I hope their Lordships will grant it.

With the highest esteem I have the honour to sign myself,

Sir,
Your most humble and most obedient Servant
[*Signed*] Holger P. Clausen jun

SL, Banks Papers, I 1:70. Autograph letter.[6]

---

[1] Sir George Steuart Mackenzie published his book *Travels in the Island of Iceland during the Summer of the Year 1810* in Edinburgh in 1811. It sold well, a second edition (with some changes) was to appear in 1812.

[2] The map of 1771 is inscribed *Nyt Carte over Island forfattet ved Professor Erichsen og Professor Schönning. Aar 1771* (New Iceland map by Professor [Jón] Eiríksson and Professor [Gerhard] Schönning. Year 1771). This map was printed in *Reise igiennem Island* (*Travels in Iceland*) by Eggert Ólafsson and Bjarni Pálsson (see Introduction, p. 21 above). Haraldur Sigurðsson is of the opinion that this map was used as the base for the map published in von Troil's *Letters on Iceland*. Further on this map, see Sigurðsson, *Kortasaga Íslands*, II, pp. 160–63. There is a map in the 1805 English edition of *Travels in Iceland* by Olafsen & Povelsen.

[3] The French government sent expeditions to Iceland at this time. See Introduction, p. 16 above.

[4] It is interesting to note that Banks is still interested in annexing Iceland, as will be seen later (Document 258).

[5] See the following document.

[6] Note in Banks's autograph, verso: 'Clausen'.

223. Petition from Holger Peter Clausen to Lords of the Privy Council, [20 December 1811][1]

To the Lords of His Majesty's most honorable Privy Council
The humble petition of Holger Peter Clausen of Iceland, merchant.
Sheweth
That the Iceland Ship *Gratierne* Thron Olsen master of about 240 Tons burthen is arrived at Leith with the following goods reported for exportation: viz 'Stockfish, salted fish or Klipfish, Shark liver oil, Wool, worsted Frocks, Stockings & Mittens, Eagle Quills, Foxskins, Eiderdown, Feathers, Dogfishskins, Household furniture.'

That the said vessel having suffered severely by the loss of the foremast & bowsprit & other damage on its present voyage, it would require a repair on its arrival at Copenhagen and that the repair would probably take so much time as to prevent its return to Leith & Iceland early enough in the ensuing year to make good its whole voyage that your petitioner therefore is desirous if the said repair should be considered to take too much time to be permitted to return to Leith with an other vessel in lieu of the *Gratierne*.

Your petitioner therefore humbly prays that he may be allowed without unloading to export the aforesaid goods & such articles as are allowed by law to be exported, to the port of Copenhagen, there to dispose of the cargo with liberty to go with this or another vessel in lieu thereof in ballast to another port in Denmark or Holstein, there to take in a Cargo of Grain, Meat Flour & pearl Barley and to proceed to a port in Great Britain to dispose of part of the cargo to pay expences & to compleat it with such goods as are allowed to be exported & to proceed to the island of Iceland and from thence to return with a cargo of Iceland produce to a port in Great Britain.[2]

Your petitioner will duly bound ever pray
[*Signed*] Holger P. Clausen j[r].

SL, Banks Papers, I 1:72. Autograph letter, a copy sent to Banks.[3]

224. Holger Peter Clausen to Banks, 26 December 1811

St. Paul's Coffeehouse

Sir!
The following statement is a translation of one sent to me, by the master[4] of the greenland ship *Freden* in the danish language.[5] I beg pardon for the incorrectness of the language I should never have ventured to translate it myself but the inaccuracy of the interpreter to whom I had delivered it, obliged me to take it back and do it myselff so good as possible.

---

[1] The Sutro catalogue dates this letter to 20 December 1811, when indeed the petition was addressed by the Board of Trade (TNA, BT 6/205, no. 37,857). According to the records of the NAS (E. 504/22/53) the ship arrived about 4 December in Leith from Iceland and continued on to Copenhagen on 10 December, doubtless travelling on the licence mentioned in Parke's letter to Wellesley (TNA, FO 40/2, 9 July 1811).

[2] This was granted on 20 January 1812 on condition that the petitioner load and export in addition articles permitted by law to be exported from Great Britain to the amount of £800 (TNA, BT 6/205, no. 37,875).

[3] Note in Banks's autograph, verso: 'Clausens Petition'.

[4] Captain Lauritz Mathiesen.

[5] This statement is enclosed, see following document.

– I beg leave to observe that it is not my statement but the masters of the *Freden*, in which I did not think myselff entittled to make any alteration, which I likely else should have done; not that I have the least doubt about the sincerity of it but I should probably have delivered it in a different a manner. I have nothing more to add, no arguments to plea for him but hope only for pity. In this hope I remain

Sir,
your most humble & most obedient servant
[*Signed*] Holger P. Clausen jr.

Wisconsin, Banks Papers, ff. 77–78. Autograph letter.[1]

225. Statement from the Master of the Greenland ship[2] *Freden* now in Leith, sent to H. P. Clausen Jr, 26 December 1811

In the year 1810 the count of Trampe came from England and brought a Licence permitting a ship to supply Greenland. I was then call'd by the Directors for the Greenland trade and ask'd if I would undertake the management of the ship *Freden* now it was resigned to the inspectors for the Greenland trade. Mr. P: Motzfeldt & Mr. N. Myhlenphort[3] together with such a quantity of Green*land* produce as the ship could carry, in order to enable them to procure the necessary food for themselves and the inhabitants, by barter, either from England or Denmark. Upon these arguments I could see no risk in fulfilling the wishes of the Directors, especially as the only intention was, to relief the innocent inhabitants from famine, which certainly had been their fate if not prevented by the arrival of this ship. – The said inspectors have been engaged in the Greenland trade. Motzfeldt in 17 & Myhlenphort in 27 years;[4] in which time it is to be presumed they may have saved so much as to be able to make this only expedition for their own account.

I made sail from Copenhagen the 17th of July 1810[5] with a cargo of provisions for Leith, from where I applied for the necessary quantity of gunpowder & shot & tobacco which graciously were permitted.[6] The 25th of August I proceeded from Leith without having the least presumption that blubber & oil not might be imported to England as those articles are the chief productions, and without the permission to export these articles it is impossible to avoid famine as the inhabitants have nothing else to pay with.[7]

Halfsunk and with the greatest trouble I arrived the 6th of November to one of the northern harbours (Egedisminde:)[8] to the greatest joy of the inhabitants who had renounced all hope of seeing any ship so late in the year. – Now some of the

---

[1] Note in Banks's autograph, verso: 'Greenland | Distressed State Children | Destroyed'.
[2] Captain Lauritz Mathiesen.
[3] The Danish inspectors in Greenland. See Document 188. The *Freden* had been temporarily 'privatised'.
[4] Motzfeldt arrived in 1794 and Myhlenphort in 1786.
[5] This is indeed correct, Trampe was able to procure British licences for the Faroe Islands and Greenland.
[6] The only Danish ship to arrive in Greenland in 1810 was the *Freden*. However, the Greenlanders were able to trade with British whalers (see Saxtorph, 'Nordgrønlands Inspektorat', pp. 27–30).
[7] The underlining in this letter is done in pencil, probably by Banks.
[8] Egedesminde (now Aasiaat) was founded in 1759 by Captain Niels Egede, son of the famous missionary Hans Egede, who named the colony (the seventh in Greenland) in memory of his father. In 1805 there were 51 inhabitants in Egedisminde itself but 218 in the whole colony.

Establishments in the Discobay¹ were relieved as they could fetch provisions upon sledges drawn by dogs, but the whole south part was entirely exposed to the greatest want and impossible to give them any relief in the winter. – The two northern Establishments Omenak² & Upemariek³ were in the same distress, and upon the later place the want was so great that they threw their newborn children for the dogs or interred them living, and as the merchant⁴ represented to them the impiety of such a deed they told him that if the family became larger they unavoidable all must perish; which they considered as a full excuse. – The want of gunpowder & lead causes extreme miseries in the winter as nothing then can be catch'd by boats, the sea being entirely frozen up. – The evening before the new years day there came a great number of greenlanders in to the merchant with large knives and demanded victuals; he had nothing himselff besides some seals=flesh which he was forced to give them, they should else have killed him. – Now he was obliged to travel with 4 sledges 200 danish miles (:about 900 english:) over the ice and large mountains, to the Diskobay, to get advice & help from the inspector. – He got so many provisions as those 4 sledges could carry and returned but was obliged to pass the winter at Omenak and live upon seals=flesh only. The want in the southern part was above any description, where there live some missionaries & other europeans.

I lay fast in the ice untill the 15th of June when I sawed the ship out to bring the southern part some little relief spared from the north, there I arrived the 26th of July. – Now I proceeded from there so quick as possible took 30 passengers on board for in the least to save them from famine. 30 more would have gone with me, but were prevented from coming to the ship by ice. Upon my home voyage I was brought in to Leith by the Engl: gunbrig the *Riffelman* where I arrived the 4th of September. – The 30th of October the ship and cargo was liberated when I paid the captors expences which were moderate. – Now I learn that the duty must be paid to which both ship and cargo not will be sufficient, because a great deal of the blubber is old and rotten, black as tar and some of it from the year 1807, but might yet give some provenue at Copenhagen to purchase corn & grains to save the rest of the inhabitants from famine, if the Brittish Government graciously would permit only the single ship to bring what relief it could to the miserable inhabitants. Formerly Greenland employ'd 8 large ships.

I have been engaged in the Greenland ships as master & mate in 21 years; in the course of that time a great number of English whale-fishers have had all succours, the Establishment could afford. – Even in this year there came an English Cap*tain* Forster to the inspector Mr. Motzfeldt and complained of want of bread, and tho' the inspector himself was convinced that he must shorten it from his own allowance he delivered him one hundred weight.

That this whole statement is the real truth I assure upon honor.

Wisconsin, Banks Papers, ff. 81–82. Autograph letter⁵

[1] Disco (now Disko) Bay, is just north of Egedesminde.
[2] Umanak (now Uummannaq), was a colony founded in 1758 but moved to its current location in 1761.
[3] Upernavik, founded in 1771, was the most northern station.
[4] Frederik Friedlieb Rosbach (1788–1821) was a merchant in Upernavik 1809–11.
[5] Notes in pencil in Banks's autograph on bottom of the page: 'Egedesminde | Discobay | Omenac | Upemaric | Ojoveng [This name does not exist. The most likely candidate is Ritenbenk on Arveprinsens Eiland north of Jakobshavn].'

226. Boulton & Baker to Banks, 2 January 1812

Chatham Place

Sir

We have just receiv'd a Letter from Messrs Orum & Wulff respecting their Iceland Trader the '*Eskefiord*',[1] for whose Safety we entertain'd serious Apprehensions from what we had heard by other Vessels lately arriv'd; They tell us however that she has been driven into Norway near <u>Bergen,</u> in a very shatterd State, so as to require a heavy & expensive Repair; for which Purpose, as well as for the Preservation of what is not wholly dammaged, the Cargoe must be taken out. Under these Circumstances we are extremely desirous to obtain Permission for her to proceed, when she can be again made fit for Sea, <u>direct</u> to Denmark, in order that the Time she will inevitably have lost (Two Months at Sea, & certainly not less than two more in repairing) and the heavy Expence incurr'd, may not be still farther encreas'd by her coming to <u>Leith</u>.

As her Licence expired on the 26 October last, we have prepared the accompanying Petition[2] for a fresh one, and entreat your kind Interference & Support, which you in former difficulties so humanely bestow'd upon us. –

If from any Circumstances, you feel averse to support our Prayer in its present Form, and would recommend any Alteration therein, or favour us with your Advice on that Subject, the Gentleman who brings it, and has occasionally had the Honor of waiting upon you from us, will receive your Instructions either now, or at any time you shall appoint; or, if you permit, our Mr. Baker (the Writer hereof) will have that Pleasure whenever it shall be agreable to you. –

We beg to subscribe ourselves very respectfully, Sir,

Y*ou*r obliged & obed*ien*t Serv*an*ts
[*Signed*] Boulton & Baker

SL, Banks MS, I 1:73. Autograph letter.[3]

227. Banks: Notes on Trade 4 January [1812][4]

The *Fredens* Licence has been granted on Condition of her Exporting a Certain Quantity of Colonial produce when she has Completed that part of her Cargo I know of no

---

[1] The *Eskefiord* received a Danish sea-licence (*söpas*) in May 1811. The vessel was permitted to sail from Newcastle instead of Leith for Iceland (TNA, BT 6/204, no. 33,962). The ship made it to Iceland, but had a very difficult return voyage and was forced into Norway. At the beginning of January 1812 the ship was in Bergen for repairs. In 1812 it proved difficult to procure a licence for Copenhagen; however, eventually after repeated petitions a licence was finally granted in August on condition the ship exported goods for £5 a ton. Thus the *Eskefiord* made it to Iceland very late in the season (TNA, BT 6/206, no. 42,655; BT 6/207, nos 45,020, 45,298).

[2] Not extant.

[3] Note in Banks's autograph, verso: 'Eskefiord' [and some pencilled notes that are so faded they are illegible].

[4] This note has been dated from the Board of Trade files where there was great difficulty in getting a licence granted for the *Freden*. On 30 December 1811 the licence was finally granted 'on condition of exporting the Articles named to the amount of not less than £400, according to the Invoice Price – and to return with a Cargo of grain meal or flour only to Leith' (TNA, BT 6/205, no. 38,008).

Obstacle to her taking on board any articles whatever that may be legaly exported & I know of no Objection to Cochineal & Vanille

Many Cargoes of /Salted/ Cod Fish Roes have been carried to the Ports of Spain both without & within the Mediterranean to be usd as bait for Catching a Fish called Sardellas[1]

[*Verso*]
Wine in the *Fredens*
Licence Why
Cochineal & Vanilla Can these be included

Polsen[2] has 24 Tons of Klipfish in Patreksfiord he
Could not Send it this year Because he had no Ship
may he send it next year

SL, Banks Papers, I 1:80. Autograph notes.

228. Rooke & Horneman to Banks, 16 January 1812

Queen Street
Cheapside

Right Honorable Sir,
Agreeable to your desire we herewith take the liberty to state what amount of goods our Iceland friends propose exporting from Leith to Denmark & Norway on their present voyage in hopes that it will meet the approbation of the Right Honorable the Lords of His Majesty's Privy Council, & that they may receive the permission of exporting this year the quantities of Shark Liver Oil & Klipfish specified. – They have requested us to add, that tho' with the assistance of their friends they have been able to offer this much, yet they would have carried these exports farther if the long detention at Leith had not in a great measure reduced their means. – We are also bound to add that Mr. Thorlacius[3] the owner of the *Torsken* and the *Bildahl*, continuing at his residence in Iceland, he, of course, has it not in his power to make personally any management with his friends in England to facilitate his operations on the present occasion, & that the present offer of export proceeds entirely from us as his /*friends*/ Agents, which we assure you is not very convenient, as we run the risk of our actions being disapproved.

By the *Norburg*,[4] Christensen master of 170 Tons burthen we offer for Mr Jess Thomsen to export at the rate of £4 per Ton say £680.[5] –

---

[1] *Sardellas* or sardelle (*Clupea aurita*), are small fish similar to a sardine.

[2] Polsen in Patreksfjörður: unidentified, possibly the factor there.

[3] Ólafur Thorlacius (1762–1815), was a native Icelandic merchant in Bíldudalur in the Western Fjords. He, along with Bjarni Sívertsen and Guðmundur Scheving, is considered one of the Icelandic pioneers engaging in trade. According to Mackenzie, correctly, he was the richest merchant in the island in 1810 (*Travels in Iceland*, p. 335).

[4] The licence was granted on condition he exported goods allowed by law to the amount of not less than £200 (TNA, BT 6/205, nos 37,607 and 37,760).

[5] New regulations regarding the neutral trade stipulated that the ships export British goods worth a specific sum.

By the *Gode Haab*,¹ Moller master of 142 Tons burthen we offer for Mr Jens Lambertsen² to export at the rate £3 per ton say £426.

By the *Jonge Goose*³ – Lund master of 100 Tons burthen we offer for Mr. J. L. Busch to export at the rate of £3 per Ton – say £300. –

By the *Torsken*,⁴ G. Clausen master of 100 Tons burthen, we offer for Mr. O Thorlacius to export at the rate of £2 per Ton say £200. –

By the *Bildahl*,⁵ Chr. Kieldsen master, of 116 Tons burthen, we offer for Mr. O. Thorlacius to export at the rate of £2 per ton say £232. –

The following are the quantities of Shark Liver Oil, which the said respective owners wish to export this year, as well as Klipfish, of which there is still a quantity aboard – viz:

Mr. Jess Thomsen⁶	15 Tons Shark liver Oil
	20 „ Klipfish
Mr. J. L. Busch⁷	30 Tons Shark Liver Oil
	10 „ Klipfish
Mr. O. Thorlacius⁸	25 Tons Shark Liver Oil
	10 „ Klipfish

independent of the Iceland produce & manufactures Mr. Clausen will no doubt have stated himself what he is able to export this voyage. –

Recommending our friends to your most tender kindness we remain with the most profound respect and the greatest esteem –

Right Honorable Sir
Your obed*ient* & humble Ser*van*ts
[*Signed*] Rooke & Horneman

Wisconsin, Banks Papers, ff. 98–99. Autograph letter.

229. Banks to William Jackson Hooker, 10 February 1812

Soho Square

My dear Sir

The Icelandic Book⁹ was Returnd from your Notary Public with a better Translation than such Gentlemen usualy give & in a very Short time I have however Found out

---

¹ This ship *Det Gode Haab* received a licence on condition of exporting goods to the amount of not less than £426, as it says in the letter (TNA, BT 6/205, no. 37,681).

² The owner was not Niels Lambertsen but Jens Lassen Busch, who also owned the *Jonge Goose*.

³ This licence was granted on the same conditions as the *Norburg*, i.e. not less than £200 (TNA, BT 6/205, no. 37,712).

⁴ This licence was granted on the same conditions as the *Norburg*, i.e. not less than £200 (TNA, BT 6/205, no. 37,760).

⁵ This licence was granted on condition that it exported an amount not less than £232 (TNA, BT 6/205, no. 37,956).

⁶ Jess Thomsen, merchant from Nordborg, merchant in Reykjavík.

⁷ Jens Lassen Busch, merchant in Ísafjörður.

⁸ Ólafur Thorlacíus, merchant in Bíldudalur.

⁹ Among the Banks papers is an abstract of the trade ordinance, dated 13 June 1787, dealing with the new trade regulations which took effect on 1 January 1788 and were in force at this time. This is a complete abstract, not

a Dane who understands English well & has some pretentions to Literature[1] I have put the Book into his hands & requested him to furnish me with a Breif abstract of the Contents & have set him to work to cop^y & Translate the Tables respecting the Trade of the Island which I think will prove usefull to your intended Publication.[2]

I Send Mr Peels Translation[3] with this in case any part of it Can be made usefull of which you are a better Judge than I can be we all go on here as usual & all desire to be kindly recommended to you.

Very Faithfully yours
[*Signed*] Jos: Banks

Kew, Hooker's Correspondence, Directors Correspondence, I, f. 35. Autograph letter.

230. Holger Peter Clausen to Banks, 23 February 1812

Leith

Right Honorable Sir!
In the moment of my departure from this country I once more venture to intrude upon your goodness by expressing my gratitude for your benevolent endeavours for the relief of the else unhappy inhabitants of Iceland. – I hope your generous mind will not refuse my

---

only the last part as printed in Hooker. Possibly the 'Forordning ang. Den islandske Handel og Skibsfart' (Ordinance regarding the Iceland trade and navigation), printed in *Lovsamling for Island,* V, pp. 417-62 (see Appendix 8). Important regulations were printed as pamphlets at the time. This is probably what is being referred to in this letter.

[1] This is doubtless Andreas Andersen Feldborg (1782–1838), a Danish writer. He had arrived in England in 1802 and travelled widely in Europe. In 1828 he was appointed a teacher of English at the University of Göttingen. He selected Danish poems for translation into English (e.g. *Poems from the Danish*) and published other works, notably *A Tour in Zealand in the year 1802; with an Historical Sketch of the Battle of Copenhagen. By a Native of Denmark*, London, 1805, a second edition being published in 1807. In 1813 his *Cursory Remarks on the Meditated Attack on Norway in 1813* was published in London. He translated various Danish works into English, e.g. on the English attack on Copenhagen in 1807, published in the *Monthly Review* 1809. He also published 'An Appeal to the English nation in behalf of Norway' in the *Pamphleteer* in 1814.

[2] Appendix F in Hooker, *A Tour in Iceland*, is entitled 'Danish Ordinances concerning the Trade of Iceland by Land and Sea; as also the Products of its Manufactories'. Three decrees were published in an English translation (Hooker, *A Tour in Iceland*, II, pp. 353–91). The first is the important 'free trade' decree of 13 June 1787 (see Appendix 8), surprisingly not the more important first two sections which spell out the ban on foreign trade but only the third section which deals with fishing, farming and manufactures (*Lovsamling for Island*, V, pp. 417-62, section III on pp. 448-62). The other two documents here are public notices (*Plakat*) as Hooker calls them. The first, dated 1 June 1792, 'Whereby sundry articles, concerning the trade to Iceland, are more specifically laid down', deals with e.g. burghership, something that had often cropped up regarding the question of who were the true Iceland merchants (printed in *Lovsamling for Island*, VI, pp. 27–9). The question of what made a merchant an Iceland burgher was a thorny question in 1807 (see Document 85). The second, dated 23 April 1793, deals with illicit trade and burghership yet again (*Lovsamling for Island*, VI, pp. 109–11). There are no tables. However, Mackenzie devoted chapter VII of his book *Travels in Iceland* to the 'State of Commerce', with tables of imports and exports.

[3] Doubtless Abraham Peel, notary public, translator of languages and custom house agent, of 3 George Yard, Lombard Street (*Holden's Annual London & Country Directory for 1811*, I, n.p.).

sincere thanks, considering that, notwithstanding the delicacy of your feelings you will allow the impossibility of a grateful mind allways to keep the expression of his gratitude in check.

As you yourself by your goodness have encouraged me I am bold enough to expose the different motives for the hoped for permission to export oil from Iceland,[1] in one view:

1. The impossibility for the inhabitants to purchase their necessities, if oil is refused by the merchants.
2. The whole quantity will in the very best year not exceed 300 Tons and consequently not have the least effect upon the brittish fisheries.
3. It is impracticable for the English to collect it round the whole island, as there is no coasting trade.

This I consider as the most important and general motives, which in their consequences are more related to the inhabitants than to the merchants. – Different from this is the exportation of Klipfish which again concerns the merchants, particularly as all the klipfish which yet is in Iceland (from the time when we knew no restrictions upon the exportation) belongs to the merchants and they will be the only sufferer if entirely prohibited to be exported, and from which will not derive the least benefit to the inhabitants as it is contrary to their way of living to consume any thing so strongly salted, and in this case it may [go] rotten in Iceland, which tho' I not was interested therein myself, I should consider as a great pity.

In case their Lordships should feel inclined to permit the exportation of some of the Klip-fish I have desired Mr. Horneman[2] to solicit on my behalf for the exportation of 30 Tons. – The following are my principal motives:

1. The fish was cured before the knowledge of any restrictions upon the exportation thereof, which sufficiently may be proved by original documents from a magistrate in Iceland, which are in the hands of Mr. Horneman.
2. The Licence for the present ship was granted on condition to export goods from England to the amount of £800. – and I have shipped to the amount of £1000. –, besides this I have ordered to be shipped to Iceland at the ensuing spring for £700. The amount of 30 Tons Klip-fish [is] at present only £540.

I hope their Lordships will make some allowance for my endeavours to make the Iceland trade so advantageous to this country as possible.

The expenses for my ship here, harbour and dock duties included, is £433 – besides my own expenses which are not less, this likewise makes a favourable balance to this country. – In case the Klip-fish exportation would be granted to me it will be no prejudice to the rest of the Iceland merchants, as those who have no fish will not be in need to desire any indulgence in that respect, and there are only three or four who have any fish left.

---

[1] The British government had permitted the export of 300 tons of klipfish from Iceland and the same amount of cod liver oil in 1811.
[2] Hans Frederik Horneman.

If you consider these arguments sufficient to plead my case I entreat you for your kind assistance. Your wish to have the best information of the Iceland business is the only excuse for my prolixity, and in hope of your forgiveness I beg leave to subscribe myself[1]

Right Honourable Sir
your most humble and most obedient Servant
[Signed] Holger P. Clausen

SL, Banks Papers, I 1:74. Autograph letter.[2]

231. Kjartan Ísfjörð to Banks, 28 February 1812

Aldgate Coffeehouse

Sir

I most humbly beg leave to inform You, that it has pleased the Board of Trade to refuse or to suspend the petitions presented by me; in consequence of which I may safely venture to assert, that my worthy relation Mr. Sivertsen and his numerous family will be utterly undone, and that the inhabitants of his district may be thereby exposed to all the horrors of famine. I understand that my Lord Bathurst has expressed an opinion that a sufficient number of vessels was employed in the Iceland trade,[3] but I beg to assure you, that scarcely a third part of the vessels engaged in that service before the War are now employed;[4] and in the case of Mr. Sivertsen's petition I must observe, that his former vessel *To Söstre* was of the burthen of 225 Tons[5] and the *Anna Dorothea* is no more than 150 tons. These matters of fact I must humbly beg leave to impress on your benevolent mind, especially as my petitions were unfortunately presented to the Lords of Trade at a moment when you were not at the Board.[6]

In immediate regard to myself I fear that I must abandon all hope of attaining the objects of my petition in the cases of the *Orion, Regina* and *Eliza*.[7] Yet I flatter myself, that the *Eliza* may be permitted to proceed from England to Denmark with a cargo of British and Colonial produce and return on a voyage to Iceland according to the regulations

---

[1] Clausen's petition was successful and the Board of Trade issued further regulations to the British consul in Iceland, John Parke, permitting the exportation of 300 tons of klipfish and fish liver oil (TNA, FO 40/2, 'Draught of Additional Instructions to H.M.'s Consul in Iceland', 7 July 1812; TNA, FO 40/2, Castlereagh to Parke, 14 July 1812).

[2] Note in Banks's autograph, verso: 'Mr Clausen | [Received] Feb*ruary* 25 – [18]12'.

[3] Consul Parke had already expressed his opinion that a sufficient number of ships were sailing to Iceland in a report to the Foreign Office and, as will be seen, Earl Clancarty, Bathurst's successor as President of the Board of Trade told Horneman that the government had 'great reluctance in admitting additional vessels into the Iceland trade' (Document 278). See also TNA, BT 6/210, October 1813, no. 54,006.

[4] The number of ships sailing to Iceland in 1811 were at least 19, though Parke estimated them to have been 18 (TNA, FO 40/2, Parke to Wellesley, 20 November 1811). During the years 1788–1807 56 merchant ships sailed on average to Iceland, though in the years preceding the outbreak of war their number declined, 41 sailing to Iceland in 1807.

[5] More like 165 tons.

[6] There is no record in TNA, BT 6.

[7] They were granted licences in June: TNA, BT 6/207, nos 42,752 (*Eliza*), 42,924 (*Regina*) and TNA, BT 6/208, no. 46,724 (*Orion*). The *Regina* also received permission to export 2 tons of iron and 50 lb of bird-shot, TNA, BT 6/205, no. 42,776. The *Orion* was back in Leith with a cargo of Iceland produce in September 1812.

established by His Britannick Majesty's Government. I must beg to notice, that the inhabitants in the neighborhood of my commercial establishment have not received any supplies from Denmark <u>since</u> the War and of course are at present greatly distressed.[1] Firmly relying on Your generous protection I have the honour to subscribe myself

Sir
Your most devoted and obedient servant
[*Signed*] K: Isfiord

Wisconsin, Banks Papers, ff.108–109. Autograph letter.

232. Jakob Aall Jr to Banks, 1 April 1812

Næs Iron Works near Brevig[2]

Pardon Right Hon*our*able Sir the Boldness with which a Stranger totally unknown to You presumes to trouble You with his Correspondence – I am however induced to take the Liberty in consequence of having an affair relating to the Sciences to submit to You and for that reason expected free Access to a Man who for the space of half a Century has shewn himself the most zealous Cultivator and Promoter of the Sciences, Men of Science all over the World consider themselves in Connexion and Friendship with each other – Wars and Contests are incapable of creating a Division amongst them – They have one and the same Object in view the Promulgation of Knowledge and the Dignity of Human Nature.

Your Right Hon*our*able Sir have particularly proved Yourself Icelands first and most generous Friend – That unfortunate Country blesses at this Hour Your Exertions to enable it to preserve a tolerable Existence – Without Your benevolent assistence it would unquestionably have been subjected to all the Misery with which a Country neglected by Nature & entirely deprived of all Connexion with the Mother Country might be menaced – The Undersigned participates with You Right Hon*our*able Sir in these warm feelings for that unfortunate Island but my Insignificance as an individual disqualifies me from effecting any thing for its Welfare. I love Iceland equally with my own native Country – Norway & Iceland from their Common Origin[3] and reciprocal Necessity for the Produce of each other are more nearly associated than any other Country in Europe – In former times this Connexion continued uninterrupted & with the greatest mutual benefit[4] – Norway is indebted to Iceland for such Lights in her Ancient History as scarcely any other Country can produce an Example of.[5]

From these Considerations I have united myself with several others of my opulent Countrymen for the purpose of founding an Establishment at Iceland having for Object the Introduction of a more liberal mode of Trade than has hitherto been in practice and to apply the Surplus to the promotion of Knowledge in the Country and other Scientific purposes – Hitherto however Licences have only been granted for Danish Ships and

---

[1] Ísfjörð's trading station was at Eskifjörður in eastern Iceland.
[2] Brevik is situated roughly half-way between Christiansand and Oslo.
[3] Most of the first settlers came from Norway.
[4] There was an active trade between Norway and Iceland during the Middle Ages, Iceland coming under the rule of Norway in 1262–4.
[5] Aall was indeed interested in Iceland: see Appendix 13.

Norway saw with pain & Sorrow richly loaded Iceland Traders filled with important Articles of Necessity pass by her to Copenhagen where such Goods are neither of so much value nor of so much Importance as in Norway I therefore venture Right Hon*oura*ble Sir to solicit Your generous Cooperation in procuring a Licence for a Norwegian Ship which may be allowed to load her Cargo in Norway after having complied with all the Conditions imposed upon Danish vessels – The Expedition will be conducted by Mr Jorgen Flood[1] the Young Man who will have the Honor of delivering this –

That the Object of this Expedition is no other than the above stated I dare venture boldly to assure You and may safely appeal to the Evidence of my Country as to my Mode of Conduct.

I take the Liberty of accompanying herewith a small Set of Norwegian and Swedish Minerals for which I solicit You right Hon*oura*ble Sir to bestow a place in Your Mineral Cabinet – In my earliest Youth was engraven in my Mind the most profound Veneration for the sacrifices You made for the Natural Sciences, of which Your Travels & Your liberal Reception of all Men of Learning will ever bear Testimony.

It will be a source of real Joy to me were this Trifle to be found worthy of Your attention and procure me the Honor of enriching Your Collection with my Northern Minerals – Amongst those now sent I may venture to call the Moraxite[2] one of the greatest rarities and in the selection of the others I have paid particular Attention to such as are here considered most worthy of notice and also to the new Species with which the North has enriched the Mineral System.

I have to lament that the accompanying Work of my productions is written in a Language unintelligible to you.[3]

I subscribe myself with the most distinguished Esteem

Right Hon*oura*ble Sir
Your mo*s*t Ob*edien*t H*um*ble Serv*a*nt
Signed/Aall Junior
Knight of Dannebrog &
Proprietor of Næs Iron Works.

Wisconsin, Banks Papers, ff. 104–105. On the right hand top corner it says 'Translated from the Danish'. The Danish signed version is in the same repository, ff. 100–101.

233. Banks to Jakob Aall Jr, [April 1812][4]

Sir

your Letter has Given me much Satisfaction & your valuable [gift?] has Enrichd Considerably the Public Collections of minerals in the British Museum I have Sir no

---

[1] This must be the same Jörgen Flood who had been Count Trampe's secretary in Iceland. He left the island with Trampe in September 1809 for England.

[2] Moraxite: Most probably morass ore (in Icelandic *mýrarauði*). The recent (proper) mineral name is goethite [$FeO(OH)$], named after the great German Johann Wolfgang von Goethe.

[3] In the British Library there are two books by J. Aall which belonged to Banks: *Om Jernmalmleier og Jerntilvirkning i Norge. Et fragmentarisk Forsög*, Copenhagen, 1806, and *Fædrelandske Ideer*, Christiansand, 1809.

[4] This letter is undated and it must have taken some time for Aall's letter to reach Banks safely. It may have been written in May, if it was Bugge's letter (next document) that Banks is referring to in the last sentence.

Collection of minerals myself & I think you will agree with me that a better use Could not be made of the Excellent Specimens you have been so good as to send to me than has been done by adding them to the Public Treasures of Science preservd here for the honor of the Countrey & the use of Students

I rejoice to hear from Norway those Just & well founded Sentiments Relative to the Feelings of men devoted to Science, which your Letter Contains, men of Science ought to place their Considerations of Each other above the hazard of interrupted [sic] by the Political Differences of their Rulers. My Literary Friends even in France are Still my unchangd well wishers, tho the map of their Population has been degraded into a State of military barbarism Scarce to be Paraleld in the history of nations. The Danes are the only people who, angry at having miscalculated the prowess of British Army to which they ought to have yielded up their Ships[1] which were no longer at their voluntary disposal but must have been transferred to one or the other of the Powerfull Contending Parties[2] have Extended their ill humer to their scientific Friends. Till I had Receivd your Letter, I had not heard from one of my Correspondents at Copenhagen Since that time I have Receivd one[3] & I hope in future that the Peace of Science which ought to be as Durable as the Existence of men will after too long an intervall of Enmity be renewd & Restord.

Wisconsin, Banks Papers, ff. 102–103. Draft.

234. Thomas Bugge[4] to Banks, 12 May 1812

Copenhagen

Sir and most illustrious colleague

It has been a long time since I have had any news of you, having been deprived of your instructive correspondence. The unfortunate war between our two countries is the cause of this. The perfidy, the injustice and the cruelty of the English ministers has ruined my poor fatherland. Our capital suffered greatly in the cruel bombardment, principally the University, which was near the cathedral,[5] which was the mark and focus of the English bombs. Almost all the houses and the colleges of our University were set on fire by the abominable fire. My house was in this quarter, it was completely burned to the ground. I lost my library of 8000 volumes, a superb collection of instruments of mathematics and physics, maps both geographical and hydrographical, all my furniture and other furnishings.[6] My loss is more than 90000 écus.[7] During these three terrible nights and days of bombardment 20 bombs fell on my house and my coach and in my garden, besides

[1] Banks had little sympathy for the Danes. See also the following letter from Bugge.
[2] The British and French.
[3] Possibly the following letter from Bugge.
[4] He titled himself State Counsellor, knight of the Dannebrog and professor of astronomy – 'Conseiller d'état, Chevalier de l'ordre de Dannebrog, Professeur en Astronomie'.
[5] The Vor Frue Kirke, Copenhagen's cathedral, was burned to the ground on the night of 5 September 1807. This is the so-called 'Second Battle of Copenhagen', 2–5 September 1807.
[6] According to the *Dansk Biografisk Leksikon*, Bugge did lose a large part of his library of 15,000 books and manuscripts, and many of his instruments.
[7] *Écu* is a French coin. It disappeared during the French Revolution but 5 franc silver coins were minted throughout the 19th century and were often still called *écu*.

the innumerable pieces of bombs, which exploded in the air and which fell like a thick and continuous hailstorm on the miserable inhabitants of Copenhagen. I give thanks to Divine Providence, that I and my large family[1] survived safe and sound from this terrible inferno, the fury and cruelty of which is so difficult to imagine without having been there. Neither I nor any member of my family lost their lives or suffered any disablement, which was the case with so many other unfortunate families. In the present situation of my poor fatherland one has to be philosophical to be without the pleasures of a happy and agreable life. I fear that the principles adopted by the two great nations will lead to despotism in government and barbarity in the sciences.

A young man among my students, Lieutenant Wormskiold,[2] who is extremely well educated in botany and astronomy, has the great courage to embark on a scientific expedition to Greenland. I ask you, Sir and most illustrious colleague, whether it is possible to make arrangements, that both my young friend and the observations and collections which he will bring to this country, are not molested by British ships-of-war and privateers. Here you are regarded as a tutelary God of Iceland and the Icelanders, and I would like to add Greenland and Mr. Wormskiold.

Thanks to Mr. Wormskiold, who will arrive in Leith in Scotland, I have the honour to transmit

1.) The new collection of the Transactions of the Royal Society of Copenhagen,[3] 9 volumes.

2.) The Transactions of the Royal Society of Sciences in Copenhagen 9 volumes and the first book of the sixth.

3.) The collection of all the geographical maps of Denmark, published up to this time, which were measured, drawn and published under my supervision.[4]

I beg that you would be so kind as to find them a place in your extensive Library and to keep them as a mark of my greatest esteem.

The Copenhagen Observatory suffered little damage in the bombardment as the roof, arches and the thick ramparts resisted the impact of the bombs.[5] As there are few things left now on the ground I have attached myself with more fervour to the heavens and I have continued my astronomical observations. I enclose some of these observations

---

[1] Bugge's wife was Ambrosia Wedseltoft (1742–95), daughter of a clergyman. They had eight children.

[2] Morten Wormskiold (1783–1845), naturalist: see Appendix 13.

[3] The Danish equivalent to the Royal Society, Det Kongelige Danske Videnskabernes Selskab, was founded in 1742. Thomas Bugge was the Secretary of the society from 1801 to his death in 1815. In this period he was responsible for publishing the annual publication *Bekiendtgiørelse fra det Kongelige Videnskabernes Selskab i Kiøbenhavn*, which first appeared in 1793. I am indebted to Katrine Hassenkam Zoref for this information.

[4] Bugge headed the surveying of Denmark. His efforts led to a breakthrough for the geographic and economic survey of the country: *Det Kongelige Danske Videnskabernes Selskabs Atlas over Kongeriget Danmark og Hertugdömmene Slesvig*, with maps from 1766 to 1841. 20 volumes (of 24) were printed before 1812 and these are the maps referred to. See Bugge, 'Landmaaling og Fremstilling af Kort under Bestyrelse af Det Kongelige Danske Videnskabernes Selskab 1761–1842', in vol. IV of *Det Kongelige Danske Videnskabernes Selskab 1742–1942. Samlinger til Selskabets Historie*. Information from the Map Department of Det Kongelige Bibliotek (Inge Uldal).

[5] This was of extreme importance to Bugge. He had, in the late 1770s, gone on a study-tour of observatories in Germany, Holland, France and England. On his return the Copenhagen observatory was rebuilt and new instruments were purchased with a royal grant.

regarding the most remarkable objects.[1] If you find them worthy of your attention you, as the president [the maitre de les communiqués] of the Royal Society can decide to include them in the Philosophical Transactions.[2]

Please accept the assurance of my highest esteem and greatest affection, with which I have the honour to be Sir and most illustrious Colleague

Your very humble and obedient servant
[signed] Thomas Bugge

University of Oxford, Bodleian Libraries, MS. Eng.hist.d.150, fols. 89–90. Autograph letter. Translated from the French.

French original:
Monseigneur et très illustre Confrère

Il ÿ a bien long tems, que j'ai eu de vos nouvelles; et et que je suis privè de Votre correspondance instructive. La malheureuse guerre entre nos deux patries en est la cause. La perfidie, l'injustice et la cruauté des ministres anglois ont abimè ma pauvre patrie. Par le bombardement cruel notre Capitale a souffert beaucoup, et principalement l'Université, qui ètoit près l'église cathedrale, qui a ètè le but et le foyer des bombes angloises. Presque toutes les maisons et les colleges de notre Université font embraser par ce feu abominable. Ma Maison ètoit dans ce quartier; elle est brulé de fond en comble; J'ai perdu ma bibliotheque de 8000 volumes, une superbe collection des instrumens de Mathématique et de Physique, des cartes geographiques et hydrographiques, toutes mes meubles et mes autres fournitures. J'ai fait une perte de plus que 90000 Ecus. Dans ces trois affreuses nuits et jours de bombardement 20 bombes sont tombèes dans ma maison et ma Carosse et dans mon jardin; outre les morceaux innombralles de bombes, qui sont sautées en l'air et qui sont tombées come une grele epaisse et continuelle sur les miserables habitans du Copenhague. Je rends grace à la Providence Divine, que moi et ma famille nombreuses, nous sommes sortis sains et saufs de ce feu infernal, dont il est difficile de s'imaginer la fureur et la cruauté, sans avoir ètè present. Ni moi ni aucun de ma familles est tué ou estropiè, ce qui est arrivè à tant d'autres familles malheureuses. Dans la présente situation de ma pauvre patrie il faut étre assez philosophe de se passer de tous les agrémens d'une vie heureuse et agrèable. J'ai crains, que les principes adoptés par deux grandes nations meneront au despotisme dans les gouvernemens et a la barbarie dans les Sciences.

Un jeune homme de mes éleves Mr: le Lietenant de Wormskiold qui est très bien instruit dans la botanique et dans l'astronomie, a le grand courage de faire un voyage scientifique à Grönlande. Je Vous prie, Monseigneur et tres illustre Confrere, s'il est possible de faire des tels arragemens, que mon jeune ami aussi bien que les observations et les collections, qu'il renvoye en sa patrie, soient respectées par les vaisseaux de guerre et par les corsaires de votre nation. On Vous regarde icei comme un Dieu tutelaire d'Islande et des Islandais, et je voudrai bien joindre ce de Grönlande et de Mr: Wormskiold.

Par l'entremise de Mr: de Wormskiold, qui doit aborder à Leight [Leith] en Ecosse, je l'ai honneur de Vous faire passer

1.) La nouvelle collection de Memoires de la Société Royale de Copenhague 9. Volumes
2.) Les Memoires de la Société Roÿale des Sciences a Copenhague 9. Volumes et le 1^{mie} Cahier de 6me
3.) La Collection de toutes les cartes geographiques de Danemark, publies jusqu'a present, qui sont mesurées, construites et publiées sous ma direction.

Je vous prie de vouloir bien leur donner une place dans Votre nombreuse bibliotheque et de les garder Comme une marque de ma plus gr*and* [torn] estime

---

[1] He was a prolific writer, his output including *Observationes astronomicæ annis 1781–83*.
[2] Bugge had already published research on astronomy in *The Philosophical Transactions* (1783).

L'Observatoire de Copenhague n'a pas recu aucune? [*torn*] dommage remarquable par le bombardement [torn: car] les voutes et les murailles epaises ont resiste [*torn*: le] choc des bombes. Comme il ÿ a très peu des choses sur la terre, je me suis attaché avec plus d'ardeur aux cieux, et j'ai continue mes observations astronomiques. Je joins quelques de ces observations sur les Objets les plus remarquables. Si Vous les trouver digner de Votre attentions Vous étéz le maitre de les communiques à la Société Royale et de les faire inserer dans les philosophical transactions.

Daignez recevoir l'assurance de la plus haute estime et du plus grande attachemnt, avec les quels j'ai l'honneur d'être
Monseigneur et très Illustre Confrere
Votre Très humble et tres obeissant Serviteur,
[*signed*] Thomas Bugge

235. Count Christian Ditlev Reventlow to Count Johan C. T. Castenschiold,[1] 12 May 1812

Copenhagen

Sir Joseph Banks having since the beginning of the war on many occasions been very actif to effectuate advantages to Iceland and to every one of its inhabitants in diminishing for them the inconveniences of the war, I consider it to be a very duty to me to require of You Sir, all communication between England and Denmark being interrupted, but not between Iceland and England to take the first possible occasion for expressing to him the gratitude, which here is felt by his noble and human deeds against that remote lands inhabitants, who used to be provided from the motherland with many things, which are of the first necessity to them, but now under the war hardly can be got from thence. – His Majesty our King is himself informed of all what Sir Joseph has done for Iceland,[2] Greenland and the island of Ferroe, and when the destroying war in Europe ceases and peace happily again knits the disjoined bands of friendship of the warient nations, then I will haste to interprete to him the gratefull sentiments of our King. I have the honour to be with the most perfect esteem Right honourable Sir,

Your humble servant
Rewentlow

Kent History and Library Centre, Banks MS, U951 Z 32/10, f. 10. Contemporary copy.
Wisconsin, Banks Papers, ff. 83–84.[3] Contemporary copy.

[1] On both copies the letter is addressed 'To the Governour of the Southern part of Iceland de Castenschiold'. Castenschiold became the governor (*stiftamtmaður*) of Iceland in 1813, but was at this time the district governor (*amtmaður*) of Southern Iceland (1810–19). This is an English version sent to Banks.
[2] For instance Count Trampe had sent reports from London. See e.g. RA, Rtk. 2214.55, Trampe to Frederik VI, 13 June 1810.
[3] Note in Banks's autograph, f. 84 verso: 'Count Reventlow | to the Stiftsamptman'.

236. Banks to [Reventlow or Bugge],[1] May 1812

our attack upon Copenhagen was an act of Foresight & wisdom & vigor Justified by the Principles of Self Defence & warranted by the Usages of war. The French made no secret of their disappointment they sufferd from their Capture of the Danish Fleet which would have been made a Peace offering to Bonaparte & Enabled him to Collect Formidable naval Forces to Exclude us from the Baltic & in proper time to issue further Loaded with French Troops to the attack of our northern provinces[2]

The danes miscalculated our power when we demanded their Ships our Force was sufficiently strong to justify them in the Eyes of all mankind for Delivering up their Ships to save their Capital but their King determind to prove his subservience to the French most unequivocaly Compelld the English to Resort to Force & to Obtain their Object the Shipping, by the Capture of Copenhagen. That the English were honest & honorable in declaring the Ships to be their only object was Clearly proved by their Evacuation of the Danish Territories the very moment their purpose was Compleated & their Conduct to the Danes during the time they occupied Copenhagen & Zeeland

Wisconsin, Banks Papers, f. 83v.[3] Draft.

237. From Morten Wormskiold, 5 June 1812

On board the vessel *Freden* in the port of Leith[4]

Sir!

The invariable kindness and the humanity, which you have always shown during the fatal hostilities, to the protection of the poor Icelanders and Greenlanders have filled me with a respect no less than the one resulting from the gratitude which you always deserve from the student of natural history. –

Sir, that is why I trust and dare hope not to offend you by begging you to permit me to have recourse to this same benevolence, should misfortune befall me.

I hope that Mr Bugge informed you about the project of my voyage in the letter he sent to Mr Borthwick[5] a few days ago. I am afraid that the regulations of the Customs at

---

[1] This is a draft written verso on the letter from Reventlow. However, as Reventlow's letter is so polite, it is not certain that this draft was meant for him. Bugge, however, in Document 234, minces no words when he describes the British destruction of Copenhagen. Had Banks received the two letters at more or less the same time as the dates suggest then, this may have been a draft meant for Bugge. However, there is no proof that it was sent (no letters to this effect are to be found in the Banks–Bugge correspondence in the Kongelige Bibliotek) and most probably permitted Banks to vent his feelings – never to be sent. It is, however, interesting as it shows very clearly Banks's opinion of the Copenhagen expedition.

[2] Northern England.

[3] This draft is not mentioned in Teigen, 'Supplementary letters of Sir Joseph Banks', the calendar of the Banks Papers, University of Wisconsin-Madison.

[4] This letter is rather difficult to translate as here we have a Dane writing in French, which is not his mother-tongue. This translation, however, fails to give this impression. The *Freden* was one of the ships engaged in the Greenland trade.

[5] Peter Borthwick, the Danish consul in Leith and a partner in Corbett, Borthwick & Co.

Leith have until now prevented me sending you the box and packet which are on board this vessel. I await your orders respecting them. –

I must further complain that the departure from Copenhagen, more sudden then I had expected, deprived me of the opportunity of having a letter of recommendation sent to you from Mr. Hornemann,[1] professor of botany in Copenhagen, who himself has enjoyed and benefitted from your loyalty and your large collections.

I owe my knowledge of botany to Mr. Hornemann having accompanied him on many voyages collecting plants for the Flora Danica[2] both in Denmark and during the summer of 1807 also in Norway. I was extremely lucky to get this wonderful opportunity to learn about my native plants. For many years now I have been preparing my project of visiting Greenland, a country which is all the more attractive to me in that it is almost totally unknown to botanists. – Now I finally have the opportunity of putting it into practice. Even though the northern regions only offer an inferior vegetation I do not doubt that I will find there enough phanerogamous plants,[3] foreign to Europe, and especially algae and other cryptogams[4] to make a tolerable collection. – Even though I am not as well versed in zoology and geography, I will use this opportunity to make observations and collections in these fields of natural history as far as my circumstances offer. I intend to spend several years so that I can traverse the whole inhabited coast. I will have to spend the winter in one of the colonies,[5] and not wishing to be idle I plan to do some geographical observations, at the very least to determine the latitude and longitude of various places, neglected until now, and which will be very necessary for the physical experiments I hope to make.

Since my arrival in Leith Mr. Borthwick has been kind enough to show me every possible civility and for the time being my only hope is that our vessel will soon receive permission to continue on its journey, because time is advancing, making me fear I will lose the summer for my project. –

Sir, I have here detailed my plan in the hope that you could perhaps honour me with some commission and please be persuaded of my eagerness to prove my respect for you by making it a rule always to obey your orders and advice.

Sir, your very humble servant,
[*Signed*] M. Wormskiold

BL, Add MS 33982, ff. 37–38. Autograph letter.[6] Translated from the French.

---

[1] Jens Wilken Hornemann (1770–1841), an eminent Danish botanist, a Linnean and university professor and eventually sole director of Copenhagen University's third botanical garden, Charlottenburg Garden, which received royal approval in 1775. He edited 18 volumes of the *Flora Danica* as well as publishing several works on botany.

[2] The *Flora Danica* was presented to Banks as a gift by Frederik VI (see Document 294).

[3] Phanerogamous plants belong to one of the two major plant divisions, including all seed-bearing plants and all flowering plants.

[4] A cryptogam is a plant that has no stamens or pistils, and therefore no true flowers or seeds: e.g. ferns, moss and fungi, the other major plant division.

[5] The trading stations in Greenland which developed into small villages were called colonies at the time.

[6] Note in Banks's autograph, bottom left-hand corner of the first page: '[Received] June 8.' Also in Banks's autograph, verso: Two lines of scrawled Latin, difficult to decipher but mentioning the Grotius Index.

French original:

Au bord du vaisseau *Freden* dans le port de Leith ce 5 Juin 1812

Monsieur!

La Bienveuillence et l'humanité inalterable par les circonstances, qui Vous ont portéer pendant tout le tems des fatales inimities, a la protection des pauvres Islandois et Grönlandois m'ont penetréer d'un respect non moindre que celui qui resulte de la reconnaissance, qu'en tout tems Vous devront les étudiant de l'histoire naturelle. –

Monsieur! C'est en quoi me confiant, que j'ose esperer de ne pas Vous offenser, en Vous suppliant de vouloir bien permettre que j'aie recours a cette meme bienveuillance, en cas de malheurs qui pourront peutetre me resulter des conjunctures presenter.

J'espere que Monsieur Bugge a bien voulu Vous informer du project de mon Voyage dans la lettre qui Vous a été remise par Monsieur Bortwick, il y a peu de jours; je plains que le reglement de la douane de Leith m'aie jusqu'ici impedié a Vous faire parvenir la caisse et le paquet y appertenant, qui se trouvent ici a bord de notre vaisseau, et a l'egard desquelles j'expecte de recevoir Vos ordres. –

Il faut encore que je plaigne, que le depart de Copenhague, plus subite que je ne me l'etoit imaginé, m'aie privé la satisfaction de Vous remettre une lettre de recommendation de la part de Monsieur Hornemann professeur en botanique a Copenhague, qui lui meme a eu le bonheur de profiter de Votre loyalité et de Vos collections immenses.

C'est a Mons*ieur* Hornemann que je dois mes connaissances en botanique; en l'accompagnant dans plusieurs voyages qu'il a fait a la recolte de plantes pour la Flora danica en Dannemark, et pendant l'été 1807 aussi en Norvege j'ai eu la plus belle occasion pour apprendre a connoitre les productions vegetales de ma patrie, et depuis plusieurs années j'avois formé le project, que maintenant j'ai le dessein d'executer, de visiter la Grönlande, pays, qui avoit d'autant plus d'attraits pour moi, qu'il est presque entierement inconnu aux botanistes. – Quoique la situation boreale ne saura permettre qu'une vegetation pauvre, je ne doute nullement et d'y trouver des plantes phanerogames etrangeres a l'Europe, et principalement des algues et autres Cryptogames de faire une assez bonne recolte. – Quoique par si bien verse dans la Zoologie et la Geografie, je ne laisserai pourtant pas de faire des observations et des collections pour les deux parties de l'histoire naturelle, tant qu'elles seront a ma portée; et comme je me suis proposé d'y sejourner plusieurs années, afin de pouvoir parcourir toute la cote habitée, il me faudra pendant l'hiver rester immobil sur une colonie quelquonque, c'est donc pour n'y etre pas oisif, que je me suis proposé a y faire des observations geographiques, du moins a la determination de /la/ latitude et longitude de plusieurs lieux, chose jusqu'ici bien negligée et qui pour quelques observations physiques que j'espere d'y pouvoir faire, me seront bien necessaires.

Depuis mon arrivée á Leith Monsieur Bortwick a eu la complaisance de me temoigner toutes les civilités possibles et pour le moment il ne me manque que l'accomplissement du souhait que notre vaisseau recoive bientot la permission de continuer son voyage, parceque sans cela le tems bien avancé me fera craindre de perdre cet éte pour mon plan. –

Monsieur! je Vous ai ici detaillé mon plan dans l'esperance que peutetre Vous m'honoreriez de quelque commission et je Vous prie d'etre persuadé de mon empressement a Vous temoigner mon respect, et de ce que toujours le regardera comme une loi, de plaire a Vos Ordres et Vos conseils.

Monsieur
Votre tres humble Serviteur
M. Wormskiold

238. Banks to William Jackson Hooker, 1 July 1812

Spring Grove

My dear Sir

Your map of the district of Iceland visited by you is Finishd & well done in my Judgment, that of the Island is also Engravd & has been Corrected Relative to the Spelling of the names by Mr Clausen[1] I expect it will be Finishd in a Few days as Soon as it is I will Send it to you with the other

Mr Clausen has brought as a Present to me all the Plans of Iceland that have been Publishd by the admiralty of Denmark, & some others Particularly a Book of harbors & views of Land if you wish to See these I will Send them with the others[2]

Dr Lind[3] to whom I Lent your Book tells me in a Letter that the hight you attribute to the waters of Geyser when I was in the Island Sept 13th is Sixty feet only & that Troills Letters State the Real hight at 90 feet[4] I have no books here So I cannot investigate the doctors asertion he however who was the man who measurd the hight of the jett Carefully having before measurd a base & prepard a fixd Quadrant to view it Found it by his Observation probably the best that has yet been made 92 feet $\frac{6}{10}$. If you Can tell me where to get one of your Books & will give me Leave I will Send it as a Present from you to Dr Lind

adieu my dear Sir
always Faithfully yours
[*Signed*] Jos: Banks

[P.S.] I see that M Verdun[5] & Some other French men who visited many northern Countries in 1774 in order to Correct the Positions of Lands by Time keeper have mentiond Iceland. These books also I will send if you wish it I would not have you ignorant of any thing that has been done Relative to our Little Island

Kew, Hooker's Correspondence, Director's Correspondence, I, f. 36. Autograph letter.
Hermannsson, 'Banks and Iceland', pp. 91–2.

---

[1] There is no map in the first edition, published in Yarmouth in 1811 though '*NOT PUBLISHED*' is stated firmly on the title page. In the second edition of 1813 there is an extremely detailed map of Iceland as a frontispiece.

[2] Thomas Bugge oversaw the surveying of Denmark (see Document 234). The whole was published under the title: *Videnskabernes Selskabs Kort over Danmark og Slesvig, under det Kongelige Videnskabernes Societets Direction ved rigtig Landmaaling optaget og ved trigonometriske samt astronomiske Operationer prövet*. The first volume was published in 1766 (the last in 1836). As Hooker wrote '*all* the plans', this is probably the work referred to. See Nörlund, *Islands kortlægning. En historisk Fremstilling*.

[3] James Lind died in 1812.

[4] On this point Lind was mistaken. Hooker states quite clearly: 'when Sir Joseph Banks visited Iceland ... the greatest elevation to which the column [of the Geyser] ascended, was ascertained to be ninety two feet' (*A Tour in Iceland*, I, p. 159). In von Troil's *Letters on Iceland*, one of the jets is measured at a height of 92 feet (p. 261).

[5] See Introduction, p. 16 above. Chapter XXI of Verdun's book deals with 'De l'Islande, des îles Féroë, & de quelques autres Islots voisins' (II, pp. 245–60), mentioning that the most recent Iceland map they had seen was made by the professors Erichsen and Schoenning, was very detailed but did not at all accord with their observations (p. 245). They then go on to discuss Horrebow's map.

239. Banks to William Jackson Hooker, 13 July 1812

Soho Square

My dear Sir

Clausen¹ whom I have seen tells me that the book from whence your account of the Iceland trade was extracted is a Collection of ordinances & Decrees of the Danish Government Respecting Iceland not Publishd together in one book but Collected by Clausen. There are in it two Mss Decrees which are of a Date I think 1791.² if you wish to have the Book I will send it for as Feldborg³ has disappointed me I have now no hopes of Procuring for you the translations I had desired him to undertake

adieu my dear Sir
in haste Faithfully yours
[*Signed*] Jos: Banks

Kew, Hooker's Correspondence, Director's Correspondence, I, f. 37. Autograph letter.

240. Magnus Stephensen to Banks, 8 August 1812

Innraholmi in Iceland

Right honourable Sir!⁴

In the month of September of that fatal year 1807 I at first time did take the liberty, with few lines⁵ to pay my due respect to the immortal & generous friend & benefactor of Iceland & of my Father, the Stiftamtmand or Governor Olav Stephensen,⁶ who now is much advanced in years,⁷ viz. the renowned friend of mankind Sir Joseph Banks. The remembrance of You, dear Sir! from my early youth, having seen your lovely countenance & kind conversation at several times in my Father's house at Svedholt, & at Bessested, Havnefiord & Reikevig,⁸ such a remembrance, joined with the universal aknowledgment of your all over the world proved generosity, where once (:and where is it not?:) Your beloved name was known, roused me, now His Danish Majesty's Counsellor of State & Chief Justice of Iceland, but then myself taken prisoner of war, & carried to Edinburgh in Scotland, in the name of my, by the war between Denmark and England, distressed country to implore your noble heart, Sir! to mediate for this unhappy island (:which is exposed to devastations of the two most contrasting elements, viz. the subterraneous fire

---

¹ Holger P. Clausen, the Iceland merchant: see Appendix 13.
² See Document 229, p. 486, n. 2.
³ Andreas Andersen Feldborg, a Danish writer: see further Document 229.
⁴ Stephensen, as was the Danish custom, always puts an umlaut over the letter y and ~ over the letter u, removed here for the convenience of the reader. His spelling in English leaves much to be desired.
⁵ Document 40, actually 17 October 1807.
⁶ The underlinings are Stephensen's. According to the renowned scholar Professor Jón Helgason 40 manuscripts were sent from Iceland in 1773, 1775 and 1777, doubtless all under the auspices of Ólafur Stephensen. In 1809 Bishop Geir Vídalín sent another (BL, Add MS 6121), see Document 201.
⁷ Ólafur Stephensen died 11 November 1812. Unfortunately, due to the state of war, his funeral was not quite up to the standard expected.
⁸ Sviðholt, Bessastaðir, Hafnarfjörður and Reykjavík.

& frozen water:) before the Brittish throne, an exception from the cruelties & hostilities of war; to mediate the release of its taken vessels, & their undisturbed trade afterwards at this island, lest its inhabitants may starve with hunger, the shipping be stopped; & our country thus soon transformed in a Desart.[1]

Such an application made a strong impression upon Your noble heart, dear Sir! Finding then a worthy cause to disembogue in the most bountiful streams, it did immediately render your tongue & your pen energetical eloquent enough, saving our in /the/ contrary case lost and unhappy island. Only your humane intercession excepted Iceland from hostilities, released its captured vessels, restored to the icelandic merchants their trifling commerce, secured our island from violence afterwards, and suchlike disturbances, as happened in the year 1809, to Your utmost indignation, only by a greedy abuse by Mr. Phelps of a mere Licence to trade granted him, being chiefly misled/*den*/ by the english rambler Savignac,[2] & assisted by the vile, foolhardy conduct of the infamous Danish traitor Jörgensen,[3] whose foolish undertakings, myself, by the gracious aid of Providence & the noble english Cap*tain* Alexander,[4] entirely did overturn on the 22 of August 1809,[5] not without great danger for my own life & wellfare, by pushing the usurpers down from their unjustly assumed Government.

A safe security from suchlike invasions here of revolutionists gives at last His Brittish Majesty's most gracious Proclamation & Order in Council of 7 Fevr*ier* 1810,[6] here strongly forbidding all Brittish Subjects to comitt any act of violence or injury against persons or properties in Iceland, a work, my country very well known & acknowledges itself indebted for, to its & the humanity's most excellent advocat & Protector, which always shall inflame its inhabitants ever to bless that eternally loved & honoured name, Sir Joseph Banks; that respected name, our country has already given a lasting remembrance, printed in its Annals,[7] & teaching our latest posterity, to whom even amongst an hostil*e* power against our principal state, we owe our liberty, our safety & our supplies of necessaries of life by our restored free commerce.

In my work Eptirmæli átjándu Aldar (:memorable Accounts of the eitheenth [*sic*] Century:) published in the icelandic tongue 1806,[8] I did myself already before quicken my

---

[1] On the whole this is not an exaggerated account of Banks's achievement for the Icelanders.

[2] That Savignac, Phelps's supercargo, was the prime mover in the Revolution is a view which was advanced by Professor Hermannsson, based on documentary evidence including this letter and one written by Captain Jones to Hooker (Kew, Hooker's Correspondence, Director's Correspondence, I, 1 September 1810). He wrote: '[Savignac] has been mentioned as the real villain in the play. And so he doubtless was. It is possible that he whispered in Jörgensen's ear before the latter left with the *Clarence* for England, that on his return he should be prepared for a revolution ... Probably the idea of a revolution matured in Savignac's mind while waiting in Iceland, and he communicated it to Jörgensen and Phelps at the time when it looked as if Count Trampe hesitated to stand by the treaty he had made with Captain Nott. Savignac was shrewd enough to push the others forward and let them take all responsibility' (Hermannsson, 'Banks and Iceland', p. 69).

[3] Jörgen Jörgensen.

[4] The Honourable Alexander Jones, youngest son of Viscount Ranelagh, commander 1806, post-captain 1811, eventually becoming a rear admiral of the Blue (*The Navy List*). Stephensen uses his Christian name as is the custom in Iceland.

[5] Stephensen is referring to the agreement of 22 August 1809. His life was hardly in danger. See Appendix 9.

[6] See Appendix 10. Stephensen is obviously under French influence (he had a smattering of French), writing *Fevr.* instead of *Feb.*

[7] For Banks in Icelandic annals, see Appendix 5.

[8] This book was in fact printed in two sizes and he is referring to the larger size. See also Document 163, p. 410, n. 7.

Countrymen's gratitude to his renowned name, as pag: 740–741,[1] but afterwards in another work of almost the same tendency, published at Copenhagen in the Danish tongue, 1808, under the title Island i det attende Aaarhundrede (:Iceland in the eitheenth [sic] Century:) dedicated & by myself personally delivered to my most gracious King, His Danish Majesty,[2] pag: 190–191 again proclaim your, dear Sir! matchless merits of my afflicted country.[3] – Both works, together with some Northern Antiquities and the Law for Kings of Denmark and Norway, entyrely engraved with costly drawings all around in great royal folio,[4] I have had the pleasure, myself, in 1809 here to deliver to Mr. Hooker, charging him with the commission, to bring these works over to you,[5] dear Sir! as a trifling mark of my gratitude. – But, doubtful as I am, either the fire of the vessel, in which Mr. Hooker first went off from hence, might have spared these works, or if they in my name ever might have been delivered to You, however thus ordered by myself, whereas I since neither can boast of the happiness to have received any answer from Your beloved hand, nor to my letter to You with Count Trampe in the same Year,[6] I do hereby venture to send another Copy of both my said works, the one unbound, with the humble wish, that your favour may grant them a place in Your most renowned library, notwithstanding the mean ligature of the one with my portrait prefixed, according to the present means of binding books at this place. – However Mr. Hooker in his very faulty Book: Tour in Iceland in the Summer of 1809 pag: 228–229,[7] has quoted such contents of my works, which may testify my own & my Countrymen's feelings of gratitude to that beloved name, Sir Joseph Banks, & pag: 222–264 made some description of his stay with me here in Borgerfiord, yet, I'll not rejoice at any thing more, than at an acknowledgment from your own dear hand, Sir!

---

[1] Indeed he does. A rough translation: 'A famous English gentleman Joseph Banks, who travelled widely around the world to observe closely nature and the customs of new countries, visited me in 1772 with Dr. Solander and Dr. Uno von Troil. They went to I Iekla, stayed here for a few weeks, leaving a reputation of affability, generosity and scholarship. Uno von Troil later published an account of their visit, *Bref rörande en Resa til Island 1772.*'

[2] King Frederik VI.

[3] *Island I det Attende Aarhundrede, historisk-politisk skildret ved Magnus Stephensen*. In his section on Banks, Stephensen heaps fulsome praise on him, his titles and exploits and gives details of the 1772 expedition. He points out that it was thanks to Banks that the Iceland ships were released and that in Iceland 'Sir Banks' is remembered for his charming and kind humanity and generosity. Finally Stephensen observes that, while Banks remains alive, humanity has not deserted every British breast.

[4] This would be the *Danske lov* (Danish law) from 1683 and the *Christian den Femtes Norske lov* (Norwegian, laws of Christian V) from 1687, the laws were separately published and collected. Stephensen would have commissioned a superb binding.

[5] Unfortunately all these books were lost when the *Margaret and Ann* went down at sea. There is no copy in the British Library, where the Banks Library is preserved.

[6] It was indeed delivered to Banks and has survived (here Document 122).

[7] Towards the end of his book Hooker discusses Stephensen's books, which he said 'treated of the most remarkable occurrences that had taken place in the later history of the country, among which it was peculiarly gratifying to me, as an Englishman to find ... how earnestly and how completely *con amore* [Stephensen] bears testimony to the noble and generous conduct of Sir Joseph Banks', and continues 'I must, however, do the Icelanders the justice to say, that there is no need of the assistance of the press to excite a stronger feeling of gratitude on their part, for the benefits that have been conferred upon them by this exalted character; for the eager enquiries that were in every place made after his welfare, by the aged, who still remember his person, and by the young, who know him from the anecdotes told by their fathers and their grandfathers, were a convincing proof of the esteem and veneration they entertain for him: so that, not unfrequently, while wandering over the wastes of Iceland, my heart has glowed, and I have felt a pride, that I should have been ashamed to dissemble, at being able to call such a man my patron and my friend' (Hooker, *A Tour in Iceland*, I, pp. 280–81).

of the receipt of these writings, which for myself & my family may be as dear a monument from a most respected friend, as, amongst several other costly presents, Your own excellent hand-gun with silver-works here in Iceland, by Yourself once presented to my old Father, in many Years was to him & myself, his eldest Son, until its wearing out made it only fit in the relics to preserve the lasting remembrance of its old noble giver. – Besides has my dear Father, now in his 83th year, lost much of his vigour, health & recollection, yet still charged me with the commission to pay his most tender & obliged respect, & to testify his due feelings of gratitude, in his own & his Country's name, to his best & most honoured very old friend, Sir Joseph Banks.

My Father and his Sons[1] will further consider as a special mark of friendship & favour to our Country, my [*sic*] it please Sir Joseph Banks to accumulate all his former benefits with the following new:

1° Not to listen to any body's dangerous contrivances against Iceland, especially of the here extremely hated Englishman, Savignac, who justly may be suspected to have seduced Mr. Phelps & Jörgensen to venture the undertaking of a revolution & to committ robbing & other acts of violence 1809 having besides afterwards, himself, offended the inhabitants of Iceland with his own injurious behaviour, wrong & violation of their territorial right & the inviolable right of their property, as well by forbidden chase upon their ground, often destroying their catching of Seadogs[2] on their own ground, where the laws only appropriate to the owners all fruits & right to such a catch, as also the most precious one of all the best icelandic produce, viz. the Eiderdown, destroying too the laying of eggs by Eiderducks & other birds on our properties, which commonly affords to the owners the most precious fruit of their ground; these costly birds, the Eiderducks, which our laws forbid to hurt, or kill anywhere & ever, or to robb or steal away their eggs and down from their just owners, under a fine and punishment as for committed violence & theft, are especially this year in great number shot, maimed and stolen away by Englishmen, Savignac himself, as leader, and some Captains of ships with their crew, from the Merchant-house Edward & C°. in London,[3] which last year did here undertake a trade, chiefly led & seduced to such a trespasses by said Savignac; some inhabitants being also violently attacked with blows and wounded in the just defence of their property, and this by Savignac himself & english Captain from one of the said ships, whose crew on several other islands, even on my own Father's Seat too, the island Vidöe,[4] and six times on my and brother's, the Bailiffs Stephen Stephensens island Andridsöe in Hvalfiord,[5] with about 60 gunshots committed said violence, robbery and theft of all the Eiderducks & eggs they there were able to get, notwithstanding all forbidding from the Government, published even on board of said ships, by the Sheriff of the town Reikiavik, by which trespasses & chiefly with the gunshots almost all Eiderducks, gathered on our island Andridsöe, even as on more other islands, are either killed and maimed, or with many gunshots at six several times frightened & quite scared away; & with them this most valuable fruit of

---

[1] On the sons see Document 45, p. 234, n. 6.

[2] Another name for the common seal.

[3] There are no reports of a merchant house of this name trading with Iceland, doubtless Everth & Hilton. In fact in the copy of this letter preserved in the papers of the Board of Trade *Everett* has been written in the margin and corrected in the copy sent to the Board of Trade.

[4] Viðey, near Reykjavík.

[5] Andriðsey in Hvalfjörður.

such a property. – Besides this has he and more Englishmen here injured some others, catching some salmons in their rivers without any Licence from the proprietors, shooting their swans & seadogs laying upon their ground, & by gunshots scaring them away.

Of such a man, as Savignac, every bad undertaking is to be expected, but if he still may enter upon the advice to deprive Iceland of its liberty, submitting it the english Kingdom, as Mr. Hooker in his Pamphlet pag: 355 seems even to point at,[1] no doubt, in consequence of Savignacs, Mr. Phelps's & the traitor Jörgensens inspirations, as he was their fellow traveller, however in that writing liing [sic] at all, this to be the wish of but a single honest Icelander, we do heartily repose all confidence in the generosity of our dearest friend, the Philantrop & immortal Patron of Iceland, Sir Joseph Banks, never to listen to such a pernicious advice against this island, nor allow it to be performed, whereas Iceland finds itself happy under the sceptre of its lawful Sovereigns, our most gracious Kings of Denmark & Norway, the most tender paternal care in every view, as to the treatment & rule, as to the granted aid & support of Iceland, being amongst them all for more than two Centuries quite hereditary; And all delightfulness in nature and politics being in the disagreeableness in both justly compared, our poor island may perhaps hardly have reason to repine at the happiness of the most Subjects in other Kingdoms. England, almost ever entangled in war, is besides, notwithstanding all her riches and power, neither able to provide our island at such a time with the first necessaries of life, as provisions, timber, tar &c: whereas her manufactures & indian trade will only be able to furnish us with wares not suitable to the most raisonnable manner of life amongst such a poor people in our hard climat, but for some better, where riches and luxurious life invite her trade to gain, nor will continuing war ever grant our poor and unpopulated island, when subjected to England, that necessary freedom, immunity from taxes & military service, as a most singular mercy of its beloved Kings.[2]

[[3]2° Whereas said english Captains, especially Storm & Smith[4] & their crew probably in the beginning misled by Savignac (·who now is returning home for a while;[5]) have, as

---

[1] Hooker stated in his appendix on the Icelandic Revolution: '[If Iceland continued under Danish rule] then will the state of the natives be more wretched than ever; unless, which I sincerely flatter myself will be the case, England should no longer hesitate about the adoption of a step to which every native Icelander looks forward as the greatest blessing that can befal/ his country, and which England herself would, I am persuaded, be productive of various signal advantages, the taking possession of Iceland and holding it among her dependencies. Iceland, thus freed from the yoke of an inefficient but presumptuous tyrant, might then, guarded by the protection of our fleets, and fostered by the liberal policy of our commercial laws, look forward to a security that Denmark could never afford, and to a prosperity that the selfishness of the Danes has also prevented: while England would find herself repaid for her generous conduct by the extension of her fisheries, the surest source of her prosperity, and by the safety which the numerous harbors of the island afford for her merchantmen against the storms and perils of the arctic ocean' (Hooker, *A Tour in Iceland*, II, pp. 62–3).

[2] This letter is important in disabusing Banks of the belief that the Icelanders would welcome a British annexation as he was convinced they would.

[3] This paragraph has been placed in brackets by pencil, probably to be transcribed and sent to a relevant governmental office, in this case the Foreign Office.

[4] Captains Smith and Storm were employees of Everth & Hilton. Magnus Stephensen had sought Consul Parke's help. The consul had promised to try to prevent a repetition of this kind of behaviour, expressing his displeasure at his fellow-countrymen and warning them that they were breaking Icelandic law. Castlereagh, the Foreign Secretary, directed Parke to lose no time in informing the British subjects in Iceland of this legal decision (TNA, FO 40/2, Castlereagh to Parke, 11 December 1812).

[5] He was returning to London and would share a cell with Jörgensen in Newgate (see Document 262).

aforesaid, committed violence, robbery & even theft of eggs and birds on my and my brothers, the Bailiff Stephen Stephensen's propriety,[1] the island Andridsöe in the Hvalfiord, at six times repeated, & suchlike crimes on other islands, viz: on my Father's, the old Governor Olav Stephensens Seat Vidöe, even as on Engöe, Thernöe, Akeröe,[2] and every where maliciously in spite and with the utmost contempt not only of the laws of our Country, but even of their own gracious Sovereign, His Brittish Majesty's Proclamation & Order in Council of the 7 Fevr*ier* 1810, which strictly forbids 'every act of depredation or violence (in Iceland) against persons, ships and goods of any of the inhabitants, and against any property in the said island respectively', having destroyed the Eiderducks & thus the most valuable Fruit of our propriety and otherwise injured its holy right, as by Mr. Clausen will farther be proved,[3] we do hereby humbly beseech your just disposition, dear Sir! that the guilty persons of said crimes & of the contempt of His Brittish Majesty's Order, to a necessary warning for others afterwards, may be strongly punished according to laws & decided to pay, at least, two hundred £ for the crimes committed on Andridsöe, as reparation for myself & my brother, for our loss by their violence & robbery on this our property, & besides justly fined, for the contempt of His Br*itish* M*ajesty*s Order.

Further, that it may be ordered for the time to come, congruously with the common right of all nations, that every Englishman, during his stay in Iceland, is amenable to the laws of said island, & consequently liable for committed crimes here to be imprisoned, fined & punished, accordingly to the laws of Iceland, even as every other its inhabitant, without farther appeal, than to the high Court of said island, lest more english travellers, as Savignac & the Captains Storm & Smith have uttered, may believe, that they with impunity here can commit every crime, as were they not amenable to the laws of Iceland, which never allows power & violence, but only right to decide all quarrels amongst the inhabitants.][4]

At last I do humbly venture generally to recommend the merchants of Iceland to all the favour & commercial liberty, which they equitably may desire, in order to their own subsistence & just gain, which may afterwards enable them, now under this most expensive time of war, to provide Iceland with the first necessaries of life, lest its inhabitants, may starve with hunger, but especially having the honour to recommend the old very honest merchants of Iceland, Mr. Jacobsen from Kieblevig,[5] who wishes to get a licence for his vessel him graciously granted, for the pursuit of his trade in Iceland, and to its better support under the present calamities of war.[6]

If any thing may excuse having thus intruded upon Your patience, dear Sir! it must only be the confidence, that your tenderness for Iceland & old friendship to my family will pardon every suit for the wellfare of this island & its recommendation to its noble Patron. –

---

[1] Stephensen doubtless meant to write *property*.
[2] Engey, Þerney and Akurey, small islands in the vicinity of Viðey, near Reykjavík.
[3] H. P. Clausen was obviously to carry the letter to England.
[4] This bracket is in pencil to identify what part of this letter was copied and sent to the Foreign Office.
[5] Christian Adolf Jacobæus of Keflavík.
[6] His vessel the *Keblevig* sailed to Iceland in 1812. In 1813 the *Skalholt* was taking timber from Norway to Iceland (RA, Rapporter fra konsulatet 1808–1816, letter from Horneman, 19 October 1813). The *Skalholt*, jointly owned by Petræus and Jacobæus, was granted a licence (TNA, BT 6/209, nos 49,502 and 50,532).

May the heaven, always rewarding true virtue bless Your respectable age with honour & prosperity in all Your noble efforts to the comfort of mankind & with the godlike joy of Your own consciousness ever to deserve the glorious name of the poor Iceland's most generous benefactor & guardian Angel.

I have the honour with utmost veneration
to remain
Right honourable Sir!
Your most obed*ien*t & humble Serv*an*t
[*Signed*] Magnús Stephensen

P.S. I do humbly request in the kindest manner to be remembered to Sir <u>John Stanley</u> Bar*one*t, even as my old Father begs himself, with much acknowledgment for his honoured letter and attention of our Family.

BL, Add MS 8100, ff. 141–144. Autograph letter.[1]

241. Count Johan C. T. von Castenschiold to Banks, 22 August 1812

Reikevig in Iceland

Sir!

His Excellency the Count Reventlow, one of his Majesty's ministers, President of the Exchequer, has directed me, as it may please You to know of the following copy of His Excellency's letter,[2] to witness to You the gratitude, which occupies the heart of every inhabitant of the islands of Iceland, Greenland and the Faroes,[3] with regard to Your successful endeavours and labours in these times of war to remove the miseries of a defenceless and suffering people.

Thus it is become my lot to express the sentiments of the inhabitants of this island, as well as those of Greenland and of the Faroes, who all without Your humane assistance doubtless were to have undergone the most dreadful starvation. But how might I here hope to succeed? How[4] could I be able to represent the feelings of the wretched housefather of [*sic*] the dispairing housemother, seeing themselves supplied with the first necessities in the moment, a horrible destruction was approaching them and their children!

Only in the country where I have the honour of being employed a magistrate,[5] more than fifty thousand inhabitants[6] appear with whom associates the populace of the other islands who all adore in You the benefactor, to whom they are indebted for their subsistence.

---

[1] Note in Banks's autograph, on the bottom left-hand corner of the first page: '[Received] Sept*ember* 18 [1812]'.
[2] Count Christian Ditlev Frederik Reventlow (1748–1827), minister of state.
[3] In the Kent copy the spelling is Faeroes throughout.
[4] Kent copy adds '& how...'.
[5] He means he was district governor of south Iceland (only in 1813 did he become governor of the island).
[6] The population in 1812 has been estimated to have been 48,317 (Jónsson and Magnússon, *Hagskinna*, p. 56).

Without Your philantropical protection all these unfortunate should have been deprived of every thing even the most necessary to support life.

Accept, gracious Sir! these lines as a small proof of the ackowledgement of Your most important benefits and of the most unlimited gratitude, the inhabitants of these islands are owing to You.

Look upon every imperfection as want of faculty not of will! For ever is Your name to find its highly merited place in the annals of Iceland.[1] – Peace and tranquillity returning, Iceland will abundantly reap the fruits it most confidently can hope from the protection it has enjoyed by You.

As at present inhabitant of this country, I myself feel and behold every day the beneficent fruits of the humane care and support with which You have favoured this country so unfortunate with respect to its situation.

I therefore most ardently, joined by every honest wish, that God allmighty still long may preserve Your for Your fellowmen so precious days. –

I remain with /the/ utmost regard and consideration
Sir
Your most obedient et[2] humble servant
[*Signed*] Castenschiold[3]

Wisconsin, Banks Papers, ff. 85–88. Autograph letter.[4]
Kent History and Library Centre, Banks MS, U951 Z32/9, f. 9.[5] Contemporary copy (amanuensis Cartlich).

242. Christian Ignatius Latrobe to Banks, 4 September 1812

Nevils Court

Sir,
Having received a letter from Mr. Frisch of Copenhagen,[6] respecting a farther supply of necessaries to the Colonists in Greenland, I take the liberty of transmitting to You an Extract of it, and feel no objection to recommend it to Your kind Consideration, as he desires, me to do, nor to the favourable attention of the Lords of Trade, after the generous disposition You have shown, to show compassion to the poor defenceless inhabitants of that barren Coast, by permitting provisions to be sent to them last Spring[7] & I am thereby

---

[1] Which it certainly did: see Appendix 5.

[2] In Denmark French was used as the language of diplomats.

[3] In Teigen's 'Supplementary letters of Sir Joseph Banks', the writer of this letter is stated to be Joachim Melchior Holten Castenschiold. Army Officer (p. 250). He was the father of the writer. Banks indeed notes that it is the governor who is writing to him.

[4] Note in Banks's autograph, on left-hand corner on first page: '[Received] Oct*ober* 2'. Note in Banks's autograph on last page: 'The Stiftsamptman of Iceland | To Sir Jos: Banks' and further verso: 'Iceland Letters | Sir Jos: Banks; and in different handwriting: 'Return'd with Lord | Bathurst's best Comp*liments*. | Many thanks for the | Perusal of the papers | Nov*ember* 18 1812'.

[5] The copy has different capital letters and slight variations in spelling and the above-mentioned notes are not on the Cartlich copy.

[6] Frisch, the director of the Royal Greenland Trade Company, had also supervised the Iceland trade.

[7] In the spring of 1811 the Privy Council for Trade and Foreign Plantations had indeed decided to send supplies to the inhabitants of the Danish settlements in the Davis Straits. The man behind this decision was not

encouraged to request, that, if consistent with Your view of the subject, his humble petition may be granted; which he seems to place with full confidence in the humanity of the British Government, entrusting even his ship *Jupiter*[1] to Your mercy by directing her to go to Leith on her return.[2] It is unnecessary for me to trouble You with many words on a subject, regarding which, I have already made such experience of Your condescending kindness, and I confidently trust, that the urgency of the case will cause the Lords of Trade again to make an exception in favour of those poor people, whose Existence depends chiefly on Your humanity, and render the sending [of] provisions to them possible, by granting the licences according to the petition.

With the most sincere Esteem
I remain ever, Sir,
Your most obliged and most faithful Serv*an*t
[*Signed*] Chr. Ign. Latrobe[3]

Wisconsin, Banks Papers, ff. 92–93. Autograph letter.[4]

243. Holger Peter Clausen to Banks, 10 September 1812

London

Right Honorable Sir!

When I had the honor to take leave of you I forgot to inquire if you had any orders for me to Copenhagen, indeed the gratitude for your generosity fill'd me so entirely that no other thought entered my mind, and hope you will not find me less worthy of your commands, tho' I omitted it then.

Since Your absence things are altered very much. Lord Clancarty is appointed President [of the Board of Trade][5] and a Mr. Robertson Vice-President.[6] – I had the hope that all was settled, but new difficulties arises.

Banks, however, but William Mellish, a London merchant who was active in the Greenland fishery. Mellish had written a letter pointing out that the inhabitants 'will be exposed to the greatest distress if not furnished with Provisions from this Kingdom'. The matter had been referred to the Lords of the Treasury and subsequently provisions were sent (TNA, BT 5/20, Minutes, 22 April and 16 May 1811).

[1] Apparently another of the 'privatized' Greenland ships.

[2] The *Jupiter* sailed in April 1812 from Bergen to Holsteinsborg (Sisimiut), apparently without a licence (see Document 246). On her return the ship had been granted a licence in mid-July (on petition from Corbett, Borthwick & Co.) to export a cargo from Leith to Trondheim (TNA, BT 6/207, no. 33,145). In October this vessel was permitted to import a return cargo from Trondheim to Dublin (TNA, BT 6/208, no. 47,565). Thus a Bergen–Greenland trade was permitted.

[3] Frisch obviously turned to Christian Latrobe, who was a Moravian brother living in London, to ask him to use his contacts to help the Greenland trade.

[4] Here the letter is torn, with a piece missing. Note in Banks's autograph on bottom left-hand corner of first page: '[Received and answered] Sept*ember* 12 – 13 [1812]'. Note in Banks's autograph, verso: 'a Licence was granted a fortnight ago to Mr. Clausen – I conclude he had | authority to Solicit it | 400 Barrels of Blubber & a Quantity of Whalebone'.

[5] The Earl of Clancarty became President of the Board of Trade in 1812 (until 1818), succeeding Earl Bathurst.

[6] Not Robertson but the Hon. Frederick John Robinson (1782–1859), later Viscount Goderich, a prominent politician and statesman. He was Vice-President of the Board of Trade from September 1812 until January 1818, eventually becoming Prime Minister briefly in 1827–8.

The 28th of August the *Tykkebay*¹ came to Leith, having been almost two years upon her voyage, and having no benefit from the exportation of the Klipfish & oil granted last year,² as she then was in Norway, the owner loaded it this year. But as the orders for Mr. [John] Parke were not arrived when he left Iceland he has no certificate,³ which is now demanded from the Privy Council. There is a Certificate from the danish authorities – that the fish is cured before they were acquainted with the prohibition of this article, but I am afraid this will be of very little use to the case. As a secondary wish, he wishes permission to touch at Norway, which indulgence has been granted this spring to the Iceland ships *Eskefiord*⁴ & *Husevig*.⁵ – I dare assure, tho' he is permitted to go with his cargo, that the exportation of this article will be far less than the granted 300 Tons.

The 27th of August the *Skalholdt*⁶ arrived at Leith. Messʳˢ Rooke & Horneman made an application the 2ᵈ of September for a Licence for her to proceed after unloading the Tallow, to Denmark and to import to Leith from thence a cargo of grain etc. with permission to go with part of the grain from Leith to Iceland but in case this not should be allowed, then to get the licence to remain in force untill the 15th of June, as the vessell will be obliged to stop during the winter in a danish port. – To day the petition is granted⁷ as far as to bring a cargo of grain to Leith; but no further and the licence only to remain in force for 4 months, which in this case will be expired in Januare, consequently in the middel of the winter, when it is impracticable to leave Copenhagen, the ships can impossibly be before the later end of May in Leith.

For the *Norburg* which was not this year in Iceland /: whose licence you had the goodness to sign the date of presentation upon:/ I applied for a Licence instead of the old, but it was refused, as too long beforehand.⁸

In this dilemma, I as usually venture to fly to your generosity for advice and assistance, if you can afford it now when you are from town.⁹ If you condescend to honor me with a

---

¹ The *Tykkebay* went through Customs on 10 September and was released for her voyage to Copenhagen on 22 September (NAS, E. 504/22/57).

² The Board of Trade had decided to permit the Iceland merchants to export 300 tons of Klipfish and 300 tons of fish liver oil during 1812.

³ The Board of Trade made out the instructions to Parke at the end of March 1813, stipulating that 300 tons of fish-liver oil would be permitted to be exported to Denmark, but no klipfish at all. If the merchants failed to obey this order their licences would become void and their ships prize. The reason given for not allowing the export was that the stock of klipfish, whose export had been 'humanely extended last year' must be exhausted by now as none had been permitted to be cured (TNA, FO 40/2, 'Draft of Additional Instructions to H.M.'s Consul in Iceland', 31 March 1813 and Castlereagh to Parke, 1 April 1813). Parke did not receive them until July (TNA, FO 40/2, Parke to Castlereagh, 13 July 1813).

⁴ The *Eskefiord* received a licence in January 1812 to proceed with a cargo of Iceland produce from Bergen to Denmark to Leith (TNA, BT 6/205, nos 38,101, 38,575, 38,597 and 38,645).

⁵ The *Husevig* was finally granted a licence to sail from Norway to Leith (TNA, BT 6/205, nos 38,575, 38,597 and 38,645).

⁶ The *Skalholt* was in Leith, leaving on 21 September for Copenhagen (NAS, E. 504/22/57, 10 September 1812).

⁷ She was permitted to land tallow in Leith and proceed to Copenhagen to return with grain etc. to Leith (TNA, BT 6/208, no. 45,927).

⁸ The Lords decided that the wishes of the *Norberg* could not be decided so long beforehand (TNA, BT 6/208, no. 46,133).

⁹ Banks was staying at this time at Revesby Abbey, his main country residence.

few lines, under the direction of Messˢ. Rooke & Horneman[1] I shall undertake nothing before I hear from you.

In the hope that you will forgive my disturbing Your repose I have the honor to subscribe myself.

Right Honorable Sir
Your most obedient and most humble Serv<i>ant</i>
Holg^r P. Clausen

Wisconsin, Banks Papers, ff. 90–91. Autograph letter.[2]

244. Christian Ignatius Latrobe to Banks, 17 September 1812

Neviles Court, Fetter Lane, London[3]

Sir,

Your very obliging letter of the 13ᵗʰ, I received with many thanks for Your condescending attention to the subject of M͟r. Frisch's letter, but was not a little alarmed at the report, of the licences for the *Hvalfisken* & *Huitfisken*,[4] which had been already granted, being 'withdrawn or returned'. M͟r: Clausen has not been quite correct in giving this information to the Board of Trade.

In consequence of this information, I went to Mr: Ballantyne,[5] whose house transacts business with the house of Corbett, Borthwick & Cº at Leith, the agents of the Danish Greenland Ships, to make Enquiry concerning M͟r: Frisch's proceedings. M͟r: Ballantyne communicated to me all the letters that had passed between the parties, from which it appeared, that M͟r: Fritsch was thankful for the licence granted to the Ship *Hvalfisken*, and intended, on the strength of it, to send that Ship to England in November, to stay During the winter at Leith and proceed to Greenland in March, 1813. No mention whatever is made by Messʳˢ. Corbett & Cº at Leith of a return of the License, & the renewal, spoken of in M͟r. Frisch's letter to me, refers only to dates, to give him the needful time to perform the Voyage.

I am not willing to admit the idea, that M͟r. Clausen should have had any sinister views in making such a representation to the Lords of Trade, to the Exclusion of M͟r. Frisch, who, (as one of our Missionaries, now in London waiting to proceed to Greenland, where

---

[1] Agents for the Iceland merchants.

[2] Note in Banks's autograph on bottom left-hand corner of first page: '[Received and answered] Sept*ember* 15–17 [1812]'.

[3] 'Neviles Court Fetter Lane' is in Banks's autograph, and is written below London and the date on the top right-hand corner of the first page.

[4] The licences for the two ships to sail with provisions for Greenland had been refused in May 1812 (TNA, BT 6/206, nos 42,105 and 42,212). Whereupon Latrobe wrote a letter himself which resulted in the *Hvalfisken* receiving a licence while the *Hvidfisken*, though issued with a licence, was banned from bringing any oil from Greenland (TNA, BT 6/206, no. 42,275). The ships sailed and at the end of July were petitioning for permission to load grain in Denmark for Greenland on their way to Leith (TNA, 6/207, no. 33,490). Thus they made the voyage.

[5] Possibly rather Bannatyne of the firm Findlay, Bannatyne & Co.

he has been Employed nearly 30 years,¹ tells me) is the regularly accredited & principal person, conducting the affairs of the Royal Danish Greenland Company. – <u>One</u> Ship surely is not able to convey a sufficient supply to the Southern Colonies, as M^r: Clausen states, and it is most desireable, that M^r. Frisch should be permitted to send out the Ships he mentions, and which are the regular traders. –

I trust, that if it is certified, that the licence for the *Hvalfisken* is not returned or withdrawn, that M^r. Frisch will not meet with unexpected difficulties, when, in reliance on its validity, he sends his Ship to Leith in November.

He certainly did an imprudent thing to venture to send the *Jupiter*² without licence from Norway to Greenland.

I should feel very uneasy to have troubled You again with so long a letter on these subjects, did I not know, with what benevolence & favour You have always considered the case of the defenceless inhabitants of Greenland & Iceland, who are so much distressed by the Effects of the War. –

With the sincerest Esteem & Gratitude
I remain ever, Sir,
Your most obliged & most devoted humb*le* Serv*ant*
[*Signed*] Chr. Ign^s Latrobe.

Wisconsin, Banks Papers, ff. 94–95. Autograph letter.

245. Banks to Christian Ignatius Latrobe, 27 September [1812]

R[evesby] A[bbey]

Dear Sir

I do not on my own Judgment Suspect Clausen of having deceivd me or the Board if he has he deserves no mercy The Licence was not Granted to him as a Favor but because it was at the time believd to be the most Effectual means of Releiving the Colony The fact of the two Licences Granted for Greenland³ being Returnd to the office was not taken upon Mr Clausens authority. I myself Calld for the Licences & saw them in the office for Trade & Could not on Enquiry learn that any Extention of time /or other further indulgence/ was asked for in Either Case⁴

The final determination of the business Seems however to me to Rest with the Danish Government all that the Board of Trade wishes to do is to Releive the miserable Colonists & I have no doubt that /*it*/ this will be done in the way most likely to Effect that purpose, if it is Pointed out to them, should /*if*/ Mr Clausen have Obtaind the Licence by any improper Conduct the Danes are the proper Persons to Punish his bad Faith if they Release & Suffer him to Proceed & it is Stated to the board that some other mode of

---

¹ This is the Moravian brother Johann Friedrich Gorke from Herrnhut who first came to Greenland in 1781 and after a visit to Europe was waiting in London for a ship to take him back to Greenland, which he managed in 1813.

² The *Jupiter* was a Greenland ship.

³ For the *Hvalfisken* and *Hvidfisken*.

⁴ This paragraph is difficult to read because of Banks many times crossing out and adding words in this draft. It is printed here as Banks intended it should be read.

proceeding is likely to be more Effectual /for the Releif of the Colony/ I have litle doubt of its being adopted to the Extent of 50 Tons of oil or Blubber the Produce of which the board at present thinks Sufficient to Supply the Colonists with the Necessary Supplies they Stand in need of

on further Consideration Mr Frisches imprudence in dispatching his Ship direct from Denmark to Greenland[1] appears greater than I apprehended it to be in my /*Last Letter*/ haste of writing my Last Letter I did not recollect that much jealousy has always been Entertaind here of a Direct Communication between the mother Country & her Colonies[2] which in Fact is the Reason /*why*/ of insisting on a final clearing out from Scotland under this Recollection it is impossible for me to hope that Mr. Frisches Ship will be well Receivd when She Comes to Leith

Wisconsin, Banks Papers, f. 96. Draft in Banks's autograph.

246. Banks to [Count Johan C. T. von Castenschiold], [November 1812]

[Soho Square]

Sir

it gives me Pain to Learn from your *Excellency's* obliging Letter[3] that the Poor assistance I have been Permitted to give to the Colonies of Denmark /*have been*/ are so much overrated by the Good Count Reventlow[4] & his Royal[5] Danish majesty[6]

it would be a Silly presumption in me to take to myself the merit of a Line of Conduct which flows from the natural Humanity of the English nation carried into effect by the wisdom & moderation of my Sovereign & his ministers; I do not wonder that the merchants who See me often on their business attribute to me more than belongs to me; in my Station as one of the Com*mittee* of *Privy Council* for Trade it is my duty to attend to Such Points as my Fellows intrust to my Charge,[7] among these are the affairs of Iceland, Greenland &c., I hear & Represent the Cases that occur but Sir it is to the Benevolence of the British Cabinet & in no Shape to me that the Regulations made in Favor of the unoffending Colonists have originated & been brought to maturity, I am merely the organ and Engine of Promulgating their Good works. To them then & not to me must the Credit of having acted with humanity wholly attach

it Greives me to be under the necessity of Destroying the Good opinion which your Gracious Sovereign & his Excellent minister have Entertain'd in my Favor but Truth Renders it necessary for me to State the Facts above mentioned, I do not however wish to Conceal from your *Excellency* that the Pleasure I feel in being allowd to be the

---

[1] He had sailed from Bergen without a British licence.

[2] In fact the cornerstone of the licensing system, there could not be a direct trade between the mother land and the colony. The trade had to be controlled by stops at British ports.

[3] This letter is Banks's reply to Document 241.

[4] Count Christian Ditlev Reventlow (1748–1827), Danish statesman: see Appendix 13.

[5] The corrections in the Wisconsin copy are made in Banks's autograph. *Royal* is in the Kent draft, but struck out in the Wisconsin fair copy. Discrepancies are few.

[6] King Frederik VI.

[7] 'Especial' is added to 'Charge' in the Kent draft.

Interpreter of the wishes of the hospitable Icelanders from whom I have Receivd so many Good offices as well as of the unfortunate inhabitants of Greenland & Feroe has been the motive that has induced me to attend diligently to their interests. This Sentiment Sir will insure a Continuation of my attention to the /*interests*/ wants of these innocent Sufferers & I beg you Sir to be assured that the Pleasure I receive from the unbounded Gratitude with which my Little services have always been Receivd, is the only Reward I wish for, & is in my judgment a most ample compensation for Exertions far more /*troublesome*/ burthensome than mine on this occasion have been

I beg you Sir to be assured that the welfare of these Suffering Colonies will Ever be present to my heart /& that the humanity of the British nation & the moral Sympathies of my Sovereign & his Ministers afford an ample security for the Continuance of the indulgences they have Receivd/.[1] That I Shall ever Recollect with pleasure the friendship which for so long [a] Period of Time united the Kingdoms of Denmark & England, a Friendship which I deem to be only for a Time Suspended & I live in hope to See that Suspension destroyed /& I Feel as/ a Certainty that Both nations will when it is Possible to do so Renew with Eagerness those ties of amity which the /*unfortunate*/ Disturbed state of the Continent has so unfortunately broken off

Wisconsin, Banks Papers, ff. 88–89. Autograph draft.
Kent History and Library Centre, Banks MS, U951 Z32/11. Contemporary draft (amanuensis, Cartlich) with corrections made by Banks.

247. Extract of a letter from Mr Frisch, Agent for the Royal Greenland Company in Copenhagen to Mr Latrobe, London, 18 December 1812

[Copenhagen]

After expressing his thanks for the Condescension of the British Government, in considering the wretched state of the Greenland Colonies & granting Licenses for Ships, going to their Relief, – he proceeds:[2]

'Mr: Clausen was sent to London by the merchants trading from hence with Iceland, to endeavour to procure a licence to bring Train oil from Iceland by way of Leith to Copenhagen, and the hope of succeeding in obtaining a similar licence for Greenland, made me commission him to petition for one for my ship *Hvalfisken*. You will easily conceive, that on account of the enormous Expence attending the voyage, it will not be possible for me to send ships to Greenland, unless I may obtain a licence to bring back from that country as much blubber, oil, whalefin[3] &c. as can be produced, without restriction as to quantity, being the only articles of trade there, by the sale of which the great expences may be covered. This I have now experienced, on the return of the *Freden*[4] a fortnight ago, as her cargo's sale fell short of her expences. I could not however but

---

[1] This insertion has been added by Banks himself and is not in the Wisconsin draft. It is, however, safe to say that it would have been added to the actual letter sent.

[2] This is a copy intended for Banks. The first paragraph is by the writer introducing the letter.

[3] 'Whalefin' is a commercial term for whalebone.

[4] The *Freden* had been sailing back and forth from Greenland since 1810. All the ships mentioned sailed to Greenland in 1813.

rejoice at her having had a successful voyage & having brought the necessaries of life, with which she was freighted, safe to the poor Colonists in South Greenland, but it is highly necessary that help be afforded them next year also (1813.) I have therefore written to the house of Corbett, Borthwick & Co. at Leith, that I wish to send the _Freden_, Capt*ain* Lauritz Matthiesen as well as the _Hvalfisken_, Capt*ain* Nils Elberg to Leith & Greenland. and I entreat you to forward a petition to Your liberal-minded Government, in behalf of this Concern, that I may obtain Licences for these two Ships, in such a form, that they may be of real service; that is, that I may be thereby enabled, to receive blubber and Whalefin, without restrictions as to quantity, by way of Leith from Greenland, and to send the needful stores & provisions to the inhabitants of that barren Coast, without loss. I most earnestly recommend this important concern to Your kind assistance.

The ship _Hvalfisken_ will load Corn /wheat/[1] at Flensburg & convey it to Leith, & go from thence with her provisions to Greenland. I trust it will be an acceptable importation. I entreat you to represent the case to the proper authorities, & hope for a favourable answer to my request.' –

Wisconsin, Banks Papers, ff. 118–119. Contemporary translation and copy.[2]

248. A project for Displanting & Dispeopling of the Danish ~~Settlements~~ Colonies in Davies Strait, [1812][3]

one or two Stout Ships to be sent there /this year/ with the necessary Supplies /*with*/ & with orders to Receive on Board as many Passengers as propose themselves & are willing to Pay for their Passage Such a Sum in Produce as they would have Paid for the Necessaries /*on board*/ brought by the Ships /which/ they must have Purchasd for their use had they Continued in the Settlement[4]

Notice to be given to the Settlers who Choose to Continue that these Ships /*would*/ will visit the Colonies next year but not afterwards & an Exhortation to Prepare themselves for Quitting the Colony when the Ships Return as the Government of England does not choose to Continue the Expence of maintaining an Enemies Colony but out of /*kindness*/ Humanity to the Individuals who Compose it are willing to provide them with a Safe Passage to Denmark

an accurate account to be taken this year of all the Europeans /who Continue to/ Reside/*nt*/ in the Colonies that /*proper*/ sufficient Shipping may be /*prepard*/ Sent out in the year 1813 for a Compleat Removal of them all

a Large /*supply*/ allowance of Powder & Shot to be supplied to the Natives this year and the Same next year with notice /to them/ that they must in Future /*be Left to*/ Rely upon their own industry for their maintenance

---

[1] Written above the word Corn.

[2] This letter was probably enclosed with Document 255.

[3] The date is derived from the contents of the letter itself viz 'this year' and '1813'.

[4] Banks was a keen advocate of extending the British Empire by creating new colonies. For example in 1783 James Matra, a fomer American colonist, had written much the same under Banks's influence in 'A Proposal for Establishing a Settlement in New South Wales' (Carter, *Sir Joseph Banks*, pp. 190, 216).

Notice of the Dispeopling of the Colony to be in Due time Given to these who use the Whale Fishery in the Strait with an exhortation to them to touch at the Different Places where Danish Settlements have been to Trade with the Indians[1] who no doubt will provide as much Trade as will Suffice to buy for them the Ammunition &c which are absolutely necessary for them Probably much more as the Trade of the Europeans when Settled there was Principaly Provided by these Indians & purchased from them

There are three moravian Settlements[2] on the Coast what must be done with these they may Surrender their Countrey[3] to the British Crown & derive advantage from the Cession

Cap*tain /Stevens/*[4] Says there are About 500 Danes in all on the Coast[5]

Wisconsin, Banks Papers, f. 79.[6] Autograph draft.

249. Banks to Magnús Stephensen, [undated, December 1812[7]]

Soho Square

Sir

The Contents of your Letter[8] gave me Pain not only on The Consideration of the ill Conduct of the British officers of merchantmen who so wantonly have trangressd the Laws of a Countrey whose Neutrality is admitted by the English nation & who Receivd them with all the Friendship of a Countrey at Peace with their nation, but on my own account also, as I Learnd for the first time from your Letter, that Swans are private Property in Iceland, & I remember with Greif to have myself violated your Laws by Shooting them under the Belief of their being Birds of the Chase; be assurd Sir that I Lament my having from ignorance inflicted an Injury on a Country to which I am attachd by the Ties of hospitality, for Abundance of which I am indebted to your Amiable Islanders; I have endeavourd & Shall Still Continue to Endeavour to Repay this Obligation with diligence & perseverance.

---

[1] These are of course the Inuit.

[2] These three settlements of the Moravian Brethren were at Ny Herrnhut, Lichtenfels and Lichtenau, all founded during the 18th century.

[3] Or possibly 'Centres'.

[4] Unidentified.

[5] This appears to be exaggerated. According to sources, 300 Europeans, including Moravian missionaries, were in Greenland at the beginning of the war (TNA, BT 5/18, Minutes, 11 May 1808). Nothing came of this project. However, many Europeans managed to leave Greenland and at the end of the war only one European missionary named Bernhard (Bernt) Hartz (1781–1822) was left in Greenland (Gad, *Grönland*, p. 205). See further on the situation in Greenland during the Napoleonic Wars, Agnarsdóttir, 'Hjálendur Danakonungs í norðri', pp. 42–9.

[6] Note in Banks's autograph, verso: '1812 | For | Displanting Greenland.'

[7] This draft is clearly written after the instructions to Parke had been given on 17 December 1812 (see below, p. 515, n. 4), but Stephensen would not have received it until the spring at the earliest.

[8] Document 240.

I laid your Letter of Complaint before My Government,[1] who understanding from Mr. Park*e*[2] that Redress has been Obtaind to the Satisfaction of the Stifts Amptman,[3] thought it Right to Leave the proceedings of Last year unnoticed, especialy as the Crimes were Committed in a Jurisdiction over which they can claim no Comptroll.

They Gave orders however to Mr Park*e* the Consul to make known to all Englishmen who visit Iceland, that they must Submit themselves to the Jurisdiction of the Countrey, and answer for all Crimes they Shall Commit in Iceland to the Laws & Courts of Justice of the Island, they moreover directed the Consul to Refuse his Clearance to every vessel on board of which any person should be, against whom any Suit Should be pending in the Courts of Iceland, unless Such Person should Quit the Ship & Remain on the Island till the issue Should be known, or should give Sufficient Security to abide by the Sentence of the Court[4]

This I trust will prove a Compleat Remedy against the Silly Licenciousness of the English Captains & a Sufficient defence for the Private Property of the Icelanders, I scarce need add that the Laws adverted to by the British Government are the Municipal Laws of the Countrey during Peace & that it is not intended to Subject English men to the operation of any Regulations or Enactment, which may have Sprung out of the unfortunate State of warfare in which the two Countries of England & Denmark are involvd, Such Laws are applicable only to a State of warfare & Cannot it is Presumed be acted upon between Two Countries at Peace with each other as Iceland & England Certainly are

Recommend me kindly to my Good old Friend your Father,[5] may health & prosperity of all kinds Enliven his Long protracted age. I myself have now nearly Reachd my 70th year & I begin to Feel my Self a Candidate to the List of Venerables, in Comparing The different Periods of my Life no one of them appear to me So Satisfactory as the Present, those Gusts of Sensibility which amidst abundant Pain were in my youth productive of the highest Gratifications have Subsided into a Calm Enjoyment of Less Rapturous delight, but I may Say of a Succession of unalloyd Pleasures unmixd with disappointments

---

[1] Banks was upset by the conduct of the Englishmen and sent this letter to the government. On 24 November 1812 the Lords of the Committee of Trade and Foreign Plantations sent an extract to the King's Advocate-General 'to report his Opinion whether British subjects are not under the terms of the Order in Council of 7 Feb. 1810 amenable to the courts in Iceland, for any Infraction of the Laws of that Island while residing there' (TNA, BT 5/22, Minutes). Robinson, Advocate-General, was 'humbly of opinion that the Icelanders may lawfully proceed against such depredators according to any Laws existing against such offences which they can enforce'. He further felt it might be 'expedient to signify to that House [Everth & Co.] H.R.H. the Prince Regent's displeasure' (TNA, BT 1/71, Robinson to Viscount Chetwynd, 30 November 1812). The decision reached was the one described by Banks here. See TNA, BT 5/22, Minutes, 9 and 15 December 1812; TNA, BT 1/72, Cooke to Chetwynd, 19 December 1812.

[2] Apparently Parke had tackled the matter. His authority had been challenged by the British sailors disobeying him. A curious note in the National Archives of Iceland states that Captains Storm and Smith (see Document 240) had appeared before the public notary, Frydensberg the treasurer, to register a protest against their countryman Parke, who had ordered them to return to England without a cargo (NAI, Bf. Rvk. A/2-1, Bréfadagbók 1806-13, 18 and 27 July 1812, nos. 3118 and 3126). They were also ordered to pay compensation to the men injured. The Icelandic chronicler Jón Espólín explained it was because of the lack of fish that they became angry and the violent incidents took place (*Íslands Árbækur*, XII, pp. 56–7).

[3] Castenschiold was the district governor at the time: see Appendix 13.

[4] Parke, who spent the winter of 1812–13 in London, was instructed to refuse clearances if necessary (TNA, FO 40/2, Castlereagh to Parke, 11 December 1812, and further letters dated 15 and 17 December). The formal instructions to this effect in the name of the Prince Regent were sent to Parke from Castlereagh on 19 December 1812 (TNA, FO 40/2).

[5] Ólafur Stephensen. He died in November 1812, before this letter was written.

or Greifs, may Such be the Lot of your Good Father and of yourself, & as your Lives have by the Strict adherence to the Laws of Virtue deservd it, I have no doubt that the Reward is Given to you.

Wisconsin, Banks Papers, ff. 73–74.[1] Draft in Banks's autograph.
Wisconsin, Banks Papers, f. 155. Contemporary copy (amanuensis Cartlich).

250. Corbett, Borthwick & Co. to Banks, 29 January 1813

Leith

To the Right Honble Sir Joseph Banks
Sir,
You have so often & humanely exerted yourself in behalf of the inhabitants of Iceland & Norway, that in every new distress, they look to you as their best friend & protector, and as such, will have too frequently occasion to encroach upon your time with statements of their difficulties. Our present application is of so much importance to the poor people in whose behalf it is made, that we have very little doubt that you will excuse us for the trouble we give & willingly support their petition. We have of late had many very urgent applications from the district of Drontheim, to state to his Majestys Government the very distressing situation of its inhabitants, in consequence of their harvest having so entirely failed in 1812, that they cannot venture to sow Barley or Oats of their own growth without exposing themselves to the almost certain failure of next years Crop also, and therefore we have petitioned the Privy Council to allow the exportation of 6 to 800[2] quarters British Barley & Oats, & now enclose a copy of our petition, most earnestly solliciting your support to it in the Council.[3]

We are so much convinced of the urgency of the case, that we shall be extremely anxious to communicate, as soon as possible, a favourable answer to our Norwegian friends; as the approach of Spring without the certainty of having the necessary seed for their next year crop, which would naturally expose them to much alarm and distress, untill they see a prospect of receiving some supply from this Country, you will therefore confer a very great favour upon them and us, in forwarding a decision upon the petition.

Should their Lordships propose any particular conditions upon which /article/ this favour can be granted, we hope that they will be communicated with as little delay as possible, as there is now little more than time to secure the arrival of the Corn within the proper Season for sowing.

We have the Honor to be most respectfully
Sir
Your most ob*edien*t humble Serv*an*t
[*Signed*] Corbett, Borthwick and Co*mp*any.

SL, Banks Papers, I 2:5. Autograph letter.

[1] They are identical except for Banks's spelling and punctuation.
[2] Difficult to decipher.
[3] Not among the Banks Papers.

251. Boulton & Baker to Banks, 1 February 1813

Chatham Place

Sir

We have receiv'd many distressing Letters on the Subject whereon our Leith Friends have, we believe, in the enclosed Letter,[1] entreated your Support: – That is, the obtaining permission from the Government to send out a little <u>Seed Corn</u> to the N. W. Coast of Norway: – say <u>Drontheim</u>[2] in particular. From <u>Tonsberg</u>[3] also we have a <u>most</u> pressing Entreaty that we would use our utmost Endeavours to obtain permission for their receiving a Supply from any Quarter, which this Government may deem least inconsistent with the present State of political Difficulties; – and pointing out in strong Colours the Hardship of their suffering so severely in a cause wherein their Situation precludes them from being concern'd![4] –

Happy indeed should we feel to be in any degree the Instrument of softening the Rigours of Warfare to People so situated. – However, this perhaps may not be the proper time for any <u>general</u> Application on that Subject. – Permit us then, Sir, only to entreat the favour of your Support to our present Petition, and to beg you will let the Bearer know when you propose attending the Board, that we cause it to be then brought forward –

We remain, with unfeigned Respect,
Sir
Y*ou*r very obed*ien*t Ser*van*t
[*Signed*] Boulton & Baker

SL, Banks Papers, I 2:3. Autograph letter

252. List. Ships clearing for Iceland 1812, 13 February 1813.

In the Year 1812 the following Ships have been cleared from here [Copenhagen] to Iceland.

Clearings No	Date	Ships Names	Masters Names	Contents of the Cargoes Ton	
966	April 17	*Orion*	J. Kettlesen	500	Barley
				225	Rye
				40	Peas
1061	April 21	*Skalholt*	P. A. Grönbeck	500	Barley
1062	April 21	*Keblevig*	H. M. Hansen	175	Rye
				250	Barley
				40	Peas

---

[1] Probably referring to Corbett, Borthwick & Co. of Leith (see previous letter).

[2] Trondheim.

[3] Tønsberg is south of Oslo, on the Oslo fjord.

[4] The situation in Norway was dire, but in 1813 a lot of licences were issued to vessels sailing to Norway (see TNA, BT 6/209).

1113	April 23	*Seyen*	H. J. Petersen	250	Barley
				125	Rye
				40	Peas
1453	May 8	*Torsken*	Peter Clausen	300	Barley
1525	May 12	*Nordstern*	N: A: Schmidt		Ballast
1547	May 12	*Eliza*	B. Hansen	500	Wheat
				200	Barley
				50	Pease
1548	May 12	*Regina*	P. Jepsen	300	Barley
				44½	Peas
				220	Wheat
1661	May 15	*Gratierne*	Thron Olsen	50	Rye
				500	Barley
1664	May 15	*Svanen*	N. Michelsen	130	Rye
1680	May 16	*Bildahl*	Chr: Kieldsen	194	Barley
				50	Rye
1841	May 26	*Helleflynderen*	Thos Christensen	250	Rye
				13	Peas
1988	June 4	*De jonge Goose*	J. H. Möller	200	Rye
2102	June 12	*Det gode Haab*	H. M. Bohne	350	Rye
				28½	Peas
2175	June 17	*Bosand*	H. A. Aas	300	Rye
2390	June 30	*Freden*	A. A. Cugson[1]	300	Barley
				100	Rye
3029	July 25	*Eskefiord*	C. W. Høyer	700	Barley
				100	Rye
				22	Peas

For all the above Vessels the Certificates of their Discharging of their Cargoes have been received, except for the *Nordstern* N. A. Schmidt, having been cleared out from here in Ballast whose Return Attest is arrivd from Øster Riisöer[2] Copenhagen Custom House 13th february 1813

Please turn over[3]

Recapitulation

Barley	barrels	3994	or Quarters	2000
Rye	do	2055	do	1028
Peas	do	271	do	139
Wheat	do	<u>720</u>	do	<u>360</u>
	Barrels	<u>7047</u>[4]	or Quarters	<u>3527</u>

SL, Banks Papers, I 1:97.

[1] This is a strange name for a Dane. In the lists of licences in the Danish National Archives (RA) the name seems to be Anders Andersen Rugsen. There the full names of the masters can be found.

[2] Presumably Øster Riisøer in Norway (visited by Mary Wollstonecraft), a privileged town in 1723, that played a role in the Napoleonic Wars in 1807–14, when Denmark-Norway joined the French camp, becoming the enemy of Norway's most important trading-partner, Great Britain.

[3] Please turn over to the next page where the 'Recapitulation' is to be found.

[4] Banks made a mistake in his calculation, the sum should be 7040.

253. From Horne & Stackhouse to Banks, 28 April 1813[1]

Liverpool

Sir

From the interest which you have uniformly been pleased to take in what concerns the Island of Iceland, we are induced to take the liberty of soliciting your attention to the Treaty between His Majesty's Government, and the Court of Denmark, which we are informed is now pending.[2]

From your kind interposition His Majesty's Government issued the Order in Council of the 7th February 1810,[3] which enabled us, in common with other British Merchants, to trade to the Island of Iceland. We have embarked considerable property in this branch of Commerce, a large proportion of which is actually in the Island.

We are induced to hope that, as the permission of a free trade with Iceland was dictated by a feeling of compassion for the wants of the Natives, so, on the conclusion of Peace with Denmark, this circumstance may operate in favor of the British Merchant, not only for the security of the Property vested in the Island & that he may be enabled to convert such property to the use design'd, but also & especially, that His Majesty's Government will stipulate for the continuance of a free trade with this part of His Danish Majesty's Possessions.[4]

The favor of your kind consideration in these matters will essentially oblige us & any suggestions you may be pleased to afford us will prove highly acceptable.

We hope for your favorable construction of the liberty we take on this occasion. And have the honor to be

Sir, your very ob*edien*t & very h*um*ble serv*an*ts
[*Signed*] Horne & Stackhouse

Wisconsin, Banks Papers, ff. 112–113. Autograph letter.[5]

---

[1] William Horne and Jonathan Stackhouse of Liverpool carried on the most extensive trade with Iceland, sending at least 3 ships in 1813, 6 in 1814, 4 in 1815, 2 in 1816 and one in 1817. This merchant house apparently began trading in 1812. The fact that the first ship they sent to Iceland was called the *Sir John Thomas Stanley* suggests that Stanley was involved in their trade, as does this very letter. See further Agnarsdóttir, 'Great Britain and Iceland', chaps VIII, 2.c, and XI.

[2] As usual Banks's aid was being sought. On 4 February 1814 Horne & Stackhouse were to petition the Privy Council 'very anxious for the security of their Property in Iceland and Faroe', hoping that in the pending treaty with Denmark free trade with these islands would continue to be secured (TNA, BT 1/84). The Foreign Office decided that as soon as diplomatic relations could be reinstated with Denmark the diplomats would take up this matter (TNA, BT 1/84, Hamilton to the Clerk of Council, 10 February 1814).

However, Article 7 of the Treaty at Kiel of 14 January 1814 simply read that trade relations between the two countries would revert to their former state before the commencement of the war.

[3] Appendix 10.

[4] Free trade between Iceland and Britain had been stipulated in the Order in Council of 7 February 1810.

[5] Note in Banks's autograph, bottom left margin: 'Queen *Stree*t Cheapside'.

254. Sir George Steuart Mackenzie to Banks, 11 May 1813

Edinburgh

Dear Sir,

The occasion of my now troubling you is my having received from London a pamphlet published by Stockdale in /*Piccadilly*/ Pall Mall, A Memoir by an Icelander,[1] which most probably you have already perused. This pamphlet, together with some other circumstances which I may hereafter be at liberty to mention to you, have roused my benevolent feelings towards the poor Icelanders to a pitch I never before felt. Our situation with Denmark is at present such as to justify some effectual measures for the relief of Iceland than those already adopted.[2] I am here surrounded by a charming family[3] with as great a store of happiness as a man could wish for; a selfishness would incline me to hold fast by what I have. But the strong feeling of benevolence which now possesses me, the desire of doing good, & a naturally active disposition prompt me, not to sit idle & enjoy the blessings of Providence, but to bestir myself in improving the condition of my fellow creatures. Re what has hitherto been done, a proper distinction has not been made between native Icelanders & Danish Merchants.[4] The latter are benefited, but I fear the former are not much better off than usual. Such however appears to me to be the present state of affairs that Iceland will be reduced to a most distressing state, if the British Government does not at once declare the Island under its protection. Now is the time for your gaining for your favourite Icelanders what they most desire; & a fine opportunity appears to be now open for your completing that wreath which the gratitude of Iceland and the admiration of your fellowmen have been twining upon your venerable brows. – I have taken the liberty to address Lord Castlereagh on the subject,[5] & to make a tender of my services to go to Iceland, where I long to exert the benevolent propensities which Providence has endowed me with. The expence of taking Iceland under our Protection would be a mere trifle; & I have some plans in view by which it would scarcely be felt.[6] Your exerting yourselff at the present moment in the cause of Iceland will be attended with certain & good effects; & should I be the happy instrument for putting the intention of Government into practice, I shall not fail to keep in mind what Iceland owes to you already. The hopes of the writer of the Memoir I do not think by any means too sanguine.

---

[1] The author is simply called 'An Icelander', the full title is *Memoir on the Causes of the Present Distressed State of the Icelanders and the Easy and Certain Means of Permanently Bettering their Condition*, London, 1813. Hermannsson believed the author to be Savignac, while scholars have also favoured Magnús Stephensen and Mackenzie himself! All three are unlikely. Though no proof has yet been found as to the author's identity, Finnur Magnússon (1781–1847), a lawyer and subsequently the Keeper of the Royal Archives to the King of Denmark, is a likely candidate. He spent a month in Scotland at the end of 1812. For a further discussion, see Agnarsdóttir, 'Hver skrifaði Íslandsbæklinginn?'

[2] Peace negotiations between Denmark and Britain were in the offing.

[3] Mackenzie (b. 1780) married in June 1802 Mary, fifth daughter of Donald MacLeod of Geanies, Sheriff of Ross-shire, and at this time five sons had been born: Alexander (1805), William (1806), George (1807), Robert (1811) and John (1813). Two further sons would be born in the following years: Donald (1815) and James (1819). He also had three daughters (no names or dates of birth). A widower, he married again in 1836, his second wife being Catherine Jardine, and his eighth son, Henry, was born in 1839.

[4] In fact this had been done in 1808: see Document 85.

[5] This letter has not been found.

[6] For Mackenzie's plans for the annexation of Iceland, see Agnarsdóttir, 'Scottish Plans for the Annexation of Iceland'.

There are various objects of industry & sources of comfort to which he has not alluded, & of which he probably never thought.

I should be most happy to learn your sentiments respecting the present state of affairs in so far as Iceland is concerned; & I have no reason to think that you will deem me unfit for the task I propose to undertake.[1] On the contrary, as I have been in the Country; & turned my attention to the means for improving the condition of the people;[2] and as I possess rank, in my own Country;[3] I have every reason to believe that you will with pleasure encourage and assist my views, which benevolence alone has dictated. I have no desire for any remuneration excepting necessary expences, & the high satisfaction of doing good. Such being my views, & pure benevolence actuating me, I need make no apology for intruding upon you; & at present I need say no more but that I am, with every sentiment of respect & esteem.

Your faithful & sincere Servant
[Signed] G. S. Mackenzie

Wisconsin, Banks Papers, ff. 130–131. Autograph letter.

255. Christian Ignatius Latrobe to Banks, 13 May 1813

Nevils Court

Sir,

I take the liberty to send You a translated Extract of Mr. Frisch's Letter to me,[4] in which he states the necessity of his having an unlimited permission to import Greenland produce. The misfortune of being captured by the *Alexandria* frigate, Capt*ain* Cathcart,[5] has rendered the *Hvalfisken*'s Voyage[6] a very disastrous one to Mr. Frisch, and as it was afterwards found, that the Licence granted by the Privy Council was perfectly correct, and ought to have protected the Ship against any Cruizer, the hardship was unmerited and encouraged me, with the more confidence, to request some indulgence in behalf of this vessel, which would also remove the very unpleasant impression made on the Swedish people, respecting the conduct of our Navy, perceiving that such an outrage is not countenanced by the British Government. Feeling as an Englishman ought, I cannot but

---

[1] Initially positive (Documents 166 and 168), by October 1810 Banks had changed his opinion, declaring to the Foreign Office that Mackenzie 'would not in my judgment have made So good a Governor as any one of the Junior Clerks of your office' (Document 177).

[2] For example the salmon fishery and the draining of bogs, a pet project of Banks's (Mackenzie, *Travels in Iceland*, pp. 205, 279).

[3] Mackenzie was the 7th baronet of Coul in the county of Ross, created in 1673.

[4] See Document 247.

[5] Captain Robert Cathcart (1774–1833) was on the Baltic station from 1811 to 1816. A veteran of the battle of the Nile on the *Bellerophon* in 1795, he commanded the *Seagull* from 1805 which was captured by a Danish brig in 1808 off the harbour in Christiansand, Norway. He was detained as a prisoner in Norway and eventually court-martialled for the loss of his ship but was honourably acquitted. At the close of 1810 he was appointed captain to the *Alexandria* frigate. See Marshall, *Royal Naval Biography*, Supplement, Part I, London, 1827, pp. 375–82. It was quite common for ships to be captured illegally.

[6] In December that year the *Hvalfisken* was at any rate granted a licence to sail to Iceland (TNA, BT6/210, no. 54,700). The *Freden* and *Hvidfisken* received the same indulgence, as did the *Albertine*, *Jubelfest* and *Sælhunden*.

read such a passage as this, in a letter from a Swedish Gentleman with grief – 'I suppose English Licences are intended only to coax ships out of their safe ports, that they may fall into Your Cruizers hands, who seem to have very little respect for the orders of Your Government. I hear also that the *Freden* and *Huitfisken*, whom You have also favoured with licences, are almost afraid to sail from Copenhagen, lest they should meet with the fate of the *Hvalfisken*.' – This will put a stop to the provisioning of the forsaken Greenland Colonies, which Mr. Frisch, the Royal Agent, says he cannot afford.

Knowing with what benevolence You have always considered the fate of the poor Colonists in Greenland and Iceland, I could not resist my impulse, to lay the Case before You.

I trust it will please God to restore You to health, and rejoice to hear, when I have called at Your house, that You are better. With the sincerest Esteem

I remain ever
Sir

Your most obedient
& most devoted humble
Serva*nt*
[*Signed*] C. I. Latrobe.

Wisconsin, Banks Papers, ff. 116–117. Autograph letter.

256. Sir George Steuart Mackenzie to Banks, 16 May 1813

Edinburgh

Dear Sir

Without waiting for your Reply to my last letter respecting Iceland,[1] or for Lord Castlereagh's determination,[2] I take up my pen to address you again. It has occurred to me that a Memorial addressed to the Prince Regent by those in this Country who have visited Iceland, might have the effect of saving the wretched Inhabitants from misery. I allude to you, Sir *John* Stanley,[3] & Mr. [William Jackson] Hooker,[4] together with myself. I feel a degree of enthusiasm & zeal in the welfare of the poor Icelanders, which renders it easy for me to sacrifice a quiet, warm fireside, & every comfort I can wish for, for the satisfaction of benefiting a country in which I was received with kindness. But it is so difficult to make disinterestedness appear in my motives for wishing to urge Government to do a good deed, that I feel exceedingly anxious. I feel so conscious that I could do a great

---

[1] Document 254, written on 11 May, only five days earlier from Scotland. Banks replied to this letter on 20 May. On 21 August 1811 Mackenzie had sent a long memorial describing the dire trade situation in Iceland and the Faroe Islands to the Committee of Council for Trade in which he mentioned he was 'well aware of the humane interest which Sir Joseph Banks has taken in the fate of the poor Icelanders' asking for a relaxation in the trade restrictions (TNA, BT 1/64). He suggested 'certain regulations', which he spelt out in detail, be implemented regarding the trade to Iceland and the Faroe Islands. He may well have addressed a similar letter to Banks or at least informed him of his intentions.

[2] Mackenzie had written to Castlereagh (Document 254). This letter does not appear to be extant.

[3] Stanley visited Iceland in 1789.

[4] Hooker visited Iceland in 1809.

deal of good, that I honestly confess I should feel greatly disappointed to see another person sent out to Iceland, however happy I should be on seeing the great point gained of it being taken under the protection of Britain. I am confident that I could fulfil the wishes of the Icelander as expressed in this sensible Memoir,[1] & that I could do even more. By the way, do you know who the author is? It will be astonishing, if he can manage so well as to conceal himself from those who he must know are well disposed towards Iceland. –

The post horse has arrived; & I have only a moment to request that you will favour me with your sentiment on the subject of my communication, & say whether you are disposed to favour the object I have in view. If anything is to be done, it ought to be done soon, as an Icelandic summer is short, & it will take some time to make the necessary arrangements. –

I am with respect

Very sincerely & faithfully

[*Signed*] Yours, G. S. Mackenzie

Wisconsin, Banks Papers, ff. 114–115. Autograph letter.

257. Corbett, Borthwick & Co. to Banks, 22 May 1813

Leith

Sir

We have to day written our agents Mess. Findlay Bannatyne & *Compan*y, to petition the Board of Trade for a renewal of the Licence granted in No*vembe*r last to the Iceland Vessell *Eliza* Cap*tain* B. Hansen,[2] and as by your recommendation a renewal has been granted, in the Licences of the other Iceland Vessells, we hope you will be kind enough to procure a favourable answer to this Petition also, the *Eliza* has been loaded with grain at Kiel, but from unavoidable circumstances, it has not been possible to get her Voyage performed during the continuance of the Licences, and as our Cruizers now bring up every vessel, whose licence is expired, however clear her destination to Britain may be, and altho the Licence is only a few days expired, it is impossible for the Cap*tain* of the *Eliza* to venture out without the renewal.

The Iceland vessell *Regina*,[3] Cap*tain* Jepsen sailed from Copenhagen for Leith about the middle of april, and by this days Mail, we are informed that on the 26 of april, she was

---

[1] Mackenzie is here referring to the *Memoir on the Causes of the Present Distressed State of the Icelanders* ... (see further Document 254).

[2] Boye Hansen was the master. The licence was granted 19 November 1812 (TNA, BT 6/208, no. 48,143). The *Eliza* eventually sailed to Iceland in the summer of 1813, leaving Leith on 7 July (NAS, E. 504/22/61). In October 1813 she was granted a licence to sail from Leith to Copenhagen under convoy, but she did not sail as Rooke & Horneman asked for a licence in lieu of the *Eliza*'s (TNA, BT 6/210, nos. 54,137 and 54,395).

[3] The detention of the *Regina* had serious consequences for the navigation to Iceland that year. If she had been taken to a British port a speedy liberation would probably have taken place. A further six to eight grain-laden Iceland ships were later captured. After some months, when the Swedes belatedly realized the grain was intended for Iceland they granted special permission for its reshipment, provided the ship and cargo were liberated by an English court. But the matter did not end there. The Board of Trade decided to use the opportunity to take steps to guard against the ever-present possibility of licence abuse and the British admiral on the Baltic station suggested that the vessels destined for the Danish North-Atlantic dependencies should be ordered to go directly

detained by the *Ulysses* Cap*tain* Fothergill,[1] and sent for Got*ten*burg.[2] Cap*tain* F*othergill* promised the Cap*tain* of the *Regina*, that the vessell should not be sent within the Swedish limits, and gave him hopes of an immediate release, the *Ulysses* is however gone out on a Cruize, and the Agents of Cap*tain* Fothergill have ordered the *Regina* to Gottenburg, the consequence of which will be, we have much reason to fear, that her Cargo of grain will be sold there, by order of the Swedish government; as the Licence obtained by us for the *Regina* was dated, (we think) 23 Nov*embe*r and to be in force for five months, it could only be expired two or three days, when the Vessell was detained on her direct Voyage to Leith, where ship & Cargo is insured by us, and neither the decisions of Sir W*illia*m Scott /nor/ (as we suppose) the wishes of the Board of Trade, seem to warrant so very strict an interpretation of the terms of the Licence, as to its continuance, and we are certain, that had the vessell been sent [to] a British Port, she would have been immediately liberated; under these circumstances, we think both the underwriters, and the Owner, have some reason to complain of the Vessells being sent to Gothenburg, when it was pretty certain, that her cargo would be immediately discharged, and sold contrary to the wishes of the owner and the express condition of the Licence, more especially as it seems to have been known to Cap*tain* Fothergill that this would immediately follow her being brought within the Swedish limits – in one of your late letters to us you mentioned that the Board of Trade, could not enforce obedience to the Licence but only recommend to the Admiralty, under whose department, the questions arising about the use of them must be decided, and therefore we are aware, that the detention of the *Regina* does not come under your cognizance as a member of the Board of Trade, but we have some hope, that you may deem it a case, in which the Board should recommend to the Admiralty, that the *Regina* with her Cargo should be ordered to Leith, there to await the decision of the Admiralty Court – we think you will be the more inclined to this, from its being partly apparent, that the improper desire of Gain on the part of the Gott*enburg* Agents, has been the cause of the Vessell being carried within the Swedish limits, that the Cargo might there be sold under their management, in place of allowing it to reach its destination as ordered in the Licence by the Board of Trade

---

from their Danish port to Vingaa (Wingo) Sound, the rendezvous for vessels bound for British and Baltic ports. There they would be convoyed to Leith. In a letter to King Charles XIII of Sweden the Board of Trade presumed 'that His Swedish Majesty moved with the same sentiments of humanity and compassion towards the inhabitants of Faroe, Iceland and Greenland, which so benevolently influenced His Majesty, will be equally desirous to afford facility to their supply from Denmark with grain, provided the same can be done without hazarding an illegitimate intercourse under cover of their licences, between the ports of Denmark and Norway' (TNA, FO 40/2, Memorandum, 2 August 1813). The King replied that he shared the compassion shown by His British Majesty towards the Icelanders and sent orders to the Swedish Navy to let these ships pass unmolested to Vingaa to join the English convoy there. However, any ship not in the convoy and discovered on the way to Norway would be seized (TNA, FO 73/85, Engeström to Douglas, 25 September 1813 [French]). Thus the British government proved instrumental in extending its protection of the Iceland trade to include the Swedes. This is a good indication of how seriously Britain took its role as protector of the Iceland trade.

[1] Captain William Fothergill (1768–1817). Fothergill had sailed to China, the Mediterranean and the West Indies and was appointed to the *Ulysses* in the summer of 1812. His ship was engaged in convoying merchantmen from Carlscrona to the Sound. He ended his career in Jamaica (*The Naval Chronicle,* vol. 38, pp. 349–436).

[2] Gothenburg in Sweden. A convention of March 1813 had given the British the right of entrepôt.

We have reason to believe that gross irregularities are committed in Gottenburg in Prize Cases, of ships detained under British Licences, and we are sorry to think, that foreigners should have so much ground to complain of the hardship imposed upon them there, because they are not always able to separate the actions of the agents there, from the intentions of the British Government or to discriminate betwixt these extra judicial proceedings, and the proper decisions of our Admiralty Court –

In the case of the *Hvalfisken* of Greenland, we cannot discover any probable grounds of detention, to justify Captain Cathcart,[1] for having sent that Vessell to Gottenburg the Captain in order to save his Cargo of Wheat, and to secure the ultimate object of his licence, a Voyage to Greenland, paid about £200 of Captors Expences, not doubting that his protest secured him legal redress, upon arrival in England, we find however, that the Captain having taken back his papers in Gottenburg, in place of allowing his ship to remain there until liberated in Doctors Commons, precluding his action against the Captor, whilst we know, that his Papers would not have been sent to Doctors Commons till this day, at least so is the case with other ships detained there to our address – this seems rather a hard limitation, as an ignorant Captain cannot in such a case be presumed to know that he thus loses all access to complaint – from the particular circumstances of the case, we intend however to attempt getting redress against Captain Cathcart in Doctors Common.

Eight vessels under British Licence, loaded in Denmark with grain to our address have lately been carried into Gottenburg, by different British Cruizers, several of them have been there for more than 2 months, and a part of the Cargoes have been sold by the Captors, and we have reason to believe, that none of the grain will now be allowed to leave Sweden; these sales have been made, and the vessels been detained from two to three months without any of their papers being sent to Doctors Commons, and without any authority whatever from the British Government, whose protection, the Licence is supposed to give the Owner, untill the proper Court finds, that an improper use had been made of it, and therefore we hope, the Board of Trade, will, when they know of the irregularity of these proceedings, put a stop to them – we have claimed the Vessells, and now endeavour to face the Captors to bring the Papers into Court.

We are most respectfully
Your obedient humble Servants
[*Signed*] Corbett, Borthwick & Company

Wisconsin, Banks Papers, ff. 126–127. Autograph letter.

---

[1] Captain Robert Cathcart, see p. 521, n. 5.

258. Banks: Memorandum, [June ?, 1813]¹

Some Notes relative to the ancient State of Iceland, drawn up with a view to explain its importance as a Fishing Station at the present time, with comparative Statements relative to Newfoundland²

A.D.
860. Iceland discovered³
Iceland was this year first discovered by Norwegian Seamen; it was uninhabited; but, from Some things met with on the Shores, the discoverers were of opinion that it had probably been visited before, though by what people has never been ascertained.⁴
874. Iceland peopled⁵
this year Iceland was peopled by Emigrants from Norway, who left their Country to avoid the oppression of their Sovereign, Harold Harfagre,⁶ they found the Island

---

¹ This memorial is undated. Hermannsson dated it to the first months of 1812 (based on the fact that Banks refers to the first edition of Mackenzie (from late 1811) and not the second (spring–April 1812)). Hermannsson wrote: 'if this edition had been out at the time of writing it would probably have been referred to, as Banks doubtless received a copy of it as soon as it was published' ('Banks and Iceland', p. 83). Actually, the page numbers Banks refers to in Mackenzie correspond more to the 1811 edition. Hermannsson believed it had been written either for Lord Liverpool, the prime minister, or for Castlereagh, the foreign secretary, 'especially as Castlereagh immediately directed his attention to the Scandinavian situation and tried to get Sweden to join the coalition against Napoleon, and in such case there was talk of separating Norway from Denmark and uniting it with Sweden' (ibid., p. 84). Mackenzie had been in contact with Banks in the summer of 1813 and persuaded him to make yet another attempt at interesting the government in annexing Iceland. According to Mackenzie in a letter dated 2 July 1813, Banks had done so, describing the Iceland fisheries as more advantageous than those of Newfoundland and adding that the possession of Iceland was of extreme importance. Thus in all probability this is the memorandum referred to and was most likely written in June 1813 (see RA, Island og Færöer, Mackenzie to Clausen, 2 July 1813; and, further on the dating, Agnarsdottir, 'Great Britain and Iceland', pp. 228–9).

² In the Royal Geographical Society of South Australia there are preserved 26 pages of copious notes in Banks's autograph regarding this memorandum, including a draft of it. He collected a lot of information but did not use all of it. The whereabouts of the original of the final version is unknown, thus the DTC copy is used here. It is a fair copy of Banks's above-mentioned draft. Where Banks's notes are significant, though not included in the fair copy, they have been footnoted.

[JB] – this denotes notes made by Banks himself in the manuscript. These have been verified as far as possible and are in general correct.

³ The dates and headings are given in separate columns in the margin in the original draft, but have been converted to subheadings here for ease of reading (see Figure 8 for the original layout).

⁴ Here Banks was going to put the following note 'as part [or parcel]' and then crossed out 'of the Dominion of the Crown of England, but not part of the Realm of England'.
According to the *Book of Icelanders* (*Íslendingabók*) by Ari 'the Learned' Þorgilsson, when the first Norse settlers arrived in c. 870 Irish hermits (anchorites) were living in Iceland. They immediately left 'taking their books and croziers with them'. Other written sources support this but there is as yet no archaeological evidence for corroboration.

⁵ In his notes Banks had written for the year 870, which he omitted in his final version, the following: 'Iceland Received a Considerable Colony of norwegians who for some Centuries Continued a free and independent Nation Conducting a Considerable Trade in the Northern Sees their Ships visiting Britain Ireland France Germany etc.'

⁶ Harald Fairhair (Haraldur hárfagri) (c. 850–930?) was the first king of Norway 872–930. It was during his reign that Iceland was settled.

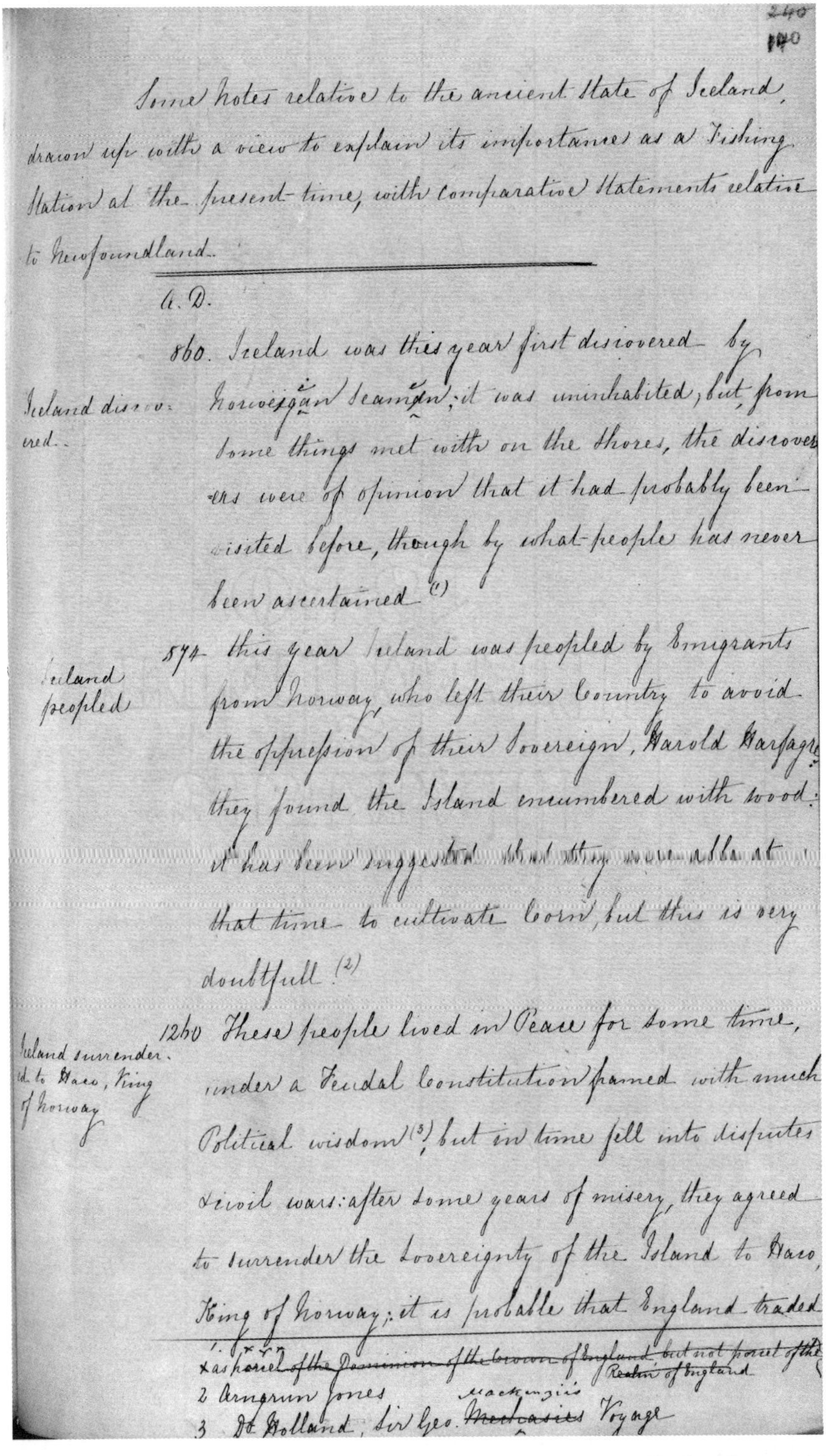

Some notes relative to the ancient State of Iceland, drawn up with a view to explain its importance as a Fishing Station at the present time, with comparative Statements relative to Newfoundland.

A.D.

Iceland discovered. — 860. Iceland was this year first discovered by Norwegian seamen; it was uninhabited, but from some things met with on the shores, the discoverers were of opinion that it had probably been visited before, though by what people has never been ascertained (¹)

Iceland peopled. — 874. this year Iceland was peopled by Emigrants from Norway, who left their country to avoid the oppression of their Sovereign, Harold Harfagr: they found the Island encumbered with wood: it has been suggested that they were able at that time to cultivate Corn, but this is very doubtfull (²)

Iceland surrendered to Haco, King of Norway. — 1260. These people lived in Peace for some time, under a Feudal Constitution framed with much Political wisdom (³), but in time fell into disputes & civil wars: after some years of misery, they agreed to surrender the Sovereignty of the Island to Haco, King of Norway; it is probable that England traded

1. ~~Iceland of the Dominion of the Crown of England~~ but not part of the Realm of England
2. Arngrim Jones Mackenzie's
3. Dr Holland, Sir Geo. ~~Mackenzie's~~ Voyage

Figure 8. First page of 'Some Notes relative to the ancient State of Iceland', 1813. These notes were 'drawn up with a view to explain its importance as a Fishing Station at the present time, with comparative Statements relative to Newfoundland'. DTC 17, 140. Courtesy of the Natural History Museum.

encumbered with wood, it has been suggested that they were able at that time to cultivate Corn, but this is very doubtfull.¹

1260. Iceland surrendered to Haco, King of Norway
These People lived in Peace for some time, under a Feudal Constitution framed with much Political wisdom,² but in time fell into disputes & civil wars: after some years of misery, they agreed to surrender the Sovereignty of the Island to Haco, King of Norway;³ it is probable that England traded with Iceland for stock fish before this Transfer took place.

1290. first notice of stockfish in England
Two hundred Stockfish were a part of the Stores embarked at Yarmouth, on board a Ship fitted out for the purpose of bringing the infant Queen of Scotland from the court of her father, the King of Norway.⁴ In Vol. p. 436 Note.⁵

1306. Toll of Stock fish at London Bridge
34th Edw*ar*d 1st. The next mention of stockfish in England is in an account of Pontage⁶ Taken at London for the repairs of London Bridge:⁷ by this regulation the Toll charg'd on a hundred Stock fish by tale was one farthing.

1312. Hull paved with Stones brought from Iceland
About this time, Kingston upon Hull, founded by Edw*ar*d the First in 1296, began to flourish. 'The first great increase of this Town (says Leland) was by passing for Fish into Iceland from whence they had the whole Trade of Stock fish into England',⁸ (Leland adds) 'in such time as all the Trade of Stockfish for England came from Iceland to Kingston. Because the Burthen of Stockfish was light, The Ships were ballasted with great Coble Stones brought from Iceland, the which in continuance, paved the Town of Kingston throughout.'⁹

---

¹ [JB]: 'Arngrim Jonas.' Banks is referring to the Icelandic scholar Arngrímur 'the Learned' Jónsson (1568–1648) and his work *Crymogæa*, written in Latin and published in Hamburg in 1609. The information here is from Chapter V. Banks was wrong to be doubtful. Corn was grown in Iceland during the warm medieval period, but Banks visited Iceland in 1772 during the Little Ice Age when corn-growing had ceased. Today in Iceland corn is only grown for experimental purposes. *Crymogæa* was the first history of Iceland, Arngrímur setting the stage for the chronology and the image of Iceland's 'Golden Period' in medieval times.

² [JB]: 'Dr. Holland, Sir Geo*rg*e Mackenzie's Voyage.'

³ King Hákon Hákonarson IV 'the Old' (c. 1204–63), reigned 1217–63.

⁴ [JB]: 'Rymer's Collections MSS, vol. 2, p. 287.' Banks is doubtless referring to the *Foedera*, a 16-volume work, published in 1704–13 by Thomas Rymer (c. 1643–1713), the historiographer royal. He consulted it often while preparing this memorandum. The princess in question was Margaret 'The Maid of Norway' (1283–90), daughter of King Eric II of Norway and Margaret, daughter of King Alexander III of Scotland. She became queen of Scotland upon her grandfather's death, at the age of three. She died en route to Orkney in 1290 and is buried in Bergen.

⁵ Halldórsson omits this note but it is in both the Adelaide draft and the DTC copy. It is clearly Banks's note to himself of the source of this piece of information. The volume number is missing and it is not clear to which work he was referring. It is possibly Rymer's *Foedera* (but that is a work of 16 volumes).

⁶ *Pontage* is a bridge-toll.

⁷ [JB]: 'Hearne, *lib. nig. Scaccarii*.' Banks is referring to Thomas Hearne's *Liber Niger Scaccarii*, published in two volumes in Oxford in 1728. In his notes the page number noted, which is correct, is: vol. 1, p. 478.

⁸ [JB]: 'Leland's *Itinerary*, vol. I, p. 49.' This is *The Itinerary of John Leland the Antiquary*, published from the original MS in the Bodleian Library by Mr Thomas Hearne, vol. 1, Oxford, 1710, fol. 53, p. 41. Leland adds: 'and partly other Fisch'.

⁹ [JB]: 'Ibid., p. 51.' In the 1710 edition this is vol. I, fol. 56, p. 43. The language has been modernized by Banks.

1339. Stockfish purchased for Edwd the 3rd
This year Edward the 3rd caused 5000 Stock fish to be purchased for his use at Boston.[1]

1357. Price of Stockfish abated by act of Parliament Statute of Herrings
in this year a statute was passed 31st Edwa*r*d 3rd ch. I for the Purpose of abating the price of Stock fish, no doubt at that time a Principal article of lenten provision; in this act it is provided that the Lord Chancellor & the Lord Treasurer of England shall, with the assistance of such Judges & Privy Councellors as they shall call to their assistance, have power to ordain remedy touching the buying & selling of stock fish at S^t. Botulfs (Boston) &c., in order that the People may have them cheaper: it is probable, that, in obedience to some ordinance issued on this occasion, John Louthen,[2] who stiled himself Fishmonger when he served the office of Lord mayor in 1348, stiled himself in 1358 when in the same office Stock fish monger.[3] This act is in fact a part of the Statute of Herrings, the first relative to Fisheries in the English code.

1380. Iceland surrendered to Denmark
in this year Norway, & its dependency, Iceland, were transferred to the Crown of Denmark.[4] On this occasion, Their Laws, revised but little alter'd by Haco, were continued to them by their new masters.[5]

1415. English ships forbidden the Iceland Fishery by proclamation Englishmen commit pyracy in Iceland
in consequence of complaints made by the king of Denmark of the ill conduct of the English in the Iceland Fishery,[6] The King, Henry the 5th, ordered proclamation to be made that 'none of our Subjects do for one year to come presume to resort to the Coasts of the Isles belonging to Denmark & Norway, more especially to the Island of Iceland, for Fishing or any other reason to the prejudice of the king of Denmark, (aliter Quam Antiquities fieri Consuevit).'[7] From this date till 1425, Many English who frequented Iceland were Pyrates as well as Fishermen; a minute account of the Enormities they committed is preserved, from whence it appears that these vessels were fitted out from Hull, Lynn & others of the Eastern ports. The names of many of

---

[1] [JB]: 'Rymer Foedera, vol. 5 p. 146; Camden's *Britannia*, p. 578.' *Britannia: or a Chorographical Description of Great Britain and Ireland, Together with the Adjacent Islands*, written in Latin by William Camden and translated by Edmund Gibson, 2nd edn, London, 1722. Banks has also added in his notes 'Olaus Magnus, 21'. This would be Olaus Magnus, *Carta marina et descriptio septemtrionalium terrarum ac mirabilium rerum in eis contentarum, diligentissime elaborata anno Dni 1539* (Marine map and description of the Northern Countries and their wonders, carefully executed in the year of our Lord 1539).

[2] 'Layton' is written above Louthen. According to a modern list this is John Lovekyn, elected Lord Mayor in 1348, 1358 and 1366 (Hope, *My Lord Mayor*, p. 183).

[3] [JB]: 'Stowe, Abridged Chronicle 8vo'. John Stow, *The Summarie of Englishe Chronicles*, London, 1567. John Lufkyn, is in 1348 called 'fishmoger' (p. 92) and in 1358 'fyshmonger' as before (p. 93).

[4] In his notes Banks has added 'this transferred the trade to Denmark'.

[5] [JB]: 'Dr Holland, Sir G. Mackenzie's *Travels* p 49.' In his notes Banks added: 'At this time the Fishery was in Great Estimation much resorted to by Englishmen as well as Foreigners.'

[6] In his notes Banks has written 'whose behaviour had in all likelihood been too boisterous to be borne by the Quiet & unoffending Icelanders'.

[7] [JB]: 'Rymer, Fæd. Tom. 9 p. 322.' Translated from the Latin except the last few words. This was a treaty between King Henry and his brother-in-law King Erik (reigned 1412–36) to the strong objections of the House of Commons (see *The Diplomatarium Islandicum* [hereafter DI], XVI, no. 79, p. 226).

these English Pyrates are preserved. This probably originated in exactions of the Danes under pretence of Toll.[1]

1425. Iceland Fishery overtraded by Bristol &c
we learn from the 'Libel of English Policy',[2] a most interesting poetical Pamphlet written about the year 1437, that, about this time the Iceland Fishery was at its greatest extent: ships of Bristol & other western ports entered into the Trade in competition with the Eastern ports, who till then had chiefly used it. In consequence of this increase of shipping, there were 12 years after this not fish enough in the Island to load the whole. (See extract from the Poem, 1437).

1430. English ships forbidden the Iceland Fisher*ies* by act of Parliament Denmark granted Licences to Fish in Iceland
An Act was passed 8th Hen*ry* 6 Ch*apter* 2 in order, as is said, to conciliate the Kings uncle, the King of Denmark, forbidding Englishmen to enter the Dominions of Denmark, except only the Town of Northberne (Bergen) in Norway, where they were to have the same Liberties & favor as the Hans[3] did enjoy; & by no means to enter into any other part of the Territories of Denmark in opposition to the King of Denmark's ordinance of prohibition. This Statute seems to have made no difference in the resort of Englishmen to Iceland, which was then on the increase, & was soon after much extended: we must therefore suppose that the King of Denmark granted Licences to our Fishermen to supersede the effects of his ordinance of Inhibition.

1436. Hen*ry* 6th grants Licences to sell &c to Iceland
a Licence was this year granted by the King, Henry 6th, to a Bishop of Hole[4] in Iceland, to engage the Master of an English Ship going to that Island to be his Proxy or attorney to visit that Bishoprick for him.[5]

1437. State of the Iceland Fishery from the Libel of English Policy
The State of the Iceland Fishery is fully set forth for this year in an interesting Poem intitled the 'Libel (libellus) of English Policy', from whence the following Lines are Transcribed.[6]
of the commodious Stock fish of Iceland p. 201
    Of Iceland to write is litle nede,
    Save of Stockfish; yet forsooth indeed
    Out of Bristowe, & Costes many one
    Men have practised by nedle & by stone
    Thider wardes within a little while
    Within twelve year and without Perill

---

[1] [JB]: '*Historia ecclesiastica Islandiæ* vol 4, p. 162.' Banks is referring to bishop Finnur Jónsson's monumental *Historia Ecclesiastica Islandiæ*, published in Copenhagen in 1772–8 in four volumes in Latin, which still awaits translation. In a footnote in the fourth volume, pp. 162–9, Jónsson details the 'pirates'.

[2] On the *Libel* see Banks's entry for 1437 and n. 6 below.

[3] The merchants of the Hanseatic League.

[4] John Bloxwich was bishop of Hólar 1435–40. He was English and never came to Iceland.

[5] [JB]: 'Foed. Tom X pp. 645, 649. [Rymer, *Foedera*].'

[6] [JB]: 'Hakluyts Voyages vol. I, p. 201.' The whole poem is to be found on pp. 187–208. Banks has 'modernized' the spelling, e.g. *yeere* becomes *year*. *The Libelle of Englyshe Polycye* was edited by George Warner and published in 1926.

Gon & come, as men were wont of old
Of Scarborough unto the Costes cold.
And now so fele Shippes this year these were
That moch losse for unfreyght they bare
Island might not make them [to] be fraught
Unto the Hawys (Househole)¹ thus much harm they caught²

1440. Hen*r*y 7ᵗʰ grants Licences to trade with Iceland³

Licenses were this year granted by King Henry the 6ᵗʰ for Ships to go to Iceland: a Bishop of Skalholt⁴ obtained one to carry corn, cloth, & Provisions, & to return with Iceland produce: the reason given for granting this favor was, that the king had been told that Iceland had neither Cloth, Corn, wine, or ale, & that the Sacrament could not be administered there, without his aid.⁵

1465. Treaty with Denmark English Ships not to visit Iceland unless with Special Licence

Oct 13⁶ of this year a treaty of alliance between England & Denmark was concluded,⁷ in which each party was to have access to the Dominions of the other; but the English were excluded from Iceland, Halgaland⁸ & Finmark: they might, however, by special Licence from Denmark resort to Iceland.⁹

1469. English men kill the Governor of Iceland for extorting grievous Tolls

That the English continued to fish in Iceland appears by a quarrel between the Kings of Denmark & England, on account of some Englishmen having killed the Governor of Iceland¹⁰ for extorting extravagant Tolls from them: on this occasion the King of Denmark seized four English Ships with their Cargoes in the Baltic.¹¹

1478. Edw*a*rd 4ᵗʰ grants Licences to Trade with Iceland

King Edward the 4ᵗʰ granted a Licence this year to Robe*r*t & Tho*ma*s Alcock [of Hull], authorising each of them to employ a Ship of 240 Tons in carrying goods not of the Staple¹² to Iceland & bringing back Iceland produce;¹³ a similar Licence was granted to the same merchants 5 years after.

---

¹ This is added by Banks.

² Banks has added in his notes for the year 1437: 'The Trade of Stockfish was overdone. Ships returned without full freights'.

³ Here Banks made a slip of the pen (copied in the DTC copy). It was, as Banks notes correctly in the text, Henry VI (reigned 1422–61) who was the king in question.

⁴ In 1440 the bishop of Skálholt was Gozewijn Comhaer, a Dutchman, bishop 1437–47.

⁵ [JB]: 'Foed. Vol 10, pp. 645; 659: 682: 711: 762' [Rymer, *Foedera*].

⁶ Actually 3 October.

⁷ The kings were: Edward IV and Christian I. For the Treaty see DI, XVI, no. 210, pp. 408–14.

⁸ Hålogaland was in the Middle Ages a petty kingdom in the very north of Norway.

⁹ [JB]: 'Foed. V. 11, P. 551: 556' [Rymer, *Foedera*].

¹⁰ Björn Þorleifsson, called 'the Rich' (1408?–67), a native governor (*hirðstjóri* was his contemporary title), was killed in 1467 not 1469.

¹¹ [JB]: 'Meursius *Historia Danica*.' Ioannes Meursius [Johannes van Meurs] (1579–1639), *Historia Danica Libri III*, published in Copenhagen, 1630–38.

¹² In medieval times trade in certain goods could only be conducted through specific towns, called staple ports. For Iceland this was Bergen in Norway.

¹³ [JB]: 'Foed: Tom 12 pp 57, 94.'

1489. The Trade of Iceland laid open to English men having Licences
a Treaty was concluded between England & Denmark,[1] in which the Iceland trade was laid open to the English, on condition of their acknowledging the Sovereignty of Denmark & renewing their Licences at the end of seven years.[2]

1494. Newfoundland discovered
Newfoundland was discovered by Sebastian Cabot.[3]

1517. Newfoundland Fishery first known in Europe
Anderson states this year as the first time that the Fishery of Newfoundland had been heard of in Europe, & that there were 50 Ships there on the Fishery; Spanish, French & Portuguese.[4]

1536. The French Fish on the banks of Newfoundland
Jacques Cartier,[5] who sailed from France on a Voyage of Discovery into the Bay of St Laurence in 1534, mentions that he saw many French Ships Fishing on the Banks of Newfoundland, but does not mention any English Ships.

1547. Fishermen of Newfoundland & Iceland freed from duties claimed by Admirals &c. Statute to consult fasting days
a Statute 2nd & third Edw*ard* 6th relieves the Fishermen of Newfoundland & Iceland from the exactions of certain duties claimed by Admirals & other persons belonging to the Admiralty; the same Parliament, ch. 19, enjoins the observation of Lent & all such days as have been accounted Fish days; in order, says the Statute, to save Flesh[6] & to set Fishers to work: see also 5 Elizabeth, 7 & 35 ch. 7, 1st James, ch. 25.

1578.[7] English Fishery in Newfoundland on the increase Iceland Fishers still much frequented
Mr. Anthony Parkhurst, in a Letter to Hackluyt this year,[8] tells him that, in the 4 years he has frequented the Newfoundland Fishery, the English Ships have increased from 30 to 50: there were besides about 100 sail of Spaniards, who made wet fish & dried it when they returned home, besides 20 or 30 that came from Biscay to kill whales for Train; of

---

[1] The treaty of 1489 between King Hans and Henry VII (DI, XVI, 444).

[2] [JB]: 'Foed: Tom 12 P 375, 381.'

[3] Sebastian Cabot (1474?–1557?), son of John Cabot, was a Venetian explorer who sailed, on behalf of Henry VII, in search of the Northwest Passage.

[4] Banks is not referring, as could have been expected as he owned a copy, to the book by Johann Anderson, *Nachrichten von Island, Grönland, und des Strasse Davis*, Hamburg, 1746, as it does not contain this information. He is probably referring to a work by Dr James Anderson (1739–1808), an economist and agriculturalist who wrote a lot about fisheries and was asked by William Pitt the Prime Minister for a survey of the fisheries of Scotland (after writing a tract on the north British fisheries, *The True Interest of Great Britain Considered*, 1783). Both this, and his *An Account of the Present State of the Hebrides and Western Coasts of Scotland* (1785), discuss briefly the Newfoundland fisheries, while *Extracts relative to the Fisheries on the North West Coast of Ireland* (1787) mentions Iceland; but all fail to yield the information Banks gives.

[5] Jacques Cartier (1491–1557), a Breton explorer who claimed Canada for France. He sailed on two voyages, the first in 1534, the second in 1536.

[6] Flesh here means 'meat'.

[7] For the year 1577 Banks notes more than once that: 'The English had 15 Ships only in N*ewfoundlan*d. The Reason given by Hackluyt is that they had so many in Iceland at that time.'

[8] Anthony Parkhurst (fl. 1561–83), merchant and explorer and a promoter of an English Newfoundland settlement, wrote the letter to Richard Hakluyt in 1578, saying that the Newfoundland fishery was not developing because the English were obtaining most of their cod from Iceland.

Portugals, about 50 sail that make all wet;[1] of French, 150 Sail. The Trade our Island has to Iceland maketh that the English are not there in such numbers as other nations.[2]

1583. an attempt to colonise Newfoundland
This year Sir Humphrey Gilbert[3] sailed to Newfoundland, to settle a Colony under a Royal Patent granted in 1578; he set up the Queen's arms, & granted Leases to many persons, Portuguese, French & Spanish, to set up Stages. They agreeing to take them; but he fail'd in his attempt to establish a Colony.

1585. Queen Elizabeth orders captures to be made in right of her Sovereignty of Newfoundland
in support of the Sovereignty of the Island established by Sir H. Gilbert two years before, The Queen sent this year a Squadron of Ships, commanded by Sir Bernard Drake,[4] who made Prizes of several Portuguese Ships laden with Fish & oil.

1595. Queen Elizabeth writes to the king of Denmark on the Iceland Fishery
Queen Elizabeth, not unmindful of the value of the Iceland Trade, wrote this year a letter to Christian, King of Denmark,[5] requesting that an English merchant of Harwich might resort to the western Isles in Iceland, for Fishing; the King returned for answer, that her Majesty's subjects had been prohibited from resorting to Iceland, because they had gone without Licence; contrary to ancient Treaties; but that they should be free to come to Iceland, the port of Westman Isles only excepted which had for a long time & still was appropriated to the use of the Danish Court.[6]

The Trade of Iceland sold by Denmark to Merchants of Hamburgh & Bremen
Before this time the trade of Iceland had been placed by the king of Denmark in the hands of the merchants of Bremen & Hamburgh;[7] this accounts for the conduct of the Danes in throwing difficulties in the way of the English who claim'd a right to use the Fishery founded on ancient Treaties.

1600. a monopoly of the Iceland trade granted to Danish Subjects
An Edict of Christian the 4th, about this Time, conveyed a monopoly of the Traffic of Iceland to certain Towns at Denmark.[8]

1615. The Iceland Fishers this year employ 150 English Vessels
Notwithstanding the difficulties which Denmark no doubt threw in the way of English Fishers in Iceland, we had this year 120 Ships & Barks employ'd in it.[9]

1616. The English commit acts of pyracy in Iceland
Harrass'd, no doubt, by interruptions from the Danes, The English again acted as pyrates as did the French also.[10]

---

[1] 'Wet' fish means that they have not been processed.

[2] [JB]: 'Hackluyt, Vol 3 p. 132.'

[3] Sir Humphrey Gilbert (1537–83), half-brother to Sir Walter Raleigh, was an English adventurer and explorer.

[4] Sir Bernard Drake (1537?–86) was an English sea-captain and explorer.

[5] King Christian IV of Denmark (1577–1648), reigned from 1588.

[6] [JB]: 'Foed. Tom. 16 P 275' [Rymer, *Foedera*]. The DTC copy says 'Coast', an obvious mistake.

[7] [JB]: 'Dr Holland, Sir G. Mackenzie p. 62.'

[8] This actually happened in 1602.

[9] [JB]: 'The Trades' increase, a Pamphlet.' *The Trade's Increase*, London, 1615, quarto, 62 pages, is said to have been written by Robert Kayll. Banks is referring to p. 295 where it says 'Iceland Voyage entertaineth one-hundred and twenty Ships and Barques'. Iceland is again mentioned on pp. 293 and 301. See also Rymer, *Foedera*, vol. 16, p. 275.

[10] [JB]: 'D^r. Holland, Sir G. Mackenzie p. 62.'

1625. The Newfoundland Fishery increase as that of the Iceland diminishes
In proportion as the Island Fishery was at this period interrupted & harassed by the Danes, that of Newfoundland increased. This year Devonshire alone had 150 ships, & carried their Fish to Spain & Italy.[1]

1635. The French permitted to dry their Fish on Shore in Newfoundland
This year 15 Ch*arles* 1st, That king, press'd for revenue by the violent struggles of his Subjects, granted to the French the liberty of curing their fish on Shore in the Island of Newfoundland, in consideration of an annual Payment of 5 per C^t. This concession was afterwards ratified by the Treaty of Utrecht, but the payment was not on that occasion noticed.

1663. The Iceland Fishery mentioned in an act of Parliament
Notwithstanding the vast increase which the Newfoundland Fishery had experienced during the civil wars, that of Iceland was not intirely forgotten. In the 15 Ch*arles* 2nd, Ch 7, an act for the encouragement of Trade, provision is made for the encouragement among others of the Iceland & Westman Isles Fishery: it was provided that no fish shall be brought from any of the Places mention'd, unless British, caught in Vessels navigated according to Law: it is also enacted in the 16 Ch. of the same Statute, that no Ship shall sail for the Fisheries of Iceland & Westman Isles untill the 18th day of March in any year.

1670. Newfoundland fisheries diminish'd Sir Jos: Childs advice to displant the Colony The Newfoundland Fishery likely soon
The Newfoundland Fishery, which in the time of the Restoration was believed to gain to England £400,000 a Year, this year diminish'd from 250 Ships to 80; Sir Josiah Child attributed this decrease to divers causes, among the rest to the advantage the Boat keepers, who were resident on the Island, had over the Ships that annually frequented it; he therefore suggested the Policy of Dispeopling & displanting the Colony, a policy which has been adhered to ever since.

The Colony, however, has not been either dispeopled or displanted, but at present abounds with inhabitants; so that it is easy to foresee that in a short time the chief benefit of the Fishery must fall to their share, as was the case long ago with the New England Fishery, which was originally managed by old English Ships.[2] These inhabitants even now purchase for the greater part of their necessary consumption from the United States, who receive their Fish in Barter, & who meet & generally undersell our merchants in the ports of the South of Europe with these very fish.

1733. Iceland Fishery monopolised by the king of Denmark English continue to fish in the Iceland seas
At this time the Trade of Iceland, which had been monopolized by the Danes since the beginning of the 17th Century, tho' they were not able till long after to exclude the English from it,[3] was carried on by the King with a capital of 4,000,000 of Dollars, or at least –

---

[1] [JB]: 'The Golden Fleece, a Pamphlet.' Banks must be referring to the *The Golden Fleece: Or the Trade, Interest and Well-Being of Great Britain considered with Remarks etc. ... mostly on wool*, published in London in 1736.

[2] [JB]: 'Sir Josiah Child.' Sir Josiah Child (1630–99) was an English politician and governor of the East India Company who wrote several books. Banks does not mention which book he is referring to but probably it is *A New Discourse of Trade*, of 1751, which mentions Newfoundland in chapter X 'Concerning plantations', pp. 154–63.

[3] The following is in his notes, but not in the fair copy: 'was conferd by Royal Charter to a company of Danish merchants who paid the King an annual Rent tho being [a] Small one for the privilege'.

nominaly so; 20 ships were sent out, but the Trade was found in due time to be a losing one.[1] He sent out annually 23 ships, some to the Flesh ports & some to the Fish ports, purchasing every thing by Barter at fix'd prices which never varied; the inhabitants were then reputed to be 50,000;[2] at what time it was that the Danes succeeded in excluding all other nations from visiting the Island does not appear; but this exclusion was continued until the year 1806,[3] when the Trade of Iceland was declar'd free to all Danes; till that time the Danish ships that us'd the Trade were nominaly King's Ships & were armed; they chased all vessels that approach'd within 4 danish, 16 English, Sea Miles off the Coast, & warn'd them off.

Since that period no notice has been taken of Strangers; some of whom, especially Dutch, have landed & traded without molestation, chiefly in the Eastern districts; when the present war broke out, abundance of Dutch ships were fishing off the coast, & some English have at all time frequented it, chiefly smacks with wells, but provided with the means of salting their fish if too long detained at sea, one of these was spoken to by Mr Clausen[4] in the year 1802.

1772. conduct of the Danes to the Icelanders Icelanders wish to become subjects of England
When Sir Jos: Banks visited Iceland this year, he found the Monopoly of the Trade there much in the same situation as Busching describes it to have been in 1733; only that the prices of the articles of Barter were by no means fix'd; the Company that traded in the King's name by their agent altered them every year in proportion to the success of the fishery, in such manner as to secure the whole produce of the Island, however large it might be, at the same charge to themselves as they were obliged to give for a small quantity: that is, just as much of European necessaries as would keep the Icelanders from starving.

This ill advised measure suppressed in the inhabitants all those energies which make men successfull. The utmost industry exerted by them produced no more advantage than was the result of ordinary exertions; of this the people grievously complained, & repeatedly solicited Sir Joseph to propose to his government the purchase of the Sovereignty of the Island from Denmark, which they thought would be sold for about £100,000. This they did not doubt very soon to make good with infinite advantage to their new masters, who, they well knew, would allow them to partake of the Blessings of British liberty.

1782. Iceland Cod more saleable in the South of Europe than any other
This year in an elaborate report of a Committee of the house of Commons on the subject of the Fisheries the following evidence is recorded; 'The Fishery in Iceland, which the people of Yarmouth carried on by the assistance of the Shetlanders in about 200 vessels

---

[1] [JB]: 'Sir G. Mackenzie p 334.'

[2] [JB]: 'Busching's *Geography*'. Anton Friderich Büsching (1724–93), German theologian and geographer, published *Géographie universelle* in Strasbourg in 1768, in six volumes. The first volume deals with Denmark, Norway, Iceland, Greenland and Sweden; Iceland is referred to on pp. 372–400. The 20 vessels referred to, are here (and not in Mackenzie) p. 358. Büsching's work, *Neue Erdbeschreibungdes ersten Theils erster Band welcher Dänemark, Norwegen und Schweden enthält*, pp. 361–87, also contains an account of Iceland: the population is mentioned on p. 371, the trade of Iceland in 1733 on p. 373.

[3] The correct year is 1788.

[4] Holger P. Clausen: see Appendix 13.

from 40 to 60 Tons; where the cod most saleable in the Spanish & Italian markets are caught, has been annihilated by the operation of the Salt Laws; several gentlemen of Yarmouth informed the Comm*itt*ee that they would revive this fishery, if it should be relieved from the oppression of these Laws.'

1799 Iceland trades to Spain & the Mediterranean Iceland a most important Fishing Station both for Sea fish & Salmon
in the years 1797, 8, & 9, a very considerable Traffic in fish, was carried on from Iceland to Spain & the Mediterranean; & this period was certainly the most favorable for the Commerce of Iceland. Mr Thorleius, a native merchant residing at Bildal,[1] speculated largely, at that time, & made a considerable fortune: at present he is esteemed the most wealthy man in Iceland.

Iceland offers the most important advantages as a Fishing Station: the facility with which fishing is carried on by the Natives is really astonishing: in the morning they go out in small Skiffs to the distance of a few miles from the Shore, and in the afternoon return with as many fine fish as their boats can contain: the rivers are frequented by [a] vast number of salmon; but these are neglected. Fish & oil are the chief articles of export, which could be extended to an indefinite amount.[2]

Conclusions, deduced from the Statements above referred to the Dates of the year under which the proofs of them are to be found;

1 That the Iceland fishery has been in former times productive to a vast extent, & must have continued as productive as formerly to the present day, had not the evil policy of Denmark, by placing a monopoly of its profits in the hands of the Crown, instead of allowing the people to take the benefit of it, prevented it from being advantageous to either party for more than two[3] centuries past.

2. That since this impolitic monopoly was abandoned by the Danes in the year 1806, Iceland-cured fish has found its way into the Mediterranean, &, in the event of Peace, will no doubt be carried there in great quantity.

3 That the seas surrounding Iceland afford at present a most extensive Fishery, sufficient, it is presumed, to supply the whole of the Catholic Countries of Europe with a sufficiency of Lenten Provisions.

4. That Iceland-cured fish is preferred in the South of Europe to that of the Newfoundland.

5. That the Newfoundland Fishery has already in part fallen into the hands of the United States of N. America, & must soon become more advantage*ou*s to them & to the Colonists settled in Newfoundland than to the mother Country.

6. That the Iceland Fishery in the hands of a nation that has at last found the means of managing it to the best advantage must soon become a very formidable rival to the United Kingdom, & in due time a successful one.

---

[1] Ólafur Þórðarson Thorlacius (1761–1815), merchant in Bíldudalur. Banks spells his name correctly in the notes but then he is using Mackenzie. Mackenzie remarks of him: 'At present he is esteemed the most wealthy man in Iceland' (1811), p. 335. Banks has this information in his notes but it is omitted in the fair copy.

[2] [JB]: 'Sir G. Mackenzie's *Travels* p. 345.' Actually Banks derives this paragraph from p. 338 in the 1811 edition, with only salmon mentioned on p. 345, while in the 1812 edition fishing is on p. 286.

[3] 'Five' has been crossed out.

7. if the People of Iceland should incline to Cede their Country to the arms of the United Kingdom, may not the following terms be accepted?

1. That private property be respected.

2. That the Ancient Laws of the Country by which they are now governed do remain in force.

3. That the Island be ceded to the Crown of the United Kingdom as a Royal fief, to be held as Alderney, Guernsey, &c. are held; The King having the same power in the Island as the King of Denmark had before its annexation.

Such an annexation of Iceland to the Crown of the United Kingdom, would place the Legislative power in the hands of the king & of course all commercial regulations in the hands of his Ministers.

The Food of the Icelanders, imported by them from Denmark, is rye, barley, oats, & pease; of these articles we can spare a sufficiency for their present reduced population; 20,000 Quarters of all Species of grain will more than suffice; of wheat the whole quantity imported in 1806 was not more than 30 quarters.

NHM, DTC 17, ff. 140–156. Contemporary copy.
Hermannsson, 'Banks and Iceland', pp. 75–83.
Adelaide, Australia, The Royal Geographical Society of South Australia. 'Notes on a brief History of Iceland' (26 pages) are preserved in Banks, Sir Joseph. Notes on Iceland, Ms 6c. Among these manuscripts is a draft of the DTC copy in the NHM, part of which is shown in Figure 8.

259. Banks: Draft re Memorandum, [June 1813][1]

The Danish officers of the Crown to be Seizd as Prisoners of war but to be sent from England to Denmark without Exchange

Kestenschiold Amptman of the South Ampt[2]
Thorersen Northern & Eastern ampt[3]
Frydensberg Landfogued[4]

Surveyors
Friis and Scheel[5]
Their Plans to be required from them also Minors Plans[6] he was a marine Surveyor & was drownd in Iceland

---

[1] This memorandum is obviously connected to the previous document. Here Banks goes into more detail on his plans for the annexation of Iceland.

[2] Johan C. T. von Castenschiold was district governor (*amtmaður*) of the Southern Amt 1810–19 and governor (*stiftamtmaður*) of Iceland 1813–19.

[3] Stefán Þórarinsson, district governor (*amtmaður*) 1783–1823.

[4] Rasmus Frydensberg the treasurer (*landfógeti*) 1804–13.

[5] The surveyors sent by the Danish government were the Norwegians Hans Frisak (1773–1834), who worked in Iceland 1803–6 and 1807–14, and Hans Jacob Scheel (1779–1815), who was in Iceland 1807–14. A scholarly account of their surveying is treated fully in Thoroddsen, *Landfræðissaga Íslands*, III, pp. 190–209.

[6] Captain Hans Erik Minor surveyed the coast of Iceland from 1776 to 1778 when he drowned in Iceland as it says in the document. His maps, that Banks wanted to get hold of, were printed by *Det Kongelige Søkort Archiv* in 1788 (Thoroddsen, *Landfræðissaga Íslands*, III, pp. 83–4, esp., p. 84, n. 1).

To Treat with Stephensen Etats Raad a native Icelander
The highest in Rank of any Native in the Island[1]
The oath of allegiance to be rendered & all who will take it to be admitted as subjects of the Crown of /Great Britain/ England.
The Island to become Subject to the Crown of England as it now is Subject to the Crown of Denmark
To Return its ancient Laws under the Kings Royal Pleasure
to be Subject to Such new Laws as the King Shall Please to Enact – of Course Subject to the Laws of Parliament by virtue of /the Kings/ Royal Assent
an English Governor to be appointed under the name of Stiftsamptman
The other offices to be filld by Icelanders During the Kings Pleasure.

all Danes who have Fishing or other Establishments to have the option of Continuing on Condition of taking the oath of allegiance to the King of England or to have three years allowd them to Remove their Property upon the same terms of Licences as are now Granted but not to Remove their warehouses but allowd to Sell them to the best bidder The Condition of their Licences from Leith to Denmark must be to Bring back a Quantity of Corn preportiond to the Cargo they are allowd to carry away in Case they Return

in Future an allowance of Corn from England to be Granted for the Support of Iceland at the Rate of Four Bushels a head of Rye that is 20,000 Quarters in case wheat is Requird one Quarter to be allowd in Exchange for Two of Rye

Military[2]
Barracks to be Erected for a Regiment of Infantry a Company of Cavalry /*To be mounted on the Island*/ & some Engineers they Should be made as Commodious as Possible in Every Particular in order to Render the Duty of this inhospitable Climate as Little disgusting as Possible

Companies of infantry one of the mounted Cavalry to be mounted on the Island where horses & Sheep & a detachment of artillery with an Engineer & Some ordnance & Stores will be an ample Garrison for the Island to insure it against any Possibility of a Ferment among the natives or against any force the Danes could Sent [sic] to Retake it Icelanders will by degrees Enter into this Corps so that it will not in the Event prove a drain to the mother Countrey

Wisconsin, Banks Papers, f. 121

---

[1] Magnús Stephensen.
[2] This is written vertically on the left-hand margin, at right angles to the main text and from bottom to top.

260. The Earl of Clancarty to Banks, 6 June 1813

Whitehall

Dear Sir Joseph,[1]

Will you be so good as to look at the accompanying papers returned here by Mr Vansittart,[2] – These potatoe bread Manufacturers, appear to be in general <u>charlatans,</u> however it is fair they should have their case examined into, tho' I should be very averse to recommend remuneration. Perhaps you could suggest the name of some person to us, who would try the different experiments mentioned in M[r]. Wheatleys papers.[3]

Upon the subject of your Islanders we have applied to the Admiralty to investigate the case of the *Regina*, & have suggested to their Lordships the propriety of a General order to their Cruizers upon future detentions, to send the Vessels to Leith instead of taking them to Swedish ports.[4]

My brother[5] desires me by letter from Galway to state that he will endeavour as soon as the season for that fishery shall arrive to send you the information you require respecting the Basking Shark, or Sunfish, that at present there is no person at Galway capable of affording him answers to your Queries. –

We sincerely hope the Gout has at length left you, & that your health will be speedily & permanently reestablished.

Yours Dear Sir Joseph
very sincerely
[*Signed*] Clancarty

BL, Add MS 56298, ff. 34–35. Autograph letter.

261. Banks to the Earl of Clancarty, 11 June 1813

[Spring Grove]

My dear Lord

When Mr. /*Whally*/ Wheatley[6] first commenced his attempt to dip his Fingers into the Public Purse, he declared that he had invented a method of extracting Flour from

---

[1] A mathematical note appears in Banks's hand, on the bottom of f. 35: 20
$3\frac{1}{2}$
$16\frac{1}{2}$

This sum is explained in the following letter.

[2] Nicholas Vansittart (1766–1851) was the Chancellor of the Exchequer, appointed on 20 May 1812.

[3] Wheatley's name is incorrectly spelled. The gentleman in question is John Whateley of Cork, Ireland. He had invented a machine 'for the speedy separation and manufacture of Farina or Flour from potatoes, and for various applications thereof to make sea-biscuits, bread and pastry', earning the 'lesser gold medal' (*Transactions of the Society of Arts*, vol. 21, 1813, p. 23). I am indebted to Will Ryan for this information.

[4] See following letter.

[5] Clancarty had four surviving younger brothers: Power (b. 1770), archbishop; William (b. 1771), rear admiral; Charles (b. 1772), archdeacon in Ardagh; and Robert (b. 1782), colonel. The brother in question is doubtless William who corresponded with Banks on the subject of the basking shark. See letter from him in NHM, DTC 18, Banks to William Le Poer Trench, undated 1813, ff. 229–230, and Trench's reply to Banks (NHM, DTC 18, 26 April 1813, ff. 231–233) in which he offers to procure further information if desired.

[6] John Whateley. See previous document.

Potatoes; he was however /*Some*/ in due time convinced, that the Substance he did extract, which he now calls Farina, is Starch & not Flour; & he ought to be aware that every Chemist in Europe & every Starch maker is as well able to make Potatoe Starch as he is, & have been in the habit of making it for a Century at least.

In the Scarce times that we Remember a few years ago, this starch was recommended & used by some /*Families*/ people to diminish the consumption of Flour in their Families; the addition of Boiled Potatoes to Flour in making bread & Puddings was then used in abundance of different ways, & still is used in some Families, but is never likely to be generally adopted.

Thus all pretentions to invention & novelty on the part of Mr. Wheatley are, I conceive, done away: if so, he can have no claim on the Public for Reward; both the improvements he suggests have been already introduced into use, & will, if the Public chuse it, be adopted on a large scale in the case of Potatoe Starch, however, we must recollect that it is subject to an excise duty, & that Penalties attach to the making of it without a Licence.

If I remember well, for I am here without Books to refer to, it requires lb 20 of Potatoes to make lb $3\frac{1}{2}$ of this Starch: the remaining lb $16\frac{1}{2}$, which also abounds in nutriment, must be employed in the food of animals: this appears to me a great waste of human Food: the cultivation of Potatoes & their use as human food increases rapidly every day; it appears to me much better Policy to encourage the use of the Potatoes as Food in their natural state, than to attempt to use them in counterfeiting any other kind of Food. I have always considered Potatoes as little Rolls prepared by nature for the use of man, who has only to bake or to boil them to make them into palatable & nutritive Food.

If any arguments are just, /*as*/ & your Lordship feels as I do that Mr. Wheatley's is an impudent attempt to impose upon Government, by calling the Starch which he has made Farina & Flour, & by preparing the use of it as Food under this false appellation, I should advise that his Son be desired to send the lb 12 of Flour he has with him to the Board, & that your Lordship refers it from thence to the Board of Excise, for information whether or not it is Starch & liable to the duties upon that article. Their answer will be a compleat Settler, I conclude, to Mr. Wheatley's Pretentions, which, I confess, I hold to be more than usually Foolish & nugatory.

I return the Papers enclosed, & with them another Letter, exaggerating, as I conceive, the Conduct of the Swedes towards the Iceland Vessels when sent in to their Ports. The real cause of complaint, however, I consider as substantiated, that of the Swedes forcing[1] the sale of the corn intended for the food of my miserable Constituents: how can these foolish people, who cannot feed themselves, wish to seize upon Norway & starve all the Population? I rejoice to hear that your Lordship has /*prepared*/ proposed so effectual /*as*/ a remedy to the Admiralty, & shall be delighted if I am fortunate enough to hear that their Lordships have adopted the measure.

NHM, DTC 18, ff. 249–251. Contemporary copy.
Chambers, *The Scientific Correspondence of Sir Joseph Banks*, vol. VI, pp. 100–101.

---

[1] The word 'forcing' is pencilled in, possibly by Dawson Turner.

262. Jörgen Jörgensen to Banks, 24 August 1813

London, Fleet Prison

Sir

I take the liberty without your previous permission, to forward you a manuscript which I have lately written in Suffolk and had put into the hands of a printer but for particular Reasons I have again withdrawn it with an Intention of postponing the publishing of it.[1] – A singular Revolution has lately taken place in my affairs, and notwithstanding the Act I committed against the Danish Crown in Iceland[2] yet my father a few hours previous to his death[3] solicited His Danish Majesty's pardon for me and the King in his own handwriting assured him a Pardon was granted me on Condition of my never setting my foot in Iceland again. These News have lately reached me, and in Addition I have seen the whole transaction inserted in the Danish Gazette (:*Dagen*:) which has also made its Appearance in Iceland. – Count Trampe in the other hand has been arrested by order of the government for having held a secret Correspondence with the Enemy.[4] – Thus the man who in this Country affected to be horror-struck at my Rebellion, has since himself /been guilty/ of Selling that province (:Drontheim:)[5] which had been entrusted to his care by his King. – Under such circumstances I do not know how far it will be prudent for me to publish my manuscript at present, it would perhaps endanger my Inheritence: yet I am anxious to preserve a work from oblivion and entire destruction which has cost me a deal of trouble and pain to compleat; – and it is therefore I now take the liberty of consigning it to your care. – I trust you will give it a place in your library, so the true and faithful account of the Icelandic Revolution may be preserved somewhere; and as I am informed that your library even after your Death will be dedicated to the public Service[6] I should be infinitely glad would you give my work a place therein. – My publication differs materially from many other accounts which have been seen upon the Subject, and should you wish to be in possession /of the documents/ on which I rest my assertions, you shall have them all.[7]

Should I not publish the manuscript on account of not giving further offence to the Danish government, I shall still endeavor to bring to light Sir George Mackenzie's

---

[1] Jörgensen had been arrested in July. He had asked Henry Jermyn (1767–1820), a Suffolk antiquary, and Hooker for assistance (BL, Eg. MS 2070, Jörgensen to Dawson Turner, 4 Aug. 1813). Jörgensen ended up in the Fleet prison where he met up again with James Savignac, confessing to Hooker that sharing a room was 'certainly not agreeable' to him 'but it answer's for cheapness Sake' (BL, Eg. MS 2070, undated (September? 1813)).

[2] That is to say by usurping the royal power and declaring Iceland an independent country.

[3] Jörgensen's father, Urban Jörgensen, watchmaker to the Danish Court, died in 1811. According to Bakewell, it was announced in the Danish gazette that the king had granted Jörgensen a pardon in response to a request from his father just before he died (*The English Dane*, p. 144).

[4] Jörgensen spread the good news to Hooker: 'Have you observed the paragraph in the newspapers from Denmark viz. "Copenhagen July 8 – The Bailiff Count de Trampe, as well as several wholesale merchants have been arrested in Norway, on suspicion of having held secret Intercourse with Sweden, and caused Corn to be exported"' (BL, Eg. MS 2070, 16 August 1813).

[5] Trampe was appointed governor of Trondheim in Norway on 23 July 1810.

[6] In accordance with Banks's wishes his library was transferred to the British Library after the death of his librarian Robert Brown (1773–1858) (see further p. 552, n. 6 below).

[7] Many of the documents are in the Department of Manuscripts in the British Library and in The National Archives, Kew.

blunders and Ignorance.¹ This however may be done in a Series of letters in the Monthly magazines. The first eighty pages of my work may not excite much public Interest, but as we advance the Subject becomes more general and therefore more interesting. I have taken a large View of a pamphlet printed about three or four months ago in the Hope of an Appeal to the British people. I have therein pointed out his erroneous propositions and shown the futility of his romantic Scheme. I have not been able to learn the name of the Author but I trust it might be known to you.² – I hope, should you not have leisure to peruse my work yourself, that you will have the goodness to direct some one else to do it, and to give /him/ you his opinion of it. – I should be happy should you then think it worthy of a place among your numerous volumes.

From a very early age, upon the particular Services in the South Seas wherein I was engaged for a considerable part of my life-time, I was taught to revere your name and I was acquainted with your virtues ere I saw you in person. – Had I published this work I should have been desirous of dedicating it to you had I not thought it would argue a great degree of /weakness/ Meanness, to force, in a manner, the Dedication of a work upon a gentleman who had so decidedly withdrawn his Countenance and favor from me. – I was ever of opinion that I should not have lost your friendship and protection had I not been grossly misrepresented to you; and some art made use of to prejudice you against me. – The non-payment of the sums of money you so generously lent me would never have induced a gentleman of your large fortune and your philantropic Disposition to forsake me had no other causes strongly operated upon your mind to produce that Effect. – I trust the attentive perusal of my work will in some degree restore your good opinion of me, which is the greatest favor you can bestow on me. –

Altho' I have no personal favor to ask you /for/ myself yet I am anxious to introduce an Object to your Notice who stands much in need of your kindness and Assistance. – I had once the pleasure and the honor to introduce to you two people who without your help must have been cast away in a strange Country, (:I mean the two Otaheitans:)³ and I should now again think myself happy might I obtrude on your goodness, and relate to you the hard case of a native of a much colder climate than that of Otaheite. The person I allude to is a young woman from Iceland under the name of Gudrun Johnsen,⁴ who came over here from Iceland in the same vessel with Mr. Parke⁵ & Mr. Savignac.⁶ I do believe she was enticed away on account of firmly believing that Mr. Savignac was an unmarried man, but to her surprise she discovered the contrary and that he had a wife in England. She is of good and decent parents⁷ and I believe she waited upon you several months ago. – She was equally astonished and grieved in hearing that Mr. Savignac was married, she found herself in this Strange land without a single friend, and what

---

¹ He is referring to Mackenzie's *Travels in the Island of Iceland*.

² The pamphlet by 'An Icelander', *Memoir on the Causes of the Present Distressed State of the Icelanders* (see Document 254).

³ See Document 39.

⁴ Guðrún Einarsdóttir Johnsen, The Dog-Day Queen, Jörgensen's girlfriend while in Iceland. See further on her Agnarsdóttir, 'Hundadagadrottningin heldur út í heim'; and Wawn, 'Hundadagadrottningin'.

⁵ John Parke, the British consul to Iceland.

⁶ James Savignac was an employee of Phelps & Co. and an active participant in the Icelandic Revolution.

⁷ Her parents were Einar Jónsson (b. 1761), a *tómthúsmaður*, a man with little or no land who made his living from the sea, and Málfríður Einarsdóttir (b. 1764), both from northern Iceland, but living at the time in Reykjavík with their three daughters.

aggravated her wretched situation, was the advantage Mr. Parker[1] took of her defenceless state, and endeavoured by all means to get her into his power and sent two Danish people to her to make such proposals as no honest woman would listen to. Upon her refusal to comply with this insidious request Mr. Parker became her Enemy and he reviled that Innocence which he was not able to corrupt. – About a month after Mr. Savignac's arrival in England he was arrested, carried to the King's Bench, and thence removed to the Fleet Prison.[2] Stranger and /un/supported as the young woman was she had no other means of Subsistence than by sharing the little Savignac, a prisoner could spare, and in order to subsist she has now been obliged to part with all her wearing apparel etc. and has only one or two gowns remaining. – During this trying and calamitous situation she has behaved with so much modesty and decorum during Savignac's Imprisonment that she is universally respected. – She sleeps in a small lodging outside this prison, and her food is the most simple: she knows nothing of England but its miseries. – She has for ten months experienced all the hardships of extreme poverty in a strange land, and this under a peculiar disadvantageous situation for I am afraid that her coming to visit Savignac in the Day Time and preparing him his meals may be construed, or has been so already, into Impropriety of conduct, tho' I can vouch for /her/ Innocence, for I do not believe there is a more ugly fellow in the World than Savignac, besides the temptation she has resisted more than anything else evince the goodness of her disposition and her correct conduct. – It is on account of these that I have dared to mention to you the situation of an unfortunate female; tho' she has hitherto preserved her virtue, yet it is hard to tel*l* how far an unprotected woman will be able to avoid the Snares which may be laid for her, and how far the lowest degree of poverty may be the cause of much Evil. – I for my part am a prisoner myself, but shall not be long so. I met Savignac and Gudrun Johnsen here by accident; had I it in my power my hand should be extended for the Relief of the young woman. I am /therefore/ not able to do anything myself and must rest satisfied by recommending her extreme case to the notice of a gentleman who was ever distinguished for his liberality and humanity. I Being a Dane Gudrun Johnsen has told me the Sorrows of her /breasts/ heart with less Reserve perhaps than she would an Englishman. – She grieves: she is desirous of returning into the bosom of her friends, nor would you believe, had you known her before, that she is the same girl which we saw in Iceland five years ago. – Her few desires, her frugality and her excellent turn of mind, do [i.e. mean], that very little will be necessary to pay for her subsistence till we can find a vessel to send her home: how far you may think proper to interfere, and to relieve the person I have mentioned from great misery, extreme poverty, and perhaps much temptation I do not know. – I have done all I could to attract your notice. I can do no more: happy should I /be/ could something be done. – I must now leave all to you. – I should be very glad should you be so good as to acknowledge the Receipt of this, I am in the Fleet-prison at present, after having last year travelled over Denmark, twice visited Portugal, once Spain, and lastly came from Gibralter.[3] – I do not know whether you ever received a printed circular letter from me? – The lodgings wherein Gudrun Johnsen lives in are in /4 Printing House

---

[1] Parke was often erroneously called Parker, particularly in Icelandic sources.

[2] The Fleet was mainly a prison for debtors and bankrupts, containing about 300 prisoners with their families. It is described by Charles Dickens in *The Pickwick Papers*.

[3] Jörgensen had been travelling in Spain and Portugal before ending up in a military hospital in Gibralter.

Square/, Coleman Street near the Bank. – When she waited upon you last her adddress was in the Adelphi, but she is now in cheaper lodgings, and to be nearer to Fleet Prison.

I have the honor to be
Your most humble and most obed*ient* Servant
[*Signed*] Jorgen Jorgensen

Wisconsin, Banks Papers, ff. 179–180. Autograph letter.

263. Hans Frederik Horneman to Banks, 28 August 1813

London

Right Honourable Sir,

I have had the honour to receive your letters of the 26th & 28th Ins*tant* – To the former I shall reply this evening or on Monday, after an interview I shall probably have to day, for the sake of ascertaining some points which you are anxious to have explained relative /to/ the unfortunate young person attached to Mr. Savignac.[1]

Agreeably to your wish I respectfully beg leave to give you my opinion on the present state of Iceland & the Document furnished by Mr. Walker.[2] That the supplies of grain last year were barely sufficient to carry the inhabitants thro' the winter I firmly believe; whether the whole import to Rykevig was only 200 Barrels as stated by M. Castenskiold,[3] I have no means of ascertaining, but I can not help entertaining strong doubts on the subject. – I think the acc*ount* Mr Parke, lately furnished to you respecting last years imports,[4] would satisfactorily explain this point. – However this may be, I feel quite satisfied, that if the seven vessels with corn from Denmark do not speedily arrive at Leith, or, if after arrival they should from some cause or other not proceed or arrive at the ports of their destination, that the situation of the Inhabitants of Iceland will be very deplorable. And I am sorry to add, that I have a letter from Mes*sieurs* Corbett Borthwick &c., this morning dated 25th Ins*tant* where not one of the 7 vessels had arrived,[5] for which Licences /were sent/ to proceed without delay to Iceland with Thursday's post sent from the Council Office. – Under these circumstances any encouragement given to British merchants to take out grain & provisions to Iceland might be /encouraged/ essential to the existence of many /human beings/ altho' I am much more inclined to consider Mr. Parkes letter[6] as correct as far as regard the fisheries, notwithstanding that other letters from Iceland of the 10th July state the Fishery to have failed. – That there is a great want of provisions every account concurs in asserting. I am convinced that the Danish

---

[1] Gudrun Johnsen: see Document 262.

[2] Possibly a George Walker of Wavertree, who, in company with Horne & Stackhouse, sent the brig *Aid* to Iceland in February 1815.

[3] Johan C. T. von Castenschiold, the Danish governor of Iceland. As almost 20 ships made the voyage to Iceland, 200 barrels of rye does not seem much.

[4] Could possibly be the list 'Ships Clearing for Iceland' (Document 252).

[5] The seven vessels were probably: *Helleflynderen, Orion, Bosand, Svanen, Gratierne, Freden* and *Eskefiord*.

[6] There is no extant letter from Parke in the Foreign Office files from these months. The letters he wrote in the summer of 1813 did not reach London until 26 October. Horneman may be referring to a letter, dated 16 May 1812, from Parke to Wellesley, on the difficult situation in Iceland (TNA, FO 40/2), a copy of which was probably sent to Banks.

Government has never refused any supplies for the Island of Iceland; but the whole is left to merchants. – I certainly do think that it could be the Duty of /the Danish/ Government to superintend that the merchants did their duty, & that for the advantage of being allowed to trade to Iceland, they should be bound to furnish the necessary supplies. – This is a subject which I shall take the liberty of expressing myself freely upon when I again write to Copenhagen, and I blame the Danish Government for not obliging the ships to sail, when they had unexpired Licences, & there was in reality no hopes of an accomodation of differences with Great Britain.

Should the said 7 vessels not arrive in time to proceed to Iceland, & it is high time now – let me intreat you to urge Government to adopt measures for the saving of so many unhappy beings from utter ruin & the calamity of a famine. – Perhaps the evil is less than represented, but I am alarmed at the thought of considering it such.

I am happy to say Mrs. Horneman is gradually recovering from her late severe indisposition. She unites with me & the children in thanks to your kind remembrance of us & /we/ beg to offer /her/ our regard to Lady Banks, Miss Banks & yourself.

With the greatest esteem I remain
Right Honourable Sir,
Your obliged & obedient Servant
[Signed] H. F. Horneman

Wisconsin, Banks Papers, ff. 132–133. Autograph letter.

264. Jörgen Jörgensen to Banks, 28 August 1813

Fleet Prison

Sir

I received your letter with a great deal of pleasure this Morning, and feel myself obliged to you that you will give my manuscript a place in your library, provided you find nothing objectionable therein.[1] – Should however that be case I can easily expunge any thing that should appear to.[2] – As a work of the nature I have written must derive a very great part of its merit from being founded on the strictest truth I should also be glad to add the original Danish and Icelandic Papers, Documents, and Receipts which I have mentioned in my narrative, all of which are in the handwriting of the persons quoted. – These are at present in the hands of Mr. Henry Jermyn of Sibton in Suffolk[3] a most intimate friend of Mr. Hooker's, and may soon be had.

---

[1] In BL, Eg. MS 2067, Jörgensen wrote (f. 368): 'By the time this my narrative will appear before the public, I am sorry to find Count Trampe will not be in England, as many may think I wish to take an advantage of his absence, but this can not be. The case, as I already long before his departure have given notice to Lord Mulgrave [Henry Phipps, 1st Earl of Mulgrave, First Lord of the Admiralty 1807–10, brother of Constantine Phipps], through a regular channel, that I would publish this small work.' It never appeared.

[2] So much for Jörgensen's self-vaunted unbiased account. In the copy he made, addressed to the Marquis Wellesley, he wrote: 'these my humble efforts to give the world a correct account of a late Revolution on the island of Iceland, your Excellency, and which can have no other merit than the truth contained therein' (BL, Eg. MS 2067, ff. 187–188, following the account of the Revolution.)

[3] Jermyn became a correspondent of Jörgensen's. See further Document 262.

I was no less pleased at your kind Intentions towards the young Icelandic Woman Gudrun Johnson. – Did I not know that she is particularly anxious to return to her native soil, and to drop every connexion that can throw a cloud on her character. I should have been the last person in the world to have ventured to recommend her to your notice and to your kindness. – Of this I am most particularly convinced that she has remained for nearly three months of this Summer with a Mr. and Mrs. Howard at Finchley,[1] who are certainly respectable people and have several children. – Gudrun Johnsen has been lately but a very short time in London. – I have also been in this prison but a short While, and the moment she saw me she immediately made known to me her wishes, her present situation, and how desirous she was to avoid scandal. – I represented to her the folly even of visiting a married man in this place, and she would not have done so had she actually not been in a very deplorable state: she has been ill for a length of time. – Her poverty shows that she is not totally abandoned, and what raises my opinion still higher of her is this, that I know that Mr. Parke previous to his departure from England several times sent gentlemen with overtures to the young woman earnestly requesting her to live with him, and he actually enclosed a ten pound note for her in one of his letters. This with her letter she returned to him without hesitation. I believe sincerely Mr. Parke is a perfect Gentleman, but men will sometimes give themselves that latitude with women, which they would not venture in other concerns of life. – I sleep in the same Room with Mr. Savignac[2] and therefore I know that at present nothing wrong is transacted, whatever may have been their former connexion, tho' I believe she is far from being fond of him, and I rather think some art has been employed to induce the woman to leave her friends and to come to England. I do assure you that Bishop Vidalinus[3] persuaded her to /to/ go to England, for Savignac told him he intended no further than /that he intended/ to have her instructed in sowing,[4] in England. But let the case be what it will, and let her have been guilty of great Indiscretion yet she is [a] good woman, and she is yet possessed of all that native and simple Innocence which so much characterises the Icelandic women. Heaven takes more pleasure in the sincere repentance of one Sinner than in ninety nine righteous who never erred. – I declare to God had I the means in my power I should apply to no one in favor of the poor Girl, but I can do nothing at present: it would be needless for me to have applied to any one but yourself, for I know none that would have gone to such great lengths to relieve misery and distress. – Mine shall be the task to prevent any ill-Effects of my advice Mr Savignac might be inclined to give her, for I have sufficient courage and spirit to keep him in order. – I shall take an opportunity to send for her to communicate the Contents of your letter to her, and sure I am that she will receive your kindness with becoming gratitude. – I am certainly vexed with Savignac that he should induce a good woman of respectable connexions to leave her country when he had not the means of support, and what makes the thing more deserving of Reproof is this that he is a married [man] to an amiable woman and he has two fine Children. – It is true I am no saint myself, nor

---

[1] Finchley was then a village just to the north of London.
[2] This was a cheaper arrangement for convicts (see Document 262, p. 541, n. 1).
[3] Geir Vídalín, the bishop of Iceland: see Appendix 13.
[4] He means sewing.

ought I to be severe upon others, for I have myself formed illicit connexions, but in so doing I have injured no female, I have no wife and no children. – My vocations have hitherto prevented me from settling any where, and it would have been imprudent in my situation of life to have married any woman whatever, what injury then that has been sustained has been inflicted on myself, and on no one else. Mr. Savignac's case was widely different and I detest a man who can act as he has done. – When I shall see Gudrun Johnsen I shall do myself the honor of writing to you later upon the Subject. –

I shall remain here about three Months and twenty days, during which time I shall have noth*in*g to do, I am desirous of employing myself in some useful manner, for the Company you meet with here is infamous and I detest them. – I should therefore wish to compile from Olavsen's voyage in Iceland,[1] from whose natural history, and from Savignac, an account of all the birds of Iceland with the manner of catching them and drawings of the Implements for that purpose.[2] I intend translating it all into English, and to send it to Mr Hooker for Correction. Yet I should not wish to waste my time on /*the*/ a Subject of that nature should it not be a new thing, and if it had already made its appearance in public. – Will you have the goodness when you shall have time to let me know if those things are published, and whether you think it will /be/ useful and interesting, I shall immediately begin my translations.

I am Sir with great truth
Your most humble & most ob*edient* Servant
[*Signed*] J Jorgensen.

Wisconsin, Banks Papers, ff. 177–178. Autograph letter.

265. Hans Frederik Horneman to Banks, Evening 28 August 1813

London

Right Honourable Sir.
I have just seen Gudrun Johnsen, whom I had requested to call on me. – She is desirous of returning to Iceland, & under the circumstances you render her an invaluable service in returning her to her parents. – I believe this young person to be possessed of /a/ sincere wish to live an honourable & virtuous life. – She has been very ill, is very poor. – Therefore I gave her £1 & promised her debt of £5 /at her lodgings/ should be paid. – She shed many tears when I mentioned to her your goodness, & the assistance she could expect from you in proceeding to Iceland. – I have her promise that she will proceed to Leith the moment the vessels are known to have arrived. – She will then want a few warm clothing for the voyage – which I have also promised her for you. – She speaks highly of Mr. Savignac's conduct to her – and I have lessened with her the unfavourable opinion she entertained of Mr. Parke. –

---

[1] He must mean the work of Eggert Ólafsson and Bjarni Pálsson, who travelled around Iceland during the years 1752–7 on an official expedition. See Introduction, pp. 20–21 above.

[2] There is no evidence that he ever got around to this.

I hope your care of her will /be/ rewarded by a virtuous life & a dutiful attachment to her parents, who by no means appear to be dissatisfied with her. – In fact, I think she really is virtuous, at least I am certain she is virtuously inclined.

Believe me with the sincerest attachment

Right Hon*ourable* Sir,
Your obliged & obed*ient* Serv*an*t
[*Signed*] H. F. Horneman

Wisconsin, Banks Papers, ff. 134–135. Autograph letter.

266. Jörgen Jörgensen to Banks, 30 August 1813

Fleet [Prison]

Sir

The moment I received your letter I sent to Gudrun Johnsen who called upon me Yesterday. No*thing* could exceed her joy on hearing that you were kind enough to say you would send her back again to her country. – She was so much more pleased as she had seen Mr. Horneman upon the same Subject on Saturday who had spoken very kindly to her. – I have not before now [seen] a smile upon the poor Girl's face since my arrival in the place, but yesterday she appeared quite happy. – She received your kindness with every mark of the sincerest Gratitude. I am sure she will be the better for it all her lifetime, and a letter from you to the Bishop[1] and any other respectable person in Iceland to recommend her, will effectively relieve her from the Apprehension of the Sneers and Insults of those women in Reckavig whose conduct Sir George Mackenzie so undhandsomely [*sic*] exposes in his work.[2] – She will be guided in every thing by Mr. Horneman, and nothing will be wanting on her part to convince you that she is not unworthy of your favors. – She is now so elated on account of the Expectation of returning to her friends and country with some degree of Credit and under your powerful protection, that should this not take place I really think she would break her heart. –

What I mentioned respecting the original documents in the Danish and Icelandic languages, relating to the Revolution, and now in the hands of Mr. Jermyn, is only true that I wish to put them in some order, and to number them, and to give an alphabetic list of the whole, stating what each document or Receipt contained, so that in case any Dane or Icelander should visit your library and be inclined to doubt or contradict the facts, that he might then turn to my list and so find the very documents in the handwriting of the persons mentioned, and hereby convince himself of the truth of my Statement past all contradiction. When you have determined to give my manuscript a place in your library I should be glad if you will let me know the same; and also if you

---

[1] Not extant.

[2] Jörgensen is referring to what Sir George Steuart Mackenzie wrote about a ball he invited the citizens of Reykjavík to in 1810: 'Several ladies, whose virtue could not bear a very strict scrutiny, were pointed out to us' (*Travels in Iceland*, p. 95).

should think it worth while to give an account of the birds in Iceland in the manner I mentioned in my last letter.¹

I am Sir:
with great Respect
Your most humble & ob*edient* Servant
[*Signed*] Jorgen Jorgensen

Wisconsin, Banks Papers, ff. 175–176. Autograph letter.²

267. Hans Frederik Horneman to Banks, 31 August 1813

London

Right Honourable Sir,
On the 28th Inst*ant* I had the honour to answer your favour, & have now the pleasure of informing you that the *Nordstern* & the *Gode Haab* have arrived at Leith, from whence they will immediately be dispatched.³ – I have now learnt how much grain each has in board. The other vessels are expected.

A person of the name of Svend Sivertsen⁴ a native of Iceland has made application to receive a Licence to purchase a vessel in Norway – & is further to carry on the trade from Iceland. – Before any petition is delivered to the Board of Trade please inform me what your opinion is about employing more vessels – I fear it can not be allowed, altho' a more open trade might be of advantage to the inhabitants.

At last I have a letter from Mr: Clausen dated Copenhagen 1 Aug. – The ship *Gratierne* will winter at Copenhagen, so will the *Swan*.⁵ I do not like this retracting or shrinking from the performance of an implied duty. –

Gudrun Johnsen has been here to day. I gave her another £1 note to purchase some linen. – I told her to be in readiness for departing – Her request was that some clothes in pawn for £5 might be taken out. – I promised to mention the subject to you. – Your decision shall be followed. – She has also delivered me the enclosed bill for lodging. – I am fearful the whole will be very expensive to you. Your answer

---

¹ The manuscripts are now in BL, Eg. MSS 2066–2070, including all letters to Hooker (BL, Eg. MS 2070).

² Note in Banks autograph, verso, where he has jotted down the following figures, the cost of something in shillings and pence:
9.8
5.6
2.6
17.8

³ *Nordstern* (registered 31 August) and *Gode Haab* (registered 4 September) (NAS, E. 504/22/61).

⁴ Svend Sivertsen (Sveinn Sigurðsson) was a factor in Eyrarbakki.

⁵ Neither ship sailed to Iceland in 1813. However, a licence granting *Svanen* (the *Swan*) to clear for Iceland was issued in August 1813 (TNA, BT 6/210, no. 53,291).

will greatly oblige me. – In the meantime allow me to call myself with the greatest esteem.

> Right Hon*ourable*. Sir
> Your obliged & ob*edient* Ser*van*t
> [*Signed*] H. F. Horneman

Wisconsin, Banks Papers, ff. 136–137. Autograph letter.[1]

268. Jörgen Jörgensen to Banks, 31 August 1813

Fleet Prison, London

Sir

Gudrun Johnsen called upon me to day after having seen Mr. Horneman: She informs me that there are now two Ships in Leith, one for the West and one for the East:[2] but as she is likewise told that Mr. Sivertsen will be here very soon, she would prefer to proceed to Iceland on his Ship, as the vessel will go straight to Reikavig,[3] where as if she should go to either part of the Country, it would be attended with much Inconvenience and Expence to a lone woman to travel overland at a late Season in the Year.[4] – The length of time she has been [in] England and the very scanty support she has had since her arrival here has reduced her to a great degree of distress, yet I find that she has preserved much frugality, and the only Sum she owes is about four pounds 10 shilling for lodgings and a couple of pounds for washing, all in her present lodgings, where she has only paid 15 Shillings per week for her board and lodging, which is the least she could pay at the present time when things are so very dear. – Her poverty has obliged her to pledge several of her clothes; I believe about 5 pounds: These of course to redeem would be very serviceable as they are certainly worth more. Upon the whole the young woman is much in want of necessaries to proceed on a voyage in a late season of the year, yet as Icelandic women by their habit wear much cheaper and coarser things than other women very little will serve her, and I think about a dozen of pounds besides what she owes would amply provide her with everything to make her comfortable on her passage. – Were I not confined and had I the means she should not want it, but situated as I am I can not advance her any thing, for I expect no Remittances till October, and for her to wait /till/ that time would frustrate her object which She has so much at heart. – However with any

---

[1] The enclosed bill was with the letter (for Gudrun Johnsen, see further in following letter):

9 Weeks Board and lodgings at 10 shillings per week   £4 – 10 – 0
To Washing   2 – 6 9
    £6 – 16 9

A Schofield Printing house Square

[2] Probably the *Nordstern* (registered 31 August; NAS, E. 504/22/61) which was bound for Ólafsvík in western Iceland and the *Gode Haab* bound for Berufjörður in the east.

[3] Bjarni Sivertsen's ship was the *Orion*.

[4] Because of the unbridged glacier rivers in south-east Iceland, travelling within Iceland was difficult and dangerous. Gyða Thorlacius, wife of one of the county magistrates at this time living in Suður-Múlasýsla in eastern Iceland, recounts in her memoirs how she travelled first on a merchant-ship from eastern Iceland to Copenhagen and then took a ship from Copenhagen to Reykjavík. This was an easier journey (Thorlacius, *Endurminningar*).

order of mine upon our house at Copenhagen (:of which after the death of my father I am a partner:) be of service I /*should*/ would with pleasure give that for the amount necessary to answer Gudrun Johnsen's purpose – Mr. Horneman has had the kindness to relieve in part her present Anxiety by advancing her on your account two pounds for which she feels very grateful: I know nothing would give her a greater pleasure than previous to her departure from England she might be permitted to thank her generous protector Sir Joseph Banks in person for his much kindness. – The hope also by going out with Mr. Sivartsen to reap a double benefit as that gentleman is so well known to you, and to be recommended to him by you would much serve her in her own Country. – The reason I am rather anxious that she should be furnished with necessaries for her voyage is this, that she has been so very ill lately during her stay in England, she has grieved very much and her constitution is not so robust as it was when she came from her own country, and therefore to be entirely without warm linens & clothes might further injure her.

I see by the Danish Newspapers that Count Trampe is arrested in part for having caused grain to be exported, and for having caused the price of grain to be raised in Norway:[1] this is precisely the charge I made against him in Iceland tho' he found means to clear himself at that time, but now I think he will not get off so cheap.

I have the honor to be Sir
Your most humble & obed*ient* Servant
[*Signed*] Jorgen Jorgensen. –

Wisconsin, Banks Papers, ff. 181–182. Autograph letter.

269. Banks to William Jackson Hooker, 2 September 1813

Spring Grove

My dear Sir

Jorgensen has opend a Correspondence with me by Sending to me a very prolix account of the attempted Revolution in Iceland which he Requests me to Deposit in my Library for the use of such as Choose to Read it my answer was that if in the Perusal I did not find any thing in my judgment improper I Would Comply with his Request[2]

on the Perusal of a Part of it for I have not yet gone through it nor do I Find any amusement in the Progress I find a larger Part of it dedicated to an Acrimonious Criticism on the Conduct of Count Trampe[3] Capt*ain* Jones[4] S*ir* [George] Mackenzie & much about himself & Little indeed of that Plain unvarnishd Narrative Which Truth best loves to be clothed in. You no doubt have Seen it I Request therefore your opinion whether I ought in any way to Concern myself with it. If you have not I will Send it to you in order that I may have the benefit of your opinion

---

[1] There was animosity between Trampe and the King's most confidential representative in Norway, General Georg Frederik von Krogh. Trampe was recalled to Copenhagen in 1812 but later reinstated. As usual Jörgensen tended to exaggerate: Trampe was not 'arrested'.

[2] See Document 262.

[3] The former governor of Iceland.

[4] Captain Jones of the *Talbot*, who brought the Icelandic Revolution to an end.

I realy think that some of the Abuse so needlessly heepd upon the unfortunate Objects of his Censure is of a nature that may be deemed Libelous & that if I allow the book to be used by those who Read in my Library I shall run the hazard of being Considerd as an Accomplice in the Publication of a Libel

he tells me that the work was Put into the hands of a Printer & Returnd but he does not tell me Why it was Sent back it may be that it was because the Printer did not think the sale of it would Pay his Expence in Paper & Press work or he may have been like me afraid of being indited as a Libeller[1]

our Old Friend Stephensen is dead[2] Vidoe is now the habitation of his son[3] Iceland has been Sadly neglected by the Danes The supply of the Last year was very inadequate to the necessities of the inhabitants. Clausen has withdrawn his two Ships from the trade[4] The Ships of this Season which ought to have been by this time ready to Sail homeward from Iceland have only begun to arrive at Leith. I was realy at one time afraid of a Famine there I hope however that this will not be the Case this year at Least as we have an account of 9 Ships of which 5 have made their appearance

The hortus Kewensis[5] will be out in a Few weeks Aiton who is very ill & goes directly to Cheltenham Expects to Receive the Last Sheet on Saturday. Mr Brown[6] is Gone to Scotland to visit his Old mother

adieu my Dear Sir
your very Faithfull H*um*ble Serv*an*t
[*Signed*] Jos: Banks

Kew, Hooker's Correspondence, the Director's Correspondence, I, 40, ff. 67–68. Autograph letter. Hermannsson, 'Banks and Iceland', p. 64.

270. Jörgen Jörgensen to Banks, 6 September 1813

Fleet Prison

Sir

Gudrun Johnsen called upon Mr. Horneman this morning: to her disappointment she learned that all the Icelandic Ships had sailed from Leith and that it was quite

---

[1] Jörgensen's account of the Icelandic Revolution was never printed.

[2] Ólafur Stephensen died on 11 November 1812. He had entertained both Banks in 1772 and Hooker in 1809.

[3] The island is Viðey and the son was Magnús Stephensen, Chief Justice of Iceland, who following his father's death had managed to procure Viðey for himself.

[4] Clausen's ships were the *Nordstern* and *Gratierne*.

[5] This refers to the 2nd edition of the *Hortus Kewensis* in 5 volumes, published in the name of William Townsend Aiton, son of William Aiton in whose name the 1st edition in 3 volumes was published in 1789, in fact largely produced and edited by Jonas Dryander until his death in October 1810, when volumes 4 and 5 were completed under his successor Robert Brown (2nd edn, 1810–13).

[6] Robert Brown (1773–1858), FRS, an eminent botanist. When Dryander died in 1810 Banks appointed Brown his successor as librarian and curator at Soho Square, a position he retained until Banks's death in 1820. He later became the librarian of the Linnean Society and keeper of the botanical collection at the British Museum. It was thanks to Brown that an independent botanical department was established in the British Museum, the first nationally owned botanical collection available to the public. He lived in Banks's house (left to him by Banks) from 1812 to his death. He was a prolific writer. In the words of his biographer William T. Stearn: 'the reputation that Brown acquired during his lifetime as one of the greatest botanists has proved well founded'(*Dictionary of Scientific Biography*, II, 517).

/un/certain when any other Vessel would arrive. The *Orion* (:Sivertsen's Ship:) had put into Norway, and it was said that he had lost his Sails & Riggings, damaged more than half of his cargo and had therefore been obliged to put into a Norwegian Port. – It is strange and singular that that same ship *Orion* has three different times met with similar accidents, and gone to Norway instead of Iceland, especially as her pretended misfortunes have always happened in the finest summer months. I am much afraid that the humane Intentions of the British Government towards Iceland are in a great measure defeated and the high premium given to Danish Ships who can succeed in bringing Corn to Norway induces the Icelandic merchant to gain a Norwegian port if possible. – Gudrun Johnsen seems quite dispirited on account of her disappointment, and her case is so much worse, as if she should remain in England the whole winter she must certainly labour under considerable Embarrassments. It was from motives of humanity that I first took the liberty of mentioning her situation to you in hopes that she might be relieved and sent home to her own Country. – It would however be a most desireable thing would you direct Mr. Horneman expressly to intimate to the young woman that all future favors must entirely depend upon her discontinuing of her visits to Mr. Savignac. – The Imputation under which Mr. Savignac lies with Respect to his Wife makes every visit to him from Gudrun Johnsen /highly improper/ and should she not entirely and for ever denounce his Society she ought not to be protected. – She has visited him once or twice under an Idea that she ought not entirely to forsake /him/ – I have told her of the Impropriety of such conduct, and should she not adhere to her promise /she has made me of staying away/ I should be much offended. – Should she not be able to go to Iceland this Winter it would be well could she be sent some few miles from London where she could live cheap and learn some useful occupation which might be of Service to her in Iceland. – I trust that Mr. Horneman's knowledge of my having written to you in behalf of the young woman may not prejudice her case and diminish your favor towards her, for tho' Mr. Horneman is a good man enough, yet the Danish houses in London hate me most cordially and have caused me more trouble and mischief than you think for. – Whatever therefore can be done in Prudence to reserve Gudrun Johnsen from misery I hope will be done from your kindness. – If you do not keep it up I am afraid she must sink for I am at a loss to guess what she can do for herself.

I should be extremely obliged to you if you could at any time request some of these Icelandic merchants who come hence direct from Copenhagen, to procure you <u>Snedorffs samtlige Skrifter</u>,[1] a work of very great merit and written with a great degree of Spirit. – Mr. Snedorff I believe was known to you, he was a Professor at the University of Copenhagen, and on his Travels in Scotland fell out of the Stagecoach and broke his head at the Age of thirty-two, to the great detriment of learning in Denmark. – His works will be of great Interest and if translated would be useful. – It is no use my asking any of the

---

[1] Frederik Sneedorff (1760–92), Danish historian. His *Samlede Skrifter* are a collection of writings, including historical lectures and letters from travels in Germany, Switzerland, France and England in the years 1791–2. In 1788 he was appointed 'professor extraordinarius' in history at the University of Copenhagen. He travelled to Oxford, Birmingham and Liverpool, and went north to Cumberland where he died in the described accident near Penrith, where he is buried.

Danes to get them for me, for they will not do it, whereas if you should be pleased to request it, they would think it an honor to comply.

I am Sir,
Your most humble & ob*edient* Servant
[*Signed*] Jorgen Jorgensen

Wisconsin, Banks Papers, ff. 185–186. Autograph letter.

271. Hans Frederik Horneman to Banks, 7 September 1813

London, Queen Street, Cheapside

Right Hon*ourable* Sir,
I have had the honour of receiving your letter of the 1ˢᵗ & 2ᵈ Ins*tant*. – Your kind & benevolent views with Gudrun Jonsen I should have gladly executed, as I think her more imprudent than otherwise in fault, but I begin to fear that it will not be possible this year. – She informs me that to travel great distances over land would be impossible for her /at/ this time /of the/ year, & that therefore only a vessel for Ryekevig direct would bring her to her home. – Now I believe from the information I have obtained, that the *Orion* is the only vessel, going to Ryekevig or near to Ryekevig, & she appears to have been obliged to run into a port in Norway to refit. – Therefore I think it not only uncertain, but improbable that the vessel will reach its destination this year. I have informed her that the moment I learn the *Orion* is arrived,[1] I shall send her word – also if the *Bosand*[2] should be destined to Rykewig, which I much doubt. – Besides, this vessel has not been heard of since she left Copenhagen. –

In the mean time she has received £2 from me as she appeared penniless & wanted a little linen, which required some time to make. – I trust you will not disapprove of this. – I must agree with you with regard to a letter to Bishop Vidalin in her behalf,[3] – Her future correct conduct will, with the attention & kindness you have shewn to her, be her surest & best recommendation. –

I have reported your opinion respecting granting a Licence for a new vessel to Mr. Sivert Sivertsen.[4]

As it will be interesting for you to know how far supplies will be received in Iceland this year I take the liberty of mentioning the various ships that have sailed & their fate – The *Eliza*[5] – has proceeded to her destination probably with a full Cargo of grain. The *Regina*[6] was at Leith. – Her cargo was sold at Gothenburg by the Captors. It is probable she will receive a fresh Cargo at Leith. –

---

[1] The *Orion* did not arrive from Copenhagen until the end of October (NAS, E. 504/22/62, and letter 26 October 1813, Hornemann to Banks, Document 277).

[2] The *Bosand* arrived in Leith in October 1813 from Copenhagen (NAS, E. 504/22/62). This ship wintered in Leith (see Document 274), leaving for Iceland in early March.

[3] Not extant. In the National Library of Iceland there is a specific file on Bishop Vidalin's correspondence from foreigners. There are no letters from Banks.

[4] Sigurður Sívertsen (b. c. 1790), the merchant son of Bjarni Sívertsen.

[5] The *Eliza*: Kjartan Ísfjörð owner, Boye Hansen master.

[6] The *Regina*: Just Ludvigsen owner, R. Jepson master.

The *Orion*¹ – run into a Norwegian port to repair.
The *Bildahl*² was at Leith, preparing to proceed in her destination.
The *Nordstern*³ now at Leith   Do   Do
*Gode Haab*⁴   Do Do Do
*Helleflynderen*⁵ put back to a Danish port to repair.
*Bosand*⁶ – has not been heard of since she left Copenhagen. –
I have the happiness to say that Mrs Horneman is now recovering, I think, entirely. She is truly sensible of your obliging inquiries, & returns her grateful thanks to Lady Banks, Miss Banks, & yourself, adding here my respects, I remain with the greatest esteem & the sincerest attachment,

Right Hon*ourable* Sir,
Your obliged & obed*ient* Serv*ant*
[*Signed*] H. F. Horneman

P.S. The enclosed was sent by Mr. Clausen's factor Mr Olsen,⁷ on his arrival at Leith with the *Nordstern* to be forwarded to you, being part of a book which he procured last year, but not in a compleat state.⁸ /H. F. H. –

Wisconsin, Banks Papers, ff. 138–139. Autograph letter.

272. Jörgen Jörgensen to Banks, 9 September 1813

[London] Fleet Prison

Sir
I am extremely vexed that I should have to write you again upon the Subject of Gudrun Johnsen and indeed it is somewhat against my Inclination so to do, thinking that you may be justly offended at my giving you so much trouble, but Gudrun Johnsen has been here again this Morning pressing me so heartily to write to you that I could not well refuse her Request. – She would write herself but she is deficient in the English language.⁹ She appears very much grieved and disappointed at not having a passage to

---

¹ The *Orion*: Bjarni Sívertsen owner, J. Ketelsen, master.
² The *Bildahl*: Ólafur Thorlacius owner, Christen Kieldsen master.
³ The *Nordstern*: H. P. Clausen owner, N. A. Schmidt master.
⁴ The *Gode Haab*: Jens Lassen Busch owner, Jörgen Hansen Möller master.
⁵ The *Helleflynderen*: Jens Lassen Busch owner, Thomas Christensen master.
⁶ The *Bosand*: Johan Gottfred Höwisch owner, Hans Andersen Aas master.
⁷ Probably Ole Christensen Olsen.
⁸ Banks had so many books dealing with Iceland that it is near impossible to suggest what book this was.
⁹ During her stay in England Gudrun Johnsen's English was to improve greatly, as her letters to Lady Stanley testify. She spelled phonetically but the meaning is quite clear: see a fascinating collection of her letters edited by Andrew Wawn, 'Hundadagadrottningin' (the letters are in English with translations into contemporary Icelandic). During the winter of 1813–14 she stayed with the Stanleys at Winnington Park in Cheshire while waiting for an Iceland-bound ship the following spring. She was invited by Lady Stanley to stay for another year: 'I wish you to consider this as your home, if you remain in England' (Bedfordshire and Luton Archives, Whitbread MSS, W1/4530, Lady Stanley to Guðrún Johnsen, 26 August 1814) but Gudrun was desperate to get back to Iceland. See further Agnarsdóttir, 'Hundadagadrottningin heldur út í heim'.

Iceland; she is much afraid that you have turned away your Countenance from her and that some one may have done her some Ill-office often with you. – I have told her that I only write to you now, upon condition that she never shows her face again within the walls of this prison unless I should send for her to communicate any News to her which could be of Service. She says she only comes here to learn if anything is done for her: and if she could but be happy enough to live in the most frugal manner any where she would be very thankful indeed. – Where she now lodges is in the house of an Umbrella maker, who /*has got a wife and*/ is a widow with a family, and are very poor people. They have been very good and kind to her, but cannot afford to give her longer credit for her board & lodgings. – She says could she discharge her present little debt to them, she might go on again with them till something could turn up in her favour. – She is in hopes, that she may find in you a generous protector who will screen her from poverty and want, and I may add, perhaps something worse than all that. – The Ships which have been in Leith are all sailed, but none of them were destined for Reikavik yet there are hopes that some will arrive bound for that place. – Should that be the Case Gudrun Johnsen may still get away.[1] – I have scarcely any Excuse to make for intruding so often upon you, but I could not well do otherwise having been so much pressed upon that account. – Gudrun Johnsen would think herself very much obliged to you, would you have the goodness to communicate your Intentions towards her through Mr. Horneman or myself. –

I am Sir
With great Respect
Your mo*st* humble & most ob*edient* Servant
[*Signed*] Jörgen Jörgensen

Wisconsin, Banks Papers, ff. 187–188. Autograph letter.[2]

273. Jörgen Jörgensen to Banks, 10 September [1813]

[London], Fleet Prison

Sir

I am perfectly ashamed at giving you any further trouble upon the same subject I have so often written to you, but Gudrun Johnsen has again applied to me tho' I do not wish to write more yet she presses so earnestly for me to do it that I could not refuse. She informs me she called upon Mr. Horneman this Morning but could learn no proper tidings relating to the ship *Orion*; and she is so perfectly destitute and distressed that nothing could exceed, I lend her three Shillings and that was more than I could well spare – Mr. Horneman told her he would write to you. – Altho' I do not wish to make my self busy upon the Subject more than I can help yet I am very glad to have an Opportunity to do justice to the character of Mr. Parke the Consul in Iceland. – I had accidentally heard that Mr. Parke was the brother of Mr. Parke a very eminent lawyer who I am not unacquainted

---

[1] Gudrun finally managed to embark for Iceland on 30 August 1814, on the *Vittoria* sent by Horne & Stackhouse from Liverpool.

[2] This letter has franked on it 'unpaid Fleet'.

with, and who is a man of the most respectable Character.¹ It struck me therefore that Mr. Parke might have been unjustly accused, and I asked Gudrun Johnsen whether it was realy true that the Consul had been guilty of such Impropriety of Conduct which I had been told. – It appeared by her confession that Mr. Parke had made some overtures to her to go with him to Iceland as housekeeper but nothing further, and the Rest was added by Mr. Savignac, who is no friend of Mr. Parkes. I had the Satisfaction to hear Gudrun Johnsen tell Mr. Savignac that he had behaved very illiberal in that respect, and she told him had she listened less to his Advice she would have been so much better off. – She now sees her folly in being guided by a man who is a lazy and idle fellow. – I am glad of having in my power from her own mouth to contradict a story which I could scarcely believe any man guilty of fabricating unless it was justly founded. – Had I in the first Instance known the connection of Mr. Parke and his true character I should not easily have given credit to such insinuations, yet I hope, that what is communicated to you will not do any harm.

Gudrun Johnsen expects anxiously to hear from you through Mr. Horneman or through some other Channel.

I am Sir
Your most humble & obedient Servant
[Signed] J. Jorgensen

Wisconsin, Banks Papers, ff. 183–184. Autograph letter.

274. Hans Frederik Horneman to Banks, 21 September 1813

[London] Queen Street, Cheapside

Right Honourable Sir,
Nothing but a wish to give you a full & ample account of every thing has prevented my answering your favour of the 10th Instant² before now. –

Gudrun Johnsen has promised me faithfully that she would never again be seen in the company of M: Savignac. – Under the circumstances I have paid her lodging & board to
A. Schofield.³   £6. 10
& previously to herself for linen   £2. 00
                           _____
                           £8. 10

for which I enclose receipts – I shall probably to morrow give her £5 to redeem her cloths – She speaks highly of her Landlady, & that she is able to provide her with some work. – She has offered to board & lodge her for 14/– a week on account of her assisting in work

---

¹ Parke's brother was James Parke (1782–1868), later created Baron Wensleydale, the first and only one (his sons all died young), an eminent lawyer and judge. *Burke's Peerage* lists James Parke's siblings, London, 1885, p. 1456. James Parke was instrumental in getting his brother appointed to the Iceland consulship (TNA, FO 40/1, James Parke to Wellesley, 6 July 1810). One of Parke's most famous cases was the trial of Queen Caroline, where he acted for the government.

² Not extant.

³ See in Document 267, p. 550, n. 1.

– I do not think this is very cheap – But Gudrun says she will charge less, when she has learnt more of the work of an umbrella maker.[1] – I think she might fill the situation of a nursery maid, & would be obliged by your opinion, also if you should have any objection to the Society of Friends of Foreigners in Distress[2] sharing in the expence of keeping her.

I now hope & trust she is half, if not wholly reclaimed, & that she will be true to her promise & avoid Savignac, who, I find, will be out of the Fleet next week. – Gudrun is poorly in health & she wishes, if you permit it, to go into the Country for a fortnight at the house of a sister of her Landlady. – I have asked her the cause of her illness – She attributes it, partly to fretting, partly to the air, which is so different from that of Iceland. –

I can not as yet agree with Adm*iral* Hope[3] regarding the unfitness of the Iceland vessels. – For the *Regina*, which it particularly applied to, has proceeded to Iceland, in a boysterous season.[4] – The *Orion* is known to be a good vessel, tho' I have my doubts about the instructions, which may have been given by Chevalier Sivertsen[5] to the master. – But time will shew. – It is thus the *Helleflynderen*[6] only to which Adm*iral* Hopes suggestion could apply, & I must say, I have heard of very severe weather in the North Sea & Cattegat.[7]

I do not indeed consider the Danish Government indifferent to the fate of the Icelanders, but I think the Icelandic merchants left too much to themselves, & not under the control, which in my opinion, they ought to be. That not a single ship should have gone with corn to Ryekewig is unpardonable. – At least I am not aware, that any of the vessels named will proceed to that port or near to it.

Messrs. Corbett & Co. write me that: 'With regard to M^r. Svend Sivertsen's application, if he is allowed to purchase a ship in Norway, & sail in Ballast to a port in Denmark, we are certain he will willingly engage to bring from thence to Iceland 200 to 250 Quarters

---

[1] Due to the windy climate few things are as useless in Iceland as umbrellas.

[2] The Society of Friends of Foreigners in Distress was founded 3 June 1806, the constitution being adopted at a general meeting on 2 April 1807. Their motto was 'Love ye the Stranger'. The Duke of Gloucester was the Patron, William Wilberforce a vice-president etc. and H. F. Horneman himself was one of the directors. The aim was 'to grant relief to indigent Foreigners here, without distinction of country or religion ... and to provide the means, to such as are desirous, to return to their own country' (p. 5). Over 1,200 had received assistance by 1814, and 17,875 before 1825. Most were Germans, with 1,157 Danes, 1,685 Swedes, 971 Norwegians and 4 from Iceland. The only stipulation was that those receiving relief had to have resided at least 6 months in the country. There is a list of donors including H. F. Horneman, Danish Consul £15.15 while Mrs Horneman contributed £1.1. The Danes A. A. Feldborg, Georg Wolff (£21), Jens Wolff and Count Reventlow were also among the contributors (*Account of the Society of Friends of Foreigners in Distress; with the Nature and Views of the Institution also The Plan and Regulations, A List of Subscribers and an Appendix containing Some of the Most Interesting Cases*, London, 1814).

[3] Rear Admiral Sir George Johnstone Hope (1767–1818). A close friend of Nelson's, he served as a captain of the fleet under Admiral Sir James Saumarez in the Baltic. He became rear admiral in 1811. He was appointed to the Board of Admiralty, becoming the confidential adviser of Viscount Melville, First Lord of the Admiralty. He is buried in Westminster Abbey. See *The Naval Chronicle*, vol. 39, p. 424.

[4] The *Regina* left from Leith for Iceland at the beginning of September (NAS, E. 504/22/61).

[5] Bjarni Sívertsen had become a knight of the Dannebrog on 11 April 1812.

[6] The *Helleflynderen* had returned to Copenhagen for repairs (see Document 271) (RA, Rapporter fra Konsulatet 1809–1816, Horneman to Banks, 7 September 1813, and from Horneman to Danish government, 19 October 1813).

[7] The Kattegat is a body of water south of Skagerrak and north of Zealand, between Denmark and Sweden.

of grain, which will be as much as he can take with a vessel of 100–120 Tons, leaving room for some necessary articles from this Country. –

I am sorry to inform you that the *Resolution*, belonging to M^r. Lambertsen, was totally wrecked at Öreback, his settlement.[1] – She had arrived in ballast from Norway, & was preparing to load – M^r. *Lambertsen* has been very unfortunate – he has not had any share in the Iceland trade during this war. – He now, thro' me, applies for an other [*sic*] Licence, which I respectfully solicit your support for.[2] –

M^{rs}. Horneman, as well as myself, return your many thanks for your kind invitation to spend another Sunday at Spring Grove before the Winter has destroyed the beauties of the Autumn. – She is now a few miles from town, on acc*ou*nt of the air, & is rapidly improving in health. – The idea of partaking of your kind invitation & spending another happy day in your company & in Lady Banks' & Miss Banks', adds to her spirits, who seem to strive to approach those pleasing moments – as well as to return to the bosom of her loving family –

With never ceasing attachment & regard I remain
Right Hon*ourable* Sir
Your obliged & obed*ient* Serv*ant*
[*Signed*] H. F. Horneman

[P.S.] The following vessels have cleared at Leith in 1813 for Iceland.[3]

Ship	Owner	grain on board		
Tykkebay	Henkel	100	Quarters	Rye
Eliza	Isefiord	199¼	D°	D°
Gode Haab	Busch	90	D°	D°
		40	D°	Barley
Bildahl	Thorlacius	110½	D°	D°
		139½	D°	Rye
Nordstern	Clausen & Olsen	100	D°	D°
		50	D°	Barley
Regina	Isefiord	308¾	D°	Rye
		= 1138.	Quarters – Grain	

The *Bosand*, with a Cargo of grain is arrived at Leith 16th Sept*ember* but the master is afraid to go so late to his harbour, which is one of the most northern in Iceland (I believe Kiblevig)[4] he will therefore probably winter at Leith.[5] –

Thus if the *Bosand*, *Orion* & *Helleflynderen* had arrived in Iceland, the supply would have been, I believe, equal to last years, which by reference to the official return of shipments made at Copenhagen may be ascertained, of which I omitted to keep a Copy.

I can not help observing, that Mr. Isefiord, the owner of the *Eliza* & *Regina*[6] has well deserved of his native Country/Iceland/ having, notwithstanding he had the Cargo of the

---

[1] Niels Lambertsen was the merchant in Eyrarbakki on the south coast of Iceland.
[2] See next document.
[3] See NAS, E. 504/22/59 and 61.
[4] Keflavík.
[5] The *Bosand* did not leave for Iceland until March 1814 (NAS, E. 504/22/63).
[6] Both ships left from Leith for Copenhagen in February 1814 (NAS, E. 504/22/63).

Regina sold at Gothenburg, shipped 500 Quarters this year, which is nearly the half of the whole quantity shipped this year. – Any favour granted him would be well deserved. –

[*Signed*] H. F. Horneman

NB. The /*Tykkebay*/ *Albertina*[1] is arrived at Leith on her return voyage. – What grain she carried to Iceland I have not learnt as yet. –

Wisconsin, Banks Papers, ff. 140–142. Autograph letter.

275. Hans Frederik Horneman to Banks, 30 September 1813

Queen Street Cheapside

Right Honourable Sir,
I am greatly obliged to you for the promise of support of Mr Sivertsen's application,[2] & I shall request his undertaking in writing, before I make /it/ /*the application*/, that he engages to carry to Iceland a Cargo of grain, if a Licence is granted to him – I shall be anxious to come to the real causes of the short supplies for Iceland this year. At present it appears the owners put off dispatching the vessels from time to time, in hope of a peace; & when the *Regina* at last sailed, her being carried into a Swedish port, where her Cargo was sold, it so alarmed the others that they dared not venture out until Adm*ira*l Hope promised, or ordered, they should be sent to British ports if detained. – By this time the season had so far advanced that it was a very great risk to sail for Iceland at all, & I have no certainty that the renewed licences, which I sent to an agent to Gothenburg to forward, have ever been received. – Those that have sailed have been cleared out in expired Licences, which is an additional risk. –

I deeply lament the state in which Iceland is thus left for next Winter & Spring, & I have always recommended, with your uniform approbation, and facility, as to the terms of the Licences granted. – Notwithstanding my urgent applications, Mr Robinson[3] has refused to grant more than 5 Months for the *Tykkebay*, lately arrived at Leith from Iceland.[4] This obliges Mr Henkel,[5] the owner, either to clear out from Copenhagen before the 21. Febr*ua*ry next, or to request a renewal. – I shall endeavour to impress on Mr Henkel the necessity of making every speed to be in Iceland next year early in Spring,[6] in order to assist with Corn as soon as possible, & I shall endeavour to persuade the other owners to be equally expeditious – supporting these solicitations by an application to the Board of Trade at Copenhagen.

---

[1] The *Albertina* belonged to the merchant Frederik Lynge who had traded in Akureyri in northern Iceland since 1788. It was granted a licence to continue to Copenhagen and return from there with grain for Iceland (TNA, BT 6/210, nos. 53,933 and 54,700).

[2] He is referring to Sveinn Sigurðsson or Sivertsen (see Document 267).

[3] The Hon. Frederick John Robinson (1782–1859) (see Document 243, p. 507, n. 6).

[4] In September 1813 the *Tykkebay* was granted a licence to proceed without unloading from Leith to Copenhagen and return to Leith (TNA, BT 6/210, no. 53,736).

[5] Henrik Henkel: see Appendix 13.

[6] Henkel did send the *Tykkebay* to Iceland the following summer. All in all, 29 application for licences (the Danish *söpas*) were made in 1814, after the war ended.

I take the liberty of enclosing a petition which my house made lately for a Licence for Mr Lambertsen to purchase a vessel instead of the *Resolution* – (which was lost at Örebæck[1] his settlement) – which was refused this morning.[2] This unfortunate Gentle*man* has had no share in the Icelandic trade, since the Danish war[3] – I therefore take the liberty of sending you the various documents & to solicit your opinion if an application might again be tried with hopes of success. –

I have paid Gudrun Johnson £6 more to enable her to take her goods out of pawn – & enclose her receipt – the Society of Friends of Foreigners in Distress[4] have placed £5 at my disposal for her benefit, & after 2 months I shall make an application for more – She hopes soon to earn something towards the payment of her board & lodging, & that, I trust, your charge of her will be greatly eased, for I confess I feared it would have become a most heavy, tho' a most pleasing one. – She promises easily, & I shall endeavour to observe her conduct, for she must not deceive us, nor do I think she will. –

My unfortunate native Country seems still blind to its dangers, & it sinks, as I foretold Chamberlain Schested[5] deaper & deaper in the gulph of destruction, if it continues to side with France. My full exertions were used, at least to /keep/ prevent D*enmark* from becoming a tool of France, & /I/ advised a decided neutrality, tho' a firm self defence, – but, alas, all in vain. –

Mrs *Horneman* as well as myself feel very grateful for the interest you kindly take in her health. I saw her on Sunday last, & the attention for the better is very great – She accepts your kind invitation with the utmost pleasure, & on Sunday week, if agreeable to you, we will have the honour of waiting on you; in the mean time we request our respectful regards to Lady Bank[s], Miss Banks & your good self –

      With the greatest respect & attachment I remain
          Right Hon*oura*ble Sir
          Your obliged & faithful Servant
          [*Signed*] H. F. Horneman

P.S. The owners are anxious I should make an application to the Swedish Ambassador,[6] for some sort of protection from His Excellency for the Icelandic vessels. – Do you approve of this?

Wisconsin, Banks Papers, ff. 106–107. Autograph letter.[7]

---

[1] Eyrarbakki.

[2] The actual text can be found in Document 277 and it was not refused because the Lords said they would 'take the matter into consideration'. Eventually, in December 'in lieu of a licence granted to the *Resolution* wrecked in Iceland' a licence was granted to import a cargo to Leith, presumably on a new ship (TNA, BT 6/210, 54,006 and 54,708).

[3] The *Resolution* did receive a *sópas* in 1813, but was wrecked. There is no record of Lambertsen sending a ship to Iceland in 1814 but in 1815 Ísfjörð sent a ship to Eyrarbakki.

[4] For 'The Society of Friends of Foreigners in Distress', see above, p. 558, n. 2.

[5] Chamberlain Ove Ramel (or Rammel) Sehestedt (1757–1838), Count Ernst Shimmelmann's successor as the head of Det kongelige General Land-Oeconomie og Commerce-Collegium – the Danish Board of Trade.

[6] The Swedish Ambassador to Great Britain was Gotthard Mauritz, Baron de Rehausen (1761–1822).

[7] This is addressed to Banks at Spring Grove, where he lay ill for so long in 1813.

276. Jörgen Jörgensen to Banks, 6 October 1813

Fleet prison

Sir

I am somewhat vexed that I am under the Necessity of troubling you with a letter at present and more so on account of the occasion of it. – When I applied to you to afford that generous Relief towards Gudrun Johnsten[1] which you have so often afforded to the distressed of every country, I did so from a motive that Johnsten should not do any thing either prejudicial to her own character or that would farther plunge her into Imprudencies. She solemnly pledged herself that she would never come to this place provided she could in any manner subsist without it. – Your immediate supply fully relieved her from her Embarrassments, and your ample weekly Allowance to her ought in conscience to support her. – I have once had occasion to send for her to go with a letter to a place where I could not trust a stranger. She came – but I am given to understand and I am convinced that she has often repeated her visits to this place and met Savignac[2] in another Room, and tho' nothing criminal may have been transacted yet his advice and his Influence over her must be highly prejudicial to her, so much more so as Savignac ought not and has no right to have any communication and less /so/ of a secret nature /that/ with any female when he has a wife of his own. It is his duty to court her friendship. – Besides this Gudrun Johnsten instead of idling away her Time ought to apply herself to learn something useful in order to diminish the Expence she is at to you, and that she may not return to her country as ignorant as she went away, and as she lives with an Umbrella-Maker she has every opportunity to learn something. – Under these considerations I think as a Gentleman and a man of honor I am bound to state what I now say to you; so that if she does not discontinue her visits she may not enjoy your Protection, and if she does that she may be frugally provided for. – Your Authority over her must put her into a right truth, and it is an easy matter to write to Mr. Horneman to inform him that you have been informed /by some person/ that Gudrun Johnsten frequently visits Mr Savignac here and she will not have the hardihood to deny it, for I know the fact to be true tho' I do [not] wish to create myself Enemies by having my name known. – It is singular with most of mankind that no sooner are they relieved from their distresses than they forget their duty. – I still hope you will never withdraw your countenance from the woman but that your fatherly admonition through Mr. Horneman may induce her to consult her own Interest and Reputation.

I am Sir
with great Respect
Your most humble & ob*edient* Servant
[*Signed*] Jorgen Jorgensen

Wisconsin, Banks Papers, ff. 189–190. Autograph letter.[3]

---

[1] Until now Jörgensen has always spelt her last name Johnsen, by which she is historically known.
[2] James Savignac, employee of Phelps & Co. and an active participant in the Icelandic Revolution.
[3] Franked: 'Fleet unpaid.'

277. Hans Frederik Horneman to Banks, 26 October 1813

London

Right Honourable Sir,

I should ere now have informed you of the result of my interview with the Earl of Clancarty, relative to the granting a new Licence for Mr Lambertsen, instead of the one which became useless by the loss of the *Resolution* on the coast of Iceland.[1] But the loss of our principal clerk & the illness of our second, added to which an inflamation in my eyes & a commi*ssio*n of forwarding about 40 Danish Invalids to their respective homes, have so unsettled me, that I had it not in my power to fulfil my Duty in a way satisfactory to myself, till now. –

His Lordship received me with his usual affability, & I conceived some hopes that my request would be granted on a fresh Memorial being presented. The following is a Copy of the answer obtained:

'My Lords have great reluctance in admitting additional vessels into the Iceland trade. Nevertheless if Mr Lambertsen is a resident inhabitant in Iceland & the fact of the *Resolution* having belonged to him, & having been wrecked as asserted shall be attested by the Certificate of Mr Parke, H. M. Consul, My Lords will take the subject into their consideration.'[2] –

This answer was forwarded to Mr Lambertsen, & I presume he is trying to comply with the conditions, where I shall again have the honour of informing you of the progress made. – Under the circumstances, I fear, there is no hope of success in an application for a Licence for Mr Svend Sivertsen,[3] who wishes to purchase a vessel & to become a regular trader to Iceland, & I have also informed him of the result of my application for Mr Lambertsen. I have informed the Danish Board of Commerce of the short supplies this year, with a view to their looking to, that supplies may be forwarded as early as possible next Spring,[4] & I am sorry to find, from information obtained from various quarters, that a dread of famine[5] is entertained in some parts of Iceland. – I shall be most gratified in learning that the few vessels, which left Leith lately, have arrived safe at their destination.

The *Orion* is arrived at Leith, after having been obliged to put into a port in Norway. – The Cargo is stated to have been much damaged & therefore sold on the spot – But M^r [Bjarni] Sivertsen will no doubt give you every information in his power, at the same time he will, I trust, not fail in his proof of /the/ necessity of fetching a Norwegian port which I think he owes himself & the general character of the Icelandic merchants.

Mr Clausen is daily expected in England, to endeavour to make arrangements for the next year.[6]

---

[1] Rooke & Horneman, on behalf of Lambertsen, petitioned for permission to purchase a vessel in a Danish port because of the loss of this ship at Eyrarbakki (TNA, BT 6/210, nos. 53,852 and 54,006). Finally in December the petition was granted (TNA, BT 6/210, no. 54,708).

[2] This is the text in TNA, BT 6/210, no. 54,006.

[3] Sveinn Sigurðsson from Eyrarbakki, factor to Niels Lambertsen.

[4] Horneman wrote a letter to the Kommercekollegiet (the Danish Board of Trade) from London on 19 October 1813 (RA, Kommercekollegiet, box 1753).

[5] In fact the population declined slightly in 1813 and again in 1814 (Jónsson & Magnússon, *Hagskinna*, p. 56).

[6] Holger Peter Clausen was the chief spokesman for the Iceland merchants. He wrote the petitions concerning the exportation of klipfish and fish-liver oil.

I am /not/ in possession of any intelligence, of a nature which I can depend upon, relative to the peaceable disposition of the Court of Copenhagen, or rather, relative to overtures, to the Allies, for of the disposition for peace with G*rea*t Britain, I entertain no doubt. –

I have, as far as my eyes would permit it, read the production of Mr Jörgensen & was going to return it /to him/ tho' only partly perused, agreeably to your request; when I accidentally found my name introduced. – He states, I should have visited the *Bahama* [prison ship] while he was a prisoner on board, under various pretences, & for the sake of preventing him going on his parole to Reading, advised him thro' a fellow prisoner, to make his escape, adding, that I knew he would be allowed to go on his parole in a few days. – To this statement he adds assertions & reflections suited to his situation, & which, I doubt not, would be deemed libellous in a Court of law.

In this stage of the business, I must confine myself with, solemnly avowing, that I did not know Mr Jörgensen would receive permission to go on his parole, & that I never entertained a thought of giving him, directly or indirectly, any advise whatsoever. – I have never indulged in further oppressing a man, so heavily oppressed already by misfortunes, but uniformly avoided talking of a subject so odious as his Revolution in Iceland. – But, where the subject has been pressed upon me, I have not hesitated to give my opinion freely. – This is generaly looked upon as a great crime in Mr Jörgensens eyes, as he still aims at proving perfect innocence. There are some other circumstances, which have made me an object of his jealousy & hatred. – I am Count Trampe's friend, I am protected by you, & I am – a Dane –

Let my conduct towards Mr Jörgensen be the most inoffensive or the most humane, it would still be in vain for me to expect to avoid totally his malicious attacks. –

I owe it to myself, in order to obtain a reputation of so foul aspersions, to request the favour of you to allow me to keep the manuscript of Mr Jörgensen, and, if the paragraph is deemed libellous, to bring it before a Court of justice; but, more especially, to convince my friends that the libel is founded in falsehood –

If you should object to my keeping possession of the manuscript, I shall certainly immediately send it to him, as I conceive to have no right to keep it back. – In this case I must wait till he endeavours to publish it.[1] –

Allow me to assure you, that I feel grateful to you for your kind attention to me on this occasion, in thus making me acquainted with Mr Jörgensen's accusations. – If you should never have read these, I am still under obligation to you in rejecting to possess a document so teeming with reproaches towards my native Country & its Government. –

Mrs Horneman returns her sincere thanks, for the truly agreeable day she spent at Spring Grove.[2] – In this I join as well as in respectful regards to Lady Banks, Miss Banck[3] & yourself –

With never ceasing sentiments of respect & attachment I remain

Right Honourable Sir,
Your obliged & obe*dien*t Ser*van*t.
[*Signed*] H. F. Horneman

Wisconsin, Banks Papers, ff. 143–144. Autograph letter.

---

[1] Jörgensen never published an account of the Icelandic Revolution.
[2] Spring Grove was Banks's residence in Middlesex.
[3] Sarah Sophia Banks, Banks's only sibling, who always resided with him.

278. Corbett, Borthwick & Co. to Banks, 9 November 1813

Leith

Sir

As we have lately heard, that some doubts have been entertained at the Board of Trade, whether the Iceland Merch*an*ts have been sufficiently anxious to supply that Island with provisions upon the Licences granted by the Hon*ourable* Board for that purpose, and aware, that after the unceasing attention paid by you to obtain for the Icelanders every possible indulgence, it is their duty to remove any suspicions as to the abuse of the Licences, and more especially to convince you, that they have not been careless in the performance of their part of the obligation to supply their countrymen with provisions,[1] after your kind exertions had given them the means, we now use the freedom to trouble you with a few remarks upon that subject.

Since the Order in Council[2] was issued, in favour of the Inhabitants of Iceland, Greenland, & Faroe, We have been Agents here, for all the vessels to which British Licences have been granted, and have had an opportunity of knowing most minutely, how these vessels have been employed during the whole time of their voyages, and we have much pleasure in being able most positively to assure you, that they could not /easily/ have deviated without our knowledge from the course pointed out in the Licences, and that we have every reason to believe, that the Iceland Merch*an*ts have used their Licences, in more strict conformity with the intentions of the Hon*ourable* Board of Trade, than has been the case with Licences granted for any other purpose whatever, and in no one instance has any ship licenced to trade with Iceland, Greenland, & Faroe either upon the voyages to or from these settlements, willingly or unnecessarily, deviated from the direct course allowed in their Licences. –

In the Year 1810 the *Orion*, *Regina* and *Eliza* were detained by one of our Cruizers, & carried into Yarmouth, the Captors alledging that they were not in the direct course to Leith, we had however the means of shewing most decidedly that the cause of detention was unfounded, and they were all restored. Since that time the *Bildal* was in like manner carried into Yarmouth on her voyage to Leith, but the proof of her being in her fair & direct course was so evident, that the agents for the Captors, was anxious to get her liberated as soon as possible – last year the Greenland ship *Hvalfisken* was detained & carried into Gottenb*ur*g, but the cause of detention seemed so unfounded, that contrary to usual practice, when the Captain takes back his papers from the Captors Sir W*illia*m Scott[3] has allowed a monition to be served upon Captain Cathcart, to answer in our claim for damages for unlawful detention. – The frequent and glaring abuses that have been practiced with other Licences probably caused their detentions, and the repeal of them may have influenced their Lordships in doubting as to the honorable use of the Iceland

---

[1] Consul Parke, for example, had written to both the Marquis Wellesley and Banks, describing the current situation and stressing the inadequate state of provisions in the island. The ships had brought little grain, but had carried off a great deal of dried fish. Some of the Danish ships had actually come in ballast; others with small cargoes, and 'not one with a sufficient supply for the district dependent upon his establishment' (TNA, FO 40/2, BT 1/66, Parke to Wellesley, 16 May 1812).

[2] The Order in Council of 7 February 1810 (Appendix 10).

[3] The King's Advocate.

Licences, but in no case has the Admiralty Court, found any, the slightest cause for a condemnation with any of the Vessells trading to Iceland Greenland & Faroe. –

With regard to the deficiency of supplies, sent to Iceland this year, and the doubts whither sufficient exertions were made by the Merch*an*ts to whom Licences were granted, we beg leave to state, that altho it is evident that the Hon*ourable* Board of Trade did not grant their Licences with any view to the Interest or Profit of the Individual Merch*an*ts to whom they were granted, but merely from a wish to alleviate the sufferings of the Inhabitants of these distant and inoffensive settlements, yet there is no doubt, that their Lordships did not expect these individuals, to supply their countrymen with provisions, under an evident risk of their own property and welfare, now in the spring of last year, our Cruizers upon the Gottenburg station, detained almost without exception, every vessel sailing with grain from Denmark, under British Licence, and carried them to Gottenburg, where the Swedish Government declared, that no grain once in a Swedish port could again go out, and therefore all these cargoes must be sold; the *Hvalfisken* was detained altho' her Licence had 4 to 5 Months to run, and the first Iceland vessell that sailed, the *Regina* was also detained, & carried into Gottenb*ur*g, and the agents for the Captors sold the Cargo of Grain, altho upon representation that it was intended for Iceland /the Swedish Government/ granted a special permission for the reshipment of it, should the ship & cargo be liberated in the English court; the ship & cargo were liberated, but the Cargo being sold, the vessell came to Leith in ballast, & then got another Cargo of Rye for Iceland.[1]

When these accounts reached Copenhagen it seemed so certain, that any Iceland ship sailing from Copenhagen with Grain, would be carried into Gotten*bur*g, that the Iceland Merch*an*ts found they would in sending out their ships expose themselves to great risk, and even with this risk would have a very small chance of getting the Grain to Iceland, it was therefore most natural for them to apply to the English Government, for more certain protection, in place of sending out the ships under a conviction, that the Grain shipped would be sold in Gottenburg, the Hon*ourable* Board of Trade & the Admiralty, granted most handsomely, all that was asked but during the correspondence, from the state of the communication betwixt the two countries, so much time was lost, that many of the ships have been unable to complete their voyages, and others foreseeing the risk of this, did not sail, that the wish to supply Iceland, with grain, was a strong motive /*with*/ all the Merch*an*ts to get their Ship out, we cannot doubt, but in addition to this, their own Interest was so much connected with the completion of the Voyage, that the loss of the whole outfit of the vessell and of the profit of a Years trade, depended upon it, and consequently any stop must have been a most serious disappointment; and with regard to the *Orion* being in Norway, we have the fullest proof that it was caused by the Perils of the Sea, and we are sorry to say, that it will cause a very considerable loss to Mr. [Bjarni]

---

[1] On 26 April the *Regina*, carrying grain from Copenhagen to Leith, was detained by the British cruiser *Ulysses*. Her licence had just expired, but instead of being sent to a British port for adjudication (where a speedy liberation would have been the likely conclusion) the *Regina* was taken to Gothenburg, where a convention of March 1813 had given the British the right of entrepôt. A further 6 to 8 grain-carrying Iceland ships were later captured. After some months, when the Swedes belatedly realized the grain was intended for Iceland, they granted special permission for its reshipment, provided the ship and cargo were liberated by an English court. This case spawned a voluminous correspondence mostly to be found in BT 6/210, FO 73/85 and FO 188/4–7. See further Agnarsdóttir, 'Great Britain and Iceland', pp. 64–5.

Sivertsen the owner, besides preventing him from getting to this country any of the Iceland goods he has laying since last year, and so anxious was he to get there, that he left Norway without the ship being fully repaired, and immediately upon arrival here, consulted with all the Iceland Cap*tain*s then here, as to the possibility of going on to Iceland, the risk however seemed so great of arriving there, the shortest day of the Year, that it was thought more adviseable to remain here, and sail as early in Spring as possible, it is well known how much the poor Norwegians were in want of Provisions, when the *Orion* lay in the harbour there, and how great the temptation was, to retain even by force this Cargo of Grain, but after selling what was entirely damaged by the leak, all that was in good order was reshipped, and is now on board the Vessell in our harbour, besides Mr. Sivertsen had the honour of being Knighted by the King of Denmark,[1] for his services in provisioning Iceland, and was appointed this Year, one of Commission to regulate the state of the Paper Currency in Iceland,[2] so that he was particularly anxious to be there – under these circumstances we are sure you will admit that no Suspicions can attach to him, as to the causes of his being in Norway – We trust you will excuse our troubling you with this letter, as we were anxious to shew that our Iceland friends were not unworthy of the Kindness you have shewn them.

We have the honour to be
your most *obedient* Serva*nt*[3]
[*Signed*] Corbett, Borthwick & *C*ompa*n*y

Wisconsin, Banks Papers, ff. 122–123. Autograph letter.[4]

279. Holger Peter Clausen to Banks, 12 November 1813

London

Right Honorable Sir!
The circumstances as yet would not permit our Government officially to confirm the power of attorney I am entrusted with by the Icela*n*d, Greenla*n*d & Færoe merchants, still I am here with the consent of our King and forwarded by a Flag of Truce to Helgoland.[5] – I am charged with the agreeable duty to assure you of the due veneration felt for Your generousity, and I am not able to do it better than by a translation of the very words, in a private letter to me from our Minister in the Foreign Department viz.:

---

[1] Bjarni Sivertsen became a knight (*Ridder*) of the Dannebrog on 11 April 1812.

[2] Inflation had hit Iceland hard. On 19 May 1813 the king appointed this commission, 'Commissorium ... ang. Forberedelse til en Lov om Pengeforandringen i Island' (Committee to prepare a law regarding the change of currency in Iceland) (*Lovsamling for Island*, VII, pp. 468, 472–4).

[3] This is a scrawl, there is little room left on the sheet of paper.

[4] Note in Banks's autograph on left margin of first page: 'Corbet[t] & Borthwick [Received] Nove*mber* 14 [18]13'.

[5] Helgoland (Heligoland), an island of 0.2 sq miles (0.6 km²) situated in the North Sea near the mouth of the Elbe, was a Danish possession from 1714 to 1814, when at the Peace of Kiel it was transferred to Great Britain. It had been seized by the British at the beginning of hostilities in 1807 and, despite its size, was of immense strategic importance and a centre for smuggling British goods into the Continent.

'I need not to tell you that we entirely rely on the continuation of Sir Joseph Banks's generous support to all which may tend to supply the Inhabitants in the Establishments of Iceland, Greenland & the Færoes. – You will know to assure Sir Joseph Banks that the high esteem, every dane, who knows his great merits of the sciences & his humanity, feels for him, is greatly augmented by the care he shew's for the relief of the inhabitants of these remote Establishments, who left to the consequences of unhappy circumstances, are worth the compassion of a humane Government.'

Our Government regrets very much the circumstances which obstructed the long wished-for peace with this country, and I am permitted, on the authority of our Minister in the Foreign Department (:the Count of Rosenkrantz:)[1] to say that our King was inclined to have made very considerable sacrifices if the cedation of Norway had been left out of the question.[2]

In hope that occurrences may take place to the accomplishment of this my most ardent wish, I do myself the Honor to subscribe me

Right Honorable Sir

Your most obedient & most humble Servant

[*Signed*] Holg^r P. Clausen jun.

Wisconsin, Banks Papers, ff. 153–155. Autograph letter.

280. Holger Peter Clausen to Banks, 12 November 1813

London

Right Honorable Sir!

According to Your kind permission I do myself the honor to lay before You a correct statement of my own trade to Iceland last year, as I consider this the truest way to answer that part of Mr Parkes report concerning the comparative prices of Iceland produce in Iceland & Denmark,[3] which I hope will convince You that the profit can not be less, considering that the merchants often are subject to lose their ships in these boisterous seas in a late season.

---

[1] Count Niels Rosenkrantz (1757–1824), Danish statesman and diplomat. At this point in his illustrious career he was the Minister for Foreign Affairs.

[2] On 14 January 1814 Denmark and Sweden were to sign a peace treaty by which Denmark ceded Norway to Sweden. The Dano-Swedish treaty was a precondition for the Anglo-Danish treaty signed the same day. By rights the North Atlantic dependencies, which were historically dependencies of the King of Norway, should have accompanied Norway. However, article 4 of the treaty stated baldly that Greenland, the Faroe Islands and Iceland were not included, quite possibly due to British pressure. The loss of Norway was a great blow to Denmark. Rosenkrantz, the Danish foreign minister, noted in the diary that he kept at the Congress of Vienna that: 'Lord Clancarty ... me dit que la Norvège auroit pu être sauvée comme l'Islande' (Lord Clancarty told me that Norway could have been saved as was Iceland) (Rosenkrantz, *Journal du Congrès*, p. 101). Clancarty was one of the British negotiators. This suggests that the English had had a hand in saving Iceland from the Swedes, which is not unlikely as it would clearly not have been in Britain's interests for Sweden to rule both the Scandinavian peninsula and the Atlantic islands. The cession of Norway to Sweden had been on the cards during the negotiations for some time.

[3] Parke was engaged in competition with the Danish Iceland merchants and naturally enough supported the British trade. He sent a lengthy report on the situation in Iceland on 20 November 1811 and further reports after that (in TNA, FO 40/2).

Mr Parkes report, would certainly have been more compleat if the state of the Exchange, at the time these prices of Iceland produce are given, were added, and it seems very singular that Mr Parke has ommitted this, considering that he attributes the higher prices given for Iceland merchandize in Iceland in the year 1812, to the worse state of the Danish paper money, which proves that this has not escaped his attention.

I find myself under the necessity to contradict the prices Mr Parke states, as they not have been [*sic*] obtained by any of the Iceland merchants in the year 1812. – My latest ship from Iceland (the *Gratiarne*) not arriving at Copenhagen before February 1813, the cargo was not sold before March, and I assure on my honor that I did /not/ get higher prices than those mentioned in the subjoined statement, and even these were the highest then obtained by any of the Iceland merchants until that time. In the same degree as our money depreciated, the goods rose, and those prices quoted by Mr Parke were not paid before June 1813, but then our exchange was about 200 dollars[1] p^r. £: and even these nominal higher prices were not obtained by any of the Iceland merchants as they were obliged to sell their cargoes, at the arrival of the ships to Copenhagen in order to be able to purchase the grain cargoes for Iceland. –

Thus Mr Parke confounds the prices of 1812 with those of 1813 which causes the apparent great advantage. – Goods have since that been much higher; but our exchange worse in the same proportion, even so low that there have been paid 400 D*ollars* for one £ sterl*ing*.

By making a calculation on the advantage the Iceland trade affords, there must be considered, 1, The merchants lose about 150 to 200 p^r C*ent* on the goods they carry to Iceland. 2, The money has in the interval the ships are fitted out and until their return, depreciated about 100 p^r C*ent*. 3, The merchants take in payment all kinds of goods the Icelanders offer and are many articles by far less profitable than Fish and oil. – 4, This trade is connected with great risk, trouble and large expences as the ships must touch Scotland twice on one voyage.

I do not see what loss it causes Iceland, tho the merchants in return for their great trouble & risk, make some fortune by the trade, it appears to me as if this might augment the hope of the Icelanders to be so much the better supplied. – Tho' the Iceland merchants gratefully acknowledge, that the trade on Iceland has been as well advantageous to them, as it has been the means to supply the inhabitants, I must on their behalf regret its not being so lucrative as Mr Parke represents it. – In this case there would be some contradiction in Mr Parke's report, when he seems to believe that the great profit should induce the merchants to drop such an advantageous trade, as I should rather thing[2] to be an encouragement for them to continue it, and which indeed is the case, as they have done every thing in their power to get the ships sent to Iceland this year. – Allready in the Month of May, the ships were ready for sea when the information reached Copenhagen that all ships without exception were detained and sent to Shwedish ports, and that the two only vessels which then had sailed (:the *Hvalfisken* for Greenland & *Regina* for Iceland:) had been sent to Gothenburgh. – Now the merchants took alarm, as it was of no use to Iceland that the ships were brought into Shwedish ports and tho' liberated, the grain sold there, they /therefore/ did not venture to dispatch their ships.

---

[1] *Rigsdaler* or rixdollar.
[2] A slip of the pen for *think*.

Through Your benevolent exertions it became possible to dispatch the Ships with safety, but this information we got by the last mail from Sweden before the stoppage of the regular intercourse between Denmark & that country took place. – In order to be convinced if regulations to that purpose were received by the Brittish Admiral, I went with a Flag of Truce to the Cattegat,[1] then to the Belt[2] where I saw Admiral Moore.[3] – He directed me to Admiral Hope at Rostock,[4] who provided us with certificates annexed to the Licences, according to which all vessels who in that late season could have any hope to reach their harbours, departed. – But some vessels destined for some of the northern & eastern ports in Iceland, where the harbours are inaccessible so late in the year were obliged to defer their voyage this year.

This is the true reason why not all the Iceland vessels are gone off, but not the fault of the merchants who would have been extremely glad, if it had been in their power to dispatch them.

I cannot ommit to assert on my honor that not a single of the Iceland vessels ever attempted to go to Norway, with the view to discharge any part of their cargo; and in respect to the charge made them of not being well fitted out for the voyage, I hope you will consider it as a proof of the reverse when I assure You that there only has been one or two instances in which the underwriters in Scotland have paid any insurance, tho' the Iceland ships frequently perform a great part of their voyage in the winter.

I venture to hope that I have explained the reasons why so many of the Iceland vessels are not sent to Iceland this year, as also that the merchants do not make any unreasonable profit by the trade, to your satisfaction, and beg /leave/ to observe that even English merchants (:Messrs Everth & Hilton:) who have traded there, and in which trade Mr Parke has been concerned complain over loss.

With the profoundest esteem I remain
Right Honorable Sir
Your most obedient & most humble Servant
[*Signed*] Holg^r P. Clausen jr.

Wisconsin, Banks Papers, ff. 124–125. Autograph letter.

281. William Horne to Sir John Thomas Stanley, [undated, early January 1814]

In the Event of a continuance of the War with Denmark, it would be highly desirable to suspend altogether the System of licensing danish Vessels to trade with Iceland & Faroe. it is unpolitick to continue them, because in many Instances, Cargoes of Grain <u>intended for the Relief of Iceland have been landed on the Coast of Norway</u>, and thus render'd the

---

[1] See Document 274, p. 558, n. 7.

[2] The Belt (*Storebælt* in Danish – the Great Belt) is a strait between Fyn and Zealand.

[3] Admiral Sir Graham Moore, GCB (1764–1843). He escorted the Portuguese royal family to Brazil in 1807 and served in the Walcheren expedition in 1809. He was the brother of General Sir John Moore and eventually became vice admiral of the Blue. At this point in his distinguished career he was serving in the North Sea Fleet, recently promoted to rear admiral in 1812.

[4] Rear Admiral Sir George Johnstone Hope (1767–1818): see Document 274, p. 558, n. 3.

Supply precarious and inadequate,[1] & enabled the danish Merchants to oppress the Natives of Iceland, by raising the Price to an enormous & unprecedented height. It is unjust to the mercantile Interests of this Country to license the Vessels of our Enemies, when British Merchants would employ their Ships in carrying on the Intercourse with Iceland & Faroe. If Opportunity is permitted <u>We will engage to supply these Islands with all the Corn wanted and we estimate to be about nine thousand Barrels Annually.</u> If the System of licensing is persever'd in we shall scarcely have an inducement to continue the Trade, on Account of the Jealousies of the danish Merchants in Iceland, who having an Interest in Vessels frequenting their Ports, not only decline trading with the English Ships, but prohibit the Natives of ever bartering their Produce for our Corn &c. at the Capital the Physician Klog[2] was influenced by the danish traders to denounce our Coffee as injurious to health, and a Conspiracy was detected at the same place (Rykavyk) for cutting the Cable of our Vessel. [*blank*] & at the Westman Islands an <u>Order</u> was made publick from a Mr. Petreus,[3] <u>prohibiting all Intercourse</u> with English Vessels, and so hostile were the Danes that we were obliged to provide our people with Arms to be prepared against their Violence.

It is now, however probable that Peace will be concluded with Denmark. & we hope that Government will not neglect those who during the War have contributed to feed the Icelanders. The near Approach of Peace renders this Matter of peculiar Importance to us. We are therefore desirous of soliciting from Government <u>a clause in the commercial Treaty with Denmark</u>, permitting British Vessels to resort to Iceland & Faroe, with the same privileges as those of the King of Denmark. We do <u>hope Government will insist on this point.</u>[4] it will be an Act of Justice to individuals, promote the Commerce of Great Britain, and ensure many Advantages to an innocent & very oppress'd race of Beings. but I wish Government would be induced to solicit Iceland a Boon or <u>take possession /of/ it</u>,[5] which might be done, without the hazard of a single Life. The little Value which has hitherto been attached to it, is more owing to the injudicious Policy with which it has been govern'd than to the Poverty of the Soil or Character of its Inhabitants. By the Improvement of its natural Advantages under the fostering Influence of the British Government the Amount of its Imports & Exports would be considerably raised. and so much so as to render the Island politically & commercially important. the productions of the two Countries are such as to render the Exchange of them mutually advantageous. consisting on our part of manufactured Goods, colonial produce, Salt & Rye. in return for Tallow, Wool, Brimstone, Skins, Oil, Salted Mutton, Salted & dried Fish &c.

The Seas on the Northwest coast abound with a Species of large Shark which yields a large proportion of Oil of Superior Quality, and there is reason to believe that a very beneficial fishery of this kind might be established. the coast is also visited by immense

---

[1] This practice is certainly often mentioned in the sources.

[2] Thomas Klog was the state physician in Iceland 1804–15.

[3] Westy Petræus.

[4] Trade was not mentioned in the treaty, but after the war the Danish government acknowledged the contribution of the English trade in provisioning the island during the hostilities by granting the British merchants trading permits for a couple of years. In September 1816 the British trade was brought to an end by a Danish ordinance (see Document 292). See further Agnarsdóttir, 'The Challenge of War on Maritime Trade in the North Atlantic: The Case of the British Trade to Iceland During the Napoleonic Wars', pp. 254–7.

[5] Yet again the question of annexing Iceland is raised.

Shoals of Herrings. these Advantages have hitherto been almost neglected. Partly from the Natives not having sufficient Means. & partly from the prohibition of foreigners, from all Intercourse with the Trade of the Island. Were Iceland an Appendage of *Great* Britain, <u>this Country</u> would <u>possess</u> nearly all the range of Coast where the <u>Fishery of Herrings could be carried on with Advantage</u> & need not fear any Competition and would Secure to herself the Monopoly of this trade. which perhaps of all others is the best Nursery for Seamen. the Natives are extremely well disposed to a Union with us.[1]

My Information respecting the Fishery of 1813 is confined to the Westman Islands & the Neighbourhood of Rykavyk, where it was <u>very unproductive</u>.[2] the Consul[3] may have ascertain'd the Produce of the fishery at the Northern Ports.

The Rye on board our Vessel was imported from Russia.[4] our Application for Licence to export Corn of British growth was occasion'd by the <u>Material</u> difference in price.

I have troubled you with the preceding Details, as the Moment is critical, and that they may be consider'd by You & Sir Joseph Banks. if he approves I shall anticipate a favorable Result as his Opinion is taken by Government on all Matters relating to the Islands.[5]

Wisconsin, Banks Papers, ff. 158–160. Contemporary copy.

282. Sir John Thomas Stanley to Banks, 13 January 1814

Winnington [Cheshire]

Dear Sir Joseph

I cannot answer your letter better than by sending you a long Extract from a Letter I received lately from Mr. Horne.[6] but first let me inform you that the Consul in Iceland has no Connection whatever with his Adventure.[7] I rather think indeed they have not agreed in Opinion in many things relating to the Trade, certainly not with respect to the productiveness of the Fisheries last year. –

Messrs. Horne & Stackhouse purchased foreign Corn when they found it impossible to export the produce of our Islands, and their Vessel saild with her Cargo for the Faroe Islands & Iceland, the day before yesterday.[8] You must own they are not wanting in Spirit in trusting their Property to /the/ such Seas at such a Season,[9] but I really believe they

---

[1] That was certainly Banks's own opinion: see e.g. Document 38.

[2] This is borne out by Icelandic sources: see e.g. Gudmundsson, *Annáll 19. aldar*, p. 169.

[3] John Parke.

[4] As was Phelps's.

[5] Eventually Horne & Stackhouse were granted permission by the Danish government to make several voyages to Iceland to close their accounts. See further p. 519, n. 3, and p. 575, n. 2 below. They also traded with the Faroe Islands and Greenland. Their principal agent was James Robb, who arrived in Reykjavík in 1813, and eventually settled in Iceland.

[6] See previous letter.

[7] 'Adventure' here means commercial enterprise.

[8] *The Sir John Thomas Stanley* was sent yet again by Horne & Stackhouse, with J. Dunstone as captain. It sailed from Liverpool to Iceland (and the Faroe Islands) on 13 January 1814, according to *The Liverpool Mercury* (21 January 1814, p. 239) though the newspaper does not mention the Faroe Islands, while this letter, a more reliable source, gives 11 January 1814. The ship carried foreign corn for Iceland and the Faroe Islands.

[9] The British were more intrepid than the Danes and Icelanders, viz. the *Clarence* sailing off in December and this ship in January. The Iceland merchants usually did not set sail until April.

have been in part stimulated, by a benevolent principle. The Accounts given to them of the Sufferings of the Icelanders were most distressing.¹

This is horrid Weather, and we are burning Coals at the Rate of a Ton a Day in the house at least. We do not appear however to have had as much Snow as you in the South. Yesterday was the worst day we have yet had. The Snow was incessant & it flew almost parallel to the ground before a fierce East Wind. it must have been drifted in places to great depths & I dread hearing Accounts of Accidents. The Cold though of long Continuance has not been very intense and we have not had above two days of Fog. and this not of the same Nature as that of London. The freezing of our Navigation & the Want of Coals have stopp'd our Salt Trade, but we expect a good year, provided the /*Custom house*/ Board of Customs abandon an Attempt made to extract a heavy Duty from the Export of the Article. We could not stand a Competition with foreign Salt, if ours was to be thrown up in price by home taxation. Surely Government gets enough by taking a Tenth of the Profits of the Trade.

I hear that Skeletons of Crocodiles have been found in our Chalk hills² and Skeletons of man in old form'd Mountains in the West Indies. do not your Philosophers want a little more knowledge of the history of the World than they have hitherto been satisfied with to account for such antidiluvian Exercise. Lady Stanley³ who came down Stairs to day for the first time since her Confinement,⁴ begs me to remember her most kindly to you & you will present both our Remembrances to your Ladies. We hope you continue to get health and believe me ever dear Sir Joseph

most sincerely yours
[*Signed*] John Thoˢ Stanley

Wisconsin, Banks Papers, f. 157. Autograph letter.

283. Hans Frederik Horneman to Banks, 25 January 1814

Queen Street, Cheapside

Right Honourable Sir,
I have had the honour to receive your letter of yesterday and of today. Mʳ [Bjarni] Sivertsen has offered Miss Gudrun Johnsen a free passage from Leith to Iceland at my particular request.⁵ He expects to sail in the middle of next month, & goes to Ryekevig direct. – She will follow your directions, & shall be provided with every thing needful for her journey

---

¹ The population of Iceland had declined. In 1813 the population was 47,805, in 1814 47,501. This is not a significant difference but in the years before the war the population had been slowly increasing (Jónsson and Magnússon, *Hagskinna*, p. 56).
² Stanley is possibly referring to an item in *The Annual Register* which stated: 'A few days ago, immediately after the late high tide, there was discovered under the cliffs between Lyme Regis and Charmouth, the complete petrification of a crocodile, seventeen feet in length, in an imperfect state' (*Chronicle*, 14 November 1812, p. 142). Banks while touring England in 1771 saw the fossilized bones of 'an Elephant found bedded in Ocre on the Mendip hills' (quoted by Beaglehole, 'Introduction', in *Endeavour Journal of Joseph Banks*, I, p. 17).
³ Lady Stanley was Maria Josepha Holroyd (1771–1863), daughter of the 1st Earl of Sheffield.
⁴ Lady Stanley gave birth to her twelfth and last child, a daugher, Octavia, late in 1813 (Adeane, *The Early Married Life*, p. 339).
⁵ This would have been on the *Orion* (NAS, E. 504/22/63).

to Leith. – Until the weather changes she does not seem inclined to depart, which is reasonable. I fear she will even then have some objection, for she appears to me to be fainthearted about her Icelandic journey. –

For the intelligence contained in your letter of this day I am truly grateful to you. – Peace was become so much a measure of necessity to Denmark, that even with the loss of Norway it is still a very desirable event.[1] – I have communicated this night the wellcome intelligence to all the Depots at which Danish prisoners of war are kept, & I figure to myself the happiness those poor captives must experience, some of which have been in their unhappy abodes for six years and upwards. – If I observe any downcast looks, it is with the brave Norwegians who are here, & who, in the moment of parting, feel thoroughly how intimately both nations were united, by custom, by kindred, by private friendship. One religion, one language. – In patience & loyalty equal. No wish ever entered their minds of independence or change of Government or master[2] – The treaty of Bergen, where an eternal union with Denmark was solemnly entered into, is now dissolved, tho' not voluntarily; and I sincerely wish Norway may be treated leniently by its new Government. – That this event will ultimately lead to an union of the three Crowns for their common safety, is my individual opinion.[3]

To Mr Clausen I have also communicated the news of peace with Denmark – By a letter from him I find, that there is no possibility of getting in to the Sound until the weather changes. –

What a gratifying reflection must it be to your benevolent mind, after so many years toil, at last to find yourself at the end of your labours, willingly, oh, and cheerfully bestowed for the sole purpose of saving an innocent people from destruction. – Sir, the palm that you have won will never wither, it will even adorn you where all other earthly possessions & acquirements sink into objects of comparative indifference, and thousand of voices will offer up their prayers to the omnipotent for your eternal happiness. –

I have the honour to be
Right Honourable Sir
Your obliged & attached Servant
[Signed] H. F. Horneman

Wisconsin, Banks Papers, ff. 161–162. Autograph letter.

---

[1] At the Peace of Kiel in January 1814 Denmark lost Norway to Sweden.

[2] Horneman was of course a Dane, but these sentiments are not unlike those expressed by Magnús Stephensen insisting the Icelanders were happy under the sceptre of the King of Denmark (see Stephensen's letter, Document 240).

[3] Horneman was wrong, though a personal union was soon formed between Norway and Sweden. At the Peace of Kiel in 1814 Denmark, in order to avoid the threatened occupation of Jutland, ceded Norway to Sweden. The three North Atlantic dependencies, however, remained under Danish rule. On 17 May 1814 the Norwegians proclaimed their independence at Eidsvoll with Christian Frederik, hereditary prince of Denmark, as King. Frederik VI had only two surviving children, both daughters, and Denmark had the Salic Law. The Swedish government did not take kindly to this and after a short conflict the Union of Sweden and Norway was proclaimed. Charles XIII (1748–1818), second son of King Adolf Frederick of Sweden and nephew of Frederick the Great, became King of Norway from 1814 until his death in 1818. In the absence of a natural heir, Jean Baptiste Bernadotte, a former marshal of Napoleon, was elected the heir-presumptive to the Swedish throne and adopted by Charles XIII. He succeeded him as King Charles XIV John (Karl XIV Johan) in 1818.

284. Banks to Horne & Stackhouse, 25 April 1814

I Thank you Gentlemen for The opportunity you have given to me of taking Leave of my old Friend Stephensen[1] & profit by it in Sending a Letter Enclosd which I beg of you to Forward to him. I wish I dare hope the Iceland Trade might in future be Left unconfind & that you would find it worth Continuing[2] but I much Fear that the illiberal & impolite Conduct of Denmark has not yet met with any amelioration.

I am Gentlemen
Your obligd H*um*ble Serv*an*t
[*Signed*] Jos: Banks

York, Kaye Collection. Autograph letter.[3]

285. Holger Peter Clausen to Banks, 28 June 1814

Copenhagen

Right Honorable Sir!

I have the honor on behalf of myself and the other Icelandic merchants humbly to present our respectful gratitude for Your humane and kind interference for the inhabitants in Iceland Greenland and the Faroe, without which they would have been unable to outlive the distressing war to which there is now happily put an end.

To express our thankfulness would require an abler pen than mine, even the most skilful writer must in this respect be far behind our feelings – My only comfort is; that your own conviction of the support you have procured the poor inhabitants on those places, for want of which they should have died the most cruel of all deaths, will render so noble a heart as Yours the most sensible joy.

With respect to myself I beg leave to inform you that I have now determined to settle constantly in Copenhagen, as my partner[4] is to perform the travelling business in Iceland.

Since the Year 1807 there has been no British consul appointed in Copenhagen, and [I] venture to aspire to the hope of not being unworthy to that situation. – My connections here as well as my situation in life, I hope, will rather promote than object this wish and if I should be entrusted with the appointment it might perhaps be in my power, through my ardour for the interest of His Britannic Majesty's subjects, to convince you of my gratitude for the benefits Iceland etc. has derived from Your gracious endeavours. – Tho' I am not entirely perfect in the English language, yet I hope to have sufficient knowledge in it, to be of use in the situation I wish for.

---

[1] Banks is not aware that Ólafur Stephensen died in 1812.

[2] The last English ship from Horne & Stackhouse, the *Providence* of 70 tons, left Iceland in 1817. In 1816 the Danish authorities had permitted a continuation of the trade so the merchants could close their accounts. The ship was sent in ballast (the ban on foreign trade was now in force again) to export Icelandic goods. The vessel arrived in Reykjavík on 11 July 1817 and sailed again on 10 August, arriving in distress in the Faroe Islands on 15 September (clearing certificate from the Custom House. Liverpool, dated 7 June 1817 and NAI, Bf. Rvk. OA/7, OA/8, OA/9. Verslun og sjóferðir 1814–18, 10 August 1817, no. 37; TNA, FO 211/12, Savignac to Foster, 1 November 1818).

[3] Note in Banks's autograph: verso: '[Received] April 26 [1814]'.

[4] Knud Schiöth (fl. 1814).

I have desired Mr Horneman to draw a petition and venture to sollicit Your kind support to the accomplishment of my wish, and have the confident hope that the knowledge you have of my character perhaps might induce you to support it.[1]

In this hope I remain
Right Honorable Sir
Your
most obedient and most humble Serv*an*t
[*Signed*] Holg^r P. Clausen[2]

Yale University Library, Joseph Banks, box 11/folder 198. Autograph letter.

286. Corbett, Borthwick & Co. to Banks, 18 March 1815

Leith

Sir

Your exertions in behalf of the inhabitants of Iceland, were in the time of their great distress, so useful, and so willingly offered to them, that we cannot doubt that you still feel, some pleasure in hearing of their connection with this country, and that you will not be unwilling to promote it, if it evidently improves their situation, whilst in so far as its limited extent allows, the trade with Iceland is also an advantageous one for our own Country – it has been represented to us, that it will be extremely difficult to continue a trade with Britain unless a limited quantity at least, of Salted Mutton, can also be shipped so as to make assorted Cargoes, and to prevent the freight being more than the Sheeps wool &c can afford to pay, we have by the request of the Iceland Merch*an*ts drawn up a Petition, but are a little uncertain whether such a Petition comes before the Treasury or Board of Trade[3] – if to the latter we have desired our friends in London to intimate to you when it is presented, and we now use the freedom to inclose a copy, under the hope, that you will still be inclined to give your support, to the Icelanders if you find, that the Interests of Britain are in no way injured and perhaps somewhat benefited by the extension of our trade with Iceland, and that consequently there is no urgent reason to reject such indulgences as they Petition for.

The quantity of Sheep wool, which can be yearly sent from Iceland is pretty considerable, and from what we have seen during the last 4 or 5 Years, we suppose it will at least average /yearly/ £9000 to £10.000 St*erling* in value, and we find that it is useful to our Manufacturers, from its being applicable to some uses, which cannot be supplied by the Wool of our own country, of similar quality, especially from its being free of Tar, or other salves, so that it can be used in the manufacture of light coloured cloths.

[1] There is no documentary evidence extant to show whether Banks put in a word. After the Peace of Kiel full diplomatic relations were re-established, the minister being Sir Augustus John Foster (1780–1848). He served in the legations of Naples and Stockholm before being appointed minister-plenipotentiary in Washington, DC, in January 1811, where he is considered to have done little to prevent the War of 1812. He was an MP 1812–14 and in May 1814 was appointed minister-plenipotentiary at Copenhagen where he remained until 1824. The secretary of legation was Edward Cromwell Disbrowe. After the war H. F. Horneman continued as consul-general in London, while in Leith the consul was Peter Borthwick (a partner in Corbett, Borthwick & Co.).

[2] Note in Banks's autograph, verso: 'Mr Clausen | [Received] July 14 – [18]14.'

[3] The petition is the following document, sent on to the Board of Trade.

You will confer a particular favour on us, as also upon the different Iceland Merchants who have had the honour of your acquaintance and protection, if you can in any way forward the object of the inclosed Petition.

We remain respectfully
Your mo*st* obe*dient* hu*m*ble Ser*vant*s
[*Signed*] Corbett, Bothwick & Co.

SL, Banks Papers, I 1: 75. Autograph letter.

287. Petition from Corbett, Borthwick & Co. to the Lords Commissioners of the Board of Trade, 18 March 1815

Unto the Right Honorable the Lords Commissioners of His Majesty's Board of Trade
The Petition of Corbett, Borthwick & Co. Leith Merchants, Agents for the Merchants in Iceland
Humbly Sheweth
That during the War with Denmark a very considerable Trade was carried on betwixt this Port and the Island of Iceland, in the course of which your Petitioners imported from thence in considerable quantities, Salted Mutton, Sheep's wool, Stockfish, Tallow, Rough Brimstone, Swan Quills, Swan Skins, Fox Skins, Sheep and Lamb Skins; and exported Cotton Goods, Rum, Coffee, Sugar and Tobacco for the consumption of that Island.
That since the Peace your Petitioners find an advantageous Trade could still be continued with Iceland, but as the Salted Mutton[1] was the only heavy Article which could be shipped for Sale in this Country, its being now prohibited renders it difficult, or almost impossible to make up the Cargoes so as to render the freights in any degree moderate, as the other articles are so bulky and light that very little fills up a vessel without heavy goods as ballast.
That the Iceland Merchants represent to your Petitioners that they cannot continue to send their ships to Leith unless Permission can be obtained to import Salted Mutton also, and therefore your Petitioners do most humbly submit to your Lordships consideration whether, all the other Articles being raw materials to be manufactured in this Country, your Lordships might not permit the Importation of at least a limited quantity of Salted Mutton under any restrictions as to exporting an equal value of British Goods or Colonial Produce, that may seem necessary, and beg leave to Petition that in this way, one thousand or fifteen hundred Barrels of Salted Mutton, might be imported yearly. It was allowed during the War duty free, but a reasonable rate of Duty might now be imposed upon it.
Your Petitioners Humbly hope that your Lordships will grant this Petition, and as in duty bound wi*ll* ever pray &c.

[*Signed*] Corbett, Borthwick & Co*mpa*ny

SL, Banks Papers, I 1:76. Autograph letter.

---

[1] During the war export of salted mutton had been permitted because of extra demand for the armies and navies.

288. Circular from Rasmus Christian Rask, 21 September 1815[1]

Leith

For a pretty long time all the Gothic Nations in emulation of one another have taken a peculiar care to preserve the Antiquities of the whole tribe of their ancestors, as being nearly connected with their National glory, and a Material branch of ancient history, very useful to keep up the public spirit and interesting to men of letters; but I wonder how they could overlook the noblest of all those sacred relicks, and carelessly leave it to all the destroying injuries of time. This most valuable remainder of Gothic Antiquity, almost the only one preserved in Iceland, is certainly the ancient general language of all the kingdoms of the North, which is still spoken throughout that Island to a truly astonishing degree of purity and elegance. This I may pretend to ascertain; for, having travelled through the kingdoms of Denmark, Sweden, and parts of Norway, in order to study the languages and the philological antiquities of the North, I have now spent these two years in travelling around the Island of Iceland, to inquire into the present state of that remarkable language; and in every corner of the country I have been able to converse with the natives in the ancient Scandinavian tongue; and I have found them reading still the old Sagas of the heroic age: nay, there are some songs of Edda celebrating the exploits of the heathen deities still understood by every peasant boy, with exception of a few difficult words. But, as the poverty of the people has increased, printing almost is fallen into disuse, and the literature and language are certainly in a declining state; the most valuable productions of wit and learning of later times existing only in Manuscripts, and in danger of being lost for ever, as, for instance, a most excellent poetical translation of Miltons Paradise Lost,[2] of which only the two first books have been published. With a view therefore to preserve the language, literature, and knowledge in the Island, I have invited all lovers of learning in the country to establish a society with annual gifts, to publish such Masterpieces, as are written in later times, and to procure such new books, as are wanting, necessary to keep up and increase learning and useful information in every way of life; and they have subscribed in considerable Number in proportion to the scanty population. But, as the Icelandic or old Scandinavian language is the source of part of the English and Scottish, and besides the Anglo Saxon (the chief Source of both) is so very nearly related

---

[1] Rasmus Christian Rask (1787–1832) was a Danish philologist whose work on Old Norse was pioneering in the field of comparative linguistics. He was the prime mover for the foundation of the Icelandic Literary Society, the aim of which was the preservation of the Icelandic language and culture. The society was founded simultaneously in Copenhagen and Reykjavík in 1816 and Rask became the first president of the Copenhagen branch. The history of the society with an English summary has been written: Líndal, *Hið íslenzka bókmenntafélag*. On his way back to Denmark from Iceland Rask visited Leith where he wrote this petition in English asking for British support. Banks did not join but Sir George Steuart Mackenzie and Ebenezer Henderson both did. The Brothers Grimm were among the foreign members of the society. See further Appendix 13. Rask puts umlauts on all his u's.

[2] Jón Þorláksson at Bægisá (1744–1819), a clergyman, had been defrocked because of fathering an illegitimate child. Reinstated after two years, he had a second child (with the same woman) only two years later. Jón was offered a position as proof-reader at the recently established Hrappsey printing-press and married the daughter of one of the owners of the press. He later became a clergyman in northern Iceland at Bægisá. He was a poet (mostly hymns) but he is best remembered for his translations, mostly from German and Danish. *Paradise Lost* was published at Hrappsey 1774. Among his other translations were Alexander Pope's *An Essay on Man* (*Tilraun um manninn*) and *Messias* by Klopstock.

to it, and in itself so difficult and confused, owing only to the incessant irruptions of the ancient Scandinavians into Great Britain, that, if I may believe my own experience at the compiling of an Anglosaxon Grammar, it will never be sufficiently extricated but through perpetual succour from the Icelandic. I thought the Britons, the most wealthy of all the Gothic Nations, ought not to be altogether unconcerned, about its conservation. Besides, although a Dane myself, I thought it an injury to the poor Icelanders and perhaps to the English generosity and zeal for every kind of knowledge, If I should not invite the British lovers of Northern antiquities and patrons of the poor but most respectable people of Iceland, to partake of this society. Any contribution, great or little, annual or not, will be received with gratitude. The Subscribers will be pleased to write beneath their names, /their/ situation place of residence and the amount and kind of assistance they intend to bestow; and the collectors of Subscribers will please to send back the lists either to the Rev*erend* Mr. Arne Helgason, Minister at the cathedral of Reykiawick, in Iceland, and president of the Icelandic Society,[1] or to me.

[*Signed*] R. Rask
under librarian to the
University of Copenhagen

NHM, DTC 19, ff. 193–195. Contemporary copy.

289. Rasmus Christian Rask to Banks, 25 September 1815

Leith

Right Honorable Sir!

I am sensible, it will appear strange, and I fear uncouth to you, to receive a letter on this from a person of whom you certainly cannot have heard so much as the name; however the character established of you in Iceland and indeed everywhere, makes me presume it. My intention with it is not only to recommend the plan inclosed to your own benevolence against science and against Iceland, but also to beg your assistance in town or in general with your acquaintance if there should be among them some friends to mankind, who would at once patronize the two, in my opinion, very worthy objects mentioned.

I have sent a copy of the plan to Sir John Stanley and to Sir George Mackenzie in Edinburgh. But I am afraid I have troubled you too much already, and so I wish, you may long enjoy the happiness, the greatest certainly in this world, to look back with satisfaction on a noble life of virtue and activity, and to see the seeds of your love of mankind thrive and shoot forth their illustrious fruits everywhere. I am, Right honorable Sir! with the greatest respect

Your most obedient humble servant
[*Signed*] R. Rask

BL, Add MS 35068, f. 135. Autograph letter.
NHM, DTC 19, f. 192. Contemporary copy.
Hermannsson, 'Banks and Iceland', pp. 94–5

---

[1] The Revd Árni Helgason (1777–1869), a prize-winning scholar, classicist and theologian. He was one of the main founders of the Icelandic Literary Society (Hið íslenzka bókmenntafélag) and president of the Reykjavík division 1816–48.

290. Sir George Steuart Mackenzie to Banks, 7 October 1815

Edinburgh

Dear Sir

My purpose in troubling you at present is to solicit your good offices in Behalf of a Society established in Iceland for the preservation of the language & of the productions of native literature.[1] I inclose a copy of a prospectus left with me by Mr. Rask, who has been exploring the antiquities & language of the North for several years. His stay here was short, & on account of his extreme modesty I saw less of him than I could have wished. The same reasons may probably have prevented his addressing you on the Subject. When you read the inclosure, I am persuaded you will use your influence to obtain some aid for the object in view. Mr. Rask is mistaken when he thinks that an annual contribution could be obtained; but if I can collect a tolerable sum, I shall inform him that he ought not to look further to this Country. I am of opinion too that a sum given at once will be of essential use in setting the business agoing at first, & in securing its future prosperity.[2] It is my intention in the course of the winter, to solicit subscriptions from all my acquaintances, & I will not refuse half a crown. – So much for Iceland.

There is yet another matter in regard to which I trust I will not solicit your attention in vain. A young man, by name Gavin White, a native of Fife, has lately invented a quadrant for the purpose of taking altitudes for the latitude, when the horizon is obscured by fog.[3] Many seamen have lamented the want of even a rude observation when surrounded by fog, while the Sun was visible & clear. The object of the present invention is to enable them in such a situation of things to find the latitude, if not with absolute precision, at least within 3 or 4 miles, which is infinitely better than no knowledge whatever. White has made, with his own hands, an instrument which I find comes within 6 miles of the truth; but it is very rudely made. I am persuaded that, even a quadrant of his construction made by a skilful artist, it would be found a most useful invention. He is poor, & cannot afford to have this done. I have been inclined therefore to mention the circumstance to you, & to request that you will state it to the Board of Longitude, with the view to solicit them to order White's quadrant to be constructed by Mr. Adie[4] of this place, under my eye; & to allow the sum of Forty or Fifty pounds for that purpose; & I will engage, if the instrument does not answer the aid of finding the latitude at sea, without the natural horizon to within 4 miles, or even less, to pay the expence myself; so much am I satisfied with the invention. Should it be found so useful as I expect, I have no doubt of the board giving White a suitable reward. I name Mr. Adie as the maker, as he is well known as a Scientific & Skilful artist, & as I should wish that both White & myself should have opportunities of suggesting what may occur as the instrument is in progress. – There can be no greater pleasure than that of bringing

---

[1] This is Hið íslenzka bókmenntafélag, see previous letter.

[2] Mackenzie donated a guinea (21 shillings or £1.05) annually.

[3] The details of the instrument were first sent to the Board of Longitude in 1817 (Cambridge University Library, RGO14/29, ff. 226–227). At a meeting of the Longitude Commissioners on 4 February 1819 it was agreed 'Mr. Gavin White is to be informed that his proposal for the observation of lunar altitudes is not considered by the Board as tending to any benefit' (Cambridge University Library, RGO14/7, f. 274).

[4] Alexander Adie, mathematical instrument maker of Edinburgh. He was in partnership with John Miller, his uncle, to whom he had been apprenticed, but Miller died in 1815 (Clifton, *Directory of British Scientific Instrument Makers,* p. 189).

humble merit from obscurity; a pleasure which you have often experienced, & will therefore forgive this attempt to enjoy it.

I am dear Sir
Yours faithfully
[*Signed*] G. S. Mackenzie

BL, Add MS 33982, ff. 92–93. Autograph letter.
NHM, DTC 19, ff. 202–204. Contemporary copy.
Hermannsson, 'Banks and Iceland', p. 95 (extract).

291. Sir George Steuart Mackenzie to Banks, 15 October 1815

Edin*burgh*

Dear Sir,

I believe that the Society in which Mr. Rask takes so much interest, is chiefly intended to convey useful knowledge to the Icelanders, while at the same time the ancient language is kept up.[1] The latter is not an object of great importance, in my opinion, any more than maintaining the language of the Scotch Highlanders. The language of our Kingdom ought as much as possible to be uniform. In the case of Iceland, it is the want of knowledge in regard to the mode in which intercourse is carried on in Europe, & of the improvements which have taken place in every thing, which has debased the minds & character of the people below the standard of Enterprise & exertion. I have written to Mr. Rask, to inquire in to the regulations of the new Society, & to know under what patronage it has been established. –

With regard to White's quadrant, I will order one to be made, & if it answers my expectations, I will send it for the inspection of the Board.[2] –

Were I to detail to you the history of Loptson,[3] from the time I first found him in a remote part of the Highlands, you would conclude that he is one of the most unworthy objects on whom charity was ever bestowed. When I took him by the hand, I resolved to put it in his power to do something for himself. I sent him to the College here, & kept him as comfortable as he could wish. I soon found however, that he was as much knave as fool – for he told falsehoods, which he might have been sure would have been detected. In Iceland I found he had been banished[4] from the country & that his return surprised the people of Reikiavik very much. His conduct became then so bad, especially in regard to his parents, that, when I was about to return to this Country, I told him that I was ready to fulfil my promise of taking him back to Scotland, but that I thought it right to inform him, my friends should no longer be the dupes of his falsehoods & that I should have nothing more to do with

---

[1] The society in question is the Hið íslenzka bókmenntafélag, see previous letters.

[2] He is referring to the Board of Longitude, see previous document.

[3] Ólafur Loptsson (b. 1783). He attended the Latin School in Reykjavík 1800–1804. He subsequently became a student of Thomas Klog, the state physician, who sent him to Copenhagen in 1807. The ship he was travelling on was captured and taken to Stornoway. There Mackenzie met him and befriended him, sending him to Edinburgh for further study. He accompanied Mackenzie to Iceland on his 1810 expedition. There his true colours, as described in this letter, became apparent to Mackenzie who dismissed him. He later went to America, where he became an assistant doctor on a ship-of-war, and from then on his life is a mystery.

[4] *Banished* is perhaps not the correct word as he had fathered four illegitimate children with four women and the law was catching up with him. He fled.

him. He preferred remaining in Iceland. On my return, I received a letter from a respectable gentleman in this neighbourhood, informing me that Loptson by some mean or other had got acquainted with a young woman a governess in his family, & that it had been settled they should be married, & inquiring what had become of him. This surprised me a good deal, & I found that he had deceived the girl & this gentleman, by saying that I had procured him a lucrative situation in the medical department of the army, & that in the mean time, my brother in law, Dr. Gregory,[1] had taken him into partnership. This was so monstrous a lie that I was astonished any body could for a moment think it within a hundred degrees of truth. This match was of course broken off. He returned to Scotland afterwards & was for sometime in Edinburgh. A considerable time /had/ passed before I heard anything more of him, when a very decent looking woman called on me, & said she was Loptson's wife, & that he had left her /& a child/ & she did not know what had become of him. She had been deceived by the fellow, & believed all his fine stories. – I understood afterwards that he had gone to America & then that he had somehow contrived to get on board a man of war as surgeon's mate. Such are a few events of his history. I should also tell you that I saw his father,[2] whom he made me believe was a substantial farmer & would be proud to furnish me with horses in return for the kindness I had shown him. He was a poor blacksmith in the service of Amtman Thoransen[3] & had not a horse in the world. He besought me to take his son away, & offered me with tears in his eyes, his whole hoard of silver dollars.[4]

The indifference of Loptson, & the extreme unwillingness he showed to go to his father's house, when we came near it, & the desire he expressed for us to hasten from the place, shocked me. I dare say he believed we had no intention of going to the eastward of Hekla. – Before we reached Iceland he spoke as if he had been in every part of it. We found he had never been anywhere except on the road between his father's house & Reikiavik – when he was servant of Dr Klog.[5] – I did not think I should have had to warn you against his impositions; for to say the truth, I expected that the Gallows would have had him long ago. – But enough on such a subject. –

I took the Liberty a short time ago, to address to your care a small packet for Mr. T. A. Knight,[6] containing a gold medal voted to him by the Caledonian Horticultural Society.[7]

I am dear Sir
Your faithfull & Ob*edient*.[8]
[*Signed*] G. S. Mackenzie

BL, Add MS 33982, ff. 94–95. Autograph letter.
NHM, DTC 19, ff. 205–208. Contemporary copy.
Hermannsson, 'Banks and Iceland', pp. 95–6 (extract).

[1] James Gregory (1753–1821), eminent professor of medicine at the University of Edinburgh. His second wife, 'a Miss McLeod' whom he married in 1796 and who bore him 11 children, was Isabella, a daughter of Donald Macleod of Geanies, sheriff of Ross-shire. Mackenzie's wife, Mary, was her sister.
[2] Loptur Ámundason, a farmer.
[3] Stefán Þórarinsson, the district governor of the Northern and Eastern Amts.
[4] Part of this sentence is illegible due to damage to the page – but here the DTC copy saves the day.
[5] State physician in Iceland.
[6] Thomas Andrew Knight (1759–1838), FRS, botanist, plant physiologist and horticulturalist. Knight was president of the Horticultural Society 1811–38. He met Banks in 1795 and they shared an extensive correspondence dating from 1795–1819, almost; 100 letters are extant.
[7] Founded in 1809, it is now the Royal Caledonian Horticultural Society based in Edinburgh.
[8] There is no room for more on the sheet of paper. Mackenzie probably would have added 'Servant'.

292. Royal Ordinance respecting an extended liberty of trade to Iceland, dated Frederiksberg 11th September 1816[1]

We Frederik the Sixth by the Grace of God King of Denmark, the Vandals and Goths, Duke of Sleswig, Holstein, Storman. Ditmarsk, Lauenburgh and Oldenburgh make known: That We have caused to be investigated and to be taken into consideration what alternative, according to the present circumstances might be necessary in the Laws concerning the trade and navigation to Our island of Iceland, in order that the trade, and the preparations already made in the Ordinances in the years 1786 and 1787 and later, might become most useful, as well for our dominions in general, as well as for Our dear and faithful subjects, in the said Our island of Iceland in particular. We therefore order and decree as follows:

1.

Our Treasury is graciously authorised, from the beginning of next year and till further notice, to issue passports or permission for a certain number of vessels to trade to Iceland, and which belong to the subjects of foreign powers.

2.

Foreigners, who expect to receive Icelandic passports must cause their petitions to be delivered at Our Treasury before the expiration of the month of January of the year for which the permission is desired. The petition must set forth the name of the vessel and its tonnage; as well as a declaration of Cargo intended to be sent to Iceland, to which in particular must be added, as far as the Cargo may consist of Timber, the volume this article will occupy in proportion to the tonnage of the vessel.

3.

On the receipt of the passport, or at any other time and place on which Our Treasury in particular cases might fix, is to be paid, in liew of Import, Export and Tonnage Duty 50 Rigsbankdollars in Silver[2] for each Last of Commerce[3] of the Tonnage of the vessel, except when the Cargo consists of Timber, in which case there is to be paid for each Last of Commerce, which this article might occupy of the ships volume, 20 Rigsbankdollars in Silver only. In this event of foreigners proving that they have freighted Danish vessels, properly manned with Danish seamen, to carry on trade to Iceland. We will grant them a reduction in the said import, to be fixed agreeably to circumstances.

---

[1] The fact that there is a translation of this ordinance among the Banks papers shows Banks's perceived interest in all things Icelandic. This is in all probability a translation by Horneman and sent to Banks.
After the Peace of Kiel the Danish government did not wish to reject the demands of the British merchants out of hand. The Russians, who hoped that the Danes would not permit the English to trade without some fitting commercial compensation, were informed that the King of Denmark wished to give the English 'un témoignage de sa satisfaction' (trans: a token of his satisfaction) for their part in provisioning Iceland during the war. This was cleverly done when the decree was finally promulgated on 11 September 1816. The decree promised the extension of liberty of commerce with Iceland. The first article began auspiciously enough, but then 17 articles follow which effectively prohibited the trade. This decree was quite simply a trade ban in disguise. The duties and licence costs were so high that they would successfully cancel any profit made by the foreign merchants, who would thus find it impossible to compete with the Danish merchants. However, Lord Castlereagh, the Foreign Secretary, used this ordinance as a pretext to appoint Thomas Reynolds (1771–1836), the Irish informer and a political embarrassment, as British consul to Iceland in 1817. See Agnarsdóttir, 'Írskur svikari'.

[2] The value of these was about 8 Rixbankdollars in Silver to the £ Sterling.

[3] $2\frac{3}{4}$ English tons are nearly equal to one Last of Commerce (*kommercelast*) Danish measurement.

4.

When the trading to Iceland has been granted to a foreign vessel, in the manner set forth, the master is bound, previous to the sailing of this vessel from the place where it has been equipped, to let the passport be countersigned by the Danish Consul at the said place, who at the same time verifies the correctness of the name of the vessel and its tonnage, as stated in the passport. In the same manner the master is to provide himself with due Certificates that the crew of the vessel is not infected by any dangerous epedemic disorder, and more in particular not by the small-pox or the measles, which Certificates previous to the departure of the vessel, are also to be exhibited to the Consul and to be countersigned by him,

5.

On arrival on the coast of Iceland the master must immediately run into a port of one of the market-towns, which are Reikevig, in the Southern District, Eskefiord and Öefiord, in the Eastern and Northern District, as well as the outport Grönnefiord situated in the Western District, which hereafter is to receive the name and privileges of a market-town, in lieu of Isefiord, which, in consequence of its less advantageous situation, for the future will have to be included among the authorised out-ports.[1] – When the master arrives at one of the said market-towns, the Icelandic passport, the before-mentioned Bill of Health, as well as the Manifest of the Cargo with the other ships papers are immediately to be delivered, at Reikevig to the Townfogd (Byfogd),[2] but in the other market-towns to the Sysselman, who is to certify on these documents, that they have been produced and found correct, and, when they have been thus found correct, they are to be returned to the master.

6.

It should appear, that the vessel on the arrival wants the said Bill of Health it is to be considered, as if not provided with the necessary Icelandic passport, and is therefore to be sent back by the magistrates, as being totally unauthorized to carry on trade, and of which the Magistrates, whom it may concern, are to make a declaration on the signing of the documents, which may be exhibited, as well as without any delay to make a report to the Amtman and carefully look to, that the vessel has no communication with the shore, which, on pain of confiscation of the ship and cargo is prohibited. When, on the contrary, the foreign vessel, agreeably to the modes prescribed; has proved itself entitled to trade in one of the market-towns of the Country, the master is then at liberty to touch at the ports of the other market-towns, and at the authorised out-ports, which latter at present are: Havnefiord, Kieblevig, Öreback and Vestmannöe in the Southern District, Berefiord, Vopnefiord, Husevig, Hofsaas and Skagestrand in the North and East District, and Reikefiord, Isefiord, Dyrefiord, Bildal, Patrikfiord, Flatöe, Stickesholm, Olafsvig, Stappen and Budenstad,[3] in the Western District.

Notwithstanding, the master is bound to exhibit the ships papers, immediately on his arrival to one of these trading ports, and before he begins any trade, to the Sysselman, who is also to provide them with his signature, certifying, that they have been exhibited accordingly.

[1] Eskifjörður, Eyjarfjörður, Grundarfjörður and Ísafjörður.
[2] *Byfogd*: *bæjarfógeti*. See Glossary.
[3] Modern Icelandic: Hafnarfjörður, Keflavík, Eyrarbakki, Vestmannaeyjar, Berufjörður, Vopnafjörður, Húsavík, Hofsós, Skagaströnd, Reykjafjörður, Ísafjörður, Dýrafjörður, Bíldudalur, Patreksfjörður, Flatey, Stykkishólmur, Ólafsvík, Stapi and Búðir.

7.

Foreign traders are in all respects to conform themselves to the Laws of the Country in general, and in particular to Our Ordinances in force respecting the speculating trade of Our subjects. It is therefore, in conformity to the placards of the 1st June 1792 and 23 April 1793 only permitted foreigners to carry on trade in retail for 4 weeks in a market-town, or in an authorised out-port, without being allowed to carry on trade in the Country by erecting sheds, buildings or tents, in the same manner as foreigners also, according to the Ordinance of the 13th June 1787 1st Chapt. 13th and 14th paragraph, are bound to furnish to the respective Magistrates a list of the merchandize, which they may have bought or sold in each of the trading ports, at which they may have traded.

8.

When a foreign trading vessel, which is provided with an Icelandic passport, arrives at an out-port, previous to its having touches at one of the ports of a market-town, and as there received the stipulated signature on the ships papers, the vessel is, by the respective Sysselman as unauthorised to trade, not to be permitted to carry on any trade, before the said signature has been produced in a lawful manner.

9.

If it is discovered that the master of a foreign vessel, or any other vessel authorised to trade to Iceland has cited in contravention of the 7th and 8th paragraph, where the rules are stated in the trade of such a vessel, he is to be fined from 50 to 100 Rixbankdollars in Silver, which fine may be recovered by the Magistrates in a lawful manner by seizing the vessel and by selling by public sale as much of the Cargo as is required for the payment of the fines. Of this fine (which, when a second time insured is paid with double the amount) the informer receives one fourth. But the remaining 3/4th are given to the funds for the poor of the place.

10.

The Cargoes of the foreign vessels must be corresponding with the declaration, which is inserted in the Icelandic passports, and with the Manifests which are exhibited to the Magistrates. If it is discovered that a foreign vessel carries other articles than those mentioned in the passport, the owner of the vessel can not expect oftener to receive permission to trade with Iceland; in the same manner as the master, in particular if it should discovered that the vessel does not at all or only in part contain Timber, which is mentioned in the passport, is to pay a fine 5 times the amount of the Import Duty, which by this incorrect declaration has been defrauded. This fine, of which the informer receives 1/4th but the other 3/4th are paid into Our Treasury, is levied in the manner prescribed.

11.

The Townfoged (Byfoged)[1] in Reikavig and the Sysselmen in whose Districts the Trading places are situated, are to have special care that foreigners act in conformity with this ordinance, as well as with others which are in force concerning the Icelandic trade, To the said Our functionaries is to be paid by each foreign trading vessel 6 Rigsbankdoll*ars* in Silver which the master pays, previous to the ships papers in each place at each place [*sic*] after their exhibition are returned to him.

---

[1] *bæjarfógeti*: see Glossary.

12.

As soon as a foreign vessel returns from Iceland the master is to deliver the Icelandic passport, which is only valid for one voyage, if at a foreign place, to Our Consul, but in Our own dominions to the inspector Officer of the Customs after which the passport is without delay sent to Our Treasury. Does the master act contrary to this, the owner of the vessel can not expect oftener to secure permission to trade to Iceland.

13.

This prohibition which is contained in the ordinance concerning the Icelandic trade of the 13th June 1787, 2d Chapter, 1st. paragraph, against the trading of merchants domiciliated in Iceland to foreign Countries, unless they have previously begun partnership with a mercantile house in Denmark or the Duchies, is, for the future to be at an end. It is therefore permitted every Icelandic Merchant, who has become Burgher and who continues for himself and his family to have an establishment to export direct to foreign places, as also to import direct from foreign places, Cargoes to Iceland with his own vessels, or in others belonging to Our subjects, which are provided with the Icelandic passport.

The 14th 15th 16th and 17th paragraphs treat solely of the trade carried on by the merchants established in Iceland.

18.

When the Merchants, who are regularly established in Iceland, for their own account and by the vessels dispatched by them from Iceland, import Cargoes direct from foreign parts, they are to enjoy the same exemptions from Duty & other Imports, which We, until further notice, have granted the trade of Our subjects to Iceland in general.

Given under our Royal Hand and Seal at the palace of Frederiksberg, the 11th September 1816

Frederik R

SL, Banks Papers, I 1:77. Contemporary copy.

293. Count Eduardo Romeo von Vargas Bedemar, 13 April 1817[1]

Copenhagen

Baron Banks,

It is not only to attainments and praise well above all eulogies we must look, to the goodness with which you have always deigned to welcome and protect the slightest effort which extends the realm of natural history. But it is above all because of what Iceland owes you which has given me the courage to offer to you with respect this little book, here enclosed.[2] Considering the scholarly research that has now made the importance

---

[1] This letter has been wrongly bound in the British Library with the year 1811 (the numbers 7 and 1 can easily be confused), when it is without doubt from 1817, when the book mentioned was published.

[2] This is *Om vulcaniske Producter fra Island* (the copy in the British Library belonged to Banks), published in Copenhagen in 1817, 57 pages in length. It is a description of a collection of volcanic minerals brought to Denmark by Governor Castenschiold. They were preserved in Prince Christian Frederik's Cabinet and the book contained descriptions of all the Icelandic minerals there (Kornerup, *Edouard Vargas Bedemar. En Eventyrers Saga*, pp. 209–13).

of this country known to Europe, and your noble and generous heart which brought a helping hand to the inhabitants when on the brink of total extinction, I take the liberty of presuming that anything that concerns this subject [Iceland] must be of interest to you. –

> I have the honour to be, Baron Banks,
> Your very humble and obedient Servant
> [*Signed*] Le Comte de Vargas Bedemar
> Chamberlain of His Danish Majesty,
> Member of the Commission of Antiquities[1]
> and of the Royal Society of Science.[2]

BL, Add MS 8100, f. 90. Autograph letter. Translated from the French.

French original:
Monsieur le Baron
Ce n'est pas seulement un mérite et tout une célébrité bien au dessus de tout éloge, ou la bonté avec laquelle Vous avez toujours daigné accueillir et protéger la moindre effort pour étendre le domaine de l'histoire naturelle, Monsieur le Baron, mais c'est surtout ce que l'Islande Vous doit, qui m'a donné le courage de Vous offrir l'hommage du petit livret ci-joint. Car après que les recherches savantes avaient fait connaître a l'Europe l'importance de ce pays, Votre coeur noble et généreux Vous fit lui porter une main sécourable au moment même que le peuple qui l'habite, fut prés de son anéantissement total. Ne dois-je donc présumer, que la moindre chose qui concerne cet objet, doive avoir quelque intéret pour Vous? –

Je profite de cette occasion pour Vous supplier d'agréer l'assurance de la profonde vénération avec laquelle j'ai l'honneur d'être,

> Monsieur le Baron
> Votre trés humble et trés obéissant Serviteur
> Le Comte de Vargas Bedemar
> Chambellan de S.M. Danoise, Membre de la Commission
> des antiquités, et de la Société Royale des Sciences.

294. From King Frederik VI to Banks, 17 September 1817

Fredricsberg[3]

Sir!
The solicitude you showed during the last war in saving our subjects in the islands of Iceland and the Faroes from the calamities and deprivation of war, which they would otherwise have been exposed to, have not escaped our notice. We know they stem from your feelings of humanity which characterise you and the knowledge you acquired while in the island of Iceland of the needs and the interesting character of this people, deservedly renowned in the annals of Northern Europe.

---

[1] Commissionen for Oldsagerne i Kiøbenhavn.

[2] Det Kongelige Danske Videnskabersselskab i Kiøbenhavn, founded in 1743. Vargas-Bedemar was indeed a member, as was Banks.

[3] Frederiksberg Palace. It was a summer palace west of central Copenhagen and a favourite residence of King Frederik VI.

Our paternal heart was touched by your generous efforts in favour of our subjects and we have now for a long time been resolved to acknowledge our gratitude.

Certain that you find in your very actions and the conviction of having eased the suffering of these people the greatest reward, we restrict ourselves to expressing our satisfaction by giving you as a mark of our goodwill and the particular esteem your literary merits have inspired in us, a copy of the botanical work Flora Danica, as well as a copy of the description of coins and medals from our Cabinet.[1] –

We pray God to keep you under his holy and worthy protection.

Your affectionate

[*Signed*] Frederic

RA, DfdUA file 817. Translation from the French. Draft[2]

French original:

Monsieur!

Les soins que Vous avez pris durant la dernière guerre pour détourner de Nos sujets dans l'isle d'Islande et dans les isles de Ferröe les calamités et les privations, auxquels ils auroient été d'ailleurs exposés, ne nous sont pas restés inconnus. Nous savons qu'ils doivent leur origine aux sentiments d'humanité qui Vous caractériscant et à la connaissance que Vous aviés acquise dans l'isle d'Islande des besoins et du caractére interessant de ce Peuple à juste titre renommé dans les annales du Nord de l'Europe.

Nôtre coeur paternal a été touché par Vos efforts genereux en faveur de Nos sujets et Nous avions depuis longtems resolu de Vous en témoigner Nôtre reconnaissance.

Persuadés que Vous trouvés dans Vos actions mêmes et dans la conviction d'avoir soulagé l'humanité souffrante la plus grande recompense, Nous bornons à Vous en exprimer Nôtre satisfaction et à Vous offrir comme une marque de Nôtre bienveillance et de l'estime particulière que Nous ont inspirées Vos mérites littéraires, un exemplaire de l'ouvrage botanique Flora Danica, ainsi qu'un exemplaire de la designation des monnoyes et médailles de Nôtre cabinet. –

Sur ce Nous prions Dieu de Vous avoir dans Sa sainte et digne garde.

Votre affectionné,

[*Signed*] Fréderic

---

[1] See Document 155. Banks's luxury copy of the *Flora Danica* cost 2,342 rixdollars and 64 skillings, while the copy sent to Sir George Steuart Mackenzie only cost 514 rixdollars and 48 skillings, but then it was uncoloured (Wagner, 'En Kongelig Gave', p. 99). The question of rewarding Banks for his help regarding Iceland was mooted by Horneman, the General-Consul in London (RA, DfdUA, 817, Rosenkrantz to the Danish ambassador in London, 3 September 1814). The King of Denmark agreed to send a gift to Banks (and Mackenzie) for their literary merits ('une marque de Sa [the King's] bienveillance et de l'estime que leurs mérites littéraires inspirent à Sa Majesté.'). This, however, took its time and in 1817 it was considered appropriate to thank Banks for his help regarding the Icelanders. Three boxes were sent to Banks in London, containing the *Flora Danica*, in colour, and *Description du Cabinet des médailles du Roi* (RA, Ges. Ark. London III indkomne skriveiser fra DfdUA 1814–1817, Rosenkrantz to Bourke, the Danish ambassador in London, 13 November 1817). Frederik read both Hooker and Mackenzie (see Wagner, 'En kongelig Gave', p. 88).

[2] 'Projet d'une lettre à adresser au nom de Sa Majesté au Chevalier Sir Joseph Banks' ('Draft of a letter in the name of His Majesty to the Knight Sir Joseph Banks'). On the draft is written *aprouvé* ('the letter is approved by the king') and below is the signature of King Frederik VI – in French.

295. John Ross to Banks, 25 July 1818

        H.M.S. *Isabella* at Sea Latitude 75° 45′ North Longitude 60.10 West

Dear Sir Joseph

I have waited for our passing the fourth and I hope the Last barrier of Ice, and untill I had finished my public Dispatches – that I might have the pleasure of giving you the latest information of our proceedings, you will be gratified to find we have arrived as far at least as could be expected by this time without any accident, or without having had a Sick person on board, – our voyage has been but a counterpart of what has been already made, except that we have found the Coast of Greenland from 68° to 75° North 100 miles further west than given in the admiralty Charts, – and that the Land called James Island, is not within 150 miles of Greenland, and therefore, must be Cumberland Island on the Coast America if every [*sic*] it was seen, but the most interesting thing is the Variation of the Compass and the extraordinary devatiation [*sic*] occasion'd by the Ships attraction, the Reports on Which I have the honor to enclose you,[1] and if you deem them worthy of being presented to the Royal Society, I request you will be pleased to offer them in my Name – I have only to add we are all in the best health and Spirits, Captain Sabine,[2] whose conduct merits my unqualified approbation, is well and joins me in every Kind wish.

    I beg to assure you of my Gratitude for your kind & valuable advice, and that I am not unmindful – you are the father of the Enterprise I have the honor to Command. – I beg you to offer my most respectful Compliments to her Ladyship – and

                I have the honor to be with the highest Respect
                          Dear Sir Joseph
                          your most obedient and
                             very humble Servant
                                [*Signed*] John Ross

P.S. I have sent you by the opportunity a quantity of the Soundings taken up here in 356 fathoms by a Machine I have invented – and which I shall send you a Drawing & Description of – and which I request you will do me the honor to present to the Royal Society.[3] This day the temperature of the mud brought up from 356 fathoms was 31° ¾ which I suppose had increased a little by the Action of the water in coming to the surface.

                                I have the honor to be
                           Your most obedient Humble Servant
                              [*Signed*] John Ross Captain

Wisconsin, Banks Papers, ff. 110–111. Autograph letter.

---

[1] Banks did indeed deposit this paper with illustrations 'On the Variation of the Compass' at the Royal Society on 21 December 1818 (see London, The Royal Society, AP/9/21). Many thanks to Fiona Keates of the Royal Society for help with locating this document.

[2] Sir Edward Sabine (1788–1883), Irish general, astronomer, geophysicist, ornithologist, explorer and the 30th President of the Royal Society. In 1818 he was invited to become an FRS and it was thanks to Banks that he was asked to take part in Ross's expedition to the Arctic, in search of the Northwest passage, as an astronomer. However, Ross and Sabine fell out, publicly, on their return.

[3] This material is not to be found in the archives of the Royal Society.

296. Count Niels Rosenkrantz to Banks, 9 September 1819

Copenhagen

Sir,

As a sequel to the Flora Danica, which the King, my master, sent to you, as mark of his special esteem,[1] he has ordered me on the basis of the same sentiment the work enclosed entitled *Testamen Hydrophytologiæ Danicæ*[2] by H. C. Lyngbye.[3]

I am glad to use the opportunity presented by this order of my Sovereign to offer you the assurance of the great consideration I have the honour to be[4]

RA, DfdUA, box 817. Draft letter. Translated from the French.

French original:
Monsieur le Chevalier
Comme suite à la Flora Danica, que le Roi, mon Maître, Vous a fait parvenir, Monsieur, comme une marque de Son estime particulière Sa Majesté m'a ordonné de Vous envoyer par l'effet de même sentiment l'ouvrage ci-joint intitulé: Testamen Hydrophytologiæ Danicæ, auctore H. C. Lyngbye /*qui répresente les desseins des plantes aquatiques Danoises*/.

Je me felicite de trouver dans l'execution de cet ordre de mon Souverain l'occasion de Vous offrir, Monsieur, l'assurance de la haute considération avec laquelle j'ai l'honneur d'être

297. Ólafur Ólafsson: Poem in celebration of Banks, [1820][5]

The Deification
of
Joseph Banks
a man
among
Great Britain's
richest
of perfection and delight
by
Icelander Ólafur Ólafsson

x x x

---

[1] See Document 294.
[2] Published in Copenhagen in 1819. Banks's gift is now in the British Library.
[3] Hans Christian Lyngbye (1782–1837), botanist and clergyman. This is the first work on Danish algae.
[4] Mackenzie also continued to receive books from the King, and in 1833 expressed his 'heartfelt acknowledgment of His Danish Majesty's condescension and kindness' (RA, DfdUA, box 817, Mackenzie to Bourke, 1 January 1833).
[5] This poem is by Ólafur Ólafsson (see Appendix 13). He sent it to his friend and relative Bjarni Thorsteinsson (1781–1876), an official in the Rentekammer, who became a district governor in Iceland in 1821. It was contained in a letter dated 24 March [1821] along with many other poems by Ólafur. Banks died on 19 June 1820 and this poem is believed to have been written in 1820. It is translated by Hjalti Snær Ægisson.

The metre is choriambic
Asclepiad with glyconic lines
Truncated and completed

x x x

Tetrastrophic tricolons
as in Odes 1–2 by Horace

Banks, the great glory of men
Among the shining stars of the English has fallen
From the sky. A man without blemish,
A learned man, a good man, a wholehearted man.

O, cry, ye Muses! Cry,
Ye youths whom he taught, the devout man.
Make your faces wet with holy
Teardrops, for your father has gone.

You, who appear gleaming with riches,
May you learn everything that he cosseted,
The wise and wealthy Banks, and use it prudently.
You will be the splendour of your nation.

When the evil and ferocious war god barred Icelanders from
Access to the sea, imposing swift death, hunger and fear,
All this was banished with a vivifying hand
By the kind Banks.

You who henceforth inhabit my fatherland,
The land Banks safeguarded, the glorious Joseph,
It would be apt for you to lament
And gloriously praise him.

Live on, kind Banks, among the deities,
Pleasantly, as you were to this former world.
You will always be a rare and lofty example
To those who dwell on the earth.

Reykjavík, National Library of Iceland, Lbs. 341 a, fol. Autograph poem.[1]

---

[1] There is also a copy in the Danish National Archives.

Latin original:

Apotheosis
Josephi Banks
Viri
inter eos
qvi
Magnæ Britanniæ
sunt
Perfectionum & amabilium
ditissimi
ab
Olavo Olavi Islando

x x x
Versus sunt Choriambici
Asclepiadei cum Glyconico
Mutilo & perfecto
x x x
Tricolon Tetrastrophon ut
Horat: Ode 1–2.

Anglorum cecidit Sidera fulgida
Inter præcipuum; Banks hominum decus,
E cælo. Sine labe
Vir, doctus, bonus, integer.

O' vos castalides plangite! Plangite
O' Vos O' juvenes qvos docuit pius
Sanctis ora rigate
Guttis, nam pater excidit.

Tu qvi divitis fulgidus appares
Discas qvod Sapiens excoluit potens
Auro Banks: bene hoc uti.
Splendor gentis eris tuæ.

Islandis pelagum mars malus & rapax
Claudendo rapida cæde, fame & metu
Explevit bonus almâ
Cuncta hæc Banks pepulit manu.

Hinc tu qvi patriam nunc habitas meam
Cui, Banks numen erat, planger te decet
Josephus gloriosis
Illum laudibus efferas.

Inter cælicolas vivito Banks bone
Illis deliciis, mundi ut eras prius,
Exemplar rarum & altum
Semper terricolis eris.

APPENDIX 1

Biographical Details of the Members of Banks's Expedition to Iceland in 1772

Asquith, John, servant of Banks 1772–3 and one of the French horn players.

Bacstrom, Sigismund (fl. 1770–99), ship's surgeon, secretary and naturalist. He probably arrived in England in 1770. He is believed to have been a Swede, though Beaglehole thought he was a Dutchman.[1] Employed by Banks in 1772, he went on the voyage to Iceland as secretary, remaining in Banks's service working in his herbarium until 1775. He was the amanuensis of many Banksian documents. He later made six voyages as surgeon in merchant ships – four to Greenland, one each to the Guinea coast and to Jamaica – and in 1780 he went on a whaling voyage to Spitzbergen. He later fell on hard times, entreating Banks's help in gaining employment. There are nine letters extant between Bacstrom and Banks from 1786–96, in which Banks showed himself willing to help him, as Bacstrom 'had always conducted himself with integrity' while in Banks's service. During his above-mentioned voyages he collected plants for Banks's herbarium.[2]

Briscoe, Peter (1747–1810), servant to Banks probably from childhood (at least 1760) on the Revesby estate in Lincolnshire. He also travelled with Banks to Newfoundland in 1767, on the *Endeavour*, and on Banks's last foreign visit to Holland. Briscoe was the first to sight the Pacific island Vahitahi[3] and is listed among the explorers of Australia.[4] He wrote a journal on the *Endeavour* voyage now in the Dixson Library, Sydney. According to Harold Carter he was 'the faithful attendant of his [Banks's] Oxford years, his London studies, the Newfoundland voyage and his county travels since'.[5] He subsequently became a grocer and his name is inscribed on a memorial tablet in St Botolph's church, Boston, Lincolnshire.[6]

Cleveley, John Jr[7] (1747–86), artist and draughtsman. Son of John Cleveley, a shipwright and painter, he was brought up near Deptford. He was a student of Paul Sandby (1731–1809) the noted landscape painter, and was also a marine painter, frequently exhibiting at the Royal Academy. He was awarded a prize for his watercolours by the

---

[1] *Endeavour Journal of Joseph Banks*, I, pp. 68–9.
[2] Dawson, *The Banks Letters*, pp. 26–7; Marshall, 'The Handwriting of Joseph Banks', p. 6; Britten 'Sigismund Bacstrom', pp. 92–7.
[3] Beaglehole, *Journals of Captain Cook*, I, p. 600.
[4] Ibid., p. 69n.
[5] Carter, *Sir Joseph Banks*, p. 64.
[6] Beaglehole, *Journals of Captain Cook*, I, p. 600.
[7] His name is often spelt as Clevely

Society of Artists. He did not accompany Constantine Phipps on his voyage to the Arctic in 1774 as has been asserted by Rauschenburg.[1]

Douvez, Antoine, the French cook.

Gore, John (1729/30–1790), captain in the Royal Navy, soldier, seaman and explorer. Gore had completed two circumnavigations before sailing with Captain Cook on the *Endeavour*, first with John Byron on HMS *Dolphin* (1764–6), and then with Samuel Wallis on the same ship (1766–8); all in all, three circumnavigations before going to Iceland. He sailed the Atlantic and went to the West Indies and the Mediterranean, eventually becoming a captain in the Royal Navy. He was, in Beaglehole's words, 'a man of commonsense and able practice – with the reputation in his maturity of being the best practical seaman in the navy'.[2] He was no less a man of action, accompanying Banks on his shooting and hunting expeditions, shooting the first kangaroo (14 July 1770), but unfortunately also a Maori for stealing a piece of cloth. He went as first lieutenant on the third Cook voyage and became its commander when Cook was murdered and Clerke died of tuberculosis. In 1780 he was made post captain at Greenwich Hospital, where he died in August 1790. He entrusted Banks with his will and the care of his young son and mistress in the event of his death.[3]

Hay, James, astronomer.

Holbrook, Robert, servant of Banks.

Lind, James (1736–1812), FRS, Scottish astronomer and physician. By 1772 Lind had already done important pioneering work on scurvy, visited China as a surgeon on an East Indiaman in 1766 and observed the transit of Venus at Hawkhill near Edinburgh in 1768. Banks did not know Lind personally, but he was an acquaintance of Solander's. On Banks's nomination Parliament agreed to make a special grant of the enormous sum of £4,000 to engage Lind. On 8 February 1772 the Council of the Royal Society recommended Lind to the Board of Longitude as 'extreamly useful ... on account of his skill and experience in his profession, and from his great Knowledge in Mineralogy, Chemistry, Mechanics, and various branches of Natural Philosophy; and also from his having spent several years in different climates, in the Indies'.[4] Subsequently, the House of Commons voted a parliamentary grant of £4,000 for the engagement of Lind on Cook's second voyage on the *Resolution*. As Cook wrote in his journal for 25 December 1771: 'The Parliament voted Four thousand pounds towards carrying on Discoveries to the South Pole, this sum was intinded for Dr Lynd of Edinburgh as an incouragement for him to embark with us, but what the discoveries were, the Parliament meant he was to make, and for which they made so liberal a Vote, I know not.'[5] When Banks had withdrawn from the voyage and Lind with him, the money was transferred to the Forsters,[6] Johann Reinhold (1727–98) the naturalist and his son, the botanist Johann Georg Adam

---

[1] 'Iceland Journals', p. 190.
[2] Beaglehole, *Journals of Captain Cook*, I, p. cxxxi.
[3] Carter, *Sir Joseph Banks*, pp. 134–5: Beaglehole, *Journals of Captain James Cook*, I, where he is frequently mentioned throughout the journal. On the other hand, he hardly figures in the Banks Iceland journal.
[4] Minutes of the Council of the Royal Society, quoted in Beaglehole, *Journals of Captain Cook*, II, p. 913.
[5] Beaglehole, *Journals of Captain Cook*, II, p. 4.
[6] For the Forsters, see above p. 47, n. 1.

(1754–94). However, Lind's 'Portable Wind Gage' (a manometer) travelled on the *Resolution*.[1] Lind was elected a Fellow of the Royal Society in 1777, continued on his travels to India and the Cape of Good Hope, and was appointed physician to the Royal Household at Windsor in 1783. His friends included James Watt and the poet Shelley. (See further on Lind in Document 15.)

Marchant, John, servant of Banks.

Miller, James (fl. 1772–82), draughtsman and engraver. He was the son of Johann Sebastian Müller, a German engraver who settled in England in 1744, calling himself John Miller (1715–90?). James Miller exhibited both at the Society of Artists and the Royal Academy.

Miller, John Frederick (fl. 1772–96), draughtsman and engraver. Brother of James Miller. He came to Banks as an 'articled draughtsman'. He drew many artefacts from the *Endeavour* voyage in 1771. In 1776 there was a falling-out with Banks. Miller had sold drawings from the Iceland expedition to Pennant and exhibited work he had done for Banks without his permission. He was dismissed, with Banks writing 'Since you have published engravings of drawings which you made for me at my expense, I desire to have no further concern with you.'[2] Like his brother he exhibited at the Society of Artists. He also illustrated Richard Weston's *Universal Botanist* (1770–77) and in 1785 he published a collection of engravings in *Various subjects of natural history wherein are delineated birds, animals and many curious plants ...*, including several which he made while in the service of Banks. He also worked on William Aiton's *Hortus Kewensis* (1789).

Mr Moreland, a gardener.[3]

Riddell, John, a member of the Riddell family, a baronetcy in Roxburghshire. He had been 'intended for the sea', as Banks says in the journal, but the Iceland voyage was apparently his first. Beaglehole adds a little bit of information, saying that this is 'presumably' the person David Hume wrote of, probably to Lind on 24 February 1772, wishing to recommend him: 'There is a young Gentleman of the name of Riddal, Grandson of Sir Walter Riddal, who goes with you in your nautical & philosophical Expedition in the Station of a Midshipman ...'[4]

Roberts, James (1752–1826) of Mareham-le-Fen in Lincolnshire, Banks's servant from an early age on the Revesby estate in Lincolnshire. In 1767 he was sent to London with horses to meet Banks on his return from Newfoundland. He became Banks's trusted servant and accompanied him, aged 16, on the *Endeavour*, and subsequently to Iceland and Holland. He attended Omai and was with him when he was inoculated against smallpox. He later became general steward of the London and country estates and, in 1818–19, when Banks was crippled by gout he was his amanuensis. He kept a journal

---

[1] Beaglehole, *Journals of Captain Cook*, II, p. 53n.

[2] Further on Miller see Britten, 'John Frederick Miller and his "Icones"', pp. 255–7.

[3] There is some dispute whether the gardener went or not. Banks's sister Sarah Sophia Banks thought not, in an annotation to her copy of the journal. But according to the journal of James Roberts he did indeed (see p.129 above). Another note by Sarah Sophia Banks says: 'James Donaldson, the Gardiner: there is a mistake in the name, as the Gardiner was Moreland' (Kent History and Library Centre, Banks MS U1590/S1/2).

[4] Beaglehole, *Endeavour Journal of Joseph Banks*, I, pp. 85–6n

of the voyage to Iceland, now preserved in the State Library of New South Wales.[1] Roberts is listed among the explorers of Australia and is commemorated on a memorial tablet in St Botolph's Church, Boston. He died on 8 July 1826 and his tombstone is in the Church of St Helen in Mareham-le-Fen, which says, 'In the year 1772 he again made a voyage with Sir J. Banks to Iceland and ascended the summit of that wonderful burning mountain Hecla.'[2]

Samarang, Alexander, a Malay. Personal servant to Banks.

Scot, Alexander, a valet.

Sidserf, Peter, servant of Banks and one of the French horn players.

Solander, Daniel Carlsson (1733–82), Swedish botanist and naturalist, FRS 1764. A protégé of Linnaeus (Carl von Linné) he went to London as his 'apostle' to promote the Linnean classification system. In 1763 he began to catalogue the natural history collections of the British Museum, becoming an assistant librarian. He met Banks in 1764 and accompanied him on the famous *Endeavour* voyage during 1768–71. After their return Solander became Banks's librarian and secretary working on cataloguing the *Endeavour* collections and the proposed *Florilegium*. He was also a close companion and friend to Banks. He was always titled Doctor or referred to as 'the doctor' but in fact he only became an honorary doctor along with Banks in 1771.[3] Solander had been due to sail on the voyage of the *Resolution* with Banks but stayed loyal to him when he withdrew, accompanying Banks to Iceland, where his language skills were a great help. He was Keeper of the Natural History Department of the British Museum from 1773. Solander's early death was a bitter blow to Banks. He left much manuscript material on the Iceland voyage, now in the Natural History Museum.[4] His correspondence has been published in English by Duyker and Tingbrand, *Daniel Solander. Collected Corrrespondence 1753–1782* (1995).

Taylor, John. There is a note by Sarah Sophia Banks in her copy of Banks's journal: 'John Taylor. Servant Boy to Doctor Lind.'[5]

Troil, Uno von (1746–1803), Swedish scholar and archbishop. Son of the archbishop of Uppsala, he was well educated and during 1770–73 he travelled to Germany, France, where he met luminaries of the French Enlightenment, and then on to England where he was introduced to Banks. When von Troil professed himself interested in the Icelandic language Banks invited him to join his Iceland expedition. Von Troil then returned to Sweden in 1773 where he was ordained, and in 1775 was appointed court chaplain of

---

[1] See pp. 115–40 above; and online at http://www.sl.nsw.gov.au/discover_collections/history_nation/voyages/discovery/endeavour/roberts/(accessed 16 June 2015).

[2] See *Cinque Ports Herald and Kent and Sussex Advertiser*, 22 July 1826, where it says: 'In 1795 he retired to Mareham House, where he spent the remainder of his days in the society of his friends.' I am indebted to Pauline Napier, Mareham-le-Fen, and Marjorie Whaler, the Society for Lincolnshire History and Archaeology, for information on James Roberts.

[3] Duyker, *Nature's Argonaut*, p. 58.

[4] On Solander see Edward Duyker's definitive biography, *Nature's Argonaut: Daniel Solander, 1733–1782: Naturalist and Voyager with Cook and Banks*.

[5] Kent History and Library Centre, Banks MS U1590/S1/2.

King Gustavus III (to whom he dedicated his book on Iceland). He enjoyed the King's favour and was eventually consecrated archbishop of Uppsala in 1786. Thus he led the clergy in the Swedish parliamentary assembly. Von Troil was a scholar. It was he who wrote about the Iceland expedition in *Letters on Iceland*, first published in Swedish 1777 and translated into German, English and French (1779–81). He also wrote a distinguished work on Swedish church history. In 1784 von Troil gave the library at Linköbing 122 volumes of Icelandic books. He wrote his autobiography, the manuscript of which is preserved in the library of Uppsala University.[1]

Walden, Frederick Herman, secretary to Banks. Next to nothing is known about him. Two letters from him to Banks are extant from the summer of 1793, written from Copenhagen regarding improvements for naval ships (29 June and 31 August 1793, BL, Add MS 8098).

Young, Nicholas, a servant to Banks. He sailed on the *Endeavour* and was the first to sight New Zealand, i.e. the Poverty Bay coast in October 1769, which Cook named Young Nick's Head, receiving a gallon of rum. He was also the first to sight Land's End on the return. He was about 12 years old, called 'Boy' in the sources, in 1769 when he had been taken on the *Endeavour* as a supernumerary. A year later he is recorded as the surgeon's servant and later entered Banks's service.[2]

---

[1] See his 'Själfbiografi'.
[2] Beaglehole, *Journals of Captain James Cook*, I, pp. 173 n, 600.

APPENDIX 2

Thomas Pennant and Banks

Thomas Pennant (1726–98) was of Welsh descent, a zoologist, geologist and mineralogist, and a renowned traveller. After studying at Oxford he travelled widely in Ireland, the British Isles and on the Continent, meeting among others Voltaire and eminent naturalists. On hearing that Banks was going to Newfoundland in 1766, he sent him a list of questions about mammals and birds he might encounter. On his return Pennant immediately wrote to Banks, very interested in the collections he had made.[1] In 1766 he published his *British Zoology* and in the same year was elected a fellow of the Royal Society. His *Arctic Zoology* was published 1784–7, drawing on Banks's collections among others.

Pennant visited Scotland in 1769, his *Tour in Scotland* appearing for the first time in 1771. When Banks set out for Iceland in 1772, first travelling to the Scottish isles he was very conscious of the fact that Pennant, on his second tour of Scotland, was also travelling in the Hebrides. Feeling that Pennant had a prior claim to that region, he wrote:

> I beleive that we passd the Ship of our freind M\\r. Pennant which I had all along carefully lookd out for not without hopes if I should be fortunate enough to meet him of tempting [him] to enlarge his plan & wishing much to spare him at all events that as I had lookd upon myself while among these Islands as treading upon ground which by prior right he has taken possession of I should communicate to him every Observation I could possibly make whenever he thought fit to publish any account of them.[2]

Banks kept his word. When Pennant, who had been frustrated in his attempt to visit Staffa because his prudent companion Mr Thompson was 'unwilling to venture in these rocky seas',[3] published his second edition *A Tour in Scotland, and Voyage to the Hebrides; 1772*, in 1774 he included with Banks's permission the 'Account of Staffa communicated by Joseph Banks, Esq.' as described in his journal. To show his appreciation Pennant penned a dedication in the form of a fulsome letter:

> To Sir Joseph Banks, Baronet.
> Dear Sir,
> I think myself so much indebted to you, for making me the vehicle for conveying to the public the rich discovery of your last voyage, that I cannot dispense with this address, the usual tribute on such occasions. You took from me all temptation of envying your superior good fortune, by the liberal declaration you made that the HEBRIDES were my ground, and yourself, as

---

[1] Carter, *Sir Joseph Banks*, p. 33, and Dawson, *The Banks Letters*, p. 660.
[2] Document 10.
[3] Pennant, *Tour of Scotland and the Hebrides,* 1790, p. 300.

you pleasantly expressed it, but an interloper. May I meet with such, in all my adventures!

In reward, the name of BANKS will ever exist with those of CLIFFORD, RALEIGH and WILLOUGHBY, on the rolls of fame, celebrated instances of great and enterprizing spirits: and the *arctic* SOLANDER must remain a fine proof that no climate can prevent the seeds of knowledge from vegetating in the breast of innate ability.

You have had justly a full triumph decreed to you by your country. May your laurels for ever remain unblighted! and if she has deigned to twine for me a civic wreath, return to me the same good wish.[4]

Included were several drawings from the Banks expedition, by John Frederick Miller and Cleveley, but engraved by others.[2] Pennant was however, not the only one to make use of Banks's Staffa account. Von Troil for example devoted letter no. XXII in *Letters on Iceland* to 'the Pillars of Basalt', with his own remarks and then the editor added Banks's account of the island of Staffa.[3]

Pennant's books on his Scottish travels were extremely popular and were published in numerous editions. Banks and Pennant would be life-long correspondents, Pennant asking Banks for many favours and commissions including loans of drawings.[4] Banks's biographer Harold B. Carter felt that Pennant had seldom properly acknowledged his debt to Banks,[5] except perhaps in the quotation above.

In May 1783 Banks learned that Pennant had bought Iceland drawings directly from John Frederick Miller. Banks threatened to prosecute if Pennant published them.[6] Pennant hastily replied that he only wished to keep them for his own amusement.[7] They patched things up and continued to correspond until 1798, the year of Pennant's death.

---

[1] Ibid., pp. i–ii. An excerpt.
[2] The 'Account of Staffa' is in Pennant, *Tour of Scotland and the Hebrides*, pp. 300–310, with six drawings from the expedition.
[3] Von Troil, *Letters on Iceland*, pp. 266–88; Banks's account follows on pp. 288–93.
[4] Dawson, *The Banks Letters*, pp. 660–64.
[5] Carter, *Sir Joseph Banks*, p. 35. And Chambers agrees with this, especially in regard to Pennant's *Arctic Zoology: Letters of Sir Joseph Banks*, p. 384.
[6] Chambers, *Scientific Correspondence of Sir Joseph Banks*, Banks to Pennant, 4 May 1783, II, p. 81.
[7] Chambers, *Scientific Correspondence of Sir Joseph Banks*, Pennant to Banks, 11 May 1783, II, p. 82.

APPENDIX 3

Sir Joseph Banks's Itinerary to Hekla, 18–29 September 1772

1.[1] fra[2] Havnefiord [Hafnarfjörður]
     s. til[3] Heder Bay [Heiðarbær]
2.   Kauresta Hraun[4]
     Thingvalle Gaa[5] [Þingvellir ravine]
     Lauger vatn Hellra[6]
     s. Laugarvatn
3.   s. Mola [Múli]
4.   Laug
     Lauga fell [Laugarfell]
     Geyse vid Haukadal [Geysir in Haukadalur]
     s. Mola
5.   Skalholt [Skálholt]
6.   Laugaras Hverer nær vid[7] Hvitaa, nær Skalholt
     a.) Miaulker hver (there boil Milk &c., dye &c) [Mjólkurhver]
     b.) Drauga hver [Draugahver]
     c.) Hildar hver [Hildarhver]
     Hvitaa [Hvítá]
     Thiorsarholt [Þjórsárholt]
     Thiorsaa [Þjórsá]
     a. Hvadeyre[8]
     s. Skard [Skarð]
7.   Leirubacke [Leirubakki]
     Næverholt [Næfurholt]
     s. Graufell [Gráfell]

---

[1] This itinerary is in Solander's hand. The numbering refers to the 11 days spent travelling to Hekla and back. This should be compared with the second part of Banks's journal.

[2] *fra* (Danish) means from.

[3] The 's.' could be short for *sedan* (Swedish): 'then'; *til* means 'to'.

[4] Probably *Kárastaðahraun*, Kárastaðir being a farm nearby. This place-name is not mentioned by Banks.

[5] *Gaa: gjá* (Icelandic) ravine.

[6] *Hellir* is cave in Icelandic. The cave near Laugarvatn.

[7] *nær vid*: near by.

[8] This place is not mentioned in Banks's journal, but is doubtless Vaðeyri, an islet in the Þjórsá river, near Þjórsárholt.

8. Hekla
    Longfell[1]
    Graufell
    Næverholt
    s. Skard
9. Thiorsaa
    Black agat.[2]
    s. Hraun giærde [Hraungerði]
10. Laugar dal [Laugardalur]
    Ølves aa [Ölfusá]
    Ingols fjæll [Ingólfsfjall]
    s Reikium [Reykir]
11. – D° – wherer[3]
    a. Badstuge where [Baðstofuhver]
    b. Fjall-where [Fjallhver] i Geyse [Geysir]
    &c
    Hellis Heyde [Hellisheiði]
    Havnefiord [Hafnarfjörður]

NHM, Solander MS, Plantae Islandicae et Notulae itinerariae, f. 2.

---

[1] The mountain Langafell, not mentioned by Banks, is near Næfurholt as is Gráfell.
[2] Banks mentions the man who obtained Icelandic agate for them on p. 108. Agate is a form of chalcedony quartz.
[3] Wherer: *hverir* (Icelandic): hot springs (phonetically spellt).

APPENDIX 4

Checklist of the Actual Titles in the Banks Library Taken Ashore in Iceland from the *Sir Lawrence* Brig, 1 September 1772.[1]

Ray, J., *Synopsis methodica anamalium Quadrupedum et Serpentini generis*, London, 1693.
Ray, J., *Synopsis methodica Avium et Piscium*, London, 1713.
Scheuchzer, J., *Agrostographia, sive Graminum, Juncorum, Cyperorum, Cyperoidum, iisque affinium methodus*, Zurich, 1719.
Micheli, P.A., *Nova plantarum genera, juxta Tournefortii methodum disposita*, Florence, 1729.
Linckius, J. H. *De Stellis marinis liber singularis; digessit C. G. Fischer*, Leipzig, 1733.
Linnaeus, C., *Flora Lapponica*, Amsterdam, 1737.
Artedi, P. *Bibliotheca Ichtyologica. Ichtyologiae. Pars 1. Philosophia Ichtyologica. Ichtyologiae.* Pars 2. *Genera Piscium. Ichotyologiae.* Pars 3. *Synonymia nominum Piscium.* Pars 4. *Descriptiones specierum Piscium.* Pars 5. Recognovit et edidit C. Linnaeus, Leyden, 1738.
Dillenius, J. J., *Historia Muscorum*, Oxford, 1741.
Anderson, J., *Nachrichten von Island, Grönland, und des Strasse Davis*, Hamburg, 1746.
Linnaeus, C., *Wästgöta resa, förrätted år 1746*, Stockholm, 1747.
Wallerius, J. G. *Mineralogia, eller mineral-riket indelt och beskrifvit*, Stockholm, 1747.
Linnaeus, C., *Hortus Upsaliensis, exhibens plantas exoticas hortu Upsaliensis academie a sese illatas ab a. 1742 in a. 1748*, Stockholm, 1748.
Linnaeus, C., *Amoenitates Academicae, seu Dissertiones variae physicae, medicae, botanicae, antehac seorsim editae, ninc collectae et auctae*, Stockholm & Leipzig, 1749.
Anderson, J., *Beschryving van Ysland, Groenland en de Straat Davis*, Amsterdam, 1750.
Linnaeus, C., *Skånska resa, förrätted år 1749*, Stockholm 1751.
Linnaeus, C., *Flora Suevica*, Stockholm, 1755.
Pontoppidan, E., *The Natural History of Norway*, London, 1755.
Horrebow, N., *The Natural History of Iceland*, London, 1758.
Linnaeus, C., *Fauna Suevica, sistens animalia Sceniae regni*, Stockholm, 1761.
Hudson, W., *Flora Anglica*, London, 1762.
Linnaeus, C., *Species plantarum*, 2nd edn, Stockholm, 1762–3.
Ström, H., *Physik og oeconomisk beskrivelse over fogderiet Söndmör, beliggende i Bergens stift i Norge*, Soröe, 1762–6.
Wallerius, J. G., *Mineralogie, uberstzt von J. D. Denso*, Berlin, 1763.

[1] Banks had also carried a select reference library with him on the *Endeavour*. A faint note at the top of the first page says: 'Books taken on shore'. The list is in the original chronological order. See Carter, *Banks. The Sources*, pp. 235–7.

Brunnich, M. T., *Ornithologia Borealis*, Copenhagen, 1764.
Linnaeus, C., *Genera plantarum. Editoio sexta ab Auctore reformata*, Stockholm, 1764.
Gunnerus J. E., *Flora Norvegica,* Trondheim, 1766.
Linnaeus, C., *Systema naturae, sive regna tria naturae systematice proposita per classes, ordines, genera, et species*, 12th edn, Stockholm, 1766–8.
Pennant, T., *British Zoology*, 4 vols., London and Chester, 1766–70.
Oeder, G. C., *Icones plantarum sponte nascentium in regnis Daniae et Norvegiae*, Vol. 3, Stockholm, 1768–70.
Linnaeus, C., *Philosophia botanica*, Vienna, 1770.
De Kerguelen, Y., *Relation d'un voyage dans la mer du nord, aux côtes d'Islande du Groenland, de Ferro, de Schettland, des Orcades et de Norvège, fait en 1767 et 1768*, Paris, 1771.
Linnaeus, C., *Mantissa plantarum altera, generum editionis 6.*, Stockholm, 1771.
Pennant, T., *Synopsis of Quadrupeds*, Chester, 1771.
Cronstedt, A. F., *An Essay Towards a System of Mineralogy, Translated by G. Von Engestrom, Revised by Em. Mendes Fa Costa*. 2nd edn, with additions and notes by M. T. Brunnich, London, 1772.
Hamilton, Sir W., *Observations on Mount Vesuvius, Mount Etna, and Other Volcanos, in a Series of Letters Addressed to the Royal Society*, London, 1772.
Olafsen, E. and Povelsen, B., *Reise igennem Island, beskreven af E. Olafsen*, Soröe, 1772.

Two further volumes are mentioned (omitted by Carter):
'Voyage to the North' is probably La Martinière, Pierre de, *A New Voyage to the North*, English edn, London 1706.
'Mellanges interesse & curieux', which might be de Surgy, Jacques-Philibert Rousselot, *Mélanges intéressans et curieux ou abrégé d'histoire naturelle morale, civile et politique, de l'Asie, l'Afrique, l'Amérique, et des Terres polaires*, 10 vols, Paris, 1766.

NHM, Solander MS, Plantae Islandicae et Notulae Itinerariae.

APPENDIX 5

Icelandic Contemporary Sources on the Banks Expedition to Iceland in 1772

These are accounts from one journal printed in Iceland in Danish and annals written in Icelandic at the time, translated into English.

Islandske Maaneds-Tidender, Hrappsey, Iceland and Copenhagen, October 1773.[1]
News from southern Iceland

Everywhere one still hears people, both high-born and low-born, expressing universal praise for the humanity and generosity of the widely celebrated and highly placed Englishmen Bank [*sic*] and Solander, who last autumn decided to visit the country. For instance it is recounted that the farmer who piloted the ship into Hafnarfjörður, where these gentlemen chose to anchor and dwell for some time, received a considerable reward. On their way to the fire-breathing mountain Hecla they are reported to have given attention with great precision to everything of importance to be found in the kingdom of nature. They were also very diligent in collecting all kinds of books and where they could be had, curious printed books and manuscripts. On the occasion of their arrival at the bishopric of Skálholt Bjarni Jónsson, the local rector of the Latin school, composed some congratulatory verses (*Gratulations vers*) in Icelandic with a Latin translation. Among them are the following ones: [Here follow nine verses of the twenty-five stanzas of the poem *Tripudium*, both in Icelandic and in Latin].[2]

The Annals[3]

Höskuldsstaðaannáll 1730–1784. Written by the Reverend Magnús Pétursson (1710–84). Published in *Annálar 1400–1800*, vol. IV, Reykjavík, 1940, pp. 545–6.

1772
After most of the merchant ships in the south of Iceland had left an English ship arrived in August in Hafnarfjörður, not very big, with cannon and well-fortified. It was said that 40 men were on board. Among them were gentlemen, elegantly dressed and their conduct

---

[1] This was the first number of this journal, which was published until 1776. In all there were 37 numbers.

[2] The complete poem is printed in Hooker, *A Tour in Iceland*, II, pp. 280–92, but only the Latin version. The editor of *Islandske Maaneds-Tidender* chose to print verses I, VII–X, XII, XV–XVI and the last XXV. See Document 6.

[3] The annals are listed according to the dates of the annalist with the oldest first.

was impeccable as were their morals. They were permitted to stay in the Danish houses or warehouses. They gave large receptions with many courses and many kinds of drinks. They invited district governors and high-ranking gentlemen. They did not conduct any trade with the inhabitants, who tolerated them. But it was said that gifts were exchanged between them and the high-ranking Icelanders. They drew and painted what they saw. Some of them travelled east to Mount Hekla and to Skálholt, bestowing gifts on high-ranking people and those who served them. District governor Ólafur [Stephensen] sent a man north to Skagafjörður on their account to obtain black lignite and black and white agate and to Hólar to seek newly printed books. They said they had sailed all around the world and had now perfected their voyage. They now wanted to go home. They sailed away at the approach of winter.

Ketilsstaðaannáll 1742–1784. Written by Pétur Þorsteinsson (1720–95). Published in *Annálar 1400–1800*, vol. IV, Reykjavík, 1940, pp. 424–5.

1773
A ship came to Hafnarfjörður in the south from England. Its commander was a lord or junker called Banks, 26 years old. He was accompanied by a Swedish doctor called Solander, with whom he was on an intimate footing. Another junker was named Troilus, an archbishop's son from Uppsala in Sweden. Accompanying them were 6 sumptuously dressed lackeys. This English gentleman was also accompanied by a few draughtsmen, who painted various things they saw. Almost 50 men in all were on the ship. They stayed for seven weeks in Hafnarfjörður and had previously sailed south under the pole and there discovered some unknown islands. Now Lord Banks also wished to explore under the north pole and received the permission of the King of Denmark to do so. Ólafur Stefánsson [Stephensen], the district governor (*amtmaður*), immediately became a friend of Mr. Banks who visited him many times with his party. These gentlemen travelled to Mount Hekla and examined it closely. They also examined the hot spring Geysir and called two Icelandic puppies they obtained here, Geysir and Hekla. They stayed at the bishopric of Skálholt with Dr. Finnur [Jónsson] giving him a silver embossed shaving razor, but the schoolmasters, who composed in Latin gratulatory odes in their honour each received a silver pocket watch. The district governor Ólafur sent men north to Hólar to obtain books. Also at his suggestion men were sent to seek natural specimens in Tindastóll Mountain. Besides which the district governor either obtained or had the major manuscripts of the Icelandic sagas copied in order to serve Mr. Banks. The district governor and Banks also exchanged many and valuable gifts. Governor Thodal also gave gifts to Mr. Banks. They took complete male and female costumes, furniture, harness and tools they had had made here back with them to England. They gave sumptuous receptions here and they conducted themselves with great modesty and generosity.

Íslands árbók 1740–1781. Written by Sveinn Sölvason (1722–82), lawman. Published in *Annálar 1400–1800*, vol. V, Reykjavík, 1955, pp. 73, 75–77.

1772
At the end of summer a rare English ship came to the south of Iceland, staying during autumn, treating our countrymen well, but neither bought nor sold any goods, but more will be written about this later.

1773
The English ship, which is mentioned for the year 1772, came to Hafnarfjörður. It is said that our countrymen had first been afraid because a ship was not expected at the time. But one man was courageous enough to row his boat over to these men and piloted the ship into the harbour. He was paid handsomely and various gifts were bestowed on him. The commander of the ship was an English junker (*jungherra*) or in their language a lord (which is said to mean the same as *lávarður* in days past). I have neither heard nor read about his given name, but he was called Bangs [*sic*]. He was very eager to explore the whole world and though he was said to be no more than 26 years of age, he had travelled far south in the world, or it was said under the southpole, he had even found some unknown islands, which I lack knowledge to write about. He wanted to sail again for the same purpose, but the Spanish objected, refusing to allow the English to conduct such spying missions in that part of the world. The ship that was being prepared for this voyage, capsized in the river Thames near London. He complained, receiving the answer: 'Either this ship, Sir, or nothing else' (this account I have from Danish newspapers). Mr Bangs was forced to change his mind and decided to explore the North Pole, to which effect he received a passport from Denmark, but whether he travelled further than to this country, I have not been able to discover. About fifty menn were aboard the ship, among them a Swedish doctor and naturalist called Solander, who lived on an intimate footing with the lord, and another junker called Troilus, son of an archbishop from Uppsala in Sweden. These men were able to talk to the people here and were his interpreters. Besides which he had six lackeys in splendid livery. He sent men north to buy Icelandic books from the printing press at Hólar and to search for rare stones in Tindastóll. Then he had bought a complete set of clothing, both male and female, as well as farming implements, harness and tools.

This English gentleman travelled himself east to Mount Hekla and immediately looked at the hot spring, which is called Geysir. He found both so remarkable that he had named two puppies after them, which he bought here. At the same time he came to Skálholt and stayed there both on his way to and from Hekla and gave the bishop a silver-embossed razor, but both the schoolmasters, who composed odes of gratulation in Icelandic and Latin in his honour, each a silver pocket watch. He paid his guides handsomely as was his custom with everyone who did him any service. On his ship he had excellent musicians, who could be heard on occasion. He also had some virtuosos, who sketched and drew many things. And as he was extremely wealthy in his fatherland, his receptions and dinners were of exceptional excellence, the likes of which have never been seen here. This ship lay here in Hafnarfjörður for seven weeks and then returned home.

Vatnsfjarðarannáll hinn yngsti 1751–1793. Written by the Reverend Guðlaugur Sveinsson (1751–93). Published in *Annálar 1400–1800*, vol. V, Reykjavík, 1961, p. 371.

1772–3
In the autumn an English ship came to Hafnarfjörður for pleasure; their commanders travelled to Bessastaðir, Skálholt and Hekla and observed here this and that; they were very generous and gave gifts to people both high and low and travelled back that same autumn.

Djáknaannálar: Úr djáknaannálum, 1731–1794. Written by Tómas Tómasson, *stúdent* (1756–1811). Published in *Annálar 1400–1800*, vol. VI, Reykjavík, 1986, pp. 175–6.

1772

In August an English ship came to Hafnarfjörður, not very large, with cannon and well fortified. The commander was Henry Banks, a young gentleman, about 26 years of age. Accompanying him were all kinds of learned men, among them were Dr Solander and a son of an archbishop from Uppsala in Sweden names Tróílus. They travelled to Skálholt and from there they climbed Hekla and observed the Geysir hot spring, drew all novelties that aroused their attention, almost 50 were on board. The district governor Ólafur [Stephensen] sent men north to Skagafjörður on their account to get black lignite and white and black agate and rare books. It was said that they got five copies of every available book. The district governor also had manuscripts copied for them. They bought complete costumes both male and female, all kinds of household utensils, tools and harness. They stayed in the Danish houses, gave large receptions with many courses and many wines, invited the district governor and other high-ranking gentlemen. They did not trade but it was said that the people at Álftanes did not lose out during their stay. It was said that the young Mr. Banks had an annual income of 24000 pounds sterling. He and Solander said they had sailed all around the world and had now completed that voyage. They now wished to return home and they sailed from here at the approach of winter. They found Hekla and Geysir novel and remarkable and therefore Banks named two puppies, which he obtained here, after them. He gave the bishop Dr Finnur [Jónsson] a silver-embossed razor but the schoolmasters each received a silver watch. He paid his guides well and all those who did him some service. On his ship were excellent musicians, who played on various occasions, also draughtsmen whom he had draw many things, for instance the daughter of the district governor who was dressed in her best apparel richly adorned with silver ornaments.[1] He gave the district governor various rare objects, among them was a silver embossed fowling-piece.

---

[1] This would have included a wrought-silver openwork belt and other decorative silver ornaments, such as a pendant and brooch. The daughter was Þórunn Ólafsdóttir and this illustration can be seen in the British Library, Department of Manuscripts, Add MS 15512, f. 9.

APPENDIX 6

Costs for Hiring the *Sir Lawrence*

1772 December [10]
[f. 259]
Joseph Banks Esquire [debtor] to James Hunter

To Joiner pr Acc^t.	£ 51.	1.	–			
To Plumer pr D°.	8.	7.	$2\frac{1}{4}$			
To Carpenter pr. D°.	6.	18.	6			
To the Baker pr D°.	28.	18.	–			
To Burrs pr D°. for Boat	35.	5.	6			
To Ship chandler pr D°. for Swivels Coullers &c	6.	13.	6			
To the Small Boat pr D°.	7.	7.	–			
To the Hearth & Copper pr. D°.	30.	10.	$11\frac{1}{2}$			
To Leather for boats Oars pr D°.	1.	–.	$9\frac{3}{4}$			
To Glazier pr D°.	1.	4.	4			
To Blocks &c for Boats	–.	16.	–			
To Smith for Ring bolts for Hearth	–.	12	$5\frac{1}{2}$	178	15	$1\frac{1}{2}$
To Coopper pr Acc^t.	5.	–.	–.			
To Brewer for Beer & Casks	18.	–.	–.			
To 13 Grose of Porter @ 6/ pr Dozⁿ,	46.	16.	–,			
To 13 Packing Casks for D°.	3.	18.	–.			
To Potatos & Bread	–.	15.	–.			
To Butcher pr Acc^t	3.	–.	5.			
To Hay & Cabage 13/6 one Bale oats 13/	1.	6.	6.			
To Small beer	1.	10.	–.			
To Oats for Hogs	2.	2.	10.			
To Boats Cable	1.	12.	2.			
To a Deepsea Line 200 f^m.	1.	2.	–.	83	– 2	11
Carried over				£261.	18	$-\frac{1}{2}$

[f. 260]

Brought Over				£261.	18	$-\frac{1}{2}$
To a Cable 120 f^m. pr Acc^t	£ 51.	7.	–.			
To Anchor 8. 3.11 @ $3\frac{1}{4}$	14.	9.	$\frac{1}{2}$	67.	10.	4
To 4 Rings 9^{sh}/4^d Stock £1. 5. –	1.	14.	4			

The above Sixty seven pound 10/4 to be charg'd Capⁿ. Riddle

[*Margin*: Port charges Lond^{on}]

To Clearing at the Custom house	£1.	11.	6			
To a Boat down the Pool	–.	5.	.			
To a Waterman attending the Ship	1.	3.	.			
To Pilotage to the Downs	5.	5.	.			
To D°. out the Needles	–.	10.	6.			
To McDonalds wagges [*sic*] pr Acct.	7.	10.	8.			
To 2 Mens D°. pr D°.	12.	18.	.			
To a Waterman returning the boat	–.	3.	.			
To 2 Boats up the Pool	–.	11.	.			
To Lights Trinity dues & Reporting	8.	10.	9.			
To Light of May & Greenwich dues	1.	4.	.			
To Searcher & clearing the Ship	–.	10.	.	40.	2.	5.
				£369.10.10		
To Ships pay from 4th July to 4th Decr. Being 5 Months @ £100				500		
				£869.10.10		
				534. 5. –.		
				335. 5.10.		

[f. 261]

Contra   P^s

By Cash.		£200.	–. –.
By D°.		100.	–. –.
By Sundries at Iceland			
By 2 firkens flower [flour?]	.	19.	.
By Cash	21.	–.	–.
By 1 Cask bread 3 Q^r. w^t.	2.	8.	–.
By 1 D°. D°.	1.	16.	–.
By 4 Grose of Bottles too Capn. Peaterson	3.	12.	–.
By 5 D°. D°.	4.	10.	–.
By Cash	200.	–.	–.
	534.	5.	–.

December 22. 1772. then examind & allowd this account & the balance being three hundred thirty five pounds five shillings & ten pence being paid the Contract of Charter party was fulfilld & became of no farther use

pr James Hunter [*signed*]

ML, Banks MS, Voluntiers, Instructions etc. for 2nd voyage. Safe 1/11 ff. 259–261.

APPENDIX 7

The Latitude and Longitude of the Danish Settlements in Davis's Streights[1]

	Latitude	Longitude
The Frau or Woman's Islands[2]	72.30	
Jacobs Bay or Omenack[3]	71	
Ritenbank[4]	70.10	
Goodhaven or Liveley[5]	69.12	52. 0.10
Jacobs haven Ice fine or river[6]	68.50	51.55
Christians hope[7]	68.50	
Egeiminden[8]	68.20	
Whale Fish Islands[9]	68.50	51.30.10
Holstein berg[10] or Weyder river[11]	67.10	52.10.10
Sugar Top[12]	65.30	52.20.10

[1] The Danish settlements, called *koloni* in Danish, often comprised both trading posts and whaling stations. They were founded during the 18th century. Very special thanks to both Dr Thorkild Kjargaard of the University of Greenland and Niels Frandsen for help with this document.

[2] *The Frau or Woman's Islands*: present-day Upernavik, the colony was founded in 1771. Woman Island or Iron Bay was the name that William Baffin gave in 1616. Its exact location is 72°47′ north latitude and 56°10′ west longitude.

[3] *Jacobs Bay or Omenack*. The colony Umanak was founded in 1761, present-day Uummannaq. Jacob's Bay is an English version of the Dutch whalers' name for the bay (Stickende Jacobs Bugt) from the 17th and 18th centuries. Its exact location is 70°40′N, 52°00′W.

[4] *Ritenbank*: Ritenbenk was founded in 1755. Its exact location is 69°45′N, 51°12′W.

[5] *Goodhaven or Liveley* (see p. 439 above): Godhavn was founded in 1773, present-day Qeqertarsuaq. Its exact location is 69°14′N, 53°31′W.

[6] *Jacobs haven Ice fine or river*: Jacobshavn was founded in 1741, present-day Ilulissat and Ilulissat Icefjord. Its exact location is 69°13′N, 53°01′W. Ilulissat is one of the world's largest ice fjords and the 'Ice fine or river' must refer to this. See further p. 611, n. 8 above.

[7] *Christians hope*: Christianshåb was founded in 1734, present-day Qasigiannguit. Its exact location is 68°49′N, 51°05′W.

[8] *Egeiminden*: Egedesminde was founded in 1759, present-day Aasiaat. Its exact location is 68°42′N, 52°45′W.

[9] *Whale Fish Islands* is the English version for the old Dutch name Walvisch Eylanden, between Disco Islands and Aasiaat. From 1778 the island has been called Kronprinsens Ejland in Danish. Its exact location is 69°10′N, 53°19′W. Captain Stevens appears to have written Hound Island in pencil on the document. Hunde Ejland [literally Dog Island] is a small island north-west of Aasiaat. In the years after 1807 this was a popular trading post visited by Englishmen.

[10] *Holstein berg*: Holsteinsborg was founded in 1755, present-day Sisimiut. Its exact location is 66°55′N, 53°40′W.

[11] *Weyders River* is an unknown place name, possibly a name used by English whalers in the meaning 'the wider river'.

[12] *Sugar Top*: Sukkertoppen was founded in 1756, present-day Maniitsoq. Its exact location is 65°24′N, 52°52′W.

Good Hope¹ in Baals River²	64.10	52. 5.10
Fisknist³	63.20	
Frederiks Hope⁴	62.30	
*Julianna Hope*⁵	61. –	50. 5.10

Cap*tain* Stevens⁶ thinks there are not more than 500 Europeans in these Colonies many if not most of these will Quit the Country, if an opportunity presents itself their chiefs however have Receivd Positive orders from home not to quit on any account. Motzfelt he thinks will stay if he is Left here.

He Speaks of a River 15 Miles⁷ wide in the Continent of Greenland opposite Disco Island the Banks of which are very high this River he says vomits out an incalcuable number of Ice Islands⁸ every year which proceed westerly to some distance & are then carried by Scottish currents along the Coast of America.

He says that he has observd a great Northerly Wind (?) which comes as he believes from beyond Spitsbergen & keeping closely [to] the Coast of Greenland proper Round Cape Farewell & then takes a Northerly course reaches almost to Disco Island where it meets the Southern current which Runs from the ? North always running on the american Shore. *T*he wood found in Iceland Left there by the Ice which every year accumulates on the N E Side of the Island he thinks must come from the Northern parts of the Russian Empire

Mr Latrobe⁹ tells me that there are 26 Europeans only attachd to the 3 Settlements of the United Brethren.¹⁰

Wisconsin, Banks Papers, f. 80. The writer of the table is unknown but the written notes are in Banks's autograph.¹¹

---

¹ *Good Hope*: Godthåb was founded in 1728, present-day Nuuk. Its exact location is 64.10′N, 35°45′W.

² *Baals River*, more commonly called Balls Revier, a name given by English explorer James Hall in 1612 to the fjords north of Godthåb. He called the two branches of the Godthåbsfjord Balls Revier and Lancaster Revier.

³ *Fisknist*: Fiskenæsset was founded in 1754, present-day Qeqertarsuatsiaat. Its exact location is 65°05′N, 50°41′W.

⁴ *Frederiks Hope*: Frederikshåb was founded in 1742, present-day Paamiut. Its exact location is 62.00′N, 49°43′W.

⁵ *Julianna Hope*: Julianehåb was founded in 1774, present-day Qaqortoq. Its exact location is 60°43′N, 46.03′W.

⁶ Unidentified.

⁷ About 3.5 miles (5–6 km) is nearer the truth.

⁸ He means icebergs. He is describing the Ilulissat Icefjord, now a UNESCO heritage site. It is the sea mouth of Sermeq Kujalleq, one of the few glaciers through which the Greenland ice cap reaches the sea. Sermeq Kujalleq is one of the fastest and most active glaciers in the world and has helped in developing our understanding of climate change.

⁹ Christian Ignatius Latrobe: see Appendix 13.

¹⁰ The Moravian settlements of Ny Herrnhut (1733), Lichtenfels (1758) and Lichtenau (1774).

¹¹ Note in Banks's autograph, verso: 'a List of the Danish Settlements in Greenland from Mr Stevens'.

APPENDIX 8

Abstract of the Trade Ordinance of 13 June 1787

Chapter 1[1]
Concerning the Iceland trade and the privileges of the merchants in general

§ 1 This trade is free to all His Majesty's subjects in Europe
„ 2 Strangers may not at all carry on any trade upon Iceland
„ 3 How to proceed with foreign ships touching Iceland in cases of necessity
„ 4 Punishment for the ship-owner who fits out a ship for such an illegal trade
„ 5 Punishment in cases where such an illegal trade is carried on by the crew
„ 6 Punishment if a foreign ship throws any thing over board when search'd
„ 7 Punishment for His Majesty's subjects who carries on such prohibited trade under the coast or in the island of Iceland
„ 8 The inhabitants and the Magistrate are bound to disclose and examine such crimes
„ 9 The accuser shall have part of the fines
„ 10 All masters going to Iceland, are to provide themselves with an Iceland passport and faithfully to behave according to the laws
„ 11 Regulation for the ships of His Majestys subjects, coming to Iceland without having [a] passport
„ 12 Regulation for the reports on goods going to Iceland
„ 13 Regulation for the direct ordered, consigned as well as those upon speculation only, outward bound cargoes
„ 14 Regulations for the homeward bound Iceland cargoes
„ 15 The merchants are exempted from paying duty
„ 16 Regulation for expeditions from Iceland to the Mediterranean
„ 17 Regulation for Expeditions from Iceland to the West-Indies
„ 18 Ships fitted out for catching whales are permitted to pass the winter in Iceland
„ 19 Ships employed in the Iceland trade are permitted to partake in the fish premiums
„ 20 Resignation of His Majesty's property to the merchants (freetraders)
„ 21 Regulation for the packet and the forwarding of letters with other ships
„ 22 The posts in Iceland shall conform themselves according to the voyage of the packet

---

[1] This is not, as is stated in the Wisconsin Papers, a 'Synopsis of a British book on trade with Iceland, outline of regulations' but a precis of every article of the trade ordinance of 13 June 1787, the so-called Free Trade Proclamation or 'Forordning ang. den islandske Handel og Skibsfart' (see Document 229). These were the trade regulations that were in force in Iceland during the Napoleonic Wars. In the ordinance all the articles were summed up in the margin. It is the marginal notes that have been translated. The document gives page numbers, but the ordinance has been published in many different works: see e.g. *Lovsamling for Island*, V, pp. 417–74.

2ᵈ. Chapter
concerning the privileges and obligations of the merchants in Iceland

§ 1 The Iceland merchants are permitted to trade with all His Majestys dominions
„ 2 None, besides those who are burghers are permitted to trade with goods imported
„ 3 The merchants may likewise carry on any other lawful trade in Iceland
„ 4 They may lawfully possess landed property
„ 5 They may have fisheries
„ 6 Encouragement for the fishery upon a large scale
„ 7 Encouragement to rise[1] the fishery specially on the north and east part of the island
„ 8 Encouragement for establishing more harbours
„ 9 How to promote the communication by sea
„ 10 Privileges for strangers setting themselves in Iceland
„ 11 The districts of the market-towns are not at all bound to trade with a certain market town
„ 12 The outharbours must always be furnish'd with a proportional quantity of necessities
„ 13 The Magistrates are oblig'd allways to procure an exact information if the island is sufficiently furnish'd with necessities
„ 14 General rules for the quality of goods imported to Iceland
„ 15 Regulations for goods damaged on the sea, specially grain
„ 16 Regulations for weight, measure and coin
„ 17 Regulations for the book of accounts, and the accounts of the inhabitants
„ 18 To what right a merchants claim is entittled
„ 19 Regulation for money set upon interrest
„ 20 Regulation for applying the laws
„ 21 In what cases the merchants are granted an extraordinary court
„ 22 The Magistrates shall assist the merchants as much as possible

3ᵈ. Chapter
Concerning the Iceland land & sea produce, and manufactories

§ 1 Regulation for the inhabitants fishery in boats
„ 2 Compensation for the proprietor whose land is used by the fishers
„ 3 Regulation for the whale-fishery
„ 4 Encouragement to catch sharks, seals, salmons, herrings and flounders
„ 5 Common rules for the quality of fish
„ 6 Rules for Iceland manufactories and a more advantageous employ thereof
„ 7 Rules for preserving the Eider-ducks
„ 8 Regulations for the sulphur & Salt manufactories
„ 9 Encouragement to bring in use more minerals
„ 10 Encouragement to make use of several other produces specially floating timber
„ 11 Advantages granted the most usefull tradesmen
„ 12 Regulation for the wool manufactors
„ 13 Encouragement for a more advantageous fabrication

---

[1] Here the English translation appears to break down; what is meant is 'increase'.

„ 14 Advantages granted the wool & linnen Weavers
„ 15 Premium granted to encourage the spinning & weaving of wool & Flax
„ 16 Dispositions made in order to instruct able apprentices in the spinning & weaving of wools & linnen

Wisconsin, Banks Papers, ff. 56–58.

APPENDIX 9

The Trading Agreements of 16 June and 22 August 1809 Made in Iceland by Captains of the Royal Navy[1]

[16 June 1809]

We undersigned

Friderich Count Trampe,
His Danish Majesty's Chamberlain
and governor of the island Iceland

Francis John Nott Esq[r 2]
Captain of His Britannic
Mayesty's Sloop *Rover*

Do hereby certify
that we have agreed to the following articles of treaty

Art. 1.

The commanding officer of His britannic Majesty's Sloop the *Rover* cannot consider the Island of Iceland as neutral during the present war between Denmark & Great Britain, yet he gives his word for himself, Officers and Ships company not to act in an hostile manner against the Island of Iceland during his present[3] Stay on the coast, and to respect private property as well as public Institutions.

Art. 2.

He also promises to protect the Fishing vessels, belonging to the island of Iceland as far as in his power.

Art. 3.

It will be granted to Vessels licenced and returning from Great Britain to trade with the merchandize they bring with them from thence. The same liberty will be admitted to brittish Trading Vessels, arriving to Iceland with the usual Bills of Lading and Clearances, and they shall enjoy the same rights and privileges, as are granted to danish Vessels.[4]

Art. 4.

It shall be allowed to brittish Subjects, if they wish, to establish themselves and to trade in the Island, to do so without taking the oath of allegiance to any other than their lawfull sovereign, the King of Great Britain, and to remain and trade one Year after certain

---

[1] Umlauts have been removed from the y (ÿ) throughout this document.
[2] Francis John Nott became a lieutenant in the Royal Navy in 1795, commander in 1801 and captain in 1810.
[3] *Present* has been added.
[4] *Ships* was first written.

intelligence of peace, between Denmark and Great Britain shall be received by the government.

Art. 5.

The brittish Subjects shall, as long as they remain in Iceland, submit to the existing Laws, conforming to which they shall enjoy the same liberties as danish Subjects.

Art. 6

English and Danish Vessels, loaden with Iceland produce, as fish &c., shall be furnished by the government with the necessary clearances to proceed to Sea.

Given under our hands and seals, at Reikevig the 16th of June 1809

    Fr. Trampe   Fr. Jho Nott
    (L. S.)   (L. S.)

The above convention has been shewn to me and I am acquainted with the contents[1]

                                                                                                                    J. Savignac

NAI, Jörundarskjöl. Original document.[2]

[The Agreement of 22 August 1809]

Agreement
between His Royal Danish Majesty's Councillor of State and Chief Justice of Iceland and His Danish Majestys Bailiff in the West County of said Island of the one part. The Honorable Alexander Jones[3] Captain of His Britannic Majestys Sloop of War the Talbot and Samuel Phelps of the City of London Esquire of the other part.

Article first

All proclamations, laws appointments, &c. &c made by Mr Jorgen Jorgensen since his arrival in this country to be abolished and totally null and void from the moment this agreement is signed.

Article second

The former government to be perfectly restored and the government to devolve upon the said Lord Chief Justice of Iceland and the said Bailiff of the West County of Iceland native Icelanders they being the next in power to the late Governor Count Trampe.

Article third

All officers under the Danish Government are at liberty to return to their offices.

Article fourth

---

[1] This is written in Savignac's hand. It was the British supercargo, employed by Phelps & Co., who had implored Nott to come to the rescue of the English trade.

[2] This is Trampe's copy. The copyist was probably his secretary, Jörgen Flood. Nott's copy of the original agreement is to be found in TNA, ADM 1/692.

[3] See p. 401, n. 2.

The Government shall be responsible for the protection of all the British Subjects and property that now is and may be on the Island and that all transgressions thefts and personal assaults committed against British Subjects or their property shall be punished with the same rigour and according to the same laws as if the property belonged to natives.

Article fifth
No Batteries to be erected (and the Battery now at Reikevig to be destroyed) no militia to be raised on the Island or the Country in any way to be fortified or armed.

Article sixth
All officers or other persons either armed or unarmed which during the late events have taken part with Mr Jorgen Jorgensen shall no longer be in employment but their persons and property in every respect (or whomsoever or of whatever nation they may be) shall be respected and protected the same as other persons or natives.

The convention between Count Trampe and Captain Nott of the 16 June last[1] shall be in full force and be published throughout this Country together with this Agreement without delay.

Article seventh
All merchants houses which are shut up in this Country shall immediately be opened and the Merchants of the Island be permitted to continue or carry on their trade as formerly.

Article eighth
All Danish property and public money to be restored.

As Witness our hands and seals this twenty-second day of August one thousand eight hundred and nine.

[*Signed*]

Magnus Stephensen
His royal danish Majest.
Counsellor of State and Chief
Justice of Iceland

Alex. Jones
Capt H· B· M; Sloop *Talbot*

Stephan Stephensen
Amtmand over Vester
Amtet i Island[2]

Saml Phelps

NAI, Jörundarskjöl. Original document.[3]

---

[1] 'of the 16 June last' has been added.
[2] Written in Gothic Danish script: district governor of Western Iceland.
[3] Jones took a copy with him, now in TNA ADM 1/1995. Published by Hooker in *A Tour in Iceland*, II, pp. 99–102; printed in *Lovsamling for Island*, VII, pp. 258–61.

APPENDIX 10

The Order in Council of 7 February 1810

At the Court of the *Queen's Palace*, the 7th of *February* 1810

PRESENT
The KING's Most Excellent Majesty in Council.

Whereas it has been humbly represented to His Majesty, that the Islands of Feroe and Iceland, and also certain Settlements on the Coast of Greenland, Parts of the Dominions of Denmark, have, since the Commencement of the War between Great Britain and Denmark, been deprived of all Intercourse with Denmark, and that the Inhabitants of those Islands and Settlements are, in consequence of the Want of their accustomed Supplies, reduced to extreme Misery, being without many of the Necessaries and of most of the Conveniencies of Life:

His Majesty, being moved by Compassion for the Sufferings of these defenceless People, has, by and with the Advice of His Privy Council, thought fit to declare His Royal Will and Pleasure, and it is hereby declared and ordered, that the said Islands of Feroe and Iceland and the Settlements on the Coast of Greenland, and the Inhabitants thereof, and the Property therein, shall be exempted from the Attack and Hostility of His Majesty's Forces and Subjects, and that the Ships belonging to Inhabitants of such Islands and Settlements, and all Goods, being of the Growth, Produce, or Manufacture of the said Islands and Settlements, on board the Ships belonging to such Inhabitants, engaged in a direct Trade between such Islands and Settlements respectively, and the Ports of London or Leith, shall not be liable to Seizure and Confiscation as Prize:

His Majesty is further pleased to order, with the Advice aforesaid, that the People of all the said Islands and Settlements be considered, when resident in His Majesty's Dominions, as Stranger Friends, under the Safeguard of His Majesty's Royal Peace, and entitled to the Protection of the Laws of the Realm, and in no Case treated as Alien Enemies:

His Majesty is further pleased to order, with the Advice aforesaid, that the Ships of the United Kingdom, navigated according to Law, be permitted to repair to the said Islands and Settlements, and to trade with the Inhabitants thereof:

And His Majesty is further pleased to order, with the Advice aforesaid, that all His Majesty's Cruizers and all other his Subjects be inhibited from committing any Acts of Depredation or Violence against the Persons, Ships, and Goods of any of the Inhabitants of the said Islands and Settlements, and against any Property in the said Islands and Settlements respectively.

And the Right Honourable the Lords Commissioners of His Majesty's Treasury, His Majesty's Principal Secretaries of State, the Lords Commissioners of the Admiralty, and the Judge of the High Court of Admiralty, and the Judges of the Courts of Vice Admiralty, are to take the necessary Measures herein as to them shall respectively appertain.

W. Fawkener.

The London Gazette, 10 to 13 February, 1810, no. 16341, front page.

APPENDIX 11

Admiralty Prize Court Bills against *Den Nye Prove*

DEN NYE PROVE – Hans Nielsen, M[r1]
To the Marshall of the Adm*iral*ty
1807
Nov. 10,:

Making Embargo on Ship at Blackwall	1. 1. –.
Painting foul Anchor & No.	3. –.
Taking Inventory of Stores & Materials & overhauling Stores	1. 4. 6.
Measuring and Mustering	1. 1. –.
Collecting and arranging Ship Papers delivering them on Oath and bringing up the Master to be produced & examined on the Standing Interrogatories	5. 5. –.
Possession Care and Management of Ship and Cargo from 16 Nov. 1807 to 20th June 1808. 216 days at 3/	32. 8. –.
Release and delivering Possession	13. 4
	£ 41.15.10.

 Received 21 June 1808 of Messrs Boulton & Baker Forty one Pounds fifteen Shillings ten Pence and Charges on Ship *Den Nye Prove* as Per Acc*oun*t for the Marshall of the Adm*iral*ty.

JNO. DEACON.

£ 41.15.10.

SL, Banks Papers, I 1:29. Contemporary copy.[2]

---

[1] *Den Nye Prove* (sometimes called *Denne Pröve* or *Nye Pröve*) was owned by the Danish merchant house Ørum & Wulff, which conducted a long-standing trade to Húsavík, Seyðisfjörður and Eskifjörður in the north and east of Iceland. The master was Hans Nielsen. The ship was captured on 26 October 1807 by the *Spencer* and brought to Yarmouth and later to London. This ship is mentioned in Documents 81, 88, 90, 91, 97 and 102. See TNA, HCA 24/166 for details concerning the capture of the ship.

These papers enumerate the cost of the proceedings in the Admiralty courts regarding the release of the Iceland ships (see Document 97). The Banks Papers in the Sutro Library have similar documents relating to the following ships: the *Bosand* (I 1:30 and 31), the *Husevig* (I 1:32 and 33) and the *Johanne Charlotte* (I 1: 34 and 35).

[2] Verso note: 'Den Nye Prove | Nielsen | Marshalls Bill | £41.15.10'.

ADMIRALTY PRIZE COURT.

[CAPTOR'S BILL]

DEN NYE PROVE –

Hans Nielsen, M*aste*r

Taken by His Majesty's Ship *Spencer*
The Hon*orab*le Robert Stopford commander
one of the Fleet under the Command of Admiral Lord Gambier and brought
to London prior to the declaration of Hostilities against Denmark.

1807	Expences on behalf of the Crown.	
	Proctors Fee retained	6. 8.
	Drawing Attestation in Verification of the Ship Papers & engrossing the same and Stamp	11. 8.
	Oath thereto and attendance	7. 8.
Nov. 18:	Attending and producing the Master as a Witness on the Standing Interrogatories, who was sworn and monished as usual by	
	Interlocutory and Act	7. 8.
	Registrars Attendance and Surrogates Fee	7. 8.
	Paid the Interpreter	3. 4
	Attending the Witness to the Exam*ination* and fixing a time and procuring his Examination to be taken	13. 4.
Dec. 2:	Attending before a Surrogate and bringing the Ship Papers	5. –.
	Praying Monition and Surrogates Fee	6. –.
	Registrars Attendance	6. 8.
	Act of Court	2. 8.
	Filing Attestation	2. 8.
	paid for Monition under Seal Stamps and Extracting	1.17. 8.
	Copy for service	6. 8.
	Paid the Marshall for Service	6. 8.
	Certificate of Service	3. 4.
	Paid the Examiner for taking the Exam*ination*s and Stamps	3. 3. 4.
	Paid him for extra labour	1. 4. –.
	Paid the Interpreter	2.11. 8.
	Praying Publication and Act	7. 8.
	Attending in the Registry inspecting the depositions and bespeaking Copy	6. 8.
	Paid for office Copy thereof, Stamps &c collating	1.14. 8.
	Extracting	6. 8.
	Perusing and Abstracting	6. 8.
	Copy for his Majestys Advocate	1. –. –.
	Attending the Translator and ordering an Abstract of the Ships Papers	6. 8.
	Paid for the Same	1.14. 6.

Attending in the Registry with the Translator and inspecting the Ship Papers & selecting such as were necessary to be translated on behalf of the Crown	6. 8.
Attending and inspecting the Translations and bespeaking Copy	6. 8.
Paid for Office Copy thereof Stamps & collating	8. 4. 6.
Extracting	6. 8.
Perusing and Abstracting	6. 8.
Copy for his Majesty's Advocate	1. -. -.
Paid for Office Copy Attestation as to Ship Papers Stamps collating & Abstracting	16. 8.
Perusing the same and Copy for His Majesty's Advocate	6. 8.
Copy Attestation of the Master as to the Property of the Ship and Clerks Fee	11. -.
Perusing and abstracting	6. 8.
Copy for His Majestys Advocate	10. -.
Copy Exhibit A: annexed thereto, Clerks Fee and perusing and Copy for His Majesty's Advocate	7. 8.
Drawing Case for the Opinion of His Majesty's Advocate thereon and fair Copy	13. 4.
Attending His Majestys Advocate therewith and with Fee paid by Claimant and afterwards carrying the same to the Claimants Proctor	6. 8.
Drawing Allegation for Condemnation of the Ship and Cargo and engrossing the same and Stamp	11. 8.
Attending and returning the Monition and giving in the Allegation when the same was admitted and Act	7. 8.
Paid filing same	4. -.
Drawing Case for the Hearing on behalf of the Crown	13. 4.
Fair Copy	6. 8.
Attending His Majesty's Advocate therewith and feeing him	6. 8.
Paid his Fee	2. 2. -.
Attending when the Cause was assigned for sentence on the 1st Assignation and Act	7. 8.
The like when the cause was assigned for Sentence on the 2d Assignation and Act	7. 8.
May 17: Attending Informations Fee when the Judge at my Petition on Motion of His Majesty's Advocate by Interlocutory decree condemned the Ship	6. 8.

	and Cargo to His Majesty as taken prior to the declaration of Hostilities against Denmark and Act	7. 8.
	Interlocutory Fees to Judge Registrar and Marshall	2.18. 8.
	Paid the Registrar for drawing the Interlocutory	6. 8.
	For various Attendances on and Consultations with His Majesty's Advocate perusing sundry Letters from His Majesty's principal Secretary of State and the Lords of the Treasury regarding this and other Iceland Vessels taking the necessary measures at various times for suspending the Proceedings and writing to His Majestys Government &c thereon, proportion thereof chargeable to this Case	13. 4.
	Attending His Majesty's Advocate in consequence of the final direction of Lord Castlereagh for releasing for releasing [sic] the Iceland Vessels and submitting the Evidence in this Case to his Consideration with Fee paid by Claimant for his Opinion whether this Vessel came within the Description of those so directed to be released	6. 8.
June:	Attending and consenting by Advice of His Majesty's Advocate to the Condemnation of this Ship and Cargo being rescinded which was done accordingly and the Ship /& Cargo/ were thereupon decreed to be released & Act	7. 8.
	Paid the Registrar for Acts Sportulage and Attendances	13. 4.
	The like for copying and dispatch	10. –.
	The Clerks	5. –.
	Officers of the Court	7. 6.
	Extra judicial Attendances & Consultation	13. 4.
	Proctors Clerks	5. –.
	Letters Messengers & sportulage	6. 8.
		£ 48. 3. –.

Den Nye Prove – Nielsen, Master

Received 20 June 1808 of Mess. Boulton & Baker Forty Eight Pounds Three Shillings – the amount of my Bill on Behalf of the Crown in the above Case

For the Kings Proctor

(Signed) JNO. FINCH

£48. 3. –.

SL, Banks Papers, I I:28. Contemporary copy.[1]

[1] Note, verso: 'Admty Prize Court Den Nye Prove Hans Nielsen Mr | Captors Bill £48.3 – Bishop [Charles Bishop] K P [The King's Proctor]'.

APPENDIX 12

Tonnage Regulations Affecting Trade between Iceland and the United Kingdom[1]

1st. The Iceland Merchants bind themselves to export Goods permitted by Law to be exported to the amount of £ [*blank*] a Ton of the Tonnage of each Vessel, at their option either to a port in the Danish Dominions, Eastwards or on their return to the Island of Iceland – Say £ [*blank*] a Ton on each Voyage, which is considered to Last altogether about 12 Months.

2d. The British Government after granting Licences permit the Iceland Merchants to export from Iceland Shark & Cod Liver Oil, Klipfish & every other article the produce or Manufacture of the said Island, without unloading or paying Duty in a British port.

3d. The Cargoes of Grain from Denmark may either be sold wholly or in part in the British port where the Vessel is permitted to go, in order to enable the Iceland merchants to make the export required.

4th. The goods from Iceland may be sold in Great Britain on paying the usual duties of His Majesty's Customs.

5th. A Merchant may export in one of his Vessels an amount exceeding £ [*blank*] a Ton of that Vessel, in which case he receives a Credit of export for his other Vessels.[2]

SL, Banks Papers, I 1:45. Contemporary copy.[3]

---

[1] This document is undated but is probably at the latest from c. August 1811. At the end of November Clausen had been complaining about the difficulties regarding the Iceland trade, and petitioned the Lords Commissioners of the Board of Trade regarding the import of klipfish and the duties on the train oil (Document 178). In December the government had decided 'that the state of the Trade of Iceland will shortly be relieved by a general regulation' (TNA, PC2/189, Fawkener to Hamilton, 18 December 1810). In August 1811 the *Orion* was granted a licence permitting the export of a cargo to Denmark 'and such addition to it at Leith as the parties may choose of permitted articles' (TNA, BT 6/204, no. 35,073). The letter from Corbett, Borthwick & Co. to Banks, dated 5 October 1811 (Document 213), demonstrates quite clearly that these tonnage regulations were now in force.

[2] Note in Banks's autograph, verso: 'Iceland proposal for Shipping English Trade.'

[3] The writer is unknown.

APPENDIX 13

Biographical Details of Banks's Correspondents

This appendix is arranged alphabetically by the surname of each of the correspondents. The biographical entry for each is brief, especially so if the man in question is well known and further information on him is easy to access. Some of the correspondents are, however, virtually unknown.[1]

Aall, Jacob (1773–1844), Norwegian politician and historian. He invested his inheritance in the immense Næs Ironworks in 1799. At the outbreak of war in 1807 he worked diligently at provisioning Norway and unsuccessfully urged his fellow Norwegians to restore, independently of Denmark, the peaceful relationship with England. In 1814 Norway declared independence and Aall was a member of the Norwegian Constituent Assembly at Eidsvoll, taking a prominent part in the framing of the constitution, later becoming a leading member of the newly founded Norwegian parliament *Storting* until 1830. His interest in Iceland was demonstrated by his financing of the publication of Werlauff's edition of the *Vatnsdæla Saga* in 1812 and the *Icelandic Lexikon* of the Reverend Björn Halldórsson in 1814. He shared Banks's concern regarding the situation in Iceland during the war, enlisting his help and sending him a collection of minerals.[2]

Anker, Peter (1744–1832), Norwegian diplomat and colonial governor in the Dano-Norwegian empire. The Anker family had become wealthy through the timber trade with England, being ennobled in 1778. Anker was sent on a six-year Grand Tour to England, Germany and France. He was very pro-English both politically and culturally. From 1773 to 1786 he was Danish consul in England, first in Hull then in London. During the American War of Independence he played an important role regarding the neutral Danish navigation. It was in 1784, when Denmark's general consul in Britain, that he came into contact with Banks regarding Iceland. In 1786 he was appointed governor of the Danish colony Tranquebar in India until 1808. In 1814 he was a member of the Meeting of Notables, which preceded the Norwegian Constituent Assembly. He travelled to England to gain British acceptance of Norwegian independence.

Baker, D. B. (fl. 1807), partner in the mercantile firm of Boulton & Baker (q.v.).

---

[1] The major sources used are the following: Andrésson, *Verzlunarsaga Íslands 1774–1806*; Benediktsson, *Sýslumannaævir*; *Dansk Biografisk Leksikon*; *Dictionary of National Biography*; *Dictionary of Scientific Biography*; Ehrencron-Müller, *Forfatterlexikon*; Erslev, *Almindeligt Forfatter=Lexicon for Kongeriget Danmark med tilhørende Bilande, fra 1814 til 1840*; *Norsk biografisk leksikon*; Ólason, *Íslenzkar æviskrár*; *Oxford Dictionary of National Biography*.

[2] Burgess, 'Ambivalent patriotism: Jacob Aall and Dano-Norwegian Identity before 1814'.

Barrow, Sir John (1764–1848), FRS, Second Secretary of the Admiralty, traveller, promoter of exploration. He served both on Lord Macartney's embassy to China (1792–4), and in the newly-acquired colony of the Cape of Good Hope in Cape Town until the French gained the colony in 1802. In 1804 he was offered the second secretaryship at the Admiralty by Dundas (then Lord Melville), which he held under Tory governments until his retirement in early 1845. Barrow was elected as a fellow to the Royal Society in 1805. He had a close relationship with Banks and his projects, being instrumental in sending off the expeditions of David Buchan, John Franklin, John Ross and William Parry. Barrow was an ardent imperialist and, with Banks's assistance, he initiated a series of expeditions to trace the course of the Niger. He was active in the African Association, interested in the search for the Northwest Passage and in Australia, advocating overseas settlement. A founder member of The Royal Geographical Society in 1830, he chaired their early meetings and later served as president (1835–7). His son of the same name, a founder member of the Hakluyt Society, visited Iceland in 1834 and wrote an account, *A Visit to Iceland, by Way of Tronyem, in the 'Flower of Yarrow' Yacht, in the Summer of 1834*, published in London in 1835 and translated into Icelandic by Haraldur Sigurðsson as *Íslandsheimsókn. Ferðasaga frá 1834*, Reykjavík, 1994.[1]

Bathurst, Henry, 3rd Earl Bathurst (1762–1834), British statesman. During this period he was President of the Board of Trade in the Portland Ministry (1807–9), and in Spencer Perceval's (1809–12). He was also Foreign Secretary for six weeks from 11 October 1809 to 6 December when Wellesley took over, during which period the aftermath of the Icelandic Revolution was under scrutiny. Thus he received the long memorandum from Count Trampe, the Governor of Iceland, castigating the revolutionaries, though it was his successor, Wellesley, who dealt with the consequences. In the Liverpool government in June 1812 Bathurst was appointed Secretary of State for War and the Colonies. He ended his career as Lord President of the Council under Wellington 1828–30.

Boddaert, Pieter (1730–95/96), Dutch physician and naturalist. He lectured on zoology at the University of Utrecht and in 1785 he published *Elenchus animalium* (Banks was one of the four scholars to whom it was dedicated), which included the first binomial names for a number of mammals. Banks met Boddaert, as he wrote in his journal, on his trip to Holland in early 1773, finding him to be 'a more liberal man than most I have found in this country'.[2] Their relationship was based on the exchange of plants and seeds and Boddaert's interest in Banks's expeditions to the north.

Bohnitze or Bohnitz, Jörgen (fl. 1810), Danish sea captain. He was master of the vessel *Jubelfest*, engaged in the Faroes trade during the Napoleonic Wars.

Boulton & Baker, London merchants, 45 Wellclose Square. Agents of the Iceland merchants, they were actively involved in relations between the British government and the Iceland trade during the Napoleonic Wars, their numerous letters to Banks a testament of their diligence. On 4 December 1807 they were entrusted with receiving public subscriptions to help the captured Danish and Norwegian masters and mates, at the time prisoners-of-war in England.

---

[1] Further see: Fleming, *Barrow's Boys*, pp. 1–12.
[2] Van Strien, 'Banks, Holland Journal', p. 174.

Bredal, Hans Georg (fl. 1807), Norwegian merchant. He was from Finnmark and a citizen of Vatneyri in western Iceland. He was the supercargo on the ship *Conferentz Raad Prætorius*, which had sailed from Bergen, and on its return from Iceland was captured on 20 September 1807 by the *Peacock* and taken to Leith. He, along with Sívertsen and Petræus, was one of the first merchants to turn to Banks for help.[1]

Bugge, Thomas (1740–1815), Danish astronomer, mathematician and surveyor, FRS (1788). He became an assistant in the Rundetaarn Observatory in 1760, and in 1761 he was sent to Trondheim to observe the Transit of Venus. He managed the first land survey and mapping of Denmark for the Danish Academy of Sciences. In 1777 he became professor of mathematics and astronomy at the University of Copenhagen and served three times as rector of the university. He travelled widely in Europe, including England, inspecting observatories. At his suggestion small observatories were built in Norway, Iceland, Greenland and Tranquebar, territories belonging to the Danish Empire. A member of numerous scientific academies, he was the Secretary of the Danish equivalent of the Royal Society (Det Kongelige Danske Videnskabernes Selskab) 1801–15 and during this period was responsible for the publication of their annual publication. He was a prolific author and he published in *The Philosophical Transactions*. In 1798 he was sent to Paris to represent Denmark in the International Commission on the Metric System. Banks and Bugge corresponded frequently on scientific matters. They also exchanged gifts, Bugge for example sending him works on Iceland.

Castenschiold, Johan Carl Thuerecht von (1787–1844), Danish governor of Iceland 1810–19. From an aristocratic family, he completed his law degree in 1806 and worked in the Rentekammer until he was appointed *amtmaður* of the Southern Amt in 1810. As such he was one of the members of the government committee instituted in 1810 in the aftermath of the Icelandic Revolution and the downfall of the Stephensen brothers, becoming governor (*stiftamtmaður*) in 1813. He left his Iceland post in 1819 and served as governor in various other Danish provinces. He gets mixed reviews in Icelandic contemporary sources. His only contact with Banks was to write him a thank-you letter for his efforts on behalf of the Icelanders.

Castlereagh, Robert Stewart (1769–1822), Viscount 1796–1821, 2nd Marquess of Londonderry 1821, FRS (1802), British statesman. He was born in Dublin and acted as Chief Secretary for Ireland 1798–1801, during which period he was involved in putting down the Irish Rebellion. While he was Secretary of State for War and the Colonies (1805–6 and 1807–9) and Secretary of State for Foreign Affairs (1812–22), Banks was often in touch with him regarding Icelandic matters, as was Sir George Steuart Mackenzie (q.v.).

Clancarty, Richard Le Poer Trench, 2nd Earl of Clancarty (1767–1837), diplomat and statesman. It was during his tenure as President of the Board of Trade (1812–18), that he and Banks corresponded regarding the detention of the Icelandic ship *Regina* in Sweden, demonstrating how seriously the British took their protection of Iceland during the wars.

---

[1] See TNA, HCA 32/985, no. 988.

Clausen, Holger Peter (1779–1825), Iceland merchant. Clausen was a wealthy merchant in Ólafsvík and Grundarfjörður from 1804 and from 1811 in Búðir, thus more or less dominating the Snæfellsnes peninsula. Sir George Steuart Mackenzie met him while in Iceland in 1810 and thought him 'remarkably intelligent' and Henry Holland found him 'an active, pleasant man, & better informed on scientific subjects than any one we have yet seen in Iceland'.[1] In London he became the leader of the Iceland merchants, as his letters to Banks testify. In 1816, when the debate regarding the freedom of the Iceland trade was raging, he wrote an interesting pamphlet defending his view that opening the trade up to foreign nations would be harmful to Iceland, *Nogle Betænkninger om det Spørgsmaal: Er en Friehandel for fremmede Nationer paa Island skadelig for Danmark og Island?* He later moved to Copenhagen, where he became a leading merchant, establishing several companies, including the Copenhagen fire insurance company (Köbenhavnsbrandforsikring), and had a seat on the Copenhagen municipal council.

Cooke, Edward (1755?–1820), government official. Cooke served as Castlereagh's Under-Secretary for War and the Colonies (1804–6) and then as Castlereagh's secretary at the Foreign Office (1807–9 and 1812–17). It was during this period that he corresponded with Banks on behalf of Castlereagh, regarding Icelandic affairs. In 1814–15 he accompanied Castlereagh to the peace negotiations in Vienna.

Corbett, Borthwick & Co., wood merchants of Dock Street, North Leith. During the Napoleonic Wars they became the agents for the vessels of Iceland, the Faroes and Greenland coming to Leith. They wrote frequently to Banks asking for his help in regard to the Iceland and Greenland trade.[2]

Denovan, James Frederick (fl. 1807–14), superintendent of police in Leith, interpreter and 'translator of languages to the High Court of Admiralty, North Leith'. According to Dr William Wright, he was 'at the Council Chamber Lieth ... interpreter, for the German, Danish & Islandic Languages' (Document 43). Denovan told Banks that he had opened a subscription for the Icelanders and generally helped them in their distress. During the war his name frequently crops up in the papers of the Board of Trade requesting licences for ships to Norway and Denmark. In the National Archives of Denmark there is a letter from a Lieutenant Falsen to the King (dated 1 October 1815) recommending him as the first man in Britain to help the unfortunate Danish prisoners and the first to begin a correspondence with Sir Joseph Banks in order to come to their aid. After the war he announced in newspapers in 1814 that he had become the Royal Danish Agent in Leith for trade.[3]

Diede, Wilhelm Christopher von, Baron zum Fürstenstein (1732–1807), Denmark's envoy to Britain. A scion of Hessian nobility, he entered the Danish civil service in 1760 (against his family's wishes). In 1763 he was appointed Denmark's envoy to Berlin, where Frederick the Great never spoke to him as he believed the Danes had been disloyal during the Seven Years War. In 1767 he was appointed 'Envoy Extraordinary of His Danish

---

[1] Mackenzie, *Travels in Iceland,* pp. 133, 177; Holland, *Iceland Journal,* pp. 147, 189–91.
[2] *The Post Office Annual Directory*, Edinburgh, 1813.
[3] *The Post Office Annual Directory*, Edinburgh, 1809, 1811 and 1813); *Holden's Directory*, 1811; Danish archives.

Majesty to the British Court' where he was very well received. He had visited England when young, was personally acquainted with George III and well connected within English society. He stayed in England until 1776. He was the man who had to hastily issue a passport to Banks when he made his sudden decision to go to Iceland.

Dryander, Jonas (1748–1810), Swedish-born botanist and curator of the Banks herbarium and library. After studying at the University of Uppsala he was employed by Banks, first as a field botanist and after Solander's death in 1782 he became Banks's librarian. He compiled the catalogue of Banks's library, published in 1798–1800 and was librarian both to the Royal Society and the Linnean Society of which he was an original Fellow, becoming Honorary Librarian and in 1796 a vice-president.

Dundas, Henry, 1st Viscount Melville (1742–1811), Scottish politician and one of Pitt's closest collaborators. He was appointed Lord Advocate in 1775 and thus came to dominate Scottish affairs, by 1796 controlling all parliamentary seats in Scotland. In 1794 he was appointed Secretary for War, an office he held until 1801 when he resigned with Pitt. It was during this tenure that Banks was approached as the Iceland expert to give his opinion on the feasibility of annexing Iceland. Dundas was, like Banks, interested in science and discovery, sending expeditions to both the Arctic and northern Australia. He was very interested in imperial matters and trade issues and was one of the organizers of the Macartney Embassy to China, in which Banks was also involved.

Everth & Hilton, London firm of fish curers and merchants. John Everth and James Hilton of Lower Thames Street ran a wholesale fish warehouse and were engaged in the herring fishery. They participated in the Iceland trade 1811–12, sending at least two ships, and had plans to conduct an Iceland fishery, which appear to have been unsuccessful.[1]

Falconer, Thomas (1738–92), classical scholar and Recorder of Chester. He was called to the Bar at Lincoln's Inn in 1760. Due to chronic ill-health he pursued interests in classical antiquities instead of law, writing several works. He was also a patron of literature and Anna Seward called him 'the Maecenas of Chester'. He knew Banks from 1767 at least and they corresponded frequently, especially about Banks's plans for exploration, Banks giving him a detailed account of his Iceland expedition in the letters in this volume.

Fenton, Perrot (fl. 1807–41), the Deputy Marshal of the Admiralty at Doctors' Commons, London.

Findlay, Bannatyne & Co., merchants of 8 New Broad Street, London.

Frederik VI (1768–1839), King of Denmark 1808–39. He was a nephew of George III, his mother being Caroline Matilda, George's youngest sister. His father, Christian VII, was mentally ill and from 1784 Frederik served as king, under the title of Crown Prince Regent. He was initially a liberal, approving the abolition of serfdom in Denmark in 1788. After the Bombardment of Copenhagen in September 1807, he turned virulently anti-British and joined Napoleon's camp. Denmark was hard hit during the Napoleonic Wars, especially by the loss of Norway in 1814. Frederik, however, was well informed on the help Banks gave to his Icelandic subjects and after the war rewarded him with a set of the *Flora Danica*.

---

[1] TNA, BT 3/11, Chetwynd to Dunsmore, 23 April 1812; *Holden's Directory, 1811*.

Frisch, Hartvig Marcus (1754–1816), director of the Royal Greenland Company. A civil servant, he first became involved in the trade of Iceland and Finmark, subsequently becoming a director of the Royal Greenland Company to which he devoted his life. He was especially interested in the possibilities of whaling for Denmark.

Halldórsson, Gunnlaugur (1772?–1814), Icelandic merchant. Educated as a clergyman, he was defrocked when he fathered a child outside wedlock. He became a merchant in Hafnarfjörður 1803–10 and in Reykjavík 1810–12. He was one of the unfortunate Icelanders to be detained in 1807 and sought Banks's help. In 1811 he fathered another illegitimate child and left for Denmark that same year. He continued to engage in trade but died in Copenhagen two years later.

Hammershaimb, Wenzel (1744–1828), the King's *landfoged* or bailiff in the Faroe Islands 1765–1815.

Hawkesbury, *see* **Jenkinson.**

Heide, Claus (d. 1774), Norwegian London-based merchant. He worked for the Norwegian firm of John Collett in Wellclose Square, timber merchants from Christiania (Oslo). When Collett died in 1759 Heide took over the firm and the position as provost for the Norwegian Church in London, becoming the leader of the Norwegian community. Thus it is no surprise that when Banks suddenly decided to go to Iceland Heide was approached to provide information about Iceland (being a dependency of Denmark-Norway), which he did as well as he could.[1]

Henkel, Henrik (d. 1817), Danish Iceland merchant. He was married to Charlotte Ísfjörð, daughter of the county magistrate Þorlákur Ísfjörð and a younger sister of Kjartan Ísfjörð (q.v.). He became the merchant in Þingeyri, in Dýrafjörður in the Western Fjords, in 1788 and was the owner of the vessel *Tykkebay*, both trading with Copenhagen and sending klipfish directly to Spain. He was a leading merchant and was one of few merchants to begin using decked vessels for fishing. He took an active part in the debate regarding the future of the Iceland trade at the end of the eighteenth century, publishing in Copenhagen no less than three pamphlets on the subject.

Holt, Andreas (1729–84), Norwegian acting councillor of the Chancery (*kancelliraad*). He had been appointed chairman of the Royal Commission of 1770 (Landsnefndin fyrri) to examine the disastrous state of the Icelandic economy and propose measures for Iceland's regeneration. The members of the commission travelled around Iceland in the summer of 1770. Holt would thus have been an ideal person to give Banks information on the current state of Iceland. When Banks consulted Claus Heide (q.v.) he had recently received a letter (Document 1) from Andreas Holt on Iceland. Solander had met Holt in London in 1766.[2] He progressed through the ranks of the Danish administration to the title of *etatsraad* (see Glossary).

Holtermann, Lorenz (fl. 1809), Norwegian merchant in Bergen.

Home, James (1760–1844), Scottish physician. He studied medicine at Edinburgh University where he met John Thomas Stanley who invited him to come to Iceland with

---

[1] Rasch, *Niels Ryberg, passim*; Polak, *Wolffs & Dorville, passim*.
[2] Gustafsson, *Mellan kung och allmoge*, p. 103.

him in 1789. Home wrote to Banks asking for advice. However, to Stanley's disappointment, he did not go and eventually succeeded his father Francis as a very successful professor of Materia Medica at Edinburgh in 1798. In 1821 he obtained the professorship of physic. Home was president of the Royal College of Physicians of Edinburgh and was elected a fellow of the Royal Society of Edinburgh in 1787.

Hooker, Sir William Jackson (1785–1865), FRS, botanist. Son of an amateur botanist, Joseph Hooker, in 1806 he inherited the estate of his godfather William Jackson, a wealthy brewer. Hooker, a member of the Linnean Society in 1806, was a protégé of Banks, who sent him to Iceland in 1809 to gather plants. Hooker travelled with the 'revolutionaries' and became a spectator, though not a participant, in the Icelandic Revolution. His book *Journal of a Tour in Iceland in the Summer of 1809* not only provides an excellent account of the events of 1809 but also includes material Banks had lent him regarding his ascent of Hekla, and the odes composed to Banks by grateful Icelanders. Jörgensen (q.v.) saved his life and Hooker remained his life-long friend, though Banks turned his back on him. The Hooker–Banks Iceland correspondence discusses Jörgensen and shows great awareness of the plight of Iceland during the Napoleonic Wars. He was appointed regius professor of botany at Glasgow University in 1820 and in 1841 was appointed the first full-time director of the Royal Botanic Gardens at Kew. Hooker published major works on the botany of bryophytes and ferns. He expanded Kew Gardens, establishing a museum of economic botany there.

Horne & Stackhouse, merchant firm in Liverpool, East-side Salthouse Dock. The owners were William Horne and Jonathan Stackhouse. They conducted a considerable trade with Iceland during the war years, as well as trading with the Faroes, and were eager to continue their trade with these islands after the war. Their pleas to this effect reached Banks through Sir John Thomas Stanley.

Horneman, Hans Frederik (fl. 1807–14), Danish merchant, later consul. In 1796 he was engaged in the Danish trade with Batavia. In 1802 he went to London and in 1809 he went into partnership with Charles Rooke and they established the firm of Rooke & Horneman (q.v.). During the war he was appointed by the Danish Board of Trade, on the recommendation of Jens Wolff, to look after the interests of the Danish prisoners-of-war. His numerous letters to Banks demonstrate the close contact he had with Banks regarding the Iceland trade, Jörgensen and Gudrun Johnsen. He was eventually appointed consul in London on 7 February 1814, but removed from the Danish consulship in June 1817 after his bankruptcy (see Rooke & Horneman). He ended up in debtor's prison shortly afterwards.[1]

Ísfjörð (Isfjord), Kjartan (1774–1845), Icelandic merchant. He was the son of a county magistrate Þorlákur Ísfjörð (1748?–81) He had been established as a merchant in Eskifjörður in eastern Iceland since 1802. His ship the *Charlotte Amalia* was the only Iceland ship condemned because it was captured the day after the formal declaration of war and he wrote several letters to Banks on the subject. In 1813 he went bankrupt, but by 1816 he was back on his feet, applying for trading passports for four ships. He eventually moved from Copenhagen to Eskifjörður, where he lived until his death.

---

[1] Polak, *Wolffs and Dorville*, p. 227.

Jenkinson, Robert Banks, 2nd Baron Hawkesbury and 2nd Earl of Liverpool (1770–1828), FRS, British statesman. He held continual senior posts in government: Foreign Secretary 1801–3, Home Secretary 1804–6 and 1807–9, Secretary for War and the Colonies, 1809–12 and then Prime Minister 1812–27. He was a close friend of Banks and he was the member of government whom Banks chose to approach with his proposals regarding the state of Iceland during the Napoleonic Wars. Hawkesbury contemplated the possibility of a British annexation of Iceland. This encouraged Banks to secure the best possible and most accurate information on Iceland and plan the country's annexation in detail in memoranda to the government.

Jensen, Hans (fl. 1811), co-director of the Greenland administration.

Jónsson, Bjarni (1725–98), Rector of Skálholt School. He studied in Copenhagen, completing an examination in philosophy and theology. He was appointed Rector by the King in 1755 and was considered a brilliant teacher. Ordained in 1781, he became a clergyman in Gaulverjabær until his death. He met Banks during his Skálholt visit in 1772 and composed an ode in his honour.

Jónsson, Teitur (1742? –1815), clergyman. He was the son of Jón Teitsson, Bishop of Hólar. He studied philosophy and linguistics at the University of Copenhagen. He obtained his baccalaureate in 1770 and was ordained in 1779. He is described as having been fond of his drink and a good Latin poet. Banks witnessed both these attributes when he visited Skálholt in 1772: Teitur was drunk, but managed to compose an ode in his honour.

Jónsson, Þorsteinn (fl. 1772–1801), farmer at Hvaleyri, near Hafnarfjörður. He was Banks's guide whilst in Iceland in 1772. In 1801 he was 70 years old and still farming, as well as fishing and working as a smith.

Jörgensen, Jörgen [or Jorgen Jorgensen, Jorgenson or Jürgensen] (1780–1814), Danish adventurer, mariner, political agent, explorer and author. He was the son of the Royal Watchmaker to the Danish Court and received a privileged early education, attending the same school as Adam Öehlenschlager, the Danish national poet. At the age of fourteen he was apprenticed on an English collier and for more than a decade he remained on British ships sailing to South Africa, Australia and the Pacific Islands. He served in the Royal Navy, as well as on whalers, sealing vessels, surveying vessels and on voyages of discovery, one in 1798 with Captain Flinders, who discovered that Van Diemen's Land (later renamed Tasmania) was a large island. He also took part in the expedition to establish a settlement, Hobart, in Van Diemen's Land. He rose to the rank of midshipman. He was a first-rate seaman and a brilliant navigator and is believed to have been the first Dane to circumnavigate the globe. He first met Banks in 1806 when he brought two Maoris and two Tahitians to London and introduced them to Banks. In 1807, at the outbreak of war, he returned to Denmark where he was given command of the largest Danish privateer, the *Admiral Juul*. The ship was captured and Jörgensen became a prisoner-of-war, on parole in London because of his status as a captain. There he became involved with Samuel Phelps who sent him on a trading venture to Iceland culminating with the Icelandic Revolution of 1809. Jörgensen took on the government of Iceland, assuming the title of 'Protector'. He was deposed by Captain Jones of the Royal

Navy in August 1809 and on his return to England was first imprisoned on the *Bahama* prison-hulk in Chatham. During this period of his life he wrote many letters to Banks. He subsequently became an author, four books being published in England including *The Copenhagen Expedition Traced to Other Causes than the Treaty of Tilsit* (1810) and *The State of Christianity in the Island of Otaheite* (1811). Despite Banks's denunciation of the Icelandic Revolution and particularly Jörgensen's role in it, Jörgensen was employed by the British government as a secret agent intermittently from 1812 to 1820 (see e.g. his *Travels through France and Germany* (1817), claiming to have witnessed the battle of Waterloo and to have met Goethe at Weimar).

Jörgensen's downfall was his compulsive gambling. He was imprisoned for debt in the Fleet where he met up with his Iceland friends (as told in his letters), and after several spells in prison, including Newgate, he was finally sentenced to death in 1822. But, as was common at the time, this was commuted to transportation for life in Van Diemen's Land. Thus he returned to Hobart. Celebrated as one of Tasmania's pioneers and first historians of the Aborigines of the island, he became an explorer, a police constable, a journalist and author, his *Aboriginal Languages of Tasmania* appearing posthumously in 1842. He died in poverty in 1841 but an obituary in the *Hobart Town Courier* described him as 'one of the most extraordinary men of our time'.[1]

Knudsen, Adser Christian (fl. 1776?–1825?), Danish merchant. He was a scion of a respected Danish merchant family and became a burgher of Reykjavík in 1799 as a factor. He set up saltworks in Akranes around 1804. In 1805 he bought the Randers commercial concern in Reykjavík. His ship, *De Fem Brödre* with him on board, was captured in Stornoway in 1807 and he was one of the Iceland merchants seeking Banks's help. He was the first Iceland merchant to obtain a licence from the Board of Trade in 1808 and he received a sizeable loan from the Danish authorities to continue his Iceland trade. He chose to sell his licensed ship the *Orion* with cargo to Count Trampe. He subsequently moved to Copenhagen, never to return to Iceland.

Konig or König, Charles Dietrich Eberhard (1774–1851), FRS, German naturalist and mineralogist. As a naturalist, Konig became an assistant to Dryander (q.v.) in Banks's library and herbarium 1801–7. He was subsequently employed by the British Museum, becoming Keeper of Natural History and Modern Curiosities in the Natural History Department in 1813. He was also quite knowledgeable about ancient Icelandic poetry, witness his letter to Banks on the subject.

Latrobe, Christian Ignatius (1758–1836), an English clergyman, musician, and composer. He attended a Moravian College in Saxony and was ordained in 1784. He was a friend and great admirer of Haydn and himself composed many works for the Moravian Church. He edited a *Selection of Sacred Music* in six volumes (1806–26), on the church music of Haydn and Mozart. His correspondence with Banks focuses on provisioning Greenland, where the Moravians had established many missions, during the Napoleonic Wars.

Lind, James (1736–1812), *see* Appendix 1.

---

[1] There are numerous books about Jörgensen: see e.g. Sprod, *The Usurper* and, one of the most recent, Bakewell, *The English Dane*.

Liverpool, *see* **Jenkinson.**

Lobnitz? Löbner, L. Lobnitz is unidentified but was probably Emilius Marius Georgius Löbner (1766–1849) who was commandant of the fort in the Faroe Islands 1796–1825 and *amtmand* (district governor) there 1816–25. The letter in this book is a translation as well as a copy and names are notoriously difficult to decipher (the copyist may have been mixing together Löbner and Captain Bohnitz of the Faroese trade).

Mackenzie, Sir George Steuart (1780–1848), FRS, chemist, mineralogist. He succeeded to the baronetcy of Mackenzie, Coul, Ross, as 7th baronet in 1796. He became interested in Iceland after meeting a young stranded Icelander, Ólafur Loptsson, in Scotland in 1807. Like Banks, he saw a British annexation of Iceland as the solution to the wartime plight of the Icelanders. He contacted Banks wishing to gain his assistance in annexing Iceland, aspiring to becoming governor. He went on a scientific expedition to Iceland in 1810, with Henry Holland and Richard Bright, publishing a magnificent book on Iceland the following year. He made valiant efforts to have Iceland annexed and when this failed he approached the Danish administration, suggesting they 'lent' Iceland to Britain during the war. This was not met with enthusiasm. He went to the Faroes in 1812. He became a Fellow of the Royal Society of Edinburgh in 1799 and a fellow of the Royal Society in 1815. Mackenzie, a Huttonian, later became interested in phrenology, publishing in 1820 *Illustrations of Phrenology*. He corresponded with many notables including Sir Walter Scott; their correspondence from 1809–31 is preserved in the National Library of Scotland. He became a member of the Icelandic Literary Society in Copenhagen.[1]

Macleay, Alexander (1767–1848), FRS, civil servant, botanist and entomologist. He was Secretary of the Transport Board 1807–17 and was in contact with Banks dealing with the Icelandic prisoners-of-war. However, he was also a naturalist, elected a fellow of the Linnean Society in 1794, and was this society's secretary 1798–1825. He was elected a fellow of the Royal Society in 1809. By 1825 he was considered to have the most extensive collection of insects of any private individual. He later became colonial secretary of New South Wales (1825). His garden in Elizabeth Bay became famous for its rare plants and he continued to send specimens to the Royal and Linnean Societies. He later entered politics.

Magnússon, Markús (1748–1825), clergyman. He studied theology at the University of Copenhagen. He was clergyman at Garðar on the Álftanes peninsula (where Bessastaðir is situated) in 1781 and where he lived all his life. He was provost (*prófastur*) for Kjalarnes. He was one of Iceland's most influential clergymen and could have become bishop had he wished. In 1809 he made the acquaintance of Jörgen Jörgensen. A letter which he wrote to a fellow cleric in the north of Iceland from that year has been preserved. In it he describes how terrified the Icelanders are of the Revolution taking place in the country.

Majendie, William Henry (fl. 1807–39), British army captain. Son of Henry William Majendie, the Bishop of Bangor, he became a captain in 1807, joining the 17th Regiment on the personal staff of General Hewitt in the East Indies. He was placed on half-pay 16

---

[1] See also Björnsson, 'Sir George Steuart Mackenzie, Bart.'

May 1816, subsequently becoming captain in the 6th (or the 1st Warwickshire) Regiment of Foot. Little else is known of him. However, in 1839 he was a high-ranking member of the Lichfield Troop of the Queen's Own Royal Yeomanry.[1]

Martini, Friedrich (1739–1812?), co-director of the Royal Greenland Trading Company.

Nolsøe, Jacob Poulsen (1775–1869), Faroese mercantile agent. Brother of Poul Poulsen Nolsøe (see p. 431, n. 2). In 1795 he was appointed to the Royal Monopoly Trade in the Faroe Islands, where he remained until 1850. His letters, of which Banks received copies, concerned the fate of his brother and the provisioning of the Faroe Islands.

Ólafsson, Ólafur (Olav Olavsen) (1753–1832). He studied law at the University of Copenhagen and architecture at the Royal Danish Academy of the Arts, where he was an award-winner. He was one of the founders of the Enlightenment-inspired Lærdómslistarfélag (The Icelandic Society of the Learned Arts) in 1779, writing many articles. He became a professor at the school of mining in Kongsberg, Norway, teaching both mathematics and law. In 1818 he was elected a member of the Norwegian parliament. He was a celebrated Latin poet, composing a celebratory ode not only to Banks but also to King Karl XIV Johan.

Pálsson, Bjarni (1719–79), state physician of Iceland. He studied medicine and natural science at the University of Copenhagen. He was sent by the Danish government with Eggert Ólafsson to explore and write an account of Iceland from 1752 to 1757. The resulting massive book, *Rejse igennem Island*, was published in Copenhagen in 1772 and subsequently translated into German, English and French (see Introduction p. 21). He graduated as a physician in 1759 and the following year was appointed the first state physician in Iceland, a life-long position. He is described as having been very intelligent, poetic and generous. It was due to his efforts that more doctors and an apothecary were appointed, doctors were appointed to each quarter of Iceland and a trained midwife sent to the island. He taught medicine. He married the daughter of Skúli Magnússon, treasurer of Iceland, one of the great men of his day. He was visited by Banks in 1772 with whom he enjoyed a memorable Icelandic dinner (see p. 19 above).

Parke, John (b. before 1782, fl. 1814), British consul to Iceland during the Napoleonic Wars. He was the son of Thomas Parke, merchant and banker of Highfield, Lancashire, and brother of James Parke, Baron Wensleydale (created 1856). He later became Consul General in Rome and Corunna. Some of his correspondence from that period is preserved in BL, Add MS 41535, 41536 and 41538.

Paus, Henrik Christian (fl. 1788–1824), Danish merchant in Iceland. He had first worked for the Royal Trading Company, then from 1788 for the Danish merchant Busch in Djúpivogur on the east coast of Iceland and finally as Busch's partner in Ísafjörður in the Western fjords, where he became a burgher, until 1824. From 1804 Paus lived in Copenhagen visiting Iceland in the summer.

Pennant, Thomas (1726–98), *see* Appendix 2.

---

[1] BL, Add MS 40472, Majendie to Sir Robert Peel, 12 November 1839, f. 220.

Petræus, Westy (1767–1829), Iceland merchant. In 1790 he had arrived in Iceland from Schleswig as agent for the Nordborg mercantile establishment in Keflavík. In 1795 he also worked for the Nordborg commercial concern in Reykjavík, owning it by 1803. In 1798 he added the Westmann Islands to his trading establishment with his partner Peter Ludvig Svane. He had two homes, one in Keflavik and one in Reykjavík, and by 1807 had become one of the leading merchants in south-western Iceland. In 1807 he was on his way to Denmark with his family but had the misfortune to be captured and taken to Leith. He was one of the two merchants sent to London in December 1807 to plead their case with Banks. There is a lot of personal information in his letters to Banks. He continued trading with Iceland until his death.

Phelps, Samuel (fl. 1809–18), British soap manufacturer and merchant. Phelps was a partner in Phelps, Troward & Bracebridge, soap manufacturers of Cupers Bridge, Lambeth. He is the key figure in the Icelandic Revolution but little is known about him. To Hooker he was 'eminent and honorable', and Mackenzie found him 'honest and good' and his character 'unimpeachable'. And Banks had initially been impressed enough with Phelps to include him in the letter of introduction he wrote to Ólafur Stephensen on behalf of Hooker (Document 118). He was the author of two books, *Observations on the Importance of Extending the British Fisheries etc.*, published in London in 1817 and reprinted 1818, and *The Analysis of Human Nature*, a two-volume work published in 1818. From the latter it can be deduced that he was well-travelled.

Preidel, John Christopher (fl. 1807–10), merchant at 13 Coleman Street, London. He appears in the documents in this volume as an agent for Icelandic merchants.

Rask, Rasmus Christian (1787–1832), Danish philologist. Rask's work on Old Norse was pioneering in the field of comparative linguistics. His aim, as is demonstrated in the documents in this volume, was to found a society for the preservation of the Icelandic language and culture. The society Hið íslenska bókmenntafélag (The Icelandic Literary Society) was founded in 1816 in Copenhagen and Reykjavík simultaneously. It is the oldest surviving literary society in the Nordic countries. Rask became the first president of the Copenhagen branch and every year he is honoured with a conference at the University of Iceland. There is no evidence that Banks considered joining this society.

Reventlow, Christian Ditlev Frederik, Count (1748–1827), Danish statesman and reformer. He attended the University of Leipzig, followed by a Grand Tour of Western Europe 1769–71, which included a visit to England. He was interested in both the arts and the sciences and is considered one of Denmark's foremost politicians. During this period in his distinguished career he was the President of the Rentekammer (the Exchequer) 1789–1813 and *gehejmestatsminister*, effectively Prime Minister. A liberal, he came into conflict with the king during the 1807–14 war and left state service in 1813, retaining, however, his seat on the Privy Council. In his letter he is ordering Castenschiold, the King's representative in Iceland, to write a thank-you letter to Banks.

Rooke & Horneman, merchants of 88 Queen Street, Cheapside, London. Rooke & Horneman was a partnership of Charles Rooke, a British merchant, and H. F. Horneman, the later Danish consul (q.v.). Charles Rooke became Horneman's partner in November 1809, investing £18,000 to Horneman's £1,180. Their partnership came to an end in

March 1816 and Horneman became bankrupt (see TNA, B 3/2243, the whole file is devoted to Horneman, Nov–Dec 1816 – Henry Frederick Horneman, of Queen Street, Cheapside, in the City of London, Merchant, Dealer and Chapman).[1]

Rosenkrantz, Niels, Count (1757–1824), was born in Norway, a Danish diplomat, privy councillor and foreign minister. After a short military career, he entered the diplomatic service in 1782. Among his appointments was the post of ambassador to the courts at Berlin and St Petersburg. He went on several missions to Napoleon and in 1810 became Foreign Minister, accompanying Frederik VI to the Congress of Vienna, and serving in that office until his death.

Ross, John (1777–1856), Scottish naval officer and Arctic explorer. Ross entered the Royal Navy at the age of nine serving in the West Indies and the Baltic. In 1808 he joined the Swedish Navy for a short period. In 1818 he commanded the *Isabella* on a mission to find the Northwest Passage, narrowly missing it. It was on this ship that he wrote to Banks giving him the latest information on the expedition. Ross was to lead further Arctic expeditions 1829–33, and in 1850 he undertook a final voyage to the Arctic in a vain attempt to rescue John Franklin, lost there with his crew in 1847.

Sívertsen, Bjarni (1763–1833), Icelandic merchant. One of the pioneers of Icelandic participation in trade, after the Danish monopoly had been lifted. He was a merchant in Hafnarfjörður (1794–1830) but also traded in Reykjavík and Keflavík. He built fishing smacks in his shipyard for his own company, the first one in 1803. One of his ships, the *Johanne Charlotte*, was captured by the British in 1807 and he came to the attention of Banks, who invited him to London to discuss the release of the ships. He returned to Iceland in 1809 and throughout the war Sívertsen was very active in trade matters and was instrumental in recovering the public money taken by the British privateer, the *Salamine*. For his efforts the King of Denmark made him a knight of the Dannebrog, the first Icelander to receive that honour. He became wealthy and eventually moved to Denmark.[2]

Smith, Culling Charles (1775–1853), British courtier and civil servant. He was the son of Charles Smith, Governor of Madras. He was appointed Under-Secretary of State for Foreign Affairs on 13 December 1809, a week after his brother-in-law, the Marquess Wellesley, became Foreign Secretary. He remained in that position until 1812, when Wellesley left the Foreign Office. He subsequently served the Royal family in various capacities. Banks wrote to him letters intended for Wellesley regarding Icelandic affairs.

Stanley, Sir John Thomas (1766–1850), 7th Baronet of Alderley Hall and in 1839 created 1st Baron Stanley of Alderley, FRS. He married Lady Maria Josepha Holroyd (1771–1863), daughter of the 1st Earl of Sheffield. He had two seats in Cheshire, Alderley Park and Winnington Hall. While at the University of Edinburgh he went on an expedition as a young but wealthy student to Iceland in 1789. The purpose of this trip was not scientific, but the three journals of his companions are a valuable source.[3] He was an MP from 1790 to 1796. During the Napoleonic Wars he appears to have engaged in the

---

[1] Polak, *Woolfs & Dorville*, p. 227.
[2] Sívertsen wrote a short autobiography mentioning Banks: 'Ágrip af æfisögu Bjarna Riddara Sivertsens'. See further Guðmundsson, *Skútuöldin* I, pp. 51–65.
[3] *The Journals of the Stanley Expedition to the Faroe Islands and Iceland in 1789*, edited by John F. West.

Iceland trade through involvement with the merchant house of Horne & Stackhouse (q.v.) and his family became much involved with Banks regarding the fate of the Icelandic girl Gudrun Johnsen.

Stephensen, Magnús (1762–1833), Chief Justice of Iceland. Born into the most prominent and wealthiest family in Iceland, the son of Ólafur Stephensen, the only native governor of Iceland (q.v.), Magnús received an exemplary education, probably the best of any of his contemporaries. Besides Latin and Greek he learnt French and German, and studied English of necessity during the Napoleonic Wars. He obtained his degree in law in 1788, but during his years of study at the University of Copenhagen he attended lectures in many other subjects, including theology, astronomy, mathematics, physics and chemistry, botany and philosophy. He was fond of dancing and learned to play the organ. While still a student he was sent to Iceland by the Danish administration to investigate the Laki eruption of 1783–4. His rise within the Danish bureaucracy of the absolute king Christian VII was swift, becoming *lögmaður* (see Glossary) in the north-west. In 1800 he became the first Chief Justice of Iceland in the High Court (Landsyfirréttur), a post he held until his death. In 1809–10 he was briefly governor of Iceland in the aftermath of the Icelandic Revolution. Stephensen was the leading Icelander of his day. He was also the leader of the Enlightenment in Iceland and owned the only printing press on the island. From 1795 to his death he published journals, many books on Enlightenment subjects, and a book of hymns, over a hundred of which he wrote himself. He was also a prolific author. He belonged to all the major scientific societies in the Danish Empire, himself heading their Icelandic equivalents. Stephensen was always a spokesman of enlightened rationalism, advocated a humane penal system and was a supporter of free trade with all countries. Though always a loyal servant of his royal master, Stephensen was an ardent patriot. It was first and foremost thanks to him that Iceland survived the Napoleonic Wars. His letter to Banks of 17 October 1807 (Document 40) was the catalyst for Banks's involvement with Iceland during the Napoleonic Wars.[1]

Stephensen, Ólafur (1731–1812), governor of Iceland. Son of a clergyman in northern Iceland, Stephensen was sent to the University of Copenhagen, where he completed his degree in law in 1751 and soon embarked on a career within the Danish royal civil service. He eventually became a district governor in 1766, which was his position when he made Banks's acquaintance in 1772. He was appointed governor of Iceland in 1790 (until 1806), the only Icelander to attain that position. He was on the whole a good government official and became extremely wealthy, not least because of an advantageous marriage with the daughter of his predecessor as district governor. He was a prolific writer especially on economic issues. Banks visited him in 1772 and Stephensen proved most helpful in fulfilling Banks's wishes. They continued to correspond for several years. He was generous and held sumptuous banquets to which he invited the British visitors. Both Hooker and Mackenzie describe the feasts in their travel books.[2]

Thielsen, Hans (fl. 1789–1814), a captain and merchant from Flensburg in Schleswig. A mercantile company from there began trading in Hafnarfjörður in the spring of 1789 and in Reykjavík from 1791. During the period 1790–95 they only had one ship sailing

---

[1] On his ideology see Sigurðsson, *Hugmyndaheimur Magnúsar Stephensens*.
[2] See his biography: Sigurðsson, *Mikilhæfur höfðingi*.

to Iceland but after that generally two, until 1807 at least. Thielsen was a citizen of Reykjavík and Christian Conrad Strube managed the trade.

Thodal, Lauritz Andreas (1719–1808), governor of Iceland 1770–85. Thodal had been on diplomatic missions to Stockholm and St Petersburg and in 1770 he was appointed *stiftamtmaður* in Iceland and the Faroe Islands, a post he retained until 1785. He was a Norwegian and was the first governor to reside in the country. Banks visited him in 1772 and describes the visit in his Journal. As Banks wrote, Thodal was an enthusiastic gardener and interested in agricultural reform. He was popular and is regarded as one of the best governors sent to Iceland. He lost everything in the British attack on Copenhagen in 1801.

Thorkelín, Grímur Jónsson (1752–1829), antiquarian, royal archivist and discoverer of *Beowulf*. Thorkelín studied philosophy, law and classics at the University of Copenhagen. He became an assistant in the royal archives in 1780 and in 1784 was awarded the title of professor. He spent the years 1786–91 in England, arriving armed with an introduction from Professor Bugge (q.v.). On 9 February 1788 Thorkelín sent Banks a long memorandum on salmon-fishing (BL, Add MS 8097, ff. 126–130). That same year he was made an Honorary Doctor of Law at the University of St Andrews, as well as acquiring various other honours. Sir John Thomas Stanley wished to have Thorkelín accompany him to Iceland in 1789, but the plan fell through. Thorkelín became so well known in London, that he was invited to become the Director of the British Museum. Instead he chose to return to Denmark, where he became the Keeper of the Royal Archives. He is perhaps best known as the discoverer and first editor of *Beowulf* in 1815. He edited a host of Old Norse texts and documents. He lost his library in the British bombardment of Copenhagen in 1807.

Trampe, Count Frederik Christopher (1779–1832), Governor of Iceland. Descended from an ancient Pomeranian noble family, he completed a law degree from the University of Copenhagen in 1798. He became a *landsdommer* (deputy judge) in Lolland and Falster in 1800. After the outbreak of hostilities in 1801 he joined the army as captain in the Regiment of the Crown Prince 1801–4. In 1804 he completed his doctorate in law (dr. juris) at the University of Kiel. In 1804 he was appointed *amtmaður* in the Western Amt of Iceland and *stiftamtmaður* in June 1806. He went on a visit to Denmark in 1807, when the ship he was sailing on was captured by the British near Kronborg, but he was released. In May 1809 he left for Iceland in the *Orion*. The Revolution, well documented in the letters in this book, took place in June and he was imprisoned on board the merchant vessel *Margaret and Ann* for its duration. On his release he went straight to England to lodge his complaint, writing a lengthy memorandum to the Foreign Secretary (Document 124). Banks, horrified by the 'atrocities' in Iceland, took his side and restitution was made. During his stay in England he travelled to Oxford and Liverpool, met Mackenzie and made the acquaintance of William, Duke of Gloucester. In the summer of 1810 he was appointed governor in Trondheim where he remained until his death in 1832. He was considered competent, hardworking and a loyal servant to the King.[1]

---

[1] See Sommerfelt's biography, 'Stiftamtmand Grev Fredrik Christopher Trampe. En Biografi'.

Troil, Uno von (1746–1803), see Appendix 1, pp. 596–7.[1]

Vargas Bedemar, Count Eduardo Romeo (1768–1847), mineralogist. Born in Magdeburg as Carl F. A. Grosse, he changed his identity to a Spanish grandee. He studied medicine at Göttingen, Halle and Berlin, but then turned to philosophical and scientific writings. He founded the Academia Italiana in Siena and, due to his friendship with the Danish diplomat Herman Schubart, he joined the Videnskabernes Selskab and secured a position as a mineralogist for the King, undertaking expeditions to Norway, the Faroes, Madeira, the Azores and Canary Islands, partly in search of Atlantis. He spent the period 1809–47 in Denmark. In 1842 he was appointed Director of the Natural History Museum in Copenhagen, having earned a good reputation. He made the acquaintance of Prince Christian Frederik (later Christian VIII) and took charge of his cabinet of natural curiosities. He published several books, among them the two-volume *Reise nach dem Hohen Norden durch Schweden, Norwegen und Lappland in den Jahren 1810, 1811, 1812 und 1814* (Frankfurt, 1819). In Copenhagen in 1817 he published *Om Vulcaniske Producter fra Island*, sending a copy to Banks (mentioned in his letter to Banks, Document 293).[2]

Vídalín, Geir Jónsson (1761–1823), bishop of Iceland. He studied the Classical languages, philosophy and theology at the University of Copenhagen. On his return to Iceland he served as the minister of Reykjavík Cathedral in 1791, being appointed bishop of Skálholt in 1797. In 1802 the two bishoprics were combined and he became the first bishop of all Iceland. Deemed intelligent and a good writer, he wrote a play and translated parts of the New Testament. His generosity led to bankruptcy in 1805 but earned him the sobriquet 'the Good' (Geir góði). He welcomed the British trade with Iceland and sent Banks some manuscripts as a token of gratitude.

Wellesley, Richard Colley Wesley, 1st Marquess (1760–1842), British statesman. He was the elder brother of the Duke of Wellington. Governor-General of India 1798–1805, he became ambassador to Spain before being appointed Foreign Secretary on 6 December 1809, where he remained until 1812, subsequently becoming Lord Lieutenant of Ireland. It was during his tenure of the Foreign Office that Wellesley became involved in Icelandic affairs; it fell to him to deal with the aftermath of the Icelandic Revolution in close collaboration with Banks.

Wellesley-Pole, William, 3rd Earl of Mornington (1763–1845), politician and younger brother of the above. He was an MP in 1790–95 and 1801–21. In 1807 he became Secretary to the Admiralty until October 1809 when he was appointed Chief Secretary to Ireland, where he was hostile to Catholic emancipation. He was out of government until 1814 when he got a seat in Liverpool's cabinet as Master of the Mint. He was involved in Icelandic affairs in 1809.

Wolff, Georg or George (1736–1828), Norwegian timber merchant. Originally from Christiania (Oslo), he moved to London and in 1766 established with his brother the

---

[1] Von Troil wrote an autobiography, 'Själfbiografi och reseanteckningar'. See also Bergström's introduction in *Brev om Island av Uno von Troil*.

[2] The main source is his biography: Kornerup, *Eduoard Vargas Bedemar*.

firm Geo. & Ernst Wolff (from 1792, Wolffs & Dorville) at 21 Wellclose Square. He opened an account at the Bank of England and the following year he was naturalized. By 1780 this firm had become the leading Norwegian timber importers in London, a position they kept until the outbreak of war in 1807. He was appointed Dano-Norwegian consul in London in 1787, as Peter Anker's (q.v.) successor. He joined the Methodist movement and became a close friend of John Wesley. He chaired the meeting on 4 December 1807 to open a public subscription to help the masters and mates of Dano-Norwegian merchant ships, captured as prisoners of war. Throughout the war he was active in helping his former fellow-countrymen and in this capacity came in contact with Banks. Wolffs & Dorville acted as de facto Dano-Norwegian consuls throughout the war although the consulate was officially closed.[1]

Wormskiold, Morten (1783–1845), Danish naturalist and botanist. He went on botanical expeditions to Norway in 1807 and was the first to visit Greenland for the same purpose in 1812–14. Bugge (q.v.) wrote to Banks asking whether he could make arrangements to ensure that Wormskiold and his collections would not be molested by British ships-of-war and privateers. Wormskiold was sailing via Leith and Bugge used the opportunity to send books to Banks. He found many new species of plants, described and illustrated in the *Flora Danica*. He subsequently joined a Russian expedition to the Pacific and the Bering Strait in 1815. However, he fell out with the expedition's leader and remained in Kamtschatka from 1816 to 1818 when he fortunately found a ship to take him back to Denmark, sailing round Asia and Africa. On his return he was hailed as the first Danish circumnavigator, though that honour may belong to Jörgen Jörgensen (q.v.). On his return he abandoned his interest in natural history.

Wright, William (1735–1819), FRS, military physician, botanical collector and naturalist. As a young man he made a voyage to Greenland in 1757. His medical career began in 1758 when he served in the Royal Navy as surgeon's second mate in the West Indies until 1763, subsequently engaging in a medical partnership in Jamaica. During this period he began collecting dried plants which he sent to Kew. He eventually became surgeon-general in Jamaica, remaining there until 1777. During his time in Jamaica Wright became a slave-owner and began a voluminous correspondence with Banks which endured for three decades. That they became firm friends is evidenced by the letters in this book. According to his biographer his 'single most important scientific ... relationship was perhaps that with Sir Joseph Banks'.[2] He became a well-known physician in Edinburgh, was elected a fellow of the Royal Society in 1778, accepted into scientific circles and elected to a host of other societies. Later, in 1796–8, he became director of military hospitals in Barbados. He adopted his nephew James Wright, who accompanied Sir John Thomas Stanley to Iceland in 1789. Banks called on his help regarding the detained Icelanders in Leith.[3]

Wulff, Jens Andreas (1772–1851), Danish merchant. He became a burgher of Reykjavík in September 1795 and soon became a partner of Niels Ørum. In 1798 they established

---

[1] Polak, *Wolffs & Dorville*, pp. 3–13, 148–9.
[2] *ODNB*, 60, p. 504.
[3] See further *Memoir of the Late William Wright, M.D*; *ODNB*.

a trading concern in Reyðarfjörður, another one in Húsavík in 1804, one in 1806 in Raufarhöfn and rented the rights to the sulphur mines near Húsavík in 1811. They also traded in Seyðisfjörður and Eskifjörður. This firm dominated the trade in the north and east of Iceland in the early decades of the nineteenth century. Two of their ships, *Den Nye Pröve* and the *Husevig*, were captured in 1807. Another of their ships was the *Eskefiord* which crops up a lot in the correspondence.[1]

---

[1] See further Fode, 'Islandshandel og fastlandsspærring', *Erhvervshistorisk Aarbog*, XXV (1974), pp. 7–34.

BIBLIOGRAPHY

Note: Icelandic names beginning with the character Þ are placed at the end of the alphabetical sequence. Where there is more than one title by the same author, they are listed in alphabetical order.

PRIMARY SOURCES

Adelaide, Australia, The Royal Geographical Society of South Australia
MS. 6c. Joseph Banks
Århus, Denmark, Statens Arkiver, Erhvervsarkivet
Købmandsfirmaet Ørum & Wulff, København, 1805–1812 kopibog
Bedford, UK, Bedfordshire and Luton Archives and Records Service
Whitbread MSS, I
Cambridge, UK, Cambridge University Library
Royal Greenwich Observatory Archives (RGO)
RGO 14/7, 29
Canberra, National Library of Australia
Papers of Sir Joseph Banks, MS 9
Copenhagen, Denmark, Rigsarkivet (hereafter RA)
Rentekammeret (Rtk)
Rtk. 2214.55, 373.121, 373.133
Departementet for de Udenlandske Anliggender (DfdUA)
DfdUA file 817, Thematically sorted files Flora Danica, Cases re presents made of Flora Danica
DfdUA file 892, Thematically sorted files Letter I [Iceland] 1741–1846. Cases concerning trade, fishing etc. in Iceland
Ges. Ark. London III 1806–1807. Alm. Korrespondancesager No. 424 Letter-books
Ges. Ark. London III indkomne skrivelser fra DfdUa 1814–1817
DfdUA, England 1807–1808, box 1920. Cases concerning compensation for the ships captured by the English
DfdUA 1771–1848, England II, Depecher 1814–15, no. 1990
Island og Færöerne 1758–1846 (Island og Færöer)
Kommercekollegiet:
Handels- og Konsulatsfagets secretariat:
Kommercekollegiet 1797–1816. Fortegnelser over de til England opbragte danske skibe, 1807, 1810–11 og udat, box 1819
Kommercekollegiet, 1797–1816. Rapporter fra Konsulatet i London, box 1753 is Rapporter fra Konsulatet 1809–1816
Efterladte Embedspapirer:
Kommercekollegiet, 1763–1813, Schimmelmannske papirer, box 2155
Kommercekollegiet 1735–1816, Etatsråd Martfelts papirer VI, box 2124

Edinburgh, Edinburgh University Library, Special Collections Department
La[ing] MS III, 379
Edinburgh, The National Archives of Scotland (NAS)
Board of Customs and Excise (CE)
CE 1/41
Exchequer and Treasury (E)
E[xchequer] 504/22/48, 52–3, 57, 59, 61–3
Edinburgh, The National Library of Scotland, Manuscript Collections
MS 682. ff. 190–193 (1814–1815)
MS 789, Letter book of Archibald Constable & Co.
London, The British Library, Department of Manuscripts
Add[itional] MS
Add MS 4857–96
Add MS 6121
Add MS 8094, 8095–98, 8100
Add MS 15509–12
Add MS 33977–8, 33981–2
Add MS 35068
Add MS 38356
Add MS 40472
Add MS 41535, 41536, 41538
Add MS 45712
Add MS 56298
Add MS 56301(7)
Egerton MS
EG MS 2066–2070
London, The British Museum, Department of Prints & Drawings
Sarah Banks's Collection of Visiting Cards: C.1–740
London, The National Archives, Kew
The Admiralty (ADM)
ADM 1/692, /1995, /3899, /4773, /4979, /5079
ADM 2/155, /657, /893, /1103, /1368
ADM 53/1103
ADM 98/114, /308
ADM 99/181, /183–5
ADM 103/579
The Board of Trade (BT)
BT 1/46, /64, /71–2, /84
BT 5/19, /20, /22
BT 6/197–210
The High Court of Admiralty (HCA)
HCA 8/14
HCA 24/165 /166
HCA 25/192
HCA 30/52
HCA 32/985, /1032, /1083, /1162, /1176, /1197, /1614, /1645, /1976
The Foreign Office (FO)
FO 40/1–2
FO 73/85
FO 83/2293

FO 95/353, /360, /371, /648
FO 188/4–7
FO 211/12
FO 366/380, /671
FO 371/23639
The Privy Council (PC)
  PC 1/3901
  PC 2/177–81, /185, /188–91
The Public Record Office (PRO)
  PRO 30/42/14/8, Notes on Proceedings in Prize Cases
The Treasury (T)
  T 1/1121, /1124
  T 4/13
  T 11/48
  T 15/1
The Treasury Solicitor and HM Procurator General (TS)
  TS 8/6, /29
The War Office (WO)
  WO 1/844, /883 /1117
London, The Natural History Museum (NHM), Botany Library
  Banks Correspondence (BC)
  MSS WRI I
  Solander MS, Flora Islandica
  Solander MS, Plantae Islandicae et Notulae itinerariae
Dawson Turner Collection (DTC)
  DTC 1, 2, 3, 4, 6, 12, 17, 18, 19
London, The Royal Botanic Gardens, Kew, Archives (Kew)
Sir Joseph Banks Papers (JBK)
  Banks Correspondence I, JBK/1/2
  Banks Correspondence I, JBK/1/4
  Banks Correspondence, II, JBK/1/7
  Hooker's Correspondence, Directors' Correspondence, I
London, The Royal Society, Carlton House Terrace (RS)
  AP/9/21
  Misc. MS 6
Madison, Wisconsin, University of Wisconsin-Madison, Memorial Library, Department of Special Collections [Wisconsin Banks Papers]
  Correspondence concerning Iceland: Written to Sir Joseph Banks. MS 3. Formerly: Banks papers: Iceland, the Danish colonies & the Polar regions, 1772–1818.
  https://uwdc.library.wisc.edu/collections/HistSciTech/BanksJ/
Maidstone, Kent, The Kent History and Library Centre
  Banks MS U 1590/S1/2
  Banks MS U 951 Z31
  Banks MSS U 951 Z 32/ 9, 10, 11, 19, 24, 25, 27
Montreal, McGill University, Blacker-Wood Library
  The Banks Iceland Journal, Manuscript B 36, 1772
New York Public Libary
  Theodorus Bailey Myers collection, 1542–1876
Nottingham Trent University
  The Sir Joseph Banks Archive Project: Transcripts by Harold B. Carter

Oxford, Bodleian Libraries, University of Oxford
MS. Eng. hist. d. 150

Reykjavík, Landsbókasafn Íslands-Háskólabókasafn [National and University Library], Department of Manuscripts
JS 93 fol.
JS 111 fol.
Lbs. 168 fol.
Lbs. 341 a, fol.
Lbs. 355 fol.
Lbs. 1344, 4to

Reykjavík, The National Archives of Iceland (NAI)
Bf. Rvk.A/2-1. Bréfadagbók 1806–13
Bf. Rvk. OA/7, OA/8, OA/9. Verslun og sjóferðir 1814–18
Jörundarskjöl
Skjalasafn landfógeta. VI. 71, p. 57
Skjalasafn stiftamtmanns III. Innkomin bréf. Nr. 267 Bréfadagbók stiftamts II, 1809
Stiftamtsjournal II. Nr. 7. 6.2.1809–22.6.1809
Stiftamtsjournal V. Nr. 10. 11.6.1811–21.2.1820

San Francisco, USA, University of San Francisco, The Sutro Library [SL, Banks Papers]
The Joseph Banks Collection, I, II

Sydney, Australia, State Library of New South Wales, Mitchell Library
Banks Papers Series 05.01, James Roberts Journal, A 1594. http://www.sl.nsw.gov.au/search-intro
Banks MS, 'Volunties, Instructions, Provisions for 2d. Voyage'

Yale University Library, USA
Joseph Banks, box 11, folder 198

York, UK, The Private Collection of Dr Patrick Kaye
Letter from Banks to Horne & Stackhouse, 25 April 1814

SECONDARY SOURCES

Printed Works

Aall, Jakob, *Fædrelandske Ideer*, Christiansand, 1809.
—, *Om Jernmalmleier og Jerntilvirkning i Norge. Et fragmentarisk Forsög*, Copenhagen, 1806.
Account of the Society of Friends of Foreigners in Distress; with the Nature and Views of the Institution also The Plan and Regulation, A List of Subscribers and an Appendix containing some of the most interesting cases, etc., London, 1814.
Adam, Alexander, *A Summary of Geography and History, both Ancient and Modern* ..., 2nd edn, Edinburgh, 1797.
Adeane, Jane H., *The Early Married Life of Maria Josepha, Lady Stanley*, London, 1899.
Agnarsdóttir, Anna, 'The Challenge of War on Maritime Trade in the North Atlantic: The Case of the British Trade to Iceland during the Napoleonic Wars' in Olaf Uwe Janzen, ed., *Merchant Organization and Maritime Trade in the North Atlantic, 1660–1815*, St John's, Newfoundland, 1998, pp. 221–58.
—, 'Gilpinsránið 1808', *Landnám Ingólfs*, 4, 1991, pp. 60–77.
—, 'Great Britain and Iceland 1800–1820', unpublished PhD thesis, London School of Economics and Political Science, 1989.
—, 'Hjálendur Danakonungs í norðri' in Guðmundur Jónsson, Helgi Skúli Kjartansson and Vésteinn Ólason, eds, *Heimtur. Ritgerðir til heiðurs Gunnari Karlssyni sjötugum*, Reykjavík, 2009, pp. 35–49.

—, 'Hundadagadrottningin heldur út í heim 1812–1814' in Anna Agnarsdóttir, Erla Hulda Halldórsdóttir, Hallgerður Gísladóttir, et al., eds, *Kvennaslóðir. Rit til heiðurs Sigríði Th. Erlendsdóttur sagnfræðingi*, Reykjavík, 2001, pp. 123–39.

—, 'Hver skrifaði Íslandsbæklinginn 1813?' in Anna Agnarsdóttir, Pétur Pétursson and Torfi H. Tulinius, eds, *Milli himins og jarðar. Maður, guð og menning í hnotskurn hugvísinda*, Reykjavík, 1997, pp. 379–93.

—, 'Iceland's "English Century" and East Anglia's North Sea World' in David Bates and Robert Liddiard, eds, *East Anglia and Its North Sea World in the Middle Ages*, Woodbridge, Suffolk, 2013, pp. 204–16.

—, 'Iceland under British Protection during the Napoleonic Wars' in Pasi Ihalainen, Michael Bregnsbo, Karin Sennefelt et al., eds, *Scandinavia in the Age of Revolution*, Farnham, 2011, pp. 255–66.

—, 'Írskur svikari, ræðismaður á Íslandi', *Ný Saga*, I, Reykjavík, 1987, pp. 4–12.

—, 'Scottish Plans for the Annexation of Iceland 1785–1813', *Northern Studies*, 29, 1992, pp. 82–91.

—, 'Sir Joseph Banks and the Exploration of Iceland' in R. E. R Banks, B. Elliott, J. G. Hawkes et al., eds, *Sir Joseph Banks: A Global Perspective*, London, 1994, pp. 31–48.

—, *'This Wonderful Volcano of Water'. Sir Joseph Banks Explorer and Protector of Iceland 1772–1820*, Hakluyt Society Annual Lecture for 2003, London, 2004.

Aiton, William Townsend, *Hortus Kewensis; or, a Catalogue of the plants cultivated in the Royal Botanic Garden at Kew*, 3 vols, London, 1789. The second edition 1810–13 had five volumes.

Alexander, Sidney, *The Complete Odes and Satires of Horace*, Princeton, 1999.

Alumni Cantabrigiensis part II, V, ed. J. A. Venn, Cambridge, 1953.

Andersen, Dan H. and Pedersen, Erik Helmer, *A History of Prices and Wages in Denmark 1660–1800*, II, Copenhagen, 2004.

Andersen, Einar, *Thomas Bugge: Et mindeskrift*, Copenhagen, 1968.

Anderson, James, *An Account of the Present State of the Hebrides and Western Coasts of Scotland*, Edinburgh, 1785.

—, *Extracts Relative to the Fisheries on the North West Coast of Ireland*, London, 1787.

—, *The True Interest of Great Britain Considered: Or a Proposal for Establishing the Northern British Fisheries*, London, 1783.

Anderson, Johann, *Frásagnir af Íslandi ásamt óhróðri Göries Peerse og Dithmars Blefkens um land og þjóð*, ed. Gunnar Þór Bjarnason and Már Jónsson, Reykjavík, 2013.

—, *Nachrichten von Island, Grönland und dem Straße Davis, zum Nutzen der Wissenschaften und der Handlung*, Hamburg, 1746.

Andrésson, Sigfús Haukur, *Verzlunarsaga Íslands 1774–1806*, I–II, Reykjavík, 1988.

Annálar 1400–1800 or *Annales islandici posteriorum sæculorum*, 6 vols, Reykjavík, 1922–87.

The Annual Register for the Year 1772, London, 1773.

The Annual Register for the Year 1811, London, 1812.

The Annual Register for the Year 1812, London, 1813.

Anonymous, 'Introduction' in Uno von Troil, *Letters on Iceland*, Dublin, 1780.

Árnason, Jón, *Dactylismus ecclesiasticus: edur Fingra-Rijm ...*, Copenhagen, 1739.

The Australian Encyclopedia in Ten Volumes, Sydney, 1958.

Babington, Charles Cardale, 'A Revision of the Flora of Iceland', *Journal of the Linnean Society of London, Botany*, 11 (53), May 1870, pp. 282–348.

Bakewell, Sarah, *The English Dane. A Life of Jorgen Jorgenson*, London, 2005.

Banks, Joseph, 'An Attempt to Ascertain the Time when the Potato (*Solanum tuberosum*) was first Introduced into the United Kingdom: with Some Account of the Hill Wheat of India', *Transactions of the Horticultural Society of London* 1, 1807, 21–5.

—, 'Journal of a Tour in Holland, 1773', ed. Kees Van Strien, Voltaire Foundation Conference Proceedings, *SVEC* 2005:01, Oxford, 2005, pp. 149–83.

—, *The Journal of Joseph Banks in the Endeavour*, ed. Averil M. Lysaght, 2 vols, facsimile edn, Guildford, Surrey, UK, 1980.

—, ed., *Reliquae Houstounianae: seu plantarum in America meridionali a Gulielmo Houstoun M.D. R.S.S. collectarum icones manu propria aere incisae; cum descriptionibus e schedis ejusdem in bibliotheca Josephi Banks, Baroneti, R.S.P. asservatis*, London, 1781.

—, *A Short Account of the Cause of the Disease in Corn, Called by Farmers the Blight, the Mildew, and the Rust*, London, 1805.

Barrow, John, *Íslandsheimsókn. Ferðasaga frá 1834*, ed. Haraldur Sigurðsson, Reykjavík, 1994.

—, *A Visit to Iceland, by way of Tronyem, in the 'Flower of Yarrow' Yacht, in the Summer of 1834*, London, 1835.

—, *Sketches of the Royal Society and Royal Society Club*, London, 1849.

Bartholin, Thomas, *Antiquitatem Danicarum de Causis Contemptæ a Danis adhuc Gentilibus Mortis. Libri Tres, ex vetustis codicibus & monumentis hactenus ineditis congesti*, Copenhagen, 1689.

Bauhin, Johann, *De plantis a divis sanctis ve nomen habentibus ... Ioanni Bauhini*, Basel, 1591.

Beaglehole, J. C., ed., *The Endeavour Journal of Joseph Banks 1768–1771*, 2 vols, Sydney, 1962.

—, 'Introduction', in *The Endeavour Journal of Joseph Banks 1768–1771*, Sydney, 1962, vol. I, pp. 1–150.

—, ed., *The Voyage of the Endeavour 1768–1771*, *The Journals of Captain James Cook on His Voyages of Discovery*, I, Cambridge, 1955.

—, ed., *The Voyage of the Resolution and Adventure 1772–1775*, *The Journals of Captain James Cook on His Voyages of Discovery*, II, Cambridge, 1959.

Benediktsson, Bogi, *Sýslumannaævir*, 4 vols, Reykjavík, 1881–1914.

Benediktsson, Jakob, ed., 'Sir Joseph Banks: Dagbókarbrot úr Íslandsferð 1772', *Skírnir*, 124, 1950, pp. 210–21.

Bergström, Ejnar Fors, 'Indledning', in Uno von Troil, *Brev om Island av Uno von Troil*, [Stockholm], 1933, pp. 1–39.

Bernier, François, *The History of the late Revolution of the Empire of the Great Mogul: together with the most considerable passages for 5 years following in that Empire ...*, London, 1671.

Biographie universelle, ancienne et moderne, II, Paris, 1843.

Björnsson, Kristbjörn Helgi, 'Auðsöfnun og áratog. Kaupmennska og útgerð í Breiðafirði á fyrstu áratugum fríhöndlunar', unpublished MA thesis in History, University of Iceland, 2014.

Björnsson, Lýður, ed., *Kaupstaður í hálfa öld 1786–1836*, Reykjavík, 1968.

Björnsson, Ólafur Grímur, 'Sir George Steuart Mackenzie, Bart.', *Náttúrufræðingurinn* 75 (1), 2007, pp. 41–50.

Björnsson, Páll, (Biornonius, D. Paulus), 'An Accompt of D. Paulus Biornonius, Residing in Iceland, Given to Some Philosophical Inquiries Concerning That Country, Formerly Recommended to Him from Hence: The Narrative being in Latine, 'tis Thus English'd by the Publisher', *Philosophical Transactions of the Royal Society*, 9, January 1, 1674, pp. 101–11, 238–40.

Black, Joseph, 'An Analysis of the Waters of some Hot Springs in Iceland', *Transactions of the Royal Society of Edinburgh*, 3, 1794, pp. 95–153.

Blefken, Dithmar, *Islandia, sive Populorum & mirabilium quae in ea Insula reperiuntur accuratior descriptio ...*, Leiden, 1607.

Bobé, Louis, 'Den grønlandske handels og Kolonisations Historie indtil 1870', *Meddelelser om Grønland*, 55, (2), Copenhagen, 1936.

Boddaert, Pieter, *Elenchus animalium*, Rotterdam, 1785.

Bondesen, P. C. B., *Slægten Bugge i Danmark og Norge*, Odense, 1891.

Booth, Christopher C., *John Haygarth, FRS (1740–1827): a Physician of the Enlightenment*, Philadelphia, 2005.

Boswell, James, *Boswell's Life of Johnson ...*, ed. G. Birkbeck Hill, I, Oxford, 1934.

Bourguignon, Henry J., *Sir William Scott, Lord Stowell*, Cambridge, 1987.

Bowden, Jean K., *John Lightfoot, His Work and Travels*, London, 1989.
Briem, Helgi P., '"King" Jörgen Jörgensen', *American Scandinavian Review*, 31, 1943, pp. 120–31.
—, *Sjálfstæði Íslands 1809*, Reykjavík, 1936.
The British Imperial Calendar, London, 1813.
Britten, James, 'John Frederick Miller and his "*Icones*"', *Journal of Botany*, 51, 1913, pp. 255–7.
—, 'Sigismund Bacstrom M.D. (fl. 1770–1799)', *The Journal of Botany*, 49, March 1911, pp. 92–7.
Buffon, Georges Louis Leclerc, *Histoire naturelle, générale et particulière*, 36 vols, Paris, 1749–88.
Bugge, Thomas, 'Determination of the Heliocentric Longitude of the Descending Node of Saturn', *Philosophical Transactions of the Royal Society*, 77, 1787, pp. 37–43.
—, 'Landmaaling og Fremstilling af Kort under Bestyrelse af Det Kongelige Danske Videnskabernes Selskab 1761–1842', *Det Kongelige Danske Videnskabernes Selskab 1742–1942. Samlinger til Selskabets Historie*, IV, Copenhagen, 1961.
—, *Observationes astronomicæ annis 1781, 1782 & 1783*, Copenhagen, 1784.
—, *Videnskabernes Selskabs Kort over Danmark og Slesvig. Under det kongelige Videnskabernes Societets Direction ved rigtig Landmaaling optaget og ved trigonometriske samt astronomiske Operationer prövet*, Copenhagen, 1766–1836.
Burgess, J. Peter, 'Ambivalent patriotism: Jacob Aall and Dano-Norwegian Identity before 1814, *Nations and Nationalism*, 4, 2004, pp. 619–37.
Burke's Peerage and Baronetcy, 106th edn, London, 1999.
Burman, Nicolaas Laurens, *Flora Indica*, Leiden, 1768.
—, *Specimen botanicum de geraniis*, Leiden, 1759.
Burrow, Edward, *A New and Compleat Book of Rates, Comprehending the Rates of Merchandize etc.*, Glasgow, 1774.
Büsching, Anton Friderich, *Géographie universelle,* 6 vols, Strasbourg, 1768.
—, *Neue Erdbeschreibungdes ersten Theils erster Band welcher Dänemark, Norwegen und Schweden enthält*, Hamburg, 1760.
Callender, R. M., *The Ancient Lead Mining Industry of Islay*, Islay Museums Trust, 1962.
Camden, William, *Britannia: or a Chorographical Description of Great Britain and Ireland, Together with the Adjacent Islands*, translated from Latin by Edmund Gibson and enlarged by Richard Gough, London, 1789.
Cameron, Hector C., *Sir Joseph Banks, KB, PRS: The Autocrat of the Philosophers*, London, 1952.
Campbell, John, *Reports of Cases Determined at Nisi Prius in the Courts of King's Bench and Common Pleas and on the Home Circuits etc. 1809–1811*, II, London, 1811.
Carter, Harold B., *His Majesty's Spanish Flock: Sir Joseph Banks and the Merinos of George III of England*, Sydney, 1964.
—, 'Introduction', in Neil Chambers, ed., *The Letters of Sir Joseph Banks. A Selection 1768–1820*, London, 2000, pp. xiii–xviii.
—, 'The Royal Society and the Voyage of HMS *Endeavour* 1768–71', *Notes and Records of the Royal Society of London*, 49 (2), 1995, pp. 245–60.
—, *The Sheep and Wool Correspondence of Sir Joseph Banks 1781–1820*, Sydney, 1979.
—, *Sir Joseph Banks 1743–1820*, London, 1988.
—, *Sir Joseph Banks (1743–1820). A Guide to Biographical and Bibliographical Sources*, London, 1987.
—, 'Sir Joseph Banks and the Plant Collection from Kew Sent to the Empress Catherine II of Russia 1795', *Bulletin of the British Museum (Natural History)*, 4(5), 1974, pp. 283–385.
Chalmers, George, *Caledonia: or an Account, Historical and Topographic, of North Britain ... with a Dictionary*, 3 vols, London, 1807–24.
—, *The Life of Mary Queen of Scots*, 2 vols, London, 1818.

Chambers, Neil, 'General Introduction' in *The Indian and Pacific Correspondence of Sir Joseph Banks*, I, London, 2008, pp. xi–xlvi.
—, 'General Introduction' in *The Scientific Correspondence of Sir Joseph Banks*, I, London, 2007, pp. ix–lix.
—, ed., *The Indian and Pacific Correspondence of Sir Joseph Banks*, I, London, 2008.
—, *Joseph Banks and the British Museum: The World of Collecting, 1770–1830*, London, 2007.
—, 'Joseph Banks, the British Museum and Collections in the Age of Empire' in R. G. W. Anderson, M. L. Caygill, A. G. MacGregor et al., eds, *Enlightening the British: Knowledge, Discovery and the Museum in the Eighteenth Century*, London, 2003, pp. 99–113.
—, 'Letters from the President: the Correspondence of Sir Joseph Banks', *Notes and Records of the Royal Society of London*, 53 (1), 1999, pp. 25–57.
—, ed., *The Letters of Sir Joseph Banks. A Selection 1768–1820*, London, 2000.
—, ed., *The Scientific Correspondence of Sir Joseph Banks 1765–1820*, 6 vols, London, 2007.
Chapuis, Albert, *Urbain Jurgensen et ses continuateurs*, Neuchatel, 1923.
Child, Josiah, *A New Discourse of Trade*, Glasgow, 1751.
Christie, Ian R., *Wars and Revolutions: Britain 1760–1815*, London, 1982.
Cicero, *De republica*, IV, London, 1948.
Cinque Ports Herald and Kent and Sussex Advertiser, 1826.
Clausen, Holger P., *Nogle Betænkninger om det Spørgsmaal: Er en Friehandel for fremmede Nationer paa Island skadelig for Danmark og Island?*, Copenhagen, 1816.
Clifton, Gloria, *Directory of British Scientific Instrument Makers, c. 1550–1851*, London, 1995.
Cobbett's Parliamentary Debates, House of Lords, London, XVII, 1810.
Coke, Lady Mary, *The Letters and Journals of Lady Mary Coke*, ed. J. A. Home, 4 vols, Edinburgh, 1889–96.
Collinge, J. M., *Foreign Office Officials 1782–1870*, London, 1979.
Collins Encyclopedia of Scotland, London, 2000.
Collins, Greenville, *Great Britain's Coasting Pilot*, London, 1693.
Collins, John, *Salt and Fishery*, London, 1682.
The Commissioned Sea Officers of the Royal Navy 1660–1815, II, London, 1954.
The Complete Peerage, 12 vols, London, 1910–59.
The Complete Sagas of Icelanders, see Hreinsson.
Cook, Andrew S., 'James Cook and the Royal Society' in Glyndwr Williams, ed., *Captain Cook: Explorations and Reassessments*, Woodbridge, Suffolk, 2004, pp. 37–55.
Cottle, A. S., trans., *Icelandic Poetry or The Edda of Saemund*, Bristol, 1797.
Dalrymple, Alexander, *A Collection of Voyages Chiefly in the Southern Atlantick Ocean*, London, 1775.
Danmarks Adels Aarbog, 1961, ed. Sven Houmøller and Albert Frabitius, Copenhagen, 1960.
Dansk Biografisk Leksikon, 3rd edn, ed. Sv. Cedergreen Bech, 16 vols, Copenhagen, 1979–84.
Dansk Biografisk Lexikon, ed. C. F. Bricka, 19 vols, Copenhagen, 1887–1905.
Dawson, Warren, *The Banks Letters. A Calendar of the Manuscript Correspondence*, London, 1958.
The Department of Foreign Affairs 1770–1848, ed. Steen M. Ousager and Hans Schultz Hansen, Copenhagen, 1997.
Det Kongelige Søkort Archiv, Copenhagen, 1788.
Dictionary of National Biography, 22 vols, Oxford, 1921–22.
Dictionary of Scientific Biography, ed. Charles Coulston Gillispie, 16 vols, New York, 1970–80.
Dillon, Richard H., ed., 'Charles Vancouver's Plan', *The Pacific Northwest Quarterly*, 41 (4), 1947, pp. 356–7.
Diment, Judith A. and Wheeler, Alwyne, 'Catalogue of the Natural History Manuscripts and Letters by Daniel Solander (1733–1782), or attributed to him, in British Collections', *Archives of Natural History*, 1984, 11 (3), pp. 468–9.

Diplomatarium Islandicum, ed. Björn Þorsteinsson, XVI, Reykjavík, 1972.
Dryander, Jonas, *Catalogus bibliothecæ historico-naturalis Josephi Banks*, 5 vols, London, 1798–1800.
Du Hamel du Monceau, H. L., and de la Marre, M., *Traité Général des Pesches, et Histoire des Poissons qu'elles fournissent, tant pour la subsistance des hommes, que pour plusieurs autres usages, qui ont rapport aux Arts et au Commerce*, Paris, 1769.
Duyker, Edward, *Nature's Argonaut: Daniel Solander, 1733–1782. Naturalist and Voyager with Cook and Banks*, Carlton South, Victoria, 1998.
— and Tingbrand, Per, eds, *Daniel Solander. Collected Correspondence 1753–1782*, Melbourne, 1995.
Edda Islandorum An. Chr. MCCXV. Islandice conscripta per Snorronem Sturlæ ... nunc primum Islandice, Danice, et Latine ex antiquissimis codicibus ... in lucem prodit opera ... P. J. Resenii, ed. Johan Peter Resen, Copenhagen, 1665.
Edinburgh Evening Courant, 28 November 1807.
Egils saga, sive Egilli Skallagrimii vita ..., ed. Guðmundur Magnússon and Grímur Thorkelín, Copenhagen, 1809.
Ehrencron-Müller, Holger, *Forfatterlexikon omfattende Danmark, Norge og Island indtil 1814*, 12 vols, Copenhagen, 1924–39.
'Eirik the Red's Saga', trans. Keneva Kunz, in *The Complete Sagas of Icelanders*, ed. Viðar Hreinsson, I, Reykjavik, 1997, pp. 1–18.
Encyclopédie, ou Dictionnaire raisonné des sciences, des arts et des métiers, par une société de gens de lettres, mis en ordre par M. Diderot de l'Académie des Sciences et Belles-Lettres de Prusse, et quant à la partie mathématique, par M. d'Alembert de l'Académie royale des Sciences de Paris, de celle de Prusse et de la Société royale de Londres, 39 vols, Berne and Lausanne, 1778–81.
Erslev, Thomas Hansen, *Almindeligt Forfatter=Lexicon for Kongeriget Danmark med tilhørende Bilande, fra 1814 til 1840 ...*, 3 vols, Copenhagen, 1843.
Espólín, Jón, *Íslands Árbækur í sögu-formi*, 12 parts, XI and XII, Copenhagen, 1843 and 1855.
'Eyrarannáll', *Annálar 1400–1800* or *Annales islandici posteriorum sæculorum*, III, Reykjavík, 1933–38, pp. 225–420.
Fabricius, Otto, *Fauna Groenlandica*, Copenhagen, 1780.
Faujas de Saint Fond, Barthélemy, *A Journey through England and Scotland to the Hebrides in 1784: A Revised Edition of the English Translation [of 1799]*, ed. Sir Archibald Geike, 2 vols, Glasgow, 1907.
Feldbæk, Ole, *Nærhed og adskillelse, 1720–1814*, Oslo and Copenhagen, 1998.
— and Justesen, Ole, *Kolonierne i Asien og Afrika*, Copenhagen, 1980.
Feldborg, Andreas Andersen, 'An Appeal to the English Nation in behalf of Norway', *The Pamphleteer*, 4(7), August 1814, pp. [233]–285.
—, *Cursory Remarks on the Meditated Attack on Norway in 1813*, London, 1813.
—, *Poems from the Danish*, London, 1815.
—, *A Tour in Zealand in the Year 1802; with an Historical Sketch of the Battle of Copenhagen. By a Native of Denmark*, London, 1804.
Fjalldal, Magnús, 'To Fall by Ambition – Grímur Thorkelín and his *Beowulf* Edition', *Neophilologus*, 92, 2008, pp. 321–32.
Fleming, Fergus, *Barrow's Boys*, New York, 1998.
Flora Danica, 51 vols, Copenhagen, 1761–1883.
Fode, Henrik, 'Islandshandel og fastlandsspærring', *Erhvervshistorisk Aarbog*, XXV, 1974, pp. 7–34.
Fraser, Antonia, *Mary Queen of Scots*, London, 1970.
Friis, Astrid, and Glamann, Kristof, *A History of Prices and Wages in Denmark 1660–1800*, I, Copenhagen, 1958.
Frydenlund, Bård, *Stormannen Peder Anker. En biografi*, Oslo, 2009.
Gad, Finn, *Grønland*, Copenhagen, 1984.
—, *The History of Greenland. I. Earliest Times to 1700*, London, 1970.

—, *The History of Greenland II: 1700 to 1782*, London, 1973.

—, 'Tasiussaq-affæren 1811', *Tidsskriftet Grønland*, 6, 1980, pp. 175–84.

Gascoigne, John, 'Banks, Sir Joseph, Baronet (1743–1820)', in *Oxford Dictionary of National Biography*, III, Oxford, 2004, pp. 691–6.

—, 'Banks, Sarah Sophia (1744–1818)' in *Oxford Dictionary of National Biography*, 3, Oxford, 2004, pp. 697–8.

—, 'Joseph Banks and the Expansion of Empire', in Margarette Lincoln, ed., *Science and Exploration in the Pacific. European Voyages to the Southern Oceans in the 18th Century*, London, 1998, pp. 39–51.

—, *Science in the Service of Empire. Joseph Banks, the British State and the Uses of Science in the Age of Revolution*, Cambridge, 1998.

—, 'The Scientist as Patron and Patriotic Symbol: the Changing Reputation of Sir Joseph Banks', in Michael Shortland and Richard Yeo, eds, *Telling Lives in Science. Essays on Scientific Biography*, Cambridge, 1996, pp. 243–66.

[Gautrek's saga, 1664] *Gothrici & Rolfi Westrogothiae regum historia lingua antiqua Gothica conscripta; quam e M.s. vetustissimo edidit, & versione notisque / illustravit Olavs Verelivs; accedunt Joannis Schefferi Argentoratensis notae politicae*, Uppsala, 1664.

'Gautrek's saga', in Hermann Pálsson and Paul Edwards, eds, *Seven Viking Romances*, London 1985.

The Gentleman's Magazine, 1772.

The Gentleman's Magazine, 1807.

The Gentleman's Magazine, 1818, part 2.

Gerard, John, *Gerard's Herbal. The History of Plants*, ed. Marcus Woodward, London 1994. Originally published as *The Herball Or Generall Historie of Plantes Gathered by John Gerarde*, London, 1597.

'Gisli Sursson's Saga', trans. Martin Regal, in *The Complete Sagas of Icelanders*, ed. Viðar Hreinsson, II, Reykjavík, 1997, pp. 1–48.

The Golden Fleece: Or the Trade, Interest and Well-Being of Great Britain considered with Remarks etc. ... mostly on wool, London, 1736.

Granville, Mary, *The Autobiography and Correspondence of Mary Granville Mrs Delany*, ed. Lady Llanover, 3 vols, London, 1862.

Greipsson, Sigurður and Anthony J. Davy, '*Leymus arenarius*. Characteristics and uses of a dune-building grass', *Búvísindi Icelandic Agricultural Sciences*, 8, 1994, pp. 41–50.

Groom, Nick, *The Making of Percy's Reliques*, Oxford, 1999.

Guðjónsson, Elsa E., 'An Icelandic Bridal Costume from about 1800', *Costume. The Journal of the Costume Society*, 23, 1989, pp. 1–21.

Guðmundsson, Gils, *Skútuöldin*, 5 vols, Reykjavík, 1944–6.

Gudmundsson, Pétur, *Annáll nítjándu aldar*, I, 1801–1830, Akureyri, Iceland, 1912–22.

Gunnarsson, Gísli, *Fiskurinn sem munkunum þótti bestur: Íslandsskreiðin á framandi slóðum 1600–1800*, Reykjavík, 2004.

—, *Monopoly Trade and Economic Stagnation. Studies in the Foreign Trade of Iceland 1602–1787*, Lund, 1983.

—, *Upp er boðið Ísaland. Einokunarverslun og íslenskt samfélag 1602–1787*, Reykjavík, 1987.

Gustafsson, Harald, *Mellan kung och allmoge – ämbetsmän, beslutprocess och inflytande på 1700 – talets Island*, Stockholm, 1985.

Hakluyt, Richard, *The Principal Navigations, Voyages, Traffiqves, and Discoveries of the English Nation*, London, 1599.

Hálfdanarson, Guðmundur, *The A to Z of Iceland,* Lanham, Md., 2010.

Halldórsson, Björn, *Lexicon Islandico-Latino-Danicum Björnonis Haldorsonii*, 2 vols, Copenhagen, 1814.

Halldórsson, Jón from Hítardalur, *Biskupasögur*, 2 vols, Reykjavík, 1903–15, specifically II, Reykjavík, 1911–15.
Hansen, Thorkild, *Arabia Felix: the Danish Expedition of 1761–1767*, New York, 1964.
Hassell, John, *A Tour of the Isle of Wight*, 2 vols, London, 1790.
Hearne, Thomas, *Liber Niger Scaccarii*, 2 vols, Oxford, 1728.
Heckscher, Eli, *The Continental System*, Oxford, 1922.
Helgason, Jón, *Árbækur Reykjavíkur 1786–1936*, 2nd edn, Reykjavík, 1942.
—, 'Finnur Magnússon', *Ritgerðakorn og ræðustúfar*, Reykjavík, 1959, pp. 171–96.
—, *Hrappseyjarprentsmiðja 1773–1794*, Copenhagen, 1928.
—, 'Íslenzk handrit í British Museum', *Ritgerðakorn og ræðustúfar*, Reykjavík, 1959, pp. 109–32.
Henchel, Ole, 'Underretning om de Islandske Svovel=Miiner samt Svovel=raffineringen sammesteds, 30 Januar 1776' in Olaus Olavius, *Oeconomisk Reyse igiennem ... Island*, II, Copenhagen, 1780, pp. 665–734.
Henderson, Ebenezer, *Iceland; or the Journal of a Residence in that Island, during the Years 1814 and 1815: Containing Observations on the Natural Phenomena, History, Literature, and Antiquities of the Island; and the Religion, Character, Manners, and Customs of its Inhabitants*, Edinburgh and London, 1818.
Henrey, Blanche, *British Botanical and Horticultural Literature: Comprising a History and Bibliography of Botanical and Horticultural Books Printed in England, Scotland, and Ireland from the Earliest Times until 1800*, 3 vols, London, 1975.
Herbert, William, *Select Icelandic Poetry Translated from the Originals with Notes in 2 Parts*, London, 1804–6.
Hermannsson, Halldór, 'Sir Joseph Banks and Iceland', *Islandica*, XVIII, Ithaca, New York, 1928.
An Historical and Genealogical Account of the Clan Maclean, from its First Settlement at Castle Duart in the Isle of Mull to the Present Period, By a Seneachie [i.e. John Campbell Sinclair].
Historical Manuscripts Commission, *Report on the Manuscripts of Earl Bathurst*, London, 1923.
Hobart Town Courier, 1841.
Hogan, James Francis, *The Convict King. Being the Life and Adventures of Jorgen Jorgenson*, London, 1891.
Holden's Annual London & Country Directory for 1811, 3 vols, London, 1811.
Holland, Henry, *The Iceland Journal of Henry Holland 1810*, ed. Andrew Wawn, The Hakluyt Society, 2nd ser., part 2, 168, London, 1987.
Holm, Sæmundur Magnússon, *Om Jordbrænderne paa Island i Aaret 1783*, Copenhagen, 1784.
Holmskjold, Theodor, *Beata ruris otia fungis Danicis a Th. H. impensa*, 2 vols, Copenhagen, 1790–99.
Hooke, Robert, *Philosophical Experiments and Observations*, Oxford, 1967.
Hooker, Sir Joseph Dalton, *Journal of the Right Hon. Sir Joseph Banks Bart., K.B., P.R.S. During Captain Cook's First Voyage in H.M.S. Endeavour in 1768–71 to Terra del Fuego, Otahite, New Zealand, Australia, the Dutch East Indies etc.*, London, 1896.
—, *Life and Letters of Sir Joseph Dalton Hooker*, ed. Leonard Huxley, London, 1918.
Hooker, Sir William Jackson, *British Jungermanniæ: Being a History and Description of Each Species of the Genus and Microscopical Analysis of the Parts,* London, 1816.
—, *Journal of a Tour in Iceland in the Summer of 1809*, 2nd edn, 2 vols, London, 1813.
Hope, Valerie, *My Lord Mayor: Eight Hundred Years of London's Mayoralty*, London, 1989.
Horrebow, Niels, *Frásagnir um Ísland*, Reykjavík, 1966.
—, *The Natural History of Iceland*, London, 1758.
—, *Tilforladelige Efterrretninger om Island*, Copenhagen, 1752.
Hortus Kewensis, see Aiton.
Hreinsson, Viðar, ed., *The Complete Sagas of Icelanders*, 5 vols, Reykjavík, 1997.
Hugason, Hjalti, *Bessastadaskolan. Ett försök til prästskola på Island 1805–1846*, Uppsala, 1983.

An Icelander, *Memoir on the Causes of the Present Distressed State of the Icelanders and the Easy and Certain Means of Permanently Bettering their Condition*, London, 1813.
Ingimundardóttir, Björk, *Skjalasafn landfógeta 1695–1904*, Reykjavík, 1986.
Islandske Maaneds-Tidender, Hrappsey, Iceland, 1774.
Íslensk bókmenntasaga, vol. III, ed. Halldór Guðmundssson, Reykjavík, 1996.
Jackson, James Grey, *An Account of the Empire of Marocco and the District of Suse*, London, 1809.
Jacobæus, H. [Halldór Jakobsson], *Fuldstændige Efterretninger om de udi Island Ildsprudende Bierge, deres Beliggende, og de Virkninger, som ved Jord=Brandene paa adskillige Tider ere foraarsagede*, Copenhagen, 1757.
Jóhannesson, Þorkell, *Saga Íslendinga 1770–1830*, Reykjavík, 1950.
Johnson, Samuel, *Johnson's Dictionary*, London, 1809.
—, *Journey to the Western Islands of Scotland*, London, 1775.
Johnston, Alexander Keith, *A General Dictionary of Geography*, London, 1882.
Jones, Evan T., 'England's Iceland Fishery in the Early Modern Period' in David J. Starkey, Chris Reid and Neil Ashcroft, eds, *England's Sea Fisheries: The Commercial Fisheries of England and Wales since 1300*, London, 2000, pp. 105–10.
Jonsell, Bengt, 'The Swedish connection' in R. E. R Banks, B. Elliott, J. G. Hawkes et al., eds, *Sir Joseph Banks: A Global Perspective*, London, 1994, pp. 23–30.
Jónsbók was printed at Hólar in 1578 under the title of *Lögbók Íslendinga*.
Jónsson, Arngrímur, *Brevis Commentarius de Islandia*, ed. Einar Sigmarsson, Reykjavík, 2008.
—, *Crymogæa: þættir úr sögu Íslands*, ed. Jakob Benediktsson, Reykjavík, 1985.
Jónsson, Bjarni, *Tractatus Historico-Criticus de Feriis Papasticis Vulgo Gagn-Dagar*, Copenhagen, 1784.
Jónsson, Bjarni frá Unnarholti, *Íslenzkir Hafnarstúdentar*, Akureyri, Iceland, 1949.
Jónsson, Björn, 'Skarðsárannáll', *Annálar 1400–1800*, I, ed. Hannes Þorsteinsson, Reykjavík, 1922–7, pp. 28–272.
Jónsson, Finnur, *Historia Ecclesiastica Islandiæ*, 4 vols, Copenhagen, 1772–8.
Jónsson, Guðmundur and Magnússon, Magnús S., eds, *Hagskinna: Icelandic Historical Statistics*, Reykjavík, 1997.
Jónsson, Klemens, *Saga Reykjavíkur*, 2nd edn, 2 vols, Reykjavík, 1944.
[Jörgensen, Jörgen], *The Copenhagen Expedition Traced to Other Causes Than the Treaty of Tilsit. By a Dane*, London, 1810.
—, *Jorgen Jorgenson and the Aborigines of Van Diemen's Land: being a reconstruction of his 'lost' book on their customs and habits, and on his role in the Roving Parties and the Black Line*, ed. N. J. B. Plomley, Hobart, Tasmania, 1991.
—, *Jorgen Jorgenson's Observations on Pacific Trade; and Sealing and Whaling in Australian and New Zealand Waters Before 1805*, ed. Rhys Richards, Wellington, NZ, 1996.
—, *Efterretning om Engelændernes og Nordamerikanernes Fart og Handel paa Sydhavet*, Copenhagen, 1807.
—, *History of the Origin, Rise, and Progress of the Van Diemen's Land Company*, London, 1829.
—, *State of Christianity in the Island of Otaheite. By a Foreign Traveller*, Reading, 1811.
—, *Travels through France and Germany, in the years 1815, 1816 & 1817*, London, 1817.
Jürgensen, Urban, *Regler for Tidens nøjagtige Afmaalning ved Uhre*, Copenhagen, 1804.
Kaempfer, Engelbertus, *Icones selectae plantarum quas in Japonica collegit et delineavit Engelbertus Kaempfer*, London, 1791.
Karlsson, Gunnar, *Iceland's 1100 Years: The History of a Marginal Society*, London, 2000.
[Kayll, Robert], *The Trades' Increase*, London, 1615. Published in the *The Harleian Miscellany*, III, London, 1809, pp. 289–315.
Keay, John and Keay, Julia, eds, *Collins Encyclopaedia of Scotland*, London, 2000.

Kerguelen Trémarec, Yves Joseph de, *Relation d'un voyage dans la mer du nord, aux côte d'Islande, du Groenland, de Ferro, de Schettland, des Orcades et de Norwége, fait en 1767 et 1768,* Paris, 1771.
Ketilsson, Magnús, *Stiftamtmenn og amtmenn á Íslandi 1750–1800,* Reykjavík, 1948.
Kjölsen, Klaus and Sjöqvist, Viggo, *Den danske udenrigstjeneste 1770–1970,* 2 vols, I, Copenhagen, 1970.
Kongelig Dansk Hof og Stats-Calender, Copenhagen, 1810.
Kornerup, Else, *Édouard Vargas Bedemar. En Eventyrers Saga,* Copenhagen, 1959.
Kristinsson, Hörður, *A Guide to the Flowering Plants and Ferns of Iceland,* Reykjavík, 2010.
Kristmannsson, Gauti, *Literary Diplomacy. The Role of Translation in the Construction of National Literatures in Britain and Germany 1750–1830,* Frankfurt, 2005.
Langford, Paul, *A Polite and Commercial People: England 1727–1783,* Oxford, 1989.
Latrobe, Christian Ignatius, *Selection of Sacred Music from the Works of Some of the Most Eminent Composers of Germany and Italy,* 6 vols, London, 1806–26.
Law Officers' Opinions to the Foreign Office 1793–1860, ed. Clive Parry, XLII, London, 1970.
Laxdæla-saga sive Historia de rebus gestis Laxdölensium ..., trans. Þorleifur Repp, Copenhagen, 1826.
Laxness, Einar, *Íslandssaga a til ö,* 3 vols, Reykjavík, 1995.
Leland, John, *The Itinerary of John Leland the Antiquary,* ed. Thomas Hearne, 9 vols, Oxford, 1710–12.
Le Maire, Jacob, *Australian Navigations Discoverd by Jacob Le Maire in the Years 1615, 1616 and 1617..., etc.* published in English by John A. J. de Villers, ed., in *The East and West Indian Mirror,* London, Hakluyt Society, 2nd ser., 18, 1906.
Lewis, Samuel, *A Topographical Dictionary of Scotland,* 2 vols, London, 1846.
Lidderdale, Thomas, *Catalogue of the Books Printed in Iceland from AD 1578 to 1880 in the Library of the British Museum,* London, 1885.
Lightfoot, John, *The Flora Scotica,* 2 vols, London, 1777.
Lincoln, Margarette, ed., *Science and Exploration in the Pacific. European Voyages to the Southern Oceans in the 18th Century,* London, 1998.
Líndal, Sigurður, *Hið íslenzka bókmenntafélag, soguágrip,* Reykjavík, 1969.
Linnaeus, see Linné, Carl von.
Linné, Carl von (Linnaeus), *Instructio peregrinatoris,* Uppsala, 1759.
—, *Systema Naturæ,* Leiden, 1735.
List of All the Officers of the Army and Royal Marines on Full and Half Pay, London, 1813.
The Liverpool Mercury, 1814.
Lloyd's Evening Post, 19–22 March 1773.
The London Gazette, 10 February 1810 and 14 August 1810.
The London Magazine or Gentleman's Monthly Intelligencer, November, 41, London, 1772.
Lovsamling for Island, ed. Oddgeir Stephensen and Jón Sigurðsson, II, Copenhagen, 1853.
Lovsamling for Island, ed. Oddgeir Stephensen and Jón Sigurðsson, III, Copenhagen, 1854.
Lovsamling for Island, ed. Oddgeir Stephensen and Jón Sigurðsson, V, Copenhagen, 1855.
Lovsamling for Island, ed. Oddgeir Stephensen and Jón Sigurðsson, VI, Copenhagen, 1856.
Lovsamling for Island, ed. Oddgeir Stephensen and Jón Sigurðsson, VII, Copenhagen, 1857.
Low, George, *A Tour through the Islands of Orkney and Schetland in 1774,* Kirkwall, Orkney, 1889.
Lyngbye, H. C., *Testamen Hydrophytologiæ Danicæ,* Copenhagen, 1819.
Lysaght, Averil M., 'Banksian Reflections', *Journal of Joseph Banks in the Endeavour,* I, Guildford, Surrey, UK, 1980, pp. 13–28.
—, 'Joseph Banks at Skara Brae and Stennis, Orkney, 1772', *Notes and Records of the Royal Society of London,* 28 (2), 1974, pp. 221–34.
—, *Joseph Banks in Newfoundland and Labrador, 1766: His Diary, Manuscripts and Collections,* London, 1971.

Lyte, Charles, *Sir Joseph Banks. 18th Century Explorer, Botanist and Entrepreneur*, Newton Abbot, 1980.

Mabey, Richard, *Flora Britannica*, London, 1996.

MacDougall, Ian, *All Men are Brethren: French, Scandinavian, Italian, German, Dutch, Belgian, Spanish, Polish, West Indian, American and Other Prisoners of War in Scotland during the Napoleonic Wars, 1803–1814*, Edinburgh, 2008.

Mackay, David, 'Agents of Empire' in David Philip Miller and Peter Hanns Reill, eds, *Visions of Empire: Voyages, Botany and Representation of Nature*, Cambridge, 1996.

Mackenzie, Sir George Steuart, *Illustrations of Phrenology*, Edinburgh, 1820.

—, *Travels in the Island of Iceland during the Summer of the Year 1810*, Edinburgh, 1811; 2nd edn, Edinburgh, 1812.

Magnus, Olaus, *Beskrivning till Carta Marina*, Stockholm, 1960.

—, *Carta marina et descriptio septemtrionalium terrarum ac mirabilium rerum in eis contentarum, diligentissime elaborata anno Dni 1539*, Venice, 1539.

Magnússon, Guðmundur and Thorkelín, Grímur, *Egils saga, sive Egilli Skallagrimii vita*, Copenhagen, 1809.

Maiden, Joseph Henry, *Sir Joseph Banks: the 'Father of Australia'*, Sydney, 1909.

Mannevillette, Jean-Baptiste d'Après de, *Description et usage d'un nouvel instrument pour observer la longitude sur mer, appelé le quartier anglais*, Paris, 1739.

'A Manuscript Inventory of the Library of Sir Joseph Banks as received by the British Museum', 2 vols, [1827] (British Library shelf-mark 460.g.1).

Marquardt, Ole, 'Change and Continuity in Denmark's Greenland Policy 1721–1870' in Eva Heinzelmann, Stefanie Robl and Thomas Riis, eds, *The Oldenburg Monarchy. An Underestimated Empire?*, Kiel, 2006.

Marsden, R. G., *A Digest of Cases Relating to Shipping, Admiralty and Insurance Law*, London, 1899.

Marshall, John, *Royal Naval Biography* Supplement Parts I, II, IX and X, London, 1828.

Marshall, John Braybrooke, 'The Handwriting of Joseph Banks, His Scientific Staff and Amanuenses', *Bulletin of the British Museum (Natural History)*, Botany Series, 6(1), 28 September 1978.

Marshall, Rosalind K., *Queen of Scots*, Edinburgh, 1986.

Martin, Martin, *A Description of the Western Islands circa 1695*, 2nd edn, London, 1716. New edn, Stirling, 1934.

—, *A Late Voyage to St Kilda*, London, 1698. New edn, Edinburgh, 1986.

Mattiolus, Pietro Andreas, *Opera quae extant omnia*, ed. Bauhin (Caspar Bauhinus), Frankfurt, 1598.

McKay, Derek, 'Great Britain and Iceland in 1809', *Mariner's Mirror*, 59(1), 1973, pp. 85–95.

McNeill, Florence Marian, *Iona: A History of the Island* (1920), 7th edn, Moffat, Scotland, 1991.

'Memoir of Sir Joseph Banks', *The New Monthly Magazine*, 1820, 14 (2), pp. 185–94.

Meursius, Joannis [Johannes van Meurs], *Historiae Danicae Libri III*, Copenhagen, 1630.

Meynell, Guy, 'Banks Papers in the Kent Archives Office, Including Notebooks by Joseph Banks and Francis Bauer', *Archives of Natural History*, 10 (1), 1981, pp. 77–88.

— and Pulvertaft, Christopher, 'The Hekla Lava Myth', *The Geographical Magazine*, 53, 1981, pp. 433–6.

Middleton, Dorothy, 'Banks and African Exploration' in R. E. R Banks, B. Elliott, J. G. Hawkes, et al., eds, *Sir Joseph Banks. A Global Perspective*, London, 1994, pp. 171–6.

Middleton, W. E. Knowles, *The History of the Barometer*, Baltimore, 1964.

Miller, David Philip, 'Joseph Banks, Empire and "Centres of Calculation" in Late Hanoverian London' in David Philip Miller and Peter Hanns Reill, eds, *Visions of Empire: Voyages, Botany and Representation of Nature*, Cambridge, 1996.

Miller, John Frederick, *Various Subjects of Natural History wherein are Delineated Birds, Animals and Many Curious Plants*, [London], 1776–82.

Miller, Philip, *The Gardener's and Florist's Dictionary*, London, 1731.

Müller, Otto Frederik, *Animalcula Infusoria Fluviatilia et Marina ...*, Copenhagen, 1786.
—, *Entomstraca seu Insecta Testacea, quae in Aquis Daniæ et Norvegiæ reperit etc.*, Copenhagen, 1785.
Münster, Sebastian, *Cosmographia*, Basel, 1628.
The National Gazetteer of Great Britain and Ireland. A Topographical Dictionary of the British Islands, 3 vols, London, 1868.
The Naval Chronicle, London, 1799–1818.
The Naval Gazetteer, Biographer and Chronologist, London, 1815.
Neri, Antonio, *The Art of Glass ... With Some Observations on the Author*, trans. and ed. Christopher Merrett, London, 1662.
Níelsson, Sveinn, *Prestatal og prófasta á Íslandi*, ed. Björn Magnússon, 2nd edn, Reykjavík, 1950.
Nordal, Guðrún, Tómasson, Sverrir and Ólason, Vésteinn, *Íslensk bókmenntasaga I*, Reykjavík, 1992.
Nörlund, N. E., *Islands kortlægning: En historisk Fremstilling*, Copenhagen, 1944.
Norman, C. B., *The Corsairs of France*, London, 1887.
Norsk biografisk leksikon, 19 vols, Kristiania [Oslo], 1923–83.
O'Byrne, William R., *A Naval Biographical Dictionary*, London, 1849.
Ólason, Páll Eggert, *Íslenzkar æviskrár*, 5 vols, Reykjavík, 1948–52.
Olafsen [Ólafsson, Eggert] and Povelsen [Pálsson, Bjarni], *Travels in Iceland Performed by Order of His Danish Majesty by Messrs. Olafsen & Povelsen*, London, 1805.
Ólafsson, Eggert and Pálsson, Bjarni, *Vice-Lavmand Eggert Olafsens og Land-Physici Biarne Povelsens Reise igiennem Island, foranstaltet af Videnskabernes Sælskab i Kiøbenhavn*, 2 vols, Copenhagen, 1772.
Ólafsson, Georg, 'Fyriraetlanir Andrew Mitchels um Grafarvog', *Ýmsar ritgerðir. Landnám Ingólfs. Safn til sögu þess*, II, Reykjavík 1936–40.
Olavius, Olaus, *Oeconomisk Reyse igiennem de nordvestlige, nordlige og nordostlige Kanter af Island*, 2 vols, Copenhagen, 1780.
Orwell, George, *Nineteen Eighty-Four*, London, 1949.
Ousager and Hansen, see *The Department of Foreign Affairs 1770–1848*.
Oxford Dictionary of National Biography, 60 vols, Oxford, 2004.
Pálsson, Bjarni see Ólafsson, Eggert.
Pálsson, Hermann and Edwards, Paul, eds, *Seven Viking Romances*, London 1985.
Pálsson, Sigurður L., 'The Journal of Magnus Stephensen 1807–1808', MA thesis, University of Leeds, 1947.
Pálsson, Sveinn, *Æfisaga Bjarna Pálssonar*, Akureyri, Iceland, 1944.
Papers of British Cabinet Ministers 1782–1900, Royal Commission on Historical Manuscripts Guides to Sources for British History, 1, London, 1982.
Park, Mungo, *Travels in the Interior Districts of Africa Performed under the Direction and Patronage of the African Association in the Years 1795, 1796 and 1797*, London, 1799.
Paykull, Gustaf von, *Fauna svecica. Insecta*, Uppsala, 1798–1800.
Pearn, John, 'Hermann Diedrich Spöring (1733–1771) Naturalist – Artist – Surgeon', http://espace.library.uq.edu.au/view/UQ:200355/Pearn_Sporing.pdf, viewed 25 May 2015.
Pennant, Thomas, *Arctic Zoology*, 2 vols and supplement, London, 1784–7.
—, *British Zoology*, 2 vols, London, 1766.
—, *A Tour in Scotland. 1769*, London, 1772.
—, *A Tour in Scotland and Voyage to the Hebrides MDCCLXXII*, 2 vols, Chester, 1774.
—, *A Tour in Scotland and Voyage to the Hebrides; MDCCLXXII*, 2nd edn, 2 vols, London, 1776.
—, *A Tour in Scotland and Voyage to the Hebrides; MDCCLXXII*, 5th edn, London, 1790.
Percy, Thomas, *Five Pieces of Runic Poetry*, London, 1763.
Pétursson, Magnús, 'Höskuldsstaðaannáll 1730–1784', *Annálar 1400–1800*, IV, Reykjavík, 1940, pp. 545–6.
Phelps, Samuel, *The Analysis of Human Nature*, 2 vols, London, 1818.

—, *Observations on the Importance of Extending the British Fisheries and of Forming an Iceland Fishing Society: ... Likewise a Short Treatise on the Quality of Salt Fit for the Fisheries; and Remarks on the Best Modes of Curing Fish, etc.*, London, 1817.

—, *A Treatise on the Importance of Extending the British Fisheries: Containing a Description of the Iceland Fisheries, and of the Newfoundland Fishery and Colony: together with Remarks and Propositions for the Better Supply of the Metropolis and the Interior, with Cured and Fresh Fish; Elucidating also the Necessity of Encouraging and Supporting Commerce and the General Industry of the Country*, London, 1818.

Philosophical Transactions of the Royal Society of London.

Phipps, Constantine John, *A Voyage towards the North Pole undertaken by His Majesty's Command, 1773*, London, 1774.

Pine, L. G., *The New Extinct Peerage 1884–1971*, London, 1972.

Pinkerton, John, *A General Collection of the Best and Most Interesting Voyages and Travels in all Parts of the World, Many of Which are now First Translated into English: Digested on a New Plan*, London, 1808–14.

Polak, Ada, *Wolffs & Dorville. Et norsk-engelsk handelshus i London under Napoleonskrigene*, Oslo, 1968.

Pontoppidan, Christian Jochum, *Geographisk Oplysning til Cartet over det sydlige Norge i trende Afdeelinger*, Copenhagen, 1785.

Ponzi, Frank, *Ísland á 18. öld/Eighteenth-century Iceland*, Reykjavík, 1980.

The Post Office Annual Directory, Edinburgh, 1809, 1811 and 1813.

Raat, Alexander J. P., *The Life of Governor Joan Gideon Loten (1710–1789)*, Hilversum, Netherlands, 2010.

Rafnsson, Sveinbjörn, 'Um eldritin 1783–1788', in *Skaftáreldar 1783–1784*, Reykjavík, 1984, pp. 243–62.

—, 'Oldsagskommissionens præsteindberetninger fra Island. Nogle forudsætninger og konsekvenser' in *Aarbøger for nordisk oldkyndighed og historie 2007*, Copenhagen, 2010, pp. 225–46.

Rasch, Aage, *Niels Ryberg 1725–1804*, Aarhus, Denmark, 1964.

Rasmussen, Stig T., ed., *Den Arabiske Rejse 1761–1767 – en dansk ekspedition set i videnskabshistorisk perspektiv*, Copenhagen, 1990.

Rauschenberg, Roy A., 'The Journals of Joseph Banks' Voyage up Great Britain's West Coast to Iceland and to the Orkney Isles July to October 1772', *Proceedings of the American Philosophical Society*, 117 (3), 1973, pp. 186–226.

Ray, John, *Catalogus Plantarum circa Cantabrigiam*, London, 1660.

—, *Synopsis Methodica Avium et Piscium*, London, 1713.

—, *Synopsis Methodica Stirpium Britannicarum*, 1690, 3rd edn, London, 1724.

Resen, Peter Johan, *Edda Islandorum An. Chr. MCCXV. Islandice conscripta per Snorronem Sturlæ ... nunc primum Islandice, Danice, et Latine ex antiquissimis codicibus ... in lucem prodit opera ... P. J. Resenii*, Copenhagen, 1665.

Reynolds, Henry, *Fate of a Free People*, Victoria, Australia, 2004.

Ring, John, *Treatise of the Gout ... and Observations on the Eau Medicinale*, London, 1811.

Róbertsdóttir, Hrefna, *Wool and Society. Manufacturing Policy, Economic Thought and Local Production in 18th century Iceland*, Gothenburg and Stockholm, 2008.

Rodger, N. A. M., *The Insatiable Earl, a Life of John Montagu, the Fourth Earl of Sandwich 1718–1792*, London, 1993.

Rosenkrantz, Niels, *Journal du Congrès de Vienne 1814–1815*, ed. Georg Nörregaard, Copenhagen, 1953.

Ross, Margaret C., *The Norse Muse in Britain 1750–1820*, Trieste, 1998.

—, ed. *The Old Norse Poetic Translations of Thomas Percy: A New Edition and Commentary*, Turnhout, Belgium, 2001.

Rottbøll, Christen Friis, 'Afhandling om en Deel enten gandske nye eller vel forhen bekiendte, men dog for os rare Planter, som i Island og Grönland ere fundne ...' in *Skrifter, som udi det Kiöbenhavnske Selskab af Lærdoms og Videnskabers Elskere ere fremlagte og oplæste ...*, X, 1770, pp. 393–462.

Rubin, Marcus, *Frederik VI's Tid fra Kielerfreden til Kongens Död*, Copenhagen, 1895.

Ryan, Lyndale, *Tasmanian Aborigines: A History since 1803*, Sydney, 2012.

Rydén, Stig, *The Banks Collection: An Episode in 18th-Century Anglo-Swedish Relations*, Stockholm, 1963.

Rymer, Thomas, ed., *Foedera*, 16 vols, London, 1704–13.

'The Saga of the Greenlanders', trans. Keneva Kunz, in Viðar Hreinsson, ed., *The Complete Sagas of Icelanders*, I, Reykjavik, 1997, pp. 19–32.

Sainty, J. C., *Admiralty Officials 1660–1870*, London, 1975.

—, *Officials of the Board of Trade 1660–1870*, London, 1974.

Saxtorph, Henrik, 'Nordgrønlands Inspektorats britiske forbindelse under englænderkrigene 1807–1814', *Arkiv*, 14, 1992, pp. 21–39.

Schou, H. H., *Beskrivelse af danske og norske Mönter 1448–1814*, Copenhagen, 1926.

Schouten, Willem Cornelisse, *Journal ou description du merveilleux voyage, fait par G. C. Schouten, natif de Horn, dans les annees 1615, 1616, 1617 (en circumnavigation le globe terrestre) ...*, Amsterdam, 1617.

The Scots Magazine, 34, Edinburgh, November 1772.

The Scots Magazine and Edinburgh Literary Miscellany, 74, Edinburgh, 1812.

Seaton, Ethel, *Literary Relations of England and Scandinavia in the Seventeenth Century*, Oxford, 1935.

Seaver, Kirsten A., 'Norse Greenland on the Eve of Renaissance Exploration in the North Atlantic' in Anna Agnarsdóttir, ed., *Voyages and Exploration in the North Atlantic from the Middle Ages to the XVIIth Century*, Reykjavík, 2000, pp. 29–44.

A Seneachie [John Campbell Sinclair], *An Historical and Genealogical Account of the Clan Maclean, from its first Settlement at Castle Duart, in the Isle of Mull, to the Present Period*, London, 1838.

Seneca, *Epistle*, I, v.

Shortland, Michael and Yeo, Richard, eds, *Telling Lives in Science. Essays on Scientific Biography*, Cambridge, 1996.

Sigurðsson, Haraldur, *Bréf frá Íslandi*, Reykjavík, 1961.

—, 'Inngangur', *Bréf frá Íslandi*, Reykjavík, 1961.

—, *Kortasaga Íslands: frá lokum 16. aldar til 1848*, II, Reykjavík, 1978.

—, *Kortasaga Íslands: frá öndverðu til loka 16. aldar*, I, Reykjavík, 1971.

—, *Ísland í skrifum erlendra manna um þjóðlíf og náttúru landsins/Writings of Foreigners Relating to the Nature and People of Iceland: A Bibliography*, Reykjavík, 1991.

Sigurðsson, Ingi, *Hugmyndaheimur Magnúsar Stephensens*, Reykjavík, 1996.

Sigurðsson, Jón, *Mikilhæfur höfðingi. Ólafur Stefánsson Stephensen stiftamtmaður og hugmyndir hans*, Reykjavík, 2011.

Simmons, Samuel Foart, ed., *The London Medical Journal*, London, 1781–90.

—, ed, *Medical Facts and Observations*, London, 1791–1800.

Smith, Edward, *The Life of Sir Joseph Banks*, London, 1911.

Sívertsen, Bjarni, 'Ágrip af æfisögu Bjarna Riddara Sivertsens', *Sunnanpósturinn*, 3, Reykjavík, March 1835, pp. 33–40.

Sneedorff, Frederik, *Samledege Skrifter*, ed. Andreas Birch, Christian Birch, and Rasmus Nyerup, 4 vols, Copenhagen, 1794–8.

Sölvason, Sveinn, 'Íslands árbók 1740–81', *Annálar 1400–1800*, V, Reykjavík, 1955, pp. 73, 75–77.

Sommerfelt, Axel, 'Stiftamtmand Grev Fredrik Christopher Trampe. En Biografi', *Trondhjemske Samlinger*, 1, 1901, pp. 1–30.

Sprod, Dan, *The Usurper. Jorgen Jorgensen and His Turbulent Life in Iceland and Van Diemen's Land 1780–1841*, Hobart, 2001.

Squibb, G. D., *Doctors' Commons: A History of the College of Advocates and Doctors of Law*, London, 1977.

[Stanley, John Thomas], *Íslandsleiðangur Stanleys 1789: Ferðabók*, ed. Steindór Steindórsson, Reykjavik, 1979.

—, *The Journals of the Stanley Expedition to the Faroe Islands and Iceland in 1789*, ed., John F. West, 3 vols, Torshavn, Faroe Islands, 1970–76.

Stanley, John Thomas, 'Stanley's Introduction to the journals', *Journals of the Stanley Expedition*, ed. John F. West, I, Torshavn, Faroe Islands, 1970, pp. xv–xvi.

Stearn, William T., *Stearn's Dictionary of Plant Names for Gardeners*, London, 1992.

Steel's Original and Correct List of the Royal Navy, London, 1809.

Steers, J. A., *The Coastline of England and Wales*, Cambridge, 1969.

Steingrímsson, Jón, *A Very Present Help in Trouble: The Autobiography of the Fire-Priest*, trans. Michael Fell, New York, 2002.

Steinþórsson, Sigurður, 'Annus Mirabilis. 1783 í erlendum heimildum', *Skírnir*, 166, 1992, pp. 133–55.

Stephensen, Magnús, *Eptirmæli Átjándu Aldar eptir Krists híngad-burd, frá Ey=konunni Íslandi*, Leirárgarðar, Iceland, 1806.

—, *Ferðadagbækur Magnúsar Stephensen 1807–1808*, ed. Anna Agnarsdóttir and Þórir Stephensen, Reykjavík, 2010.

—, *Island i det Attende Aarhundrede, historisk-politisk skildret*, Copenhagen, 1808.

—, *Kort Beskrivelse over den nye Vulcans Ildspudning i Vester=Skaptefields=Syssel paa Island i Aaret, 1783*, Copenhagen, 1785.

—, *Philosophisches Schilderung der gegenwärtigen Verfassung von Island: nebst Stephensens zuverlässiger Beschreibung des Erdbrandes im Jahre 1783 ...*, ed. C. U. D. Eggers, Altona [Holstein], 1786.

—, 'Varnarrit' [Defence, Apologia] of 19 September 1815, in *Ísafold*, 9, 1882, nos. 2 and 4, pp. 5–8, 13–15.

Stephensen, Ólafur, 'Um not af nautpeningi' [On the use of cattle], *Rit þess (konunglega) íslenzka Lærdómslistfélagsins*, 6, 1785, pp. 20–96.

Stow, John, *The Summarie of Englishe Chronicles*, London, 1567.

Strien, Kees van, ed, 'Joseph Banks, Journal of a Tour in Holland, 1773', *SVEC* [Studies on Voltaire and the Eighteenth Century] 2005:01, Oxford, 2005, pp. 83–183.

Sutherland, Lucy, *The East India Company in Eighteenth-Century Politics*, Oxford, 1952,

Sveinsson, Guðlaugur, 'Vatnsfjarðarannáll hinn yngsti 1751–1793', *Annálar 1400–1800*, V, Reykjavík, 1961, p. 371.

Svenskt Biografiskt Lexicon, 23 vols, Uppsala, 1842–76.

Tacitus, *Annals*, 15, viii.

'The Tale of the Mountain-Dweller (Bergbúa þáttur)', trans. Marvin Taylor, in *The Complete Sagas of Icelanders*, ed. Viðar Hreinsson, II, Reykjavik, 1997, pp. 444–8.

Teigen, Philip M., ed., 'Supplementary Letters of Sir Joseph Banks, third series', *Journal of the Society for the Bibliography of Natural History*, 7 (3), 1975, 249–57.

Thorkelín, Grímur, *De Danorum Rebus Gestis Secul. III & IV: Poëma Danicum dialecto Anglosaxonica / ex Bibliotheca Cottoniana Musaei Britannici edidit versione lat. et indicibus auxit Grim. Johnson Thorkelin* [Beowulf, first edn], Copenhagen, 1815.

Thorlacius, Gyða, *Endurminningar frú Gyðu Thorlacius frá dvöl hennar á Íslandi 1801–1815*, ed. Victor Bloch, Reykjavík, 1947.

Thoroddsen, Þorvaldur, *Landfræðissaga Íslands*, ed. Gísli Már Gíslason and Guttormur Sigbjarnason, 5 vols, 2nd edn, Reykjavík, 2003–9.

Thunberg, Carl Peter, *Flora Capensis*, 2 vols, Uppsala, 1807, and Stuttgart, 1823.

—, *Prodromus plantarum Capensium*, Uppsala, 1794–1800.
—, *Flora Japonica*, Leipzig, 1784.
Tómasson, Tómas, 'Djáknaannálar: Úr djáknaannálum, 1731–1794', *Annálar 1400–1800*, VI, Reykjavík, 1986, pp. 175–76.
Torfæus, Thormodus [Þormóður Torfason], *Series dynastarum et regum Daniae*, IV, Copenhagen, 1702.
The Trade's Increase, London, 1615. Republished in *The Harleian Miscellany*, III, London, 1809, pp. 289–315.
The Transactions of the Royal Society of Edinburgh, 111, 1794.
Transactions of the Society of Arts, 21, 1813.
Troil, Uno von, *Brev om Island av Uno von Troil*, Stockholm, 1933.
—, *Bref rörande en resa til Island MDCCLXXII*, Uppsala, Sweden, 1777.
—, *Letters on Iceland*, Dublin, 1780.
—, 'Självbiografi och reseanteckningar' in Henrik Schück and Oscar Levertin, eds, *Svenska memoarer och bref, etc.*, I, Stockholm, 1900, pp. 159–238.
Turner, Dawson, *Fuci, or Coloured Figures and Descriptions of the Plants Referred to by Botanists to the Genus Fucus*, 4 vols, London, 1808–19.
Turton, William. *A General System of Nature through the Three Grand Kingdoms of Animals, Vegetables, and Minerals*, 7 vols, London, 1806.
Tusser, Thomas, *A Hundreth Good Points of Husbandrie*, London, 1557.
—, *Five Hundreth Good Points of Husbandrie*, London, 1573.
—, *Thomas Tusser. His Good Points of Husbandry*, ed., Dorothy Hartley, Bath, 1931.
Tving, Rasmus, *Træk af Grønlandsfartens Historie*, Copenhagen, 1944.
Vaccari, E., 'The organized traveller: scientific instructions for geological travels in Italy and Europe during the eighteenth and nineteenth centuries' in Patrick N. Wyse Jackson, ed., *Four Centuries of Geological Travel: The Search for Knowledge on Foot, Bicycle, Sledge and Camel*, Geological Society, London, Special Publications, no. 287, 2007, pp. 7–17.
Vargas Bedemar, Edvard Romeo von, *Om vulcaniske Producter fra Island*, Copenhagen, 1817.
Verdun de la Crenne, Jean René A., de Borda, Jean-Charles Chevalier and Pingré, Alexandre Guy, *Voyage fait par ordre du roi en 1771–1772, en diverses parties de l'Europe, de l'Afrique et de l'Amérique ...*, Paris, 1778.
Verne, Jules, *Journey to the Centre of the Earth*, Oxford, 2008.
Vídalín, Jón, *Húspostilla*, Hólar, 1718.
Villers, John A. J. de, ed., 'Australian Navigations Discovered by Jacob Le Maire in the Years 1615, 1616 and 1617 ...', in *The East and West Indian Mirror*, London, 1906.
Von Troil, see Troil.
Wagner, Peter, 'En Kongelig Gave', *Fund og forskning i det Kongelige Biblioteks Samlinger*, 40, Copenhagen, 2001, pp. 81–101.
Warner, George F., ed., *The Libelle of Englyshe Polycye: a Poem on the Use of Sea-power, 1436*, Oxford, 1926.
Wawn, Andrew, *The Anglo Man. Þorleifur Repp, Philology and Nineteenth-Century Britain*, Reykjavík, 1991.
—, 'The Enlightenment Traveller and the Idea of Iceland: the Stanley Expedition of 1789 Reconsidered', *Scandinavica*, 28, 1989, pp. 5–16.
—, 'Hundadagadrottningin. Bréf frá Íslandi: Guðrún Johnsen og Stanleyfjölskyldan frá Cheshire, 1814–16', *Saga*, 23, 1985, pp. 99–133.
—, 'John Thomas Stanley and Iceland: the Sense and Sensibility of an Eighteenth-Century Explorer', *Scandinavian Studies*, 53 (1), 1981, pp. 52–76.
—, *The Vikings and the Victorians*, Cambridge, 2000.
Weld, Charles Richard, *A History of the Royal Society*, 2 vols, London, 1868.

Werlauff, E. C., *Vatnsdølernes Historie og Finnboge hiin Stærkes Levnet*, Copenhagen, 1812.
West, John F., *Faroe: the Emergence of a Nation*, London, 1972.
—, 'Introduction' in *The Journals of the Stanley Expedition to the Faroe Islands and Iceland in 1789*, I, Torshavn, Faroe Islands, 1970, pp. v–xii.
—, *The Journals of the Stanley Expedition to the Faroe Islands and Iceland in 1789*, 3 vols, Torshavn, Faroe Islands, 1970–76.
—, 'Stanley's Introduction to the Journals', *The Journals of the Stanley Expedition to the Faroe Islands and Iceland in 1789*, I, Torshavn, Faroe Islands, 1970, pp. xv–xvi.
Weston, Richard, *The Universal Botanist and Nurseryman*, 4 vols, London, 1770–77.
Worm, Ole, *Seu Danica Literatura Antiqvissima, Vulgo Gothica dicta, luci reddita, opera O. Wormiii; acc. de prisca Danorum poesi dissertation*, Amsterdam, 1636.
Worm=Müller, Jacob S., *Norge gjennem nødsaarene. Den norske regjeringskommisssion 1807–1810*, Christiania [now Oslo], 1918.
Wright, William, *Memoir of the late William Wright, M.D.*, Edinburgh, 1828.
Wyse Jackson, Patrick N. ed., *Four Centuries of Geological Travel: The Search for Knowledge on Foot, Bicycle, Sledge, and Camel*, Geological Society of London Special Publication, no. 287, London 2007.
Young, Arthur, *Observations on the Present State of the Waste Lands of Great Britain. Published on Occasion of the Establishment of a New Colony on the Ohio*, London, 1773.
Þórarinsson, Sigurður, *Heklueldar* (Reykjavík, 1968).
Þorgilsson, Ari or Ari the learned [*Ari fróði*], *Íslendingabók* [The Book of Icelanders], ed., Jakob Benediktsson, Reykjavík, 1968.
Þorkelsson, Jón, *Saga Jörundar hundadagakóngs*, Copenhagen, 1892.
Þorláksson, Helgi, 'Útflutningur íslenskra barna til Englands á miðöldum', *Sagnir*, 4, 1983, pp. 47–53.
—, *Sjórán og siglingar. Ensk-íslensk samskipti 1580–1630*, Reykjavík, 1999.
Þorsteinsson, Björn, *Enska öldin í sögu Íslendinga*, Reykjavík, 1970.
—, 'Henry VIII and Iceland', *Saga-Book of the Viking Society for Northern Research*, London, 1957–9, pp. 67–101.
—, *Tíu þorskastríð 1415–1976*, Reykjavík, 1976.
— and Jónsson, Bergsteinn, *Íslandssaga til okkar daga*, Reykjavík, 1991.
Þorsteinsson, Hannes, *Guðfræðingatal: stutt æfiágrip þeirra guðfræðinga íslenzkra, er tekið hafa embættispróf við Kaupmannahafnarháskóla 1707–1907*, Reykjavík, 1910.
Þorsteinsson, Pétur, 'Ketilsstaðaannáll 1742–1784', *Annálar 1400–1800*, IV, Reykjavík, 1940, pp. 424–5.

Websites

http://www.measuringworth.com
http://www.biographi.ca [*Dictionary of Canadian Biography* online]

INDEX

Aalborg, Jutland, 307
Aall, Jacob, Jr, 489–90, 490–91, 625
Aas, Hans Andersen, 280, 287, 289, 296, 518, 555
Aasiaat, 481n, 610n
Académie des Sciences, 50n, 156n
Adam, Alexander, 331
Adam, Johann Georg, 47n, 594
Addison, Joseph, 181
Adie, Alexander, 580
Admiral Juul, 338n, 362n, 402n, 405, 632
Adventure [various ships of that name], 10, 45n, 47, 432, 436n, 467
Æðey, 314
Africa (see also West Africa), 28, 154n, 186, 216, 468, 641
African Association, 28, 468, 626
Aiton, William, 552, 595
Aiton, William Townsend, 552, 595
Akurey, 504n
Akureyri, 312n, 560n
Albert, Prince, 67n
Albertina, 521n, 560
d'Alembert, Jean le Rond, 10, 13, 50n
Alexander, 224n, 225
Alexander, Captain D.T., 432
Alexandria, 521
Almannagjá, 93–5n
Alnwick Castle, 112
Alströmer, Johan, 209
Althing – see Alþingi
Alþingi, xxiii–xxv, 12, 14, 18, 93n, 235n, 242n, 257n
America (see also United States), 28, 31, 51, 78n, 152, 154n, 187n, 219, 228, 260, 319, 337, 399, 401, 409, 414, 431n, 432n, 513n
Amsterdam, 16n, 22, 166n, 187n
Ámundason, see Loptur Ámundason
Anderson, James, 532n
Anderson, Johann, 12n, 532n
Andriðsey, 502
Angerstein, Anna Elisabet, 193
Anglo-Danish relations, 84, 235, 255, 564, 567–8, 571–2, 574
Anker, Peter, 204–5, 206–7, 219n, 625, 641
Anna Dorothea, 310n, 436n, 443n, 488

Anna Maria, 438
Arabia, 16, 104
Ardtornish Point, 65n
Argyll, 57n, 67
Argyll, Duke of, 5, 65n
Argyllshire, 65n
Ari Guðlaugsson, 132n
Ari the Learned, see Ari Þorgilsson
Ari Þorgilsson, 83n, 94n, 259n
Ármannsfell, 95n
Armstrong, Thompson & Co., 358
Arnarnesvogur, 91
Árnessýsla, 313n
Arngrímur Jónsson, 12
Árni Helgason, 579
Arnold, John, 52, 203n, 207
Arran, 79n, 183n, 412n
Artaurinish, 65, 124n
Ásmundsdóttir, see Jórunn Ásmundsdóttir
Asp, Per Olof von, 184
Asquith, John, 48, 156, 593
Augusta, 50
Austfirðir (the Eastern fjords), 313n
Australia, 5, 8, 20, 35, 36, 593, 596, 626, 629, 632
Austur-Skaftafellssýsla, 313n

Bacstrom, Sigismund, 7, 42, 46, 102n, 118, 156, 593
Baðstofuhver, 108n, 601
Baffin, William, 610
Bagot, Lewis, 182
Bahama, 362, 371n, 377n, 390n, 564, 632
Baine, John, 212n, 326n
Baker, D. B., 288, 289, 625
Balls Revier, 611n
Baltic, 154n, 229n, 244, 251, 303n, 307n, 373n, 418n, 427, 428, 434, 449, 495, 524n, 531, 558n,
Banks, (Sir) Joseph,
 Expedition to Iceland 1772, 7–10, 17–25, 42–3, 113–14, 176–9, 226, 593–7, 600–601, 602–3, 604–7, 608–9
 Icelanders ask Banks for help, 228–9, 239–40, 245–8, 248–9, 250–51, 265–6, 268, 269, 270–74, 274–5, 278–9, 280–81, 283–4, 285, 285–6, 287–91, 307–8, 381–2, 385–7, 391

663

images, Frontispiece, Plates 1, 12, 16
journals (Iceland expedition), 45–140
life, 1–7, 8–29
plans for annexations, 29–34, 233–5, 236, 240–42, 243–5, 256–7, 257–65, 275–6, 503, 526–37, 537–8
Protector of Iceland, 29–30, 229–30, 230–31, 231–2, 233, 234, 236–7, 238–9, 251–2, 253, 267, 275, 277, 281, 282, 294–7, 297–300, 456–7, 489, 494, 499–500, 503, 505–6, 511–12, 587–8, 590–91
travels, 3, 4–7, 25–6.
see also Royal Society, Icelandic Revolution
Banks, Lady (Dorothea Hugessen), 27, 210n
Banks, Sarah Sophia, 27, 39–40, 52n, 156n, 157n, 564, 595n, 596
Banterer, 280n
Bankes, Captain, 432n
Barðastrandarsýsla, 314n
Barrow, Sir John, 31, 300n, 326–7, 394, 626
Bartholin, Thomas, 328
Básendar, 313
Bassant, see *Bosand*
Bathurst, Henry, 3rd Earl, 25, 33, 283n, 294–7, 309n, 311n, 318n, 320–22, 335–56, 359–60, 368n, 377n, 388n, 405n, 416n, 454, 464, 467n, 471, 477, 488, 506n, 626
Battine, William, 394n
Bauhin, Gaspard, 175
Bayly, William, 46
Beachy Head, 50, 118
Beaglehole, John Cawte, 8, 37, 39, 593, 594, 595
Bedemar, Eduardo Romeo von Vargas, 586–7, 640
Bedre Tider, 237, 439n,
Beinn a' Chaolais, 60n
Benedikt Gröndal, 241n
Benediktsson, see Jakob Benediktsson
Benners, Isaac S., 212n
Bentinck, Margaret, Dowager Duchess of Portland, 23n
Bergen, 110n, 154n, 158n, 269, 358n, 388, 430, 483, 507n, 508n, 511n, 528n, 530, 531n, 574, 627, 630,
Bergman, Torbern Olof, 194
Bergmann, Lorentz Andersen,
Berkshire, 3, 274,
Bernier, François, 181
Bernstorff, Andreas Peter von, Count, 15
Bernstorff, Johann H. E., Count, 85n
Berufjörður, 314, 550, 584
Bessastaðir (Bessested), 82, 128, 130, 132n, 136, 176n, 221n, 222n, 301, 331n, 349n, 395, 606, 634
Bickerton, Sir Richard, 300n

Bjering (Biering), Hans Peter, 352, 369n
Bildahl (*Bildal*), 257, 267n, 376, 396, 428, 433n, 434, 443, 484, 485, 518, 555, 559, 565
Bildal, see *Bildahl*, Bíldudalur
Bishop, Charles (King's Proctor), 238n, 262n, 268n, 270n, 271n, 275, 280n, 282
Bíldudalur (Bildal), 246n, 261n, 315, 450, 484n, 485n, 536, 584
Bjarnason, see Páll Bjarnason
Bjarni Halldórsson, 90n
Bjarni Helgason, 104n
Bjarni Jónsson, 18, 19, 20n, 25, 42, 102n, 108n, 159, 160–62, 188–91, 197, 604, 632
Bjarni Pálsson (Povelsen), 19, 19n, 20, 21, 22, 42, 164–6, 202, 547n, 635
Bjarni Sívertsen, xxiv, 227n, 229, 230, 233n, 236, 238, 239n, 243, 244, 245–8, 254n, 261n, 265–6, 266, 267, 271–4, 278, 278–9, 279n, 280, 283–4, 291n, 293n, 295, 307–8, 309, 310, 332n, 336n, 339n, 344n, 364, 396n, 418, 437, 463, 465, 484n, 550, 553n, 558, 563, 567, 573, 637, Plate 11
Björn Halldórsson, 221n, 625
Björn Jónsson (apothecary), 164
Björn Jónsson (clergyman), 169n
Björn Stephensen, 234n, 267
Björn Þorleifsson (the rich), 531n
Björnsen, Ole, 344n
Björnsson, see Finnbogi Björnsson, Páll Björnsson
Black, Joseph, 24n, 217n, 531n
Black Mountains, 61n
Blagden, Charles, 26, 203
Blefken, Dithmar, 12
Bloxwich, John, 530n
Board of Agriculture [British], 28, 208n, 400n
Board of Longitude, [British], 28, 46, 580, 581n, 594
Board of Trade [British], 32, 33, 283n, 309n, 320n, 323n, 335n, 359, 365n, 371, 372, 375, 382n, 412n, 424, 426–8, 435, 436, 437, 439, 440, 441, 442, 443, 447, 451, 452, 454, 455–6, 462, 463, 464, 466, 470, 471, 472, 473, 476, 477, 478, 480n, 483n, 488, 502n, 507, 508n, 509, 510, 523, 524, 525, 549, 565, 566, 576, 577, 624n, 626, 627, 628, 631, 633
Boddaert, Peter, 187–8, 626
Bohne, H. M., 518
Bohnitze, or Bohnitz, Jörgen, 432–3, 438, 440, 441, 626, 634
Bombardment of Copenhagen, 30, 419, 466n, 491–2, 495, 629, 639
Bomore, see Bowmore
Bonaparte, Napoleon, 27, 30, 31, 401, 495, 526n, 574n
Booshala Island, 69

INDEX

Borda, Jean-Charles Chevalier de, 16n
Borgarfjarðarsýsla, 162, 163, 312
Borgarfjörður, 244n, 398n
Borrowstounness [Borrowstones], 111
Borthwick, Peter, 495, 576n
Bothwell, James Hepburn, Earl of, see Hepburn
Bosand, 243, 257, 267n, 280, 287, 289, 290n, 295, 296, 302, 303, 304, 373n, 518, 544n, 554, 555, 559, 620n
Boston, United Kingdom, 430, 529, 593, 596
Boston, United States, 181, 414n, 433
Boswell, James, 6, 8n, 69n, 78n, 111n
botany, xvii, 1, 2, 6n, 8, 9, 10, 21, 23n, 26, 29, 42–3, 50, 51, 52, 60, 61, 77, 86, 91, 102, 108, 155, 187–8, 193–6, 201, 211–15, 216, 445, 492, 496, 631, 637
Bougainville, Louis-Antoine Comte de, 6
Boulton & Baker, 280–81, 282, 287–8, 289–90, 290, 290–91, 291, 302–3, 304, 304–6, 306, 308–9, 309, 310, 359–60, 366–7, 372–3, 373, 430, 483, 517, 626–7
Bounty, 28, 187n
Bowmore, 57, 77, 120n
Boyesen, Mr, 288, 289
Bracebridge, Abraham, 301n, 322n
Bredal, Hans Georg, 230n, 238, 243, 245–8, 271–4, 297–8, 627
Brevik, 489
Briem, see Helgi P. Briem
Bright, Richard, 24, 604
Briscoe, Peter, 3, 4, 5, 7, 46n, 48, 110, 119, 121, 122, 123, 124, 126, 139, 140, 156, 593
Bristol, 11, 55, 530,
British and Foreign Bible Society, 24
British Fisheries Society, 67n
British licences to trade (see also Iceland trade), 249, 284, 285, 302, 304–5, 308, 309, 310, 358, 359–60, 420–21, 422, 430,
British Museum, 2, 3, 21, 22, 28, 34n, 35, 72, 325, 490, 596, 633, 639
Brougham, Henry, 79n
Broughton, Charles Rivington, 394
Broussonet, Pierre Marie Auguste, 204
Brown, Robert, 27, 35, 541n, 552
Brown, Rogers & Brown, 432n
Browning, Thomas, 393
Brownrigg, Robert, 445n
Brúará, 98
Buchan, Alexander, 4, 39
Búðir, 584, 628
Buffon, Georges-Louis Leclerc, Comte de, 50
Bugge, Thomas, 203, 207–8, 491–4, 495, 498n, 627
Buller, James, 300n
Bülow, Johan, 216

Burman, Nicolaas Laurens, 187, 188
Busch, Jens Lassen, 387, 416, 485, 555n, 559, 635
Büsching, Anton Friderich, 535
Byron, John, 594
Byron, Lord, 328n

Cabot, John, 532n
Cabot, Sebastian, 532
Cader Idris, 61
Cadiz, 52n, 372n
Cairn na Burgh Beag, 75n
Cairn na Burgh Mòr, 75n
Calcutta, 29
Caledonian Horticultural Society, 582
Cambridge, 2, 175n
Camelford, 65n
Campbell, Daniel, 57n, 58n, 59n
Campbell, Donald, 61
Campbell, John Lorne, 78n
Campbells [clan Campbell], 64n
Campbeltown, 57n
Canada, 16, 401n, 532n
Canna, 78, 172, 173
Cape Cornwall, 54
Cape of Good Hope, 5, 6n, 7, 206n, 403n, 595, 626
Cape Town, 5, 626
Carcass, 26n
Carlscrona, 307, 310, 524n
Carmarthan Van, 61n
Carmarthenshire Beacons, 61n
Caroline Matilda, Queen, 17, 84n
Carter, Harold B., 35, 37, 40–41, 90, 111, 113n, 603
Cartier, Jacques, 532
Castenschiold, Johan Carl Thuerecht von, 410n, 413n, 494, 505–506, 511–12, 537n, 544, 586n, 627
Castle Artaurinish, 65
Castle Duart, 65, 124,
Castlereagh, Lord, Robert Stewart, Viscount Castlereagh, 33, 267, 275n, 279, 281, 283n, 286, 324, 503n, 520, 526n, 583n, 627, 628
Cathcart, Captain Robert, 521, 525, 565
Catherine II of Russia, 187n, 208n
Ceylon, 4n, 8n, 29, 179n, 445
Chalmers, George, 412
Chambers, Neil, 36, 37
Charlotte Amalia, 257n, 294n, 298, 299, 386, 387, 391, 440, 631
Charles I of England, 12, 412n, 534
Charles II of England, 412n, 534
Charles XIII of Sweden, 524n, 574n
Charles XIV John of Sweden, 574n
Charlotte of England, Queen, 6, 7n, 192n

665

Chelsea Physic Garden (Apothecaries garden), 2, 18, 21
Cheshire, 3, 326n, 555n, 637
Chester, 629
Child, Sir Josiah, 534
Childe, Gordon, 20n
China, 28, 594, 626, 629
Christensen, J, 230n
Christensen, master of the *Norburg*, 484
Christensen, Thomas, 518, 555n
Christian I of Denmark, 11, 110n, 531n
Christian IV of Denmark, 533
Christian VII of Denmark, 17, 84, 330n, 638
Christiansand, 341, 489n, 521
Christianshåb, 610n
Choiseul, Étienne François de, 16
Christina (or Christine) Maria, 257, 267n
Christine Henriette, 402, 405
Clancarty, Richard Le Poer, 2nd Earl, 488n, 507, 539, 539–40, 563, 568n, 627–8
Clarence, 301, 311, 318, 319, 320, 321, 325n, 337, 338, 340, 348, 361n, 377, 378, 388, 398, 409n, 500n, 572n
Clausen [Claussen], Holger Peter, 32, 351, 352, 375, 387–8, 396, 397, 426–8, 429–30, 433, 434, 435–6, 440, 459, 463, 476n, 479, 480, 486–8, 498, 499, 504, 507–9, 510, 512, 535, 549, 552, 555, 559, 563, 567–8, 568–70, 574, 575–6, 624n, 628
Cleveley, John Jr, 7, 21, 22n, 46, 48, 62n, 65n, 73n, 76n, 77n, 78n, 83n, 84n, 118, 124, 130, 132, 133, 138, 139n, 156, 593–4, 599, Plates 2, 3, 4, 5, 6, 7
Cleveley, John Sr, 593
Clio, 426n
Cochrane, Honourable John, 205n, 206n, 218n, 222n, 223n, 324n
Coeverden, Louisa Josephina van, 400n
Coke, Edward, Viscount, 6
Coke, Sir John, Secretary of State to Charles I of England, 12
Coke, Lady Mary, 5
Coll, Isle of, 78
Collins, Greenvile, 55
Colonsay, 62n, 170
Columba (Saint), 62n, 69n, 73n, 74, 75, 78n
Columbus, 31, 260
Comhaer, Gozewijn, 531n
Conferentz [Conference] raad Prætorius, 230, 243, 246, 250, 267n, 297, 627
Cook, Captain James, xv, 3n, 4, 5, 6, 10, 20, 26, 27, 45n, 180n, 181n, 209n, 400n, 594, 597
Cooke, Edward, 275n, 277, 279, 281–2, 283, 394–5, 628

Corbett, Borthwick & Co, 374–6, 388n, 417–18, 428n, 432n, 435, 436–9, 440–41, 442–3, 446–8, 451–2, 453–5, 455–6, 462–3, 463–4, 466, 469–71, 472–3, 476–7, 478, 495n, 507n, 509, 513, 516, 517n, 523–5, 544, 565–7, 576–7, 577, 628
Cork, 232n, 237n, 257, 261n, 539n
Cornwall, 119n
Corryvreckan, 63n, 64, 123
County Antrim, 67
County Down, 55n
Cowes, 51, 52, 118
Craignure, 64n
Crawford, John F., 212
Crickett, John (Marchal of the High Court of Admiralty), 262n
Croÿ, Anne Emmanuel Duc de, 50

Dalasýsla, 315
Davis Straits (Davis Strait), 292, 374n, 439, 506n, 610
Dayes, Edmund, 326n
Deal, 5, 50
Delany, Mrs, 23
Denmark, 33, 84, 96n, 128, 133, 134, 155, 164n, 204n, 205, 206, 218n, 222, 223, 224, 227, 232, 233, 234, 235, 240 241, 242, 246, 247, 249, 250, 251, 253, 255, 260n, 261, 264, 266, 267, 270, 272, 274, 275, 276, 279, 280, 285, 287, 289, 294, 295, 298, 302, 304, 306, 309, 310, 311n, 319, 322, 323, 342, 359, 360, 362, 367, 368, 369, 372, 373, 374, 375, 376, 378, 379, 380, 383, 385, 387, 389, 390, 391, 392, 398, 403, 413, 416, 418, 419, 424, 425, 429, 430, 431, 434, 435, 438, 440, 442, 447, 448, 449, 450, 451, 453, 454, 455, 456, 457, 458, 459, 460, 461, 462, 463, 465, 466, 470, 471, 472, 473, 476, 477, 480, 481, 483, 484, 488, 489, 492, 494, 496, 498, 499, 508, 511, 512, 513, 515, 519, 520, 525, 529, 530, 531, 532, 533, 535, 536, 537, 538, 543, 544, 553, 558, 561, 566, 568, 570, 571, 574, 575, 577, 578, 583, 586, 588, 605, 606, 615, 616, 618, 621, 623, 624, 625, 627, 628, 629, 630, 632, 636, 637, 639, 640
Denovan, James Frederick, 232, 233, 238–9, 628
Deptford, 276, 593
Diderot, Denis, 10, 13
Diede, Wilhelm Christopher von, 9, 48, 155–8, 628–9
Diment, Judith A., 42, 43n
Disbrowe, Edward Cromwell, 576n
Dodman Point, 54n
Dolphin, 4n, 6, 594

Donaldi, Anna (abbess), 126
Donaldson, James, 156, 595n
Domett, William, 300n
Dorlton, George, 5
Douglas, John, 181n
Douvez or Douez, Antoine, 48, 133, 156, 594
Dover, 50, 117
Downpatrick, 55n
Drake, Sir Bernard, 533
Draugahver, 102, 600
Drontheim, see Trondheim
Dryander, Jonas, 27, 202n, 203–4, 212, 552n, 629, 633
Dublin, 22, 55, 120, 384, 507n, 627
Dunbar, 112
Dundas, Henry, 1st Viscount Melville, 111, 212n, 218–24, 626, 629
Dundas, Lawrence, 139
Dundee, 302n, 308n, 367
Dundram, 120
Dundrum, 55
Dunkirk, 381, 386
Dunstone, J., 572n
Dýrafjörður, 315, 450, 584, 630

East India Company, 47, 52n, 534n
Eastern fjords, see Austfirðir
Eckersberg, Christoffer Wilhelm, Plate 9
Eddystone, 53
Edgar, 440, 442
Edgcumbe, 53
Edgcumbe, George, 53n
Edgcumbe, Richard, 53n
Edinburgh, 20, 46n, 93n, 111, 112, 116, 140, 156, 157n, 168, 237, 324, 331n, 386, 579, 580n, 582, 594, 641
Edward I of England, 528
Edward III of England, 529
Edward IV of England, 11, 531
Edward VI of England, 532
Egede, Hans, 481n
Egede, Niels, 481n
Egedesminde, 452n, 481, 482n, 610
Egeria, 388n
Eggers, C. U. D., 206n, 222n
Eggert Ólafsson (Olafsen), 19n, 21, 22, 106, 479n, 547, 635
Egypt, 16
Eigg, 78
Einar Jónsson, 542n
Einarsdóttir, see Guðrún Einarsdóttir Johnsen, Málfríður Einarsdóttir
Einarsson, see Hálfdan Einarsson, Ísleifur Einarsson

Eiríksson, see Jón Eiríksson
Ekmansson, Friderik, 22n
Elbe, 377n, 395, 409, 410, 411n, 433n, 460
Elberg, Nils or Jepsen, 447, 513
Eliza, 440n, 442, 443n, 447, 466, 488, 518, 523, 554, 559, 565
Elizabeth I of England, 50n, 280n, 400, 532, 533
Elizabeth, Princess of England, 6, 39
Endeavour, xv, 4–6, 7, 8, 20, 22, 23, 27, 29, 37, 39, 41, 86n, 176n, 180n, 186n, 187n, 190n, 193n, 201n, 226n, 593, 594, 595, 596, 597, 602n, Plate 16
Engey, 504
Esja, 92, 163
Eskefiord, 430, 483, 508, 518, 544n, 642
Eskifjörður, 287n, 298, 312, 366n, 367n, 466n, 489n, 584, 620n, 631, 642
Espólín, see Jón Espólín
Etna, 24, 177n, 182, 260
Eton, 1, 2
Everth & Hilton, 467, 502, 503n, 515n, 570, 629
Everth, John, 467, 629
Eyjarfjarðarsýsla (Vaðlasýsla), 314
Eyjafjörður, 314
Eyrarbakki (Öreback), 313, 350n, 352, 353n, 450, 459, 549n, 559, 561, 563n, 584

Fabricius, Otto, 208
Falconer, Thomas, 8n, 119, 168–74, 174–6, 176–9, 101–2, 105–6, 629
Falmouth, 119
Faroe Islands: Trade and licences, 374–6, 431–3, 438, 440–41, 443–4, 446, 451, 454–6, 461, 462, 463, 470, 472–3
Faujas de Saint Fond, Barthélemy, 18n, 27n
Fawkener, William Augustus, 302, 619
Faxaflói, 83
Feldborg, Andreas Andersen, 486n, 499, 558n
Fell, Michael, 395n, 410, 419n
Fem Brödre, 229n, 243, 257n, 270n, 324n, 633
Fenton, Perrot, 230–31, 629
Filippus Gunnarsson, 338n
Findlay, Bannatyne & Co, 441, 453, 509, 629
Findoe, Jurgen, 438
Fingal (Fuihn), 18n, 65, 69, 125, 170
Fingals Cave, 67n, 70–72, 125, 170n
Finnbogi Björnsson, 345, 379
Finnsson, see Hannes Finnsson
Finnur Jónsson, 18, 101, 108n, 134, 190n
Finnur Magnússon, 216n, 311n, 520
Fiskenæsset, 611
Fjallhver, 108n, 601
Fladda, 75
Flatey, 261n, 315, 584

Flensburg, 248n, 249, 249n, 286, 345, 369n, 420, 422, 513, 638
Flinders, Matthew, 28, 632
Flinn, Lieutenant, 250n, 253
Flint, Andreas, Plate 10
Flood, Jörgen, 490, 616n
Flora, 327n, 395n, 433n
Florence, 10
Florida, 67n, 173n
Foreign Office [British], 33, 34, 364n, 365, 371–2, 376–7, 377–81, 382–3, 393–4, 403, 412, 415, 416n, 420n, 424, 460n, 464–5, 467n, 503n, 504n, 519n, 521n, 544n, 628, 637, 640
Forner, Mr, 217
Forster, Captain, 482
Forster, Johann Reinhold, 47, 594
Forster, Johann Georg Adam, 47, 594–5
Forsvaag, S., 297n
Fortuna, 268, 269
Forward, 432n
Foster, Sir John Augustus, 576n
Fothergill, Captain William, 524
Fowey, 119
France, 16, 63, 117, 201n, 215n, 293, 317, 374, 431n, 491, 479, 492n, 498, 526n, 532, 553n, 561, 596, 625, 640
Franklin, Benjamin, 10, 25
Franklin, John, 28, 468n, 626, 637
Freden, 366, 367, 373n, 416, 430, 435n, 438–9, 447, 452, 453, 463, 469–71, 472, 476–7, 478, 480–81, 481–2, 483, 484, 512, 513, 518, 521n, 522, 544n
Frederik II of Denmark, 189
Frederik IV of Denmark, 184n
Frederik V of Denmark, 16
Frederik VI of Denmark, 33, 206n, 330n, 331n, 356n, 370n, 384n, 466n, 494, 496n, 501n, 511, 574n, 583–6, 587–8, 629
Frederiksberg, 587
Frederikshåb, 611
Freebairn, Charles, 59, 60, 61n, 76, 122, 123n
Freeport, 59n, 61, 76n, 123
Frisch, Hartvig Marcus, 446–8, 451n, 506, 507n, 509–10, 511, 512–13, 521–2, 630
Frisak, Hans, 537
Fruhling, 438
Frydensberg, Rasmus, 263n, 267, 312n, 318, 321, 340, 345n, 347, 350, 369n, 370, 380n, 400n, 410n, 515n, 537
Fuihn, see Fingal
Fury, 431n, 444n

Gahn, Henrik, 193
Gallions Reach, 117

Galliot, 438
Galloway, 120
Gambier, Admiral Lord, 287, 621
Garðahraun, 91
Garðar, Álftanes, 634, 84n, 132, 301n
Garthshore, Maxwell, 212
Geirfuglasker, 80, 87
Geir Jónsson Vídalín, 241n, 244, 263n, 327n, 349, 354, 456–7, 499n, 546, 554, 640
geology, 83, 84, 86 8, 91, 93, 94, 95, 96, 98, 104, 106–7, 108, 152–3, 260
George III of England, 4, 6, 7n, 17, 26, 29, 39, 52n, 84n, 192n, 211n, 629
Gerard, John, 2
Geysir (Great Geysir, Geyser), 18, 21, 23, 41, 42n, 97, 132n, 133–4, 137, 177, 180n, 181–2, 183, 185, 217, 226, 326, 329, 498, 600, 601, 605, 606, 607, Plate 7
Geysers (hot springs), 9, 18, 19, 22, 24n, 34, 97, 98, 101–2, 104, 108, 132, 601n
Giant's Causeway, 67, 69n, 175
Gibbs, Sir Vicary, 394n
Gibson, Mr, 431
Gilbert, Humphrey, 533
Gilpin, Thomas (Gilpin's Raid), 318, 319, 320n, 321, 331, 332n, 336, 350, 364n, 369, 376n, 379, 383n
Gísli Magnússon, 202n
Gísli Símonsen, 338n, 344n
Glasgow, 130
Gloucester, Duke of, 558n, 639
Gode Haab, 485, 518, 549, 550n, 555, 559
Godhavn, 439n, 453n, 469n, 610
Godthåb, 611
Goethe, Johann Wolfgang von, 490n, 633
Gore, John, 7, 48, 118, 119, 124, 156, 594
Gorgie, 46
Gorke, Johann Friedrich, 510n
Gothenburg, 229, 359, 386, 431, 442, 444, 524–5, 540, 554, 560, 565–6, 569
Grafarvogur, 92
Graham, Mr, 60
Gratierne, 459, 475–6, 480, 518, 544n, 549, 552n
Gravesend, 10, 48, 49, 155n
Gráfell, 42n, 106, 600, 601n
Great Britain, 32, 33, 180, 258, 293, 304, 308, 310, 329, 333, 356, 359, 371, 404, 423, 424, 425, 426, 427, 428, 437, 440, 465, 480, 538, 545, 564, 571, 572, 579, 590, 615–16, 618, 624
Greenland,
  administration, 447
  Danish settlements, 610–11
  Displanting and dispeopling, 513–14

trade and licences, 292–3, 375, 435, 438–9,
446–8, 451–2, 453–4, 462–3, 466, 469–73,
476–7, 480–82, 492, 495, 506–7, 509–511,
512–13, 521–2, 525, 629, 634
  fishery, 374–5, 424
Greenwich, 28, 117n, 323, 609
Gregory, James, 582
Greville, Charles Francis, 25, 180n
Grimm, Brothers, 578n
Grímur Pálsson (Poulsen), 351, 370
Grindavík, 11, 313
Grönbeck, P. A., 517
Gröndal, see Benedikt Gröndal
Grundarfjörður, 312, 315, 584, 628
Guðbrandur Þorláksson, 12, 21
Guðlaugsson, see Ari Guðlaugsson
Guðlaugur Þorgeirsson, 84, 90n, 132n
Guðmundsson, see Þorleifur Guðmundsson
Guðmundur Pétursson, 419
Guðmundur Runólfsson, 18n, 83n, 130n
Guðmundur Scheving, 261n, 484n
Guðrún Einarsdóttir Johnsen, 542, 543, 544n, 546,
547, 548, 549, 550, 551, 552, 553, 555–6, 557,
573, 631, 637
Gullbringusýsla, 87, 163, 221n, 301n, 312, 395
Gunnar Högnason, 96n
Gunnar Pálsson, 22
Gunnarsson, see Filippus Gunnarsson
Gunnlaugur Halldórsson (Haldorsen), 239–40, 630
Gustavus III of Sweden, 180, 210n, 597
Gyða Thorlacius, 550n

Haabet, 267n
Haakon IV, King of Norway, see Hákon Hákonarson
IV
Hákon Hákonarson, 344n
Hafnarfjörður (Hafnefiord, Havnefiord), xi, 13, 16,
17, 18, 19, 21, 42n, 83, 90n, 91n, 92n, 108n,
109, 132n, 158, 176n, 198n, 236n, 248n, 261n,
286n, 299, 301n, 310n, 313, 320n, 321n, 337n,
339n, 377n, 395, 398n, 459n, 499, 584, 600,
601, 604, 605, 606, 607, 629, 632, 637, 638
The Hague, 25
Hakluyt, Richard, 12, 532
Hákon Hákonarson IV (Haco), King of Norway, 528
Halesworth, 406n
Hálfdán Einarsson, 22, 90n
Hall, James, 611n
Hall, John, 112n
Halldór Guðmundsson Thorgrímsen, 458
Halldórsson, see Bjarni Halldórsson, Björn
   Halldórsson, Gunnlaugur Halldórson
Hamilton, William, 177, 180n
Hammershaimb, Wenzel, 432–3, 630

Hampshire, 51, 400n
Hannes Finnsson, 22, 101n, 190n
Hanseatic merchants, 12, 530
Hansen, Boye, 518, 523, 554n
Hansen, H. M., 517
Hansen, Phillip Christian, 268, 269
Harboe, Ludvig, 104n
Harald Hairfair, 328, 526
Harriet, 403n
Harrison, George, 238n, 268n
Harrow, 1
Hartman, Mr, 137
Hartz, Bernhard, 514n
Hassell, John, 50, 51n
Hastings, Warren, 406n
Haukadalur, 101, 183n, 600
Havkalven, 419n
Hawkesbury, Lord, see Jenkinson
Hawkesworth, Dr John, 4n, 39
Hawley, Sir David Henry, 40, 174n
Hay, James, 156, 594
Hayaden, see *Najaden*
Haygarth, John, 175, 176
Hebrides, 55n, 57n, 60n, 62n, 64n, 69n, 75n, 78n,
110n, 111n, 115, 117, 170n, 598
Hecla, 28
Heide, Claus, 9, 152–3, 153–5, 259n, 630
Heiðarbær, 94, 600
Heimaey, 313
Hekla (Hecla, Heckla), 9, 17, 18, 19, 20, 23, 24, 26,
40, 42n, 93n, 96, 102n, 104–8, 116, 133,
134–6, 137, 154, 155, 157n, 160, 167, 177, 188,
189, 197, 226, 331n, 408, 501n, 582, 596, 600,
601, 604, 605, 606, 607, 631, Plates 4, 12, 13
Helga Sigurðardóttir, 160n, 197n
Helgason, see Árni Helgason, Bjarni Helgason
Helgi P. Briem, 254n
Helgoland, 567
Helleflynderen, 416n, 518, 544n, 555, 558, 559
Hellisheiði, 88, 109, 601
Helston, 54n
Henderson, Ebenezer, 24, 578n
Hendrichs (Henrich, Henricksen), Boy, 454
Henkel, Henrik, 448–51, 559, 560, 630
Henry III of England, 129n
Henry V of England, 529
Henry VI of England, 530, 531
Henry VII of England, 531, 532n
Henry VIII of England, 12, 50n, 51n, 52n, 412n
Hepburn, James, Earl of Bothwell, 412
Herbert, Henry, 327
Herbert, William, 327
Herschel, William, 29
Hertford, Lady, 6

Hewitt, Sir George, 474, 634
Hibolt, Captain, 399
Hið íslenska bókmenntafélag (the Icelandic Literary Society), 578–9, 580, 581n
Hildarhver, 102, 600
Hnappadalssýsla, 315
Hodges, William, 47
Hodgkinson, William Banks, 1
Hofsós, 584, 287n, 296, 303n, 314
Holbrook, Robert, 48, 156, 594
Hole, Lewis, 388n
Holland, 25, 26, 39, 50n, 51, 178, 179, 180, 187n, 193n, 201n, 381, 386, 492n, 593, 595, 626
Holland, Henry, 24, 259n, 628, 634
Holm, Captain, 350
Holmskiold, Johan Theodor, 201, 216
Holt, Andreas, 9, 152–3, 154, 184, 200–202, 630
Holtermann, Lorentz, 358, 630
Holroyd, Maria-Josepha, Lady Stanley, 555n, 573
Holyhaven, 367n
Hooke, Robert, 12n
Hooker, Joseph, 630
Hooker, Joseph Dalton, 5n, 187n, 405n, 468n
Hooker, Sir William Jackson, xii–xiii, xxi, 23, 24, 25, 32, 202n, 203n, 324n, 326, 329, 331, 334, 362, 396, 401, 405–6, 407, 408, 409, 409–10, 411, 413–15, 445–6, 468–9, 479, 485–6, 498, 499, 501, 503, 522, 541n, 545, 547, 551–2, 588n, 630–31, 636, 638
Hólar (Holum), 12, 21, 22, 84n, 90n, 107, 137n, 199n, 202, 220n, 241, 242, 244, 331, 530, 605, 606, 632
Hólm, Sæmundur Magnússon, 203n
Hólmshraun, 94
Holst, Boy Petersen, 286
Holstein, 480, 583
Holsteinsborg, 439n, 452n, 507n, 610
Holt, Andreas, 9, 154, 184, 630
Hölter, Diðrik, 237n, 270n
Hölter, Jóhanna Margrét, 270n
Hölter, Katrín Margrét, 237n
Hölter, Marta María, 237n
Holtermann, Lorenz, 630
Home, Sir Everard, 212
Home, George, 229, 247n, 251n
Home, James, 210–11, 630
Hompesch, Baron Charles von, 318, 319, 321, 363, 379, 444
Hope, Rear-Admiral Sir George Johnstone, 558, 560, 570
Hope, John, 2nd Earl of Hopetoun, 111
Hope, William Johnstone, 300n
Hopetoun House, 111

Horne, William, 570–72, 631
Horne & Stackhouse, 519, 544n, 556n, 572, 575n, 631
Horneman, Hans Frederik, 387n, 396–7, 418, 433, 433–4, 448–51, 487, 544–5, 547–8, 549–50, 554–5, 557–60, 560–61, 563–4, 573–4, 588n, 631
Hornemann, Jens Wilken, 496
Horrebow, Niels, 13, 15, 22, 47n, 80, 84n, 498n, Plate 13
hot springs, see geysers, Geysir
Howard, Mr and Mrs, 546
Höwisch, Johan Gottfred, 280n, 287, 296, 303, 304n, 555n
Howth Head, 55
Hoy, 109
Høyer, C. W., 518
Hrappsey, 159n, 578n, 604
Hraungerði, 108, 601
Hugessen, Dorothea (Lady Banks), 27, 40n, 210n, 216, 545, 555, 559, 584
Hull, 302n, 374n, 418, 435n, 473, 528, 529, 531, 625
Hume, David, 595
Hunter, Captain James, 10, 48, 117, 118, 128, 132, 156, 608, 609
Húnavatnssýsla, 314
Húsavík, 287n, 296, 314, 366n, 584, 620n, 642
Husevig, 280, 287, 289, 290n, 302, 308, 309, 366, 367, 368, 373, 376, 433n, 434, 459n, 508, 620n, 642
Huskisson, William, 238n
Hutchinson, Captain, 364
Hvaleyri, Plate 5, 83n, 86, 195n, 199, 632
Hvalfisken, 447, 462n, 509, 510, 512, 513, 521, 522, 525, 565, 566, 569
Hvalfjörður, 162, 502n
Hvalseyjar, 315
Hveragerði, 108n
Hvide Biorne, 438
Hvidfisken (*Hvidfischen*), 436n, 438, 509n, 510n, 521n
Hvítá, 104, 108n, 312, 600

Iceland,
  Anglo-Icelandic relations until 1772, 11–13, 526–36
  Anglo-Icelandic relations after 1772, 228, 502–4, 514–15
  animal husbandry, 14, 97, 137, 221, 260
  annexation (proposals for annexation of Iceland by Great Britain), 29–34, 222–4, 254–6, 256–7, 257–65, 292–3, 317, 318–20, 321–2, 324–5, 425, 474–5, 479, 520, 522–3, 536–8, 571

INDEX

British trade with Iceland during the Napoleonic
  Wars (see also Icelandic Revolution), 318–19,
  322, 323, 326–7, 457–60, 467, 570, 570–2,
  572–3, 575, 615–17, 624
climate, 220, 258–9, 264, 279
costume, 82, 87, 88, 89, 132, 137, Plate 6, Figures
  2, 3
crops, 85, 90, 114, 132, 220–21, 241, 242, 244,
  258
fishery, 205, 206, 218–19, 260, 279, 374, 572
fortification and arms, 221, 241, 242, 257, 323,
  538
geology, 91, 93–5, 104, 106, 114, 130, 203–4
  (Laki eruption)
geography, 220, 258, 312–315, Plate 13
government, 13–14, 15, 241, 242, 244, 257,
  262–3, 267, 321
Iceland in 1772–1815, 13–16
Icelanders' attitudes towards Banks, 226, 229, 242,
  255, 322, 329, 330, 333, 409–10, 424–3, 535,
  575
language, 578–9, 580, 581
literature and manuscripts, 21, 102, 137, 183–4,
  189, 191, 198–200, 227, 325, 327–9 578–9,
  580
minerals, 162–4, 191–2
music, 137
poetry, odes, annals, 159–60, 167, 408, 590–92,
  604–7
population, 14, 221, 227, 269
religion, 84, 90, 101, 128, 261
ships (see under individual names)–society, 14,
  264–5
trade (including imports and exports, British
  licences), 14–15, 154, 222, 227, 240, 241, 243,
  244, 249, 260–62, 273, 292–3, 294–7, 302–3,
  304–5, 306, 315–17, 318–20, 360, 366–7,
  367–8, 396–7, 416–20, 423, 42–5, 426–8, 429,
  433–4, 436–7, 440, 442–3, 448–451, 458–9,
  463–4, 464–5, 475–6, 480, 483–9, 504, 507–9,
  517–18, 523–4, 544–5, 549–50. 552. 554–5,
  558–60, 560–61, 565–7, 568–70, 576–7, 577,
  583–6, 612–14
Icelandic Revolution of 1809, xi, xiii, xv, 32, 229n,
  301n, 329n, 332–3, 335–356, 357–8, 358n,
  360n, 361, 364–6, 376–7, 386–371, 371–2,
  377–81, 388–90, 391–4, 395n, 397–404, 404,
  404–5, 408, 409n, 500, 502, 541, 542, 551, 562,
  564, 627, 631, 632, 633, 638, 640
Ilulissat, 610n
India, 8n, 30, 212, 223n, 595, 625, 640
Ingólfsfjall, 108, 601
Investigator, 28
Iona (Y Columb Kill), 69, 73, 74n, 125–6

Ireland, 54n, 55n, 57, 69n, 72, 120, 122, 126, 154n,
  170, 171, 172, 177, 220, 384, 458, 465, 526n,
  598, 627, 640
Ireland, Bennett, 467n
Ireland's Eye, 55
Isabella, 636
Ísafjarðarsýsla, 314
Ísafjörður, 312, 315, 450, 485n, 584, 635
Ísfjörð or Isfiord, see Kjartan Ísfjörð, Þorlákur Ísfjörð
Islay (Ilay, Ila), 57–8, 59, 60, 62n, 63, 76, 77, 120,
  122, 123, 169
Isle of Man, 55, 120
Isle of Wight, 51, 118
Ísleifur Einarsson, 241n, 267, 318, 321n, 347, 369n,
  377n, 410n

Jackson, Captain George, 301n, 337, 338, 377, 378
Jackson, James Grey, 468
Jacobsen, Jacob, 268, 269
Jacobshavn, 610
Jacobæus, Christian Adolf, 396, 459, 465, 504
Jacobæus, P, see Páll Jakobsson
Jacquin, Nicolaus Joseph von, 212
Jakob Benediktsson, 39
Jakobsson, see Páll Jakobsson
Jamaica, 29, 211, 339n, 524n, 593, 641
James I of England, (VI of Scotland), 189n, 412
James III of Scotland, 110n
James, Earl of Arran, 412n
James Island, 589
Jardine, James, 426n
Java, 25
Jenkinson, Charles, 1st Baron Hawkesbury and 1st Earl
  of Liverpool, 25
Jenkinson, Robert Banks, 2nd Baron Hawkesbury and
  2nd Earl of Liverpool, 25, 30, 33, 233n, 236,
  254n, 257–65, 275–6, 278n, 279, 322n,
  368–71, 631–2
Jensen, Hans, 446–8, 632
Jepsen, P., 518
Jergensen or Jerghensen, see Jörgensen
Jermyn, Henry, 541n, 545, 548
Jessen, Captain, 376, 438
Johanne Charlotte, 229, 243, 257, 261, 267n, 295,
  307, 310, 620, 637
Johnson, Samuel, 6, 8, 69n, 78n, 111n, 468
Johnston, Sir Alexander, 469
Jones, Captain Alexander Montgomery, 332, 352,
  353, 354, 355, 357, 361n, 366n, 370, 380, 390,
  393, 401, 402, 406n, 408n, 414, 500, 551,
  616–17, 632
Jones, Charles Wilkinson, 4th Viscount Ranelagh,
  410n
Jonge Goose, 376, 416n, 433n, 434, 485, 518

Jón Eiríksson, 479n
Jón Espólín, 332n, 347n, 515n
Jón Jónsson (Arakot), 344n
Jón Matthíasson, 137
Jón Steingrímsson, 17
Jón Vídalín, 166
Jón Þorláksson, 578n
Jónsson, see Arngrímur Jónsson, Bjarni Jónsson, Björn Jónsson (apothecary), Björn Jónsson (clergyman), Einar Jónsson, Finnur Jónsson, Jón Jónsson, Sigfús Jónsson, Steingrímur Jónsson, Teitur Jónsson, Þorsteinn Jónsson
Jörgensen, Jörgen (Jorgen Jorgensen, Jorgenson or Jürgensen), 32, 224–6, 255n, 291–2, 292–3, 301–2, 311, 311–17, 318n, 319n, 320n, 325n, 329n, 332, 333, 335–58, 338, 340, 341, 342, 360–64, 371, 372n, 376n, 377–81, 388–90, 391–2, 393n, 397–405, 406, 407, 408, 409, 414–15, 420, 445n, 468, 500, 502, 503, 541–4, 545–7, 548–9, 550–57, 562, 564, 616–17, 631, 632–3, 634, 641, Plates 9, 13
Jórunn Ásmundsdóttir, 195
Jubelfest (*Jubelfast*), 376, 432n, 433n, 435n, 438, 440, 441, 452, 455, 461n, 462, 463, 521n, 626
Juin, Nicolas, 53
Julianehåb, 611
Jupiter, 436n, 438, 507, 510
Justitia, 301n, 318n, 320n, 337n, 344n, 378n
Jura, 60, 63, 64n, 78, 123, 169, 170
Jura Point, 123
Jutland, 249n, 307, 310, 319, 386n, 428n, 450, 574n
Juul, Carsten Christensen, 298, 386, 387
Jurgensen or Jürgensen, see Jörgensen
Jürgensen, Urban (brother), 226
Jürgensen, Urban (father), 541n

Kaempfer, Engelbert, 215n
Kallström, Andreas, 194
Kapelluhraun, 91, 92n
Kárastaðahraun, 42n, 600
Karl, see Charles
Kattegat, 428, 558
Kayll, Robert, 533n
Keblevig, 459n, 504n, 517
Keflavík, 232, 243, 313, 396n, 434n, 459n, 504, 559, 584, 635, 637
Kelp, 66, 77, 78n, 138n, 170–71, 174–5, 176, 244, 262
Kent, 27, 50n
Kerguelen-Trémarec, Yves Joseph de, 16
Ketelsen, J., 555n
Kew Gardens (Royal Botanic Gardens at Kew), 6, 7, 18, 21, 24, 26, 28, 29, 36, 211n, 405n, 468n, 631, 641

Kiel, 523
Kieldsen, Christian, 376, 428, 485, 518, 555n
Kilarrow, 58, 59, 121
King-Hele, Desmond, xv–xvi, 36
King's Lynn, 11, 529
Kirkwall, 110, 139
Kjartan Ísfjörð (Isfiord), 298–300, 381, 385–7, 391, 440, 442–3, 466n, 488–9, 554n, 630, 631
Kjósarsýsla, 163, 301n, 312
Klog, Thomas, 324n, 349n, 571, 581n, 582
Klow, Anna Helene, 158n
Klow, Miss, 130n
Knatchbull, Sir Edward, 40n
Knatchbull-Hugessen, Edward, Lord Brabourne, 35, 40n
Knight, Thomas Andrew, 582
Knudsen, Adser Christian, 229, 270–71, 298–9, 341, 435, 369, 371, 379, 398, 633
Knudsen, Lars, 309, 338n
Koefoed, Hans Wölner, 318, 321, 337n, 339n, 340, 341, 350, 370, 380n, 400n,
Kollafjörður, 92
Kolvig, Frederik, 299, 381, 386
Kommercekollegiet (Danish Board of Trade), 85n, 184n, 205n, 387n, 563n, 631
Kongelige Danske Videnskabernes Selskab (Danish Royal Society), 21, 207n, 492n, 587, 627
Konig, Charles Dietrich Eberhard, 327–9, 633
König, Johan Gerhard, 8, 15, 187n, 201n, 211
Kópavogur, 91
Krísuvík, 312
Krogh, Georg Frederik von, 551n
Kronprinsens Ejland, 610n
Krossavík, 419

Labrador, 2, 3, 4, 26, 39, 193
Lack, Thomas, 323, 467n, 471
Lady Nelson, 224, 225
Lágafell, 267n
Laggan, 59, 77
Laki, 17, 203–4, 205n, 220n, 260n, 331n, 638
Lambertsen, Niels, 350–51, 352–3, 459, 465, 485, 559, 561, 563
Lambeth, 301n, 636
Landeyjar, 313
Land's End, 54, 597
Langafell, 601
Langavatn, 315
Lapland, 3
La Revanche, 398n
La Syrène, 398n
Latrobe, Christian Ignatius, 506–7, 509–10, 510–11, 512–13, 521–2, 611, 633

Laugarás, 102
Laugaráshver, 102, 600
Laugardalur, 96, 601
Laugarfell, 600
Laugarnes, 42n, 101, 132n, 600
Laugarvatn, 18, 96n, 97
Laugarvatnshellar, 96, 600n
Lauraguais, Louis Léon Félicité, Comte de, 6n, 50, 117
Lavinia, 268
Leach, Mr, 67, 124, 126, 171n
League of Armed Neutrality, 30, 218n, 223n
Le Poer Trench, William, 539n
Leiden, 22, 175
Leirubakki, 108, 600
Leith, 20, 109n, 111, 140, 227, 228, 229, 230, 231, 232n, 233236, 237n, 238, 241, 242, 243, 245n, 247n, 250, 251n, 253, 256n, 257, 271, 275, 278n, 283n, 285, 297n, 302n, 308n, 309, 323, 326n, 333n, 342, 359n, 362n, 371, 375, 376n, 384, 388, 396, 397, 398, 416, 418, 419, 420, 421, 422, 423, 424n, 426, 427, 428, 430, 431, 432, 433, 435, 437, 438, 441, 442, 443n, 447, 448, 451, 452, 453, 454n, 455, 456, 458, 461n, 462n, 463, 464, 466, 469, 470, 471, 473, 476, 478, 480, 481, 482, 483, 484, 488n, 492, 495n, 496, 507, 508, 509, 510, 511, 512, 513, 517, 523, 524, 538, 539, 544, 547, 549, 550, 552, 554, 555, 556, 558n, 559, 560, 561n, 563, 565, 566, 573, 574, 577, 578n, 618, 624, 627, 628, 635, 641
Leland, John, 528
Le Maire, Jacob, 166
Lerwick, 319n, 444n
Lawson, Mr., 111
Lichtenau, 514n, 611n
Lichtenfels, 514n, 611n
Lightfoot, Dr John, 2, 26, 61n
Limehouse, 117
Lincolnshire, 1, 108, 593, 595
Lind, James, 7, 20, 23n, 46, 48, 61, 78, 86, 90, 93n, 106, 107, 116, 117, 118, 119, 121, 122, 124, 128, 133, 140, 156, 178, 181, 182–3, 185, 198, 498, 594–5, 596
Lindberg, Jörgen Hansen, 447n
Lindegren, Carl, 185
Linköping, 597
Linnean Society, 6n, 28, 212n, 552n, 629, 631, 634
Linnaeus (Carl von Linné), 3, 5n, 8, 21, 28, 29, 79, 175n, 180n, 185, 187, 193n, 201, 209, 213n, 217, 596
Liston, Captain John, 335n, 340, 341, 342, 353, 354, 355, 377n, 378, 395n, 402, 409, 411, 419
Literary Club, 6n

Litli-Dímon, 96
Liverpool, 215n, 307n, 319, 320, 323, 361n, 375n, 376n, 396n, 410, 418n, 432n, 440n, 463n, 553n, 572n, 631, 639
Liverpool, Lord, see Jenkinson 573, 576,
Livingstone, David, 67n
Lizard Point, 54
Löbner, Emilius Marius Georgius, 431n
Lobnitz, 432–3
Loch Don, 64, 124
Loch Indaal, 57, 63, 77n, 120
Loch na Gaul, 71
Loch Sunart, 76
London, xii, 2, 4n, 8, 9, 10n, 20, 22, 23, 26, 27, 29, 32, 47n, 48n, 52, 76, 86n, 93n, 109n, 112, 117, 152n, 154n, 168, 169, 179, 184, 184n, 185n, 188, 189n, 191, 192, 193n, 194n, 198, 200, 212n, 216, 217n, 224, 226, 230, 232, 234, 238, 240, 243, 253, 254n, 256, 257, 265n, 268, 271, 273, 274n, 280n, 285, 287, 291n, 298, 299, 305n, 306n, 307, 308, 311n, 332, 336, 340, 342, 361n, 362, 363, 366, 370n, 374n, 375, 381, 384, 386n, 388, 390n, 391, 396, 398, 399, 403, 417, 419n, 420, 426, 427, 431, 434n, 435n, 439, 440, 443, 444, 445, 452, 458, 459, 460n, 462, 463, 464, 470, 472, 476, 494n, 502, 507n, 509, 510n, 512, 515n, 520, 528, 546, 553, 593, 595, 596, 606, 609, 616, 618, 621, 625, 626, 628, 629, 630, 631, 632, 636, 637, 639, 641,
Loptsson, see Ólafur Loptsson
Loptur Amundason, 582n
Loten, John Gideon, 179
Louis XV of France, 16
Louis XVI of France, 10
Louisiana, 16
Low, George, 110n
Ludvigsen, Just, 299n, 554n
Lufkyn, John, 539n
Lund (Sweden), 156
Lund, Master, 485
Lunga, 124
Lyngbye, Hans Christian, 590
Lynge, Frederik, 560n
Lyons, Israel, 2
Lysaght, Averil Margaret, 2, 8, 39, 41

Macartney, George, 28, 31, 46n, 626, 629
MacCoul, Fhinn (see Fingal), 69
McDonald, King, 59
McDougall, Mr., 123
Macdougall, Alexander, 62n
Macduffies, 62
McFees, 62
Maclean (Maclane), Sir Allan, 67, 124, 171

McLeay, Alexander, 251–2, 253, 275n, 276–7, 284n, 364, 415n, 634
Mackay, David, 29
Macleod, Donald, 520n, 582n
MacLeod, Mary, 520n
MacNeal, 124
McNeil, Alexander, 62n, 123
Mackenzie, Sir George Steuart, 24, 137n, 163n, 323–5,325, 334, 365n, 381–2, 386, 395–6, 409n, 410, 411, 412, 413, 414, 415, 418420n, 424, 425, 426n, 429n, 459n, 468, 479, 486n, 520–21, 522–3, 526, 536n, 541, 542, 548, 551, 578n, 579, 580–81, 581–2, 588n, 590n, 627, 628, 634, 636, 638, 639
MacPherson, James, 65n, 74
Madeira, 5, 86n, 640
Maelström, 63
Magnus, Olaus, 529n
Magnús Pétursson, 604
Magnús Stephensen, xxv, 226–9, 233, 234, 237, 241n, 247n, 248, 254–6, 258n, 325n, 329n, 330–33, 346n, 355, 358, 379n, 383n, 395, 406n, 409, 414, 499–505, 514–16, 520n, 538, 552, 574n, 617, 638, Plate 10
Magnús Sæmundsson, 95n
Magnúsdóttir, see Sigríður Magnúsdóttir, Þórunn Magnúsdóttir
Magnússon, see Finnur Magnússon, Gísli Magnússon, Markús Magnússon
Majendie, Henry William, 474n, 634
Majendie, William Henry, 474–5, 475, 634–5
Malmquist, Peter, 349n
Málfríður Einarsdóttir, 542n
Maniitsoq, 610n
Mannevillette, Jean-Baptiste d'Après de, 55
Marchant, John, 48, 156, 595
Margaret of Norway (The Maid of Norway), 528
Margaret and Ann, 357, 368, 369, 370, 371, 377, 378, 398, 405, Plate 15
Maria Bonaventura, 419
Markús Magnússon, 301–2, 634
Marta María Stephensen, 237n
Martfelt, Christian, 85
Martin, Martin, 73n, 174
Martini, Friedrich, 446–8, 635
Mary, Queen of Scots, 412
Masson, Francis, 7, 28, 206n
Matthíasson, see Jón Matthíasson
Mathiesen, Lauritz, 438, 452n, 453, 480n, 481n
Mattiolus, Pietro Andreas, 175
Mauritz, Gotthard, Baron de Rehausen, 561n
Mellish, William, 435n, 473, 507n
Melville, 1st Viscount, see Dundas.
Mendelssohn, Felix, 67n

Merioneth, 61n
Merkurhraun, 108
Merrett, Christopher, 175
Mevagissey, 54n
Michelsen, N. 376, 518
Miller, David Philip, 29
Miller, James, 7, 18, 21, 22n, 46, 48, 58n, 64n, 118, 124, 130, 132, 156, 595
Miller, John, 580n
Miller, John Frederick, 7, 18, 21, 22n, 42n, 46, 48, 58n, 59n, 65n, 118, 121, 122, 123, 124, 130, 132, 138, 156, 202, 595, 599
Miller, Philip, 2
Minor, Hans Erik, 537
Mitchell, Andrew, 318, 320n, 321, 358n
Mjólkurhver, 102, 600
Möller, Jörgen Hansen, 485, 518, 555n
Moltke, Count, Plate 14
Montagu, John, The Earl of Sandwich, 4, 6, 7, 40n, 45, 46, 47n, 50
Montgomery, Richard, 401
Montgomery, Sarah, 401n
Moore, Sir Graham, 570
Moreland, James, 48, 156n, 595
Mornington, see Wellesley-Pole
Mortimer, John Hamilton, 4n
Morvern (Morven), 65, 67, 124, 170
Motzfeldt, Peter Hanning, 439, 447, 470, 481, 482
Mulgrave, see Phipps, Constantine and Henry
Munk, Mr., 454
Mull (Isle of, Sound of), 64, 65, 67, 69, 75, 124, 125, 126, 170, 171, 172
Mull of Cantyre, 55
Mull of Chanach, 122
Murray, Johan Anders, 180, 185, 212
Murray, Patrick, Lord Elibank, 112
Múlasýslur, 313
Múli, 42n, 98, 600
Müller, Johann Sebastian, 595
Müller, Otto Frederik, 208
Münster, Sebastian, 483n
Myhlenphort, Marcus Nissen, 454, 481
Mýrasýsla, 315

Næfurholt, 104, 600
Nagle, Sir Edmund, 326, 401
Naiad, 401n,
Najaden (*Hayaden*), 438, 441, 452, 454, 455, 461n, 462n
Napoleonic Wars, xv, 1, 24, 29, 31, 33, 223n, 224n, 319n, 436n, 626, 628, 631, 633, 635, 637, 638
Natural History Museum (London), xvi, 22, 35, 36, 41, 42, 179n, 596
Navy Board, 7, 26, 45

INDEX

Nelson, Horatio, 558n, 26n, 223n
Nepean, Sir Evan, 206
Nes (Seltjarnanes), 165
Newcastle, 112, 181n, 358,430
Newfoundland, 2, 3, 4, 12, 26, 39, 205, 206, 218, 219, 235, 374, 424, 532–4, 536, 593, 595, 598
Newport, 51
Newton, Isaac, 27
New Zealand, 5, 39, 176, 187, 597
Neyborough, Mr, 137
Nicholl, John, 231n, 270n, 297n
Nielsen, Hans, 280, 287n, 289n, 296, 620, 621
Niger, 3, 53n
Niger River, 28, 186, 468n, 626
Nikulasdóttir, see Þuríður Nikulásdóttir
Nissen, Niels, 438
Nolsøe, Jacob, 194, 431–2, 443–4, 446, 461
Nolsøe, Poul Poulsen, 431, 444
Norberg, 397, 458, 508n,
Nordborg (Norberg), 319
Nordstern, 518, 549, 550n, 552n, 555, 559
North Star, 431n, 444n
Norður-Múlasýsla, 313n, 419n
Norður-Þingeyjarsýsla, 314n
North America, 17n, 65n, 93n, 206n, 225n, 414n, 536, 581n, 582, 589, 611
Norfolk, 11
North Pole, 26, 179, 180, 182, 605, 606
Norway, 11, 36, 55n, 63, 64n, 74, 82, 96n, 133, 155, 170, 204n, 206, 207, 227, 242, 243, 244, 246, 260n, 269, 280, 285, 287, 293, 297n, 303n, 304, 305–6, 308, 309, 322, 323, 337, 342, 358n, 359–60, 360, 388, 397, 397–9, 403n, 414n, 416n, 418, 419–20, 422, 430, 432, 434n, 448, 450–51, 459, 483, 484, 489–91, 496, 504n, 508, 510, 516, 517, 518n, 521n, 524n, 526, 528, 529, 530, 531n, 535n, 540, 541n, 549, 551, 553, 554, 558, 559, 563, 566, 567, 568, 570, 574, 578, 625, 627, 628, 629, 635, 637, 640, 641
Nott, Captain Francis John, 339–42, 355, 378, 383n, 398–9, 500n, 615–16, 617
Nuuk, 611n
Ny Herrnhut, 514n, 611n,
Nye Pröve, 243n, 257, 280, 287, 289, 290n, 295, 296, 302, 308n, 620–23, 642

Ólafsvík, 315, 352n, 459, 550n, 584, 628
Ólafsson, see Eggert Ólafsson, Ólafur Ólafsson.
Ólafur Loptsson, 324–5, 581–2, 634
Ólafur Ólafsson, 590–92, 635
Ólafur Stephensen, xii, 19, 24, 29, 30, 88n, 94n, 109n, 130n, 132n, 162–4, 183–4, 191–2, 198–200, 222n, 227, 234n, 254–5, 312n, 318n, 325–6, 329–30, 395n, 414, 460, 499n, 515n, 552, 575, 636, 638
Ólafur Magnússon, 244
Ólafur Þórðarson Thorlacius, 261n, 465, 484, 485, 536, 555n
Ölfusá, 108, 601
Óli Sandholt, 346n
Olsen, Gottsche Hans, 300
Olsen, Ole Christensen, 555
Olsen, Thron, p. 475, 480, 518
Önundarfjörður, 450
Oransay, 62, 75, 76, 123, 126, 169, 170
Order in Council of 7 February 1810, 33, 284n, 333n, 374n, 375, 376n, 383–5, 392n, 395–6, 407n, 423, 424, 426–27, 431, 435, 437, 439, 447, 462, 464, 470–71, 472–3, 476–7, 478, 500, 504, 519, 565, 618–19
Öreback, see Eyrarbakki
Orion (Sívertsen's ship), 308, 440n, 442, 443n, 463, 464, 467, 488, 517, 544n, 550n, 553, 554, 555, 556, 559, 563, 565, 566, 573n, 624
Orion (Trampe's ship), xiii, 333n, 341, 354, 361n, 365n, 371n, 378, 380, 398, 399, 402, 404, 406, 415, 633, 639, Plate 15
Orkneys, 20, 22n, 40, 93n, 109, 110n, 115, 116, 117, 138, 139, 140, 168, Plate 8
Ørum, Niels, 287, 296, 641
Ørum and Wulff, 280, 287, 288, 306n, 419n, 452, 459, 483, 620n
Orwell, George, 64n
Ossian, 63, 66, 69
Ostend, 127
Øster Riisøer, 518
Otway, Admiral William Albany, 4/3
Öxará, 95n
Oxford, 1, 2, 6, 215n, 176, 553n, 593, 598, 639

Paamiut, 611
Páll Bjarnason, son of Bjarni Jónsson, 160n
Páll Björnsson, Reverend, 13
Páll Jakobsson, 102n, 160n
Pallas, Peter Simon, 187
Palmerston, Viscount (Lord Palmerston), 47n, 300n
Pálsson, Bjarni, 19, 20, 21, 42, 164, 479, 635
Pálsson, Grímur, see Grímur Pálsson
Pálsson, Gunnar (Reverend), see Gunnar Pálsson
Papin, Denis, 97
Paris, 10, 156, 174, 209, 627
Park, Mungo, 28, 186n, 468
Parke, James (brother of John Parke), 557n, 635
Parke, John (British Consul in Iceland), 33, 393n, 396n, 410, 412, 416n, 450, 457–8, 458–60, 460–61, 463, 464–5, 467–8, 480, 488n, 503n,

508, 514n, 515, 543n, 544, 546, 547, 556–7, 563, 565n, 568–70, 635
Parke, Thomas (father of John Parke), 635
Parkhurst, Anthony, 532
Parkinson, Sydney, 4, 39, 179
Parry, William (artist), 26
Parry, William Edward (naval officer), 28, 626
Pasch, Lorens, 209
Paterson, Captain, 137
Patreksfjörður, 16, 315n, 398n, 450n, 484, 584n
Paus, Henrik Christian, 387–8, 635
Paykull, Gustaf von, 217
Peacock, 227n, 228n, 239n, 250, 297n, 627
Peel, Abraham, 486
Pelican, 298, 386n
Pembroke, 54n
Pennant, Thomas, 2, 3n, 23, 40, 55n, 57n, 59n, 60n, 62n, 64n, 67n, 68n, 69n, 70n, 71n, 72n, 73, 74n, 75n, 76n, 78n, 79, 126n, 170, 171, 173, 174, 178, 182n, 183, 202, 595, 598–9, 635
Percy, Thomas (Bishop), 328
Petersen (Master of the *Husevig*), 280, 287n
Petersen, H. J. (Master of the *Seyen*), 351, 352, 376, 518
Petersen, Peter Lorentz, Captain, 452n
Petræus, Westy, 245–8, 250–51, 252, 271–4, 635–6
Pétur Þorsteinsson, 605
Pétursson, Guðmundur, see Guðmundur Pétursson, Magnús Pétursson
Phelps, Samuel, 301n, 311n, 320n, 321n, 322, 323, 324n, 326, 327n, 329, 337n, 361n, 375n, 396n, 424, 433, 460, 542, 572n, 636, see also Phelps, Troward & Bracebridge
and Icelandic Revolution: 32, 326n, 332–3, 335n, 340–46, 349–57, 364–6, 369–71, 376, 377–81, 382n, 388–90, 393, 395, 397–406, 408n, 409–11, 412–15, 419–22, 500, 502–3, 616–17, 632
Phelps, Troward & Bracebridge, 301n, 311n, 320n, 322n, 636
Phillips, Thomas, xi
Philp, Philps, see Phelps
Phipps, Constantine (Captain), 2, 8, 26, 28, 53, 179, 180n, 182, 183, 186, 187n, 545n, 594
Phipps, Henry, 1st Earl of Mulgrave, 545n
Phoebe (frigate), 398
Pierie, Lieutenant, 212n
Pingré, Alexandre Guy, 16
Pitcairn, David, 211, 212
Plumer, Thomas (Sir), 394n
Plymouth, 47, 52, 118
Pocock, Nicholas, 326n
Polsen (Merchant), 484
Pomona, 138–9

Pontoppidan, Carl (merchant), 158n,
Pontoppidan, Christian Jochum, 207n
Portland, 118
Portland, 3rd Duke of, William Henry Cavendish Cavendish-Bentinck, 320n, 626
Portland, Dowager Duchess of, see Bentinck, Margaret
Preidel, John Christopher, 249, 268, 268–9, 269–70, 274n, 284n, 285–6, 420–21, 422, 636
Pringle Stoddard (Commander), 268
Privy Council (British), 28, 29, 30, 32, 33, 230, 298n, 302, 304n, 308n, 309n, 310n, 323n, 359n, 373n, 383n, 384, 416, 421, 422, 423, 428n, 430, 431n, 444n, 446n, 464n, 465, 476, 479, 480–81, 481–2, 484, 506n, 511, 516, 519n, 521, 618
Privy Council (Danish), 216n, 412, 636
Prize ships, 62n, 154, 285, 293, 333n, 352, 354, 378, 380, 394, 406, 419n, 426, 440, 444n, 471, 477, 508n, 525, 533, 618, 620–23, 641
Problus, 268, 269
Providentia, 319, 358
Providence, 575
Providence, Rhode Island, 414
Ptarmigan, 18, 60, 61, 114, 133
Pulten(e)y, Dr Richard, 3n, 212

Qaqortoq, 611
Qasigiannguit, 610n
Qeqertarsuaq, 610n
Qeqertarsuatsiaat, 611n

Racehorse, 26n
Rafn Þorgrímsson Svarfdalín, Plate 11
Ragnheiður Stephensen, 130n
Raleigh, Walter, 533n, 599
Ramsay, Robert, 110
Ramsden, Jesse, 86n
Rangárvallarsýsla, 96n, 104n
Rask, Rasmus Christian, 578–9, 579, 580, 581, 636
Ratisbon, 7n
Rauðöldur, 106n
Rauschenberg, Roy Anthony, 39, 41, 45n, 47n, 50n, 53n, 54n, 60n, 88n, 93n, 95n, 96n, 98n, 99n, 100n, 101n, 102n, 106n, 107n, 109n, 155n, 156n
Ray, John, 175, 602
Reading, 274, 403, 564
Reckawick, see Reykjavík
Regina, 257, 267n, 299, 303n, 309, 440n, 442, 443n, 488, 518, 523–4, 539, 554, 558, 559, 560, 565–6, 569, 627
Rehausen, Baron de see Mauritz, Gotthard
Reikavik, see Reykjavík

INDEX

Rennthier, 249, 268, 269, 270n, 272n, 274n, 285, 410n, 417, 420, 422–3, 428, 429n, 433n
Reventlow, Christian Ditlev Frederik, Count, 494, 495, 505, 511, 558n, 636
Revesby Abbey, 1, 2, 3, 4, 29, 155, 204n, 361, 365, 425, 468, 508n, 593, 595
Reyðarfjörður, 314, 642
Reykir, 42n, 108n, 601
Reykjafjörður, 584n
Reykjarfjörður, 315n
Reykjanes, 87n, 314n
Reykjavellir, 104n
Reykjavík (Reikavik), 129n, 132, 162n, 165n, 199n, 220n, 241n, 243n, 244n, 248n, 249n, 254n, 261n, 267n, 286n, 312n, 318n, 319n, 329n, 331n, 339n, 341n, 358n, 380n, 384n, 395n, 410n, 457n, 459n, 485n, 499, 504n, 542n, 548n, 550n, 570n, 575n, 578n, 579n, 581n, 629, 633, 635, 636, 637, 638, 640, 642, 646, Plate 14
Reynolds, Joshua, frontispiece, 6n, 26n
Reynolds, Thomas, 583n
Resolution (Commanded by Captain James Cook) 6–7, 10, 26, 45–7, 116, 153n, 594–5, 596
Resolution (belonging to Niels Lambertsen), 559, 561, 563
Richard, Arthur, 352n
Richmond, Thomas, 5
Riddel(l), John, 7, 48, 156, 595
Riffleman, 170, 182
Ritonbonk, 182n, 610n
Robb, James, 570n
Roberts, James, 5, 7, 13n, 17, 18n, 20n, 23, 26, 41, 46n, 48, 91n, 156, 158n, 595–6
  Iceland journal, 115–40, Plate 8
Robinson, Sir Christopher, 287n, 297n, 376, 382n, 384n, 394n, 471n, 515n
Robinson, Frederick John (Viscount Goderich), 507n, 560
Rød Øldur, see Rauðöldur
Roebuck, John, 111
Rogers, Woodes, 399
Rojndin Freya, see *Royndin Friða*
Rome, 7n, 635
Rooke, Charles, 387n, 631, 636
Rooke & Horneman, 387, 391, 418–20, 484–5, 508–9, 523n, 563n, 631, 636
Rosbach, Frederik Friedlieb, 482n
Rose, George, 309n, 464
Rosenkrantz, Niels, Count, 568, 590
Roslin, Alexander, 209
Ross, John, 589, 626, 636–7
Rottbøll, Christen Friis, 201n
Rousseau, Jean-Jacques, 10

Rover, 323n, 339, 357n, 378, 615
Roy, William, 207n
Royal Academy of Arts, London, 47, 593, 595
Royal Botanic Garden, Copenhagen, 201
Royal Botanic Garden, Edinburgh, 111n
Royal Botanic Gardens, Kew, 7, 26, 36, 211, 468n, 631
Royal Caledonian Horticultural Society, Edinburgh, 582
Royal Chartered General Trading Company of Copenhagen, 219n, 261n
Royal Clarence Baths, 53n
Royal College of Physicians, Edinburgh, 630
Royal Danish Academy of Arts, 635
Royal Danish Academy of Sciences, 13, 21, 208n, 492, 627
Royal Danish Agricultural Society (Landhusholdnings Selskab)
Royal Exchange, London, 291n, 397
Royal Geographical Society, 28, 626
Royal Navy (British), 4, 30, 31, 32, 615, 632; see also Icelandic Revolution
Royal Observatory at Greenwich, 28
Royal Society of Edinburgh, 24n, 111n, 382n, 630, 634
Royal Society (London), 2, 3, 4, 12–13, 25, 27–8, 35, 175n, 177n, 203n, 207, 208, 381–2, 493, 589, 594–5, 598, 626, 629, 634, 641, Plate 1
Royal Society of Sciences, Uppsala, 217
Royndin Friða, 443n, 444
Rugsen, Anders Andersen, 318n
Rùm, 78, 173
Runólfsson see Guðmundur Runólfsson
Russia, 160n, 187, 208n, 417n, 572, 583n, 611, 641
Ryberg, Niels, 154n
Rymer, Thomas, 528n, 529n
Rödefiord, 230n, 267n, 387n, 388n

Sælhunden, 436n, 521n
Sæmundsson, see Magnús Sæmundsson
Sæmundur Magnússon Hólm, 203n
St Columba, 62n, 69n, 73n, 74, 75, 78n
St David's Head, 54
St Helena, 5
St John, 74
St John's, Newfoundland, 3n
St Kilda, 76, 78, 79, 173, 174
St Magnus Cathedral, 110n
St Martin, 74
St Michael's Mount, 119n
St Oran, 74–5
St Petersburg, 85n, 130, 187n, 208, 638
St Vincent, 29

677

Salamine, 317n, 318n, 319n, 331, 336n, 363n, 444n, 637
Samarang, Alexander, 156, 596
Sandown Castle, 50
Sandby, Paul, 593
Sandholt, see Óli Sandholt
Sandwich, Earl of, see Montagu, John
San Francisco, 36
Sappho, 338n, 362n, 402
Saunders, Mr, 139
Savignac, James, 301n, 319n, 320n, 337, 338, 339n, 340–41, 345, 350, 353, 358n, 377–80, 397, 409n, 411, 414n, 500, 502–4, 520n, 541n, 542–3, 544, 546–7, 553, 557–8, 562, 616
Savory, 124n
Scalholt, see Skálholt
Scania, 11
Scarba, 63, 123, 170, 174
Scarborough, 11, 531
Scheel, Hans Jacob, 537
Scheving, see Guðmundur Scheving
Schimmelmann, Ernst, 205n, 206n
Schiöth, Knud, 575n
Schmidt, N. A., 518, 555n
Schofield, A., 550n, 557
Schönning, Gerhard, 479n
Schouten, Willem of Hoorn, 166
Schrøter, Johan Henrik, 432n
Schumacher, Hans Christian Friedrich, 216
Scopoli, Giovanni Antonio, 187
Scoresby, William Jr, 28
Scotland, 6n, 9, 20, 23, 183, 205n, 227, 228, 238, 271, 272, 273, 353, 465, 468, 473n, 499, 532n, 570, 578, 598–9, 629. See also individual towns and islands, Fingal's Cave
  agriculture, 57, 59–60, 63, 76–7
  Banks's descriptions of, 55–79, 109–12, 169–74
  education, 78
  fishing, 12, 64–5, 532n
  houses, 55, 58, 59, 67, 73, 77
  industries, 59, 60, 64n, 65–6, 76–7, 78n, 139n, 170–76, 465
  literature, 65n
  maps, 49, 56
  monarchs, 59, 64n, 69n, 74, 126, 412n, 528
  people, 58, 62, 73, 77–8
  religion, 57, 58, 74–5
  Roberts's descriptions of, 120–27, 138–40
  trade, 12, 384, 420–22, 428, 431, 433, 438n, 458, 459n, 473n, 511, 569
Scott, Alexander, 117, 118, 119, 156
Scott, Sir Walter, 67n, 78n, 634
Scott, William (Sir), 382n, 406n, 524, 565
Sehestedt, Ove Ramel, 561n

Seien, see *Seyen*
Selhun, 438
Selkirk, Alexander, 399n
Seltjarnarnes, 19, 165
Sergel, Johan Tobias, 217n
Seyen, 229n, 232n, 246, 261n, 283, 351, 352n, 370n, 376, 433, 436n, 437, 518
Seyðisfjörður, 287n
Shannon, 268, 434n
Shawfield, 57n, 58n, 59n
Shetland, 110n, 258, 374, 437, 462, 535
Shooter's Hill (Shuter's Hill), 117
Sibthorp, Humphrey, 2
Sidserf, Peter, 48, 156, 596
Sigfús Jónsson, 344
Sigríður Magnúsdóttir, 85n, 130n, Plate 6
Sigurðsson, see Sveinn Sigurðsson
Sigurðardóttir, see Helga Sigurðardóttir
Sigurður Sívertsen, 554
Silfverhjelm, Göran Ulrik, 217
Simmons, Samuel Foart, 204
Símonsen see Gísli Símonsen
Sinclair, Sir John, 208
Sir Lawrence, 7n, 10, 13, 17, 48, 156, 608–9
Sisimiut, 507n, 610n
Sívertsen, see Bjarni Sívertsen, Sigurður Sívertsen
Skaftafellssýsla (Skaptafells Syssel), 203n, 313
Skagafjarðarsýsla, 162n, 314n, 584n
Skagaströnd, 270n, 314n
Skalholt, 396, 459n, 504n, 508, 517
Skálholt (Scalholt), 18, 22n, 42, 98n, 101, 102, 104n, 107n, 108n, 113, 134, 157n, 159, 160, 166, 177, 203, 220, 222n, 241, 242, 301n, 331n, 531, 600, 604, 605, 606, 607, 632, 640
Skaptafells Syssel, see Skaftafellssýsla
Skarð, 42n, 104, 600
Skye (Skie), 78, 172, 173
Slade, Sir Thomas, 53
Slate Islands, 64, 170
Smeaton, John, 53n
Smith, Adam, 15
Smith, Captain, 503, 504, 515n
Smith, Culling Charles, 371–2, 376–7, 382, 391, 393, 393–4, 412–13, 415–16, 424–5, 425–6, 637
Smith, James Edward, 6n, 212
Sneedorff, Frederik, 215, 216, 553
Snowdon, 61
Snæfellsjökull, 80n, 450n
Snæfellsnes, 80n, 628
Snæfellsnessýsla, 315n
Society of Apothecaries, 18, 21
Society of Antiquaries, 2
Society of Dilettanti, 6n

Society of Friends of Foreigners in Distress, 558, 561
Soho Square, 27, 240, 245n, 430n, 552n
Solander, Daniel Carl, 3, 4, 5, 6–7, 8, 10, 17, 19, 20, 22, 25, 26, 27, 39, 42, 46, 48, 67, 82, 86, 91n, 93n, 107, 108n, 116–24, 126, 128–30, 132–3, 136, 138–40, 153, 154, 155–6, 159, 160n, 164, 166, 182n, 183, 185, 190, 191, 193, 197, 198, 201n, 209, 226, 501n, 594, 596, 599, 600n, 604–7, 629, 630
Sölvason, see Sveinn Sölvason
Somerled, 74n
Sotheby, Wilkinson & Hodge, 35
South America, 399
Spencer, 280n, 620n, 621
Spitzbergen, 180n, 186, 593, 611,
Spöring, Herman Diedrich, 4, 39
Sprightly, 232n, 238
Spring Grove, 26n, 430n, 559, 561n, 564
Stackhouse, Jonathan, 519n, 631, see also Horne & Stackhouse
Staffa, 10n, 18n, 20n, 22n, 23, 41, 55n, 67–73, 75, 104, 116, 124–5, 171–3, 175, 185, 193, 202n, 598–9
Staffordshire, 1
Stanley, Sir John Thomas, 24, 29, 210, 211n, 212n, 213n, 217, 232, 236, 245, 265n, 326, 329n, 330, 505, 519n, 522, 555n, 570–72, 572–3, 579, 630, 631, 637, 639, 641
Stanley, Lady Maria, see Holroyd, Maria-Josepha
Stanley, Octavia, 573n
Stapi, 315n, 584n
Stefán Stephensen, 130n, 234n, 237n, 241n, 245, 254n, 262n, 263, 267n, 333n, 410n
Stefán (Þórarinsson) Thorarensen, 241n, 244n, 262n, 263, 410n, 537n, 582
Stefán Þórðarson (Thorderson), 83n, 128n
Steingrímsson, see Jón Steingrímsson
Stenness, 138n, 139n, 140n
Stephensen, *see* Björn Stephensen, Magnús Stephensen, Marta María Stephensen, Ólafur Stephensen, Ragnheiður Stephensen, Stefán Stephensen, Þórunn Stephensen
Stevens, Captain, 514, 611
Stevenson, Robert Louis, 67n
Stewart, Robert, see Castlereagh
Stewart, Thomas, Lieutenant, 333n, 352n, 393n
Stockholm, 4n, 85n, 194, 208, 576n
Stoddard, Captain Pringle, 268
Stonehenge, 23
Stopford, Captain Robert, 621
Stóri-Dímon, 96
Storm, Captain, 503, 504, 515n
Stornoway, 243, 257, 271n, 324n, 581n, 633

Stow, John, 529n
Strandasýsla, 314n
Strien, Kees van, 39
Stromness, 20, 138, 139, 381n
Stuart, William (Lord), 268
Strube, Christian Conrad, 319, 338n, 345, 369, 379n, 638
Struensee, Johan Friedrich, 84n
Stykkishólmur, 315n, 584n
Suður-Múlasýsla, 313n, 550n
Suðuroy, 432n
Sugar Top, see Sukkertoppen
Suhm, Peter Frederik, 216
Sukkertoppen, 610n
sulphur, xxii, 15, 97, 101, 107, 109, 133, 136, 181, 205n, 219–20, 223n, 242, 312, 314, 316, 331n, 613, 642
Svane, Peder Ludvig, 244, 250, 635
Svanen (*Swanen, Swan*), 257, 267n, 373n, 376, 433n, 434, 437, 473, 518, 544n, 549
Svarfdalín, see Rafn Þorgrímsson Svarfdalín
Sveinsson, see Guðlaugur Sveinsson
Sveinn Sigurðsson (Svend Sivertsen), 350n, 549n, 560n, 563n
Sveinn Sölvason, 605
Sviðholt, 94n, 132n, 222n, 227n, 499n
Swabey, H. B., 394n
Sweden, Swedish, 3, 8, 10, 15n, 46n, 82n, 133n, 159, 180n, 184n, 185n, 193, 194n, 208, 209, 210n, 217n, 239, 285, 294, 295, 303n, 307, 310, 359, 360n, 373n, 381, 402, 418n, 430, 451, 490, 521–2, 523n, 524–5, 526n, 535n, 539, 540, 541n, 558n, 560, 561, 566, 568n, 570, 574n, 578, 597, 627, 629, 639
Royal Society of Sciences, Uppsala, 217
Royal Swedish Academy of Sciences, 209n
Royal Swedish Navy, 208, 307n, 524n, 636
Sydney, 28, 35, 41, 593
Syria, 16

Talbot, 332, 333n, 352, 353, 357, 362n, 370, 380, 390, 393, 401n, 402n, 406n, 551n, 616–17
Tahiti, 4, 5, 6, 19, 26, 28, 176, 180, 185, 187n, 224, 632
Taylor, James, 238
Taylor, John, 156, 596
Teitur Jónsson, 166–8, 632
Tersmeden, Magdalena Elisabet, 195n
Thames, 48n, 117n, 366, 386n, 388, 606
Thames, 398n
Thielsen, Hans, 247n, 248–9, 258n, 268, 269, 274–5, 276–7, 284n, 286n, 417–18, 419, 420n, 421n, 422, 423, 428, 429n, 433–4, 638
Thiorsaae, see Þjórsá

Thodal, Laurits Andreas (Governor of Iceland)), 85n, 90n, 130n, 132, 134n, 158–9, 177n, 202, 220, 222n, 605, 638
Thompson, Mrs, 111, 140
Thomsen, Jesse, 319, 338, 345, 397, 419, 420n, 458–9, 484, 485
Thomsen, Nicholi, 459
Thorarensen, see Stefán Þórarinsson Thorarensen
Thorderson, see Stefán Þórðarson
Thorgrímsen, see Halldór Guðmundsson Thorgrímsen
Thorkelín, Grímur Jónsson, 27, 207, 210n, 215, 215–16, 228n, 411–12, 638–9
Thorlacius, see Gyða Thorlacius, Ólafur Þórðarson Thorlacius
Thorshavn (Thorshaven), 461
Thurot, Francois, 62, 169
Tierra del Fuego, 5
Tindastóll, 162, 605, 606
Tiree (Tirey), 78
Titherington & Allanson, 373n
Tobermory, 67, 126, 171, 173n
Torsken, 484, 485, 518
Townsend, A. C., 90n
Troward, Richard, 322n, 402, 404, 406, see also Phelps, Troward & Bracebridge
Thunberg, Carl Peter, 217
Trampe, Count Frederik Christopher, 241, 242, 244, 254n, 256, 262n, 267, 276, 318n, 332, 333, 361n, 362, 363n, 365, 372, 383n, 384n, 386, 388n, 396, 398–400, 403, 413, 418, 420, 422, 435n, 437, 438, 442, 462–3, 470–71, 481, 490n, 494n, 541, 545n, 551, 564, 633, 639, Fig. 6
and Icelandic revolution, 335–56, 357–8, 364, 368, 370, 371, 377–80, 389–90, 391–2, 393–4, 400–401, 405–6, 408, 410n, 414, 419, 500n, 615–17
Tranquebar, 8n, 625, 627
Transport Board, 251–3, 274, 276–7, 284, 360, 362, 364, 388, 390, 403, 634
Transport Office, Leith, 250, 253, 271n, 272
Treasury [British], 231, 238, 268–70, 278n, 283n, 290, 322–3, 375, 390n, 396n, 507n, 576, 623
Treshnish Isles, 75n, 124n
Troil, Uno von (Troilus), 9, 10, 13n, 18, 19, 22, 23, 48, 70n, 80n, 86, 90, 102, 106n, 109n, 117, 118, 119, 121, 122, 124, 128, 133, 140, 156, 166, 179–80, 183, 184–5, 190n, 193–6, 202n, 203, 208–210, 217–18, 226n, 259n, 408, 479n, 498, 501n, 596–7, 599n, 605, 606, 607, 639
Trondheim (Drontheim), 301n, 318n, 320n, 393n, 420, 507n, 516, 517, 541
Turner, Dawson, 23, 35, 174n, 254n, 334

Tusser, Thomas, 468
Tvende Söstre (*Two Sisters*, *To Söstre*), 227, 229n, 230, 231n, 233, 234, 236, 239, 243, 244n, 245n, 246, 250, 257, 265, 267n, 278n, 283, 284n, 295, 297n, 307n, 309n, 310n, 418, 488
Tykkebay, 350n, 369n, 437, 508, 559, 560, 630

Ulysses, 524, 566n
United States, 17n, 36, 187n, 400n, 534, 536
Upernavik, 439n, 469n, 482n, 610
Uppsala, 3, 22, 156, 190n, 193n, 217n, 596–7, 605, 606, 607, 629, 639
Úthlíðarhraun, 98
Utrecht, 179n, 534, 626
Uummannaq, 482n, 610n

Vaðeyri, 42n, 600n
Vahl, Martin, 216
Valgerður Þórðardóttir, 132n
Vancouver, Charles, 400, 401
Vancouver, George, 28, 400n
Vansittart, Nicholas, 539
Vargas Bedemar, Count Eduardo Romeo von, see Bedemar
Vashon, Admiral James, 228n, 231, 238
Vatneyri, 246n, 627
Verdun de la Crenne, Jean-René Antoine Marquis de, 16, 498
Verne, Jules, 67n, 80
Versailles, 10, 16
Vestfirðir (the Western fjords), 16, 246n, 314n, 398n, 450n, 484n, 630, 635
Vestmannaeyjasýsla, 313
Vestmannaeyjar (Westman Islands), 11, 243n, 244n, 312n, 313, 348, 351, 370n, 379, 533, 534, 571, 572, 584, 635
Vesuvius, 24, 177n, 182, 185, 203, 260
Victoria, Queen, 67n
Vídalín, see Geir Vídalín, Jón Vídalín
Viðey, 312n, 325n, 329n, 395n, 460n, 502n, 552n, Plate 14
Vigilant, 414n, 433n
Virginia, 261n
Vittoria, 556n
Voltaire, 363, 598
Vopnafjörður, 314n, 419n, 584n
Vrow Christiana, 91n, 130, 138

Walden, Frederick Hermann, 7, 46, 48, 118, 119, 130, 138, 139n, 156, 597
Wales, 3, 26, 54n, 55n, 61n, 77, 119
Wales, William, 46
Walker, George, 544
Wallis, Samuel, Captain, 4n, 6, 594

Ward, Robert, 300n
Ward, William, 298n, 386n
Wedseltoft, Ambrosia, 492n
Wellesley, Arthur, Duke of Wellington, 402, 403
Wellesley, Richard, Marquess Wellesley, 33, 335n, 357–8, 368, 371n, 372, 376n, 377–81, 382n, 392, 393, 394n, 404n, 416n, 424, 457, 458, 464–5, 565n, 626, 637, 640
Wellesley-Pole, William, 3rd Earl of Mornington, 300, 320n, 361n, 640
Wellington, see Wellesley
West, Benjamin, 6n
Western Fjords see Vestfirðir
Western Isles, 13, 22n, 23, 40, 56 (Map 2), 59n, 117, 157n, 168n, 169, 174, 178
West Indies, 28, 30, 218n, 223, 228n, 431n, 573, 612
Westman Islands, see Vestmannaeyjar
Whateley, John, 539n
Wheeler, Alwyne, 42
White, Gavin, 580, 581
White River, see Hvítá
Wilberforce, William, 391, 443, 558n
William of Orange, 53n
Winnington Park, 555n, 637
Wolff, Georg(e), 152, 247n, 253, 272n, 363n, 431–2, 443–4, 446, 461, 558n, 640–41
Wolff, Jens, 247n, 271n, 272n, 558n, 631, 641
Wolffs & Dorville, 306n, 473, 640–41
Wormskiold, Morten, 492, 495–7, 641
Wright, James, 213, 641
Wright, Dr William, 211–13, 229–30, 231–2, 233, 236–7, 240, 641–2
Wulff, Jens Andreas, , 287n, 296, 302, 303, 304–5, 306, 308, 360, 366, 367–8, 416–17, 430, 452n, 641–2, see also Ørum & Wulff

Yarmouth (Great), 11, 232n, 249, 268, 280n, 285, 286, 287, 296, 298n, 303n, 334n, 386n, 396, 403, 420, 434n, 442, 528, 535–6, 565, 620n
Yarmouth, Isle of Wight, 52
Young, Nicholas, 46n, 48, 156, 597

Zealand, 11, 230
Zoffany, Johann, 7, 10, 46, 47

Þerney, 504n
Þingvallahraun, 94n
Þingvallavatn, 94n, 96n, 313
Þingvellir (Thingvalle), 18, 42n, 93n, 94n, 95n, 96n, 108n, 262, 600
Þjórsá, 42n, 104n, 108n, 600
Þjórsárholt, 42n, 134n, 600
Þórarinn Þorsteinsson, 195n
Þórðardóttir, see Valgerður Þórðardóttir
Þórðarson, see Stefán Þórðarson
Þorgeirsson, see Guðlaugur Þorgeirsson
Þorgilsson, see Ari Þorgilsson
Þórhallsson, see Þorlákur Þórhallsson
Þorláksson, see Guðbrandur Þorláksson, Jón Þorláksson
Þorlákur Ísfjörð, 381n, 630, 631
Þorlákur Skúlason, 189n
Þorlákur Þórhallsson, 101n
Þorleifsson, see Björn Þorleifsson (the rich)
Þorleifur Guðmundsson, 244n, 250n, 256n, 257n
Þorsteinsson, see Pétur Þorsteinsson, Þórarinn Þorsteinsson
Þorsteinn Jónsson, xi, 90n, 198, 199, 632, Plate 5
Þórunn Ólafsdóttir, 607n
Þórunn Stephensen, 130n
Þuríður Nikulasdóttir, 344n